CLINICAL PHYSIOLOGY
OF ACID-BASE
AND ELECTROLYTE DISORDERS

CLINICAL PHYSIOLOGY OF ACID-BASE AND ELECTROLYTE DISORDERS

Second Edition

Burton David Rose, M.D.

Director, Division of Nephrology
The St. Vincent Hospital
Worcester, Massachusetts
Professor of Medicine and Physiology
University of Massachusetts Medical School

McGraw-Hill Book Company

New York   St. Louis   San Francisco   Auckland
Bogotá   Guatemala   Hamburg   Johannesburg
Lisbon   London   Madrid   Mexico   Montreal
New Delhi   Panama   Paris   San Juan   São Paulo
Singapore   Sydney   Tokyo   Toronto

CLINICAL PHYSIOLOGY OF ACID-BASE AND ELECTROLYTE DISORDERS

Copyright © 1984, 1977 by McGraw-Hill, Inc. All rights reserved. Printed in the United States of America. Except as permitted under the United States Coypright Act of 1976, no part of this publication may be reproduced or distributed in any form or by any means, or stored in a data base or retrieval system, without the prior written permission of the publisher.

3 4 5 6 7 8 9 0 DOC DOC 8 9 8 7 6 5

ISBN 0-07-053622-8

This book was set in Times Roman by University Graphics, Inc. The editors were Beth Ann Kaufman and Steven Tenney; the production supervisor was Avé McCracken.
R. R. Donnelley & Sons Company was printer and binder.

Library of Congress Cataloging in Publication Data

Rose, Burton David.
    Clinical physiology of acid-base and electrolyte
disorders.
    Bibliography: p.
    Includes index.
    1. Acid-base imbalances. 2. Water-electrolyte
imbalances. 3. Kidneys. 4. Water-electrolyte balance
(Physiology) I. Title [DNLM: 1. Acid-base imbalance.
2. Kidney—Physiology. 3. Water-electrolyte balance.
4. Water-electrolyte imbalance. WD 220 R795c]
RC630.R67 1984   616.6′1   83-22227
ISBN 0-07-053622-8

For Gloria, Emily, Anne, and Daniel

CONTENTS

PART FOUR PHYSIOLOGIC APPROACH TO ACID-BASE AND ELECTROLYTE DISORDERS

PREFACE

This book integrates the essentials of renal and electrolyte physiology with the common clinical disorders of acid-base and electrolyte balance. Its underlying premise is that these clinical disturbances can be best approached from an understanding of basic physiologic principles. Thus, Chapters 1 to 8 review the physiology of the body fluids, how the kidney functions, and the effects of hormones on the kidney. This is followed by a discussion of the renal and extrarenal factors involved in the normal regulation of volume (sodium), water, acid-base, and potassium balance (Chapters 9 to 13). In addition to providing the foundations for understanding how disease states can overcome these regulatory processes, the initial chapters can also be used by first-year medical students studying renal physiology.

The material presented in these chapters represents the core of information that, in my opinion, the clinician should possess. It is not meant to be an exhaustive review; in those areas where controversy exists, I have chosen to note the presence of uncertainty and to refer the interested reader to appropriate references, instead of extensively reviewing each theory. Since the primary purpose of this book is to teach the reader how to approach clinical problems, the physiologic discussions are correlated with situations in clinical medicine whenever possible.

Part 4 (Chapters 14 to 28) contains a separate chapter on each of the major acid-base and electrolyte disturbances. In addition to discussing etiology, symptoms, diagnosis, and treatment, each chapter begins with a short summary of the pathophysiology of the specific disorder with cross-references to more complete discussions in the earlier chapters. Although this leads to a certain amount of repetition, it has the advantage of allowing each clinical chapter to be read independently of the other parts of the book, making the book easier to use by a physician dealing with an acutely ill patient.

The second edition, which has been completely rewritten, differs from the first edition in several important ways. First, new information that has accumulated over the past seven years has been included. Some of this information challenges previously held concepts on how the kidney functions and what changes occur with fluid and electrolyte disorders. Second, my continuing exposure to patients, medical students, and house officers has helped me to refine and clarify some of the discussions, allowing them to be more easily understood and applied. Third, a new chapter (Chapter 14) has been added on the meaning and clinical applications of commonly ordered urinary tests, including the sodium, potassium, and chloride concentrations, pH, and osmolality. Since these parameters are used for diagnostic and therapeutic reasons in a variety of disorders, it seemed appropriate to summarize their significance in one place.

In both the physiologic and clinical sections, problems are presented at the end of most of the chapters. These problems are intended not only to test understanding but also to emphasize important concepts frequently misunderstood by physicians dealing wth these disorders. The answers to the problems are presented in Chapter 29. Finally, at the request of many students and house officers who have used this book, Chapter 30 contains a summary of important equations and formulas that are useful in the clinical setting.

ACKNOWLEDGEMENTS

This book could not have been written without the assistance of many people. Although many residents at Saint Vincent Hospital and medical students at the University of Massachusetts Medical School made valuable contributions, I would like to particularly note the efforts of Robert Black and Gordon Saperia who reviewed the entire manuscript. Of equal importance was the outstanding organizational and secretarial work of Sheila Putnam.

Burton David Rose

CLINICAL PHYSIOLOGY
OF ACID-BASE
AND ELECTROLYTE DISORDERS

ONE

WATER AND ELECTROLYTE PHYSIOLOGY

PHYSIOLOGY OF BODY FLUIDS

INTRODUCTION

The cells are the major functioning units in the body. Their ability to function normally is dependent upon a variety of factors including the delivery of nutrients, the removal of waste products, and the maintenance of a stable physicochemical environment in the extracellular fluid. The kidneys play a major role in the last two processes, excreting some of the waste products of metabolism (such as urea) as well as electrolytes and water to keep the composition of the extracellular fluid relatively constant. Before discussing how the kidneys accomplish these goals in the following chapters, in this chapter we will review the volume and composition of the different fluid compartments plus the factors determining the distribution of water among them. As an introduction, a brief review of basic chemical terms and concepts will be presented.

Units of Solute Measurement

The concentration of a given solute can be expressed in milligrams per deciliter† (mg/dL), millimoles per liter (mmol/L), milliequivalents per liter (meq/L), or milliosmoles per liter (mosmol/L). For sodium ion (Na^+), 2.3 mg/dL (or 23 mg/L), 1 mmol/L, 1 meq/L, and 1 mosmol/L all refer to the same concentration of Na^+. Since these units are used clinically, it is important to understand their meaning.

Atomic weight and molarity Table 1-1 lists the atomic weights of the most important elements in the body. The atomic weight is an assigned number which allows comparison of the relative weights of the different elements. By definition, one atom of oxygen is assigned a weight of 16, and the atomic weights of the other elements are determined in relation to that of oxygen. In a molecule, i.e., a substance containing two or more atoms, the molecular weight is equal to the sum of the atomic weights of the individual atoms (Table 1-1). For example, the molecular weight of water (H_2O) is 18 ($2 \times 1 + 16$).

One mole (mol) of any substance is defined as the molecular (or atomic) weight of that substance in grams. Similarly, one millimole (mmol) is equal to one-thousandth of a mole or the molecular (or atomic) weight in milligrams. Since the atomic weight of Na^+ is 23, 23 mg is 1 mmol and 23 mg of Na^+ in 1 liter of water represents a Na^+ concentration (written [Na^+]) of 1 mmol/L. The concept of molarity is important because, from Avogadro's law, 1 mol of any nondissociable substance contains the same number of particles (approximately 6.02×10^{23}). Thus, 1 mmol of Na^+ contains the same number of atoms as 1 mmol of Cl^- even though the former weighs 23 mg and the latter weighs 35.5 mg. However, 1 mmol of NaCl (58.5 mg) largely dissociates into Na^+ and Cl^- ions and therefore contains almost twice as

---

†Milligrams per deciliter is also referred to by many laboratories and in many journals and texts as "milligrams per 100 milliters" or "milligrams percent."

Table 1-1 Atomic and molecular weights of physiologically important substances

Substance	Symbol or formula	Atomic or molecular weight
Calcium ion	Ca^{2+}	40.1
Carbon	C	12.0
Chloride ion	Cl^-	35.5
Hydrogen ion	H^+	1.0
Magnesium ion	Mg^{2+}	24.3
Nitrogen	N	14.0
Oxygen	O	16.0
Phosphorus	P	31.0
Potassium ion	K^+	39.1
Sodium ion	Na^+	23.0
Sulfur	S	32.1
Ammonia	NH_3	17.0
Ammonium ion	NH_4^+	18.0
Bicarbonate ion	HCO_3^-	61.0
Carbon dioxide	CO_2	44.0
Glucose	$C_6H_{12}O_6$	180.0
Phosphate ion	PO_4^{3-}	95.0
Sulfate ion	SO_4^{2-}	96.1
Urea	NH_2CONH_2	60.0
Water	H_2O	18.0

many particles. As will be seen, these relationships are important in understanding electrochemical equivalence and in the measurement of osmotic pressure.

Although the concentration of uncharged molecules, e.g., glucose and urea, also can be measured in millimoles per liter, they are more commonly measured in milligrams per deciliter. For example, the molecular weight of glucose is 180. Consequently, a glucose concentration of 180 mg/L (or 18 mg/dL) is equal to 1 mmol/L. To convert from milligrams per deciliter to millimoles per liter, the following formula can be used:

$$mmol/L = \frac{mg/dL \times 10}{molecular\ weight} \tag{1-1}$$

(The multiple of 10 is used to convert milligrams per deciliter into milligrams per liter.)

Electrochemical equivalence Positively charged particles are called cations, and negatively charged particles are called anions. When cations and anions combine, they do so according to their ionic charge (or valence) and not according to their weight. Electrochemical equivalence refers to the combining power of an ion. One equivalent is defined as the weight in grams of an element that combines with or replaces 1 g of hydrogen ion (H^+). Since 1 g of H^+ is equal to 1 mol of H^+ (6 $\times$

10^{23} particles), 1 mol of any univalent anion (charge equals $1-$, also 6×10^{23} particles) will combine with this H^+ and is equal to one equivalent (eq). For example,

$$1 \text{ mol } H^+ + 1 \text{ mol } Cl^- \rightarrow 1 \text{ mol } HCl$$
$$(1 \text{ g}) \qquad (35.5 \text{ g}) \qquad (36.5 \text{ g})$$

By similar reasoning, 1 mol of a univalent cation (charge equals $1+$) also is equal to 1 eq since it can replace H^+ and combine with 1 eq of Cl^-. For example,

$$1 \text{ mol } Na^+ + 1 \text{ mol } Cl^- \rightarrow 1 \text{ mol } NaCl$$
$$(23 \text{ g}) \qquad (35.5 \text{ g}) \qquad (58.5 \text{ g})$$

In contrast, ionized calcium (Ca^{2+}) is a divalent cation (charge is $2+$). Consequently, 1 mol of Ca^{2+} will combine with 2 mol of Cl^- and is equal to 2 eq:

$$1 \text{ mol } Ca^{2+} + 2 \text{ mol } Cl^- \rightarrow 1 \text{ mol } CaCl_2$$
$$(40 \text{ g}) \qquad (71 \text{ g}) \qquad (111 \text{ g})$$

The body fluids are relatively dilute, and most ions are present in milliequivalent quantities (one-thousandth of 1 eq equals 1 meq). To convert from units of millimoles per liter to milliequivalents per liter, the following formulas can be used:

$$\text{meq/L} = \text{mmol/L} \times \text{valence} \tag{1-2}$$

or from Eq. (1-1),

$$\text{meq/L} = \frac{\text{mg/dL} \times 10}{\text{mol wt}} \times \text{valence} \tag{1-3}$$

There are two advantages in measuring ionic concentrations in milliequivalents per liter. First, it emphasizes the principle that ions combine milliequivalent for milliequivalent, not millimole for millimole or milligram for milligram. Second, to maintain electroneutrality, there is an equal number of milliequivalents of cations and anions in the body fluids. As we shall see in later chapters, the need to preserve electroneutrality is an important determinant of ion transport in the kidney and ion movement between the cells and the extracellular fluid. This obligatory relationship could not be appreciated if the ionic concentrations were measured in millimoles per liter or in milligrams per deciliter (Table 1-2).

It should be noted that not all ions can be easily measured in milliequivalents per liter. The total calcium (Ca^{2+}) concentration in the blood is approximately 10 mg/dl. From Eq. (1-3),

$$\text{meq/L of } Ca^{2+} = \frac{10 \times 10}{40} \times 2 = 5 \text{ meq/L}$$

However, roughly 50 to 55 percent of plasma Ca^{2+} is bound by albumin and, to a much lesser degree, citrate so that the physiologically important ionized (or unbound) Ca^{2+} concentration is only 2.0 to 2.5 meq/L.

There is a different problem with phosphate since it can exist in different ionic forms, as $H_2PO_4^-$, HPO_4^{2-}, or PO_4^{3-}, and an exact valence cannot be given. We can

Table 1-2 Normal plasma electrolyte concentrations

Electrolyte	meq/L	mmol/L
Cations		
Na^+	142.0	142.0
K^+	4.3	4.3
Ca^{2+}†	2.5	1.25
Mg^{2+}†	1.1	0.55
Total	149.9	148.1
Anions		
Cl^-	104.0	104.0
HCO_3^-	24.0	24.0
$H_2PO_4^-$, HPO_4^{2-}	2.0	1.1
Proteins	14.0	0.9
Other‡	5.9	5.5
Total	149.9	135.5

†The values for Ca^{2+} and Mg^{2+} include only the ionized form of these ions.
‡This includes SO_4^{2-} and organic anions such as lactate.

estimate an approximate valence of minus 1.8 since roughly 80 percent of extracellular phosphate exists as HPO_4^{2-} and 20 percent as $H_2PO_4^-$ (see page 208). If the normal serum phosphorus concentration is 3.5 mg/dL (phosphate in the blood is measured as inorganic phosphorus), then

$$\text{meq/L of phosphate} = \frac{3.5 \times 10}{31} \times 1.8 = 2 \text{ meq/L}$$

Similarly, only an average valence can be given for the polyvalent protein anions. If the plasma protein concentration is 0.9 mmol/L and the average valence is minus 15, then

$$\text{meq/L of protein} = 0.9 \times 15 = 14 \text{ meq/L}$$

Passive and Active Solute Transport

Water and solutes are continually exchanged between the different fluid compartments. This occurs by passive and active mechanisms. The movement of particles is passive if it develops spontaneously and does not require a supply of metabolic energy. In contrast, particle movement is active if it is dependent upon energy derived from metabolic processes. In humans, solute movement occurs by both passive and active mechanisms whereas all water movement is passive† (Table 1-3).

†For a complete review of the principles of active and passive transport, see Refs. 1 and 2.

Table 1-3 Mechanisms of solute transport

Passive
  Simple and facilitated diffusion
  Coupled transport
  Solvent drag

Active
  Primary active transport
  Secondary active transport
  Endocytosis

Membrane permeability The primary barrier to the movement of solutes in the body is the cell membrane. Although small solutes can pass through most cell membranes, this occurs much more slowly than in water. For example, the permeability of the skeletal muscle cell membrane to K^+ is only 10^{-7} times the K^+ permeability of a similar thickness of water [3].

The ability of a solute to cross a membrane is dependent in part upon its solubility in the lipoprotein matrix of the membrane. Substances which are lipid-soluble, such as gases (oxygen and carbon dioxide) and urea, can pass directly through the membrane, utilizing virtually its entire surface area. In contrast, ions and glucose are water-soluble (hydrophilic) and must cross the membrane in some other way. This may occur through water-filled channels created by membrane proteins which traverse the lipid bilayer [1]. Since only this part of the membrane is available for the transport of hydrophilic solutes, ions and glucose will move through the membrane more slowly than oxygen, carbon dioxide, or urea.

In addition, solute permeability is inversely related to solute size. Thus, larger macromolecules such as proteins either cannot cross the membrane or tend to do so more slowly than smaller molecules such as sodium and glucose (see Fig. 3-3). This size constraint is presumably imposed by the width of the channels through which solute transport occurs.

Simple diffusion Net passive movement of a solute across a membrane generally requires the presence of a favorable concentration or electrical gradient. The particles in a solution are in constant random motion and, therefore, are as likely to move in one direction as any other. Suppose two urea-containing solutions are separated by a membrane which is permeable to urea. As a result of random movement, urea particles on both sides of the membrane will enter and cross the membrane (Fig. 1-1). However, the number of particles randomly entering the membrane will be proportional to the number of particles in the solution, i.e., to the concentration of urea. If the urea concentration is higher on side 1 than on side 2, then the random movement (or diffusion) of urea from side 1 to side 2 will exceed that in the opposite direction. This will continue until the urea concentration is the same on both sides of the membrane. At this time, termed the *equilibrium state,* urea will still move across the membrane but the rate will be the same in both directions. The net effect

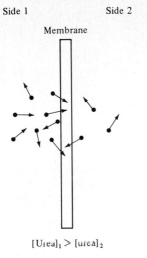

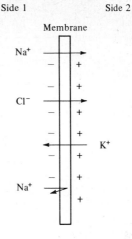

Figure 1-1 header: Side 1 ... Side 2, Membrane, [Urea]₁ > [urea]₂ rendered as $[\text{Urea}]_1 > [\text{urea}]_2$

Figure 1-1 Random movement of urea particles across a membrane by diffusion. Since the urea concentration is higher on side 1 than on side 2, there will be net movement of urea from side 1 to side 2 until the urea concentration is the same on both sides.

Figure 1-2 Effect of electrical forces on the movement of ions across a membrane. The movement of Na^+ from side 1 to side 2 makes side 2 electrically positive with respect to side 1. This electrical gradient favors the movement of an anion (Cl^-) from side 1 to side 2 or of a cation (K^+) from side 2 to side 1. However, the positive charges on side 2 retard the further movement of a cation (Na^+) into that compartment.

is the movement of urea down a concentration gradient from an area of high concentration to one of low concentration.

Electrical forces In addition to concentration gradients, the passive movement of ions (but not uncharged particles such as glucose and urea) is affected by the electrical potential across the membrane. Since like charges repel and opposite charges attract, positively charged particles tend to move to the negative side of the membrane and negatively charged particles tend to move to the positive side of the membrane.

The possible effects of electrical forces are illustrated in Fig. 1-2. If Na^+ moves across a membrane from side 1 to side 2 either by diffusion or active transport, side 2 will become electrically positive in relation to side 1. The creation of this electrical potential can have three different effects on ion transport: (1) An anion such as Cl^- can move to the positive side of the membrane, thereby following Na^+; (2) a cation such as K^+ can move to the negative side of the membrane, moving in the opposite direction from Na^+; and (3) if no other ion movement occurs, the positive charges on side 2 will repel Na^+, acting to retard further Na^+ transport across the membrane. In the first two instances, electroneutrality is maintained as an anion follows Na^+ or a cation is exchanged for Na^+. In these settings, the electrical potential gen-

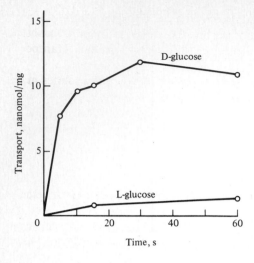

Figure 1-3 Accelerated passive transport down a concentration gradient of D-glucose in comparison to L-glucose due to D-glucose transport protein from human erythrocytes. *(Adapted from M. Kasahara and P. C. Hinkle, J. Biol. Chem., 252:7384, 1977.)*

erated by the movement of Na^+ is dissipated by the transport of other ions. However, in the third situation, there is a persistent separation of charges across the membrane (see "Generation of the Membrane Potential" below).

Facilitated diffusion Although solutes can freely diffuse across the cell membrane down a concentration or electrical gradient, it now seems that this transport occurs in many circumstances by facilitated diffusion. In this process, solute-specific carriers are present in the cell membrane that allow the solute to cross the membrane more rapidly than it would by simple diffusion [1]. For example, a carrier which is specific for D-glucose has been identified in red blood cells. As a result, D-glucose can enter the red cell at a much faster rate than L-glucose even though both compounds have the same molecular size (Fig. 1-3). Carrier-mediated facilitated diffusion has also been demonstrated in the kidney for many solutes, including sodium and bicarbonate [4,5]. This transport is passive since net solute movement occurs only in the presence of a favorable electrochemical gradient.

The manner in which the combination of solute with its carrier promotes passive transport is incompletely understood. Since non-lipid-soluble substances (such as glucose and sodium) penetrate the lipid phase of the membrane slowly, the presence of carrier proteins which traverse the membrane may provide a pathway that facilitates the transport of these solutes [1].

Coupled transport Coupled transport refers to transport involving a carrier that recognizes two solutes and promotes the movement of both across the membrane. This process appears to be particularly important in the renal proximal tubule where there are sodium-glucose, sodium-phosphate, and sodium–amino acid carriers present in the brush border [4]. The net effect is mutual enhancement of solute movement from the tubular lumen into the cell (cotransport). This may result from the attachment of one solute to the carrier increasing the affinity of the carrier for the other solute. Thus, sodium enhances the proximal reabsorption of organic solutes, and the presence of organic solutes promotes the reabsorption of sodium [4,6]. Although coupled transport may appear to be similar to simple facilitated diffusion, it may differ in one important respect; namely, that one of the solutes may actually be transported uphill against an electrochemical gradient (see "Secondary Active Transport" below).

Coupled transport can also result in *countertransport* in which two solutes move in opposite directions across the membrane. A major example of this phenomenon occurs in the proximal tubule where a sodium-hydrogen carrier promotes the movement of sodium from the lumen into the cell and hydrogen from the cell into the lumen [7].

Solvent drag Solute transport may also be passively coupled to that of water. When water moves across a membrane because of an osmotic pressure gradient (see below), frictional forces between the solvent (water) and solutes it contains result in membrane-permeable solutes being carried along with the water. This phenomenon, called *solvent drag*, is independent of concentration or electrical gradients for the solutes.

Active transport of solutes When solutes move *against* concentration and/or electrical gradients, energy is required. An example of active transport is shown in Fig. 1-4, which depicts the relationship between a muscle cell and its extracellular environment, the interstitial fluid. Because of the different concentrations in these fluid compartments, Na^+ tends to move passively into the cell and K^+ to move out of the cell. This is prevented by the active transport of Na^+ out of the cell and K^+ into the cell against these gradients. The activity of this Na^+-K^+ pump, which traverses the cell membrane, appears to be regulated by an enzyme, sodium-potassium-activated adenosine triphosphatase (Na^+-K^+-ATPase) [8–10]. This enzyme catalyzes the hydrolysis of the high-energy compound ATP into the lower-energy compound adenosine diphosphate (ADP). The energy released by this reaction is then utilized for the active transport of Na^+ and K^+. In general, three Na^+ ions are pumped

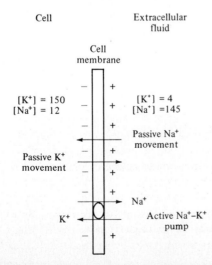

Cell   Extracellular fluid

Cell membrane

$[K^+] = 150$
$[Na^+] = 12$

$[K^+] = 4$
$[Na^+] = 145$

Passive Na^+ movement

Passive K^+ movement

Na^+

K^+

Active Na^+-K^+ pump

Figure 1-4 Forces involved in the maintenance of the steady-state Na^+ and K^+ concentrations across the cell membrane. Although there are passive gradients favoring the movement of Na^+ into the cell and K^+ out of the cell, this is balanced by the active transport of Na^+ out of the cell and K^+ into the cell against these gradients.

outward and two K^+ ions inward for each molecule of ATP that is hydrolyzed [8–10].

The Na^+-K^+ pump plays an important role in the maintenance of the ionic composition and volume of the intracellular fluid and in the entry of nutrients into the cell (see below). In the renal tubular cells (and other transporting cells such as those of the intestinal mucosa), active transport is also essential for the movement of solutes *across the cells* (see Chap. 5) [10]. In addition, the Na^+-K^+ pump participates in the generation of the membrane potential that is essential for neuromuscular transmission.

Generation of the membrane potential To varying degrees, charges are separated across all cell membranes in the body, resulting in a persistent membrane potential. Two factors play an important role in the generation and maintenance of this potential: the Na^+-K^+ pump, which maintains a high concentration of Na^+ in the extracellular fluid and K^+ in the cells, and the relative membrane permeability (roughly 100 times more permeable to K^+ than to Na^+ in skeletal muscle)† [9].

Since the cell K^+ concentration is 150 meq/L and the extracellular fluid K^+ concentration is 4 meq/L, K^+ diffuses out of the cell down a concentration gradient (Fig. 1-4). As a result, the extracellular side of the membrane becomes electrically positive with respect to the interior of the cell. Anions do not follow K^+ out of the cell because almost all the cell anions (primarily proteins and organic phosphates) are unable to cross the cell membrane. Sodium, the principal extracellular cation (see Table 1-6), enters the cell relatively slowly down this electrical gradient since the membrane is less permeable to Na^+ than to K^+. Thus, the diffusion of K^+ out of the cell will continue until the extracellular side of the membrane becomes sufficiently positive to prevent further diffusion. This relationship at equilibrium can be expressed by the following equation [9]:

$$E_m = -61 \log \frac{r[K^+]_{cell} + 0.01 \, [Na^+]_{cell}}{r[K^+]_{ecf} + 0.01 \, [Na^+]_{ecf}}$$

where E_m is the membrane potential, r is the Na^+/K^+ active transport ratio of 3:2, 0.01 is the relative permeability of Na^+ to K^+, and the subscript ecf refers to the extracellular fluid. If the normal concentrations of Na^+ and K^+ (Fig. 1-4) are substituted in this equation:

$$E_m = -61 \log \frac{\dfrac{3}{2}(150) + 0.01\,(12)}{\dfrac{3}{2}(4.4) + 0.01\,(145)}$$

$$= -88 \text{ mV} \qquad \text{(cell interior negative)}$$

---

†Relative and absolute ion permeabilities may vary in different organs. For example, transporting epithelia such as the renal tubular cells have a relatively high Na^+ permeability at the luminal membrane (see below).

This role of K^+ in the generation of the membrane potential is important clinically since the life-threatening changes of hyperkalemia and hypokalemia, e.g., cardiac arrhythmias and muscle paralysis, are due in part to alterations in this potential (see Chap. 26).

Although the development of the membrane potential means that electroneutrality is not strictly maintained, the quantity of charges that is separated across the membrane is extremely small and not detectable clinically. For example, a membrane potential of 90 mV requires the separation of only 10^{-9} meq of K^+ per cm^2 of membrane or about one one-hundred-thousandth of the intracellular K^+ pool [9].

Secondary active transport Carrier-mediated cotransport can result in the movement of one of the solutes against a concentration gradient without the usual direct requirement for the utilization of energy for uphill or active transport. In the renal proximal tubule, for example, sodium in the lumen moves passively into the tubular cell across the luminal membrane since the membrane is relatively permeable to sodium and the luminal concentration of sodium (similar to that in the plasma, or about 150 meq/L) far exceeds that of the cell (about 10 meq/L). The energy generated by this downhill sodium movement in some way allows another solute (such as glucose) to be cotransported into the cell *against* its concentration gradient (cell glucose concentration greater than that in the tubular lumen) [4]. The energy for this process is indirectly provided by the Na^+-K^+-ATPase pump at the peritubular membrane (Fig. 1-5). By actively extruding sodium from the cell, this pump maintains the low intracellular sodium concentration that is necessary for passive sodium entry into the cell at the luminal membrane.

Endocytosis Endocytosis is a form of active transport that permits the cellular uptake of particles that are too large to cross the cell membrane directly. In this process, a portion of the membrane initially invaginates around the macromolecule and then becomes completely pinched off, existing as an intracellular membrane-bound vesicle which then releases the transported material. In the kidney, endocytosis is the mechanism by which the proximal tubule reabsorbs large polypeptides such as insulin [11]. The energy for endocytosis is supplied by the hydrolysis of ATP.

Summary Figure 1-5 shows the different types of active and passive solute transport that can be present in one cell, in this case a proximal tubular cell. Note that the orientation of the cell is important, since the transport mechanisms differ at the luminal and peritubular aspects of the cell membrane. The central process is primary active transport by Na^+-K^+-ATPase in the *peritubular membrane* which pumps Na^+ out of and K^+ into the cell. This results in a low cell Na^+ concentration which allows cotransport across the *luminal membrane* of Na^+ (down its concentration gradient) and glucose (by secondary active transport *against* its concentration gradient). These solutes are then returned to the systemic circulation at the peritubular membrane: Na^+ by active transport and glucose probably by facilitated diffusion. The luminal membrane is also the site both of countertransport of Na^+ and H^+ (a process essential for bicarbonate reabsorption and acidification of the urine; see Chap. 12) and of endocytosis of large polypeptides.

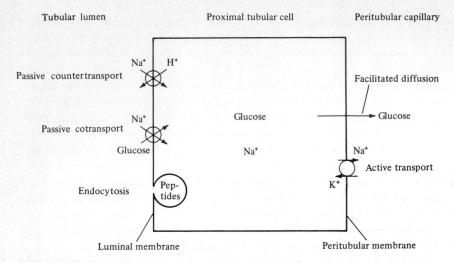

Tubular lumen   Proximal tubular cell   Peritubular capillary

Figure 1-5 Schematic representation of some of the mechanisms involved in proximal tubular solute transport. Polypeptides that enter the cell by endocytosis are hydrolyzed by lysosomal enzymes into amino acids, which then move across the peritubular membrane into the peritubular capillary, perhaps by facilitated diffusion.

Water Movement, Osmosis, and Osmotic Pressure

The movement of water between the different fluid compartments is determined by two forces: hydrostatic pressure and osmotic pressure. If two solutions are separated by a membrane permeable to water, then the application of mechanical pressure to one of the fluid compartments will push water across the membrane into the other compartment. In the body, the highest hydrostatic pressures are in the vascular space, being generated by the contraction of the heart. Consequently, in the capillary bed, this pressure tends to push the plasma water out of the vascular space into the interstitium. Fortunately, the plasma volume is preserved by the counterbalancing effect of the osmotic forces generated by the plasma proteins (see below).

The concepts of osmosis and osmotic pressure can be easily understood from the simple experiment in Fig. 1-6. Suppose distilled water in a beaker is separated into

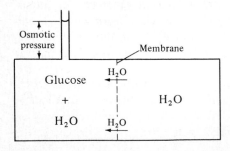

Figure 1-6 Effect of adding an impermeable solute such as glucose to the fluid on one side of a membrane. As water moves into the glucose compartment, a hydrostatic pressure is generated (measured by the height of the column of water above the glucose compartment) which at equilibrium will be equal to the osmotic pressure of the solution.

two compartments by a membrane that is permeable to water but not to solutes and that the solute glucose is added to the fluid on one side of the membrane. Water molecules exhibit random motion and can move across a membrane by a mechanism similar to that for diffusion of solutes. When solutes are added to water, the random movement (or activity) of the water molecules is reduced [2]. Since water moves from an area of high activity to one of lower activity, water will flow into the compartment containing glucose. In theory, this movement of water, called *osmosis,* should continue indefinitely because the activity of water is always reduced in the glucose compartment. However, since the compartment is rigid, the increase in volume will result in an increase in hydrostatic pressure, causing the fluid column above the compartment to rise. This hydrostatic pressure tends to push water back into the solute-free compartment. Equilibrium will be reached when the hydrostatic pressure (as measured by the height of the column) is equal to the forces pulling water across the membrane. This hydrostatic pressure which opposes the osmotic movement of water is called the *osmotic pressure* of the solution.

The osmotic pressure of a solution is proportional to the *number of particles per unit volume of solvent,* not to the type, valence, or weight of the particles. The unit of measurement of osmotic pressure is the osmole. One osmole (osmol) is defined as one gram molecular weight (1 mol) of any nondissociable substance (such as glucose) and contains 6.02×10^{23} particles. In the relatively dilute fluids in the body, the osmotic pressure is measured in milliosmoles (one-thousandth of an osmole) per kilogram of water (mosmol/kg). Since most solutes are measured in the laboratory in units of millimoles per liter, milligrams per deciliter, or milliequivalents per liter, the following formulas must be used to convert into mosmol/kg:

$$\text{mosmol/kg} = n \times \text{mmol/L} \qquad (1\text{-}4)$$

or from Eqs. (1-1) and (1-2),

$$\text{mosmol/kg} = n \times \frac{\text{mg/dL} \times 10}{\text{mol wt}} \qquad (1\text{-}5)$$

$$= n \times \frac{\text{meq/L}}{\text{valence}} \qquad (1\text{-}6)$$

where n is the number of dissociable particles per molecule. When $n = 1$, as for Na^+, Cl^-, Ca^{2+}, urea, and glucose, 1 mmol/L will generate a potential osmotic pressure of 1 mosmol/kg. However, if a compound dissociates into two or more particles, 1 mmol/L will generate an osmotic pressure greater than 1 mosmol/kg. For example, at the concentrations present in the body, approximately 75 percent of NaCl dissociates into Na^+ and Cl^-. Thus, for each 1 mmol/L of NaCl, there will be 0.75 mmol/L each of Na^+ and Cl^- and 0.25 mmol/L of NaCl or 1.75 mosmol/kg (Table 1-4) [12].

In addition to the number of particles, the osmotic pressure is proportional to the temperature of the solution. These relationships can be expressed by van't Hoff's law (which is similar to Boyle's law for gases):

$$\text{Osmotic pressure} = nCRT \text{ (in atmospheres)} \qquad (1\text{-}7)$$

Table 1-4 Relationship between various units of measurement†

Substance	Atomic or molecular weight, mg	mmol	meq	mosmol
Na$^+$	23	1	1	1
Cl$^-$	35.5	1	1	1
NaCl	58.5	1	2	1.75‡
Ca^{2+}	40.0	1	2	1
CaCl$_2$	111.0	1	4 (Ca^{2+}, 2Cl$^-$)	~3‡ (Ca, 2Cl)
Glucose	180.0	1	...	1
Urea	60.0	1	...	1

†Adapted from R. Hays, *Clinical Disorders of Fluid and Electrolyte Metabo-lism*, 3d ed., M. H. Maxwell and C. R. Kleeman (eds.), McGraw-Hill, New York, 1980.

‡The calculation of milliosmoles for NaCl and CaCl$_2$ reflects the incomplete dis-sociation of these compounds in the body fluids.

where C is the total solute concentration in moles per unit volume of solvent; n is the number of dissociable particles per molecule; R is a constant with the same value as the gas constant per mole (0.082); and T is the absolute temperature (in kelvins). At body temperature (310 K), the osmotic pressure of 1 mosmol of solute (that is, 1 mmol of any solute where $n = 1$) in 1 kg of water is

$$\text{Osmotic pressure} = 0.001 \times 0.082 \times 310 = 2.54 \times 10^{-2} \frac{\text{atm}}{\text{mosmol/kg}}$$

Since there is 760 mmHg/atm at sea level,

$$\text{Osmotic pressure} = 2.54 \times 10^{-2} \times 760 = 19.3 \frac{\text{mmHg}}{\text{mosmol/kg}}$$

Let us now consider what would happen in a beaker, similar to that in Fig. 1-6, if a solute which was able to cross the membrane were added to one compartment (Fig. 1-7). For example, urea is lipid-soluble and able to cross most cell membranes. If urea is added to the fluid on one side of the membrane, it will move down a con-centration gradient into the solute-free compartment. The equilibrium state will be

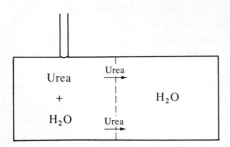

Figure 1-7 Effect of adding a permeable solute such as urea to the fluid on one side of a membrane. In this setting, equilibrium is reached by urea equilibration across the membrane rather than water movement into the urea compartment. Consequently, no osmotic pressure is generated.

characterized by equal urea concentrations in each compartment and not by water movement into the urea compartment. As a result, no osmotic pressure will be generated at equilibrium. Thus, Eq. (1-7) must be modified to take into account the degree to which the solute crosses the membrane, i.e., its permeability:

$$\text{Osmotic pressure} = a(nCRT) \tag{1-8}$$

where a refers to the permeability coefficient of the solute. If the membrane is impermeable to the solute, e.g., glucose, or the solute cannot equilibrate across the membrane for some other reason, then a equals 1. For example, the cell membrane is permeable to Na^+ but the cell Na^+ concentration is maintained at low levels by the active transport of Na^+ out of the cell (Fig. 1-4). Thus, the membrane is effectively, although not actually, impermeable to Na^+. In these settings, an osmotic pressure is generated, and these solutes are called *effective osmoles*. If the solute is able to cross the membrane and reach equal concentrations in both compartments, e.g., urea, then a equals 0. Since no pressure is generated, urea is an *ineffective osmole*. If several solutes are present, some of which can cross the membrane and some of which cannot, then a will have a net value between 0 and 1.

In the laboratory, the osmotic concentration of a solution is measured not as an osmotic pressure but according to other properties of solutes such as their ability to depress the freezing point or the vapor pressure of water. Solute-free water freezes at $0°C$. If 1 osmol of any solute (or combination of solutes) is added to 1 kg of water, the freezing point of this water will be depressed by $1.86°C$. For example, the freezing point of the plasma water is normally about $-0.521°C$. This represents an osmolality of 0.280 osmol/kg (0.521/1.86) or 280 mosmol/kg. However, the experiment with urea shows that *there is an important difference between the osmotic pressure of a solution and its osmolality a measured by the freezing-point depression.* When the freezing-point depression is measured, all solutes contribute in relation to their concentration, including those that are ineffective osmoles. In contrast, ineffective osmoles do not contribute to the osmotic pressure of the solution. For example, the plasma osmolality of 280 mosmol/kg represents a potential osmotic pressure of 5404 mmHg (280 × 19.3). However, almost all the plasma solutes (primarily Na^+ salts) are ineffective since Na^+ salts are able to cross the capillary wall separating the plasma from the interstitial fluid. The net effective plasma osmolality is only 1.3 mosmol/kg (generating an osmotic pressure of 25 mmHg) and is due to the plasma proteins which are restricted to the vascular space (see below).

Osmolality, osmolarity, and specific gravity The terms osmolality and osmolarity frequently are confused. *Osmolality* refers to the number of osmoles per kilogram of water. As a result, the total volume will be one liter of water plus the relatively small volume occupied by the solutes. *Osmolarity* refers to the number of osmoles per liter of solution. In this situation, the volume of water is less than one liter by an amount equal to the solute volume. Thus, osmolality is measured in milliosmoles per kilogram of water (abbreviated mosmol/kg) and osmolarity in milliosmoles per liter (mosmol/ L). In practice, this difference is negligible because of the low solute concentrations in the body fluids.

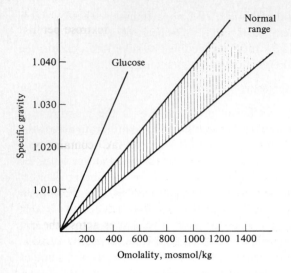

Figure 1-8 Relationship between the specific gravity and osmolality of the urine from normal subjects who have neither glucose nor protein in the urine. For comparison, the relationship between the specific gravity and osmolality for glucose solutions is included. *(Adapted from B. Miles, A. Paton, and H. deWardener, Br. Med. J., 2:904, 1954. By permission of the British Medical Journal.)*

The solute concentration of a solution also can be estimated by measuring its *specific gravity* which is defined as the weight of the solution compared with that of an equal volume of distilled water. For example, plasma is approximately 0.8 to 1.0 percent heavier than water and therefore has a specific gravity of 1.008 to 1.010. Since the specific gravity is proportional to the weight, as well as to the number, of particles in the solution, its relationship to the osmolality of the solution is dependent upon the molecular weights of the solutes. As illustrated in Fig. 1-8, in a normal urine whose main solutes are urea, Na^+, Cl^-, K^+, NH_4^+, and $H_2PO_4^-$, the specific gravity varies in a predictable way with the osmolality. However, if larger molecules, such as glucose, are present in high concentrations, there will be a disproportionate increase in the specific gravity as compared with the osmolality. An interesting clinical example occurs after the administration of radiographic dyes (molecular weight approximately 550). As the dye is excreted in the urine, the urine specific gravity may exceed 1.040 while the urine osmolality is elevated to a much lesser degree. The antibiotic carbenicillin, which is frequently given in doses of 24 to 36 g/day, can also produce this effect [13]. With these exceptions in mind, the specific gravity can be used to estimate the urine osmolality if the latter cannot be directly measured. The clinical importance of measuring the urine osmolality or specific gravity is summarized in Chap. 14.

Isotonic and isosmotic solutions An experiment similar to that in Fig. 1-6 can be produced if red blood cells are suspended in distilled water. Water will move into the solute-containing compartment (in this case, the erythrocytes), resulting in cell swelling and hemolysis. In contrast, if a solution is used with the same effective osmolality as that of the cell (280 mosmol/kg), there will be no osmotic gradient and no change in cell volume. Such a fluid which maintains the cell volume is called *isotonic*. In comparison, fluids which tend to cause cell swelling are called *hypotonic* and those which cause cell shrinkage are called *hypertonic* since they possess a higher osmotic pressure than the cell. For example, 5% dextrose in water (5 g of dextrose per 100 mL of water) is an isotonic solution. That it has the same osmolality as the cell, i.e., isosmotic with the cell, can be appreciated from Eq. (1-5):

$$\text{mosmol/kg} = \frac{5000 \text{ mg/dL} \times 10}{180} = 278$$

Similarly, 0.45% NaCl, 0.9% NaCl, and 3% NaCl represent hypotonic, isotonic, and hypertonic solutions, respectively. It should be noted that 0.9% NaCl contains Na^+ in a concentration of 154 meq/L, the same Na^+ concentration as in the plasma water (see Table 1-6).

Although isotonic fluids are isosmotic, the converse may not be true. For example, hemolysis occurs if erythrocytes are suspended in an isosmotic urea solution. This results from water following urea into the cell as urea equilibrates across the cell membrane (in contrast to dextrose and NaCl, which do not). Thus, a urea solution is physiologically equivalent to distilled water.

The clinical implications of these experiments are extremely important. Although a variety of fluids can be safely infused intravenously, distilled water and pure urea solutions should never be used since the ensuing hemolysis can result in kidney failure and death. The infused fluids do not have to be isotonic (in the appropriate setting, hypotonic or hypertonic solutions may be indicated), but they must contain some osmotically active solute to prevent cell lysis.

Summary Solutes which are unable to cross a membrane, i.e., effective osmoles, generate an osmotic pressure tending to pull water into the compartment containing the solutes. In terms of osmolality, *water moves from an area of relatively low osmolality to one of higher osmolality*. Although this sounds as if water is moving against a concentration gradient, it must be remembered that the activity (or random motion) of water is *inversely* proportional to solute concentration. Thus, water moves appropriately from an area of relatively high activity to one of lower activity.

DISTRIBUTION AND COMPOSITION OF BODY FLUIDS

The total body water (TBW) accounts for 45 to 50 percent of the body weight in adult females and 55 to 60 percent of the body weight in adult males. Approximately 50 percent of this water is in muscle, 20 percent in the skin, 10 percent in the blood, and the remainder in the other organs. The remaining 40 to 55 percent of the body weight comprises solids (particularly the skeleton) and adipose tissue. The individual and sexual variability in fractional water content is largely due to variations in the percentage of the body weight comprising bone or adipose tissue. Thus, the lower water content in females is primarily due to their increased percentage of adipose tissue relative to males.

The TBW is contained in two major compartments which are divided by the cell membrane: the *cell water* and the *extracellular water* (Fig. 1-9). The extracellular

Total body water $\cong$ 60 percent body weight

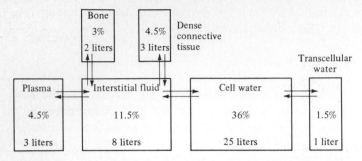

Figure 1-9 Distribution of total body water in an average 70-kg male. The percentages refer to the fraction of the body weight contained in each compartment. (*Adapted from I. Edelman and J. Leibman, Am. J. Med., 27:256, 1959, and D. M. Woodbury, in Physiology and Biophysics, T. C. Ruch and H. D. Patton (eds.), Saunders, Philadelphia, 1974.*)

water is subdivided into several compartments, the largest of which are the *interstitial-lymph water,* constituting the extracellular environment in which the cells function, and the rapidly circulating *plasma water.* These fluids are in equilibrium across the capillary wall. In addition, there are three smaller extracellular fluid compartments. The first is the water found in dense connective tissues such as cartilage and tendons. The second consists of water bound in bone matrix which is not much altered under physiologic conditions. The third is called the *transcellular water* and is composed of epithelial secretions such as the digestive secretions, sweat, and cerebrospinal, pleural, synovial, and intraocular fluids.

Measurement of Body Water

Since the volume of the different water compartments cannot be measured directly, indirect methods must be utilized. These methods, called dilution techniques, are based on the relationship between the concentration of a substance in solution, its volume of distribution, and the quantity of the substance present:

$$\text{Quantity} = \text{volume} \times \text{concentration} \qquad (1\text{-}9)$$

This relationship assumes that the concentration is uniform throughout the volume of distribution. For example, if 5 g of glucose is added to a beaker with an unknown volume of distilled water and the glucose concentration after mixing is 1 g/dL (or 10 g/L), then

$$\text{Volume} = \frac{5 \text{ g}}{10 \text{ g/L}} = 0.5 \text{ liter}$$

The same principle can be used to measure the TBW. A substance is required that will distribute rapidly and uniformly through the same volume of distribution as water. In humans two isotopes of water, deuterium oxide (D_2O) and tritiated water, and the drug antipyrine have been used [14,15]. Equation (1-9) must be altered slightly to account for excretion of the indicator:

$$\text{Volume of distribution} = \frac{\text{quantity infused} - \text{quantity excreted}}{\text{concentration}} \quad (1\text{-}10)$$

For example, suppose 100 mL of D_2O as an isotonic NaCl solution is infused into a 70-kg male. If, after an equilibratory period of 2 h, the plasma D_2O concentration is 0.0025 mL/mL and 0.5 mL is lost in the urine and by vaporization from the skin and respiratory tract, then

$$\text{Volume of distribution} = \frac{100 \text{ mL} - 0.5 \text{ mL}}{0.0025 \text{ mL/mL}} = 39{,}800 \text{ mL}$$

Thus, the TBW is approximately 39.8 liters or 57 percent of the body weight.

Similar methods have been applied to the measurement of the volumes of the extracellular fluid and the plasma [14,15]. An indicator is required that is confined to the compartment to be measured and that distributes evenly through that compartment.† Sulfate and nonmetabolizable sugars such as inulin and mannitol have been used to measure the extracellular volume; radioiodinated albumin and Evans blue (a dye which binds to albumin) have been used to measure the plasma volume. Since the cell water is not accessible for analysis, the volume of the cell water must be estimated from the TBW minus the extracellular fluid volume. As might be expected, there are significant differences between individuals according to age, weight, and sex as well as variations with time in a given individual. Figure 1-9 summarizes the water distribution that might be found in a normal 70-kg male. The TBW is approximately 42 liters (60 percent of body weight), of which 25 liters (60 percent of the TBW) is in the cells and 17 liters (40 percent of the TBW) is in the extracellular fluid. The extracellular water is divided between the interstitial-lymph (8 liters), plasma (3 liters), dense connective tissue (3 liters), bone (2 liters), and transcellular (1 liter) compartments.

Measurement of Body Solutes

Ions constitute roughly 95 percent of the solutes in the body water. Although the total quantity of an ion in the body can be determined only by the analysis of cadavers, the exchangeable (or available) pool of the ion can be measured in a living subject by using the isotope dilution techniques described above (e.g., with ^{42}K). The results of the studies for Na^+ and K^+, the predominant cations in the body water, are shown in Table 1-5. Several points deserve emphasis. First, there is a marked difference in distribution, Na^+ being primarily in the extracellular fluid whereas K^+

---

†There are certain limitations to the measurement of the extracellular volume since the markers that are used may enter cells to a slight degree, may be secreted in the gastrointestinal tract, or may be unable to enter some of the extracellular areas in dense connective tissue and bone. Thus, the results vary to a moderate degree, depending upon the indicator used. The values presented here represent mean values from different studies.

Table 1-5 Distribution of Na^+ and K^+ in humans†

Compartment	Na^+	K^+
	meq/kg body weight	
Total body	58.0	53.8
Extracellular	52.8	2.5
Intracellular	5.2	51.3
Total exchangeable	41.0	52.8

†Data from D. M. Woodbury, in *Physiology and Biophysics,* 20th ed., T. C. Ruch and H. D. Patton (eds.), Saunders, Philadelphia, 1974.

is almost entirely in the cells. Second, essentially all the body K^+ is in the exchangeable pool. In contrast, only 70 percent of the body Na^+ is exchangeable, the remainder being bound in bone. This difference is not limited to Na^+. Almost all the Ca^{2+} (in bone) and most of the Mg^{2+} (in bone and in cells) in the body are nonexchangeable. The distinction between exchangeable and nonexchangeable solutes is physiologically important since only the exchangeable solutes are osmotically active (see below).

Although hydrogen ions (H^+) play an essential role in the regulation of endogenous metabolism, they are not included in this discussion since they are present in very low concentrations, being measured in nanoequivalents per liter, that is, 10^{-6} meq/L. The physiologic importance of H^+ will be considered separately in Chap. 11.

Ionic Composition of Extracellular Fluid

Plasma water In the extracellular fluid, Na^+ is quantitatively the only important cation, with Cl^- and HCO_3^- being the major anions (Table 1-6). In addition, the plasma proteins, which are essentially limited to the vascular space (see below), constitute an important fraction of the plasma anions.

The ionic composition of the plasma is measured in the laboratory as milliequivalents per liter of plasma. However, this is not an accurate measure of the composition of the plasma water since only about 930 mL of each liter of plasma is water [16,17]. The remaining 70 mL is occupied by the plasma proteins and to a lesser degree lipids. Thus, to convert from milliequivalents per liter of plasma to the *physiologic* measurement of milliequivalents per liter of plasma water (the electrolytes being only in the aqueous phase), the concentrations measured in the laboratory must be divided by 0.93. For example,

$$[Na^+] = 142 \text{ meq/L plasma} \div 0.93 = 153 \text{ meq/L plasma water}$$

$$[Cl^-] = 104 \text{ meq/L plasma} \div 0.93 = 112 \text{ meq/L plasma water}$$

In the presence of hyperlipemia or hyperproteinemia, the plasma water content may be less than 93 percent. In these disorders, the plasma Na^+ concentration as measured in the laboratory may be reduced since there is less water and consequently

Table 1-6 Ionic composition of the body water compartments†

Ion	Plasma, meq/L	Plasma water,‡ meq/L	Interstitial fluid,§ meq/L	Skeletal muscle cell, meq/L
Cations				
Na^+	142.0	152.7	145.1	12.0
K^+	4.3	4.6	4.4	150.0
Ca^{2+} (ionized)	2.5	2.7	2.4	4.0
Mg^{2+} (ionized)	1.1	1.2	1.1	34.0
Total	149.9	161.2	153.0	200.0
Anions				
Cl^-	104.0	111.9	117.4	4.0
HCO_3^-	24.0	25.8	27.1	12.0
$HPO_4^{2-}, H_2PO_4^-$	2.0	2.2	2.3	40.0
Proteins	14.0	15.0	0.0	54.0
Other	5.9	6.3	6.2	90.0††
Total	149.9	161.2	153.0	200.0

†Adapted from D. M. Woodbury, in *Physiology and Biophysics,* 20th ed., T. C. Ruch and H. D. Patton (eds.), Saunders, Philadelphia, 1974.

‡Plasma water content assumed to be 93 percent of plasma volume.

§Gibbs-Donnan factors (see below) used as multipliers are 0.95 for univalent cations, 0.90 for divalent cations, 1.05 for univalent anions, and 1.10 for divalent anions.

††This largely represents organic phosphates such as ATP.

less Na^+ in each liter of plasma. For example, if the plasma water Na^+ concentration is normal at 153 meq/L but the water content of plasma is only 80 percent, the measured plasma Na^+ concentration will be 122 meq/L (153 × 0.80) (see Chap. 23).

Interstitial fluid The ionic concentrations in the interstitial fluid are different from those in the plasma, owing to the effects of the *Gibbs-Donnan equilibrium.* In the section on osmotic pressure, the effect of solute concentration on water movement across a membrane permeable to water but not to solutes was described. However, the capillary wall is different in that it is permeable to all the ions in the plasma with the exception of the plasma protein anions. A simplified example of the effects of a membrane permeable to Na^+ and Cl^- but not proteins is illustrated in Fig. 1-10. On one side of the membrane (compartment 1) are nine Na^+ and nine Cl^- ions; on the

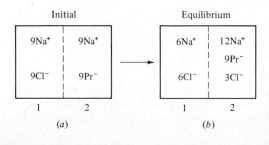

Figure 1-10 Effect of nondiffusible protein anions (Pr^-) on the distribution of diffusible ions across the membrane. The equilibrium state in (b) is called the Gibbs-Donnan equilibrium.

other side (compartment 2) are nine Na^+ and nine Pr^- ions. As described above, ions can move passively across a membrane down concentration or electrical gradients. Since the membrane is permeable to Cl^-, this ion will move across the membrane from compartment 1 to compartment 2 down a concentration gradient, with Na^+ following to maintain electroneutrality. However, this will increase the number of Na^+ ions in compartment 2, creating a gradient for the movement of Na^+ back into compartment 1, a gradient opposite in direction from that of Cl^-. How these ions are distributed at equilibrium, called the Gibbs-Donnan equilibrium, can be expressed by the following equation:

$$[Na^+]_1 \times [Cl^-]_1 = [Na^+]_2 \times [Cl^-]_2 \qquad (1\text{-}11)$$

(A simple derivation of this equation, which holds true for any pair of univalent cations and anions that can diffuse freely across the membrane, is presented in the Appendix at the end of this chapter.)

In this example, if the number of Na^+ and Cl^- ions moving from compartment 1 to compartment 2 is each equal to x, then $[Na^+]_1 = [Cl^-]_1 = 9 - x$; $[Na^+]_2 = 9 + x$; and $[Cl^-]_2 = x$. Thus,

$$(9 - x)(9 - x) = (9 + x)(x)$$

$$81 - 18x + x^2 = 9x + x^2$$

$$x = 3$$

Consequently, the equilibrium state will be characterized by six Na^+ and six Cl^- ions in compartment 1 and twelve Na^+, three Cl^-, and nine Pr^- ions in compartment 2.

Three important points about this equilibrium state should be noted. First, electroneutrality is maintained, as the number of cations is equal to the number of anions in each compartment. Second, because of the presence of nondiffusible anions in compartment 2, there is an unequal distribution of diffusible ions as the cation concentration is higher and the diffusible anion concentration lower in this compartment. Third, the osmolality generated by the diffusible ions is *always less* in the protein-free compartment than in the protein-containing compartment. If one takes all numerical combinations whose product is a given number (36 in this example), the sum of the two numbers in the combination will be *least* when the square root is used. In terms of Fig. 1-10, since $[Na^+]_1$ and $[Cl^-]_1$ are the same (6 is the square root of 36), the number of Na^+ and Cl^- particles in the protein-free compartment $(6 + 6)$ is less and therefore the osmolality is less than that in the protein-containing compartment $(12 + 3)$.

The physiologic equivalents of this experiment occur across the capillary wall separating the plasma, with its protein anions, from the interstitial fluid and across the glomerular capillary wall in the kidney separating the plasma from the glomerular filtrate (see Chap. 3). From the Gibbs-Donnan equilibrium, the cation concentrations in the relatively protein-free interstitial fluid (and glomerular filtrate) are lower than those in the plasma water. At the electrolyte and protein concentrations present in the plasma, the conversion factor is 0.95 for univalent cations and 0.90

Table 1-7 Mean electrolyte content of the transcellular fluids†

Fluid	Na$^+$, meq/L	K$^+$, meq/L	Cl$^-$, meq/L	HCO$_3^-$, meq/L
Saliva	33	20	34	0
Gastric juice‡	60	9	84	0
Bile	149	5	101	45
Pancreatic juice	141	5	77	92
Ileal fluid	129	11	116	29
Cecal fluid	80	21	48	22
Cerebrospinal fluid	141	3	127	23
Sweat	45	5	58	0

†Adapted from A. Arieff, in *Clinical Disorders of Fluid and Electrolyte Metabolism,* 2d ed., M. H. Maxwell and C. R. Kleeman (eds.), McGraw-Hill, New York, 1972.

‡The Cl$^-$ concentration exceeds the Na$^+$ + K$^+$ concentration by 15 meq/L in gastric juice. This largely represents the secretion of H$^+$ by the parietal cells.

for the ionized fractions, i.e., nonbound, of the divalent cations, Ca^{2+} and Mg^{2+} (see Appendix at the end of this chapter). On the other hand, the anion concentrations are higher than those in the plasma by a factor of 1.05 for univalent anions and 1.10 for divalent anions. Thus (Table 1-6),

$$\text{Interstitial [Na}^+] = 0.95 \times \text{plasma water [Na}^+] = 145 \text{ meq/L}$$

$$[\text{Cl}^-] = 1.05 \times \text{plasma water [Cl}^-] = 117 \text{ meq/L}$$

In addition to these changes in ion concentration, the osmotic pressure of plasma water is greater than that of the interstitial fluid. This is due to two factors: the osmotic effect of the plasma proteins and the increased number of ions in the plasma due to the Gibbs-Donnan effect. The physiologic importance of this increase in osmotic pressure in holding water within the vascular space will be discussed below.

Transcellular fluid The transcellular fluids largely represent secretions from epithelial cells and have different ionic compositions from the plasma and interstitial fluids (Table 1-7). It should be noted that the concentrations in Table 1-7 are only mean values and that there may be wide variations within a given individual with time. Although only a small volume of water is present in the transcellular space at any one time, a much larger volume is secreted into this space each day. This is particularly true of the gastrointestinal secretions which average 3 to 6 L/day and are then reabsorbed. In some disease states, e.g., vomiting and diarrhea, the loss of these transcellular secretions from the body can produce serious abnormalities in fluid and electrolyte homeostasis.

Ionic Composition of Intracellular Fluid

The volume and composition of the cellular fluid from a particular organ must be measured indirectly from the difference between the total tissue and extracellular

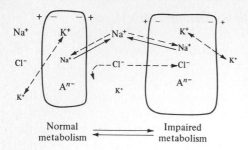

Normal metabolism ⟶ Impaired metabolism

Figure 1-11 Schematic representation of a normal cell, on the left, and a swollen cell, on the right, resulting from temporary inhibition of energy metabolism. The changes are depicted as reversible. A^{n-} refers to the nondiffusible anions in the cell. *(From A. Leaf, Am. J. Med., 49:291, 1970.)*

quantities of water and ions. The composition of a typical skeletal muscle cell is shown in Table 1-6. Cells from other organs have some differences in composition, e.g., different functioning proteins, but the basic pattern is similar. It can be seen that there are major differences between the cellular and extracellular fluids. In contrast to the extracellular fluid, the principal intracellular electrolytes are K^+, Mg^{2+}, organic phosphates, e.g., adenosine triphosphate, and proteins. The high cell K^+ concentration and low Na^+ concentration are maintained not by an impermeability of the cell membrane to these ions but by the active Na^+-K^+ pump in the membrane which transports Na^+ out of, and K^+ into, the cell.

In addition to maintaining intracellular electrolyte concentrations, the Na^+-K^+ pump plays an important role in Na^+ and K^+ handling by the kidney (see Chaps. 5 and 13) and in the regulation of the cell volume. The high concentration of nondiffusible anions in the cell sets up a Gibbs-Donnan equilibrium across the cell membrane. As a result, there tend to be more particles and a higher osmotic pressure within the cells than in the interstitial fluid. This should pull water into the cell and produce cell swelling and eventual cell dysfunction. However, this sequence is prevented by the active transport of Na^+ out of the cell. If the pump is inhibited, Na^+ will enter the cell down a concentration gradient. To maintain electroneutrality, either Cl^- will follow Na^+ into the cell or K^+ will leave the cell. To the extent that the former occurs, there will be an increase in cell osmolality, resulting in cell swelling (Fig. 1-11). The net effect is an increase in the cell Na^+ and Cl^- concentrations and volume and a decrease in the cell K^+ concentration [18].

The ionic concentrations of the intracellular fluid listed in Table 1-6 are mean values, and it should not be assumed that the distribution of these ions within the cell is uniform. The cell is a heterogeneous structure containing several different functioning units, e.g., the mitochondria, the endoplasmic reticulum, and the nucleus, and recent experimental data indicate that the compositions of these organelles differ from one another. For example, the H^+ concentration within the cell compartments may vary as much as twofold (see Chap. 11) [19].

EXCHANGE OF WATER BETWEEN CELLULAR AND EXTRACELLULAR FLUIDS

Osmotic forces are the prime determinant of water distribution in the body. Since water can freely cross almost all cell membranes, the body fluids are in osmotic equilibrium as the osmolalities of the intracellular and extracellular fluids are the same [20]. Consequently, the distribution of the total body water between the cells and the extracellular fluid is determined by the number of osmotically active particles in each compartment.† Na^+ salts are the principal extracellular osmoles and act to hold water in the extracellular space. Conversely, K^+ salts account for almost all the intracellular osmoles (most of the cell Mg^{2+} being bound and osmotically inactive) and act to hold water within the cells. Although the cell membrane is permeable to Na^+ and K^+, these ions are able to act as effective osmoles because they are restricted to their respective compartments by the Na^+-K^+ pump in the cell membrane. As a result, the volumes of the extracellular and intracellular fluids are determined by the volume of the TBW and by the ratio of the exchangeable Na^+ to the exchangeable K^+. In contrast to Na^+ and K^+, urea is an ineffective osmole which distributes evenly through both compartments and therefore does not affect the distribution of water.

Under normal circumstances, the water and electrolyte content in the body is maintained within relatively narrow limits as variations in dietary intake are matched by appropriate variations in urinary excretion (see Chaps. 9 and 10). Nevertheless, it is important to understand the potential physiologic effects of changes in solute or water balance since these disturbances can occur in the clinical setting. If the osmolality of one fluid compartment is changed, water will move across the cell membrane to reestablish osmotic equilibrium. How this affects water distribution and solute concentrations can be appreciated from the following examples (Fig. 1-12). For the sake of simplicity, let us assume that the osmolality of the body fluids is 280 mosmol/kg and is due entirely to 140 meq/L of Na^+ salts in the extracellular fluid and to 140 meq/L of K^+ salts in the cells, i.e., we are assuming that Na^+ and K^+ salts dissociate completely into cations and anions. As depicted in Fig. 1-12*a*, an average 70-kg male might have a TBW of 42 liters (or 42 kg) of which 25 liters is intracellular and 17 liters is extracellular. What will happen if 420 meq of NaCl (420 mosmol) without water is added to the extracellular fluid (Fig. 1-12*b*)? Since the NaCl remains extracellular, there will be an increase in the extracellular fluid osmolality, resulting in water movement out of the cells down an osmotic gradient. The following sequence can be used to calculate the characteristics of the new equilibrium state:

1. Initial total body solute = 280 mosmol/kg $\times$ 42 kg = 11,760 mosmol
2. Initial extracellular solute = 280 mosmol/kg $\times$ 17 kg = 4760 mosmol
3. New total body solute = 11,760 + 420 = 12,180 mosmol
4. New osmolality of body fluids = 12,180 mosmol $\div$ 42 kg = 290 mosmol/kg
5. New extracellular solute = 4760 + 420 = 5180 mosmol
6. New extracellular volume = 5180 mosmol $\div$ 290 mosmol/kg = 17.9 kg
7. New intracellular volume = 42 $-$ 17.9 = 24.1 kg
8. New extracellular $[Na^+]$ = osmolality/2 = 145 meq/L

---

†Although there are more ions in the cells than in the extracellular fluid (Table 1-6), there is no difference in osmolality since many of the cell ions are bound, e.g., to proteins, and are osmotically inactive.

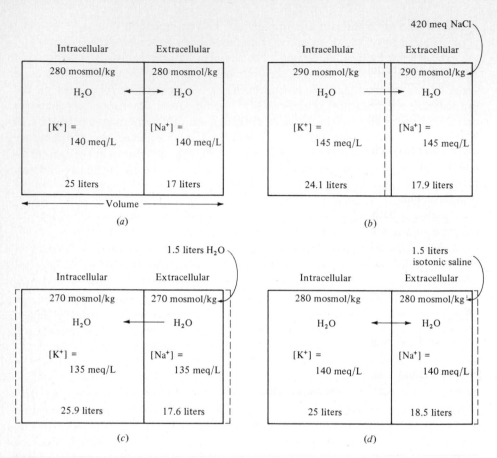

Figure 1-12 Osmolality of the body fluids and the distribution of the total body water between the intracellular fluid and the extracellular fluid in the control state (*a*) and after the addition of NaCl (*b*). H_2O (*c*), or isotonic NaCl and H_2O (*d*) to the extracellular fluid. For simplicity, it is assumed that Na^+ salts account for all the extracellular osmoles and K^+ salts for all the intracellular osmoles. See text for details. (*Adapted from L. E. Earley, Clinical Disorders of Fluid and Electrolyte Metabolism, 2d ed., M. H. Maxwell and C. R. Kleeman (eds.), McGraw-Hill, New York, 1972.*)

Thus, increasing the quantity of extracellular solute results in the movement of 900 mL of water from the cells into the extracellular fluid. *The net effect is an increase in the osmolality of both compartments even though the added solute is restricted to the extracellular space.*

A different sequence occurs if 1.5 liters of solute-free water is added to the extracellular fluid, e.g., by ingestion. This reduces the extracellular fluid osmolality, creating an osmotic gradient favoring the entry of water into the cells (Fig. 1-12*c*). To calculate the new steady state, steps 1 and 2 are similar to those above:

1. Initial total body solute = 11,760 mosmol
2. Initial extracellular solute = 4760 mosmol

3. Initial intracellular solute = 11,760 − 4760 = 7000 mosmol
4. New total body water = 42 + 1.5 = 43.5 kg
5. New osmolality of body fluids = 11,760 mosmol ÷ 43.5 kg = 270 mosmol/kg
6. New extracellular volume = 4760 mosmol ÷ 270 mosmol/kg = 17.6 kg
7. New intracellular volume = 7000 mosmol ÷ 270 mosmol/kg = 25.9 kg
8. Ratio of intracellular volume to TBW = 25.9 ÷ 43.5 = 60%
9. New extracellular $[Na^+]$ = osmolality/2 = 135 meq/L

Since there is no change in the ratio of intracellular to extracellular solute, the fractional composition of the TBW is unchanged (cell water is 60 percent of TBW). However, the TBW is increased, resulting in expansion and dilution of both compartments.

Finally, if both NaCl and water are given as 1.5 liters of isotonic NaCl, there will be no change in osmolality and consequently no water movement across the cell membrane (Fig. 1-12*d*). Since the administered NaCl will remain in the extracellular space, the only effect will be a 1.5-liter increase in the extracellular fluid volume.

These experiments illustrate an important and often misunderstood concept, that *the plasma Na$^+$ concentration is a measure of concentration and not of volume.* In each instance, the extracellular fluid volume is increased, because of an increase in either the TBW and/or the total exchangeable Na$^+$, yet the plasma Na$^+$ concentration is, respectively, increased, decreased, and unchanged. This occurs because the plasma sodium concentration reflects the *ratio* of the amounts of solute and water present *not* the absolute amount of either solute or water. Thus, *there is no necessary correlation between the plasma Na$^+$ concentration and the extracellular fluid volume.* For example, these parameters change in a parallel direction when Na$^+$ is administered (Fig. 1-12*b*) but in an opposite direction (low plasma Na$^+$ concentration, high extracellular fluid volume) when water retention occurs (Fig. 1-12*c*).

Relation of Plasma Sodium Concentration to Osmolality

The osmolality of the plasma (P_{osm}) is equal to the sum of the osmolalities of the individual solutes in the plasma. Most of the plasma osmoles are Na$^+$ salts (Table 1-6), with lesser contributions from other ions, glucose, and urea. The osmotic effect of the plasma ions can be estimated from 2 × plasma Na$^+$ concentration. The validity of this approximation results from the interplay of several factors. First, NaCl dissociates into roughly 1.75 particles (0.75 Na$^+$, 0.75 Cl$^-$, and 0.25 NaCl) [12], so that the plasma Na$^+$ concentration must be multiplied by 1.75. Second, the plasma Na$^+$ concentration must be divided by 0.93 to arrive at the physiologically important plasma water Na$^+$ concentration (Na$^+$ being present only in the aqueous phase of plasma). Thus,

$$\text{Osmolality of Na}^+ \text{ salts} = \frac{1.75}{0.93} \times \text{plasma } [Na^+]$$

$$= 1.88 \times \text{plasma } [Na^+]$$

The remaining $0.12 \times$ plasma Na^+ concentration is equal to 17 mosmol/kg (0.12×140) which fortuitously is the approximate osmotic pressure generated by K^+, Ca^{2+}, and Mg^{2+} salts.

The osmotic contributions of glucose and urea, both of which are measured in milligrams per deciliter, can be calculated from Eq. (1-5). The molecular weights of glucose (mol wt 180) and the two nitrogen atoms in urea (mol wt 28, since urea is measured as blood urea nitrogen or BUN) are used as the basis for these calculations.

$$\text{mosmol/kg of glucose} = \frac{[\text{glucose}] \times 10}{180} = \frac{[\text{glucose}]}{18}$$

$$\text{mosmol/kg of urea nitrogen} = \frac{BUN \times 10}{28} = \frac{BUN}{2.8}$$

Thus, the P_{osm} can be estimated from

$$P_{osm} \cong 2 \times \text{plasma } [Na^+] + \frac{[\text{glucose}]}{18} + \frac{BUN}{2.8} \qquad (1\text{-}12)$$

The *effective plasma* (and extracellular fluid) *osmolality* is defined as those osmoles acting to hold water within the extracellular space. Since urea is an ineffective osmole.

$$\text{Effective } P_{osm} \cong 2 \times \text{plasma } [Na^+] + \frac{[\text{glucose}]}{18} \qquad (1\text{-}13)$$

The normal values for these parameters are

$$\text{Plasma } [Na^+] = 137\text{--}145 \text{ meq/L}$$

$$[\text{Glucose}] = 60\text{--}100 \text{ mg/dL, fasting}$$

$$BUN = 10\text{--}20 \text{ mg/dL}$$

$$P_{osm} = 275\text{--}290 \text{ mosmol/kg}$$

$$\text{Effective } P_{osm} = 270\text{--}285 \text{ mosmol/kg}$$

Under normal circumstances, glucose accounts for only 5 mosmol/kg, and Eq. (1-13) can be simplified to

$$\text{Effective } P_{osm} \cong 2 \times \text{plasma } [Na^+] \qquad (1\text{-}14)$$

Thus, in most conditions, the plasma Na^+ concentration is a reflection of the P_{osm}, a finding consistent with the fact that Na^+ salts are the principal extracellular osmoles.

Since the body fluids are in osmotic equilibrium,

$$\text{Effective } P_{osm} = \text{effective osmolality of total body water}$$

$$= \frac{\text{extracellular solute} + \text{intracellular solute}}{TBW}$$

As described above, exchangeable Na^+ (Na_e^+) salts are the primary effective extra-cellular solutes, and exchangeable K^+ (K_e^+) salts are the primary effective intracellular solutes. Therefore,

$$\text{Effective } P_{osm} \propto \frac{2 \times Na_e^+ + 2 \times K_e^+}{TBW} \qquad (1\text{-}15)$$

(The multiple 2 is used to account for the osmotic contribution of the anions accompanying Na^+ and K^+.) If we now combine Eqs. (1-14) and (1-15) [12],

$$\text{Plasma } [Na^+] \propto \frac{Na_e^+ + K_e^+}{TBW} \qquad (1\text{-}16)$$

As illustrated in Fig. 1-13, this relationship holds over a wide range of plasma Na^+ concentrations in humans. The importance of these variables on the plasma Na^+ concentration can be appreciated from the examples in Fig. 1-12. Increasing the Na_e^+ elevates the plasma Na^+ concentration (Fig. 1-12b); increasing the TBW decreases the plasma Na^+ concentration (Fig. 1-12c); and increasing the Na_e^+ and TBW proportionately has no effect on the plasma Na^+ concentration (Fig. 1-12d). The effect of K_e^+ is less apparent but can be important clinically [21,22]. For example, if K^+ is lost from the extracellular fluid because of diarrhea, there will be a fall in the plasma K^+ concentration. This will create a concentration gradient favoring the movement of K^+ from the cells into the extracellular fluid. To preserve electroneutrality, Na^+ will move into the cells, lowering the plasma Na^+ concentration.

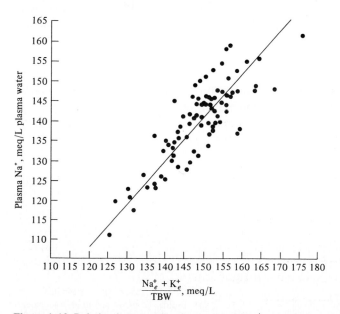

Figure 1-13 Relation between the plasma water Na^+ concentration and the ratio of $(Na_e^+ + K_e^+)/$ TBW. *(Adapted from I. Edelman, J. Leibman, M. O'Meara, and L. Birkenfeld, J. Clin. Invest., 37:1236, 1958, by copyright permission of The American Society for Clinical Investigation.)*

The clinical application of these concepts occurs in patients with hyponatremia (low plasma Na^+ concentration) or hypernatremia (high plasma Na^+ concentration) (see Chaps. 22 to 24). From these relationships, we can see that hyponatremia usually represents hypoosmolality and can be produced by Na^+ and K^+ loss or water retention. Conversely, hypernatremia represents hyperosmolality and can be produced by Na^+ gain or water loss. The toxicity of hyperkalemia (high plasma K^+ concentration) prevents the retention of enough K^+ to cause an important elevation in the plasma Na^+ concentration.

EXCHANGE OF WATER BETWEEN PLASMA AND INTERSTITIAL FLUID

The supply of nutrients to the cells and the removal of waste products from the cells occur across the capillary wall by the diffusion of solutes and gases (O_2 and CO_2) between the plasma and the interstitial fluid which constitutes the immediate extracellular environment of the cell. Equally important is the maintenance of a proper distribution of water between these compartments. Although osmotic forces contribute to the distribution of water across the capillary wall, the situation differs from that across the cell membrane. Since the capillary is permeable to Na^+ salts and glucose, there are few effective osmoles in the interstitial fluid (see below). In contrast, the plasma proteins move across the capillary wall only to a limited degree. Therefore, they are effective osmoles and act to pull water from the interstitial fluid into the vascular space. This osmotic pressure generated by the plasma proteins is called the *colloid osmotic pressure* or the *plasma oncotic pressure*. Fluid does not continuously move into the capillary because the oncotic pressure is balanced by the capillary hydrostatic pressure, generated by the propulsion of blood from the heart, which tends to push water out of the vessels into the interstitium. In addition, oncotic and hydrostatic pressures present in the interstitium contribute to the regulation of fluid exchange between the plasma and the interstitial fluid (Fig. 1-14).

The relationship between net filtration from the vascular space into the interstitium and the hydrostatic and oncotic pressures can be expressed by Starling's law [23–27]:

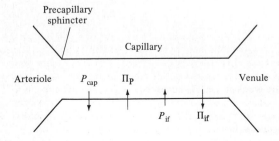

Precapillary sphincter

Capillary

Arteriole P_{cap} Π_P Venule

P_{if} Π_{if}

Figure 1-14 Schematic representation of the factors controlling fluid movement across the capillary wall between the plasma and the interstitial fluid.

Table 1-8 Postulated average values for Starling's forces in humans†

Hydrostatic pressures	
P_{cap} (mean)	17
P_{if}	−4
Mean gradient ($P_{cap} - P_{if}$)	21
Oncotic pressures	
Π_P	26
Π_{if}	5.3
Mean gradient ($\Pi_P - \Pi_{if}$)	20.7
Net gradient	+0.3
+ = filtration	
− = absorption	

†Units are millimeters of mercury (mmHg).

Net filtration = K_f (hydrostatic pressure gradient − oncotic pressure gradient)

$$= K_f[(P_{cap} - P_{if}) - (\Pi_P - \Pi_{if})] \qquad (1\text{-}17)$$

where K_f refers to the net permeability of the capillary wall, a function of the surface area available for filtration as well as the unit permeability (or porosity) of the membrane; P_{cap} and P_{if} are the capillary and interstitial fluid hydrostatic pressures; and Π_P and Π_{if} are the plasma and interstitial fluid oncotic pressures. The latter is derived from filtered plasma proteins and to a lesser degree mucopolysaccharides in the interstitium.

The normal values for these parameters in humans in the resting state is uncertain, largely because of difficulties in measurement of the capillary and interstitial hydrostatic pressures (Table 1-8)† [25–27]. For example, it has been suggested that the interstitial hydrostatic pressure is actually slightly negative, thereby promoting filtration [26,27]. This negative pressure may be generated by the removal of interstitial fluid by the lymphatic vessels. Regardless of the exact values, there is general agreement that the capillary hydrostatic pressure and plasma oncotic pressure are quantitatively the most important and that there is a small net gradient favoring filtration of 0.3 to 0.5 mmHg.

†These values are obtained primarily from peripheral capillaries (such as skeletal muscle) and do not necessarily apply to all visceral capillaries [28]. In the kidney, for example, the glomerular capillary hydrostatic pressure is substantially higher, resulting in an appropriately greater tendency to filtration (see Table 3-1). In the liver, on the other hand, the hepatic sinusoids are highly permeable to proteins, thereby tending to abolish the oncotic pressure gradient, i.e., Π_{if} almost equals Π_P. The net effect is a relatively high gradient favoring filtration as the hydrostatic pressure gradient is unopposed. This occurs even though the capillary hydrostatic pressure is lower than that in skeletal muscle since a substantial part of hepatic blood flow comes from a low pressure system, the portal vein.

Plasma Oncotic Pressure

From van't Hoff's law [Eq. (1-8)],

$$\text{Osmotic pressure} = a(nCRT)$$

we would expect the osmotic pressure generated by a solute to be a linear function of solute concentration since a, n, R, and T are all constants. Although this holds true for most solutes, it does not for the plasma proteins. As depicted in Fig. 1-15, the oncotic pressure generated by the plasma proteins is greater than that predicted on the basis of protein concentration from van't Hoff's law. This difference is due in part to the Gibbs-Donnan equilibrium since more particles are present in the protein-containing compartment. For example, let us return to Fig. 1-10 and assume that Na^+ and Cl^- are the only diffusible ions. If compartment 1 represents the interstitium, then $[Na^+]_{if} = [Cl^-]_{if} = 145$ meq/L (Table 1-6). In compartment 2, which now represents the plasma water, the plasma Na^+ concentration will exceed the plasma Cl^- concentration by 15 meq/L which is the charge on the plasma proteins (Table 1-6). Thus,

$$[Na^+]_{if} \times [Cl^-]_{if} = [Na^+]_P \times [Cl^-]_P$$

$$145 \times 145 = ([Cl^-]_P + 15) \times [Cl^-]_P$$

$$[Cl^-]_P = 137.7 \text{ meq/L}$$

$$[Na^+]_P = 152.7 \text{ meq/L}$$

As a result, the total number of milliequivalents per liter in the plasma water (137.7 + 152.7 = 290.4) exceeds that in the interstitial fluid (145 + 145 = 290) by 0.4 meq/L or 0.4 mmol/L. Although this difference appears small, it should be remembered that the plasma protein concentration is only 0.9 mmol/L (Table 1-2). Consequently, since proteins are nondissociable, the net osmotic effect of the plasma pro-

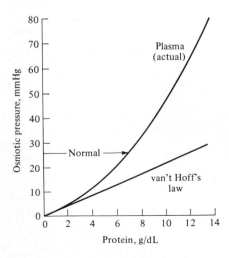

Figure 1-15 Relationship between protein concentration and osmotic pressure as predicted from van't Hoff's law and as actually occurs with plasma. (*Adapted from E. Landis and J. R. Pappenheimer, in Handbook of Physiology, sec 2, Circulation, vol. II, W. F. Hamilton and P. Dow (eds.), American Physiological Society, Washington, D.C., 1963.*)

teins is increased from 0.9 mosmol/kg (0.9 mmol/L equals 0.9 mosmol/kg) to 1.3 mosmol/kg by the Gibbs-Donnan effect. Since 1 mosmol/kg generates an osmotic pressure of 19.3 mmHg, this effect increases the capillary oncotic pressure from 17.4 mmHg (0.9 × 19.3) to 25 mmHg (1.3 × 19.3).†

As described above, the concentration of diffusible ions in the interstitial fluid differs from that in the plasma because of the effects of the negative charges on the plasma proteins. At the electrolyte and protein concentrations normally present in the plasma, the ratio of the interstitial to plasma concentrations should be 0.95 for univalent cations and 1.05 for univalent anions. This example illustrates how these values can be derived since the ratio of the $[Na^+]_{if}$ to the $[Na^+]_p$ is 0.95 (145/152.7), and that of the $[Cl^-]_{if}$ to the $[Cl^-]_p$ is 1.05 (145/137.7).

Capillary Hydrostatic Pressure

In humans, the average capillary hydrostatic pressure is approximately 25 mmHg at the arterial end and falls to 10 mmHg by the venous end [26,27]. This pressure is determined by three factors: the arterial pressure, which has a mean value of 85 to 95 mmHg; the resistance at the precapillary sphincter (Fig. 1-14); and the venous pressure. The sphincter resistance determines the degree to which the arterial pressure is transmitted to the capillary. This is important because the ability to vary sphincter resistance allows the capillary hydrostatic pressure to be held relatively constant in the presence of changes in the arterial pressure. How this occurs can be appreciated from the relationship between resistance (R), the pressure drop across the resistance (ΔP), and the blood flow (Q):

$$\Delta P = Q \times R \tag{1-18}$$

Thus, an increase in resistance elevates the ΔP, and a decrease in resistance reduces the ΔP. For example, if the arterial pressure is increased, a rise in precapillary resistance by constriction of the sphincter will increase the ΔP, thereby preventing an increment in the capillary pressure (and in capillary blood flow since the elevations in pressure and resistance balance out). If this did not occur, then every patient with high blood pressure would tend to develop edema (defined as a palpable swelling due to expansion of the interstitial fluid volume) since the increase in capillary hydrostatic pressure would act to push water out of the vascular space into the interstitium. Although neural and humoral factors may contribute, capillary resistance is largely under local control, e.g., by stretch receptors in the sphincter wall and local metabolic factors, a process that is called *autoregulation* [31]. Autoregulation also occurs in the glomerular capillaries in the kidney, where it plays an important role in the maintenance of renal blood flow (see Chap. 3).

In contrast to these events at the arterial end of the capillary, the resistance at the venous end of the capillary is less well regulated. Consequently, alterations in

---

†In addition to this Gibbs-Donnan effect, other poorly understood factors contribute to the discrepancy between the actual and predicted osmotic pressures produced by the plasma proteins. For a more complete discussion, see Refs. 29 and 30.

venous pressure produce parallel changes in capillary hydrostatic pressure (see below).

Net Filtration

As described above, there is normally a small net gradient favoring filtration of 0.3 to 0.5 mmHg (Table 1-8). However, this gradient is not uniform throughout the capillary (Fig. 1-16). At the arterial end, the capillary hydrostatic pressure is at its peak (about 25 mmHg). As a result, the hydrostatic pressure gradient exceeds the oncotic pressure gradient by 7 to 9 mmHg [25,26], thereby promoting the movement of fluid into the interstitium. By the venous end of the capillary, the capillary hydrostatic pressure has fallen to a level (as low as 10 mmHg) such that the oncotic pressure gradient now exceeds the hydrostatic pressure gradient. In addition, the surface area and membrane porosity are higher on the venous side [25]. Thus, most of the filtered fluid is absorbed back into the capillary. The small amount of fluid and solutes that does not reenter the capillary is removed by the lymph vessels and then returned to the systemic circulation.

Safety factors Since the mean gradient only slightly favors filtration, it might be assumed that a small increase in capillary hydrostatic pressure (due to an elevated

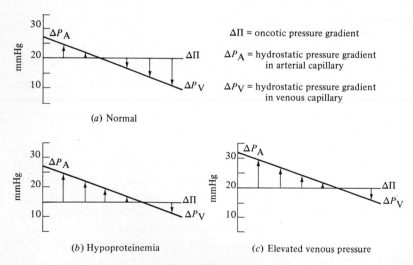

(a) Normal

(b) Hypoproteinemia

(c) Elevated venous pressure

Figure 1-16 Relationship between the hydrostatic pressure gradient at the arterial (ΔP_A) and the venous (ΔP_V) ends of the capillary, the oncotic pressure gradient ($\Delta\Pi$), and the movement of fluid out of (arrows pointing upward) or into (arrows pointing downward) the capillary. In the normal subject (a), there is a gradient ($\Delta P_A - \Delta\Pi$) favoring filtration at the arterial end of the capillary. As the capillary hydrostatic pressure falls, a gradient develops ($\Delta\Pi - \Delta P_V$) which promotes the absorption of fluid back into the capillary. The net effect is a small amount of filtration which is then removed by the lymph vessels. In the presence of hyproproteinemia (b) or elevated venous pressure (c), there is an increase in the gradient favoring fluid movement out of the capillary, predisposing toward the formation of edema. *(Adapted from C. A. Wiederheilm, J. Gen. Physiol., 52:29s, 1968.)*

venous pressure) or a small decrease in plasma oncotic pressure (due to hypoprotein-emia) would lead to fluid accumulation in the interstitium (Fig. 1-16) and ultimately to clinically apparent edema. However, experimental and clinical observations indicate that edema does not occur until there are more marked alterations in these parameters [25,26,28,32]. Two factors appear to be responsible for this. First, lymphatic flow is able to increase so that the excess filtrate can initially be carried away. Second, as fluid moves into the interstitium, the oncotic pressure falls (both by dilution and by removal of interstitial proteins by the enhanced lymphatic flow) and the hydrostatic pressure may rise. Both of these changes reduce the gradient promoting filtration and retard further fluid accumulation. Thus, edema will not occur until these protective mechanisms are overcome; in general, this requires at least a 10- to 15-mmHg increase in capillary hydrostatic pressure or reduction in plasma oncotic pressure [28]. For example, edema due to hypoalbuminemia usually does not occur until the plasma albumin concentration falls below 2.5 to 3.0 g/dL (normal equals 4 to 5 g/dL) (see Chap. 16) [32].

BALANCE STATE

Despite wide variations in dietary intake, the volume and composition of the body fluids are maintained in an extremely narrow range as excretion is adjusted to match intake. Thus the amount of a substance added to the body each day by dietary ingestion or endogenous production is equal to the amount eliminated from the body by excretion or endogenous utilization. This is referred to as the *balance state* or the *steady state*. As we will see in Chaps. 9 to 13, variations in urinary excretion are the major means by which the body maintains water and electrolyte homeostasis. Although urinary excretion can effectively correct an excess of a substance, it should be noted that a deficit can be corrected only by increased intake.

These principles can be illustrated by a brief review of how the body maintains water balance (Table 1-9). There are three sources of water addition to the body: (1) drinking; (2) water contained in food, e.g., meat is roughly 70 percent water and fruits and vegetables are almost 100 percent water; and (3) water generated by the oxidation of carbohydrates, fats, and proteins. In a person on a normal diet, the latter two sources are relatively constant whereas the amount of water ingested by drinking is partially controlled by the hypothalamus which affects the sensation of thirst.

Table 1-9 Typical daily water balance in a normal human

Water intake, mL/day		Water output, mL/day	
Source		Source	
Ingested water	1400	Urine	1500
Water content of food	850	Skin	500
Water of oxidation	350	Respiratory tract	400
		Stool	200
Total	2600	Total	2600

Water is eliminated from the body in the urine and stool and by vaporization from the skin and respiratory tract. The latter losses are relatively constant in contrast to urinary water excretion, which is highly variable, being largely dependent upon the presence or absence of antidiuretic hormone (ADH). ADH is secreted from the posterior lobe of the pituitary gland in response to an increase in the effective plasma osmolality and acts to reduce renal water excretion (see Chaps. 6 and 8).

Water balance is maintained by regulating both water intake via thirst and water excretion via ADH (see Chap. 10). These effects are mediated by hypothalamic osmoreceptors which are so sensitive that the plasma osmolality usually does not vary more than 1 to 2 percent. After the ingestion of a water load, for example, the ensuing reduction in the plasma osmolality (see Fig. 1-12c) decreases ADH secretion, resulting in an appropriate increase in urinary water excretion. On the other hand, a water deficit raises the plasma osmolality, which stimulates both ADH secretion and thirst. The former minimizes further water loss while the latter restores water balance by enhancing water intake.

APPENDIX: GIBBS-DONNAN EQUILIBRIUM

The Gibbs-Donnan equation can be derived from the following thermodynamic principles. An ion, for example, Na^+, can move passively across a membrane down electrical and chemical (concentration) gradients. Thus, the net flux of Na^+ (U_{Na+}) across the membrane can be expressed as [3]

$$U_{Na+} = \text{electrical gradient} + \text{chemical gradient} \qquad (1\text{-}19)$$

The effect of the electrical forces can be measured from the term ZFE_m, where Z is the valence of Na^+ ($1+$), F is Faraday's constant, and E_m is the membrane potential. The chemical gradient can be expressed as $RT \ln ([Na^+]_1/[Na^+]_2)$, where R is the gas constant, T is temperature in kelvins, and $[Na^+]_1$ and $[Na^+]_2$ represent the Na^+ concentrations on the two sides of the membrane. Thus,

$$U_{Na+} = Z_{Na+}FE_m + RT \ln \frac{[Na^+]_1}{[Na^+]_2} \qquad (1\text{-}20)$$

Similarly,

$$U_{Cl-} = Z_{Cl-}FE_m + RT \ln \frac{[Cl^-]_1}{[Cl^-]_2} \qquad (1\text{-}21)$$

By definition, both U_{Na+} and U_{Cl-} are zero in the equilibrium state:

$$0 = Z_{Na+}FE_m + RT \ln \frac{[Na^+]_1}{[Na^+]_2} \qquad (1\text{-}22)$$

$$0 = Z_{Cl-}FE_m + RT \ln \frac{[Cl^-]_1}{[Cl^-]_2} \qquad (1\text{-}23)$$

If we add these equations, the terms for the electrical gradient cancel out since Z_{Na^+} is $1+$ and Z_{Cl^-} is $1-$:

$$0 = RT \ln \frac{[Na^+]_1}{[Na^+]_2} + RT \ln \frac{[Cl^-]_1}{[Cl^-]_2} \tag{1-24}$$

or

$$RT \ln \frac{[Na^+]_1}{[Na^+]_2} = -RT \ln \frac{[Cl^-]_1}{[Cl^-]_2} \tag{1-25}$$

Since $-\ln (a/b) = \ln (b/a)$,

$$RT \ln \frac{[Na^+]_1}{[Na^+]_2} = RT \ln \frac{[Cl^-]_2}{[Cl^-]_1} \tag{1-26}$$

or

$$\frac{[Na^+]_1}{[Na^+]_2} = \frac{[Cl^-]_2}{[Cl^-]_1} \tag{1-27}$$

$$[Na^+]_1 \times [Cl^-]_1 = [Na^+]_2 \times [Cl^-]_2 \tag{1-28}$$

This is called the Gibbs-Donnan equation and holds for all univalent ions that distribute passively across the membrane. For example, across the capillary membrane separating the plasma (P) from the interstitial fluid (if)

$$\frac{[Na^+]_P}{[Na^+]_{if}} = \frac{[K^+]_P}{[K^+]_{if}} = \frac{[Cl^-]_{if}}{[Cl^-]_P} = \frac{[HCO_3^-]_{if}}{[HCO_3]_P} \tag{1-29}$$

In humans, the Gibbs-Donnan factor for univalent cations is 0.95 ($[Na^+]_{if}$ is 0.95 $[Na^+]_P$) and for anions is 1.05 ($[Cl^-]_{if}$ is 1.05 $[Cl^-]_P$).

The divalent cations, Ca^{2+} and Mg^{2+} are significantly protein-bound. Therefore, the Gibbs-Donnan equation can be written only for the free (or nonbound) concentrations of these ions. Since Z_{Cl^-} is $1-$ and $Z_{Ca^{2+}}$ is $2+$, Eqs. (1-22) and (1-23) must be rewritten in the following way to allow the terms for the electrical gradient to cancel out:

$$0 = Z_{Ca^{2+}}FE_m + RT \ln \frac{[Ca^{2+}]_1}{[Ca^{2+}]_2} \tag{1-30}$$

$$0 = 2Z_{Cl^-}FE_m + 2RT \ln \frac{[Cl^-]_1}{[Cl^-]_2} \tag{1-31}$$

If we add these equations and then solve in a manner similar to that above,

$$2 \ln \frac{[Cl^-]_2}{[Cl^-]_1} = \ln \frac{[Ca^{2+}]_1}{[Ca^{2+}]_2} \tag{1-32}$$

Since $2 \ln (a/b) = \ln (a/b)^2$,

$$\frac{[Ca^{2+}]_1}{[Ca^{2+}]_2} = \left[\frac{[Cl^-]_2}{[Cl^-]_1}\right]^2 = \left[\frac{[Na^+]_1}{[Na^+]_2}\right]^2 \qquad (1\text{-}33)$$

Since the Gibbs-Donnan factor for univalent cations is 0.95, it is 0.90 (0.95 $\times$ 0.95) for divalent cations. Similarly, the factor for divalent anions is 1.10 (1.05 $\times$ 1.05).

PROBLEMS

1-1 Thirst is stimulated by an increase in the effective P_{osm} (see Chap. 10). What effect will the following have on thirst?

    (*a*) An increase in the plasma Na^+ concentration

    (*b*) An increase in the blood urea nitrogen (BUN)

1-2 What is the relationship between the plasma Na^+ concentration and the plasma osmolality? Between the plasma Na^+ concentration and the extracellular volume?

1-3 If glucose is added to the extracellular fluid, what will happen to the following?

    (*a*) The plasma osmolality

    (*b*) The extracellular volume

    (*c*) The intracellular volume

    (*d*) The plasma Na^+ concentration

1-4 Laboratory tests for a patient provide the following values:

$$P_{osm} \qquad = 290 \text{ mosmol/kg}$$

$$\text{Plasma } [Na^+] = 125 \text{ meq/L}$$

$$\text{BUN} \qquad = 28 \text{ mg/dL}$$

If glucose is the only other osmole in the extracellular fluid, calculate the plasma glucose concentration in mg/dL.

1-5 If the plasma Cl^- concentration is 104 meq/L, the Cl^- concentration in the relatively protein-free glomerular filtrate will be approximately 117 meq/L. What are the two factors responsible for this difference?

1-6 What effect will the following have on the distribution of fluid between the vascular space and the interstitium?

    (*a*) An increase in arterial pressure

    (*b*) A decrease in venous pressure

    (*c*) An increase in the plasma albumin concentration

1-7 What is the difference between the factors determining osmolality and osmotic pressure?

1-8 Is Na^+ an effective or ineffective osmole across the capillary membrane? Across the cell membrane?

REFERENCES

1. Hays, R. M.: Dynamics of body water and electrolytes, in *Clinical Disorders of Fluid and Electrolyte Metabolism,* 3d ed., M. H. Maxwell and C. R. Kleeman (eds.), McGraw-Hill, New York, 1980.

2. Brown, A. C.: Passive and active transport, in *Physiology and Biophysics,* vol. II, T. C. Ruch and H. D. Patton (eds.), Saunders, Philadelphia, 1974.

3. Woodbury, J. W.: The cell membrane: ionic and potential gradients and active transport, in *Medical Physiology and Biophysics,* T. C. Ruch and J. F. Fulton (eds.), Saunders, Philadelphia, 1960.

4. Kinne, R., and I. L. Schwartz: Isolated membrane vesicles in the evaluation of the nature, localization and regulation of renal transport processes, Kidney Int., 14:547, 1978.

5. Cohen, L. H., A. Mueller, and P. R. Steinmetz: Inhibition of the bicarbonate exit step in urinary acidification by a disulfonic stilbene, J. Clin. Invest., 61:981, 1978.

6. Burg, M., C. Patlak, N. Green, and D. Villey: Organic solutes in fluid absorption by renal proximal convoluted tubules, Am. J. Physiol., 231:627, 1976.

7. Kinsella, J. L., and P. S. Aronson: Properties of the Na^+-H^+ exchanger in renal microvillus membrane vesicles, Am. J. Physiol., 238:F461, 1980.

8. Sweadner, K. J., and S. M. Goldin: Active transport of sodium and potassium ions: mechanism, function, and regulation, N. Engl. J. Med., 302:777, 1980.

9. DeVoe, R. D., and P. C. Maloney: Principles of cell homeostasis, in *Medical Physiology,* V. B. Mountcastle (ed.), 14th ed., Mosby, St. Louis, 1980.

10. Katz, A. I.: Renal Na-K-ATPase: its role in tubular sodium and potassium transport, Am. J. Physiol., 242:F207, 1982.

11. Carone, F. A., and D. R. Peterson: Hydrolysis and transport of small peptides by the proximal tubule, Am. J. Physiol., 238:F151, 1980.

12. Edelman, I. S., J. Leibman, and M. P. O'Meara: Interrelations between serum sodium concentration, serum osmolarity and total exchangeable sodium, total exchangeable potassium and total body water, J. Clin. Invest., 37:1236, 1958.

13. Zwelling, L. A., and J. E. Balow: Hypersthenuria in high-dose carbenicillin therapy, Ann. Intern. Med. 89:225, 1978.

14. Woodbury, D. M.: Physiology of body fluids, in *Physiology and Biophysics,* vol. II, T. C. Ruch, and H. D. Patton (eds.), Saunders, Philadelphia, 1974.

15. Pitts, R. F.: *Physiology of the Kidney and Body Fluids,* Year Book, Chicago, 1974.

16. Eisenman, A. J., L. B. MacKenzie, and J. P. Peters: Protein and water content of serum and cells of human blood, with a note on the measurement of red blood cell volume, J. Biol. Chem., 116:33, 1936.

17. Albrink, M., P. M. Hold, E. B. Man, and J. P. Peters: The displacement of serum water by the lipids of hyperlipemic serum. A new method for the rapid determination of serum water, J. Clin. Invest., 34:1483, 1955.

18. Leaf, A.: Regulation of intracellular fluid volume and disease, Am. J. Med., 49:291, 1970.

19. Adler, S.: The simultaneous determination of muscle cell pH using a weak acid and weak base, J. Clin. Invest, 51:256, 1972.

20. Maffly, R. H., and A. Leaf: The potential of water in mammalian tissues, J. Gen. Physiol., 42:1257, 1959.

21. Laragh, J. H.: The effect of potassium chloride on hyponatremia, J. Clin. Invest., 33:807, 1954.

22. Fichman, M. P., H. Vorherr, C. R. Kleeman, and N. Telfer: Diuretic-induced hyponatremia, Ann. Intern. Med., 75:853, 1971.

23. Starling, E.: On the absorption of fluids from the connective tissue spaces, J. Physiol. (London), 19:312, 1896.

24. Landis, E., and J. R. Pappenheimer: Exchange of substances through the capillary walls, in *Handbook of Physiology,* sec. 2, Circulation, vol. II, W. F. Hamilton and P. Dow (eds.), American Physiological Society, Washington, D.C., 1963.

25. Wiederhielm, C. A.: Dynamics of transcapillary fluid exchange, J. Gen. Physiol., 52:29s, 1968.

26. Guyton, A. C.: *Textbook of Medical Physiology,* Saunders, Philadelphia, 1981, chap. 30.

27. Brace, R. A.: Progress toward resolving the controversy of positive vs. negative interstitial fluid pressure, Circ. Res., 49:281, 1981.

28. Taylor, A. E.: Capillary fluid filtration: Starling forces and lymph flow, Circ. Res., 49:557, 1981.

29. Scatchard, G., A. C. Batchelder, and A. Brown: Chemical, clinical, and immunological studies on

the products of human plasma fractionation. VI. The osmotic pressure of plasma and of serum albumin, J. Clin. Invest., 23:458, 1944.

30. Reiff, T. R.: Osmotic water loss in aging: a hypothesis, Hosp. Pract., 17(4):196, 1982.
31. Braunwald, E.: Regulation of the circulation, N. Engl. J. Med., 290:1420, 1974.
32. Darrow, D. C., E. B. Hopper, and M. K. Cary: Plasmapheresis edema. I. The relation of reduction of serum proteins to edema and the pathological anatomy accompanying plasmapheresis, J. Clin. Invest., 11:683, 1932.

RENAL PHYSIOLOGY

INTRODUCTION TO RENAL FUNCTION

INTRODUCTION

The kidney performs a variety of essential functions:

1. It participates in the maintenance of the constant extracellular environment that is required for adequate functioning of the cells. This is achieved by excretion of some of the waste products of metabolism (such as urea, creatinine, and uric acid) and by specifically adjusting the urinary excretion of water and electrolytes to match intake and endogenous production. As will be seen, the kidney is able to regulate individually the excretion of water and solutes such as sodium, potassium, and hydrogen, largely by changes in tubular reabsorption or secretion.
2. It secretes hormones that participate in the regulation of systemic and renal hemodynamics (renin, prostaglandins, and bradykinin), red blood cell production (erythropoietin), and calcium, phosphorus, and bone metabolism (vitamin D) (see Chaps. 3 and 8).
3. It performs such miscellaneous functions as catabolism of peptide hormones [1] and synthesis of glucose (gluconeogenesis) in fasting conditions [2,3].

In this chapter, we will review briefly the morphology of the kidney and the basic processes of reabsorption and secretion. The regulation of renal hemodynamics, the specific functions of the different nephron segments, and the relationships between hormones and the kidney will then be discussed in the ensuing chapters.

RENAL MORPHOLOGY

The basic unit of the kidney is the nephron, with each kidney in humans containing approximately 1.0 to 1.3 million nephrons. Each nephron consists of a glomerulus, which is a tuft of capillaries interposed between two arterioles (the afferent and efferent arterioles), and a series of tubules lined by a continuous layer of epithelial cells (Fig. 2-1). The glomeruli are located in the outer part of the kidney, called the cortex, whereas the tubules are present in both the cortex and the inner part of the kidney, the medulla (Figs. 2-1 and 2-2).

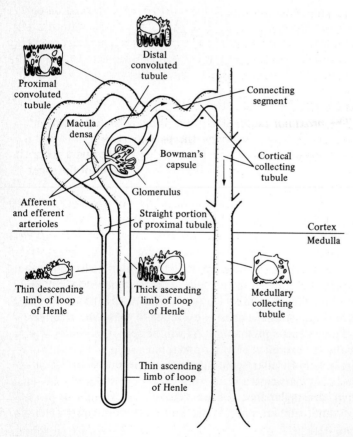

Figure 2-1 Relationships of the component parts of the nephron. (*Adapted from R. Vander, Renal Physiology, 2d ed., McGraw-Hill, New York, 1980.*)

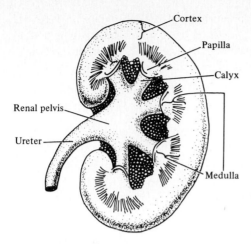

Cortex

Papilla

Calyx

Renal pelvis

Ureter

Medulla

Figure 2-2 Section of a human kidney. The outer portion (the cortex) contains all the glomeruli. The tubules are located in both the cortex and the medulla with the collecting tubules forming a large portion of the inner medulla (the papilla). Urine leaving the collecting tubules drains sequentially into the calyces, renal pelvis, ureter, and then the bladder. *(Adapted from R. Vander, Renal Physiology, 2d ed., McGraw-Hill, New York, 1980.)*

The initial step in the excretory function of the nephron is the formation of an ultrafiltrate of plasma across the glomerulus. This fluid then passes through the tubules and is modified in two ways: by *reabsorption and secretion.* Reabsorption refers to the removal of a substance from the filtrate; secretion refers to the addition of a substance to the filtrate. As will be seen, the different tubular segments make varying contributions to these processes.

Fluid filtered across the glomerulus enters Bowman's space and then the proximal tubule (Fig. 2-1). The *proximal tubule* is composed anatomically of an initial convoluted segment and a later straight segment, the pars recta, which enters the outer medulla. The *loop of Henle* begins abruptly at the end of the pars recta. It generally includes a thin descending limb and thin and thick segments of the ascending limb. This hairpin configuration plays a major role in the excretion of a hyperosmotic urine (see Chap. 6). It is important to note that the length of the loops of Henle is not uniform (Fig. 2-3). Approximately 40 percent of nephrons have short loops which penetrate only the outer medulla or may even turn around in the cortex [4]. The remaining 60 percent have long loops which course through the medulla and may extend down to the papilla (the innermost portion of the medulla). The length of the loops is largely determined by the cortical location of the glomerulus (Fig. 2-3): those glomeruli in the outer cortex (about 30 percent) have only short loops; those in the juxtamedullary region (about 10 percent) have only long loops; and those in the midcortex may have either short or long loops. The functional significance of these differences will be discussed in Chap. 6.

The thick ascending limb also has a cortical segment which returns to the region of the parent glomerulus. It is in this area, where the tubule approaches the afferent glomerular arteriole, that the specialized tubular cells of the *macula densa* are located (Fig. 2-4). The juxtaglomerular cells of the afferent arteriole and the macula densa compose the juxtaglomerular apparatus, which is important in the control of renin secretion (see Chap. 3).

After the macula densa, there are three cortical segments (Fig. 2-3): the *distal*

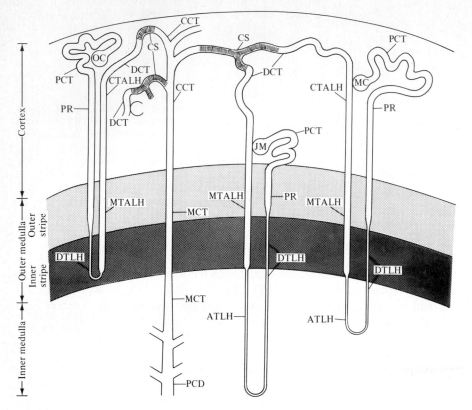

Figure 2-3 The anatomic relationships of the different nephron segments according to location of the glomeruli in the outer cortex (OC), midcortex (MC), or juxtamedullary area (JM). The major nephron segments are labeled as follows: proximal convoluted tubule (PCT); pars recta (PR); descending and ascending thin limbs of the loop of Henle (DTLH and ATLH); medullary and cortical aspects of the thick ascending limb of the loop of Henle (MTALH and CTALH); distal convoluted tubule (DCT); connecting segment (CS); cortical collecting tubule (CCT); and medullary collecting tubule (MCT) and its terminal portion, the papillary collecting duct (PCD). *(Adapted from H. R. Jacobson, Am. J. Physiol., 241:F203, 1981.)*

convoluted tubule, the *connecting segment* (previously considered part of the late distal tubule), and the *cortical collecting tubule* [5,6]. Note that the connecting segments of many nephrons drain into a single collecting tubule. Fluid leaving the cortical collecting tubule flows into the *medullary collecting tubule* and then drains sequentially into the calyces, the renal pelvis, the ureters, and the bladder (Fig. 2-2).

The segmental subdivision of the nephron is based upon different permeability and transport characteristics that translate into important differences in function. The proximal tubule, for example, is a relatively permeable epithelium. This property enables the proximal tubule to reabsorb approximately 70 percent of the filtrate in a composition that is similar to that of the plasma. In contrast, the collecting tubules are relatively impermeable to passive solute and water movement. As a

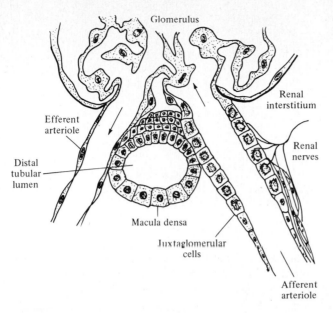

Figure 2-4 Diagram of the juxtaglomerular apparatus. The juxtaglomerular cells in the wall of the afferent arteriole secrete renin into the lumen of the afferent arteriole and the renal lymph. Stretch receptors in the afferent arteriole, the sympathetic nerves ending in the juxtaglomerular cells, and the composition of the tubular fluid reaching the macula densa all may contribute to the regulation of renin secretion. *(Adapted from J. O. Davis, Am. J. Med., 55:333 1973.)*

result, these segments can establish and maintain steep concentration gradients between the tubular fluid and the plasma since passive back-diffusion is limited. This ability allows the collecting tubules to make the final small changes in urinary composition that permit solute and water balance to be maintained.

REABSORPTION AND SECRETION

The rate of glomerular filtration averages 135 to 180 L/day in a normal adult. Since this represents a volume that is more than 10 times that of the extracellular fluid and approximately 60 times that of the plasma, it is evident that almost all this fluid must be returned to the systemic circulation. This process is called *tubular reabsorption* and occurs in two steps: The substance to be reabsorbed is first transported from the tubular lumen into the cell, usually across the luminal aspect of the cell membrane; it then moves across the peritubular aspect of the cell membrane into the interstitium and then the capillaries that surround the tubules† (Fig. 2-5).

†The spatial orientation of the cells is important since the luminal and peritubular aspects of the cell membrane (which are separated by the tight junction) have different permeability and transport properties. For example, active Na^+ transport occurs only at the peritubular membrane since this is the site of the Na^+-K^+-ATPase pump (see Fig. 1-5).

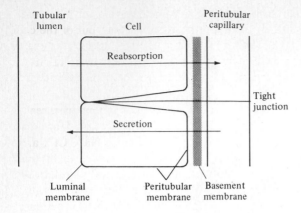

Figure 2-5 Schematic representation of reabsorption and secretion in the nephron.

Although most reabsorbed solutes are returned to the systemic circulation intact, some are metabolized within the cell, particularly low-molecular-weight proteins in the proximal tubule (see Chap. 5). Solutes can also move in the opposite direction, from the peritubular capillary through the cell and into the urine; this process is called *tubular secretion* (Fig. 2-5).

Filtered solutes and water may be transported by one or both of these mechanisms. For example, Na^+, Cl^-, and H_2O are reabsorbed; hydrogen ions are secreted; K^+ and uric acid are both reabsorbed and secreted; and filtered creatinine is excreted virtually unchanged since it is not reabsorbed and only a small amount is added to the urine by secretion.

Reabsorption and secretion may involve both active and passive mechanisms. As

Table 2-1 Summary of the net daily reabsorptive work performed by the kidney†

	Filtered	Excreted	Percent net reabsorption
Water	180 liters	0.5–3.0 liters	98–99
Na^+	26,000 meq	100–250 meq	>99
Cl^-	21,000 meq	100–250 meq	>98
HCO_3^-	4,800 meq	0	100
K^+	800 meq	40–120 meq	85–95‡
Urea	54 g	27–32 g	40–50

†These values are for a normal adult male. The glomerular filtration rate, and therefore the filtered load of solutes and water, is approximately 25 percent lower in women.

‡Although almost all of the filtered K^+ is reabsorbed, the net reabsorption of K^+ is only 85 to 95 percent. This is due to the secretion of K^+ into the lumen, primarily in the cortical collecting tubule (see Chap. 13).

an example, the reabsorption of Na^+ occurs by the passive movement of Na^+ from the tubular lumen into the cell and then the active transport of Na^+ out of the cell into the peritubular capillary (see Chap. 5). On the other hand, K^+ is secreted from the cortical collecting tubule cell into the lumen. Although the movement of K^+ from the cell into the lumen is passive, the uptake of K^+ by the cell from the peritubular capillary is active (see Chap. 13).

The tubular cells perform these functions in an extremely efficient manner, reabsorbing almost all the filtrate to maintain the balance between intake and excretion. On a normal diet, more than 98 to 99 percent of the filtered H_2O, Na^+, Cl^-, and HCO_3^- is reabsorbed (Table 2-1). In comparison, only 40 to 50 percent of the filtered urea (a product of protein metabolism) is reabsorbed, and consequently a relatively large amount of urea is present in the urine. As we will see in later chapters, the renal handling of these substances can be regulated in an individual manner. Although this process of filtration and almost complete reabsorption may seem inefficient, a high rate of filtration is required for the excretion of those waste products of metabolism (such as urea and creatinine) that enter the urine primarily by glomerular filtration.

Composition of Urine

The composition of the urine differs from that of the extracellular fluid in two important ways. First, although the composition of the extracellular fluid is maintained within narrow limits, the quantity of solutes and water in the urine is highly variable, being dependent upon the intake of these substances. A normal subject appropriately excretes more Na^+ on a high-salt diet than on a low-salt diet. In both instances, the steady state in maintained as output equals intake. Similarly, the urine volume is greater after a water load than after water restriction. This relation to intake means that *there are no absolute "normal" values for urinary solute or water excretion*. We can describe a normal range which merely reflects the range of dietary intake, for example, 100 to 250 meq/day for Na^+. Second, whereas ions compose 95 percent of the extracellular fluid solutes, the urine has high concentrations of uncharged molecules, particularly urea.

Summary of Nephron Function

The following chapters in Part Two will describe the roles of the different nephron segments in the regulation of solute and water homeostasis. These functions are summarized in Table 2-2. As can be seen, there are marked differences in segmental function, a finding consistent with the differences in segmental histology (Fig. 2-1) and permeability and transport characteristics [4]. In addition, multiple sites participate in the regulation of the rates of excretion of the different substances in the filtrate. This diversity provides the flexibility that allows the kidney to maintain solute and water balance even in the presence of major changes in dietary intake.

Table 2-2 Summary of the contribution of the different nephron segments to solute and water homeostasis

Nephron segment	Major Functions
Glomerulus	Forms an ultrafiltrate of plasma
Proximal tubule	Reabsorbs isosmotically 70 percent of the filtered NaCl and H_2O Reabsorbs K^+, glucose, amino acids, calcium, phosphate, magnesium, urea, uric acid, and bicarbonate (by H^+ secretion) Secretes H^+, ammonia, and organic acids and bases
Loop of Henle	Countercurrent multiplier; reabsorbs NaCl in excess of H_2O Major site of active regulation of magnesium excretion
Distal tubule and connecting segment	Reabsorb a small fraction of filtered NaCl Major site of active regulation of calcium excretion
Collecting tubules	Site of final modification of the urine Reabsorb NaCl; urine NaCl concentration can be reduced to less than 1 meq/L Reabsorb H_2O and urea relative to the ADH concentration present, allowing a dilute or concentration urine to be produced Secrete K^+ in cortical collecting tubule, the major source of urinary K^+; reabsorb or secrete K^+ in medullary collecting tubule Secrete H^+, ammonia; urine pH can be reduced to 4.5 to 5.0

PROBLEM

2-1 What is the relationship between the amount of water excreted, the amount filtered, and the amount reabsorbed?

REFERENCES

1. Carone, F. A., and D. R. Peterson: Hydrolysis and transport of small peptides by the proximal tubule, Am. J. Physiol., 238:F151, 1980.
2. Owen, O. E., P. Felig, A. P. Morgan, J. Wahren, and G. F. Cahill, Jr.: Liver and kidney metabolism during prolonged starvation, J. Clin. Invest., 48:574, 1969.
3. Burch, H. B., R. G. Narins, C. Chu, S. Fagioli, S. Choi, W. McCarthy, and O. H. Lowry: Distribution along the rat nephron of three enzymes of gluconeogenesis in acidosis and starvation, Am. J. Physiol., 235:F246, 1978.
4. Jacobson, H. R.: Functional segmentation of the mammalian nephron, Am. J. Physiol., 241:F203, 1981.
5. Imai, M.: The connecting tubule: a functional subdivision of the rabbit distal nephron segments, Kidney Int., 15:346, 1979.
6. Morel, F., D. Chabardès, and M. Imbert: Functional segmentation of the rabbit distal tubule by microdetermination of hormone-dependent adenylate cylase, Kidney Int., 9:264, 1976.

THREE

RENAL CIRCULATION AND GLOMERULAR FILTRATION RATE

INTRODUCTION

The blood flow to the kidneys averages 20 percent of the cardiac output. In terms of flow per 100 g weight, the renal blood flow (RBF) is four times greater than that to the liver or exercising muscle and eight times that of coronary blood flow. Blood enters the kidney through the renal arteries and passes through serial branches (interlobar, arcuate, interlobular) before entering the glomeruli via the afferent arterioles. The portion of the plasma not filtered across the glomerular capillary wall then leaves the glomeruli via the efferent arterioles and enters the postglomerular capillaries. In the cortex, these capillaries run in apposition to the adjacent tubules, although not necessarily the tubule segments from the same glomerulus [1]. In addition, branches from the efferent arterioles of the juxtamedullary glomeruli enter the medulla and form the vasa recta capillaries (Fig. 3-1). Blood returns to the systemic circulation through veins similar to the arteries in name and location.

The renal circulation affects urine formation in the following ways:

1. The rate of glomerular filtration is an important determinant of solute and water excretion.
2. The peritubular capillaries in the cortex modulate proximal tubular reabsorption and secretion (see Chap. 5).

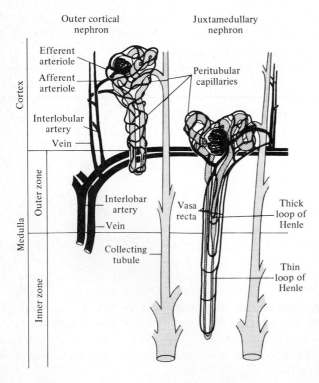

Figure 3-1 Comparison of the anatomy and blood supplies of outer cortical and juxtamedullary nephrons. Note that the efferent arterioles from the juxtamedullary nephrons not only form peritubular capillaries around the convoluted tubules but enter the medulla and form the vasa recta. (*Adapted from R. F. Pitts, Physiology of the Kidney and Body Fluids, 3d ed. Copyright © 1974 by Year Book Medical Publishers, Inc., Chicago. Used by permission.*)

3. The vasa recta capillaries participate in the countercurrent mechanism, permitting the conservation of water by the excretion of a hyperosmotic urine (see Chap. 6).

The remainder of this chapter will review glomerular function, the factors responsible for the regulation of the glomerular filtration rate (GFR) and renal plasma flow, and the clinical methods used to measure these parameters.

GLOMERULUS

The glomerulus consists of a tuft of capillaries that is interposed between the afferent and efferent arterioles. Each glomerulus is enclosed within an epithelial cell capsule (Bowman's capsule) that is continuous both with the epithelial cells that surround the glomerular capillaries and with the cells of the proximal convoluted tubule (Fig. 3-2). Thus, the glomerular capillary wall, through which the filtrate must pass, consists of three layers: the endothelial cell, the glomerular basement membrane, and the epithelial cell. The epithelial cells are attached to the basement membrane by discrete foot processes. The pores between the foot processes (slit pores) are closed by a thin membrane called the slit diaphragm (Fig. 3-2).

The major function of the glomerulus is to allow the filtration of small solutes (such as sodium and urea) and water while restricting the passage of larger molecules (Fig. 3-3). Solutes up to the size of inulin (mol wt 5200) are freely filtered. In contrast, myoglobin (mol wt 17,000) is filtered less completely than inulin whereas albumin (mol wt 69,000) is filtered only to a limited degree. Filtration is also limited for ions or drugs which are bound to albumin, such as roughly 40 percent of the circulating calcium.

The basement membrane appears to be the primary barrier to the filtration of larger molecules, although the slit diaphragms between the foot processes of the epithelial cells may play a contributory role [2,3]. The size limitation to filtration suggests that functional pores (perhaps reflecting the molecular organization of proteins) exist within the basement membrane.

Molecular *charge* has also been shown to be an important determinant of filtration across the glomerular capillary [3]. As illustrated in Fig. 3-4, neutral dextrans are filtered to a greater degree than anionic dextran sulfates of similar molecular sizes. This inhibitory effect of charge appears to be due to electrostatic repulsion by anionic sialoproteins present in the basement membrane and surrounding the epithelial cell foot processes. This effect is clinically important since albumin is a polyanion in the physiologic pH range. As with dextran sulfate, albumin is filtered to a much lesser degree (about 5 percent) than neutral dextran of the same size. Thus, charge as well as size limits the filtration of albumin. Both experimental and clinical observations have suggested that the loss of these negatively charged sialoproteins may be responsible for the increased filtration of albumin seen in certain glomerular diseases [4,5].

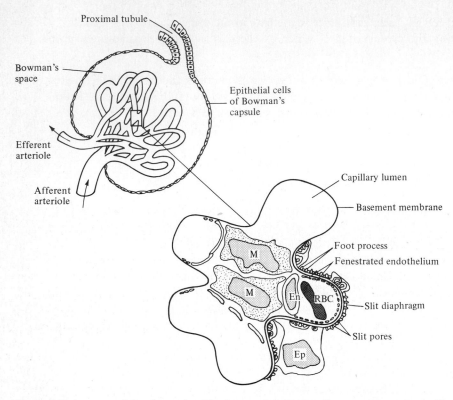

Proximal tubule

Bowman's space

Epithelial cells of Bowman's capsule

Efferent arteriole

Afferent arteriole

Capillary lumen

Basement membrane

M

M

Foot process

Fenestrated endothelium

En RBC

Slit diaphragm

Slit pores

Ep

Figure 3-2 Anatomy of the glomerulus. The bottom drawing is a diagram of part of a capillary tuft with the mesangial cells (M) in the middle surrounded by capillaries. The capillary wall has three layers composed of the fenestrated endothelial cells (En), the basement membrane, and the epithelial cells (Ep) which attach to the basement membrane by discrete foot processes. Between the foot processes are slit pores which are closed by a thin membrane, the slit diaphragm. *(Adapted from R. Vander, Renal Physiology, 2d ed. McGraw-Hill, New York, 1980, and H. Latta, in Handbook of Physiology, sec. 8, Renal Physiology, vol. I, J. Orloff, R. W. Berliner, and R. Geiger (eds.), American Physiological Society, Washington, D.C., 1973.)*

Albumin is the main component of the plasma oncotic pressure, which serves to hold fluid within the vascular space (see page 32). Thus, the normal glomerular impermeability to albumin helps to maintain the plasma volume by preventing albumin loss in the urine. The importance of this can be appreciated from patients with the nephrotic syndrome who have an abnormal increase in glomerular permeability, resulting in albuminuria and hypoalbuminemia. The reduction in the plasma oncotic pressure favors the movement of fluid from the vascular space into the interstitium and the development of edema (see Chap. 16).

The glomerular cells also have synthetic, phagocytic, and endocrine functions. The epithelial cells are thought to be responsible for the synthesis of the basement membrane and for the removal of macromolecules that pass through the basement membrane [6]. Macromolecules that enter the capillary wall but are unable to cross

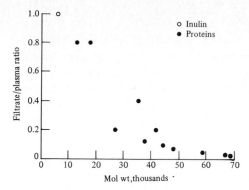

Figure 3-3 Glomerular permeability to proteins. As the molecular weight falls below 70,000, there is a progressive increase in the filtrate-to-plasma ratio. The ratio of 1.0 for inulin indicates complete filtration of this substance. *(Adapted from E. Renkin and R. Robinson, N. Engl. J. Med., 290:785, 1974. By permission from the New England Journal of Medicine.)*

the basement membrane are phagocytosed by circulating macrophages that move in and out of the mesangium and perhaps by the mesangial cells in the central part of the glomerular tuft (Fig. 3-2) [7]. In addition, at least two hormones are secreted by the glomerulus: renin and prostaglandins [8]. The latter participate in the regulation of both renal blood flow and renin release (see below).

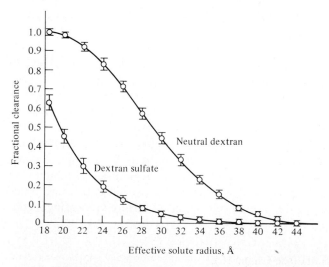

Figure 3-4 Fractional clearances (the ratio of the filtration of a substance to that of inulin, which is freely filtered) of tritiated neutral dextran and dextran sulfate as a function of molecular radius. The decreased filtration of dextran sulfates of varying sizes is suggestive of a charge limitation to filtration. As a reference, the effective molecular radius of albumin is about 36 Å. *(From B. M. Brenner, C. Baylis, and W. M. Deen, Kidney Int., 12:229, 1977. Reprinted by permission from Kidney International.)*

Renin-Angiotensin System

The afferent arteriole of each glomerulus contains specialized cells, called the juxtaglomerular cells (see Fig. 2-4), which secrete the proteolytic enzyme renin.† Renal hypoperfusion, produced by hypotension or volume depletion, and increased sympathetic activity are the major stimuli to renin release and initiate the sequence shown in Fig. 3-5 [9–11]. Renin cleaves a decapeptide angiotensin I from renin substrate (angiotensinogen), an α_2-globulin produced in the liver. Angiotensin I is then converted into an octapeptide, angiotensin II.‡ This reaction is catalyzed by an enzyme called converting enzyme, which is located in the lung, the luminal membrane of vascular endothelial cells, and other organs [11].

Actions of angiotensin II Angiotensin II has two major systemic effects, both of which act to reverse the hypovolemia or hypotension that is usually responsible for the stimulation of renin secretion (Fig. 3-5) [9–11]. First, it promotes renal Na^+ and H_2O retention and therefore expansion of the plasma volume. This is mediated both by enhanced secretion of aldosterone from the adrenal cortex (see Chap. 8) and by direct stimulation of tubular reabsorption by angiotensin II [13]. Second, it produces arteriolar vasoconstriction, which, by elevating systemic vascular resistance, increases the systemic blood pressure. Enhanced sensitivity to norepinephrine as well as the direct effect of angiotensin II may contribute to this response [14,15]. These hemodynamic effects tend to complete a negative feedback loop as the resulting elevations in blood volume and blood pressure, as well as a direct inhibitory effect of angiotensin II on the juxtaglomerular cells, reduce further renin secretion [16].

Angiotensin II also has a dual effect on the renal circulation. Direct arteriolar constriction tends to decrease renal blood flow. However, this effect is partially attenuated because angiotensin II also promotes the renal secretion of vasodilator prostaglandins (see below) [8].

Since the renal and particularly the vascular effects of angiotensin II act to increase the systemic blood pressure, *angiotensin II contributes importantly to the maintenance of blood pressure in all circumstances in which renin secretion is enhanced and circulating angiotensin II levels are high.* This is true in the hypertension associated with renal artery stenosis (in which renal ischemia stimulates renin release) as well as *in normotensive states* associated with effective circulating volume depletion (see Chap. 9) such as true volume depletion, heart failure, and hepatic cirrhosis [17–20]. For example, the administration of an angiotensin II inhibitor to a normotensive patient with hepatic cirrhosis can lower the blood pressure by as much as 25 mmHg [20]. In contrast, renin release and therefore circulating angio-

---

† A precursor of renin, called inactive renin, is also present in the kidney and may be secreted into the systemic circulation [8a]. The physiologic role of this compound is uncertain.

‡ Angiotensin II may be further metabolized to a heptapeptide, angiotensin III. In humans, angiotensin III possesses only 15 to 30 percent of the activity of angiotensin II, is present in much lower concentrations, and is of uncertain physiologic significance [12].

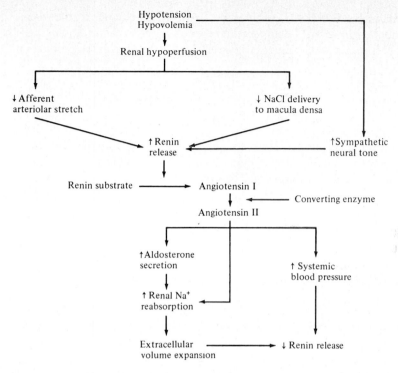

Figure 3-5 Renin-angiotensin-aldosterone system.

tensin II levels are relatively low in normal subjects on a regular diet [17,21]. As a result, angiotensin II does *not* play an important role in the maintenance of blood pressure in this setting.

Control of renin secretion In normal subjects, the major determinant of renin secretion is Na^+ intake: a high intake expands the extracellular volume and decreases renin release whereas a low intake (or fluid loss from any site) leads to mild volume depletion and stimulation of renin secretion (see Fig. 8-12). The associated changes in angiotensin II and aldosterone production then allow Na^+ to be excreted with volume expansion and retained with volume depletion.

These changes in volume are primarily sensed at one or more of three sites, leading to the activation of effectors which govern the release of renin (Fig. 3-5): (1) baroreceptors (or stretch receptors) in the wall of the afferent arteriole, which are stimulated by a reduction in renal perfusion pressure and whose effect appears to be mediated by the local production of prostaglandins, particularly prostacyclin (see Chap. 8) [22–24];† (2) the cardiac and arterial baroreceptors which regulate sympathetic neural activity and the level of circulating catecholamines, both of which enhance renin secretion via the β_1-adrenergic receptors [26,27]; and (3) the cells of the macula densa in the early distal tubule (see Fig. 2-4) which appear to be stimulated by a reduction in sodium chloride delivery. As with the baroreceptors, the macula densa effect is mediated primarily by alterations in prostaglandin production [28,29].

The interaction of these receptors can be appreciated from the response to hypovolemia. The decrease in volume initially lowers the blood pressure which diminishes the stretch in the afferent arteriole, increases sympathetic activity, and reduces flow to the macula densa (in part by enhancing proximal reabsorption). Each of these changes promotes renin secretion. This response can be largely abolished by inhibiting its mediators with a combination of indomethacin (an inhibitor of prostaglandin synthesis) and propranolol (a β-adrenergic blocker) [30].

DETERMINANTS OF GLOMERULAR FILTRATION RATE

The initial step in urine formation is the separation of an ultrafiltrate of plasma across the wall of the glomerular capillary. As with other capillaries, fluid movement across the glomerulus is governed by Starling's forces, being proportional to the permeability of the membrane and to the balance between the hydrostatic and oncotic pressure gradients (see Chap. 1):

GFR $= K_f$ (hydrostatic pressure gradient $-$ oncotic pressure gradient)

$$= K_f[(P_{GC} - P_{BS}) - (\Pi_P - \Pi_{BS})]$$

where K_f is the ultrafiltration coefficient of the membrane,‡ a function of both the area of the capillary tuft available for filtration and the unit permeability (or porosity) of the membrane; P_{GC} and P_{BS} are the hydrostatic pressures in the glomerular capillary and Bowman's space; and Π_P and Π_{BS} are the oncotic pressures in the plasma and Bowman's space. Since the filtrate is essentially protein-free, Π_{BS} is zero and

$$\text{GFR} = K_f(P_{GC} - P_{BS} - \Pi_P) \tag{3-1}$$

The GFR in normal adults is approximately 95 to 120 mL/min. This degree of filtration is, per weight, more than 1000 times that in muscle capillaries. Two factors

---

†The interaction between renin and renal prostaglandins is complex and potentially confusing since each stimulates the secretion of the other [8,22,23] and they induce opposing vascular actions—angiotensin II is a vasoconstrictor and most renal prostaglandins are vasodilators (see Chap. 8). However, angiotensin II is a systemic vasoconstrictor whereas the effect of renal vasodilator prostaglandins tends to be limited to the kidney because they are rapidly metabolized in the lungs if they enter the systemic circulation. Thus, the net effect of simultaneous secretion of renin and renal prostaglandins is that angiotensin II causes systemic vasoconstriction and raises the blood pressure while the prostaglandins minimize the degree of renal vasoconstriction, thereby maintaining renal blood flow [25,25a].

‡K_f in this setting refers to the permeability of the membrane to water and small solutes (such as sodium, urea, and glucose) whose filtration is not limited by molecular size (see Fig. 3-3).

Table 3-1 Comparison of Starling's forces governing fluid movement across the glomerulus and a muscle capillary†

	Glomerulus (primate)		Muscle capillary (human)
	Afferent arteriole	Efferent arteriole	
Hydrostatic pressures			
P_{GC} or P_{cap}‡	46	45	17
P_{BS} or P_{if}	10	10	−4
Mean gradient	36	35	21
Oncotic pressures			
Π_P	23	35	26
Π_{BS} or Π_{if}	0	0	5.3
Mean gradient	23	35	20.7
Net gradient	+13	0	+0.3
+ = filtration			
− = absorption	(Mean = +6 mmHg)		

†Units are millimeters of mercury (mmHg).
‡The subscripts "cap" and "if" refer to the capillary and interstitial forces in the muscle capillary.

account for this difference: (1) the K_f of the glomerulus is 50 to 100 times that of a muscle capillary [31,32]; and (2) the capillary hydrostatic pressure and therefore the mean gradient favoring filtration ($P_{GC} - P_{BS} - \Pi_P$) is much greater in the glomerulus than in a muscle capillary (Table 3-1) [33,34]. Although almost all of the filtered electrolytes and water is reabsorbed, the high GFR is required to allow the filtration and subsequent excretion of a variety of metabolic waste products such as urea and creatinine (see below).

Changes in the GFR can be produced by alterations in any of the factors in Eq. (3-1): capillary permeability, capillary hydrostatic pressure, the hydrostatic pressure in Bowman's space, and the plasma oncotic pressure [35]. In addition, the renal plasma flow (RPF) is also an important determinant of the GFR.

Capillary Permeability

The factors controlling the K_f are incompletely understood, but it appears that this parameter remains relatively constant in most normal conditions [32,35]. Furthermore, small changes in K_f will not affect the GFR since it is *the hydrostatic and oncotic pressures, not the capillary permeability,* that normally limit the filtration of small solutes and water (see Fig. 3-7) [35]. Although a variety of hormones, including angiotensin II, antidiuretic hormone, and prostaglandins, can affect the K_f, the physiologic significance of these effects remains to be defined [36–38]. However, in disease states such as glomerulonephritis, the K_f may be substantially reduced, a change which contributes to the fall in GFR that frequently occurs in these disorders [39].

Capillary Hydrostatic Pressure

The glomerular capillaries are uniquely interposed between two arterioles in a manner that permits rapid regulation of the GFR through changes in the P_{GC}. In this situation, P_{GC} is determined by the aortic pressure and the resistance at the afferent and efferent arterioles. How this occurs can be appreciated from the relationship between resistance (R), the pressure drop across the resistance (ΔP), and blood flow (Q):

$$\Delta P = Q \times R \qquad (3\text{-}2)$$

The resistance at the afferent arteriole determines the degree to which the aortic pressure is transmitted to the glomerulus. For example, basal afferent arteriolar resistance must be less than that in the precapillary sphincter in skeletal muscle to account for the substantially higher hydrostatic pressure in the glomerulus (Table 3-1).

In addition, P_{GC} and therefore the GFR will be affected by changes in arteriolar tone from its resting level. Constriction of the afferent arteriole reduces the P_{GC} and therefore the GFR since less of the systemic pressure is transmitted to the glomerulus; dilation of the afferent arteriole, on the other hand, enhances both P_{GC} and GFR (Fig. 3-6). In comparison, constriction of the *efferent* arteriole retards fluid movement from the glomerulus into the efferent arteriole, resulting in an increase in P_{GC} and GFR; dilation of the efferent arteriole facilitates fluid entry into the efferent arteriole, diminishing P_{GC} and GFR (Fig. 3-6).

Arteriolar tone also affects the RPF, from which the glomerular filtrate is derived. In the kidney, the resistance to flow across the arterioles constitutes 85 percent of renal vascular resistance, the remaining 15 percent coming from the peritubular capillaries and renal veins [40]. If Eq. (3-2) is written for the renal circulation, then

$$RPF = \frac{\text{aortic pressure} - \text{renal venous pressure}}{\text{renal vascular resistance}} \qquad (3\text{-}3)$$

where the difference between aortic and renal venous pressures represents the ΔP across the kidney. This relation shows that an increase in tone at either end of the glomerulus will elevate total renal resistance and reduce RPF. Thus, GFR and RPF are regulated in parallel at the afferent arteriole, e.g., constriction decreases both, and inversely at the efferent arteriole, e.g., constriction reduces RPF but may augment P_{GC} and GFR. As a result, alterations in efferent (but not afferent) arteriolar tone affect the ratio of the GFR to the RPF since these parameters tend to change in opposite directions. This ratio is called the *filtration fraction* and in humans is normally about 20 percent (GFR is 125 mL/min; RPF is 625 mL/min).

The opposing effects of efferent arteriolar tone on P_{GC} and RPF also means that the direct relationship between this resistance and the GFR described above (Fig. 3-6) must be modified since the RPF is an independent determinant of GFR (see below). For example, although efferent arteriolar constriction increases the P_{GC}, the

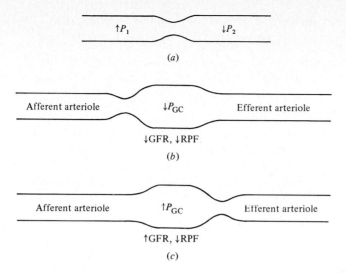

Figure 3-6 Relationship between arteriolar resistance, GFR, and RPF. (*a*) If flow is constant, constriction of a vessel results in a rise in pressure proximally (P_1) and a fall distally (P_2). (*b*) Constriction of the afferent arteriole reduces P_{GC} and GFR. (*c*) Constriction of the efferent arteriole increases P_{GC} and GFR. Since constriction of either arteriole also increases renal vascular resistance, RPF will fall in both (*b*) and (*c*). Arteriolar vasodilation has the opposite effects.

concomitant elevation in renal vascular resistance will reduce RPF, which will tend to lower the GFR. Depending upon the magnitude of efferent constriction, the net effect may be an increase, no change, or, if RPF is sufficiently reduced, even a fall in GFR.

Arteriolar resistance is partially under intrinsic myogenic control but also can be influenced by the renal sympathetic nerves, angiotensin II, and renal prostaglandins (see below).

Hydrostatic Pressure in Bowman's Space

Alterations in P_{BS} do not appear to play an important role in the physiologic regulation of GFR [41,42]. An obvious exception occurs with ureteral obstruction where the resistance to urine flow results in an increase in P_{BS}, reducing the hemodynamic gradient favoring filtration and therefore the GFR [43].

Plasma Oncotic Pressure

The plasma oncotic pressure is determined by the plasma protein concentration, particularly of albumin. Under normal conditions, the plasma albumin concentration is maintained within narrow limits. However, in certain clinical states, the plasma albumin concentration and Π_P are abnormal, and this can result in changes in GFR. For example, patients with volume depletion due to vomiting or diarrhea may develop hemoconcentration and an increase in the plasma protein concentration. This increases Π_P, contributing to the decrease in the GFR seen with these disorders.

Renal Plasma Flow

The RPF is another major determinant of the GFR. The mechanism by which this occurs is as follows. Experimental studies in rats and primates have demonstrated that the gradient favoring filtration falls to zero by the efferent arteriole (Table 3-1). This has been called *filtration equilibrium* and is due to the removal of the protein-free filtrate from the glomerulus which results in a progressive increment in the protein concentration (and therefore the oncotic pressure) in the fluid remaining in the glomerulus [35,41,44]. Since the hydrostatic pressures in the glomerulus and Bowman's space remain constant, this elevation in the plasma oncotic pressure (from 23 to 35 mmHg) ultimately abolishes the hydrostatic pressure gradient favoring filtration (Fig. 3-7), and filtration ceases. In the primate, this occurs after the filtration of 20 percent of the RPF, a filtration fraction similar to that seen in humans. Further filtration *at the same RPF* cannot occur, i.e., the GFR cannot exceed 20 percent of the RPF, without an increase in glomerular hydrostatic pressure or a reduction in plasma oncotic pressure.

However, the GFR will change if the RPF changes [35]. If, for example, the RPF is reduced with no alteration in glomerular hydrostatic pressure, then filtration equilibrium will still be reached after the filtration of 20 percent of the RPF. Thus, the GFR falls in proportion to the decrement in RPF, e.g., a 15 percent reduction in RPF induces a 15 percent reduction in GFR. Conversely, a 15 percent increase in RPF leads to a 15 percent increase in GFR.

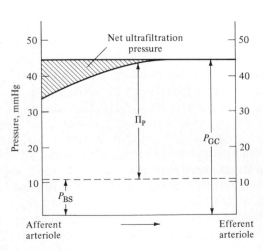

Figure 3-7 Depiction of the hemodynamic forces along the length of the primate glomerular capillary. The dotted line represents the hydrostatic pressure in Bowman's space P_{BS}. The plasma oncotic pressure is added to this so that the middle solid line represents the sum of the forces retarding filtration: $P_{BS} + \Pi_P$. The upper solid line represents the glomerular hydrostatic pressure (P_{GC}), and the shaded area depicts the net gradient favoring filtration, $P_{GC} - P_{BS} - \Pi_P$, which is $+13$ mmHg at the afferent arteriole. As a result of ultrafiltration, Π_P increases until the filtration gradient is abolished and filtration ceases. This is in contrast to muscle capillaries where filtration is limited by a decline in capillary hydrostatic pressure (see Chap. 1). *(Adapted from D. A. Maddox, W. M. Deen, and B. M. Brenner, Kidney Int., 5:271, 1974, and W. M. Deen, C. R. Robertson, and B. M. Brenner, Am. J. Physiol., 223:1178, 1972. By permission from Kidney International.)*

Note that the oncotic pressure of the fluid leaving the efferent arteriole and entering the peritubular capillary is determined by the plasma protein concentration and the degree to which the plasma proteins are concentrated by the removal of the protein-free filtrate, i.e., by the filtration fraction. As will be seen, the filtration fraction and the peritubular capillary oncotic pressure are important determinants of proximal tubular reabsorption (see page 95).

REGULATION OF GLOMERULAR FILTRATION RATE AND RENAL PLASMA FLOW

Regulation of renal hemodynamics is primarily achieved via changes in arteriolar resistance which can affect both RPF (from Eq. 3-3) and GFR (by altering P_{GC} and RPF). In contrast, the other determinants of GFR—K_f, P_{BS}, and Π_P—are relatively constant in normal subjects and are not readily altered.

Neurohumoral Influences

A variety of vasoactive hormones, including angiotensin II, norepinephrine (either circulating or released from the renal sympathetic nerves), prostaglandins, and kinins, can affect renal hemodynamics by their effects on arteriolar resistance [36]. Angiotensin II and norepinephrine constrict the efferent arteriole and, to a much lesser degree, the afferent arteriole [45, 45a]. The enhanced arteriolar tone decreases the RPF. However, the GFR is initially well maintained since the forces which tend to reduce the GFR—afferent arteriolar constriction and diminished RPF—are counteracted by the preferential efferent arteriolar constriction which increases the P_{GC} (Fig. 3-6) [45].

Renal vasodilator prostaglandins act to modify these effects of the vasoconstrictors. Both angiotensin II and norepinephrine stimulate the renal production of prostaglandins [8,46,47], which then decrease the arteriolar constriction, thereby minimizing the renal ischemia [25,48,49]. To a lesser degree, increased secretion of vasodilator kinins by the kidney also may act to preserve renal perfusion in this setting [50,51].

The physiologic importance of these effects can be illustrated by the hemodynamic response to volume depletion. In this setting, angiotensin II and norepinephrine secretion are enhanced. The ensuing systemic arteriolar constriction tends to maintain the blood pressure and to decrease musculocutaneous, splanchnic, and renal blood flow, which allows the preferential maintenance of coronary and cerebral perfusion. However, renal blood flow and therefore the GFR are diminished to a lesser degree because of the protective effect of the renal prostaglandins and kinins [25,48–51]. This permits the kidney to maintain its excretory functions despite the presence of hypovolemia. If, in this situation, a prostaglandin synthesis inhibitor such as indomethacin is administered, the action of the vasoconstrictors is relatively unopposed, resulting in marked reductions in RPF and, because of its plasma-flow dependence, GFR (Fig. 3-8) [25,48,49]. This effect has important clinical implications since pros-

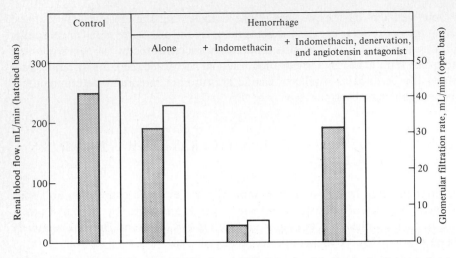

Figure 3-8 Renal blood flow (hatched bars) and GFR (open bars) in the dog in the control state and during hypotensive hemorrhage alone, with the use of indomethacin, or with indomethacin plus denervation and the administration of an angiotensin II antagonist. RBF and GFR are relatively well maintained following hemorrhage because prostaglandins antagonize the vasoconstrictive effects of the sympathetic nerves and angiotensin II. The administration of indomethacin to inhibit prostaglandin synthesis in this setting results in marked renal ischemia. However, this effect is not seen if vasoconstriction is blocked by denervation and the administration of an angiotensin antagonist since the vasoconstrictive stimulus to prostaglandin release has been eliminated. *(Adapted from W. L. Henrich, T. Berl, K. M. McDonald, R. J. Anderson, and R. W. Schrier, Am. J. Physiol., 235:F46, 1978.)*

taglandin synthesis inhibitors, which are widely used in the treatment of arthritis and other disorders, can lead to the development of renal failure when given to patients with diseases associated with effective circulating volume depletion such as heart failure or hepatic cirrhosis (see Chap. 16) [52,53]. (Although inhibitors of kinin synthesis, such as aprotinin, could have a similar effect [50,51], these drugs are rarely given clinically.)

In contrast to their effect in hypovolemic states, angiotensin II, norepinephrine, prostaglandins, and kinins do not appear to play an important role in the regulation of renal hemodynamics in normal subjects in whom there is a relatively low rate of secretion of these hormones [46,48,50,54]. In this setting, GFR and RPF are controlled primarily by autoregulation and tubuloglomerular feedback, which are intrarenal, not systemic, processes.

Autoregulation

Since P_{GC} is an important determinant of GFR, it might be expected that small variations in arterial pressure could induce large changes in GFR. However, over a wide range of mean arterial pressures the GFR and RPF remain roughly constant (Fig. 3-9) [55]. This phenomenon, which is also present in other capillaries [56], is intrinsic to the kidney, occurring in denervated, perfused kidneys, and has been termed *autoregulation*. Since GFR and RPF are maintained in parallel, autoregulation is primarily mediated by changes in afferent arteriolar resistance (Fig. 3-6) [42,57]. As systemic pressure rises, an increase in afferent arteriolar tone prevents the increase in pressure from being transmitted to the glomerulus, allowing P_{GC} and GFR to remain unchanged. The enhanced arteriolar resistance also increases total renal vascular resistance and, from Eq. (3-3), can prevent any change in the RPF. Conversely, as blood pressure decreases, afferent arteriolar dilation will initially protect both GFR and RPF. However, the afferent arteriole becomes maximally dilated at a mean arterial pressure of approximately 70 to 80 mmHg in normotensive subjects. As a result, autoregulation is lost at pressures below this level. In this setting, GFR and RPF fall in proportion to the drop in blood pressure and the GFR ceases when the systemic pressure reaches 40 to 50 mmHg.

When the arterial pressure is increased, autoregulation is important in preventing the filtered load from exceeding tubular reabsorptive capacity. Although proximal, loop, and distal tubular reabsorption can increase in proportion to changes in GFR (the phenomenon of glomerulotubular balance, see Chap. 5), the ability of the collecting tubules to augment reabsorption is limited. If collecting tubule load became excessive due to an inappropriate increase in GFR, potentially serious losses of Na^+ and water could occur.

The mechanism by which autoregulation is mediated is incompletely understood. The simplest hypothesis is that myogenic stretch receptors in the wall of the afferent arteriole are of primary importance, similar to the role of the precapillary sphincter in the muscle capillary (see page 35) [56]. A reduction in renal perfusion pressure, for example, will decrease the degree of stretch, which will then promote arteriolar dilation. However, renal autoregulation is substantially more complex than the myogenic theory as evidenced by the observation that the autoregulation of RPF occurs earlier than, and can be dissociated from, that of GFR [58,59]. Thus, factors in addition to myogenic properties of the afferent arteriole must play a contributory role.

Angiotensin II and tubuloglomerular feedback are particularly important in this regard. The renin-angiotensin system is activated as the renal perfusion pressure is

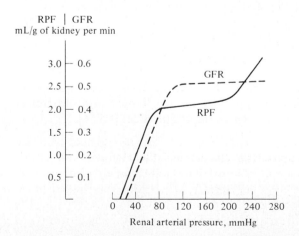

Figure 3-9 Autoregulation of GFR and RPF in the dog. (*Adapted from R. E. Shipley and R. S. Study, Am. J. Physiol., 167:676, 1951.*)

lowered. Its role in autoregulation has been demonstrated by the administration of an angiotensin antagonist as hypotension is induced. In this settting, the autoregulation of RPF is intact but not that of the GFR which begins to fall [59]. This finding indicates that stretch receptor–induced *afferent* arteriolar dilation maintains the RPF but only partially protects the GFR. It is the ensuing increase in angiotensin II production that returns the GFR to normal by preferentially *constricting* the *efferent* arteriole, thereby raising the glomerular capillary hydrostatic pressure.

Although this sequence is reversed when the renal perfusion pressure is elevated, angiotensin II levels are low in the basal state and it is unlikely that any further reduction is responsible for the maintenance of GFR. There is, however, substantial evidence for the role of tubuloglomerular feedback in this setting.

Tubuloglomerular Feedback

It has generally been thought that glomerular filtration is the primary event with tubular reabsorption responding to changes in the filtered load. However, it now appears that alterations in tubular function can affect the GFR. This phenomenon, termed *tubuloglomerular feedback* (TGF), seems to be mediated by the macula densa segment of the early distal tubule which senses changes in the delivery and subsequent reabsorption of sodium, chloride, or other solutes [57,60,61]. If flow to the macula densa is enhanced, there is a reduction in GFR (Fig. 3-10) which returns distal flow to normal. This effect may be mediated both by afferent arteriolar constriction and by decreased capillary permeability (due perhaps to contraction of the mesangium which will diminish the surface area available for filtration) [60,62].

The mechanism by which TGF occurs is unknown. Angiotensin II appears to have only a permissive role [63] and it seems likely that some other vasoconstrictor produced within the kidney is at least in part responsible for this process† [60,64]. This agent is released when there is an increase in macula densa flow [60]; this is in contrast to the secretion of *renin* which is enhanced by a reduction in macula densa flow [29].

The physiologic role of TGF is also uncertain. It has been proposed that TGF may be responsible for autoregulation [57]. If renal perfusion pressure rises, the ensuing elevation in capillary hydrostatic pressure will tend to increase the GFR. The augmentation in filtered load will be sensed by the macula densa, which will return the GFR, RPF, and macula densa flow toward normal. However, if TGF is abolished by the luminal perfusion of furosemide (which inhibits NaCl reabsorption in the macula densa), there is only a partial impairment in autoregulation [65]. Thus, the role of TGF in autoregulation appears to be contributory with the stretch receptors in the afferent arteriole also having an important effect [65,66].

---

†One possibility is adenosine which is generated from the utilization of adenosine triphosphate (ATP) [64]. If the GFR is increased, more NaCl is filtered and then actively reabsorbed in the loop of Henle and macula densa. This step requires the catabolism of ATP, and the ensuing generation of adenosine could be responsible for the afferent arteriolar and mesangial constriction that return the GFR and filtered load toward normal [64].

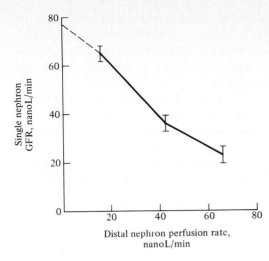

Figure 3-10 Relationship of single nephron GFR to distal nephron (macula densa) perfusion rate in dogs. As the perfusion rate increases (via the insertion of a micropipette into the late proximal tubule), there is a progressive reduction in GFR. *(From L. G. Navar, Am. J. Physiol., 234:F357, 1978.)*

It may be that the primary role of TGF is to prevent excessive salt and water losses (by reducing the filtered load) when the GFR is increased or tubular fluid reabsorption in the more proximal segments is reduced [60]. For example, carbonic anhydrase inhibitors like acetazolamide are proximally acting diuretics which produce a fall in GFR even if fluid losses are replaced. This effect may be due to TGF since there is no decrease in GFR if the diuretic-induced increase in macula densa flow is prevented by experimentally perfusing the loop of Henle at a constant rate [67]. Furthermore, the reduction in GFR seen in the intact animal is exactly sufficient to return macula densa flow to baseline levels.

Thus, it may be that *it is macula densa flow itself, not the GFR, that is being regulated* [67a]. This would be appropriate physiologically since the bulk of the filtrate is reabsorbed in the proximal tubule and loop of Henle; the final *qualitative* changes in the urine are made in the distal nephron which has a more limited total reabsorptive capacity that could be overwhelmed if delivery were substantially increased. Thus, TGF and autoregulation are intrarenal mechanisms which maintain distal fluid delivery at a relatively constant level so that minor changes in urinary excretion (according to changes in intake) can be made in the distal nephron.

Distribution of Renal Blood Flow

In normal circumstances, approximately 80 percent of renal blood flow goes to the outer cortex, 10 to 15 percent to the inner cortex (the site of the juxtamedullary nephrons; see Fig. 2-3), and the remaining 5 to 10 percent to the medulla. The low rate of medullary flow, due in part to the relatively high resistance in the vasa recta capillaries, plays an important role in the countercurrent mechanism and urinary concentration (see Chap. 6).

This distribution may be altered in a variety of conditions. In particular, there is a marked reduction in outer cortical flow with a preferential increase in perfusion to the inner cortex in disorders associated with renal Na^+ retention such as heart failure and hypotension [68–70]. These changes are reversible if circulatory hemodynamics can be normalized [69]. The mechanism by which they occur is unknown as afferent arteriolar stretch receptors, angiotensin II, circulating catecholamines, and prostaglandins have all been implicated [70–73].

The physiologic effect of these changes is also uncertain. It had been postulated that increasing inner cortical flow might promote Na^+ retention because the juxtamedullary nephrons with their loops of Henle have a greater reabsorptive surface and therefore could reabsorb Na^+ more efficiently than those in the outer cortex. However, the significance of the intrarenal shunting of blood flow remains unclear since Na^+ retention can occur with normal outer cortical perfusion [72,74], and redistribution of blood flow is not necessarily associated with redistribution of glomerular filtration [75].

Summary

The GFR is normally maintained within relatively narrow limits to prevent inappropriate GFR-induced fluctuations in solute and water excretion. Regulation of the GFR is primarily achieved by alterations in arteriolar tone which influence both the hydrostatic pressure in the glomerular capillary and renal blood flow. In normal subjects, the GFR is maintained by autoregulation, a phenomenon which is mediated by at least three factors: stretch receptors in the afferent arteriole, angiotensin II, and tubuloglomerular feedback.

In certain conditions, however, the effects of autoregulation are overridden. Severe hypovolemia or hypotension, for example, results in marked increases in the release of angiotensin II and norepinephrine. The net effect is renal vasoconstriction (not autoregulatory vasodilation), a reduction in renal blood flow, and a lesser fall or no change in GFR (since efferent constriction increases the glomerular capillary hydrostatic pressure). This response is appropriate since it allows renal excretory function (via the GFR) to be maintained while permitting the preferential shunting of blood to the coronary and cerebral circulations. To some degree, the renal ischemia is minimized by a vasoconstrictor-induced increase in the renal secretion of vasodilator prostaglandins.

CLINICAL EVALUATION OF RENAL CIRCULATION

Concept of Clearance and Measurement of GFR

Estimation of the GFR is an essential part of the evaluation of patients with renal disease. Since the total kidney GFR is equal to the sum of the filtration rates of each of the functioning nephrons, the total GFR can be used as an *index of functioning renal mass*. For example, the loss of one-half of the functioning nephrons should initially lead to roughly a 50 percent decrease in the GFR.† Thus, *the GFR can be used to document the presence, estimate the severity, and follow the course of kidney disease.* A reduction in GFR implies either progression of the underlying disease or the development of a superimposed and potentially reversible problem such as diminished renal perfusion due to volume depletion. An increase in GFR, on the other hand, indicates improvement or possibly hypertrophy in the remaining nephrons.

Measurement of the GFR is also helpful in determining the proper dosage of those drugs that are excreted by the kidney by glomerular filtration. When the GFR falls, drug excretion will be reduced, resulting in an increase in plasma drug levels and potential drug toxicity. To prevent this, drug dosage must be lowered in proportion to the decrease in GFR [76].

How can the GFR be measured? Consider a compound, such as the fructose polysaccharide inulin (not insulin), with the following properties:

1. Able to achieve a stable plasma concentration
2. Freely filtered at the glomerulus
3. Neither reabsorbed, secreted, synthesized, nor metabolized by the kidney

In this situation,

$$\text{Filtered inulin} = \text{excreted inulin}$$

The filtered inulin is equal to the GFR times the plasma inulin concentration (P_{in}), and the excreted inulin is equal to the product of the urine inulin concentration (U_{in}) and the urine volume (V, in milliliters per minute or liters per day). Therefore,

$$\text{GFR} \times P_{in} = U_{in} \times V \tag{3-4}$$

$$\text{GFR} = \frac{U_{in} \times V}{P_{in}}$$

The term $(U_{in} \times V)/P_{in}$ is called the clearance of inulin and is an accurate estimate of the GFR. The inulin clearance, in milliliters per minute, refers to that volume of plasma cleared of inulin by renal excretion. For example, if 1 mg of inulin is excreted per minute ($U_{in} \times V$) and the P_{in} is 1.0 mg/dL (or to keep the units consistent, 0.01 mg/mL), then the clearance of inulin is 100 mL/min; that is, 100 mL of plasma has been cleared of the 1 mg of inulin that it contained.

Despite its accuracy, the inulin clearance is rarely performed clinically because

---

†Chronically, the GFR will return to about 70 percent of the previous baseline because of hypertrophy and an increase in nephron filtration rate in the remaining nephrons [75a, 75b]. To the degree that this occurs, the GFR will actually provide an *over*estimate of the number of functioning nephrons. Although this adaptation is at first beneficial by maintaining a high GFR, there is substantial evidence that it is ultimately harmful, leading to glomerular sclerosis and renal failure [75b]. This deleterious effect of nephron hyperfiltration may contribute to the progressive nature of many kidney diseases. Both the hyperfiltration and perhaps the long-term decline in renal function can be prevented by limiting dietary protein intake [75b].

it involves both an intravenous infusion of inulin and an assay for inulin that is not available in most laboratories. The most widely used method to estimate the GFR is the endogenous creatinine clearance [77–79]. Creatinine is derived from the metabolism of creatine in skeletal muscle and is released into the plasma at a relatively constant rate. As a result, the plasma creatinine concentration (P_{cr}) is very stable, varying less than 10 percent per day in serial observations in normal subjects. Like inulin, creatinine is freely filtered across the glomerulus and is neither reabsorbed nor metabolized by the kidney. However, a small amount of creatinine enters the urine by tubular secretion in the proximal tubule [77]. Due to tubular secretion, the amount of creatinine excreted exceeds the amount filtered by 10 to 20 percent. Therefore, the creatinine clearance (C_{cr}),

$$C_{cr} = \frac{U_{cr} \times V}{P_{cr}} \tag{3-5}$$

will tend to exceed the inulin clearance by 10 to 20 percent. Fortuitously, this is balanced by an error of almost equal magnitude in the measurement of the P_{cr}. The most commonly used method involves a colorimetric reaction after the addition of alkaline picrate. The plasma, but not the urine, contains noncreatinine chromagens (acetone, proteins, ascorbic acid, pyruvate) which account for approximately 10 to 20 percent of the normal P_{cr} [77]. Since both the U_{cr} and P_{cr} are elevated to roughly the same degree, the errors tend to cancel out and the C_{cr} is a reasonably accurate estimate of the GFR [77–79], particularly if the GFR is greater than 40 mL/min (normal equals 95 to 120 mL/min) [77]. However, as renal failure progresses and the GFR falls, less creatinine is filtered and proportionately more of the urinary creatinine is derived from tubular secretion. As a result, the $U_{cr} \times V$ is much higher than it would be if creatinine were excreted only by glomerular filtration, and the C_{cr} can exceed that of inulin by 10 to 40 percent or more [77]. This error, however, does not substantially detract from the clinical usefulness of the C_{cr}. For example, a C_{cr} of 35 mL/min indicates the presence of moderately severe renal disease. The fact that the GFR (as measured by the inulin clearance) may actually be only 25 mL/min is not so important since knowledge of the exact GFR is usually not necessary.†

The C_{cr} is usually determined in the following way. Venous blood is used for the P_{cr}. Urinary creatinine excretion $U_{cr} \times V$ is concomitantly measured on a 24-h collection since shorter collections tend to give less reliable results [82]. For example, a 30-year-old female who weighs 60 kg is being evaluated for possible kidney disease and the following results are obtained:

$$P_{cr} = 1.5 \text{ mg/dL}$$

$$U_{cr} = 100 \text{ mg/dL}$$

$$V = 1080 \text{ mL/day}$$

---

†It has been suggested that the GFR can be more accurately determined by measuring the clearance of radioactively labeled iothalamate or vitamin B_{12} [80,81]. However, for clinical purposes, the C_{cr} remains a satisfactory method that is simple to perform and available in almost all laboratories.

And

$$1080 \text{ mL/day} \div 1440 \text{ min/day} = 0.75 \text{ mL/min}$$

Thus:

$$C_{cr} = \frac{U_{cr} \times V}{P_{cr}}$$

$$= \frac{100 \times 0.75}{1.5} = 50 \text{ mL/min}$$

Since this is roughly one-half the normal C_{cr}, this patient has lost approximately one-half of her GFR.

The major error involved in the determination of the C_{cr} is an incomplete urine collection. For this reason, it is important to know the normal values for creatinine excretion. In adults under the age of 60, daily creatinine excretion should be 20 to 25 mg/kg lean body weight in males and 15 to 20 mg/kg in females [77,83]. From the ages of 60 to 90, there is a progressive 50 percent reduction in creatinine excretion (from 20 to 10 mg/kg in males), probably due to a decrease in skeletal muscle mass [83]. If creatinine excretion is found to be much less than these values, an incomplete collection should be suspected. In the patient described above, creatinine excretion was 18 mg/kg per day (1080 mg/60 kg), suggesting that a complete collection was obtained.

P_{cr} and GFR

Changes in, and estimation of, the GFR also can be ascertained from measurement of the P_{cr}, a simpler test to perform than the C_{cr}. In a subject in the steady state,

$$\text{Creatinine excretion} = \text{creatinine production} \quad (3\text{-}6)$$

Creatinine excretion is roughly equal to the amount of creatinine filtered (GFR $\times$ P_{cr}) whereas the rate of creatinine production is relatively constant. If these substitutions are made in Eq. (3-6), then

$$\text{GFR} \times P_{cr} = \text{constant} \quad (3\text{-}7)$$

Thus, *the P_{cr} varies inversely with the GFR*. If, for example, the GFR falls by 50 percent, creatinine excretion also will be reduced. As a result, newly produced creatinine will accumulate in the plasma until the filtered load again equals the rate of production. This will occur when the P_{cr} has doubled,

$$\text{GFR}/2 \times 2P_{cr} = \text{GFR} \times P_{cr} = \text{constant}$$

In adults, the normal P_{cr}† is 0.8 to 1.3 mg/dL in males and 0.6 to 1.0 mg/dL in females [77].

---

†The P_{cr} should be measured when the patient is fasting, since cooked meat and its broth contain enough creatinine to acutely raise the P_{cr} by as much as 1.0 mg/dL [84]. In a subject with normal renal function, this can result in a doubling of P_{cr} and an apparent 50 percent reduction in GFR.

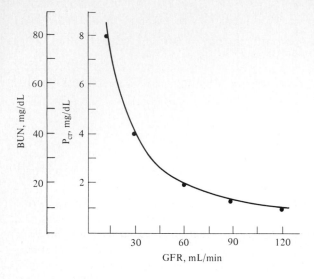

Figure 3-11 Steady-state relationship between the plasma creatinine concentration (P_{cr}), blood urea nitrogen (BUN), and the GFR.

The reciprocal relationship between the GFR and the P_{cr} is depicted in Fig. 3-11. There are three important points to note about this relationship. First, this curve is valid *only in the steady state*. If a patient develops acute renal failure with a sudden drop in the GFR from 120 to 12 mL/min, the P_{cr} on day 1 will still be normal since there will not have been time for creatinine to accumulate in the plasma. After 7 to 10 days, the P_{cr} will stabilize roughly at 10 mg/dL, a level consistent with the reduced GFR.

It should be remembered that the steady state can be disturbed by changes in creatinine production as well as in GFR. When creatinine production is acutely increased, as with severe muscle breakdown, the P_{cr} can increase out of proportion to any change in GFR [85].

Second, it is important to note the *shape of* the curve. In a patient with normal renal function, *an apparently minor increase in the P_{cr} from 1.0 to 2.0 mg/dL can represent a marked fall in the GFR from 120 to 60 mL/min.* In contrast, in a patient with advanced renal failure, a marked increase in the P_{cr} from 6.0 to 12.0 mg/dL reflects a relatively small reduction in the GFR from 20 to 10 mL/min. Thus, the initial elevation of the P_{cr} represents the major loss in GFR.

Third, the relationship between the GFR and the P_{cr} is dependent upon the rate of creatinine production, which is largely a function of muscle mass. In Fig. 3-11, a normal GFR of 120 mL/min is associated with a P_{cr} of 1.0 g/dL. Although this may be true for a 70-kg male, a similar GFR in a 50-kg female might be associated with a P_{cr} of only 0.6 mg/dL. In this setting, a P_{cr} of 1.0 mg/dL is not normal and reflects a 40 percent fall in GFR.

To account for the effects of body weight, age, and sex on muscle mass, the following formula has been derived to estimate the C_{cr} from the P_{cr} in the steady state in adult males [86]:

$$C_{cr} = \frac{(140 - \text{age}) \times \text{lean body weight}}{P_{cr} \times 72}$$

This value should be multiplied by 0.85 in females since a lower fraction of the body weight is composed of muscle. The units of measure used in this formula are: C_{cr}, mL/min; age, years; lean body weight, kg; and P_{cr}, mg/dL.

The results obtained with this formula appear to correlate fairly well with a simultaneously measured C_{cr}. Its usefulness can be illustrated by the observation that a P_{cr} of 1.4 mg/dL represents a C_{cr} of 101 mL/min in an 85-kg, 20-year-old man

$$C_{cr} = \frac{(140 - 20) \times 85}{1.4 \times 72}$$

but only a C_{cr} of 20 mL/min in a 40-kg, 80-year-old woman,

$$C_{cr} = \frac{(140 - 80) \times 40}{1.4 \times 72} \times 0.85$$

This example calls attention to the danger of overdosing elderly patients who have seriously impaired renal function despite a relatively normal P_{cr}. The use of this simple formula, while not absolutely accurate, will help to avoid this problem.

In summary, the P_{cr} varies inversely with the GFR in the steady state. Because of this relationship, serial measurements of the P_{cr} can be used to look for disease progression in patients with kidney dysfunction. The loss of functioning nephrons usually is associated with a reduction in the GFR and should result in an increase in the P_{cr}. On the other hand, if the P_{cr} remains stable (and there has been no loss of muscle mass), it is reasonably safe to assume that the GFR has not significantly changed.

Blood Urea Nitrogen and GFR

Changes in the GFR also can be detected by changes in the concentration of urea in the blood, measured as the blood urea nitrogen (BUN). Like creatinine, urea is excreted primarily by glomerular filtration and the BUN tends to vary inversely with the GFR (Fig. 3-11).

However, two factors can alter the BUN without change in the GFR or P_{cr}. First, urea production may not be constant. Urea is formed by the hepatic metabolism of amino acids not utilized for protein synthesis. As amino acids are deaminated, ammonia is produced. The development of toxic levels of ammonia in the blood is prevented by the conversion of ammonia (NH_3) into urea in a reaction that can be summarized by the following equation:

$$2NH_3 + CO_2 \rightarrow H_2N - \overset{\overset{\displaystyle O}{\displaystyle \|}}{C} - NH_2 + H_2O$$
$$\text{Urea}$$

Thus, urea production and the BUN are increased when more amino acids are metabolized in the liver. This may occur with a high-protein diet, enhanced tissue breakdown (due to trauma, gastrointestinal bleeding, or the administration of corticosteroids), or decreased protein synthesis (due to tetracycline) [87]. On the other hand, urea production and the BUN are reduced by severe liver disease or a low protein intake.

Second, urea excretion is not determined solely by glomerular filtration. Approximately 40 to 50 percent of the filtered urea is normally reabsorbed by the tubules. The reabsorption of urea tends to follow passively that of sodium. Thus, in states of volume depletion in which sodium reabsorption is increased, urea reabsorption also is enhanced. The net result is reduced urea excretion and an elevation in the BUN that is not due to a fall in GFR [87]. Under most conditions, the ratio of the BUN to the P_{cr} is 10 to 15:1. When this ratio exceeds 20:1, one of the conditions associated with enhanced urea production or tubular reabsorption should be suspected [87].

In summary, a reduction in the GFR results in elevation in both the BUN and P_{cr}. Because of the variability in urea production and reabsorption, the P_{cr} is a more reliable reflection of the GFR. For similar reasons, the urea clearance is not an accurate estimate of the GFR. Since urea is reabsorbed and the degree of reabsorption is variable, the quantity of urea excreted is much less than the amount filtered. As a result, the urea clearance is only 50 to 70 percent that of inulin [88]. Thus, the C_{cr} is the preferred clinical method for measuring the GFR with one exception. As described above, renal failure is associated with an increase in the percent of excreted creatinine derived from tubular secretion [77,79]. Thus, the C_{cr} can exceed the inulin clearance by 10 to 40 percent or more while the urea clearance will continue to be less than C_{in}. Consequently, in patients with advanced renal failure (P_{cr} greater than 4 mg/dL), the GFR is best estimated by taking the average of the urea and creatinine clearances [89]:

$$GFR = \frac{C_{cr} + C_{urea}}{2}$$

Measurement of Renal Plasma Flow

The principles of clearance also can be used to measure the RPF. Paraaminohippuric acid (PAH) is an easily measured indicator that enters the urine by glomerular filtration and by the organic acid secretory pathway in the proximal tubule. By filtration and secretion, PAH is almost entirely removed from the plasma in a single pass through the kidney. Therefore,

$$PAH \text{ presented to the kidney} = PAH \text{ excreted}$$

$$RPF \times P_{PAH} = U_{PAH} \times V$$

$$RPF = \frac{U_{PAH} \times V}{P_{PAH}} = C_{PAH}$$

If the hematocrit (Hct) is known, then the renal blood flow can be calculated from

$$RBF = \frac{C_{PAH}}{1 - Hct}$$

The normal RPF and RBF in humans are roughly 625 mL/min and 1100 mL/min, respectively. Since only 85 to 90 percent of the PAH actually is removed from the circulation in a single pass, both RPF and RBF will be underestimated by 10 to 15 percent.

Whereas serial measurements of the plasma creatinine concentration or creatinine clearance may be helpful in assessing the course of a patient with renal insufficiency, measurement of the RPF provides little useful information and is rarely performed in clinical situations.

PROBLEMS

3-1 A 68-year-old man with acute renal failure is admitted to the hospital. The following plasma creatinine concentration values are obtained:

Day	[Creatinine], mg/dL
1	1.0
2	3.0
3	5.0

If the patient weighs 60 kg, what would you estimate the GFR to be on day 2?

3-2 If the left renal artery is constricted, producing a decrease in renal perfusion, what will happen to (*a*) renin secretion from the left kidney; (*b*) the arterial blood pressure; and (*c*) renin secretion from the right kidney?

3-3 A creatinine clearance test is performed on an 80-kg male. The following results are obtained:

$$\text{Plasma [creatinine]} = 3.5 \text{ mg/dL}$$

$$\text{24-h urine volume} = 800 \text{ mL}$$

$$\text{Urine [creatinine]} = 125 \text{ mg/dL}$$

(*a*) Calculate the creatinine clearance.
(*b*) Is this an accurate estimate of the GFR?

3-4 A patient becomes mildly volume-depleted, leading to activation of the renin-angiotensin system. What will happen to (*a*) the RPF; (*b*) the GFR (in relation to the change in RPF); (*c*) the filtration fraction; and (*d*) the concentration of albumin in the peritubular capillary?

REFERENCES

1. Beeuwkes, R., and J. V. Bonventre: Tubular organization and vascular-tubular relations in the dog kidney, Am. J. Physiol., 229:695, 1975.
2. Venkatachalam, M. A., and H. G. Rennke: The structure and molecular basis of glomerular filtration, Circ. Res., 43:337, 1978.
3. Brenner, B. M., T. H. Hostetter, and H. D. Humes: Molecular basis of proteinuria of glomerular origin, N. Engl. J. Med., 298:826, 1978.
4. Bennett, C. M., R. J. Glassock, R. L. S. Chang, W. M. Deen, C. R. Robertson, and B. M. Brenner: Permselectivity of the glomerular capillary wall: studies of experimental glomerulonephritis in the rat using dextran sulfate, J. Clin. Invest., 57:1287, 1976.

5. Robson, A. M., J. Giangiacomo, R. A. Kienstra, S. T. Naqvi, and J. R. Inglefinger: Normal glomerular permeability and its modification by minimal change nephrotic syndrome, J. Clin. Invest., 54:1190, 1974.

6. Sharon, Z., M. M. Schwartz, B. U. Pauli, and E. J. Lewis: Kinetics of glomerular visceral epithelial cell phagocytosis, Kidney Int., 14:526, 1978.

7. Kreisberg, J. I., and M. J. Karnovsky: Glomerular cells in culture, Kidney Int., 23:439, 1983.

8. Schlondorff, D., S. Roczniak, J. A. Satriano, and V. W. Folkert: Prostaglandin synthesis by isolated rat glomeruli: effect of angiotensin II, Am. J. Physiol., 239:F486, 1980.

8a. Hsueh, W. A., E. J. Carlson, and V. J. Dzau: Characterization of inactive renin from human kidney and plasma. Evidence of a renal source of circulating inactive renin, J. Clin. Invest., 71:506, 1983.

9. Peart, W. S.: Renin-angiotensin system, N. Engl. J. Med., 292:302, 1975.

10. Haber, E.: The renin-angiotensin system and hypertension, Kidney Int., 15:427, 1979.

11. Laragh, J. H., R. L. Soffer, and D. B. Case: Converting enzyme, angiotensin II and hypertensive disease, Am. J. Med., 64:147, 1978.

12. Carey, R. M., E. D. Vaughan, Jr., and M. J. Peach: Activity of [des-aspartyl]-angiotensin II and angiotensin II in man, J. Clin. Invest., 61:20, 1978.

13. Hall, J. E., A. C. Guyton, N. C. Trippodo, T. E. Lohmeier, R. E. McCaa, and A. W. Cowley, Jr.: Intrarenal control of electrolyte excretion by angiotensin II, Am. J. Physiol., 237:F424, 1979.

14. Malik, K. U. and A. Nasjletti: Facilitation of adrenergic transmission by locally generated angiotensin II in rat mesenteric arteries, Circ. Res., 38:26, 1976.

15. Spertini, F., H. R. Brunner, B. Waeber, and H. Gavras: The opposing effects of chronic angiotensin-converting enzyme blockade by captopril on the responses to exogenous angiotensin II and vasopressin vs. norepinephrine in rats, Circ. Res. 48:612, 1981.

16. Shade, R. E., J. P. Davis, J. A. Johnson, and R. W. Gotshall: Mechanism of action of angiotensin II and antidiuretic hormone on renin secretion, Am. J. Physiol., 224:926, 1973.

17. Gavras, H., H. R. Brunner, E. D. Vaughan, Jr., and J. H. Laragh: Angiotensin-sodium interaction in blood pressure maintenance of renal hypertensive and normotensive rats, Science, 180:1369, 1973.

18. Watkins, L., Jr., J. A. Burton, E. Haber, J. R. Cant, F. W. Smith, and A. C. Barger: The renin-angiotensin-aldosterone system in congestive failure in conscious dogs, J. Clin. Invest., 57:1606, 1976.

19. Curtiss, C., J. N. Cohn, T. Vrobel, and J. A. Franciosa: Role of the renin-angiotensin system in the systemic vasoconstriction of chronic heart failure, Circulation, 58:763, 1978.

20. Schroeder, E. T., G. H. Anderson, S. H. Goldman, and D. H. P. Streeten: Effect of blockade of angiotensin II on blood pressure, renin and aldosterone in cirrhosis, Kidney Int., 9:511, 1976.

21. Noth, R. H., S. Y. Tan, and P. J. Mulrow: Effects of angiotensin II blockade by saralasin in normal man, J. Clin. Endocrinol. Metab., 45:10, 1977.

22. Data, J. L., J. G. Berber, W. J. Crump, J. C. Frolich, J. W. Hollifield, and A. S. Nies: The prostaglandin system: a role in canine baroreceptor control of renin release, Circ. Res., 42:454, 1978.

23. Berl, T., W. L. Henrich, A. L. Erickson, and R. W. Schrier: Prostaglandins in the beta-adrenergic and baroreceptor-mediated secretion of renin, Am. J. Physiol., 236:F472, 1979.

24. Beierwaltes, W. H., S. Schryver, E. Sanders, J. Strand, and J. C. Romero: Renin release selectively stimulated by prostaglandin I_2 in isolated rat glomeruli, Am. J. Physiol., 243:F276, 1982.

25. Oliver, J. A., J. Pinto, R. R. Sciacca, and P. J. Cannon: Increased renal secretion of norepinephrine and prostaglandin E_2 during sodium depletion in the dog, J. Clin. Invest., 66:748, 1980.

25a. Henrich, W. L., T. Berl, K. M. McDonald, R. J. Anderson, and R. W. Schrier: Angiotensin II, renal nerves, and prostaglandins in renal hemodynamics during hemorrhage, Am. J. Physiol., 235:F46, 1978.

26. Kopp, U., and G. F. DiBona: Interaction of renal β_1-adrenoreceptors and prostaglandins in reflex renin release, Am. J. Physiol., 244:F418, 1983.

27. Himori, N., A. Izum, and T. Ishimori: Analysis of β-adrenoreceptors mediating renin release produced by isoproterenol in conscious dogs, Am. J. Physiol., 238:F387, 1980.

28. Gerber, J. G., R. D. Olson, and A. S. Nies: Interrelationship between prostaglandins and renin release, Kidney Int., 19:816, 1981.

29. Francisco, L. L., J. L. Osborn, and G. F. DiBona: Prostaglandins in renin release during sodium deprivation, Am. J. Physiol., 243:F537, 1982.
30. Henrich, W. L., R. W. Schrier, and T. Berl: Mechanisms of renin secretion during hemorrhage in the dog, J. Clin. Invest., 64:1, 1979.
31. Pappenheimer, J. R., E. M. Renkin, and L. M. Borrero: Filtration, diffusion, and molecular sieving through peripheral capillary membrane: a contribution to the pore theory of capillary permeability, Am. J. Physiol., 167:13, 1951.
32. Deen, W. M., J. L. Troy, C. R. Robertson, and B. M. Brenner: Dynamics of glomerular ultrafiltration in the rat. IV. Determination of the ultrafiltration coefficient, J. Clin. Invest., 52:1500, 1973.
33. Maddox, D. A., W. M. Deen, and B. M. Brenner: Dynamics of glomerular ultrafiltration. VI. Studies in the primate, Kidney Int., 5:271, 1974.
34. Guyton, A. C.: *Textbook of Medical Physiology,* Saunders, Philadelphia, 1981, chap. 30.
35. Brenner, B. M., and H. D. Humes: Mechanics of glomerular ultrafiltration, N. Engl. J. Med., 297:148, 1977.
36. Dworkin, L. D., I. Ichikawa, and B. M. Brenner: Hormonal modulation of glomerular function, Am. J. Physiol. 244:F95, 1983.
37. Ausiello, D. A., J. I. Kriesberg, C. Roy, and M. J. Karnovsky: Contraction of cultured rat glomerular cells of apparent mesangial origin after stimulation with angiotensin II and arginine vasopressin, J. Clin. Invest., 65:754, 1980.
38. Ichikawa, I., and B. M. Brenner: Evidence for glomerular actions of ADH and dibutyryl cyclic AMP in the rat, Am. J. Physiol., 233:F102, 1977.
39. Maddox, D. A., C. M. Bennett, W. M. Deen, R. J. Glassock, D. Knetson, and B. M. Brenner: Control of proximal tubule fluid reabsorption in experimental glomerulonephritis, J. Clin. Invest., 55:1315, 1975.
40. Renkin, E., and R. Robinson: Glomerular filtration, N. Engl. J. Med., 290:785, 1974.
41. Brenner, B. M., J. L. Troy, T. M. Daugharty, W. M. Deen, and C. R. Robertson: Dynamics of glomerular ultrafiltration in the rat. II. Plasma-flow dependence of GFR, Am. J. Physiol., 223:1184, 1972.
42. Robertson, C. R., W. M. Deen, J. L. Troy, and B. M. Brenner: Dynamics of glomerular ultrafiltration. III. Hemodynamics and autoregulation, Am. J. Physiol., 223:1191, 1972.
43. Yarger, W. E., H. S. Aynedjian, and N. Bank: A micropuncture study of postobstructive diuresis in the rat, J. Clin. Invest., 51:625, 1972.
44. Deen, W. M., C. R. Robertson, and B. M. Brenner: A model of glomerular ultrafiltration in the rat, Am. J. Physiol., 223:1178, 1972.
45. Myers, B. D., W. M. Deen, and B. M. Brenner: Effects of norepinephrine and angiotensin II on the determinants of glomerular ultrafiltration and proximal tubule fluid reabsorption in the rat, Circ. Res., 37:101, 1975.
45a. Edwards, R. M.: Segmental effects of norepinephrine and angiotensin II on isolated renal microvessels, Am. J. Physiol., 244:F526, 1983.
46. Dunn, M. J., and V. L. Hood: Prostaglandins and the kidney, Am. J. Physiol., 233:F169, 1977.
47. Dunham, E. W., and B. G. Zimmerman, Release of prostaglandin-like material from dog kidney during nerve stimulation, Am. J. Physiol., 219:1279, 1970.
48. Schor, N., I. Ichikawa, and B. M. Brenner: Glomerular adaptations to chronic dietary salt restriction or excess, Am. J. Physiol., 238:F428, 1980.
49. Henrich, W. L., T. Berl, K. M. McDonald, R. J. Anderson, and R. W. Schrier: Angiotensin II, renal nerves, and prostaglandins in renal hemodynamics, Am. J. Physiol., 235:F46, 1978.
50. Johnston, P. A., D. B. Bernard, N. S. Perrin, L. Arbeit, W. Lieverthal, and N. G. Levinsky: Control of rat renal vascular resistance during alterations in sodium balance, Circ. Res., 48:728, 1981.
51. Johnston, P. A., N. S. Perrin, D. B. Bernard, and N. G. Levinsky: Control of rat renal vascular resistance at reduced perfusion pressure, Circ. Res., 48:734, 1981.
52. Walsh, J. J., and R. C. Venuto: Acute oliguric renal failure induced by indomethacin: possible mechanism, Ann. Intern. Med., 91:47, 1979.
53. Zipser, R. D., J. C. Hoefs, P. F. Speckhart, P. K. Zia, and R. Horton: Prostaglandins: modulators

of renal function and pressor resistance in chronic liver disease, J. Clin. Endocrinol. Metab., 48:895, 1979.

54. Hollenberg, K., G. H. Williams, K. J. Taub, I. Ishikawa, C. Brown, and D. Adams: Renal vascular response to interruption of the renin-angiotensin system in normal man, Kidney Int., 12:285, 1977.

55. Forster, R. P., and J. P. Maes: Effect of experimental neurogenic hypertension on renal blood flow and glomerular filtration rates in intact denervated kidneys of unaesthetized rabbits with adrenal glands demedullated, Am. J. Physiol., 150:534, 1947.

56. Braunwald, E.: Regulation of the circulation, N. Engl. J. Med., 290:1420, 1974.

57. Navar, L. G.: Renal autoregulation: perspectives from whole kidney and single nephron studies, Am. J. Physiol., 234:F357, 1978.

58. Jackson, T. E., A. C. Guyton, and J. E. Hall: Transient response of glomerular filtration rate and renal blood flow to step changes in arterial pressure, Am. J. Physiol., 233:F396, 1977.

59. Hall, J. E., A. C. Guyton, T. E. Jackson, T. G. Coleman, T. E. Lohmeier, and N. C. Trippodo: Control of glomerular filtration rate by renin-angiotensin system, Am. J. Physiol., 233:F366, 1977.

60. Wright, F. S., and J. P. Briggs: Feedback regulation of glomerular filtration rate, Am. J. Physiol., 233:F1, 1977.

61. Briggs, J. P., J. Schnermann, and F. S. Wright: Failure of tubule fluid osmolarity to affect feedback regulation of glomerular filtration, Am. J. Physiol., 239:F427, 1980.

62. Ichikawa, I.: Direct analysis of the effector mechanism of the tubuloglomerular feedback system, Am. J. Physiol., 243:F447, 1982.

63. Ploth, D. W., and R. N. Roy: Renal and tubuloglomerular feedback effects of [Sar^1, Ala^8] angiotensin II in the rat, Am. J. Physiol., 242:F149, 1982.

64. Spielman, W. S., and C. I. Thompson: A proposed role for adenosine in the regulation of renal hemodynamics and renin release, Am. J. Physiol., 242:F423, 1982.

65. Moore, L. C., J. Schnermann, and S. Yarmizu: Feedback mediation of SNGFR autoregulation in hydropenic and DOCA- and salt-loaded rats, Am. J. Physiol., 237:F63, 1979.

66. Young, D. K., and D. J. Marsh: Pulse wave propagation in rat renal tubules: implications for GFR autoregulation, Am. J. Physiol., 240:F446, 1981.

67. Tucker, B. J., R. W. Steiner, L. C. Gushwa, and R. C. Blantz: Studies on the tubulo-glomerular feedback system in the rat: the mechanism of reduction in filtration rate with benzolamide, J. Clin. Invest., 62:993, 1978.

67a. Moore, L. C.: Interaction of tubuloglomerular feedback and proximal nephron reabsorption in autoregulation, Kidney Int., 22(suppl 12):S-173, 1982.

68. Kilcoyne, M. M., D. H. Schmidt, and P. J. Cannon: Intrarenal blood flow in congestive heart failure, Circulation, 47:786, 1973.

69. Sparks, H. V., H. H. Kopald, S. Carriere, J. E. Chimoskey, M. Kimoshita, and A. C. Barger: Intrarenal distribution of blood flow with chronic congestive heart failure, Am. J. Physiol., 223:840, 1972.

70. Grandchamp, A., R. Veyrat, E. Rosett, J. R. Scherrer, and B. Truniger: Relationship between renin and intrarenal hemodynamics in hemorrhagic hypotension, J. Clin. Invest., 50:970, 1971.

71. Stein, J. H., S. Boonjarern, R. C. Maux, and T. F. Ferris: Mechanism of the redistribution of renal cortical blood flow during hemorrhagic hypotension in the dog, J. Clin. Invest., 52:39, 1973.

72. Mimran, A., L. Guiod, and N. K. Hollenberg: The role of angiotensin in the cardiovascular and renal response to salt restriction, Kidney Int., 5:348, 1974.

73. Data, J. L., L. C. T. Chang, and A. S. Nies: Alteration of canine renal vascular response to hemorrhage by inhibitors of prostaglandin synthesis, Am. J. Physiol., 230:940, 1976.

74. Westenfelder, C., J. A. L. Arruda, R. Lockwood, S. Boonjarern, L. Nascimento, and N. A. Kurtzman: Distribution of renal blood flow in dogs with congestive heart failure, Am. J. Physiol., 230:537, 1976.

75. Bruns, F. J., E. A. Alexander, A. L. Riley, and N. G. Levinsky: Superficial and juxtamedullary nephron function during saline loading in the dog, J. Clin. Invest., 53:971, 1974.

75a. Deen, W. M., D. A. Maddox, C. R. Robertson, and B. M. Brenner: Dynamics of glomerular ultrafiltration in the rat. VII. Response to reduced renal mass, Am. J. Physiol., 227:556, 1974.

75b. Brenner, B. M.: Hemodynamically mediated glomerular injury and the progressive nature of kidney disease, Kidney Int., 23:647, 1983.

76. Coggins, C. H., W. M. Bennett, and I. Singer: Drugs and the kidney, in *Pathophysiology of Renal Disease,* B. Rose (ed)., McGraw-Hill, New York, 1981.

77. Doolan, P. D., E. L. Alpen, and G. B. Theil: A clinical appraisal of the plasma concentration and endogenous clearance of creatinine, Am. J. Med., 32:65, 1962.

78. Tobias, G. J., R. F. McLaughlin, Jr., and J. Hopper, Jr.: Endogenous creatinine clearance: a valuable clinical test of glomerular filtration and a prognostic guide in chronic renal disease, N. Engl. J. Med., 266:317, 1962.

79. Bennett, W. M., and G. A. Porter: Endogenous creatinine clearance as a clinical measure of glomerular filtration rate, Br. Med. J., 4:84, 1971.

80. Elmwood, C. M. and E. M. Sigman: The measurement of glomerular filtration rate and effective renal plasma flow in man by iothalamate I and iodopyracet I, Circulation, 26:441, 1967.

81. Breckenridge, A., and A. Metcalfe-Gibson: Methods of measuring glomerular filtration rate: a comparison of inulin, vitamin B_{12}, and creatinine clearance, Lancet, 2:265, 1965.

82. Dodge, W. F., L. B. Travio, and C. W. Daeschner: Comparison of endogenous creatinine clearance with inulin clearance, Am. J. Dis. Child., 113:683, 1967.

83. Siersbaek-Nielsen, K., J. M. Hansen, J. Kampmann, and M. Kristensen: Rapid evaluation of creatinine clearance, Lancet, 1:1133, 1971.

84. Jacobsen, F. K., C. K. Christensen, C. E. Mogensen, F. Andreasen, and N. S. C. Heilskov: Pronounced increase in serum creatinine after eating cooked meat, Br. Med. J., 1:1049, 1979.

85. Hamilton, R. W., L. B. Gardner, A. S. Penn, and M. Goldberg: Acute tubular necrosis caused by exercise-induced myoglobinuria, Ann. Intern. Med., 77:77, 1972.

86. Cockcroft, D. W., and M. H. Gault: Prediction of creatinine clearance from serum creatinine, Nephron, 16:13, 1976.

87. Dossetor, J. B.: Diagnosis and treatment, creatininemia versus uremia: the relative significance of blood urea nitrogen and serum creatinine concentrations in azotemia, Ann. Intern. Med., 65:1287, 1966.

88. Smith, H. W., W. Goldring, and H. Chasis: The measurement of the tubular excretory mass, effective blood flow and filtration rate in the normal human kidney, J. Clin. Invest., 17:263, 1938.

89. Lubowitz, H., E. Slatopolsky, S. Shankel, R. E. Rieselbach, and N. S. Bricker: Glomerular filtration rate: determination in patients with chronic renal disease, J. Am. Med. Assoc., 199:252, 1967.

FOUR

EVALUATION OF MICROPUNCTURE DATA

MICROPUNCTURE
ISOLATED TUBULAR SEGMENTS

The following chapters in Part Two will present our current understanding of how the kidney regulates ion and water homeostasis. Although we may know that Na^+ excretion increases after a Na^+ load and decreases with Na^+ restriction, these changes in urinary composition do not give much information on how Na^+ is handled by the kidney and which sites in the nephron are involved in the regulation of Na^+ excretion. Since most of the information on the function of the different nephron segments has come from micropuncture experiments and the in vitro study of isolated tubular segments in animals, this chapter will briefly describe the principles involved in these techniques.

MICROPUNCTURE

Micropuncture experiments are performed in the following way. Under the microscope, the glomeruli of the outer cortical nephrons of the rat (or other experimental animal) are identified. A nonreabsorbable dye, such as lissamine green, is then injected into Bowman's space. As the dye passes through the nephron, the proximal tubule and subsequently the superficial distal cortical nephron can be seen. By puncturing these segments with a micropipette, specimens of the tubular fluid can be obtained and analyzed. The juxtamedullary nephrons and the medullary segments (see Fig. 2-3) are not generally accessible to micropuncture.

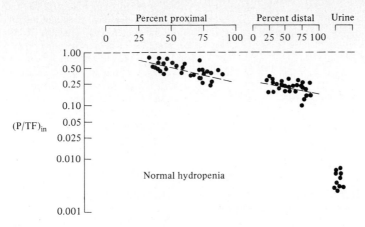

Figure 4-1 Summary of water handling by the kidney in the rhesus monkey as a function of proximal and distal tubular length. Each point represents a separate sample. The plasma to tubular fluid inulin concentration ratio, $(P/TF)_{in}$, is equal to the fraction of the filtered water remaining in the tubule at the particular puncture site. *(Adapted from C. M. Bennett, B. M. Brenner, and R. W. Berliner, J. Clin. Invest., 47:203, 1968, by copyright permission of The American Society for Clinical Investigation.)*

Let us consider the fate of the fluid filtered at the glomerulus. At any point in the tubule, the fraction of the filtered water (equal to the GFR) remaining in the tubule will be

$$\% \text{ filtered water in the lumen} = \frac{\text{tubular fluid volume } (V) \times 100}{\text{GFR}} \tag{4-1}$$

The GFR can be measured by the inulin clearance (C_{in}) (see Chap. 3).

$$C_{in} = \frac{\text{tubular fluid (TF) or urine [inulin]} \times \text{volume}}{\text{plasma [inulin]}}$$

Thus, from Eq. (4-1)

$$\% \text{ filtered water in the lumen} = \frac{V}{TF_{in} \times V/P_{in}} \times 100$$

$$= \frac{P_{in}}{TF_{in}} \times 100 \tag{4-2}$$

Using this formula, we can evaluate segmental water reabsorption in the rhesus monkey from the micropuncture data in Fig. 4-1.† At the end of the proximal tubule, 30 percent of the filtered water remains in the lumen $(P_{in}/TF_{in} = 0.30)$, meaning that 70 percent of the filtered water has been reabsorbed in this segment. As the filtrate

---

†Percent distal in the diagrams in this chapter refers to the superficial distal cortical nephron and may include the distal convoluted tubule, connecting segment, and initial cortical collecting tubule (see Fig. 2-3). It is likely that it is function in the cortical collecting tubule that is primarily being measured.

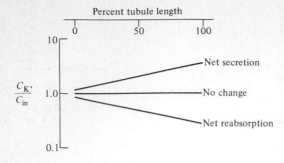

Figure 4-2 Hypothetical results of micropuncture experiments, depicting the effects of net secretion or net reabsorption of K^+ on the ratio of the clearance of K^+ (C_{K+}) to the clearance of inulin (C_{in}). Since inulin is neither secreted nor reabsorbed, an increase in C_{K+}/C_{in} represents the addition of K^+ to the lumen (tubular secretion), and a decrease in C_{K+}/C_{in} represents the removal of K^+ from the lumen (tubular reabsorption).

flows through the tubules, more than 99 percent is ultimately reabsorbed as evidenced by the P_{in}/U_{in} being less than 0.01 in the final urine.

Similar methods can be used to evaluate solute transport. For example, at any point in the tubule, the fraction of the filtered K^+ remaining in the tubule will equal

$$\% \text{ filtered } K^+ \text{ in the lumen} = \frac{K^+ \text{ in the lumen}}{\text{filtered } K^+} \times 100 \qquad (4\text{-}3)$$

Since the amount of K^+ in the lumen is equal to the *tubular fluid* K^+ concentration times the urine volume and the amount of K^+ filtered is equal to the GFR times the *plasma* K^+ concentration,

$$\% \text{ filtered } K^+ \text{ in the lumen} = \frac{\text{tubular fluid } [K^+] \times \text{volume}}{\text{GFR} \times \text{plasma } [K^+]} \times 100 \qquad (4\text{-}4)$$

Since the clearance of K^+ (C_{K+}) is equal to (tubular fluid $[K^+] \times$ volume)/plasma $[K^+]$ and the GFR is equal to the inulin clearance, this equation can be simplified to

$$\% \text{ filtered } K^+ \text{ in the lumen} = \frac{C_{K+}}{C_{in}} \times 100 \qquad (4\text{-}5)$$

By measuring C_{K+}/C_{in},† one can determine whether there has been reabsorption or secretion of K^+ (Fig. 4-2). If C_{K+}/C_{in} is falling, then K^+ must have been removed from the tubule by reabsorption. If C_{K+}/C_{in} is rising, then K^+ must have been added to the tubule by secretion.

Another indicator of tubular function is to compare the tubular fluid (TF) and plasma (P) concentrations of K^+. For example, the tubular cells can generate concentration gradients between the tubular fluid and the plasma by active transport. This can be detected by the demonstration of a diminishing TF/P $[K^+]$ ratio.

Figure 4-3 illustrates the results of micropuncture experiments in the rat. From

---

† Since the volume terms cancel out,

$$C_{K+}/C_{in} = \frac{(TF/P) \ [K^+]}{(TF/P) \ [in]}$$

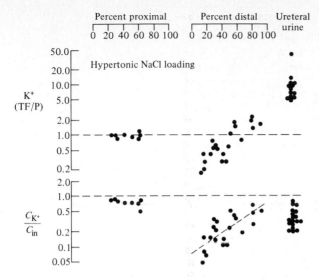

Figure 4-3 Summary of K^+ handling by the rat kidney as a function of proximal and distal tubular length. Each point represents a separate sample. The TF/P ratios reflect concentration gradients between the tubular fluid and the plasma. Changes in C_{K^+}/C_{in} represent net reabsorption from or secretion into the lumen. (*Adapted from G. Malnic, R. M. Klose, and G. Giebisch, Am. J. Physiol., 211:529, 1966.*)

this diagram, the following conclusions can be drawn regarding K^+ transport by the rat kidney:

1. K^+ is reabsorbed in the proximal tubule (diminishing C_{K^+}/C_{in}). Since there is no change in the K^+ concentration (TF/P [K^+] equals 1), K^+ has been reabsorbed with water in a concentration similar to that in the plasma.
2. K^+ is secreted into the lumen in the distal nephron (rising C_{K^+}/C_{in}), most of which occurs in the cortical collecting tubule (see Chap. 13).

Although the loops of Henle and the deeper aspects of the collecting tubules are not accessible to micropuncture, their role in ion transport can be estimated, respectively, by the difference between late proximal and early distal samples and between late distal and ureteral urine samples. Thus, from Fig. 4-3,

3. K^+ is reabsorbed in the loop of Henle (comparison of late proximal and early distal samples shows a decrease in C_{K^+}/C_{in}).
4. There is neither net reabsorption nor secretion of K^+ in the medullary collecting tubules (C_{K^+}/C_{in} is constant). The increase in TF/P [K^+] between the late distal sample and ureteral urine represents water reabsorption in the collecting tubules and not K^+ addition to the tubular fluid.

A similar analysis of Na^+ transport reveals (Fig. 4-4):

1. Na^+ is reabsorbed throughout the nephron (C_{Na^+}/C_{in} falls progressively from the proximal tubule to the distal tubule to the ureteral urine) with roughly 70 percent of the filtered Na^+ being reabsorbed in the proximal tubule (C_{Na^+}/C_{in} equals 0.3 at the end of the proximal tubule).

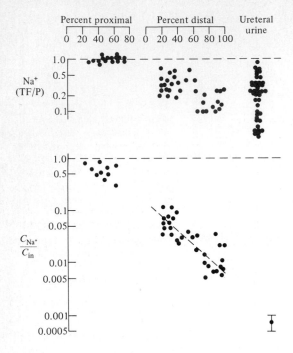

Figure 4-4 Summary of Na$^+$ handling by the rat kidney as a function of proximal and distal tubular length. *(Adapted from G. Malnic, R. M. Klose, and G. Giebisch, Am. J. Physiol., 211:529, 1966.)*

2. There is no change in the Na$^+$ concentration in the proximal tubule (TF/ P [Na$^+$] = 1), but reabsorption occurs against progressively steeper gradients in the distal nephron. In the ureteral urine, the TF/P [Na$^+$] may be less than 0.01 during volume depletion.

It should be noted that these experiments have been performed only on the outer cortical nephrons whose glomeruli are accessible to micropuncture. The juxtamedullary nephrons, which are inaccessible to micropuncture, differ anatomically and, in some ways, physiologically from those in the outer cortex (see Fig. 2-3) [1,2].

ISOLATED TUBULAR SEGMENTS

The study of isolated tubular segments involves the dissection of the tubular segments away from the intact kidney and then the perfusion or incubation of these tubules with solutions of varying ionic and hormonal composition (Fig. 4-5) [3]. These experiments have helped to elucidate the ionic and water permeabilities, transport characteristics, and hormone responsiveness of these nephron segments and, in conjunction with micropuncture data, have given a better understanding of nephron function [4].

In addition to being able to more carefully study segmental function, this technique has two advantages over micropuncture. First, those segments not accessible to micropuncture can be examined, including the juxtamedullary nephrons, the pars

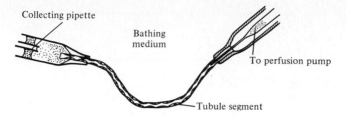

Figure 4-5 Diagrammatic illustration of the method by which isolated tubular segments are perfused in vitro. Fluid is periodically sampled from the collecting pipette. *(Adapted from J. P. Kokko, J. Clin. Invest., 49:1838, 1970, by copyright permission of The American Society for Clinical Investigation.)*

recta of the proximal tubule, the loop of Henle, and the deep collecting tubules. Second, the characteristics of human nephron segments have been able to be studied for the first time with preliminary results suggesting that the human nephron behaves in a manner similar to that in animal models [5].

PROBLEMS

4-1 If only water is removed from the tubule, what will happen to (*a*) the ratio of the tubular fluid K^+ concentration to the plasma K^+ concentration (TF/P [K^+]), and (*b*) the C_{K^+}/C_{in}?

4-2 From Fig. 4-3, it can be seen that the TF/P [K^+] increases in the distal tubule and in the collecting tubules (changes in the collecting tubules can be estimated from the difference between the late distal tubule and ureteral urine samples). What is responsible for these changes in the TF/P ratio in each segment?

REFERENCES

1. Kawamura, S., M. Imai, D. W. Seldin, and J. P. Kokko: Characteristics of salt and water transport in superficial and juxtamedullary straight segments of proximal tubules, J. Clin. Invest., 55:1269, 1975.
2. Jacobson, H. R.: Characteristics of volume reabsorption in rabbit superficial and juxtamedullary proximal convoluted tubules, J. Clin. Invest., 63:410, 1979.
3. Burg, M. B., J. J. Grantham, M. Abramow, and J. Orloff: Preparation and study of fragments of single rabbit nephrons, Am. J. Physiol., 210:1293, 1966.
4. Jacobson, H. R., and J. P. Kokko: Symposium on isolated perfused tubule, Kidney Int., 22:415, 1982.
5. Chabardès, D., M. Gagnan-Brunette, M. Imbert-Teboul, O. Gontcharevskaia, M. Montégut, A. Clique, and F. Morel: Adenylate cyclase reponsiveness to hormones in various portions of the human nephron, J. Clin. Invest., 65:439, 1980.

FIVE

PROXIMAL TUBULE

Fluid filtered at the glomerulus enters the proximal tubule where about 65 to 70 percent of the filtrate is normally reabsorbed [1]. The primary event in proximal tubular function is the active transport of Na^+, which then allows water and many of the other filtered solutes to be reabsorbed *passively* in a reabsorbate that is isosmotic to plasma. In addition, some solutes are secreted rather than reabsorbed in this segment, including hydrogen ions (H^+) and organic acids and bases.

Although the proximal tubule plays a major role in solute reabsorption, the degree of reabsorption of individual solutes is not uniform. For example, almost all of the filtered glucose and amino acids are reabsorbed in this segment but only about 90 percent of the HCO_3^-, 70 percent of the Na^+, and 60 percent of the Cl^- [1]. This chapter will review the basic aspects of how the proximal tubule selectively performs these functions as well as some of the clinical implications of the relationship between the reabsorption of Na^+ and that of the other solutes in the filtrate.

ANATOMY

The proximal tubule has a convoluted segment which begins at the glomerulus and then a straight segment (pars recta) which ends in the outer medulla in the descending limb of the loop of Henle (see Fig. 2-3). However, closer examination has revealed the presence of *three* distinct proximal segments with different cell types: S_1 in the early convoluted segment; S_2 in the late convoluted segment and early pars recta; and S_3 in the remainder of the pars recta (Fig. 5-1) [2]. To some degree, these cell types are associated with different functional characteristics. In particular, tubular secretion by the organic acid and base secretory pumps is most prominent in the S_2 segment [2].

CELL MODEL FOR PROXIMAL TRANSPORT

The anatomy of the proximal tubule is similar to that of other transporting epithelia (Fig. 5-2) [3]. The cells have two membranes with different permeability and transport characteristics: the luminal membrane, which separates the cell from the tubular lumen; and the peritubular (or basolateral) membrane, which separates the cell from the interstitium and peritubular capillary. In addition, there is an intercellular

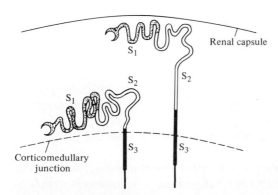

Figure 5-1 Schematic representation of the distribution of the S_1, S_2, and S_3 segments defined by cell type in the proximal tubules from outer cortical and juxtamedullary nephrons. *(From P. B. Woodhall, C. C. Tisher, C. A. Simonton, and R. R. Robinson, J. Clin. Invest., 61:1320, 1978, by copyright permission of the American Society for Clinical Investigation.)*

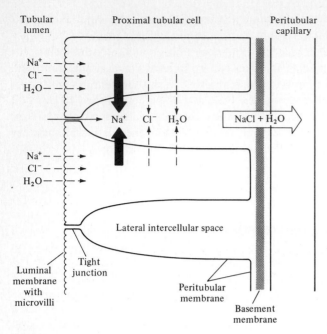

Figure 5-2 Schematic representation of active NaCl and H_2O reabsorption in the proximal tubule. Na^+ enters the cell passively (dashed arrow) and then is actively transported into the intercellular space (dark solid arrow). Cl^- and H_2O follow the movement of Na^+ down passive electrical and osmotic gradients. Some of this may occur through the tight junction as well as across the cell membranes. The NaCl and H_2O that enter the intercellular space can either move into the capillary and be returned to the systemic circulation or leak back into the lumen across the tight junction.

space between adjacent cells which is open both at the capillary end and to a lesser degree at the luminal end across the tight junction.

The proximal tubule normally reabsorbs more than 100 L/day (65 to 70 percent of a daily filtration rate of 150 to 180 liters). It is well suited for this task because of a series of adaptations, each of which facilitates net reabsorption of the filtrate (see below):

1. The luminal membrane has microvilli which serve to increase the surface area available for reabsorption. In addition, the microvilli have a brush border which contains specific carrier proteins as well as an enzyme, carbonic anhydrase, that plays an important role in HCO_3^- reabsorption.
2. The tight junction is relatively "leaky" in comparison to other nephron segments. This allows some proximal reabsorption to occur passively down concentration or osmotic gradients through the tight junction. This paracellular pathway is a low-resistance route in comparison to having to traverse both the luminal and peritubular membranes [4].
3. The activity of the Na^+-K^+-ATPase pump in the peritubular membrane, which is responsible for active Na^+ transport, is higher than in most other nephron segments [5].

In general, filtered Na^+ passively enters the cell across the luminal membrane and is then actively transported into the intercellular space. Cl^- and H_2O follow either across the cell membranes or through the tight junction down the electrical and osmotic gradients created by the movement of Na^+ (Fig. 5-2) [3]. The reabsorbate that accumulates in the intercellular space may enter the capillary and be returned to the systemic circulation or may leak back into the lumen across the tight junction. As will be seen, net proximal Na^+ and water reabsorption is affected by multiple factors including filtered solutes which are reabsorbed with Na^+, peritubular capillary hemodynamics, and neurohumoral factors such as norepinephrine and angiotensin II [3].

Cell Entry

Figure 5-3 depicts the electrical and chemical gradients present in a typical proximal tubular cell. The low cell Na^+ concentration, the cell negative potential difference across the luminal membrane, and the relatively high Na^+ permeability of the luminal membrane all promote the passive entry of Na^+ into the cell. To preserve electroneutrality, this Na^+ movement is associated either with Cl^- reabsorption or

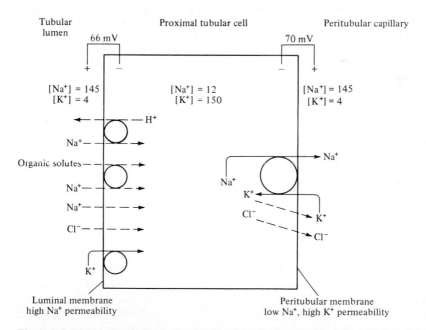

Figure 5-3 Electrical and chemical properties of a typical proximal tubular cell. Na^+ entry into the cell is passive, occurring in part by cotransport with organic solutes or countertransport with H^+. To maintain electroneutrality, the movement of Na^+ is linked to passive Cl^- entry into the cell or to H^+ secretion from the cell into the lumen (see Fig. 12-1). Active transport of Na^+ occurs across the peritubular membrane by a process mediated by Na^+-K^+-ATPase. The net effect is the reabsorption of NaCl, which creates an osmotic gradient for water movement across the cell. The solid lines depict active transport; the dashed lines indicate passive movement. The units are milliequivalents per liter.

the secretion of H^+ from the cell into the lumen, a process that results in HCO_3^- reabsorption (see Fig. 12-1).

Some of Na^+ entry into the cell occurs by *cotransport*† as specific Na^+-glucose, Na^+–amino acid and Na^+-phosphate carrier proteins are present in brush border vesicles in the luminal membrane [1,6]. The net result is mutual enhancement of solute transport, an effect that may be mediated by the attachment of one solute to the carrier, increasing the affinity of the carrier for the other solute [6–8]. Thus, the removal of glucose or amino acids from the luminal fluid impairs proximal Na^+ reabsorption [7,8]. Na^+ entry into the cell may also occur by *countertransport*† with H^+, as the carrier promotes both Na^+ reabsorption and H^+ secretion [9]. It should be noted that these passive processes are forms of *secondary active transport* since the solutes cotransported with Na^+ move into the cell against a concentration gradient, e.g., the phosphate concentration in the cell exceeds that in the lumen [6]. This uphill movement can occur passively because the downhill movement of Na^+ into the cell in some way generates enough energy to pull the cotransported solute along.

Movement into the Intercellular Space

At the peritubular membrane, Na^+ in the cell must be transported into the intercellular space against electrical and concentration gradients (Fig. 5-3). The energy required for this process is derived from the hydrolysis of ATP by the Na^+-K^+-ATPase pump. This pump transports Na^+ out of the cell and K^+ into the cell in a 3:2, not a 1:1, ratio [10,11]. Thus, the net effect is the transfer of one Na^+ ion into the intercellular space with an anion ultimately following to maintain electroneutrality.

In addition to promoting Na^+ reabsorption, the Na^+-K^+-ATPase pump also is responsible for maintaining the low cell Na^+ concentration. This is extremely important since Na^+ entry into the cell and consequently that of the cotransported (or countertransported) solutes is dependent upon the high Na^+ concentration gradient across the luminal membrane. As a result, net proximal transport can be virtually abolished by inhibition of the Na^+-K^+-ATPase pump [12]. (This is also true of other segments as the Na^+-K^+-ATPase pump is the active transporter of Na^+ throughout the nephron [5,10].)

Solutes other than Na^+ move passively across the peritubular membrane down concentration or electrical gradients. This is thought to occur primarily by facilitated diffusion, a process mediated by solute-specific carriers in the membrane (see Fig. 1-3) [9,13]. Since this final step is passive, *it is entry into the cell across the luminal membrane that is the main regulatory step in the proximal transport of these solutes.* For example, proximal phosphate reabsorption is appropriately increased in states of phosphate depletion. This effect is mediated by enhanced uptake by the Na^+-phosphate carrier in the luminal membrane (see below) [14].

---

†The different types of passive and active solute transport are discussed in detail in Chap. 1.

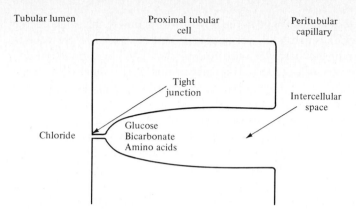

Figure 5-4 Schematic representation of the differences in solute composition between the lumen and the intercellular space in the later segments of the proximal tubule. In comparison to the plasma and intercellular space, the lumen has a very low concentration of glucose, bicarbonate, and amino acids but a higher chloride concentration.

Passive Transport

Although active transport by the Na^+-K^+-ATPase pump has been thought to be responsible for essentially all of proximal tubular NaCl and water transport, it now appears that *passive mechanisms account for about one-third of proximal fluid reabsorption* [15–18]. The mechanism by which this occurs is as follows. The early proximal convoluted tubule reabsorbs most of the filtered glucose, amino acids, and HCO_3^- (by coupled transport with Na^+) but a lesser amount of Cl^-. Water is then reabsorbed down an osmotic gradient. The net effect is that the tubular fluid has an osmolality similar to that of the plasma, a *higher chloride concentration,* but little if any glucose, bicarbonate, or amino acids. In contrast, the intercellular spaces in the later segments of the proximal tubule have solute concentrations similar to that of the plasma since they are in equilibrium with fluid in the peritubular capillary (Fig. 5-4). If the tight junction which separates the lumen from the intercellular space were equally permeable to all solutes, there would be no net fluid movement since the effective osmolalities of the two solutions would be similar. However, the permeability to Cl^- exceeds that to the other solutes, particularly HCO_3^- [19,20].

In this setting, passive fluid reabsorption can occur across the tight junction into the intercellular space by two mechanisms:†

1. Chloride can traverse the tight junction down its concentration gradient, with sodium and water then following down respective electrical and osmotic gra-

---

†This process appears to occur only in the outer and midcortical nephrons which comprise 85 to 90 percent of all nephrons (see Fig. 2-3). In contrast, active Na^+ transport is responsible for almost all NaCl reabsorption in the juxtamedullary proximal tubules which are not preferentially permeable to chloride [17,20].

dients. (Bicarbonate does not move in the opposite direction to the same degree since the tight junction is much less permeable to bicarbonate.) The presence of primary Cl^- transport has been demonstrated by the finding of a potential difference that is *lumen positive* (due to the reabsorption of the anion Cl^-) in the late proximal tubule [17]. The potential difference would be *lumen negative* if Na^+ transport were primary (see Fig. 13-9) [17].

2. Water can move across the tight junction down an osmotic gradient, with sodium chloride following both by solvent drag (see Chap. 1) and by diffusion since the loss of water raises the solute concentrations in the lumen [3]. This movement of water occurs because the tight junction is more permeable to Cl^- than to the solutes in the intercellular space. Thus, Cl^- is a relatively *ineffective* osmole (see Chap. 1). As a result, the *effective osmolality* in the intercellular space exceeds that in the lumen (thereby promoting water reabsorption), even though the total osmolality is the same in both compartments.

It is likely that HCO_3^- is the most important of the solutes that promote passive transport since it is present in the highest concentration (24 mmol/L versus only 5 mmol/L [90 mg/dL] for glucose). A clinical example of the effect of HCO_3^- is seen in the response to the administration of acetazolamide. This proximally acting diuretic is a carbonic anhydrase inhibitor that diminishes the reabsorption of bicarbonate (see Chap. 12). It also produces a substantial reduction in proximal NaCl reabsorption even though it has no known direct action on Na^+ or Cl^- transport [21]. This effect is presumably due to diminished passive reabsorption, resulting from the decrease in HCO_3^- transport. A similar reduction in proximal NaCl reabsorption occurs in metabolic acidosis, a disorder in which the plasma HCO_3^- concentration is decreased (see Chap. 19) [22]. In this setting, less HCO_3^- is filtered (because of the low plasma level) and therefore less is available for proximal reabsorption.

Capillary Uptake

The movement of the reabsorbate from the intercellular space into the peritubular capillary (derived from the efferent arteriole) is governed by Starling's forces (Fig. 5-5) (see Chap. 1):

Capillary uptake $= K_f$ (oncotic pressure gradient $-$ hydrostatic pressure gradient)

$$= K_f ([\Pi_{ptc} - \Pi_{if}] - [P_{ptc} - P_{if}])$$

where K_f is the peritubular capillary permeability coefficient; Π_{ptc} and Π_{if} are the oncotic pressures in the capillary and interstitium, and P_{ptc} and P_{if} are the hydrostatic pressures in the capillary and interstitium.

The approximate normal values for the hydrostatic and oncotic pressures in the peritubular capillary are depicted in Fig. 5-5. The mean hydrostatic pressure is much less than arterial pressure due to the resistances at the glomerular arterioles. In contrast, the oncotic pressure is higher than arterial pressure because of the removal of

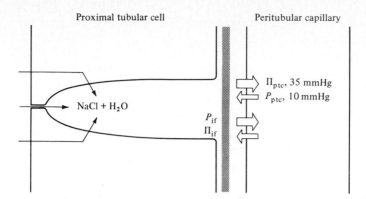

Figure 5-5 Role of Starling's forces in the uptake of the reabsorbate by the peritubular capillary. Approximate values for the capillary hydrostatic and oncotic pressures are included and show a gradient favoring fluid movement into the capillary. The interstitial pressures are more difficult to measure, and normal values are not known.

protein-free filtrate in the glomerulus (see Table 3-1). The net effect is a relatively large gradient ($\Pi_{ptc} - P_{ptc} = 25$ mmHg) favoring the movement of the reabsorbate from the intercellular space into the capillary. In addition, the transport of NaCl and H_2O from the cell or lumen into the intercellular space leads to an elevation in interstitial hydrostatic pressure and a reduction in interstitial oncotic pressure (by dilution), both of which further promote fluid uptake by the capillary.

However, these values are not constant since they are influenced by glomerular arteriolar tone, which is itself under neurohumoral control (see page 65). For example, the degree to which the systemic blood pressure is transmitted to the peritubular capillary is dependent upon glomerular arteriolar resistance. Arteriolar constriction increases the pressure drop across the glomerulus, thereby reducing capillary hydrostatic pressure; arteriolar dilation, in comparison, allows the capillary pressure to rise toward the arterial pressure.

The peritubular capillary oncotic pressure, on the other hand, is affected by the amount of the RPF that is filtered (called the filtration fraction, GFR/RPF). The filtration of protein-free fluid acrosss the glomerulus raises the protein concentration and, therefore, the oncotic pressure in the fluid remaining in the vascular space (see Fig. 3-7). If relatively more fluid is filtered, i.e., if the filtration fraction is increased, there will be a greater than usual elevation in the protein concentration in the fluid leaving the glomerulus and entering the peritubular capillary. Changes in the filtration fraction are primarily induced by changes in resistance at the *efferent* arteriole. For example, efferent arteriolar constriction will tend to raise the GFR (by increasing glomerular hydrostatic pressure), lower the RPF (because of elevated renal vascular resistance), and therefore increase the filtration fraction (see Chap. 3).

Thus, efferent arteriolar constriction (induced by angiotensin II or norepinephrine, both of which preferentially act at this site) [23] alters peritubular capillary hemodynamics in such a way—increased oncotic pressure, reduced hydrostatic

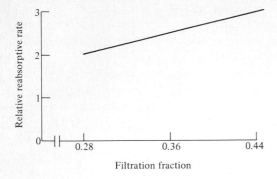

Figure 5-6 Relationship between the filtration fraction, altered within the physiologic range, and the relative rate of proximal reabsorption in the rat. *(Redrawn from J. E. Lewy and E. E. Windhager, Am. J. Physiol., 214:943, 1968.)*

pressure—as to promote capillary uptake and net proximal reabsorption. The physiologic importance of this effect is illustrated in Fig. 5-6, which demonstrates that proximal reabsorptive rate varies directly with the filtration fraction.

A possible clinical example of the role of peritubular capillary hemodynamics may occur in patients with congestive heart failure. In this disorder, the filtration fraction [24] and proximal reabsorption [25] are frequently elevated, thereby contributing to the low rate of Na$^+$ and water excretion that is commonly seen (see Chap. 16). These changes may be mediated by the increases in angiotensin II production and sympathetic neural tone that are induced by the reduction in cardiac output. (In addition to its hemodynamic effects, norepinephrine also may directly stimulate proximal reabsorption.) [26].

Backflux Across the Tight Junction

The mechanism by which changes in peritubular capillary hemodynamics influence proximal reabsorption is incompletely understood [3]. In a variety of conditions in which net proximal transport is reduced—increased peritubular capillary hydrostatic pressure, decreased peritubular capillary oncotic pressure, and volume expansion—enhanced movement of raffinose or sucrose from the peritubular capillary into the lumen has been demonstrated [27–29]. Since these sugars do not enter cells, this movement must reflect increased permeability of the tight junction, and suggest that the reduction in proximal reabsorption may be mediated by increased backflux from the intercellular space into the lumen. For example, if capillary uptake is impeded by a fall in peritubular capillary oncotic pressure, the fluid that accumulates in the intercellular space as active Na$^+$ transport continues will preferentially move back into the lumen across the tight junction.

There are, however, two problems with this passive backflux theory [3]. First, an increase in permeability of the tight junction would be expected to *enhance* not reduce Cl$^-$ reabsorption because of the favorable concentration gradient (see Fig. 5-4). Second, backleak would be expected to involve all the solutes in the reabsorbate. In contrast, microperfusion studies have demonstrated that NaCl reabsorption is diminished but not that of glucose or HCO$_3^-$ [3,30]. How NaCl transport is selectively affected by altered peritubular capillary forces is not known.

Glomerulotubular Balance

The efficiency with which proximal transport is regulated can be appreciated from the phenomenon of glomerulotubular balance. The urinary excretion of Na^+ and water is equal to the difference between the amount filtered across the glomerulus and the amount reabsorbed by the tubules. To maintain the extracellular fluid volume, it is important that tubular reabsorption varies with the changes that can occur in the GFR. For example, a normal adult male filters approximately 180 L/day (125 mL/min) and then reabsorbs 178 to 179 liters, resulting in a daily urine output of 1 to 2 liters. If there were a slight increase in the GFR to 183 L/day and no change in tubular reabsorption, there would be a 3-liter increase in the urine output and a serious reduction in the extracellular fluid volume. Fortunately, this does not occur since, over a wide range of spontaneous and experimental variations in the GFR, there is a proportional change in tubular reabsorption [31–33]. Thus, if there is a 1.5 percent increase in the GFR (from 180 to 183 L/day), there will be a similar increase in tubular reabsorption, resulting in only a small elevation in the urine output. This response, in which the absolute level of tubular reabsorption is directly related to the filtration rate, is called *glomerulotubular balance*. Notice that, at all levels of GFR in Fig. 5-7, approximately 60 percent of the filtrate is reabsorbed in the proximal tubule. Similarly, the more distal nephron segments reabsorb a constant fraction of the load delivered to them from the proximal tubule [32]. This is another way to define glomerulotubular balance: that the fractional tubular reabsorption remains roughly constant despite changes in the GFR.

The mechanism by which glomerulotubular balance is mediated in the proximal tubule is incompletely understood, but both peritubular and luminal factors are thought to contribute [34–37]. For example, if the GFR increases while RPF remains constant, the protein concentration in the plasma leaving the glomerulus will rise due to the loss of more protein-free filtrate. The ensuing elevation in the oncotic pressure in the peritubular capillary will then promote increased net proximal reabsorption.

Another mechanism contributing to glomerulotubular balance is the presence of

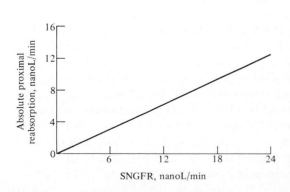

Figure 5-7 Phenomenon of glomerulotubular balance in the proximal tubule. Since fractional Na^+ and water reabsorption remains constant, absolute proximal reabsorption in a nephron is directly proportional to the single nephron filtration rate (SNGFR). A similar relationship between absolute Na^+ reabsorption and the amount of Na^+ delivered to the segment is present in the loop of Henle and the distal tubule. (*Adapted from A. Spitzer and M. Brandis, J. Clin. Invest., 53:279, 1974, by copyright permission of The American Society for Clinical Investigation.*)

factors in the filtrate which enhance Na^+ and H_2O reabsorption [34,36]. As described above, bicarbonate, glucose, and amino acids augment Na^+ reabsorption both by cotransport [7,8] and by creating gradients for passive reabsorption (see Fig. 5-4). An elevation in GFR will augment the filtered load of these (and other) solutes, and their subsequent reabsorption could then promote glomerulotubular balance for Na^+ and H_2O [34].

Glomerulotubular balance in the proximal tubule, loop of Henle, and distal tubule is one of three intrarenal mechanisms that act to prevent fluid delivery from exceeding the limited total reabsorptive capacity of the collecting tubules [33]. The others are *autoregulation,* which keeps the GFR relatively constant despite variations in renal arterial pressure, and *tubuloglomerular feedback,* which lowers the GFR if the load to the macula densa segment of the early distal tubule is increased (see Chap. 3). Thus, one view of nephron function is that the proximal tubule and loop of Henle are responsible for the reabsorption of the bulk of the filtrate, with the distal nephron (particularly the collecting tubules) making small variations in electrolyte and water excretion in accordance with changes in intake [38]. This process operates most efficiently if the distal delivery of filtrate is kept at a nearly constant level.

It should be noted that the relationship between glomerular filtration and tubular reabsorption is not constant and may be reset at a different level when there are changes in the effective circulating volume (see Chap. 9). *The fraction of the filtered Na^+ and water reabsorbed in the proximal tubule tends to be increased by volume depletion and decreased by volume expansion* [39–41]. These changes are appropriate, however, since in these conditions the maintenance of constant fractional Na^+ reabsorption is not desirable. For example, enhanced reabsorption leading to Na^+ and water retention is a proper response to volume depletion. As described above, hypovolemia-induced increases in sympathetic activity and angiotensin II production may contribute to this process by their effects on arteriolar resistance and, secondarily, peritubular capillary hemodynamics. These hormones also may directly enhance proximal Na^+ reabsorption [26,42].

Summary

The proximal tubule reabsorbs isosmotically about 65 to 70 percent of the filtrate. In the early proximal tubule, this occurs in three steps: entry into the cell across the luminal membrane, movement across the peritubular membrane into the intercellular space, and uptake by the peritubular capillary. Despite the large amount of reabsorption that occurs, the only major active (energy-requiring) step in this process is mediated by the Na^+-K^+-ATPase pump in the peritubular membrane. In addition to directly promoting Na^+ reabsorption (with Cl^- and H_2O following down the electrical and osmotic gradients that result from Na^+ transport), this pump maintains the low cell Na^+ concentration that allows passive Na^+ entry into the cell. The reabsorption of many other solutes (such as glucose, phosphate, amino acids, and bicarbonate) occurs by carrier-mediated coupled transport with Na^+ across the luminal membrane. Furthermore, the preferential reabsorption of these solutes with Na^+ in the early proximal tubule creates osmotic and concentration gradients that permit about one-third of total proximal Na^+ and H_2O reabsorption to occur passively through the tight junction (see Fig. 5-4).

Uptake by the peritubular capillary of fluid transported into the intercellular space is regulated by Starling's forces. Depending upon the magnitude of these forces, which can be influenced by vasoactive hormones, the reabsorbate either enters the capillary and is returned to the systemic circulation or leaks back into the lumen across the tight junction. Modulations in net proximal tubular reabsorption appear to be influenced by luminal, peritubular capillary, and neurohumoral factors.

PRIMACY OF SODIUM TRANSPORT IN PROXIMAL TUBULAR FUNCTION

Na^+ reabsorption creates respective electrical, osmotic, and concentration gradients for passive Cl^-, H_2O, and urea reabsorption. In addition, the proximal transport of most other solutes also is influenced by that of Na^+. Thus, in hypovolemic states, there is increased proximal reabsorption of bicarbonate, glucose, calcium, and uric acid as well as Na^+. Similarly, glomerulotubular balance can be demonstrated for HCO_3^- and glucose and probably is related to glomerulotubular balance for Na^+ [43,44]. The mechanisms responsible for these relationships are incompletely understood, but linkage to Na^+ by carrier-mediated cotransport at the luminal membrane plays at least a contributory role.

The remainder of this chapter will review the different transport systems for these solutes (both in the proximal tubule and other nephron segments) and the potential clinical implications of their relationship to Na^+ reabsorption. In many cases, the changes which occur are at the expense of the homeostatic requirements for these solutes.

Bicarbonate

Ninety percent of the filtered HCO_3^- is reabsorbed in the proximal tubule and the remainder in the distal tubule and collecting tubules. HCO_3^- reabsorption is accomplished by the active transport of H^+ from the cell into the lumen (see Chap. 12).

Relation of $T_{mHCO_3^-}$ to sodium transport The term $T_{mHCO_3^-}$ refers to the maximum tubular reabsorptive capacity for HCO_3^-. To measure the $T_{mHCO_3^-}$, $NaHCO_3$ can be infused intravenously to raise the plasma HCO_3^- concentration and, thereby, the filtered load. Tubular reabsorption, measured in milliequivalents of HCO_3^- reabsorbed per minute, can then be calculated from

$$\text{Tubular reabsorption} = \text{filtered load} - \text{urinary excretion}$$

$$= C_{in} \times \text{plasma } [HCO_3^-] - \text{urine } [HCO_3^-] \times \text{volume}$$

where C_{in} is the inulin clearance (an estimate of the GFR) in liters per minute.

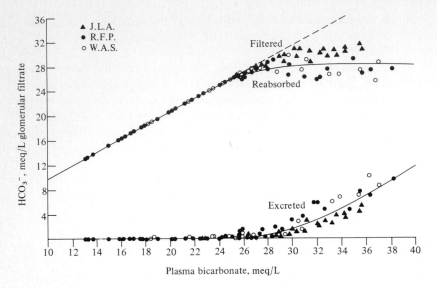

Figure 5-8 Filtration, reabsorption, and excretion of bicarbonate as functions of plasma concentration in normal humans. *(From R. F. Pitts, J. Ayer, and W. Shiess, J. Clin. Invest., 28:35, 1949, by copyright permission of the American Society for Clinical Investigation.)*

The results of such an experiment are illustrated in Fig. 5-8. There appears to be a maximum HCO_3^- reabsorption of 26 to 28 meq/L of glomerular filtrate.† This would be an appropriate mechanism by which the kidney prevents the plasma HCO_3^- concentration from exceeding the normal value of 22 to 26 meq/L, since the extra HCO_3^- would be excreted in the urine. However, the reabsorption of HCO_3^- (by H^+ secretion) is linked to Na^+ transport, and the infusion of $NaHCO_3$ expands the extracellular volume, a stimulus known to diminish proximal Na^+ reabsorption. Figure 5-9 depicts the results of two HCO_3^- titration experiments in a patient with moderate renal failure. When $NaHCO_3^-$-induced volume expansion was allowed to occur, HCO_3^- reabsorption reached a plateau when the plasma HCO_3^- concentration was 28 meq/L. When volume expansion was minimized by prior volume depletion, HCO_3^- reabsorption continued to rise even when the plasma HCO_3^- concentration had risen to 36 meq/L. In the rat, if volume expansion were prevented, no $T_{mHCO_3^-}$ could be demonstrated even if the plasma HCO_3^- concentration were greater than 60 meq/L [45].

Thus, *there is no absolute T_m for HCO_3^-, since the reabsorptive capacity for HCO_3^- varies directly with the fractional reabsorption of Na^+* (Fig. 5-10)‡ [45,46]. This assumes clinical importance in patients with metabolic alkalosis (high plasma

---

†Since the $T_{mHCO_3^-}$ is measured in milliequivalents reabsorbed per minute, the maximum reabsorption of 28 meq/L of glomerular filtrate must be corrected for the GFR. If the GFR were 125 mL/min (or 0.125 L/min), then the $T_{mHCO_3^-}$ would be 3.5 meq/min (28 meq/L × 0.125 L/min).

‡HCO_3^- reabsorption also is affected by other factors, including changes in the arterial pH and plasma K^+ concentration. The regulation of HCO_3^- reabsorption will be discussed in Chap. 12.

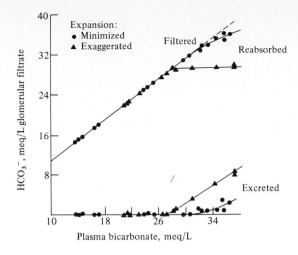

Figure 5-9 HCO$_3^-$ titration curves obtained from a patient with a GFR of 37 mL/min studied under conditions of minimized and exaggerated expansion of extracellular fluid volume. *(From E. Slatopolsky, P. Hoffsten, M. Purkerson, and N. S. Bricker, J. Clin. Invest., 49:988, 1970, by copyright permission of the American Society for Clinical Investigation.)*

HCO$_3^-$ concentration, alkaline arterial pH) and volume depletion (see Chap. 18). The normal response to an increase in the plasma HCO$_3^-$ concentration is to excrete the excess HCO$_3^-$ in the urine. However, HCO$_3^-$ reabsorption is increased, in the presence of volume depletion, resulting in the retention of the excess HCO$_3^-$ and perpetuation of the alkalosis. HCO$_3^-$ excretion will increase only if the stimulus to Na$^+$ retention is removed by the restoration of normovolemia.

Glucose

Under normal conditions, all the filtered glucose is reabsorbed in the proximal tubule and returned to the systemic circulation via the peritubular capillaries [47]. This process occurs in two steps. Filtered glucose enters the cell by passive cotransport with Na$^+$ (even though glucose moves uphill against a concentration gradient) and then leaves the cell at the peritubular membrane, probably by facilitated diffusion [6,48].

Studies in subjects in normal NaCl balance have shown a T_m for glucose of

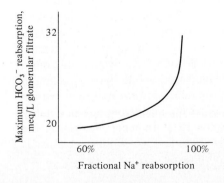

Figure 5-10 Relationship between HCO$_3^-$ reabsorption and fractional Na$^+$ reabsorption in patients with renal failure. A similar relationship in which HCO$_3^-$ reabsorption varies with that of Na$^+$ can also be demonstrated in normal subjects. *(Adapted from E. Slatopolsky, P. Hoffsten, M. Purkerson, and N. S. Bricker, J. Clin. Invest., 49:988, 1970, by copyright permission of the American Society for Clinical Investigation.)*

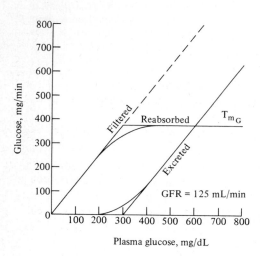

Figure 5-11 Filtration, reabsorption, and excretion of glucose as functions of plasma concentration in normal humans. The curves for reabsorption and secretion are drawn in two ways: (1) as idealized, sharply breaking curves and (2) as rounded curves more descriptive of the true relationships. With a T_m for glucose of 375 mg/min and a GFR of 125 mL/min, glucose excretion should not begin until the plasma glucose concentration is greater than 300 mg/dL (sharply breaking curve). However, due to tubular heterogeneity (see text), there is "splay" in the glucose titration curve (rounded curves) and glucosuria begins when the plasma glucose concentration exceeds 180 to 200 mg/dL, well before the saturation of tubular reabsorptive capacity. The relative lack of splay in the HCO_3^- titration curve (Fig. 5-8) may be due to the ability of distal HCO_3^- reabsorption to compensate for changes in proximal reabsorption. This is in contrast to glucose, which is reabsorbed entirely in the proximal tubule. *(From R. F. Pitts, Physiology of the Kidney and Body Fluids, 3d ed., Copyright © 1974 by Year Book Medical Publishers, Inc., Chicago. Used by permission. Adapted from H. R. Wright, H. F. Russo, H. R. Sheggs, E. A. Patch, and K. H. Beyer, Am. J. Physiol., 149:130, 1947.)*

approximately 375 mg/min (Fig. 5-11). As a result, glucose should not appear in the urine until the filtered glucose load exceeds this value. If the GFR is 125 mL/min, glucosuria should not begin until the plasma glucose concentration is greater than 300 mg/dL [125 mL/min × 3 mg/mL (or 300 mg/dL) = 375 mg/min]. (The normal plasma glucose concentration is 60 to 100 mg/dL, fasting.)

However, glucose excretion usually begins when the plasma glucose concentration exceeds 180 to 200 mg/dL. This deviation from the T_m is called *splay* and has been ascribed to heterogeneity in the relationship between glomerular size and proximal tubular length within individual nephrons [49]. A nephron with a large glomerulus, i.e., high filtered load, or a relatively short proximal tubule, i.e., low reabsorptive capacity, will spill glucose in the urine at a lower plasma glucose concentration than predicted from the T_m for the whole kidney. Clinically, glucosuria is most commonly seen when the filtered load is increased due to hyperglycemia in uncontrolled diabetes mellitus. Less often, there is a defect in proximal reabsorption that may be selective as in renal glucosuria [50], or part of a generalized abnormality in proximal transport as in the Fanconi syndrome [51]. In renal glucosuria, the appearance of glucose in the urine at a normal plasma glucose concentration is thought to be due either to a decreased number of glucose carriers or to a reduction in the affinity of the carriers for glucose [52].

Urea

Urea is lipid-soluble and able to cross most cell membranes by passive diffusion. The reabsorption of water in the proximal tubule increases the tubular fluid urea concentration and allows urea to be reabsorbed passively down a concentration gradient. Urea also is reabsorbed in the more distal nephron segments, the importance of which will be discussed in Chap. 6. In general, about 40 to 50 percent of the filtered urea is reabsorbed.

The urea concentration in the blood is measured as the blood urea nitrogen (BUN). The BUN tends to vary inversely with the GFR, a reflection of the importance of glomerular filtration in urea excretion (see Chap. 3). Thus, an elevation in the BUN is often due to a fall in GFR. There are, however, two important exceptions: conditions associated with enhanced urea production such as gastrointestinal bleeding or a high-protein diet, and volume depletion in which the increase in proximal Na^+ and H_2O reabsorption results in increased urea reabsorption and consequently a rise in the BUN. This is referred to as *prerenal azotemia* since the elevation in BUN is not due to renal disease.

The plasma creatinine concentration is a more accurate estimate of the GFR since creatinine production is relatively constant and its excretion is not influenced by Na^+ transport. In the presence of volume depletion, the plasma creatinine concentration will be elevated only to the degree that the GFR is diminished. A reduction in GFR increases both the BUN and the plasma creatinine concentration, and the normal BUN/creatinine ratio of 10:1 is maintained. In contrast, when urea reabsorption is increased due to hypovolemia, the BUN rises without change or with less change in the plasma creatinine concentration, and the ratio can exceed 20 to 30:1 [53].

Calcium and Phosphate

Na^+, K^+, and Cl^- are almost completely absorbed in the gastrointestinal tract, and their steady-state concentrations in the extracellular fluid are maintained primarily by changes in urinary excretion. In contrast, calcium and phosphate absorption is incomplete, and variations in intestinal absorption and calcium phosphate release from bone as well as in urinary excretion contribute to the regulation of calcium and phosphate balance (see Chap. 8).

Calcium Approximately 40 percent of the plasma Ca^{2+} is bound to albumin and is not filtered at the glomerulus. Of the remaining 60 percent, 50 percent exists as the physiologically important ionic (free) Ca^{2+}, and 10 percent is bound to citrate, bicarbonate, and phosphate. The filtered Ca^{2+} is reabsorbed throughout the nephron (except for the late cortical and all of the medullary collecting tubule), with about 5 percent being excreted on a regular diet [54]. Although 85 percent of the filtered Ca^{2+} is reabsorbed in the proximal tubule and medullary loop of Henle, most of this transport is passive, following gradients established by NaCl and water reabsorption [54]. The regulation of Ca^{2+} excretion according to physiologic needs appears to occur primarily in the cortical distal nephron, including the cortical thick ascending limb of the loop of Henle, the distal tubule, and the connecting segment (previously considered part of the late distal tubule; see Fig. 2-3) [54,55]. In these segments, parathyroid hormone (PTH) stimulates Ca^{2+} reabsorption, apparently by activating a hormone-specific adenyl cyclase system [55–57]. Thus, if Ca^{2+} absorption from the gut is increased, the ensuing small elevation in the plasma Ca^{2+} concentration diminishes PTH secretion,† resulting in reduced distal Ca^{2+} reabsorption and appropriately increased Ca^{2+} excretion.

The passive reabsorption of most of the filtered Ca^{2+} in the proximal tubule and loop of Henle means that Ca^{2+} transport in these segments will be affected by changes in net NaCl transport. Thus, on a constant Ca^{2+} intake, variations in Na^+ reabsorption (due to diet, drugs, or changes in the effective circulating volume) will alter both Na^+ and Ca^{2+} excretion [58,59].

These characteristics are useful in the therapy of hypercalcemia and of nephrolithiasis due to hypercalciuria [59]. Hypercalcemia can be corrected by increasing Ca^{2+} excretion. This can be achieved by decreasing Na^+ reabsorption in the proximal tubule with a high Na^+ intake and in the loop of Henle by the use of a diuretic which inhibits loop Na^+ reabsorption, such as furosemide (see Fig. 16-9). Conversely, lowering calcium excretion may reduce the frequency of stone formation in patients with idiopathic hypercalciuria [60,60a]. This can be done by increasing Na^+ and, secondarily, Ca^{2+} reabsorption in the proximal tubule and loop of Henle by inducing volume depletion with a low Na^+ intake and a diuretic. The diuretic must act distal to the medullary ascending limb, e.g., hydrochlorothiazide, so that it can enhance Na^+ excretion without also increasing Ca^{2+} excretion [59,61].

Phosphate Eighty to ninety-five percent of the filtered phosphate is normally reabsorbed, almost all of this occurring in the proximal tubule [62,63]. Filtered phosphate initially moves from the lumen into the cell by cotransport with Na^+ [6,14], and then diffuses across the peritubular membrane into the peritubular capillary. Proximal phosphate transport is primarily regulated by the plasma phosphate concentration and PTH (which decreases phosphate reabsorption), both of which affect uptake by the Na^+-phosphate carrier [14,64–67]. In states of phosphate depletion, phosphate uptake by this carrier is increased [14]. This effect, which is independent of PTH [14,65], plays a major role in the ability of the kidney to markedly reduce phosphate excretion in the presence of phosphate depletion [14]. The mechanism by which this occurs is uncertain [68].

In contrast, phosphate uptake by the carrier is diminished after a phosphate load [68], resulting in an appropriate increase in urinary excretion. Although this in part represents a direct effect of an elevation in the plasma phosphate concentration, increased secretion of PTH appears to be of primary importance [65,67,68]. Phosphate loading initially raises the plasma phosphate concentration. This induces a small elevation in the calcium-phosphate product, which reduces the plasma Ca^{2+} concentration, probably as a result of calcium phosphate precipitation in bone and

---

†PTH secretion varies inversely with the plasma calcium concentration. The role of PTH in the maintenance of calcium and phosphate balance is discussed in detail in Chap. 8.

soft tissues [69]. This minor degree of hypocalcemia stimulates the secretion of PTH [70], which then allows the excess phosphate to be excreted. PTH also enhances bone resorption, resulting in the release of Ca^{2+} from bone and correction of the hypocalcemia.

Magnesium

Circulating Mg^{2+} is partially protein-bound so that only about 70 to 80 percent is filtered across the glomerulus. In general, about 3 percent of the filtered Mg^{2+} escapes reabsorption in humans and is excreted [71,72]. This value is appropriately increased after a Mg^{2+} load and falls to very low levels with Mg^{2+} depletion. In contrast to other solutes and water, most of the filtered Mg^{2+} (60 to 65 percent) is reabsorbed in the loop of Henle, not in the proximal tubule (20 to 30 percent) [71,72]. Furthermore, alterations in Mg^{2+} excretion are primarily due to changes in loop reabsorption. The factors involved in the regulation of Mg^{2+} excretion are poorly understood, but a direct effect of the plasma Mg^{2+} concentration on tubular function may be involved [71].

Uric Acid

Uric acid is formed from metabolism of purine nucleotides. With a pK_a of 5.75, the reaction

$$Uric\ acid \rightleftharpoons H^+ + urate^-$$

is shifted far to the right at the normal arterial pH of 7.40 (see Chap. 11). As a result, most uric acid circulates as the urate anion. Filtered urate is handled entirely in the proximal tubule where three separate processes are involved [73,74]: reabsorption of almost all of the filtered urate in the early proximal tubule; tubular secretion by the organic acid secretory pathway in the midproximal tubule of an amount normally equal to about 50 percent of the filtered load; and postsecretory reabsorption of most of the secreted urate in the late proximal tubule. The net effect is the excretion of 6 to 12 percent of the amount filtered. Alterations in excretion according to urate homeostasis are thought to be mediated by changes in the rate of tubular secretion.

Net urate reabsorption also varies directly with proximal Na^+ transport [75] and, in the presence of volume depletion, both Na^+ and urate excretion are reduced. This is responsible for the increase in the plasma urate concentration (hyperuricemia) frequently seen in patients on diuretic therapy. If the diuretic-induced Na^+ and water losses are replaced, hyperuricemia does not develop since there is no stimulus to Na^+ retention [76].

Proteins

Filtered proteins are largely reabsorbed in the proximal tubule, the mechanism by which this occurs being dependent upon their size [77]. Amino acids enter the cell by cotransport with sodium and then leave the cell by facilitated diffusion across the peritubular membrane [77,78]. There are several different amino acid carriers, each of which recognizes different groups of amino acids.

Larger proteins are handled differently [77]. Small peptides, such as angiotensin II, are hydrolyzed by brush border peptidases, and the amino acids are then reabsorbed. Larger compounds, such as insulin and lysozyme, enter the cell by endocytosis, are metabolized into amino acids within the cell, and then are returned to the systemic circulation. The net effect of these processes is twofold: the preservation of nitrogen balance by minimizing urinary losses, and participation in hormonal homeostasis by being a major site of metabolism for polypeptide hormones such as insulin, gastrin, and glucagon [77].

SECRETORY PATHWAYS

In addition to their reabsorptive functions, the proximal tubular cells secrete hydrogen ions (see Chap. 12) and organic acids and bases [79,80]. The last two processes occur primarily in the S_2 segment of the proximal tubule (see Fig. 5-1), which has the highest number of secretory pumps [2,81]. In general, tubular secretion occurs in three steps: movement of the organic solute from the peritubular capillary into the interstitium by diffusion, active transport of the solute into the cell by secretory pumps in the peritubular membrane, and passive diffusion into the lumen across the luminal membrane down the concentration gradient created by solute transport into the cell [79,80].

The organic acids circulate predominantly as their anionic salts and are secreted in that form. Most organic anions, both endogenous (urate, hippurate, ketoacid anions) and exogenous (penicillins, cephalosporins, salicylates, diuretics, and radiocontrast media) compete for a common secretory pathway. Therefore, the presence of one organic anion can inhibit the secretion of another. For example, fasting subjects frequently develop hyperuricemia. It is thought that this is secondary to the ketonemia of fasting which would diminish urate secretion.

Organic anion secretion can also be inhibited by the drug probenecid. This property of probenecid is useful clinically, as it has been given in conjunction with penicillin therapy. By reducing penicillin secretion (and excretion), higher blood levels of the antibiotic can be achieved.

Similarly, the organic bases such as creatinine and certain drugs (including quinidine, procainamide, and cimetidine) are secreted by a common pathway [80].

The fact that these pathways are relatively nonspecific and are able to secrete foreign substances makes them well adapted for a major role in the elimination of a variety of foreign compounds from the body. This is particularly important for the many drugs and chemicals that are highly protein-bound and therefore cannot be excreted by glomerular filtration. In addition, tubular secretion is essential for the action of diuretics such as furosemide since these drugs are effective only at the luminal surface and are not well filtered because of protein-binding [80].

Other transport processes may also be involved in the renal handling of organic acids and bases. As described above, urate is both reabsorbed and secreted in the proximal tubule [73]. In addition, these substances may undergo passive reabsorption or secretion depending upon the urine pH [80]. Salicylic acid, for example, exists both as the intact acid, and the organic anion:

$$\text{salicylic acid} \rightleftharpoons H^+ + \text{salicylate}^-$$

The intact acid, but not the organic anion, can freely diffuse across cell membranes because it is nonpolar. This difference makes salicylate excretion pH-dependent. Raising the urine pH will shift the above reaction to the right. The ensuing fall in the urinary salicylic acid concentration will allow cellular salicylic acid to diffuse into the lumen, thereby increasing total drug excretion. Thus, elevating the urine pH is an important component of the treatment of salicylate intoxication (see Chap. 19).

PROBLEMS

5-1 If there is no change in the extracellular volume, what effect will an increase in the GFR have on (a) fractional proximal Na^+ reabsorption; (b) absolute proximal Na^+ reabsorption; (c) absolute proximal HCO_3^- reabsorption?

5-2 What factors determine the peritubular capillary oncotic pressure? What effect should a reduction in the peritubular capillary oncotic pressure have on net proximal Na^+ reabsorption?

5-3 A subject with a previous BUN of 10 mg/dL and plasma creatinine concentration of 1.0 mg/dL reports 3 days of diarrhea and poor appetite. On physical examination the patient appears to be volume-depleted. Repeat blood tests reveal BUN equal to 40 mg/dL, creatinine concentration of 1.0 mg/dL, and plasma urate concentration of 10.1 mg/dL (normal is 4 to 8 mg/dL). Ketones are noted in the urine and are believed to reflect the ketosis associated with fasting.

    (a) Why has the BUN increased?

    (b) Has there been any change in the GFR?

    (c) What factors may be responsible for the hyperuricemia?

5-4 Patients with the proximal form of renal tubular acidosis have an impaired ability to reabsorb HCO_3^- in the proximal tubule (see Chap. 19). What effect should this have on proximal NaCl reabsorption?

REFERENCES

1. Cogan, M. G.: Disorders of proximal nephron function, Am. J. Med., 72:275, 1982.
2. Woodhall, P. B., C. C. Tisher, C. A. Simonton, and R. R. Robinson: Relationship between para-aminohippurate secretion and cellular morphology in rabbit proximal tubules, J. Clin. Invest., 61:1320, 1978.
3. Rector, F. C., Jr.: Sodium, bicarbonate, and chloride absorption by the proximal tubule, Am. J. Physiol., 244:F461, 1983.
4. Andreoli, T. E., J. A. Schafer, S. L. Troutman, and M. L. Watkins: Solvent drag component of Cl⁻ flux in superficial proximal straight tubules: evidence for a paracellular component of isotonic fluid absorption, Am. J. Physiol., 237:F455, 1979.
5. Garg, L. C., M. A. Knepper, and M. B. Burg: Mineralocorticoid effects on Na-K-ATPase in individual nephron segments, Am. J. Physiol., 240:F536, 1981.
6. Kinne, R., and I. L. Schwartz: Isolated membrane vesicles in the evaluation of the nature, localization and regulation of renal transport processes, Kidney Int., 14:547, 1978.

7. Burg, M., C. Patlak, N. Green, and D. Villey: Organic solutes in fluid absorption by renal proximal convoluted tubules, Am. J. Physiol., 231:627, 1976.

8. Kokko, J. P.: Proximal tubule potential difference: dependence on glucose, HCO_3^- and amino acids, J. Clin. Invest., 52:1362, 1973.

9. Kinsella, J. L., and P. S. Aronson: Properties of the Na^+-H^+ exchanger in renal microvillus membrane vesicles, Am. J. Physiol., 238:F461, 1980.

10. Katz, A. I.: Renal Na-K-ATPase: its role in tubular sodium and potassium transport, Am. J. Physiol., 242:F207, 1982.

11. Grantham, J. J.: The renal sodium pump and vanadate, Am. J. Physiol., 239:F97, 1980.

12. Burg, M. B., and N. Green: Role of monovalent ions in the reabsorption of fluid by isolated perfused proximal renal tubules of the rabbit, Kidney Int., 10:221, 1976.

13. Grinstein, S., R. J. Turner, M. Silverman, and A. Rothstein: Inorganic anion transport in kidney and intestinal brush border and basolateral membranes, Am. J. Physiol., 238:F452, 1980.

14. Kempson, S. A., S. V. Shah, P. G. Werness, T. Berndt, P. H. Lee, L. H. Smith, F. G. Knox, and T. P. Dousa: Renal brush border membrane adaptation to phosphorus deprivation: effects of fasting versus low phosphorus diet, Kidney Int., 18:36, 1980.

15. Neumann, K. H., and F. C. Rector, Jr.: Mechanism of NaCl and water reabsorption in the proximal convoluted tubule of rat kidney: role of chloride concentration gradients, J. Clin. Invest., 58:1110, 1976.

16. Kiil, F.: Renal energy metabolism and regulation of sodium reabsorption, Kidney Int., 11:153, 1977.

17. Jacobson, H. R.: Characteristics of volume reabsorption in rabbit superficial and juxtamedullary proximal convoluted tubules, J. Clin. Invest., 63:410, 1979.

18. Green, R., J. H. V. Bishop, and G. Giebisch: Ionic requirements of proximal tubular sodium transport. III. Selective luminal anion substitution, Am. J. Physiol., 236:F268, 1979.

19. Andreoli, T. E., and J. A. Schafer: Effective luminal hypotonicity: the driving force for isotonic proximal tubular fluid absorption, Am. J. Physiol., 236:F89, 1979.

20. Holmberg, C., J. P. Kokko, and H. R. Jacobson: Determination of chloride and bicarbonate permeabilities in proximal convoluted tubules, Am. J. Physiol., 241:F386, 1981.

21. Mathisen, O., T. Monclair, H. Holdaas, and F. Kiil: Bicarbonate as mediator of proximal tubular NaCl reabsorption and glomerulotubular balance, Scand. J. Lab. Clin. Invest., 38:7, 1978.

22. Cogan, M. G., and F. C. Rector: Proximal reabsorption during metabolic acidosis in the rat, Am. J. Physiol., 242:F499, 1982.

23. Edwards, R. M.: Segmental effects of norepinephrine and angiotensin II in isolated renal microvessels, Am. J. Physiol., 244:F526, 1983.

24. Heller, B. I., and W. E. Jacobson: Renal hemodynamics in heart disease, Am. Heart J., 39:188, 1950.

25. Bell, N. H., H. P. Schedl, and F. C. Bartter: An explanation for abnormal water retention and hypoosmolality in congestive heart failure, Am. J. Med., 36:351, 1964.

26. Bello-Reuss, E.: Effect of catecholamines on fluid reabsorption by the isolated proximal convoluted tubule, Am. J. Physiol., 238:F347, 1980.

27. Hayslett, J. P.: Effect of changes in hydrostatic pressure in peritubular capillaries on the permeability of the proximal tubule, J. Clin. Invest., 52:1314, 1973.

28. Imai, M., and J. P. Kokko: Effect of peritubular protein concentration on reabsorption of sodium and water in isolated perfused proximal tubules, J. Clin. Invest., 51:314, 1972.

29. Boulpaep, E. L.: Permeability changes of the proximal tubule of Necturus during saline loading, Am. J. Physiol., 222:517, 1971.

30. Berry, C. A., and M. G. Cogan: Influence of peritubular protein on solute absorption in the rabbit proximal tubule. A specific effect on NaCl transport, J. Clin. Invest., 68:506, 1982.

31. Glabman, S., H. S. Aynedjian, and N. Bank: Micropuncture study of the effect of acute reductions in glomerular filtration rate on sodium and water reabsorption by the proximal tubules of the rat, J. Clin. Invest., 44:1410, 1965.

32. Schrier, R. W., and M. H. Humphreys: Role of distal reabsorption and peritubular environment in glomerulotubular balance, Am. J. Physiol., 222:379, 1972.

33. Kunau, R. T., Jr., H. L. Webb, and S. C. Borman: Characteristics of sodium reabsorption in the loop of Henle and distal tubule, Am. J. Physiol., 227:1181, 1974.

34. Häberle, D. A., and H. Von Baeyer: Characteristics of glomerulotubular balance, Am. J. Physiol., 244:F355, 1983.

35. Brenner, B. M., and J. L. Troy: Postglomerular vascular protein concentration: evidence for a causal role in governing fluid reabsorption and glomerulotubular balance by the renal proximal tubule, J. Clin. Invest., 50:336, 1971.

35a. Ichikawa, I., J. R. Hoyer, M. W. Seiler, and B. M. Brenner: Mechanism of glomerulotubular balance in the setting of heterogeneous glomerular injury. Preservation of a close functional linkage between individual nephrons and surrounding microvasculature, J. Clin. Invest., 69:185, 1982.

36. Häberle, D. A., T. T. Shiigai, G. Maier, H. Schiffl, and J. M. Davis: Dependency of proximal tubular fluid transport on the load of glomerular filtrate, Kidney Int., 20:18, 1981.

37. Wright, F. S.: Flow-dependent transport processes: filtration, absorption, secretion, Am. J. Physiol., 243:F1, 1982.

38. Moore, L. C.: Interaction of tubuloglomerular feedback and proximal nephron reabsorption in auto-regulation, Kidney Int., 22 (suppl 12):S-173, 1982.

39. Dirks, J. H., W. J. Cirksena, and R. W. Berliner: The effect of saline infusion on sodium reabsorption by the proximal tubule of the dog, J. Clin. Invest., 44:1160, 1965.

40. Stein, J. H., R. W. Osgood, S. Boonjarern, J. W. Cox, and T. F. Ferris: Segmental sodium reabsorption in rats with mild and severe volume depletion, Am. J. Physiol., 227:351, 1974.

41. Weiner, M. W., E. J. Weinman, M. Kashgarian, and J. P. Hayslett: Accelerated reabsorption in the proximal tubule produced by volume depletion, J. Clin. Invest., 50:1379, 1971.

42. Hall, J. E., A. C. Guyton, N. C. Trippodo, T. E. Lohmeier, R. E. McCaa, and A. W. Cowley, Jr.: Intrarenal control of electrolyte excretion by angiotensin II, Am. J. Physiol., 232:F538, 1977.

43. Bennett, C. M., P. D. Springberg, and N. R. Falkinburg: Glomerular-tubular balance for bicarbonate in the dog, Am. J. Physiol., 228:98, 1975.

44. Kwong, T., and C. M. Bennett: Relationship between glomerular filtration rate and maximum tubular reabsorption rate of glucose, Kidney Int., 5:23, 1974.

45. Purkerson, M., H. Lubowitz, R. White, and N. S. Bricker: On the influence of extracellular fluid volume expansion on bicarbonate reabsorption in the rat, J. Clin. Invest., 48:1754, 1969.

46. Kurtzman, N. A.: Regulation of renal bicarbonate reabsorption by extracellular volume, J. Clin. Invest., 49:586, 1970.

47. Kurtzman, N. A., and V. G. Pillay: Renal reabsorption of glucose in health and disease, Arch. Intern. Med., 131:901, 1973.

48. Aronson, P. S., and B. Sacktor: The Na^+ gradient-dependent transport of D-glucose in renal brush border membranes, J. Biol. Chem., 250:6032, 1975.

49. Oliver, J., and M. MacDowell: The structural and functional aspects of the handling of glucose by the nephrons and the kidney and their correlation by means of structural-functional equivalents, J. Clin. Invest., 40:1093, 1961.

50. Elsas, L. J., and L. E. Rosenberg: Familial renal glycosuria: a genetic reappraisal of hexose transport by kidney and intestine, J. Clin. Invest., 48:1845, 1969.

51. Roth, K. S., J. W. Foreman, and S. Segal: The Fanconi syndrome and mechanisms of tubular transport dysfunction, Kidney Int., 20:705, 1981.

52. Wolff, L. I., B. L. Goodwin, and C. E. Phelps: T_m-limited renal tubular reabsorption and the genetics of renal glucosuria, J. Theor. Biol., 11:10, 1966.

53. Dossetor, J. B.: Diagnosis and treatment, creatininemia versus uremia: the relative significance of blood urea nitrogen and serum creatinine concentrations in azotemia, Ann. Intern. Med., 65:1287, 1966.

54. Ng, R. C. K., R. A. Peraino, and W. N. Suki: Divalent cation transport in isolated tubules, Kidney Int., 22:492, 1982.

55. Bourdeau, J. E., and M. B. Burg: Effect of PTH on calcium transport across the cortical thick ascending limb of Henle's loop, Am. J. Physiol., 239:F121, 1980.

56. Chabardès, D., M. Gagnan-Brunette, M. Imbert-Teboul, O. Gontcharevskaia, A. Montégut, A. Clique, and F. Morel: Adenylate cyclase responsiveness to hormones in various portions of the human nephron, J. Clin. Invest., 65:439, 1980.

57. Costanzo, L. S., and E. E. Windhager: Effects of PTH, ADH, and cyclic AMP on distal tubular Ca and Na reabsorption, Am. J. Physiol., 239:F478, 1980.

58. Kleeman, C. R., J. Bohannan, D. Bernstein, S. Ling, and M. H. Maxwell: Effect of variations in sodium intake on calcium excretion in normal humans, Proc. Soc. Exp. Biol. Med., 115:29, 1964.

59. Martinez-Maldonado, M., G. Eknoyan, and W. N. Suki: Diuretics in nonedematous states, Arch. Intern. Med., 131:797:1973.

60. Coe, F. L.: Treated and untreated recurrent calcium nephrolithiasis in patients with idiopathic hypercalciuria, hyperuricosuria, or no metabolic disorder, Ann. Intern. Med., 87:404, 1977.

60a. Brocks, P., C. Dahl, H. Wolf, and I. Transbøl: Do thiazides prevent recurrent idiopathic renal calcium stones? Lancet, 2:124, 1981.

61. Breslau, N., A. M. Moses, and I. M. Weiner: The role of volume contraction in the hypocalciuric action of chlorothiazide, Kidney Int., 10:164, 1976.

62. Dennis, V. W., and P. C. Brazy: Divalent anion transport in isolated renal tubules, Kidney Int., 22:498, 1982.

63. Knox, F. G., H. Osswald, G. R. Marchand, W. S. Spielman, J. A. Haas, T. Berndt, and S. P Youngberg: Phosphate transport along the nephron, Am. J. Physiol., 233:F261, 1977.

64. Agus, Z. S., J. B. Puschett, D. Senesky, and M. Goldberg: Mode of action of parathyroid hormone and cyclic adenosine 3′,5′-monophosphate on renal tubular phosphate reabsorption in the dog, J. Clin. Invest., 50:617, 1971.

65. Steele, T. H.: Renal response to phosphorus deprivation: effect of the parathyroids and bicarbonate, Kidney Int., 11:327, 1977.

66. Dennis, V. W., E. Bello-Reuss, and R. R. Robinson: Response of phosphate transport to parathyroid hormone in segments of rabbit nephrons, Am. J. Physiol., 233:F29, 1977.

67. Hruska, K. A., S. Klahr, and M. R. Hammerman: Decreased luminal membrane transport of phosphate in chronic renal failure, Am. J. Physiol., 242:F17, 1982.

68. Kempson, S. A., G. Colon-Otero, S. Y. Lise Ou, S. T. Turner, and T. P. Dousa: Possible role of nicotinamide adenine dinucleotide as an intracellular regulator of renal transport of phosphate in the rat, J. Clin. Invest., 67:1347, 1981.

69. Herbert, L. A., J. Lemann, Jr., J. R. Petersen, and E. J. Lennon: Studies of the mechanism by which phosphate infusion lowers serum calcium concentration, J. Clin. Invest., 45:1886, 1966.

70. Reiss, E., J. M. Canterbury, M. A. Bercovitz, and E. L. Kaplan: The role of phosphate in the secretion of parathyroid hormone in man, J. Clin. Invest., 49:2146, 1970.

71. Dirks, J. H.: The kidney and magnesium regulation, Kidney Int., 23:771, 1983.

72. Carney, S. L., N. L. M. Wong, and G. A. Quamme: Effect of magnesium deficiency on renal magnesium and calcium transport in the rat, J. Clin. Invest., 65:180, 1980.

73. Boss, G. R., and J. E. Seegmiller: Hyperuricemia and gout: classification, complications and management, N. Engl. J. Med., 300:1459, 1979.

74. Fanelli, G. M., Jr., and I. M. Weiner: Pyrazinoate excretion in the chimpanzee: relation to urate disposition and the actions of uricosuric drugs, J. Clin. Invest., 52:1946, 1973.

75. Weinman, E. G., G. Eknoyan, and W. N. Suki: The influence of the extracellular fluid volume on the tubular reabsorption of uric acid, J. Clin. Invest., 55:283, 1975.

76. Steele, T. H., and S. Oppenheimer: Factors affecting urate excretion following diuretic administration in man, Am. J. Med., 47:564, 1969.

77. Carone, F. A., and D. R. Peterson: Hydrolysis and transport of small peptides by the proximal tubule, Am. J. Physiol., 238:F151, 1980.

78. Schafer, J. A., and D. W. Barfuss: Membrane mechanisms for transepithelial amino acid absorption and secretion, Am. J. Physiol., 238:F335, 1980.

79. Grantham, J. J.: Studies of organic anion and cation transport in isolated segments of proximal tubules, Kidney Int., 22:519, 1982.

80. Irish, J. M., III, and J. J. Grantham: Renal handling of organic anions and cations, in *The Kidney*, B. M. Brenner and F. C. Rector, Jr. (eds.), 2d ed., Saunders, Philadelphia, 1981.

81. Shimomura, A., A. M. Chonko, and J. J. Grantham: Basis for heterogeneity of *para*-aminohippurate secretion in rabbit proximal tubules, Am. J. Physiol., 240:F430, 1981.

LOOP OF HENLE AND THE COUNTERCURRENT MECHANISM

INTRODUCTION

The 30 to 35 percent of the filtrate that is not reabsorbed in the proximal tubule enters the loop of Henle, which has a characteristic hairpin configuration in the medulla (see Fig. 2-3). This segment has two major functions. First, it reabsorbs approximately 15 to 20 percent of the filtered NaCl, primarily in the thick ascending limb [1]. In general, Na^+ and Cl^- passively enter the cell in the thick ascending limb by carrier-mediated cotransport down a favorable concentration gradient for Na^+ [2]. As in the proximal tubule, the Na^+-K^+-ATPase pump plays a central role in this process by transporting Na^+ out of the cell into the peritubular capillary, thereby maintaining a low cell Na^+ concentration [2,3]. The previous suggestion that reabsorption in the loop of Henle occurs by primary active Cl^- transport [4] has not been substantiated [2].

The total amount of NaCl reabsorbed in the loop tends to vary directly with the load delivered from the proximal tubule [5,6], thereby allowing the loop of Henle to participate in glomerulotubular balance (see Chap. 5). This flow dependence may be related to changes in the tubular fluid NaCl concentration [6]. The ascending limb normally reabsorbs NaCl without water, thereby lowering the luminal concentration of these ions (see below). If more fluid is delivered to the ascending limb but NaCl reabsorption initially remains constant, there will be less of a reduction in the tubular fluid NaCl concentration. Thus, the fluid in the more distal parts of the ascending limb will have a *higher* NaCl concentration which will promote passive NaCl entry into the cell [6a]. Neither hormones nor peritubular capillary hemodynamics appear to be important in this process [7].

The second major function of the loop of Henle is to reabsorb NaCl in excess of water, an effect that is essential for the excretion of urine with an osmolality different from that of the plasma. This characteristic of loop function is dependent upon the varying permeability properties of the different loop segments—the descending limb and the thin and thick ascending limbs.

COUNTERCURRENT MECHANISM

Fluid leaving the proximal tubule is isosmotic to plasma. However, the excretion of an isosmotic urine is usually not adequate to meet the homeostatic requirements of the body. After a water load, for example, water must be excreted in excess of solute. This requires the excretion of urine that is hypoosmotic to plasma. Conversely, water must be retained and a hyperosmotic urine excreted after a period of water restriction. The formation of a dilute (hypoosmotic to plasma) or concentrated (hyperosmotic to plasma) urine is achieved via the countercurrent mechanism which includes the loop of Henle, the cortical and medullary collecting tubules, and the blood supply to these segments.

Before discussing these processes in detail, it is useful to summarize their basic aspects. The excretion of a *concentrated* urine involves two major steps:

1. The medullary interstitium is made hyperosmotic by the reabsorption of NaCl without water in the ascending limb of the loop of Henle.
2. As the urine enters the medullary collecting tubule, it equilibrates osmotically with the interstitium, resulting in the formation of a concentrated urine. Antidiuretic hormone (ADH), released from the posterior pituitary (see Chap. 8), plays an essential role in this process by increasing collecting tubule permeability to water, which is very low in the basal state.

In addition, two modifying factors are important in the *maintenance* of medullary hyperosmolality:

1. Water equilibrating in the medullary collecting tubule dilutes the interstitium. To minimize this effect, the volume of urine presented to this segment is markedly reduced in the cortex by water reabsorption in the cortical collecting tubule.
2. Medullary blood flow in the vasa recta is arranged in a hairpin configuration to minimize removal of the excess interstitial solute.

Urinary *dilution* also has two basic steps, the first of which is the same as in urinary concentration:

1. NaCl reabsorption without water in the ascending limb of the loop of Henle decreases the osmolality of the tubular fluid at the same time as it increases the osmolality of the interstitium.
2. The urine remains dilute if water reabsorption in the collecting tubules is minimized by keeping these segments poorly permeable to water. This requires the relative absence of ADH.

Countercurrent Multiplication: Loop of Henle

In humans, the maximum urine osmolality that can be attained is 900 to 1400 mosmol/kg. Since the urine becomes concentrated by equilibrating with the medullary interstitium, this means that a similar osmolality must be achieved in the interstitium. The process by which the interstitial osmolality is increased from 285 mosmol/kg (isosmotic to plasma) to 900 to 1400 mosmol/kg is called *countercurrent multiplication.* (Countercurrent refers to the opposite directions of flow in the descending and ascending limbs that result from the hairpin configuration of the loop.)

The exact mechanism of countercurrent multiplication is uncertain as several different theories have been proposed [8–11]. Nevertheless, it is generally agreed that the descending and ascending limbs of the loop of Henle have different physiologic properties that are essential to this process. The descending limb is permeable to water and to a lesser degree solutes whereas both the thin and thick segments of the ascending limb are *impermeable to water* but are able to transport NaCl into the interstitium. Active NaCl reabsorption in the thick ascending limb is *the primary step in countercurrent multiplication.* In contrast, NaCl transport in the *thin* ascending limb is primarily passive.† However, since the net effect, i.e., NaCl reabsorption without water, is the same as that in the *thick* ascending limb, the following discussion will assume that the ascending limb functions in a homogeneous manner.

It should be noted that countercurrent multiplication primarily occurs in the 60 percent of nephrons which have long loops of Henle (see Fig. 2-3). The glomeruli of these nephrons are located in the juxtamedullary area and midcortex. In contrast, there is little contribution from the outer cortical nephrons which have short loops that turn around in the outer medulla or even the inner cortex.

Generation of medullary interstitial hyperosmolality If one could start at a hypothetical time zero, the fluid in the descending and ascending limbs and in the interstitium would be isosmotic to plasma, similar to that delivered from the proximal

---

†The mechanism by which passive NaCl reabsorption occurs is as follows. Fluid delivered out of the proximal tubule has a Na^+ concentration of 150 meq/L, similar to that in the plasma water. As this fluid enters the descending limb, water moves down an osmotic gradient into the hyperosmotic interstitium (see below). As a result, the Na^+ concentration rises markedly to a level greater than that in the interstitium. The subsequent movement of this fluid into the thin ascending limb, which has a relatively high Na^+ permeability, then allows Na^+ to be reabsorbed passively down its concentration gradient [8,9].

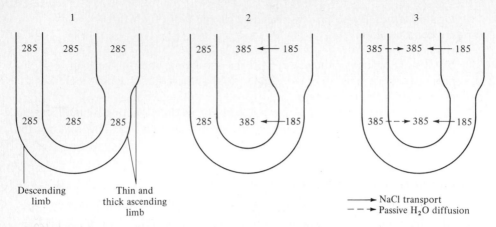

Descending limb

Thin and thick ascending limb

⟶ NaCl transport
– – ⟶ Passive H₂O diffusion

Figure 6-1 Role of active NaCl transport in initiating countercurrent multiplication. In step 1, at time zero, the fluid in the descending and ascending limbs and the interstitium is isosmotic to plasma. In step 2, NaCl is transported out of the ascending limb into the interstitium to a gradient of 200 mosmol/kg. In step 3, the fluid in the descending limb equilibrates osmotically with the hyperosmotic interstitium, primarily by water movement out of the tubule. Dilution of the interstitium by this water movement is prevented by continued NaCl transport out of the ascending limb. The result is the creation of an osmotic gradient between the ascending limb and the relatively hyperosmotic descending limb and interstitium.

tubule (Fig. 6-1). The initial and essential step in countercurrent multiplication is the transport of NaCl from the ascending limb of Henle into the interstitium to a maximum gradient of 200 mosmol/kg. Since the ascending limb is impermeable to water, this results in an increase in interstitial osmolality from 285 to 385 mosmol/kg. The fluid in the descending limb then equilibrates osmotically with the interstitium, primarily by water movement out of the tubule. As water enters the interstitium, interstitial osmolality is maintained by continued NaCl transport out of the ascending limb. The net effect is the establishment of a 200 mosmol/kg gradient between the fluid in the ascending limb (185 mosmol/kg) and that in the interstitium and descending limb (385 mosmol/kg).

As urine flows through the tubules, and NaCl transport in the ascending limb continues, the initial step in Fig. 6-1 is *multiplied,* resulting in the generation of a much higher interstitial osmolality. An example of how this might occur is shown in Fig. 6-2. For the sake of simplicity, steps 1 to 8 are depicted as discrete instants in time, even though ion transport and urine flow occur simultaneously in the intact organism. In steps 1 and 2, a 200-mosmol/kg gradient is established between the fluid in the ascending limb and that in the descending limb and interstitium. In step 3, urine moves through the tubules, with the hyperosmotic fluid in the descending limb flowing into the ascending limb. As NaCl is again pumped into the interstitium to a gradient of 200 mosmol/kg in step 4, the osmolality of the inner medullary interstitium is now 485 mosmol/kg, as compared with 385 mosmol/kg in step 2.

These steps illustrate the basic aspects of countercurrent multiplication: NaCl transport out of the ascending limb makes the interstitium and descending limb

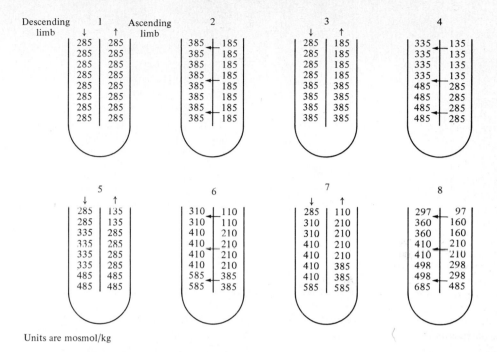

Units are mosmol/kg

Figure 6-2 Principle of countercurrent multiplication of concentration based on the assumption that at any level along the loop of Henle a concentration gradient of 200 mosmol/kg can be established between ascending and descending limbs by the active transport of ions. The osmolality of the interstitium is the same as that in the descending limb and has been omitted from the diagram. *(Adapted from R. F. Pitts, Physiology of the Kidney and Body Fluids, 3d ed., Copyright © 1974 by Year Book Medical Publishers, Inc., Chicago. Used by permission.)*

hyperosmotic; the hyperosmotic fluid in the descending limb then flows in a countercurrent fashion into the ascending limb; the combination of a higher tubular fluid osmolality in the inner medullary ascending limb (385 mosmol/kg in step 3 versus 285 mosmol/kg in step 1) and reestablishment of the 200-mosmol/kg gradient between the ascending limb and interstitium results in a further elevation in interstitial osmolality.

As the sequence in Fig. 6-2 goes on (steps 5 to 8), the osmolality continues to rise, being highest in the tubule at the hairpin turn and in the interstitium at the papillary tip (the inner medulla). The osmolality at these sites is directly proportional both to the length of the loops and to the gradient achieved between the ascending limb and the interstitium. In humans, the maximum osmolality at the papillary tip can reach 900 to 1400 mosmol/kg (Fig. 6-3).† This is relatively inefficient in comparison with other mammals. For example, the desert rat, which infrequently comes

---

†Only about one-half of the papillary solute is NaCl, with urea accounting for most of the remainder (see "Role of Urea" below).

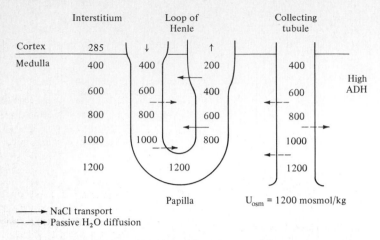

	Interstitium	Loop of Henle	Collecting tubule	
Cortex	285	↓ ↑		
Medulla	400	400 200	400	
	600	600 400	600	High ADH
	800	800 600	800	
	1000	1000 800	1000	
	1200	1200	1200	

Papilla U_{osm} = 1200 mosmol/kg

⟶ NaCl transport
---⟶ Passive H_2O diffusion

Figure 6-3 Countercurrent multiplication and the excretion of a concentrated urine. The transport of NaCl from the ascending limb results in the formation of an interstitial osmolal gradient from 285 mosmol/kg in the cortex to 1200 mosmol/kg at the papillary tip. In the presence of ADH, the urine becomes concentrated as it equilibrates with the interstitium in the medullary collecting tubule. The contribution of urea to the concentrating process is discussed in the text and has been omitted from the diagram. The collecting tubule is also the site of active Na^+ transport, the importance of which is discussed in Chap. 7.

in contact with water, has relatively long loops of Henle and can attain interstitial and urine osmolalities in the range of 5000 mosmol/kg.

Notice also that the osmolality of the tubular fluid leaving the ascending limb is hypoosmotic to plasma (Fig. 6-3). This fluid is further diluted by NaCl reabsorption without water in the cortical aspect of the thick ascending limb. As a result, the osmolality of the urine leaving the loop of Henle is approximately 100 mosmol/kg. If the collecting tubules are impermeable to water (ADH absent), this dilute urine is excreted relatively unchanged. In contrast, if the collecting tubules are permeable to water (ADH present), the urine equilibrates with the interstitium and a concentrated urine is excreted (Fig. 6-3). Thus, *the final osmolality of the urine is determined by the water permeability of the collecting tubules, not by events in the loop of Henle.*†

Collecting Tubules

As with the loops of Henle, the cortical and medullary collecting tubules possess distinct permeability characteristics, being in the basal state relatively impermeable to the passive transport of water, NaCl, and, with the exception of the inner medullary collecting tubule, urea [8,9,12]. The impermeability to NaCl is important in

†Although it is primarily those nephrons with long loops of Henle that create the interstitial osmolal gradient, urine from all nephrons drains into the collecting tubules and reaches osmotic equilibrium with the medullary interstitium in the presence of ADH (see Fig. 2-3).

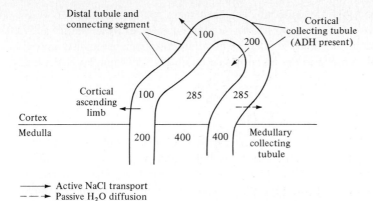

Active NaCl transport
---- Passive H_2O diffusion

Figure 6-4 Representation of the role of the cortical collecting tubule in the concentrating process. As a result of water reabsorption in this segment, the hypoosmotic urine formed in the ascending limb of the loop of Henle is made isosmotic to plasma and is much reduced in volume.

that it permits the high NaCl concentration in the interstitium to act as an effective osmotic gradient between the tubular fluid and interstitium.

ADH promotes urinary concentration primarily by increasing the water permeability of the collecting tubules [12–14]. In the medullary collecting tubule, this allows the tubular fluid to reach osmotic equilibrium with the hyperosmotic interstitium. The reabsorbed water then returns to the systemic circulation via the capillaries of the vasa recta.

The *cortical collecting tubule* also plays an essential role in the concentrating process. The maximum urine osmolality attained in the medulla cannot exceed that in the interstitium at the papillary tip. As water leaves the medullary collecting tubule, it decreases the interstitial osmolality by dilution, thereby reducing the maximum urine osmolality that can be achieved. This effect is minimized because the volume of fluid presented to the medullary collecting tubule is markedly reduced by ADH-induced water reabsorption in the cortical collecting tubule† [8,9,13]. In the presence of ADH, the hypoosmotic fluid in the cortical collecting tubule equilibrates with the cortical interstitium which is isosmotic to plasma (Fig. 6-4). If the osmolality of the tubular fluid entering this segment is 100 mosmol/kg, then osmotic equilibration will result in the reabsorption of almost two-thirds of water that has been delivered. This reduction in volume permits concentration of the urine to proceed in the medulla with minimum dilution of the medullary interstitium. Since cortical blood flow is more than 10 times maximum urine flow, the water reabsorbed in the cortex is rapidly returned to the systemic circulation without dilution of the cortical interstitium.

---

†Water reabsorption is probably minimal in the distal tubule and connecting segment, which, like the ascending limb, are relatively impermeable to water and show essentially no response to ADH [13,15].

In the absence of ADH, the collecting tubules remain poorly permeable to water. As a result, much less water is reabsorbed and a dilute urine is excreted. Since active NaCl transport continues in these segments, the minimum urine osmolality may be reduced from 100 mosmol/kg in the distal tubule to 50 to 75 mosmol/kg in the final urine.

Countercurrent Exchange: Vasa Recta

The capillaries of the vasa recta, derived from the efferent arterioles of the juxta-medullary glomeruli (see Fig. 3-1), return the NaCl and water reabsorbed in the loop of Henle and collecting tubules to the systemic circulation. These vessels are well adapted for such a role since Starling's forces in the capillary are much in favor of fluid uptake (see Chap. 1): the oncotic pressure which promotes uptake is approximately 26 mmHg whereas the hydrostatic pressure which pushes fluid out of the capillary is only about 9 mmHg [16].

The vasa recta also play an integral role in the maintenance of the medullary osmolal gradient. Since they are permeable to solutes and water, the vasa recta reach osmotic equilibrium with the interstitium. In the descending limb of the capillary (Fig. 6-5), solute enters and water leaves the capillary as the plasma osmolality

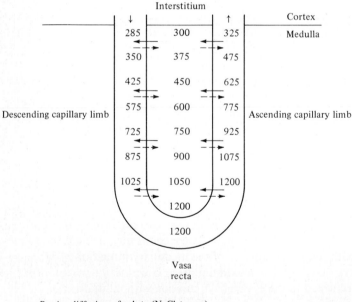

—————▶ Passive diffusion of solute (NaCl + urea)
— — —▶ Passive diffusion of H$_2$O

Figure 6-5 Principle of countercurrent exchange in the vasa recta capillaries. In the descending capillary limb, solute enters and water leaves the capillary down concentration gradients, tending to reduce interstitial osmolality. These processes are reversed in the ascending capillary limb, thereby preserving the interstitial osmolal gradient. *(Adapted from R. F. Pitts, Physiology of the Kidney and Body Fluids, 3d ed., Copyright © 1974 by Year Book Medical Publishers, Inc., Chicago. Used by permission.)*

reaches 1200 mosmol/kg at the papillary tip. If the vasa recta left the kidney at this point, the combination of solute removal and water addition would reduce medullary osmolality. However, medullary hyperosmolality is maintained because the vasa recta, like the loops of Henle, turn around at the papillary tip and return to the cortex. The solute removed from the interstitium in the descending limb is returned to the interstitium, i.e., exchanged, in the ascending limb down a concentration gradient from the lumen to the interstitium. Similarly, water added to the interstitium in the descending limb reenters the capillary in the ascending limb. Allowing for a lag in equilibration, the blood returning to the cortex is only slightly hyperosmotic to plasma (325 mosmol/kg).

The low rate of medullary blood flow (6 percent of renal blood flow) also contributes to the maintenance of interstitial hyperosmolality. If medullary blood flow is increased, more blood at 325 mosmol/kg will leave the medulla, and a significant washout of medullary solute can occur, with a reduction in interstitial osmolality [17]. This last effect will also diminish water reabsorption in the descending limb of the loop of Henle since there will be a lesser osmotic gradient between the tubular fluid and the interstitium (see Fig. 6-3). A clinical example where each of these changes occurs is an *osmotic diuresis* in which a large amount of nonreabsorbed solute is present in the urine. This may be seen with glucosuria in uncontrolled diabetes mellitus (see Chap. 25) or after an intravenous infusion of mannitol. In these settings, medullary blood flow is enhanced by an unknown mechanism [17], resulting sequentially in a decrease in papillary osmolality [18], and an elevation in urine output, primarily due to a fall in descending limb water reabsorption [19].

Role of Urea

Of the 1200 mosmol of solute per kilogram present at the papillary tip during antidiuresis, about half is NaCl and half is urea [20]. The high interstitial concentration of urea is produced by the diffusion of urea down a concentration gradient from the inner medullary collecting tubule into the interstitium (Fig. 6-6). ADH, acting in both the cortex and the medulla, plays a central role in this process by increasing the water permeability of the collecting tubules. As water is reabsorbed in the cortex and outer medulla, the urea concentration in the tubular fluid rises markedly since these segments are essentially impermeable to urea [12]. In contrast, permeability to urea in the *inner* medullary collecting tubule is relatively high in the basal state [12] and may increase further with ADH [21]. These effects allow urea to passively diffuse into the interstitium at this site.

In addition to ADH, urea accumulation in the medulla is also indirectly dependent upon active NaCl transport in the ascending limb. By making the tubular fluid dilute and the interstitium concentrated, NaCl reabsorption creates the osmotic gradients that allow water reabsorption to occur in the collecting tubules, thereby raising the tubular fluid urea concentration.

Some of the urea in the interstitium reenters the tubule in the ascending limb and, to a lesser degree, the descending limb [8,9]. The net effect of this urea recycling is that the quantity of urea in the early distal tubule is the same as or slightly exceeds the amount filtered even though 60 to 70 percent of the filtered urea has

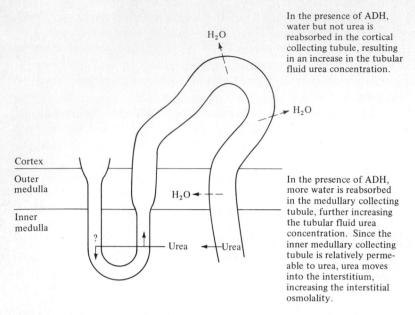

In the presence of ADH, water but not urea is reabsorbed in the cortical collecting tubule, resulting in an increase in the tubular fluid urea concentration.

In the presence of ADH, more water is reabsorbed in the medullary collecting tubule, further increasing the tubular fluid urea concentration. Since the inner medullary collecting tubule is relatively permeable to urea, urea moves into the interstitium, increasing the interstitial osmolality.

Figure 6-6 Mechanism by which urea achieves high concentrations in the medullary interstitium.

been reabsorbed in the proximal tubule [22]. Thus, both urinary and interstitial urea concentrations are maintained at high levels in the presence of ADH.†

The importance of urea in the concentrating process can be appreciated from studies in rats with hereditary diabetes insipidus who secrete no ADH [23]. In the absence of ADH, urea accumulation in the interstitium is minimized since the lack of water reabsorption in the cortical and outer medullary collecting tubules prevents the increase in the tubular fluid urea concentration that is necessary for urea diffusion. As a result, papillary osmolality is much less than normal, due to the almost complete inhibition of the papillary buildup of urea (Fig. 6-7). However, Na^+ accumulation in the papilla, derived from NaCl reabsorption in the medullary ascending limb, is not directly ADH-dependent, and papillary Na^+ content is normal. *Since the maximum urine osmolality (U_{osm}) cannot exceed that in the interstitium, the ability to conserve water by excreting a highly concentrated urine is reduced when there is no papillary accumulation of urea.* Thus, even after the administration of ADH for 3 days, the maximum U_{osm} in rats with diabetes insipidus is less than one-half that seen in normal subjects [23]. The more chronic administration of ADH for 28 days returns papillary urea and osmolality to normal and restores normal concentrating ability.

†The volume of the medullary interstitium is very small. For example, the weight of both kidneys in humans is approximately 350 g, most of which is composed of nephron segments, tubular fluid, and blood vessels. Thus, the attainment of high concentrations of urea in the interstitium requires only small amounts of urea. Since approximately 27 to 32 g of urea is excreted per day (see Table 2-1), the interstitial accumulation of urea does not importantly affect total urea excretion.

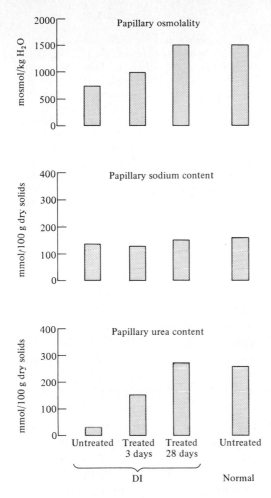

Figure 6-7 Simultaneous renal papillary osmolality, sodium content, and urea content in rats with diabetes insipidus (DI) before and after treatment with daily injections of ADH and in normal antidiuretic rats. In untreated DI rats, there is a marked reduction in papillary osmolality due to a fall in papillary urea content. Both osmolality and urea content slowly return to normal with ADH therapy. No significant difference in papillary sodium content was found among these groups. *(Adapted from A. R. Harrington and H. Valtin, J. Clin. Invest., 47:502, 1968, by copyright permission of the American Society for Clinical Investigation.)*

These data demonstrate that *papillary osmolality is not constant but varies with the availability of ADH*. A similar situation probably exists in humans. In patients with central diabetes insipidus or those who shut off ADH secretion by chronic water loading (see Chap. 24), the ability to concentrate the urine is impaired, presumably because of this reduction in medullary osmolality. As illustrated in Fig. 6-8, a mild defect in urinary concentration can be induced with only 3 days of overhydration.

Summary

The countercurrent mechanism permits the kidney to excrete urine with an osmolality that varies in humans from a minimum of 50 mosmol/kg to a maximum of 900 to 1400 mosmol/kg. The primary event in this process is active NaCl transport out of the thick ascending limb of the loop of Henle into the medullary interstitium, producing dilution of the tubular fluid and concentration of the interstitium. Because

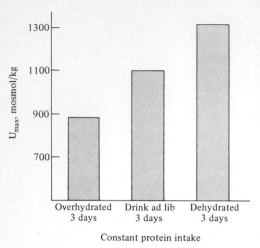

Figure 6-8 Effect of bodily hydration on maximum urinary concentration (U_{max}). Normal subjects, who served as their own controls, were put on either high water intake, low water intake, or ad lib intake for 3 days. On the fourth day, the U_{max} was measured after 12 h of water restriction and the administration of exogenous ADH to assure maximum antidiuretic effect. The reduction in U_{max} seen with overhydration, which suppresses endogenous ADH secretion, may have been due to a reduction in medullary urea content and osmolality. (*Adapted from F. H. Epstein, C. R. Kleeman, and A. Hendrikx, J. Clin. Invest., 36:629, 1957, by copyright permission of the American Society for Clinical Investigation.*)

of the different permeability characteristics of the descending and ascending limbs, this first step results in countercurrent multiplication in which a medullary osmolal gradient is created which reaches its maximum at the papillary tip. Other important factors in the determination of the urine osmolality include the vasa recta capillaries, which have a hairpin configuration that minimizes washout of the interstitial solute; ADH, which increases the water permeability of the collecting tubules; and, in the presence of ADH, the cortical collecting tubule in which water reabsorption reduces the volume of fluid presented to the medulla, the medullary collecting tubule in which the tubular fluid equilibrates osmotically with the interstitium, and urea which diffuses into the inner medullary interstitium and contributes to its high osmolality.

The osmolal changes that may occur as the tubular fluid moves through the nephron are summarized in Fig. 6-9. The isosmotic urine delivered from the proximal tubule becomes hyperosmotic in the descending limb as it equilibrates with the medullary interstitium and then becomes hypoosmotic in the ascending limb as NaCl is reabsorbed without water. *The final osmolality of the urine is determined in the collecting tubules in a manner dependent upon ADH.* In the presence of ADH, collecting tubule water permeability is increased, allowing osmotic equilibration of the tubular fluid with the isosmotic interstitium in the cortex and then the hyperosmotic interstitium in the medulla. The result is the excretion of a concentrated urine. ADH also contributes to the high medullary osmolality by permitting urea entry into the interstitium. In the absence of ADH, the hypoosmotic urine leaving the loop of Henle does not equilibrate with the interstitium, and a dilute urine is excreted.

It should be noted that these effects of ADH are not all-or-none but are dose-related. This is important because normal daily needs usually do not require maximal dilution or concentration of the urine. For example, to remain in the steady state, a normal subject may excrete 800 mosmol of solute and 2 liters of water daily. The excretion of urine with an average osmolality of 400 mosmol/kg requires a submaximal ADH response.

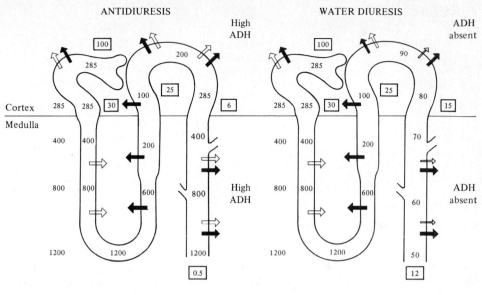

ANTIDIURESIS   WATER DIURESIS

Figure 6-9 Summary of NaCl and H₂O transport throughout the nephron during an antidiuresis and a water diuresis. The tubular fluid and interstitial concentrations are expressed in milliosmoles per kilogram (mosmol/kg); the large, boxed numbers represent the percentage of the glomerular filtrate remaining in the tubule at each site. Note that the composition and volume of the tubular fluid are essentially the same at the end of the loop of Henle as the excretion of a concentrated or dilute urine is determined primarily in the collecting tubules.

PROBLEMS

6-1 Explain the roles of NaCl reabsorption in the medullary ascending limb, urea accumulation, and the vasa recta in the production and maintenance of the hypertonicity of the medullary interstitium.

6-2 What is the role of NaCl reabsorption without water in the medullary ascending limb on (a) the excretion of a concentrated urine; and (b) the excretion of a dilute urine? What is the contribution of NaCl reabsorption without water in the cortical ascending limb and distal tubule to these processes?

6-3 In addition to osmotic diuretics, other diuretics are available which increase the urine output by inhibiting active NaCl reabsorption (see Chap. 16). What would be the likely site of action of a non-osmotic diuretic if it (a) inhibited both concentration and dilution; (b) inhibited dilution but not concentration?

6-4 What is the mechanism by which water is reabsorbed in the descending limb of the loop of Henle? Why is water reabsorption in the descending limb reduced during an osmotic diuresis? What might happen to descending limb water reabsorption in an untreated patient with central diabetes insipidus (absence of ADH)?

6-5 What effect will a low-protein diet (urea is an end product of protein metabolism) have on concentrating ability?

REFERENCES

1. Bennett, C. M., B. M. Brenner, and R. W. Berliner: Micropuncture study of nephron function in the rhesus monkey, J. Clin. Invest., 47:203, 1968.
2. Burg, M. B.: Thick ascending limb of Henle's loop, Kidney Int., 22:454, 1982.
3. Garg, L. C., M. A. Knepper, and M. B. Burg: Mineralocorticoid effects on Na-K-ATPase in individual nephron segments, Am. J. Physiol., 240:F536, 1981.
4. Burg, M. B., and N. Green: Function of the thick ascending limb of Henle's loop, Am. J. Physiol., 224:659, 1973.
5. Kunau, R. T., Jr., H. L. Webb, and S. C. Borman: Characteristics of sodium reabsorption in the loop of Henle and distal tubule, Am. J. Physiol., 227:1181, 1974.
6. Wright, F. S.: Flow-dependent transport processes: filtration, absorption, secretion, Am. J. Physiol., 243:F1, 1982.
6a. Wright, F. S., and J. Schnermann: Interference with feedback control of glomerular filtration rate by furosemide, triflocin, and cyanide, J. Clin. Invest., 53:1695, 1974.
7. Diezi, J., M. Nenninger, and G. Giebisch: Distal tubular function in superficial rat tubules during volume expansion, Am. J. Physiol., 239:F228, 1980.
8. Jamison, R. L., and R. H. Maffly: The urinary concentrating mechanism, N. Engl. J. Med., 295:1059, 1976.
9. Kokko, J. P.: Renal concentrating and diluting mechanisms, Hosp. Pract., 14(2):110, 1979.
10. Bonventre, J. V., and C. Lechene: Renal medullary concentrating process: an integrative hypothesis, Am. J. Physiol., 239:F578, 1980.
11. Jamison, R. L., and C. R. Robertson: Recent formulations of the urinary concentrating mechanism: a status report, Kidney Int., 16:537, 1979.
12. Rocha, A. S., and J. P. Kokko: Permeability of medullary nephron segments to urea and water: effect of vasopressin, Kidney Int., 6:379, 1974.
13. Imai, M.: The connecting tubule: a functional subdivision of the rabbit distal nephron segments, Kidney Int., 15:346, 1979.
14. Bleich, H. I., E. S. Boro, and R. M. Hays: Antidiuretic hormone, N. Engl. J. Med., 295:659, 1976.
15. Woodhall, P. B., and C. C. Tisher: Response of the distal tubule and cortical collecting duct to vasopressin in the rat, J. Clin. Invest., 52:3095, 1973.
16. Sanjana, V. M., P. A. Johnston, W. M. Deen, C. R. Robertson, B. M. Brenner, and R. L. Jamison: Hydraulic and oncotic pressure measurements in inner medulla of mammalian kidney, Am. J. Physiol., 228:1921, 1975.
17. Thurau, K.: Renal hemodynamics, Am. J. Med., 36:698, 1964.
18. Goldberg, M., and M. A. Ramirez: Effects of saline and mannitol diuresis on the renal concentrating mechanism in dogs: alterations in renal tissue solutes and water, Clin. Sci., 32:475, 1967.
19. Seely, J. F., and J. H. Dirks: Micropuncture study of hypertonic mannitol diuresis in the proximal and distal tubule of the dog kidney, J. Clin. Invest., 48:2330, 1969.
20. Ullrich, K. J., K. Kramer, and J. W. Boylan: Present knowledge of the countercurrent system in the mammalian kidney, Prog. Cardiovasc. Dis., 3:395, 1961.
21. Rocha, A. S., and L. H. Kudo: Water, urea, sodium, chloride, and potassium transport in the in vitro isolated perfused papillary collecting duct, Kidney Int., 22:485, 1982.
22. Lassiter, W., M. Mylle, and C. W. Gottschalk: Net transtubular movement of water and urea in saline diuresis, Am. J. Physiol., 206:669, 1964.
23. Harrington, A. R., and H. Valtin: Impaired urinary concentration after vasopressin and its gradual correction in hypothalamic diabetes insipidus, J. Clin. Invest., 47:502, 1968.

SEVEN

DISTAL NEPHRON

INTRODUCTION

The distal nephron begins at the macula densa at the end of the cortical thick ascending limb and consists of four segments: the distal tubule, the connecting segment (previously considered part of the late distal tubule), the cortical collecting tubule, and the medullary collecting tubule (see Fig. 2-3). These segments perform different functions and can be separated by histologic appearance and hormone responsiveness (Table 7-1) [1–3].

The distal nephron, particularly the collecting tubules, is the site at which the final qualitative changes in urinary excretion are made. For example, the Na^+ concentration is usually 30 to 50 meq/L in the fluid leaving the loop of Henle but can be appropriately reduced to less than 1 meq/L by the end of the medullary collecting

Table 7-1 Hormone responsiveness of the distal nephron segments

Segment	Hormone responsiveness
Distal tubule	Parathyroid hormone, calcitonin
Connecting segment	Parathyroid hormone
Cortical collecting tubule	Antidiuretic hormone, aldosterone, ?natriuretic hormone
Medullary collecting tubule	Antidiuretic hormone

tubule in states of volume depletion. This steep concentration gradient can be maintained because the distal nephron is relatively impermeable to the passive movement of water (in the absence of antidiuretic hormone) and solutes. Consequently, the gradient generated by active Na^+ transport is not dissipated by passive back diffusion from the plasma into the tubular fluid. In contrast, proximal Na^+ reabsorption occurs without change in the tubular fluid Na^+ concentration (see Fig. 4-4) since the proximal tubule is a highly permeable epithelium and the development of a tubular fluid-to-plasma concentration gradient for Na^+ is limited by the backflux of Na^+ into the lumen across the tight junction (see Chap. 5).

Although the collecting tubules can generate steep concentration gradients, their *total* reabsorptive capacity is limited. In terms of active Na^+ transport, this is exemplified by a lower level of Na^+-K^+-ATPase activity than is present in other nephron segments (except for the descending and thin ascending limbs of the loop of Henle, where transport is essentially passive) [4]. Therefore, the collecting tubules function most efficiently when the bulk of the filtrate is reabsorbed in the proximal tubule and loop of Henle, and distal delivery is held relatively constant. As described in Chaps. 3 and 5, three intrarenal processes minimize changes in distal delivery in normal subjects: *autoregulation* which maintains the GFR in the presence of variations in renal arterial pressure; *glomerulotubular balance* in which proximal, loop, and distal tubular reabsorption increase if there is an elevation in GFR; and *tubuloglomerular feedback* which lowers the GFR if the load to the macula densa is enhanced. These processes are important since an inappropriate increase in distal delivery could overwhelm reabsorptive capacity, leading to potentially serious losses of NaCl and water.

This chapter will briefly review the major functions of the distal nephron. Their role in the maintenance of ion and water balance will be discussed in detail in Chaps. 9 to 13.

DISTAL TUBULE AND CONNECTING SEGMENT

Sodium and Water

The distal tubule and connecting segment reabsorb about 5 to 10 percent of the filtered Na^+, with Cl^- following passively [5]. As with the loop of Henle, distal Na^+ reabsorption varies directly with Na^+ delivery and therefore participates in glomerulotubular balance [6]. Thus, an increase in delivery results in a proportionate increase in segmental Na^+ reabsorption. This effect is independent of aldosterone

[7] and peritubular capillary hemodynamics [8] and may, as in the loop of Henle, be related to changes in the tubular fluid Na^+ concentration [9]. If more Na^+ is delivered to the distal tubule, the associated elevation in the luminal Na^+ concentration favors passive Na^+ entry into the tubular cell.

In contrast, water reabsorption in these segments is probably minimal since they are poorly permeable to water in the basal state and do not respond to antidiuretic hormone [1,7,10].

Calcium

The early cortical distal nephron (including the cortical thick ascending limb as well as the distal tubule and connecting segment) is the major regulatory site of urinary Ca^{2+} excretion (see Chap. 5) [11]. This is primarily mediated by parathyroid hormone, which, by activating a hormone-specific adenyl cyclase, promotes the reabsorption of Ca^{2+} in these segments without affecting that of Na^+ or water [11,12]. The physiologic significance of calcitonin, which also stimulates adenyl cyclase in the distal tubule [1], is uncertain.

Potassium and Hydrogen

The distal tubule secretes H^+ and the connecting segment can secrete both H^+ and K^+ [3,10]. However, the collecting tubules play a much more important role in the regulation of H^+ and K^+ excretion (see below).

CORTICAL AND MEDULLARY COLLECTING TUBULES

Water

As discussed in the preceding chapter, the final urine osmolality is determined in the collecting tubules. In the absence of antidiuretic hormone (ADH), the water permeability of the collecting tubules is low and a dilute urine is produced (minimum urine osmolality equals 50 to 75 mosmol/kg). In the presence of ADH, water permeability is increased, resulting in the osmotic equilibration of the tubular fluid with the hyperosmotic interstitium and the excretion of a concentrated urine (maximum urine osmolality equals 900 to 1400 mosmol/kg).

Sodium

The collecting tubules usually reabsorb 5 to 7 percent of the filtered Na^+ [5], and variations in Na^+ reabsorption in these segments are probably the major determinant of fluctuations in daily Na^+ excretion [13,14]. This effect is at least in part mediated by aldosterone, which promotes Na^+ reabsorption in the cortical collecting tubule [4,7]. For example, a reduction in Na^+ intake enhances aldosterone secretion via activation of the renin-angiotensin system (see Chap. 8). This results in increased

Na^+ reabsorption in the cortical collecting tubule and an appropriate fall in Na^+ excretion. The opposite sequence occurs with a Na^+ load as aldosterone secretion is diminished. In states of chronic Na^+ loading, a natriuretic hormone may also contribute to the decrease in cortical collecting tubule Na^+ reabsorption (see Chap. 8) [15,16].

The factors governing Na^+ transport in the medullary collecting tubule are not known [17,18] as aldosterone does not appear to act at this site [19]. Active NaCl transport occurs in the inner medullary (or papillary) segment, a process that undoubtedly contributes to the very low urine Na^+ concentration attained in hypovolemic states [18].

Chloride Chloride is generally reabsorbed with Na^+ throughout the nephron. However, studies in animals given a high Na^+–low Cl^- diet have demonstrated that Cl^- can be conserved independently of Na^+ [19a]. Active Cl^- reabsorption in the inner medullary collecting tubule may play an important role in this setting [19a].

Potassium

Most of the excreted K^+ is derived from secretion from the cortical collecting tubular cell into the lumen [20] (see Chap. 13). This process is in part regulated by aldosterone. A K^+ load, for example, directly stimulates the release of aldosterone which then promotes the secretion and subsequent excretion of the excess K^+.

The medullary collecting tubule, which does not respond to aldosterone [19], plays a contributory role in the maintenance of K^+ balance as a high-potassium diet leads to K^+ secretion and a low-potassium diet to net reabsorption [18,20–22]. These effects may be directly mediated by alterations in the plasma K^+ concentration. For example, K^+ depletion lowers the plasma and then the renal tubular cell K^+ concentration [22]. The latter change may promote net K^+ reabsorption.

Hydrogen

After an acid load, the urine pH can be reduced to a minimum of 4.5 to 5.0 in the collecting tubules by H^+ secretion. This represents secretion against a plasma-to-tubular fluid concentration gradient for H^+ of almost 1:1000. (The relationship between pH and H^+ concentration is discussed in Chap. 11.) The absolute amount of H^+ secreted in the collecting tubules depends upon multiple factors, including systemic acid-base, NaCl, and K^+ balance, aldosterone, and the availability of urinary buffers (see Chap. 12).

RENAL PELVIS, URETERS, AND BLADDER

Minor modifications in the composition of the urine can occur after the urine has left the tubules. The renal pelvis is permeable to urea but not water or other solutes. As a result, urea may diffuse from the pelvis back into the medullary interstitium, thereby contributing to medullary hyperosmolality [23]. The ureters and bladder, on the other hand, allow some diffusion of solutes and water such that compositional changes of as much as 10 to 15 percent may occur in low-flow states when contact time is prolonged [24]. Thus, when a concentrated urine is produced, the urine osmolality may fall due both to water movement into and urea diffusion out of the urine.

REFERENCES

1. Imai, M.: The connecting tubule: a functional subdivision of the rabbit distal nephron segments, Kidney Int., 15:346, 1979.
2. Imai, M., and R. Nakamura: Function of distal convoluted and connecting tubules studied by isolated nephron fragments, Kidney Int., 22:465, 1982.
3. Kaissling, B.: Structural aspects of adaptive changes in renal electrolyte excretion, Am. J. Physiol., 243:F211, 1982.
4. Garg, L. C., M. A. Knepper, and M. B. Burg: Mineralocorticoid effects on Na-K-ATPase in individual nephron segments, Am. J. Physiol., 240:F536, 1981.
5. Bennett, C. M., B. M. Brenner, and R. W. Berliner: Micropuncture study of nephron function in the rhesus monkey, J. Clin. Invest., 47:203, 1968.
6. Kunau, R. T., Jr., H. L. Webb, and S. C. Borman: Characteristics of sodium reabsorption in the loop of Henle and distal tubule, Am. J. Physiol., 227:1181, 1974.
7. Gross, J. B., M. Imai, and J. P. Kokko: A functional comparison of the cortical collecting tubule and the distal convoluted tubule, J. Clin. Invest., 55:1284, 1975.
8. Diezi, J., M. Nenninger, and G. Giebisch: Distal tubular function in superficial rat tubules during volume expansion, Am. J. Physiol., 239:F228, 1980.
9. Wright, F. S.: Flow-dependent transport processes: filtration, absorption, secretion, Am. J. Physiol., 243:F1, 1982.
10. Jacobson, H. R.: Functional segmentation of the mammalian nephron, Am. J. Physiol., 241:F203, 1981.
11. Ng, R. C. K., R. A. Peraino, and W. N. Suki: Divalent cation transport in isolated tubules, Kidney Int., 22:492, 1982.
12. Costanzo, L. S., and E. E. Windhager: Effects of PTH, ADH, and cyclic AMP on distal tubular Ca and Na reabsorption, Am. J. Physiol., 239:F478, 1980.
13. Stein, J. H., R. W. Osgood, S. Boonjarern, N. W. Cox, and T. F. Ferris: Segmental sodium reabsorption in rats with mild and severe volume depletion, Am. J. Physiol., 227:351, 1974.
14. Stein, J. H., R. W. Osgood, S. Boonjarern, and T. F. Ferris: A comparison of the segmental analysis of sodium reabsorption during Ringer's and hyperoncotic albumin infusion in the rat, J. Clin. Invest., 52:2313, 1973.
15. deWardener, H. E., and E. M. Clarkson: The natriuretic hormone: recent developments, Clin. Sci., 63:415, 1982.
16. Fine, L. G., J. J. Bourgoignie, K. H. Hwang, and N. S. Bricker: On the influence of the natriuretic factor from patients with chronic uremia on the bioelectric properties and sodium transport of the isolated collecting tubule, J. Clin. Invest., 58:590, 1976.
17. Stokes, J. B.: Ion transport by the cortical and outer medullary collecting tubule, Kidney Int., 22:473, 1982.
18. Rocha, A. S., and L. H. Kudo: Water, urea, sodium, chloride, and potassium transport in the in vitro isolated perfused papillary collecting duct, Kidney Int., 22:485, 1982.
19. Stokes, J. B., M. J. Ingram, A. D. Williams, and D. Ingram: Heterogeneity of the rabbit collecting tubule: localization of mineralocorticoid hormone action to the cortical portion, Kidney Int., 20:340, 1981.
19a. Luke, R. G.: Effect of adrenalectomy on the adrenal response to chloride depletion in the rat, J. Clin. Invest., 54:1329, 1974.

20. Wright, F. S. : Potassium transport by successive segments of the mammalian nephron, Fed. Proc., 40:2398, 1981.
21. Stetson, D. L., J. B. Wade, and G. Giebisch: Morphologic alterations in the rat medullary collecting duct following potassium depletion, Kidney Int., 17:45, 1980.
22. Linas, S. L., L. N. Peterson, R. J. Anderson, G. A. Aisenbrey, F. R. Simon and T. Berl: Mechanism of renal potassium conservation in the rat, Kidney Int., 15:601, 1979.
23. Bonventre, J. V., R. J. Roman and C. Lechene: Effect of urea concentration of pelvic fluid on renal concentrating ability, Am. J. Physiol., 239:F609, 1980.
24. Levinsky, N. G., and R. W. Berliner: Changes in composition of urine in ureter and bladder at low urine flow, Am. J. Physiol., 196:549, 1959.

EFFECTS OF HORMONES ON RENAL FUNCTION

The preceding chapters discussed the reabsorptive and secretory functions of the individual nephron segments. These processes are affected by a variety of hormones, some of which are produced within the kidney such as renin, vitamin D, prostaglandins, and kinins. As will be seen, these hormones play an important role in the maintenance of fluid and electrolyte balance since they allow the *individual regulation* of the rate of excretion of the different solutes and water. In addition, the kidney secretes erythropoietin, a hormone that promotes red cell production by the bone marrow.

ANTIDIURETIC HORMONE AND WATER BALANCE

Antidiuretic hormone (the human form is called arginine vasopressin) is a polypeptide synthesized in the supraoptic and paraventricular nuclei in the hypothalamus (Fig. 8-1). Secretory granules containing antidiuretic hormone (ADH) migrate down the axons of the supraopticohypophyseal tract into the posterior lobe of the pituitary where they are stored and subsequently released after appropriate stimuli. In addition, some of the secretory granules produced in the paraventricular nuclei enter the

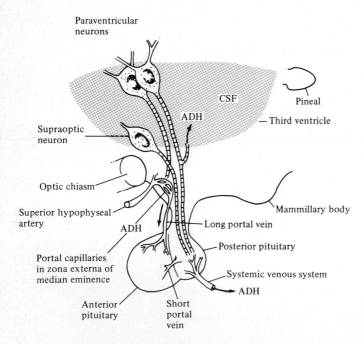

Figure 8-1 Diagram of the mammalian hypothalamus and pituitary gland showing paths for the secretion of antidiuretic hormone (ADH). The hormone is formed in the supraoptic and paraventricular nuclei, transported in granules along their axons, and then secreted at three sites: the posterior pituitary gland, the portal capillaries of the median eminence, and the cerebrospinal fluid (CSF) of the third ventricle. *(Adapted from E. A. Zimmerman and A. G. Robinson, Kidney Int., 10:12, 1976. Reprinted by permission from Kidney International.)*

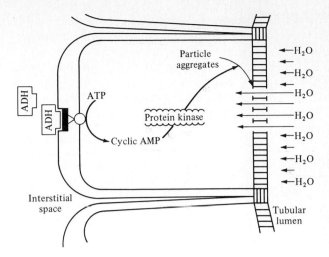

Figure 8-2 Scheme of the proposed mechanism by which ADH alters the water permeability of the luminal membrane of the collecting tubule cells. Cytoplasmic particle aggregates which fuse with the luminal membrane appear to represent or contain the channels through which water movement occurs. *(Adapted from T. P. Dousa, H. Sands, and O. Heckter, Endocrinology, 91:757, 1972. Permission granted from Lippincott.)*

cerebrospinal fluid or the portal capillaries in the median eminence (Fig. 8-1) [1]. The latter effect probably accounts for the observation that lesions of the posterior pituitary or tract below the median eminence do not usually lead to permanent diabetes insipidus (ADH-lack) since ADH produced in the hypothalamus still has access to the systemic circulation.

ADH is rapidly metabolized in the liver and kidney with a half-life in the circulation of only 15 to 20 min.

Actions

ADH is a prime determinant of renal water excretion (see Chap. 6). By augmenting the water permeability of the cortical and medullary collecting tubules [2], ADH increases renal water reabsorption, resulting in an elevation in the urine osmolality. In the absence of ADH, water reabsorption is reduced in these segments and large volumes of dilute urine are produced.

The mechanism by which ADH increases the water permeability of the collecting tubules is illustrated in Fig. 8-2. The first step is the attachment of ADH to a specific receptor on the peritubular membrane of the cell. This hormone-receptor complex activates adenyl cyclase,† resulting in the generation of cyclic adenosine 3′,5′-monophosphate (cyclic AMP) from ATP. Cyclic AMP then activates a protein kinase which initiates a sequence of events resulting in cytoplasmic, membrane-bound particle aggregates moving to and fusing with the luminal membrane (Fig. 8-3) [3,4]. These aggregates are thought to represent or contain the channels for water movement across the membrane.

---

†Adenyl cyclase and cyclic AMP also mediate the effects of many other hormones. Each system has a specific cell receptor that is activated only by the appropriate hormone. For example, an adenyl cyclase system is present in the proximal tubule that is sensitive to parathyroid hormone (PTH). There is no cross-activation between the PTH and ADH systems [5].

Figure 8-3 ADH-induced intramembranous particle aggregates (arrows) on luminal membrane of the toad bladder, an epithelium similar to the mammalian collecting tubule. Aggregates are seen between and near the bases of microvilli (MV). These changes correlate specifically with enhanced water permeability. *(From W. A. Kachadorian, S. D. Levine, J. B. Wade, V. A. DiScala, and R. M. Hays, J. Clin. Invest., 59:576, 1977, by copyright permission of the American Society for Clinical Investigation.)*

Extrarenal effects ADH is an arterial vasoconstrictor (hence the name vasopressin) which tends to raise the systemic blood pressure. This may be physiologically important since ADH secretion is markedly increased in hypovolemic or hypotensive states (see Fig. 8-6) and may contribute to the maintenance of blood pressure in these settings [6,7]. In general, however, the renin-angiotensin and sympathetic nervous systems play the primary role in blood pressure regulation (see Chap. 9) [8].

ADH has several other actions of uncertain physiologic significance. These include a reduction in the volume of salivary, sweat, and gastric secretions [9], a possible role as a corticotropin-releasing factor by hormone secreted into the portal capillaries in the median eminence [10], and a possible role in memory consolidation from hormone secreted into the cerebrospinal fluid (Fig. 8-1) [11].

Control of ADH Secretion

The major stimuli to ADH secretion are an increase in the plasma osmolality (P_{osm}) and a decrease in the effective circulating volume. This is appropriate, since ADH promotes renal water reabsorption, reducing the P_{osm} and increasing volume (Fig. 8-4).

Plasma osmolality An increase in the P_{osm} enhances ADH secretion and thirst (see below), resulting in water retention and a reduction in the P_{osm} toward normal. Conversely, water loading lowers the P_{osm}, which inhibits ADH release. The ensuing reduction in collecting tubule water reabsorption lowers the urine osmolality, thereby allowing the water load to be excreted and the P_{osm} to return to normal. Since the half-life of ADH in the circulation is 15 to 20 min, the maximum diuresis after a water load is delayed for 90 to 120 min, the time required for the metabolism of the previously circulating ADH.

The location of the osmoreceptors governing ADH release was demonstrated by the classic experiments of Verney, in which local infusions of hypertonic saline,

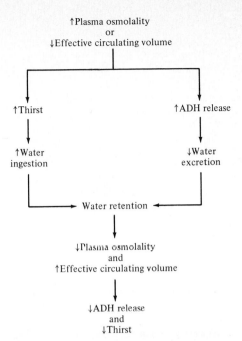

Figure 8-4 Feedback loop for the stimulation of ADH release and thirst.

which increase the local P_{osm} without affecting the systemic P_{osm}, were used [12]. Such an infusion into the carotid artery, but not the femoral artery, resulted in ADH release and an antidiuresis. This suggested that the receptors, which are separate from the hormone-producing cells [13], are located in the brain and not peripherally.

The increment in plasma osmolality is perceived in the hypothalamus as an effective osmotic gradient between the plasma and the receptor cell, inducing water movement out of the cell. This reduction in receptor cell volume (or an increase in the concentration of some intracellular solute) seems to be the signal that stimulates ADH secretion.

Since Na^+ salts are the major extracellular osmoles (see Chap. 1), *the plasma Na^+ concentration is the primary osmotic determinant of ADH release* [14]. If the plasma Na^+ concentration rises, for example, the associated elevation in P_{osm} promotes the secretion of ADH. In contrast, *an increase in the plasma concentration of urea does not stimulate ADH release* because this solute readily crosses cell membrane (whereas Na^+ cannot). Consequently, there is no decrease in osmoreceptor cell size as osmotic equilibrium is reached by urea entry into the cell rather than water movement out of the cell, i.e., urea is an ineffective osmole. Although glucose (the other quantitatively important extracellular solute) generally moves into cells slowly, i.e., it is an effective osmole like Na^+, elevations in the plasma glucose concentration do *not* enhance ADH secretion [14,15], suggesting that glucose is able to rapidly enter the osmoreceptor cells.

The osmoreceptors are extremely sensitive, responding to changes in osmolality of as little as 1 percent [14,16]. In humans, ADH release begins when the P_{osm} exceeds 275 to 285 mosmol/kg and increases progressively as the effective P_{osm} is

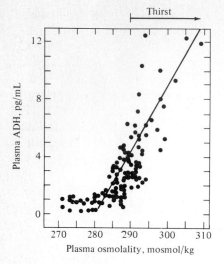

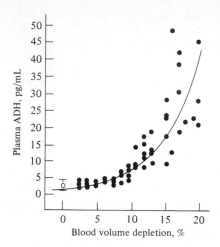

Figure 8-5 Relationship of plasma ADH concentration to plasma osmolality in normal humans in whom the plasma osmolality was changed by varying the state of hydration. Notice that the osmotic threshold for thirst is 5 to 10 mosmol/kg higher than that for ADH. *(Adapted from G. L. Robertson, P. Aycinena, and R. L. Zerbe, Am. J. Med., 72:339, 1982.)*

Figure 8-6 Relationship of ADH levels to isosmotic changes in blood volume in the rat. Notice that much higher ADH levels can occur with hypovolemia than with hyperosmolality. *(From F. L. Dunn, T. J. Brennan, A. E. Nelson, and G. L. Robertson, J. Clin. Invest., 52:3212, 1973, by copyright permission of the American Society for Clinical Investigation.)*

elevated (Fig. 8-5) [17]. This system is so efficient that the P_{osm} usually does not vary by more than 1 to 2 percent despite wide fluctuations in water intake.

Effective circulating volume Patients with effective volume depletion, as with vomiting, hypotension, or heart failure (see Chap. 9), may secrete ADH, even in the presence of a low plasma osmolality [13,17–19]. These findings indicate the existence of nonosmolal, volume-sensitive receptors for ADH release (Fig. 8-6). Parasympathetic afferents in the carotid sinuses are of primary importance in this response. Changes in the rate of afferent discharge affect the activity of the vasomotor center in the medulla and subsequently in vasomotor neurons in the region of the supraoptic nucleus [13]. Although low-pressure, left-atrial receptors contribute to this volume effect in some animal species, they appear to be less important in primates and humans [20,21].

The carotid sinuses are actually pressure receptors. However, they are able to function indirectly as volume receptors. How this occurs can be appreciated from the formula relating pressure, cardiac output, and vascular resistance:†

Mean arterial pressure = cardiac output × systemic vascular resistance

---

†This product of cardiac output and systemic vascular resistance actually equals the change in pressure across the circulation, i.e., the mean arterial pressure minus mean venous pressure. However, the mean venous pressure (1 to 7 mmHg) is normally so much lower than the mean arterial pressure that only a small error results from ignoring the venous pressure.

Thus, if volume depletion decreases cardiac output, there tends to be a decrease in blood pressure, which can be sensed by the carotid sinus receptors. As an example, acute thoracic inferior vena cava constriction results in pooling of blood in the venous system and a reduction in the venous return to the heart. This lowers the cardiac output and, secondarily, arterial blood pressure. This experimental model of volume depletion results in an increase in U_{osm} and a fall in free-water clearance (a measure of the volume of solute-free water excreted per minute; see Chap. 10) (Fig. 8-7). The abolition of these urinary responses by prior hypophysectomy suggests that they are mediated by an increase in ADH secretion. That they also are abolished by carotid baroreceptor denervation suggests that the increment in ADH release is governed by baroreceptor afferents.

The carotid baroreceptors also sense primary changes in pressure. As a result, ADH release varies inversely with acute fluctuations in systemic arterial pressure, appropriately increasing with hypotension and decreasing with hypertension [22–24]. These changes are blocked by denervation of the carotid sinuses.

The sensitivity of the volume receptors is different from that of the osmoreceptors. Although a relatively large volume change is required before ADH is released, ADH secretion increases exponentially once this threshold is reached (Fig. 8-6) [14,19].

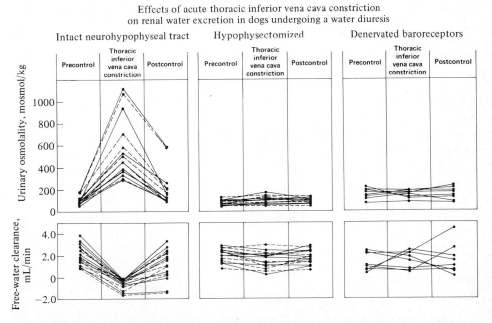

Figure 8-7 Effect of thoracic inferior vena cava constriction on urine osmolality (above) and free-water clearance (below) in intact (left), hypophysectomized (middle), and baroreceptors-denervated animals (right). The renal nerves play no role in this response as evidenced by the equivalent results in intact and denervated kidneys (solid and dashed lines, respectively). *(From R. J. Anderson, P. Cadnapaphornchai, J. A. Harbottle, K. M. McDonald, and R. W. Schrier, J. Clin. Invest., 54:1473, 1974, by copyright permission of the American Society for Clinical Investigation.)*

Importance of thirst The maintenance of water balance is importantly related to water intake as well as the ADH effect on water excretion. Water intake to meet homeostatic requirements is regulated by the sensation of thirst which is controlled centrally in an area of the hypothalamus that overlaps the region governing ADH release [25]. In general, thirst is affected by those stimuli determining the secretion of ADH, *osmolality* and *volume* [15,25], although the osmotic threshold for thirst is 5 to 10 mosmol/kg higher than that for ADH (Fig. 8-5) [17]. As hyperosmolality or volume depletion increases thirst, the ensuing water retention acts to return the plasma osmolality or volume toward normal. The hypovolemic stimulus to thirst may be mediated in part by the renin-angiotensin system which also is activated by volume depletion (see Fig. 3-5) [25].

Although thirst is regulated centrally (including cortical areas which influence nonessential or social drinking), it is sensed peripherally usually as the feeling of a dry mouth [26]. Similarly, the cessation of thirst (satiety) is initially mediated peripherally by oropharyngeal mechanoreceptors or by a feeling of gastric distention [26,27]. The net effect is that satiety occurs before much of the ingested water has been absorbed and therefore before the hyperosmolality or volume depletion has been corrected. This is an appropriate response since there is a 30- to 60-min delay before ingested water is completely absorbed. Thus, in a hyperosmolal subject, water intake would be excessive if thirst continued until the plasma osmolality returned to normal since there would still be a substantial amount of nonabsorbed water remaining in the gastrointestinal tract.

Interactions of the osmotic and volume stimuli The hormone-producing cells in the supraoptic and paraventricular nuclei receive input from both the osmotic and volume receptors, resulting in positive or negative interactions [13]. For example, volume depletion potentiates the ADH response to hyperosmolality (Fig. 8-8) but can prevent the inhibition of ADH release normally induced by a fall in the P_{osm} [13,17–19].

In the clinical setting, volume depletion is a common cause of water retention and hypoosmolality (see Chap. 23). This occurs in part because the presence of ADH

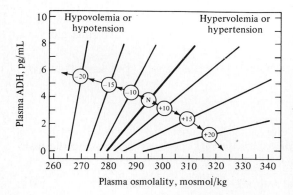

Figure 8-8 The influence of hemodynamic status on the osmoregulation of ADH in otherwise healthy humans. The numbers in the center circles refer to the percentage change in volume and/or pressure; *N* refers to the normovolemic normotensive subject. Notice that the hemodynamic status affects both the slope of the relationship between the plasma ADH and osmolality and the osmotic threshold for ADH release. (*Adapted from G. L. Robertson, R. L. Shelton, and S. Athar, Kidney Int., 10:25, 1976. Reprinted by permission from Kidney International.*)

Table 8-1 Factors influencing ADH secretion

Stimuli	Inhibitors
Hyperosmolality	Hypoosmolality
Hypovolemia	Hypervolemia
Stress, e.g., pain	Ethanol
Nausea	Phenytoin
Hypoglycemia	
Hypoxemia	
Nicotine	
Morphine	
Other drugs (see Table 23-3)	

prevents the excretion of the water load taken in because of the hypovolemic stimulus to thirst.

Other factors affecting ADH secretion In addition to osmolality and volume, ADH release can be influenced by a variety of other factors that are not directly related to osmolal or volume balance (Table 8-1) [28–35]. Neither the physiologic role of these effects nor the mechanisms by which they occur is well understood. Nevertheless, they may become clinically important in certain circumstances. In surgical patients, for example, elevated levels of ADH may persist for several days after the operation [28], a stress response that appears to be mediated by pain afferents [29]. If a large amount of free water is given in this setting, water retention and hyponatremia may ensue since the ADH prevents excretion of the excess water.

ALDOSTERONE

The steps involved in adrenal cortical steroid synthesis are illustrated in Fig. 8-9. These hormones are synthesized in different areas of the adrenal cortex: aldosterone in the zona glomerulosa, and glucocorticoids (particularly cortisol), androgens, and estrogens in the zona fasciculata and reticularis. The zona glomerulosa is well adapted for the production of aldosterone [36]. It has a low concentration of 17α-hydroxylase, the enzyme necessary for cortisol and androgen synthesis. More importantly, the two-step conversion of corticosterone to aldosterone, by the addition of an aldehyde at the 18-carbon position [37], can be achieved only in the zona glomerulosa [36].

Aldosterone is converted to inactive metabolites in the liver and, to a much lesser degree, in the kidney.

Actions

Aldosterone acts primarily on the cortical collecting tubule cell to increase the reabsorption of Na^+ and Cl^- and the secretion of K^+ and H^+ (Fig. 8-10) [38–40]. After aldosterone is administered, there is a 90-min latent period before electrolyte excre-

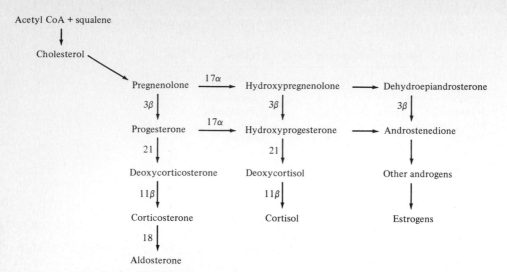

Acetyl CoA + squalene

Cholesterol

Pregnenolone —17α→ Hydroxypregnenolone ——→ Dehydroepiandrosterone

3β ↓   3β ↓   3β ↓

Progesterone —17α→ Hydroxyprogesterone ——→ Androstenedione

21 ↓   21 ↓

Deoxycorticosterone   Deoxycortisol   Other androgens

11β ↓   11β ↓

Corticosterone   Cortisol   Estrogens

18 ↓

Aldosterone

Figure 8-9 Schematic pathways of adrenal steroid biosynthesis. The numbers at the arrows refer to specific enzymes: 17α = 17α-hydroxylase; 3β = 3β-hydroxysteroid dehydrogenase; 21 = 21-hydroxylase; 11β = 11β-hydroxylase; 18 refers to a two-step process resulting in the addition of an aldehyde at the 18-carbon position. Deficiencies in any of these enzymes can lead to abnormal mineralocorticoid (as well as glucocorticoid and androgen) production, resulting in disturbances in Na^+, K^+, and H^+ balance (see Chaps. 18, 27, and 28).

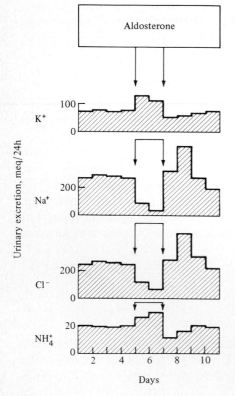

Figure 8-10 Effect of aldosterone on the daily urinary excretion of K^+, Na^+, Cl^-, and H^+ (as NH_4^+; see Chap. 12) in a normal subject maintained on a constant dietary intake. Note that the quantitatively most prominent effect is the marked reduction in NaCl excretion. *(From G. W. Liddle, Arch. Intern. Med., 102:998, 1958. By permission of the American Medical Association, copyright 1958.)*

Capillary   Cortical collecting tubule cell   Tubular
                                                                                 lumen

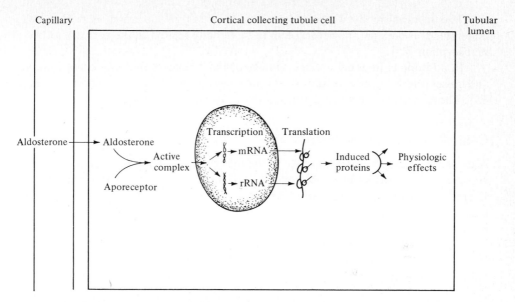

Figure 8-11 General model for the proposed mechanism of aldosterone action. The hormone enters the cell by diffusion and combines with an aporeceptor in the cytoplasm to generate an active complex which then translocates to the nucleus. Within the nucleus, the steroid-receptor complex interacts with the chromatin to enhance messenger RNA (mRNA) and ribosomal RNA (rRNA) transcription which, in turn, leads to increased translation of induced proteins. The synthesis of induced proteins ultimately mediates the physiologic effects of the hormone. *(From D. Feldman, J. W. Funder, and I. S. Edelman, Am. J. Med., 53:545, 1972.)*

tion is affected. The sequence of events responsible for this delay is depicted in Fig. 8-11. Aldosterone diffuses into the cell and attaches to a specific receptor in the cytoplasm. The hormone-receptor complex then moves into the nucleus where it interacts with specific sites on the nuclear chromatin to enhance messenger RNA and ribosomal RNA transcription. This in turn is translated into new protein synthesis that ultimately mediates at least some of the physiologic effects of the hormone. Enhanced Na^+ transport, for example, is linearly related to the increment in messenger RNA synthesis [41].

The alterations in electrolyte transport produced by aldosterone are mediated by two changes in the cortical collecting tubule cell: (1) the permeability of the luminal membrane to Na^+ and K^+ is enhanced; and (2) there is increased activity of the Na^+-K^+-ATPase pump in the peritubular membrane [39,42–44]. The net result is that Na^+ reabsorption is augmented since both Na^+ diffusion from the lumen into the cell and the active transport of Na^+ out of the cell into the capillary are facilitated (see Fig. 5-3). Similarly, the combination of increased K^+ transport from the capillary into the cell by the Na^+-K^+-ATPase pump and enhanced K^+ permeability of the luminal membrane favors K^+ secretion from the cell into the lumen (see Chap. 13). The increase in luminal membrane permeability appears to be the primary effect of aldosterone-induced protein with the ensuing enhanced entry of luminal Na^+ into the cell secondarily leading to the elevation in Na^+-K^+-ATPase activity [43,44].

The mechanism by which aldosterone enhances H^+ secretion is not as well understood. It does not, however, seem to be directly related to the increase in Na^+ reabsorption [45].

In addition to its renal actions, aldosterone also reduces the Na^+ concentration and raises the K^+ concentration in salivary, sweat, and colonic secretions [44,46–48]. These changes are generally of limited physiologic importance.

Control of Aldosterone Secretion

Aldosterone plays an important role in the maintenance of volume and K^+ balance via its effects on NaCl and K^+ excretion [49]. Thus, it is appropriate that angiotensin II (the production of which varies inversely with volume) and an elevation in the plasma K^+ concentration are the major stimuli of aldosterone secretion. Angiotensin II and hyperkalemia act on the zona glomerulosa, promoting the conversion of cholesterol to pregnenolone and, more importantly, of corticosterone to aldosterone (see Fig. 8-9) [50,51]. Adrenocorticotropic hormone (ACTH) and hyponatremia also can enhance aldosterone release, but these effects are physiologically less important.

Renin-angiotensin system The volume stimulus to aldosterone secretion is primarily mediated by the renin-angiotensin system (see Fig. 3-5) [38,52]. In normal subjects, both the plasma renin activity and aldosterone release vary inversely with dietary Na^+ intake (Fig. 8-12). An increase in Na^+ intake, for example, initially expands

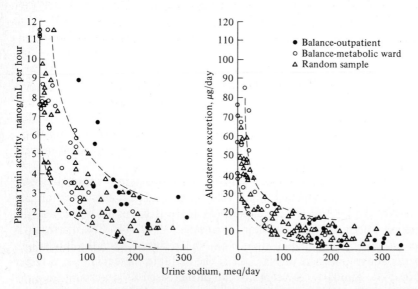

Figure 8-12 Relation of both plasma renin activity and 24-h urinary excretion of aldosterone to the concurrent daily rate of sodium excretion, used as an estimate of daily Na^+ intake. For these normal subjects, the data indicate an inverse relationship between dietary Na^+ intake and renin and aldosterone secretion. *(From J. H. Laragh, I. Baer, H. R. Brunner, F. R. Bühler, J. E. Sealey, and E. D. Vaughan, Jr., Am. J. Med., 52:633, 1972.)*

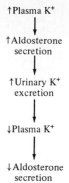

↑Plasma K⁺

↑Aldosterone
secretion

↑Urinary K⁺
excretion

↓Plasma K⁺

↓Aldosterone
secretion

Figure 8-13 Role of aldosterone in potassium homeostasis.

the extracellular volume, resulting in reductions in renin and aldosterone production. The latter effect then allows the excess Na⁺ to be excreted.

Conversely, a reduction in the effective circulating volume will enhance the secretion of renin and therefore that of aldosterone. The ensuing Na⁺ retention returns the volume toward normal. The importance of renin in this sequence has been demonstrated by the loss of the hypovolemic stimulus to aldosterone secretion in nephrectomized patients [53,54].

Plasma K⁺ concentration Aldosterone secretion varies directly with the plasma K⁺ concentration [55]. This represents a direct effect on the zona glomerulosa [50] and is extremely sensitive, as increments in the plasma K⁺ concentration of as little as 0.1 to 0.2 meq/L can induce a significant rise in aldosterone secretion [55]. The resultant increase in K⁺ excretion then returns the plasma K⁺ concentration toward normal (Fig. 8-13).

ACTH ACTH, released from the anterior pituitary, regulates adrenal glucocorticoid and androgen synthesis and release. It is of less importance in aldosterone secretion, causing only a transient rise in hormone secretion [56]. This limited response may be due to overproduction of deoxycorticosterone, an ACTH-dependent steroid with aldosteronelike activity (see Fig. 8-9). The ensuing fluid retention will diminish renin secretion and secondarily that of aldosterone [57].

Although ACTH is not of primary importance in the regulation of aldosterone release, it may play a *permissive* role in the response to volume depletion. Hypophysectomized animals, for example, respond to hypovolemia with a much less than normal increment in aldosterone secretion [58]. This effect, however, may not be due to diminished ACTH secretion since it is not reproduced with cortisol-induced ACTH suppression [58]. It is possible that lack of other pituitary hormones, such as β-lipotropin [59,60], may be responsible for the submaximal increase in aldosterone release in this setting.

Hyponatremia Hyponatremia (plasma Na⁺ concentration less than 125 to 130 meq/L) is another stimulus to the secretion of aldosterone [54,61]. It is not likely,

Table 8-2 Interrelationships between aldosterone and Na$^+$ and K$^+$ balance

Clinical state	Aldosterone secretion	Proximal or loop Na$^+$ reabsorption	Distal Na$^+$ delivery	Change in urinary excretion Na$^+$	K$^+$
K$^+$ load	↑	↓	↑	0	↑
K$^+$ depletion	↓	↑	↓	0	↓
Na$^+$ load	↓	↓	↑	↑	0
Na$^+$ depletion	↑	↑	↓	↓	0

however, to be physiologically important since large changes in the plasma Na$^+$ concentration do not occur in normal subjects due to the effects of ADH (see above). Even when hyponatremia is present, its effect on aldosterone is frequently overridden by concomitant changes in the effective circulating volume [62]. Thus, aldosterone secretion is increased in the hyponatremic patient who is volume-depleted, but may be reduced in a patient who is volume-expanded, e.g., due to the retention of water [62].

Maintenance of Sodium and Potassium Balance

Since aldosterone affects both Na$^+$ and K$^+$ handling, it might be expected that the regulation of the excretion of one ion would interfere with that of the other. This does not occur because of two additional effects: (1) K$^+$ secretion is highly dependent upon the rate of fluid delivery to the cortical collecting tubule (see Fig. 13-8); and (2) K$^+$ balance influences Na$^+$ reabsorption in the proximal tubule or perhaps the loop of Henle by an unknown mechanism [63,63a].

How these factors allow Na$^+$ and K$^+$ excretion to be regulated independently is depicted in Table 8-2. A K$^+$ load, for example, increases aldosterone secretion and therefore K$^+$ excretion. K$^+$ also reduces proximal and/or loop Na$^+$ and water reabsorption. Thus, the increase in cortical collecting tubular Na$^+$ reabsorption induced by aldosterone is counteracted by the augmented Na$^+$ and water delivery from the more proximal segments. The net effect is enhanced excretion of K$^+$ with little change in that of Na$^+$. On the other hand, effective volume depletion increases aldosterone secretion (by stimulating renin release), thereby reducing Na$^+$ excretion. K$^+$ excretion is not importantly affected since hypovolemia also enhances proximal Na$^+$ reabsorption. The ensuing decrease in distal fluid delivery diminishes K$^+$ secretion, counteracting the effect of the increase in aldosterone [64]. This explains why untreated patients with heart failure (who are effectively volume-depleted due to a low cardiac output) do not spontaneously develop K$^+$ wasting and hypokalemia even though aldosterone secretion is frequently elevated.

Aldosterone Escape

If aldosterone is given to a normal subject on an adequate NaCl intake, NaCl and water retention and K$^+$ loss will be seen initially, resulting in a weight gain of 2 to

3 kg after several days. Then, a spontaneous diuresis ensues secondary to the volume expansion [65,66]. This phenomenon has been called *aldosterone escape* but does *not* represent aldosterone resistance since urinary K^+ loss continues [65], the Na^+ concentration remains low in other aldosterone-sensitive secretions such as in sweat and saliva [66], and the cortical collecting tubule continues to respond to aldosterone [67]. Thus, it appears that the escape phenomenon is due to decreased Na^+ reabsorption in some aldosterone-independent nephron segment, perhaps the loop of Henle [66]. How this occurs is not known, but a natriuretic hormone may be involved [68]. The clinical correlate of this effect occurs in patients with autonomous overproduction of aldosterone (primary hyperaldosteronism; see Chap. 27) who present with hypokalemia but not edema since continued fluid retention is prevented.

NATRIURETIC HORMONE

Expansion of the extracellular volume with a Na^+ load results in an appropriate increase in Na^+ excretion. Initially, it was felt that this was due to an increase in the GFR (first factor) or a decrease in aldosterone secretion (second factor). However, aortic constriction to lower the GFR and the administration of high doses of aldosterone do not prevent the excretion of the Na^+ load [69]. The factor(s) responsible for this decrease in tubular reabsorption have been termed *third factor*. It is now clear that third factor is itself composed of several factors including intrarenal hemodynamics and, probably, a *natriuretic hormone* (see Chap. 9) [70,71].

An example of a disorder in which a natriuretic hormone may be particularly important is chronic renal failure. In this setting, patients who have lost as much as 90 percent of their GFR (with a consequent marked reduction in the filtered Na^+ load) may still be able to maintain Na^+ balance on a normal Na^+ intake by reducing tubular Na^+ reabsorption. Other circumstances that also require a decrease in tubular Na^+ reabsorption include a high Na^+ intake and aldosterone escape (see above).

In each of these three conditions, circulating substances have been identified which inhibit Na^+ transport both in vivo and in vitro [68,71–73]. That the release of this "hormone" is induced by volume expansion is supported by the following observations in renal failure. The excretion of Na^+ is determined by the difference between the filtered load and tubular reabsorption. In renal failure, the GFR is reduced and Na^+ excretion will fall if there is no change in tubular reabsorption. As a result, Na^+ retention and volume expansion will ensue if dietary Na^+ intake is held constant. In this setting, in which tubular Na^+ reabsorption must be reduced to maintain Na^+ balance, a circulating natriuretic factor can be demonstrated [74]. However, if Na^+ intake is restricted so that Na^+ retention and volume expansion do not occur, a circulating natriuretic hormone cannot be found [74]. This experiment supports the existence of a natriuretic hormone that is released specifically in response to persistent hypervolemia.†

---

†The relative roles of aldosterone, natriuretic hormone, and other factors in the regulation of Na^+ excretion will be discussed in Chap. 9.

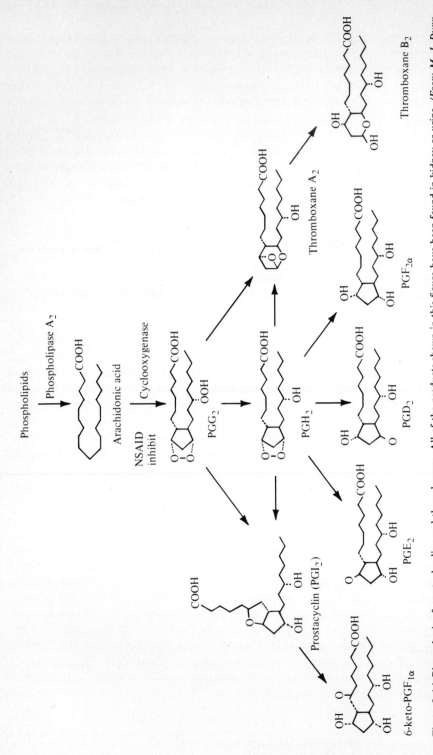

Figure 8-14 Biosynthesis of prostaglandins and thromboxane. All of the products shown in this figure have been found in kidney or urine. *(From M. J. Dunn and E. J. Zambraski, Kidney Int., 18:609, 1980. Reprinted by permission from Kidney International.)*

Table 8-3 Renal actions of prostaglandins and possible complications with nonsteroidal anti-inflammatory drugs

Effect of prostaglandins	Potential complication of nonsteroidal anti-inflammatory drugs
Maintain renal blood flow and GFR by antagonizing vasoconstrictive effects of angiotensin II and norepinephrine	Renal ischemia and renal failure in conditions associated with increased angiotensin II and/or norepinephrine such as hepatic cirrhosis, heart failure, nephrotic syndrome, and volume depletion
Increase renin secretion	Hyperkalemia due to hyporeninemic hypoaldosteronism
Antagonize effect of ADH on water permeability	Potentiates effect of ADH. However, water retention and hyponatremia have not been reported, perhaps because the initial water retention will reduce further ADH secretion
May increase sodium excretion in hypovolemic states	May promote further fluid retention in edematous states

The characteristics of this natriuretic hormone (or possibly hormones) are incompletely understood. The site of origin and afferent signal governing its release are not known although it has been postulated that the sensor is in the cardiopulmonary circulation [71,75] and that the hormone is secreted from the brain, probably in the area of the hypothalamus† [76,77]. Once released, the natriuretic hormone may act by decreasing Na^+ reabsorption in the collecting tubules [68,72], an effect that may be mediated by a reduction in Na^+-K^+-ATPase activity in these segments [71,77].

PROSTAGLANDINS

A variety of prostaglandins are synthesized in the kidney (Fig. 8-14) [78]. Glomerular and vascular endothelium, the medullary collecting tubules, and the renomedullary interstitial cells have been identified as sites of prostaglandin production. In general, PGE_2 is the primary metabolite produced in the medulla, whereas many prostaglandins, including PGE_2, PGI_2 (prostacyclin), and $PGF_{2\alpha}$, are produced in the cortex [79].

The functions of the renal prostaglandins have become better understood [78,79]. They have a variety of renal effects but little systemic activity since most of these hormones are rapidly metabolized in the pulmonary circulation. The intrarenal effects may have important clinical implications (Table 8-3) in view of the increasing

---

†Another possibility is that changes in volume perceived by atrial receptors can lead to the release of a natriuretic hormone directly *from the atria* [77a].

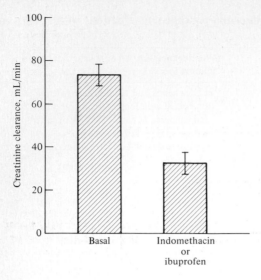

Figure 8-15 Reduction in creatinine clearance from a mean of 73 mL/min to 32 mL/min following the administration of the nonsteroidal anti-inflammatory drugs, indomethacin or ibuprofen, to 12 patients with hepatic cirrhosis and ascites. Urinary PGE excretion in the patients in the basal state was 11 times higher than in normals. *(From R. D. Zipser, J. C. Hoefs, P. F. Speckhart, P. K. Zia, and R. Horton, J. Clin. Endocrinol. Metab., 48:895, 1979.)*

use of the many nonsteroidal anti-inflammatory drugs (NSAID) which are inhibitors of the cyclooxygenase enzyme that converts arachidonic acid to PGG_2† (Fig. 8-14).

Actions

Renal hemodynamics Renal prostaglandins are primarily vasodilators (except for $PGF_{2\alpha}$ and thromboxane). They appear to play no role in the regulation of renal perfusion in the basal state when their secretion rate is relatively low. However, prostaglandin synthesis is increased (probably within the glomeruli) by the potent vasoconstrictors, angiotensin II and norepinephrine [78,80,80a]. The ensuing prostaglandin-induced vasodilation partially counteracts the vasoconstriction, thereby minimizing the degree of renal ischemia (see page 65) [78,81]. As a result, NSAID have no effect on renal perfusion in normal subjects but *can lead to renal ischemia and renal insufficiency* in hypovolemic states in which angiotensin II and norepinephrine secretion are increased. These include true volume depletion (see Fig. 3-8) [81] and edematous states associated with effective circulating volume depletion such as hepatic cirrhosis, heart failure, and the nephrotic syndrome (Fig. 8-15) [82–84].

---

†One exception appears to be sulindac, a NSAID which does not inhibit renal prostaglandin synthesis and would not be expected to lead to the complications in Table 8-3 [79a]. Sulindac is the only NSAID which has no active drug excreted with urine, a characteristic that probably explains its lack of renal effect.

Renin secretion The stimulation of renin secretion induced by baroreceptors in the afferent glomerular arteriole and by the macula densa cells in the distal tubule appears to be mediated by prostaglandins, particularly prostacyclin (see page 59) [85,85a]. This response is blocked by NSAID, and, in some patients, these drugs can lead to hyperkalemia due to hyporeninemic hypoaldosteronism (since angiotensin II is the major stimulus to aldosterone secretion) [86].

Antagonism of ADH effect Prostaglandins (particularly PGE_2) impair the ability of ADH to increase the urine osmolality [87]. This is in part due to a reduction in the ADH-induced generation of cyclic AMP by the collecting tubule cell [88]. In addition, PGE_2 has effects that interfere with the medullary interstitial accumulation of urea and NaCl that are essential for the production of a concentrated urine (see Chap. 6). These include a decrease in the reabsorption of urea in the inner medullary collecting tubule and possibly that of NaCl in the thick ascending limb of Henle's loop [87,89].

The physiologic significance of these prostaglandin effects is uncertain. It has been suggested that a short negative feedback loop may be present in which ADH stimulates local PGE_2 synthesis by the collecting tubule cells and/or the renomedullary interstitial cells, thereby preventing an excessive elevation in the urine osmolality [87,90,91]. However, such a direct regulatory role appears unlikely in humans since the effect of ADH on PGE_2 release seems to be due to its pressor, not its antidiuretic properties [80]. Thus, the effect of ADH may be similar to that of angiotensin II or norepinephrine, stimulating prostaglandin synthesis to maintain renal perfusion in the presence of renal vasoconstriction [80,80a].

Regardless of the exact relationship between these hormones, the administration of NSAID leads to an enhanced effect of ADH as evidenced by an increase in the urine osmolality that may exceed 200 mosmol/kg [90,92]. Although this could lead to inappropriate water retention and hyponatremia, this complication has not been described, probably because the initial fall in the plasma Na^+ concentration will diminish ADH secretion (see Fig. 8-5).

Sodium excretion The effect of renal prostaglandins on sodium excretion is controversial. Although prostaglandins have been reported to decrease NaCl reabsorption in the thick ascending limb of Henle's loop and the collecting tubules [93], these observations have not been confirmed by others [94]. If prostaglandins do promote Na^+ excretion, they would do so only when their production is increased. Thus, these hormones would have little effect in the basal state when their secretion rate is low, but they could play a role in hypovolemic or edematous states where prostaglandin synthesis is increased, at least in part, by angiotensin II and norepinephrine [78,95]. (Once again, prostaglandins are acting as modulating hormones, preventing excessive Na^+ retention.) In these settings, NSAID could lead to reduced Na^+ excretion both by increasing tubular reabsorption and decreasing the GFR. In edematous states, this could increase the degree of edema.

HORMONAL REGULATION OF CALCIUM AND PHOSPHATE BALANCE

The regulation of calcium and phosphate homeostasis involves changes in intestinal, bone, and renal function. In contrast to Na^+, Cl^-, and K^+, which are completely absorbed in the intestinal tract, the absorption of calcium and phosphate is incomplete. On a normal adult diet of 1000 mg of calcium per day, roughly 700 mg may be absorbed. However, 600 mg of calcium from digestive secretions may be lost in the feces so that the net absorption may be only 100 mg. In the steady state, this 100 mg is excreted in the urine.

Within the body, most of the calcium and much of the phosphate are found in bone as hydroxyapatite, $Ca_{10}(PO_4)_6(OH)_2$, the main mineral component of bone [96]. Phosphate also is present in high concentrations in the cells. Of the calcium in the extracellular fluid, roughly 40 percent is bound to albumin, 10 percent is complexed with citrate, bicarbonate, or phosphate, and 50 percent exists as the physiologically important ionized (or free) Ca^{2+}. The phosphorus in the blood consists of phospholipids, ester phosphates, and inorganic phosphates. The latter are completely ionized, existing in the extracellular fluid primarily as HPO_4^{2-} and $H_2PO_4^-$ in a ratio of 4:1 (see page 208).

Although only a small fraction of the total body calcium and phosphate is located in the plasma, it is the plasma concentrations of ionized Ca^{2+} and inorganic phosphate that are under hormonal control. This function is mediated primarily by parathyroid hormone and vitamin D, which affect intestinal absorption, bone formation and resorption, and urinary excretion. The possible physiologic roles of other hormones such as calcitonin [97] in the regulation of calcium and phosphate balance are not clear at this time.

Parathyroid Hormone

Parathyroid hormone (PTH) is a polypeptide secreted from the parathyroid glands in response to a decrease in the plasma concentration of ionized Ca^{2+}. It acts to increase the plasma Ca^{2+} concentration in three ways (Fig. 8-16) [98]:

1. In the presence of permissive amounts of vitamin D, it stimulates bone resorption, resulting in the release of calcium phosphate.
2. It enhances intestinal Ca^{2+} and phosphate absorption by promoting the formation of 1,25-dihydroxycholecalciferol (1,25-D) the major active metabolite of vitamin D (see below).
3. It augments renal Ca^{2+} reabsorption.

These effects are reversed by a small elevation in the plasma Ca^{2+} concentration, which lowers PTH secretion.

PTH also affects phosphate balance (Fig. 8-16). By its effects on bone and the intestinal tract, more phosphate enters the extracellular fluid. However, PTH also reduces proximal tubular phosphate reabsorption, resulting in enhanced excretion. The last effect, which is due at least in part to inhibition of phosphate uptake by the

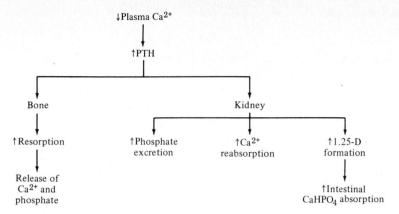

Figure 8-16 Effect of PTH on Ca^{2+} and phosphate metabolism. The net effect is an increase in the plasma Ca^{2+} concentration with no change or a decrease in the plasma phosphate concentration.

sodium-phosphate carrier protein in the brush border of the proximal tubular cell (see Chap. 5), usually predominates as PTH tends to lower the plasma phosphate concentration.

PTH affects renal calcium and phosphate excretion by activating specific adenyl cyclase systems in different nephron segments, particularly the proximal tubule and the early cortical distal nephron [99,100]. This is manifested by an increase in active distal Ca^{2+} reabsorption [101] and a reduction in proximal phosphate reabsorption [102].

PTH also has proximal actions independent of calcium and phosphate balance. In the early proximal tubule, PTH decreases the reabsorption of bicarbonate, Na^+, Ca^{2+}, amino acids, and water [102–106]. These changes are largely prevented if bicarbonate transport is inhibited prior to the administration of PTH [107,108], suggesting that the primary effect of PTH at this site is to reduce bicarbonate reabsorption. This could then secondarily affect the transport of other solutes and water since early proximal bicarbonate reabsorption results in the generation of passive gradients for net fluid reabsorption in the later aspects of the proximal tubule (see page 93).

These changes in proximal transport, however, are not necessarily reflected in the final urine. Na^+ reabsorption is enhanced distally so that an increase in Na^+ excretion is not usually seen [102]. Similarly, the PTH-induced stimulation of distal Ca^{2+} reabsorption overcomes the proximal effect, resulting in a net fall in Ca^{2+} excretion† [101,103]. In contrast, bicarbonate and amino acid excretion are elevated since these solutes are primarily reabsorbed proximally (see Chap. 5) [104–106]. However, these last effects are small and of minor clinical importance.

---

†Although PTH tends to augment net tubular Ca^{2+} reabsorption, it also enhances the filtered Ca^{2+} load by increasing the plasma Ca^{2+} concentration. This can ultimately override the direct effect on tubular reabsorption, producing an increase in Ca^{2+} excretion. However, enhanced excretion becomes appropriate once the plasma Ca^{2+} concentration rises.

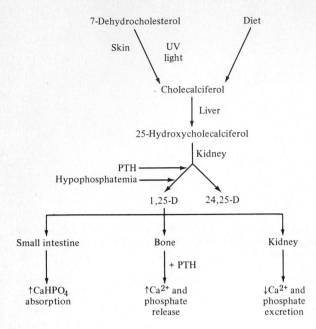

Figure **8-17** Metabolic activation of vitamin D and its effects on calcium and phosphate homeostasis. The result is an increase in the plasma Ca^{2+} and phosphate concentrations.

Vitamin D

Vitamin D (cholecalciferol) is a fat-soluble steroid which is present in the diet and also can be synthesized in the skin from 7-dehydrocholesterol in the presence of ultraviolet light. Vitamin D then undergoes a series of metabolic conversions, beginning with 25-hydroxylation in the liver (Fig. 8-17) [109,110]. Further hydroxylation occurs, mostly in the renal proximal convoluted tubular cells [111], either to 1,25-dihydroxycholecalciferol (1,25-D), the most active form of vitamin D, or to 24,25-dihydroxycholecalciferol (24,25-D), whose physiologic role is not well defined [112,113]. The formation of 1,25-D is primarily stimulated by PTH and hypophosphatemia [109,110].

The main action of 1,25-D is to enhance the availability of calcium and phosphate for both new bone formation and the prevention of symptomatic hypocalcemia and hypophosphatemia. This is primarily achieved by increases in intestinal absorption and bone resorption (Fig. 8-17) [109,110]. The latter effect is mediated in part by PTH as permissive amounts of 1,25-D are necessary for the PTH effect on bone (but probably not for its effect on the kidney) [114]. 25-Hydroxycholecalciferol may contribute to these responses [112,115], but it is much less active than 1,25-D [109,110]. Vitamin D also reduces the urinary excretion of calcium and phosphate (if the plasma concentrations and therefore the filtered load remain constant), although this effect is probably of lesser clinical importance [116].

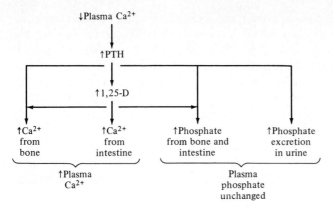

Figure 8-18 Physiologic sequence of events following the development of hypocalcemia. Because of the PTH interaction with vitamin D, the end result is a rise in the plasma Ca^{2+} concentration, with little or no change in the plasma phosphate concentration. *(Redrawn from H. DeLuca, Am. J. Med., 58:39, 1975.)*

Regulation of Plasma Calcium and Phosphate Concentrations

Figures 8-18 and 8-19 depict a model for the roles of PTH and vitamin D in the maintenance of the plasma Ca^{2+} and phosphate concentrations. The plasma Ca^{2+} concentration, as routinely measured in the laboratory (normal equals 8.5 to 10.5 mg/dL), includes all the Ca^{2+} in the plasma, of which only about 50 percent is the physiologically important, ionized Ca^{2+}. In general, measuring the total plasma Ca^{2+} concentration is sufficient, since changes in this parameter usually are associated with similar changes in the ionized Ca^{2+} concentration. An exception occurs in

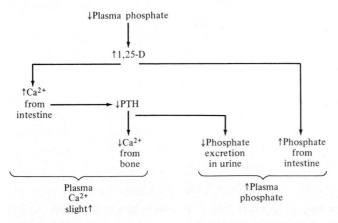

Figure 8-19 Sequence of events following the stimulation of 1,25-D formation by hypophosphatemia. The net effect is an increase in the plasma phosphate concentration with only a slight increase in the plasma Ca^{2+} concentration. *(Redrawn from H. DeLuca, Am. J. Med., 58:39, 1975.)*

patients with hypoalbuminemia in whom the total plasma Ca^{2+} concentration is reduced without change in the ionized Ca^{2+} concentration. To correct for this, the measured plasma Ca^{2+} concentration should be increased by 0.8 mg/dL for each 1.0 g/dL fall in the plasma albumin concentration (normal plasma albumin concentration equals 3.5 to 5.0 g/dL). For example, if the plasma Ca^{2+} concentration is 7.5 mg/dL and the plasma albumin concentration is 2.0 g/dL (roughly 2.0 g/L less than normal), then the corrected plasma Ca^{2+} concentration would be 7.5 + (2 × 0.8) = 9.1 mg/dL, which is normal.

In the presence of hypocalcemia, PTH secretion is enhanced, promoting the formation of 1.25-D (Fig. 8-18). PTH increases calcium phosphate release from bone and urinary phosphate excretion. 1,25-D augments intestinal calcium phosphate absorption. The net effect is an increase in the plasma Ca^{2+} concentration with little change in the plasma phosphate concentration. This sequence is reversed with hypercalcemia as PTH secretion and 1,25-D formation are reduced.

The normal plasma phosphate concentration, measured in the laboratory as the plasma inorganic phosphorus concentration, i.e., the concentration of phosphorus contained in the inorganic phosphates, is 2.5 to 4.5 mg/dL. If hypophosphatemia develops, 1,25-D synthesis is directly enhanced, increasing intestinal calcium phosphate absorption (Fig. 8-19). The small increase in the plasma Ca^{2+} concentration produced by 1,25-D suppresses PTH secretion, reducing both calcium phosphate release from bone and urinary phosphate excretion. The net effect is an increase in the plasma phosphate concentration with only a slight increment in the plasma Ca^{2+} concentration. The hormonal response to hyperphosphatemia is discussed in the next section.

Calcium and Phosphate Metabolism in Renal Failure

Although a complete review of the clinical disorders of calcium and phosphate balance is beyond the scope of this chapter, the changes that occur in patients with renal failure provide an interesting example of how the homeostatic mechanisms governing calcium and phosphate balance can be impaired due to alterations in PTH and vitamin D metabolism [117].

Phosphate balance in renal failure Renal failure is characterized by a decrease in the functioning renal mass and, therefore, in the total GFR. With the initial fall in GFR, there is a reduction in the filtered phosphate load and consequently in phosphate excretion. If intake remains constant, there will be an increase in the plasma phosphate concentration, which will drive the reaction

$$Ca^{2+} + HPO_4^{2-} \rightleftharpoons CaHPO_4$$

to the right [118], resulting in the precipitation of $CaHPO_4$ in bone and soft tissues and in hypocalcemia. This initiates the sequence illustrated in Fig. 8-20. The fall in the plasma Ca^{2+} concentration stimulates the secretion of PTH which, by increasing calcium release from bone and phosphate excretion in the urine, returns both the plasma Ca^{2+} and phosphate concentrations to normal. The net effect is an increased

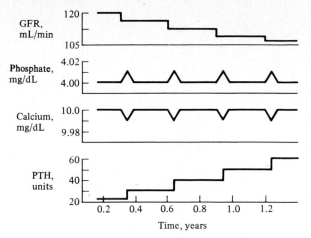

Figure 8-20 Hypothetical model for the pathogenesis of secondary hyperparathyroidism in advancing chronic renal disease. *(From E. Slatopolsky and N. S. Bricker, Kidney Int., 4:141, 1973. Reprinted by permission from Kidney International.)*

baseline level of PTH required to maintain calcium and phosphate balance. With each succeeding decrement in GFR, this process is repeated with gradually increasing secretion of PTH. The central role of phosphate retention in the genesis of the secondary hyperparathyroidism of renal failure has been shown by the demonstration that hypersecretion of PTH does not occur if phosphate retention is prevented by reducing phosphate intake in proportion to the decrease in GFR (Fig. 8-21) [119,120].

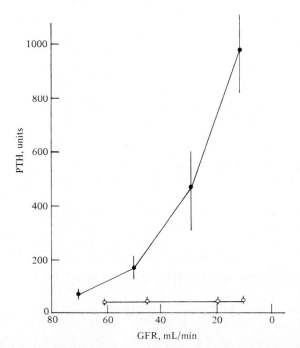

Figure 8-21 The relationship between PTH levels and GFR in two groups of dogs: those maintained on a 1200-mg/day phosphorus diet (closed circles) and those maintained on a diet containing less than 100 mg of phosphorus per day (open circles). *(From E. Slatopolsky, S. Caglar, J. P. Pennell, D. D. Taggart, J. M. Canterbury, E. Reiss, and N. S. Bricker. J. Clin. Invest., 50:492, 1971, by copyright permission of The American Society for Clinical Investigation.)*

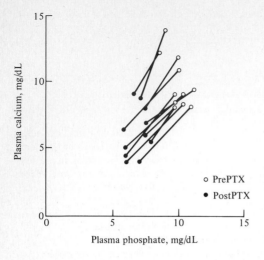

Figure 8-22 Changes in total plasma calcium and phosphate levels observed in 11 uremic patients before and following subtotal parathyroidectomy (PTX) for severe secondary hyperparathyroidism. *(From S. G. Massry, J. W. Coburn, M. M. Popovtzer, J. H. Shinaberger, M. H. Maxwell, and C. R. Kleeman, Arch. Intern. Med., 124:431, 1969. Copyright 1969, American Medical Association.)*

However, PTH loses its ability to maintain normophosphatemia when the GFR falls below 30 mL/min. Because of the inhibitory effect of PTH on proximal phosphate reabsorption, the fraction of the filtered phosphate that is reabsorbed can fall from the normal 80 to 95 percent to as low as 15 percent in severe renal failure [121]. At this point, PTH is unable to further increase phosphate excretion, resulting in persistent hyperphosphatemia if intake is not diminished. In addition, since PTH continues to enhance the release of phosphate from bone, the net effect of PTH in this setting is to increase the plasma phosphate concentration. Thus, if a subtotal parathyroidectomy is performed, both the plasma *calcium and phosphate* concentrations are reduced (Fig. 8-22). This is due to diminished bone resorption and to the deposition of calcium phosphate in bones previously demineralized by chronic hyperparathyroidism.

Calcium balance in renal failure Hypocalcemia almost always occurs at some time in patients with renal failure [122]. In addition to hyperphosphatemia, impaired synthesis of 1,25-D and resistance to the effects of both vitamin D and PTH contribute to this problem. In general, plasma 1,25-D levels remain normal until the GFR falls below 30 mL/min [123]. Both the reduction in functioning renal mass and the concurrent development of hyperphosphatemia [124] may play a role in the decreased renal generation of 1,25-D.

Although 1,25-D levels are normal in mild renal failure (GFR 30 to 50 mL/min), these patients already have diminished intestinal Ca^{2+} absorption, decreased mineralization of bone, and resistance to the skeletal effects of PTH (which requires the presence of vitamin D) [125,126]. These abnormalities are compatible with decreased vitamin D effect and suggest that there is end-organ resistance to vitamin D since circulating hormone levels are still normal in early renal failure. How this occurs is not known, but the administration of exogenous 1,25-D can at least partly overcome this resistance and reverse the skeletal and intestinal dysfunction [125–128].

Clinical consequences The clinical consequences of these changes in calcium and phosphate homeostasis are threefold. First, both hyperparathyroidism and 1,25-D deficiency can produce bone disease, causing osteitis fibrosa and osteomalacia, respectively [122,125,129].

Second, calcium phosphate may precipitate out of the plasma and be deposited in arteries, soft tissues, joints, and the viscera, e.g., the myocardium [117,129,130]. This complication is called *metastatic calcification* and appears to be dependent upon the *solubility product* of calcium phosphate. This is defined as the maximum product of the plasma concentrations of calcium and phosphate that can exist in solution without precipitating. When measured in milligrams per deciliter (mg/dL), the solubility product for calcium phosphate is approximately 60 [118]. When this value is exceeded, metastatic calcification is likely to occur, particularly if the product is greater than 72, as in patients with severe hyperparathyroidism who have marked hyperphosphatemia and a relatively normal plasma Ca^{2+} concentration [117,129,131].

The signs and symptoms associated with metastatic calcifications are dependent upon the location of the deposits and include ischemic necrosis due to arterial calcification, heart failure, and cardiac arrhythmias [117,129]. It has also been suggested that renal calcifications may contribute to the progression of the renal failure [132,133].

Third, occasional patients may develop the symptoms of hypocalcemia, including tetany, paresthesias, and carpopedal spasm.

CATECHOLAMINES

Catecholamines, released from the sympathetic nerves and the adrenal medulla (norepinephrine and epinephrine), can importantly influence renal function.† Norepinephrine is a potent vasoconstrictor, acting to reduce renal blood flow in hypovolemic states. Increased sympathetic activity also enhances proximal Na^+ reabsorption, an effect that contributes to the compensatory renal Na^+ retention seen with volume depletion [134–136]. It is unclear, however, whether this is due to direct stimulation of proximal Na^+ transport [137] or to altered peritubular capillary hemodynamics resulting from the increase in arteriolar resistance (see page 96). In addition, the sympathetically induced activation of the renin-angiotensin-aldosterone system indirectly enhances net Na^+ reabsorption (see Chap. 3) [138].

Another catecholamine, dopamine, appears to be released from dopaminergic nerve endings in the kidney [134] and has opposite effects from norepinephrine and epinephrine. Dopamine is a *vasodilator* which directly decreases Na^+ reabsorption; it acts in the *straight* segment of the proximal tubule, not the convoluted segment which is the site of action of norepinephrine [139]. The release of dopamine is

---

†The hemodynamic and renal effects of the sympathetic nervous system are discussed in detail in Chap. 9.

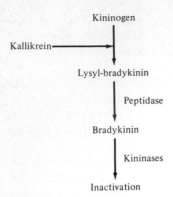

Figure 8-23 Kallikrein-kinin system.

enhanced with volume expansion, e.g., due to a high NaCl intake, a situation in which net sympathetic activity is generally reduced [140,141]. Thus, dopamine could act as a natriuretic hormone, contributing to the appropriate natriuresis seen with volume expansion. However, the physiologic importance of this effect is as yet unproven.

KININS

Kinins are another set of hormones produced in the kidney [142,143]. The process begins with the secretion of the enzyme kallikrein by the cells in the distal nephron, particularly those in the distal tubule and connecting segment (see Fig. 2-3) [143a]. This enzyme catalyzes the conversion of inactive kininogen (a plasma protein) into lysyl-bradykinin and then, in the presence of a peptidase, into bradykinin (Fig. 8-23). Kinin generation probably occurs both in the tubular lumen and, since secreted kallikrein appears to reach the vascular compartment, in the vascular space.

The physiologic role of the renal kinins is incompletely understood. Like the renal prostaglandins, they are vasodilators which may act to minimize renal ischemia in hypovolemic states in which angiotensin II and norepinephrine secretion are increased [144,145]. Renal kinins may also be involved in renin and prostaglandin release and in the regulation of Na^+ excretion [143,146]. It is not likely, however, that they are important circulating hormones since bradykinin is rapidly metabolized by kininases, one of which is converting enzyme, the same enzyme that converts angiotensin I into angiotensin II (see Fig. 3-5).

ERYTHROPOIETIN

Erythropoietin is the major stimulus to erythropoiesis, acting directly on red cell precursors in the bone marrow. The kidney is the primary site of erythropoietin production [148,149]. However, it has been unsettled as to whether the kidney synthesizes the intact hormone or an inactive compound (called erythrogenin) which then cleaves a plasma protein to release erythropoietin. Although recent studies have demonstrated that the kidney is able to produce erythropoietin directly [150,151], this does not preclude a contribution from the second mechanism. The site of hormone synthesis within the kidney is uncertain, with at least the glomerular capillary cells being involved [149].

Hypoxemia is the primary stimulus to erythropoietin release, resulting in an appropriate increase in red cell production and therefore oxygen-carrying capacity [149]. The importance of erythropoietin has been demonstrated in patients with renal failure. Anemia is common in this setting and in many patients is due primarily to the associated impairment in erythropoietin production [152,153].

MISCELLANEOUS

Cortisol, an adrenal cortical hormone (see Fig. 8-9), has several effects on renal function. It has weak aldosteronelike activity and also contributes to the maintenance of renal blood flow and the GFR by systemic actions which influence the cardiac output and arterial blood pressure [154]. These effects of cortisol are primarily permissive since variations in cortisol secretion are not necessary for normal renal function. However, cortisol deficiency is associated with a variety of hemodynamic and renal disturbances, including reductions in blood pressure, cardiac output, renal blood flow, and GFR [154]. Furthermore, the combination of a low GFR and a hypotension-induced increase in ADH secretion limits the ability to excrete water, producing a tendency to water retention and the development of hyponatremia (see Chap. 23) [154].

In addition to cortisol, a variety of other hormones can affect renal function, including calcitonin, estrogens, glucagon, insulin, and prolactin [116,155,156]. However, the physiologic significance of these effects is unknown.

SUMMARY

Variations in hormone secretion are referred to as primary, i.e., nonphysiologic, or secondary, i.e., physiologic. For example, aldosterone secretion is appropriately increased by volume depletion; this is called secondary hyperaldosteronism. Conversely, the autonomous hypersecretion of aldosterone, e.g., due to an adrenal adenoma, is referred to as primary hyperaldosteronism. The importance of hormones in the regulation of water and ion homeostasis can be appreciated from the findings in patients with primary changes in hormone secretion.

The excessive secretion of ADH results in increased water reabsorption in the collecting tubules. This is called the syndrome of inappropriate ADH secretion and is characterized by water retention, hyponatremia, and hypoosmolality (see Chap. 23). In contrast, water reabsorption is reduced in the absence of ADH as large volumes (up to 15 to 20 L/day) of dilute urine are produced. This may be due to deficient hormone secretion (central diabetes insipidus) or to renal resistance to the effects of ADH (nephrogenic diabetes insipidus) (see Chap. 24). Despite the polyuria, patients with either form of diabetes insipidus tend to remain in near normal water balance by increasing water intake to match the urine output.

Primary hyperaldosteronism is associated with enhanced K^+ and H^+ secretion in the cortical collecting tubule, resulting in hypokalemia and metabolic alkalosis (see Chap. 27). As described above, the increase in Na^+ reabsorption induced by aldosterone does not produce edema, because of the phenomenon of aldosterone escape. On the other hand, hypoaldosteronism is characterized by variable degrees of salt wasting, hyperkalemia, and metabolic acidosis due to the reductions in Na^+ reabsorption and K^+ and H^+ secretion (see Chap. 28).

PTH increases bone resorption and the urinary excretion of phosphate. As a result, hypercalcemia and a normal or reduced plasma phosphate concentration are produced by the primary hypersecretion of PTH. In contrast, hypoparathyroidism is associated with hypocalcemia and hyperphosphatemia.

Vitamin D increases the availability of calcium and phosphate by augmenting bone resorption and the gastrointestinal absorption of $CaHPO_4$. Consequently, vitamin D deficiency is characterized by hypocalcemia and hypophosphatemia, and vitamin D excess (due to the chronic administration of high doses of vitamin D) by hypercalcemia and a normal or increased plasma phosphate concentration. Since adequate levels of calcium and phosphate are necessary for bone mineralization, vitamin D deficiency may be associated with rickets in children and osteomalacia in adults.

PROBLEMS

8-1 Aldosterone secretion can be increased by an autonomous adrenal adenoma or in the presence of volume depletion. What will the plasma renin activity be in these two conditions?

8-2 Patients with renal failure have an elevated P_{osm} due to the increase in the BUN, yet do not have persistent ADH release. Why?

8-3 If patients make no ADH, they will excrete large volumes of dilute urine. What will protect them against the development of negative water balance leading to symptomatic hyperosmolality?

8-4 Licorice contains a steroid, glycyrrhizic acid, which has aldosteronelike activity. Patients ingesting large amounts of licorice may have hypokalemia and alkalosis (due to increased renal H^+ and K^+ loss) but not edema. Why?

8-5 A patient with renal failure has hypocalcemia and hyperphosphatemia. What factors are responsible for the hypocalcemia? Would you start vitamin D in an attempt to correct the hypocalcemia?

8-6 Patients with hypercalciuria due to a defect in renal Ca^{2+} reabsorption usually do not become hypocalcemic. Why?

8-7 ADH secretion is increased by effective circulating volume depletion. How does ADH act to increase the effective circulating volume? What effect does ADH have on renal Na^+ reabsorption?

8-8 Patients with which of the following conditions are likely to develop a decline in renal function following the administration of a nonsteroidal anti-inflammatory drug?
- (*a*) Low-salt diet
- (*b*) High-salt diet
- (*c*) Untreated hypertension
- (*d*) Heart failure
- (*e*) Severe vomiting

REFERENCES

1. Zimmerman, E. A., and A. G. Robinson: Hypothalamic neurons secreting vasopressin and neurophysin, Kidney Int., 10:12, 1976.
2. Rocha, A. S., and J. P. Kokko: Permeability of medullary nephron segments to urea and water: effect of vasopressin, Kidney Int., 6:379, 1974.
3. Muller, J., W. A. Kachadorian, and V. A. DiScala: Evidence that ADH-stimulated intramembrane particle aggregates are transferred from cytoplasmic to luminal membranes in toad bladder epithelial cells, J. Cell Biol., 85:83, 1980.
4. Levine, S. D., W. A. Kachadorian, D. N. Levin, and D. Schlondorff: Effects of trifluoperazine on function and structure of toad urinary bladder, J. Clin. Invest., 67:662, 1981.
5. Lewis, C. R., and G. D. Aurbach: Renal adenyl cyclase: anatomically separate sites for parathyroid hormone and vasopressin, Science, 159:545, 1968.
6. Aisenbrey, G. A., W. A. Handelman, P. Arnold, M. Manning, and R. W. Schrier: Arginine vasopressin during fluid deprivation in the rat, J. Clin. Invest., 67:961, 1981.
7. Andrews, C. E., Jr., and B. M. Brenner: Relative contributions of arginine vasopressin and angiotensin II to maintenance of systemic arterial pressure in the anesthesized water-deprived rat, Circ. Res., 48:254, 1981.
8. Johnston, C. I., M. Newman, and R. Woods: Role of vasopressin in cardiovascular homeostasis and hypertension, Clin. Sci., 61(suppl.):129s, 1981.
9. Wakim, K. G.: Reassessment of the source, mode, and locus of action of antidiuretic hormone, Am. J. Med., 42:394, 1967.
10. Aizawa, T., N. Yasuda, M. A. Greer, and W. H. Sawyer: In vivo adrenocorticotropin-releasing activity of neurohypophyseal hormones and their analogs, Endocrinology, 110:98, 1982.
11. Weingartner, H., P. Gold, J. C. Ballenger, S. Smallberg, R. Summers, D. R. Rubinow, R. M. Post, and F. K. Goodwin: Effects of vasopressin on human memory functions, Science, 211:601, 1981.
12. Verney, E. B.: Absorption and excretion of water: the antidiuretic hormone, Lancet, 2:781, 1946.
13. Schrier, R. W., and D. G. Bichet: Osmotic and non-osmotic control of vasopressin release and the pathogenesis of impaired water excretion in adrenal, thyroid, and edematous disorders, J. Lab. Clin. Med., 98:1, 1981.
14. Robertson, G. L.: Thirst and vasopressin function in normal and disordered states of water balance, J. Lab. Clin. Med., 101:351, 1983.
15. Zerbe, R. L., and G. L. Robertson: Osmoregulation of thirst and vasopressin secretion in normal subjects: effect of various solutes, Am. J. Physiol., 244:E607, 1983.
16. Leaf, A., and A. R. Mamby: The normal antidiuretic mechanism in man and dog: its regulation by extracellular fluid tonicity, J. Clin. Invest., 31:54, 1952.
17. Robertson, G. L., P. Aycinena, and R. L. Zerbe: Neurogenic disorders of osmoregulation, Am. J. Med., 72:339, 1982.
18. Leaf, A., and A. R. Mamby: An antidiuretic mechanism not regulated by extracellular fluid tonicity, J. Clin. Invest., 31:60, 1952.
19. Dunn, F. L., T. J. Brennan, A. E. Nelson, and G. L. Robertson: The role of blood osmolality and volume in regulating vasopressin secretion in the rat, J. Clin. Invest., 52:3212, 1973.
20. Gilmore, J. P., and I. H. Zucker: Failure of left atrial distention to alter renal function in the nonhuman primate, Circ. Res., 42:267, 1978.
21. Goetz, K. L., G. C. Bond, and W. E. Smith: Effect of moderate hemorrhage in humans on plasma ADH and renin, Proc. Soc. Exp. Biol. Med., 145:277, 1974.
22. Berl, T., P. Cadnapaphornchai, J. A. Harbottle, and R. W. Schrier: Mechanism of stimulation of vasopressin release during beta adrenergic stimulation with isoproterenol, J. Clin. Invest., 53:857, 1974.
23. Anderson, R. J., P. Cadnapaphornchai, J. A. Harbottle, K. M. McDonald, and R. W. Schrier: Mechanism of effect of thoracic inferior vena cava constriction on renal water excretion, J. Clin. Invest, 54:1473, 1974.
24. Berl, T., P. Cadnapaphornchai, J. A. Harbottle, and R. W. Schrier: Mechanism of suppression of

vasopressin during alpha-adrenergic stimulation with norepinephrine, J. Clin. Invest., 53:219, 1974.

25. Fitzsimons, J. T.: The physiologic basis of thirst, Kidney Int., 10:3, 1976.

26. Rolls, B. J., R. J. Wood, E. T. Rolls, H. Lind, W. Lind, and J. G. G. Ledingham: Thirst following water deprivation in humans, Am. J. Physiol., 239:R476, 1980.

27. Thrasher, T. N., J. F. Nistal-Herrera, L. C. Keil, and D. J. Ramsay: Satiety and inhibition of vasopressin secretion after drinking in dehydrated dogs, Am. J. Physiol., 240:E394, 1981.

28. Moore, F. D.: Common patterns of water and electrolyte change in injury, surgery, and disease, N. Engl. J. Med., 258:277, 1958.

29. Ukai, M., W. Moran, Jr., and B. Zimmerman: The role of visceral afferent pathways on vasopressin secretion and urinary excretory patterns during surgical stress, Ann. Surg., 168:16, 1968.

30. Rowe, J. W., R. L. Shelton, J. H. Helderman, R. E. Vestal, and G. L. Robertson: Influence of the emetic reflex on vasopressin release in man, Kidney Int., 16:729, 1979.

31. Baylis, P. H., R. L. Zerbe, and G. L. Robertson: Arginine vasopressin response to insulin-induced hypoglycemia in man, J. Clin. Endocrinol. Metab., 53:935, 1981.

32. Anderson, R. J., R. G. Pluss, A. S. Berns, J. T. Jackson, P. E. Arnold, R. W. Schrier, and K. M. McDonald: Mechanism of effect of hypoxia on renal water excretion, J. Clin. Invest., 62:769, 1978.

33. Rowe, J. W., A. Kilgore, and G. L. Robertson: Evidence in man that cigarette smoking induces vasopressin release via an airway-specific mechanism, J. Clin. Endocrinol. Metab., 51:170, 1980.

34. Eisenhofer, G., and R. H. Johnson: Effects of ethanol ingestion on thirst and fluid conservation in humans, Am. J. Physiol., 244:R568, 1983.

35. Moses, A. M., and M. Miller: Drug-induced dilutional hyponatremia, N. Engl. J. Med., 291:1234, 1974.

36. Ganong, W. F., L. Coulton, A. B. Alpert, and T. C. Less: ACTH and the regulation of adrenocortical secretion, N. Engl. J. Med., 290:1006, 1974.

37. Ulick, S.: Diagnosis and nomenclature of the disorders of the terminal portion of the aldosterone biosynthetic pathway, J. Clin. Endocrinol. Metab., 43:92, 1976.

38. Knochel, J. P., and M. G. White: The role of aldosterone in renal physiology, Arch. Intern. Med., 131:876, 1973.

39. Garg, L. C., M. A. Knepper, and M. B. Burg: Mineralocorticoid effects on Na-K-ATPase in individual nephron segments, Am. J. Physiol., 240:F536, 1981.

40. Stokes, J. B., M. J. Ingram, A. D. Williams, and D. Ingram: Heterogeneity of the rabbit collecting tubule: localization of mineralocorticoid hormone action to the cortical portion, Kidney Int., 20:340, 1981.

41. Edelman, I. S.: Receptors and effectors in hormone action on the kidney, Am. J. Physiol., 241:F333, 1981.

42. Hierholzer, K., and M. Wiederholt: Some aspects of distal tubular solute and water transport, Kidney Int., 9:198, 1976.

43. Petty, K. J., J. P. Kokko, and D. Marver: Secondary effect of aldosterone on Na-K-ATPase activity in the rabbit cortical collecting tubule, J. Clin. Invest., 68:1514, 1981.

44. Will, P. C., J. L. Lebowitz, and U. Hopfer: Induction of amiloride-sensitive sodium transport in the rat colon by mineralocorticoids, Am. J. Physiol., 238:F261, 1980.

45. Al-Awqati, Q., L. H. Norby, A. Mueller, and P. R. Steinmetz: Characteristics of stimulation of H^+ transport by aldosterone in turtle urinary bladder, J. Clin. Invest., 58:351, 1976.

46. Lauler, D., R. B. Hickler, and G. Thorn: The salivary sodium-potassium ratio. A useful "screening" test for aldosteronism in hypertension, N. Engl. J. Med., 267:1136, 1962.

47. Grand, R. J., P. A. di Saint'Agnese, R. C. Talamo, and J. C. Pallavicini: The effects of exogenous aldosterone on sweat electrolytes, J. Pediatr., 70:346, 1967.

48. Charron, R. C., C. E. Leme, D. R. Wilson, T. S. Ing, and O. M. Wrong: The effect of adrenal steroids on stool composition, as revealed by in vivo dialysis of faeces, Clin. Sci., 37:151, 1969.

49. Young, D. B., R. E. McCaa, Y. Pan, and A. C. Guyton: Effectiveness of the aldosterone-sodium and -potassium feedback control system, Am. J. Physiol., 231:945, 1976.

50. Williams, G. H., and L. M. Braley: Effects of dietary sodium and potassium intake on acute stim-

ulation of aldosterone output by isolated human adrenal cells, J. Clin. Endocrinol. Metab., 45:55, 1977.

51. Aguitera, G., and K. J. Catt: Loci of action of regulators of aldosterone biosynthesis in isolated glomerulosa cells, Endocrinology, 104:1046, 1978.

52. Ames, R., A. Borkowski, A. Sicinski, and J. Laragh: Prolonged infusions of angiotensin II and norepinephrine on blood pressure, electrolyte balance, and aldosterone and cortisol secretion in normal man and in cirrhosis with ascites, J. Clin. Invest., 44:1171, 1965.

53. Cooke, C. R., D. S. Gann, P. K. Whelton, T. H. Hsu, T. Bledsoe, M. A. Moore, and W. G. Walker: Hormonal responses to acute volume changes in anephric subjects, Kidney Int., 23:71, 1983.

54. McCaa, R. E., J. D. Bower, and C. S. McCaa: Relative influence of acute sodium and volume depletion on aldosterone secretion in nephrectomized man, Circ. Res., 33:555, 1973.

55. Himathongham, T., R. Dluhy, and G. H. Williams: Potassium-aldosterone-renin interrelationships, J. Clin. Endocrinol. Metab., 41:153, 1975.

56. Crabbe, J., W. Reddy, E. Ross, and G. Thorn: The stimulation of aldosterone secretion by adrenocorticotropic hormone, J. Clin. Endocrinol. Metab., 19:1185, 1959.

57. Aguitera, G., K. Fujita, and K. J. Catt: Mechanisms of inhibition of aldosterone secretion by adrenocorticotropin, Endocrinology, 108:522, 1981.

58. Williams, G. H., L. I. Rose, R. G. Dluhy, J. F. Dingman, and D. P. Lauler: Aldosterone response to sodium restriction and ACTH stimulation in panhypopituitarism, J. Clin. Endocrinol. Metab., 32:27, 1973.

59. Mulrow, P. J.: Glucocorticoid-suppressible hyperaldosteronism: a clue to the missing hormone?, N. Engl. J. Med., 305:1012, 1981.

60. Matsuoka, H., P. J. Mulrow, R. Franco-Saenz, and C. Hao Li: Effects of β-lipotropin and β-lipotropin-derived peptides on aldosterone production in the rat adrenal gland, J. Clin. Invest., 68:752, 1981.

61. Davis, J. O., J. Urquhart, and J. T. Higgins, Jr.: The effects of alterations of plasma sodium and potassium concentration on aldosterone secretion, J. Clin. Invest., 42:597, 1963.

62. Bartter, F. C., G. W. Liddle, L. E. Duncan, Jr., J. K. Barber, and C. Delea: The regulation of aldosterone secretion in man: the role of fluid volume, J. Clin. Invest., 35:1306, 1956.

63. Tannen, R. L.: Relationship of renal ammonia production and potassium homeostasis, Kidney Int., 11:453, 1977.

63a. Stokes, J. B.: Consequences of potassium recycling in the renal medulla. Effects on ion transport by the medullary thick ascending limb of Henle's loop, J. Clin. Invest., 70:219, 1982.

64. Finn, A., and L. Welt: Effect of aldosterone administration on electrolyte excretion and GFR in the rat, Am. J. Physiol., 204:243, 1963.

65. August, J. T., D. H. Nelson, and G. W. Thorn: Response of normal subjects to large amounts of aldosterone, J. Clin. Invest. 37:1549, 1958.

66. Knox, F. G., J. C. Burnett, Jr., D. E. Kohan, W. S. Spielman, and J. C. Strand: Escape from the sodium-retaining effects of mineralocorticoids, Kidney Int., 17:263, 1980.

67. Schwartz, G. J., and M. B. Burg: Mineralocorticoid effects on cation transport by cortical collecting tubules in vitro, Am. J. Physiol., 235:F576, 1978.

68. Sealey, J., J. Kirshman, and J. Laragh: Natriuretic activity in plasma and urine of salt-loaded man and sheep, J. Clin. Invest., 48:2210, 1969.

69. deWardener, H. E., I. H. Mills, W. F. Clapham, and C. J. Hayter: Studies on the efferent mechanism of the sodium diuresis which follows the administration of intravenous saline in the dog, Clin. Sci., 21:249, 1961.

70. deWardener, H. E.: Natriuretic hormone, Clin. Sci., 53:1, 1977.

71. deWardener, H. E., and E. M. Clarkson: The natriuretic hormone: recent developments, Clin. Sci., 63:415, 1982.

71a. Bourgoignie, J. J., K. H. Hwang, C. Espinel, S. Klahr, and N. S. Bricker: A natriuretic factor in the serum of patients with chronic uremia, J. Clin. Invest., 51:1514, 1972.

72. Fine, L. G., J. J. Bourgoignie, K. H. Hwang, and N. S. Bricker: On the influence of the natriuretic factor from patients with chronic uremia on the bioelectric properties and sodium transport of the isolated collecting tubule, J. Clin. Invest., 58:590, 1976.

73. Clarkson, E. M., S. M. Raw, and H. E. deWardener: Two natriuretic substances in extracts of urine from normal man when salt-depleted and salt-loaded, Kidney Int., 10:381, 1976.

74. Schmidt, R. W., J. J. Bourgoignie, and N. S. Bricker: On the adaptation in sodium excretion in chronic uremia: the effects of "proportional reduction" of sodium intake, J. Clin. Invest., 53:1736, 1974.

75. Epstein, M., N. S. Bricker, and J. J. Bourgoignie: Presence of a natriuretic factor in urine of normal men undergoing water immersion, Kidney Int., 13:152, 1978.

76. Kaloyanides, G. J., M. B. Balabanian and R. L. Bowman: Evidence that the brain participates in the humoral natriuretic mechanism of blood volume expansion in the dog, J. Clin. Invest., 62:1288, 1978.

77. Haupert, G. T., Jr., and J. M. Sancho: Sodium transport inhibitor from bovine hypothalamus, Proc. Nat. Acad. Sci. U.S.A. 76:4658, 1979.

77a. Currie, M. G., D. M. Geller, B. R. Cole, J. G. Boylan, W. Yusheng, S. W. Holmberg, and P. Needleman: Bioactive cardiac substances: potent vasorelaxant activity in mammalian atria, Science, 221:71, 1983.

78. Dunn, M. J., and V. L. Hood: Prostaglandins and the kidney, Am. J. Physiol., 233:F169, 1977.

79. Dunn, M. J., and E. J. Zambraski: Renal effects of drugs that inhibit prostaglandin synthesis, Kidney Int., 18:609, 1980.

79a. Bunning, R. D., and W. F. Barth: Sulindac. A potentially renal-sparing nonsteroidal anti-inflammatory drug, J. Am. Med. Assoc., 248:2864, 1982.

80. Nadler, J., R. D. Zipser, R. Coleman, and R. Horton: Stimulation of renal prostaglandins by pressor hormones in man: comparison of PGE_2 and prostacyclin (6 keto $PGF_{1\alpha}$), J. Clin. Endocrinol. Metab., 56:1260, 1983.

80a. Scharschmidt, L. A., and M. J. Dunn: Prostaglandin synthesis by rat glomerular mesangial cells in culture. Effects of angiotensin II and arginine vasopressin, J. Clin. Invest., 71:1756, 1983.

81. Oliver, J. A., J. Pinto, R. R. Sciacca, and P. J. Cannon: Increased renal secretion of norepinephrine and prostaglandin E_2 during sodium depletion in the dog, J. Clin. Invest., 66:748, 1980.

82. Zipser, R. D., J. C. Hoefs, P. F. Speckhart, P. K. Zia, and R. Horton: Prostaglandins: modulators of renal function and pressor resistance in chronic liver disease. J. Clin. Endocrinol. Metab., 48:895, 1979.

83. Walsh, J. J., and R. C. Venuto: Acute oliguric renal failure induced by indomethacin: possible mechanism, Ann. Intern. Med., 91:47, 1979.

84. Arisz, L., A. J. M. Donker, J. R. H. Brentjens, and G. K. van der Hem: The effect of indomethacin on proteinuria and kidney function in the nephrotic syndrome. Acta Med. Scand., 199:121, 1976.

85. Beierwaltes, W. H., S. Schryver, E. Sanders, J. Strand, and J. C. Romero: Renin release selectively stimulated by prostaglandin I_2 in isolated rat glomeruli, Am. J. Physiol., 243:F276, 1982.

85a. Francisco, L. L., J. L. Osborn, and G. F. DiBona: Prostaglandins in renin release during sodium deprivation, Am. J. Physiol., 243:F537, 1982.

86. Tan, S. Y., R. Shapiro, R. Franco, H. Stockard, and P. J. Mulrow: Indomethacin-induced prostaglandin inhibition with hyperkalemia: reversible cause of hyporeninemic hypoaldosteronism, Ann. Intern. Med., 90:783, 1979.

87. Stokes, J. B.: Integrated actions of renal medullary prostaglandins in the control of water excretion, Am. J. Physiol., 240:F471, 1981.

88. Beck, N. P., T. Kaneko, U. Zor, J. B. Field, and B. B. Davis: Effects of vasopressin and prostaglandin E on the adenyl cyclase, cyclic 3', 5' adenosine monophosphate system of the renal medulla of the rat, J. Clin. Invest., 50:2461, 1971.

89. Roman, R. J., and C. Lechene: Prostaglandin E_2 and $F_{2\alpha}$ reduce urea reabsorption from the rat collecting duct, Am. J. Physiol., 241:F53, 1981.

90. Kramer, H. J., K. Glänzer, and R. Düsing: Role of prostaglandins in the regulation of renal water excretion, Kidney Int., 19:851, 1981.

91. Kirschenbaum, M. A., A. G. Lowe, W. Trizna, and L. G. Fine: Regulation of vasopressin action by prostaglandins. Evidence for prostaglandin synthesis in the rabbit cortical collecting tubule, J. Clin. Invest., 70:1193, 1982.

92. Berl, T., A. Raz, H. Wald, J. Horowitz, and W. Czaczkes: Prostaglandin synthesis inhibition and the action of vasopressin: studies in man and rat, Am. J. Physiol., 232:F529, 1977.

93. Kokko, J. P.: Effect of prostaglandins on renal epithelial electrolyte transport, Kidney Int., 19:791, 1981.

94. Fine, L. G., and M. A. Kirschenbaum: Absence of direct effects of prostaglandins on sodium chloride transport in the mammalian nephron, Kidney Int., 19:797, 1981.

95. Rathaus, M., E. Podjarny, E. Weiss, M. Ravid, S. Bauminger, and J. Bernheim: Effect of chronic and acute changes in sodium balance on the urinary excretion of prostaglandins E_2 and $F_{2\alpha}$ in normal man, Clin. Sci., 60:405, 1981.

96. Rasmussen, H., and P. Bordier: The cellular basis of metabolic bone disease, N. Engl. J. Med., 289:25, 1973.

97. Austin, L. A., and H. Heath, III: Calcitonin: physiology and pathophysiology, N. Engl. J. Med., 304:269, 1981.

98. Aurbach, G. D., and D. A. Heath: Parathyroid hormone and calcitonin regulation of renal function, Kidney Int., 6:331, 1974.

99. Chabardès, D. M. Gagnan-Brunette, M. Imbert-Teboul, O. Gontcharevskaia, M. Montégut, A. Clique, and F. Morel: Adenylate cyclase responsiveness to hormones in various portions of the human nephron, J. Clin. Invest., 65:439, 1980.

100. Imai, M.: The connecting tubule: a functional subdivision of the rabbit distal nephron segments, Kidney Int., 15.346, 1979.

101. Ng, R. C. K., R. A. Peraino, and W. N. Suki: Divalent cation transport in isolated tubules, Kidney Int., 22:492, 1982.

102. Agus, Z. S., J. B. Puschett, D. Senesky, and M. Goldberg: Effects of parathyroid hormone on renal tubular reabsorption of calcium, sodium, and phosphate, Am. J. Physiol., 224:1143, 1973.

103. Widrow, S. H., and N. G. Levinsky: The effect of parathyroid extract on renal tubular calcium reabsorption in the dog, J. Clin. Invest., 41:2151, 1962.

104. Crumb, C. K., M. Martinez-Maldonado, G. Eknoyan, and W. Suki: Effects of volume expansion, purified parathyroid extract, and calcium on renal bicarbonate absorption in the dog, J. Clin. Invest., 54:1287, 1974.

105. Muldowney, F. P., D. V. Carroll, J. F. Donohue, and R. Freaney: Correction of renal bicarbonate wastage by parathyroidectomy, Q. J. Med., 40:487, 1971.

106. Muldowney, F. P., R. Freaney, and D. McGeeney: Renal tubular acidosis and aminoaciduria in osteomalacia of dietary or intestinal origin, Q. J. Med., 37:517, 1968.

107. Dennis, V. W.: Influence of bicarbonate on parathyroid hormone-induced changes in fluid absorption by the proximal tubule, Kidney Int., 10:373, 1976.

108. Knox, F. G., J. A. Haas, and C. P. Lechene: Effect of parathyroid hormone on phosphate reabsorption in the presence of acetazolamide, Kidney Int., 10:216, 1976.

109. Haussler, M. R. and T. A. McCain: Basic and clinical concepts related to vitamin D metabolism and action, N. Engl. J. Med., 297:974, 1041, 1977.

110. DeLuca, H. F.: Vitamin D metabolism and function, Arch. Intern. Med., 138:836, 1978.

111. Kawashima, H., J. A. Kraut, and K. Kurokawa: Metabolic acidosis suppresses 25-hydroxy-vitamin D_3-1α-hydroxylase in the rat kidney. Distinct site and mechanism of action, J. Clin. Invest., 70:135, 1982.

112. Bordier, P., H. Rasmussen, P. Marie, L. Miravet, J. Gueris, and A. Ryckewart: Vitamin D metabolites and bone mineralization in man. J. Clin. Endocrinol. Metab., 46:284, 1978.

113. Corvol, M. T., M. F. Dumontier, M. Garbedian, and R. Rappaport: Vitamin D and cartilage, II: Biological activity of 25-hydroxycholecalciferol and 24,25-dihydroxycholecalciferol on cultured growth plate chondrocytes, Endocrinology, 102:1269, 1978.

114. Gerblich, A. A., S. M. Genuth, and J. G. Haddad: A case of idiopathic hypoparathyroidism and dietary vitamin D deficiency: the requirement for calcium and vitamin D for bone, but not renal responsiveness to PTH, J. Clin. Endocrinol. Metab., 44:507, 1977.

115. Recker, R., P. Schoenfeld, J. Letteri, E. Slatopolsky, R. Goldsmith, and A. Brickman: The efficacy of calcifediol in renal osteodystrophy, Arch. Intern. Med., 138:857, 1978.

116. Puschett, J. B., W. S. Beck, Jr., A. Jelonek, and P. C. Fernandez: Study of the renal tubular interactions of thyrocalcitonin, cyclic adenosine 3′, 5′-monophosphate, 25-hydroxycholecalciferol, and calcium ion, J. Clin. Invest., 53:756, 1974.

117. Rose, B. D.: *Pathophysiology of Renal Disease,* McGraw-Hill, New York, 1981, chap. 9.

118. Herbert, L. A., J. Lemann, Jr., J. R. Petersen, and E. J. Lennon: Studies of the mechanism by which phosphate infusion lowers serum calcium concentration, J. Clin. Invest., 45:1886, 1966.

119. Slatopolsky, E., S. Caglar, J. P. Pennell, D. D. Taggart, J. M. Canterbury, E. Reiss, and N. S. Bricker: On the pathogenesis of hyperparathyroidism in chronic experimental renal insufficiency in the dog, J. Clin. Invest., 50:492, 1971.

120. Rutherford, W. E., P. Bordier, P. Marie, K. Hruska, H. Harter, A. Greenwalt, J. Blondin, J. Haddad, N. S. Bricker, and E. Slatopolsky: Phosphate control and 25-hydroxycholecalciferol administration in preventing experimental renal osteodystrophy in the dog, J. Clin. Invest., 60:332, 1977.

121. Slatopolsky, E., A. M. Robson, I. Elkan, and N. S. Bricker: Control of phosphate excretion in uremic man, J. Clin. Invest., 47:1865, 1968.

122. David, D. S.: Calcium metabolism in renal failure, Am. J. Med., 58:48, 1975.

123. Slatopolsky, E., R. Gray, N. D. Adams, J. Lewis, K. Hruska, K. Martin, S. Klahr, H. DeLuca, and J. Lemann: Low serum levels of 1,25 $(OH)_2D_3$ are not responsible for the development of secondary hyperparathyroidism in early renal failure, Kidney Int., 14:733, 1978.

124. Larkin, R. G., J. W. Colston, L. S. Galante, S. J. MacAuley, I. M. A. Evans, and I. MacIntyre: Regulation of vitamin-D metabolism without parathyroid hormone, Lancet, 2:289, 1973.

125. Massry, S. G., D. A. Goldstein, and H. M. Malluche: Current status of the use of 1,25 $(OH)_2D_3$ in the management of renal osteodystrophy, Kidney Int. 18:409, 1980.

126. Goldstein, D. A., R. E. Horowitz, R. Petit, B. Haldimann, and S. G. Massry: The duodenal mucosa in patients with renal failure: response to 1,25$(OH)_2D_3$, Kidney Int., 19:324, 1981.

127. Healy, M. D., H. Malluche, D. Goldstein, F. Singer, and S. Massry: Effects of long-term therapy with calcitriol in patients with moderate renal failure, Arch. Intern. Med., 140:1030, 1980.

128. Massry, S. G., S. Tuma, S. Dua, and D. A. Goldstein: Reversal of skeletal resistance to parathyroid hormone in uremia by vitamin D metabolites; evidence for the requirement of 1,25$(OH)_2$ D_3 and 24,25$(OH)_2$ D_3. J. Lab. Clin. Med., 94:152, 1979.

129. Katz, A. I., C. L. Hampers, and J. P. Merrill: Secondary hyperparathyroidism and renal osteodystrophy in chronic renal failure: analysis of 195 patients with observations on the effects of chronic dialysis, kidney transplantation and subtotal parathyroidectomy, Medicine, 48:333, 1969.

130. Alfrey, A. C., C. C. Solomons, J. Ciricillo, and N. L. Miller: Extraosseus calcification: evidence for abnormal pyrophosphate metabolism in uremia, J. Clin. Invest., 57:692, 1976.

131. Velentzas, C., H. Meindok, D. G. Oreopoulos, H. E. Meema, S. Rabinovich, D. Sutton, and R. Ogilvie: Detection and pathogenesis of visceral calcification in dialysis patients and patients with malignant disease, Can. Med. Assoc. J., 118:45, 1978.

132. Ibels, L. S., A. C. Alfrey, L. Haut, and W. E. Huffer: Preservation of function in experimental renal disease by dietary restriction of phosphate, N. Engl. J. Med., 298:122, 1978.

133. Ibels, L. S., A. C. Alfrey, W. E. Huffer, P. W. Craswell, and R. Weil, III: Calcification in end-stage kidneys, Am. J. Med., 71:33, 1981.

134. Moss, N. G.: Renal function and renal afferent and efferent nerve activity, Am. J. Physiol., 243:F425, 1982.

135. Schrier, R. W., M. H. Humphreys, and R. C. Ufferman: Role of cardiac output and the autonomic nervous system in the antinatriuretic response to acute constriction of the thoracic superior vena cava, Circ. Res., 29:490, 1971.

136. Gill, J. R., Jr., and A. G. T. Casper: Role of the sympathetic nervous system in the renal response to hemorrhage, J. Clin. Invest., 48:915, 1969.

137. Bello-Reuss, E.: Effect of catecholamines on fluid reabsorption by the isolated proximal convoluted tubule, Am. J. Physiol., 238:F347, 1980.

138. Haber, E.: The renin-angiotensin system and hypertension, Kidney Int., 15:427, 1979.

139. Bello-Reuss, Y. Higashi, and Y. Kaneda: Dopamine decreases fluid reabsorption in straight portions of rabbit proximal tubule, Am. J. Physiol., 242:F634, 1982.

140. Alexander, R. W., J. R. Gill, Jr., H. Yamabe, W. Lovenberg, and H. R. Keiser: Effects of dietary sodium and of acute saline infusion on the interrelationship between dopamine excretion and adrenergic activity in man, J. Clin. Invest., 54:194, 1974.

141. Carey, R. M., G. R. van Loon, A. D. Baines, and E. M. Ortt: Decreased plasma and urinary dopamine during dietary sodium depletion in man, J. Clin. Endocrinol. Metab., 52:903, 1981.

142. Levinsky, N. G.: The renal kallikrein-kinin system, Circ. Res., 44:441, 1979.

143. Carretero, O. A., and A. G. Scicli: The renal kallikrein-kinin system, Am. J. Physiol., 238:F247, 1980.

143a. Proud, D., M. A. Knepper, and J. J. Pisano: Distribution of immunoreactive kallikrein along the rat nephon, Am. J. Physiol., 244:F510, 1983.

144. Johnston, P. A., D. B. Bernard, N. S. Perrin, L. Arbeit, W. Lieverthal, and N. G. Levinsky: Control of rat renal vascular resistance during alterations in sodium balance, Circ. Res., 48:728, 1981.

145. Johnston, P. A., N. S. Perrin, D. B. Bernard, and N. G. Levinsky: Control of rat renal vascular resistance at reduced perfusion pressure, Circ. Res. 48:734, 1981.

146. Sealey, J. E., S. A. Atlas, and J. H. Laragh: Linking the kallikrein and renin systems via activation of inactive renin: new data and an hypothesis, Am. J. Med., 65:994, 1978.

147. Zusman, R. M., and H. R. Keiser: Prostaglandin biosynthesis by rabbit renomedullary interstitial cells in tissue culture: stimulation by angiotensin II, bradykinin, and arginine vasopressin, J. Clin. Invest., 60:215, 1977.

148. Erslev, A. M.: Renal biogenesis of erythropoietin, Am. J. Med., 58:25, 1975.

149. Fisher, J. W.: Control of erythropoietin production, Proc. Soc. Exp. Biol. Med., 173:289, 1983.

150. Sherwood, J. B., and E. Goldwasser: Extraction of erythropoietin from normal kidneys, Endocrinology, 103:866, 1978.

151. Fried, W., J. Barone-Varelas, and M. Berman: Detection of high erythropoietin titers in renal extracts of hypoxic rats, J. Lab. Clin. Med., 97:82, 1981.

152. Desforges, J. F., and J. P. Dawson: The anemia of renal failure, Arch. Intern. Med., 101:326, 1958.

153. Brown, R.: Plasma erythropoietin in chronic uraemia, Br. Med. J., 2:1036, 1965.

154. Linas, S. L., T. Berl, G. L. Robertson, G. A. Aisenbrey, R. W. Schrier, and R. J. Anderson: Role of vasopressin in impaired water excretion of glucocorticoid deficiency, Kidney Int., 18:58, 1980.

155. Symposium on Hormones and the Kidney, S. E. Bradley (ed.), Kidney Int., 6:261, 1974.

156. DeFronzo, R. A., C. R. Cooke, R. Andres, G. R. Faloona, and P. J. Davis: The effect of insulin on renal handling of sodium, potassium, calcium, and phosphate in man, J. Clin. Invest., 55:845, 1975.

REGULATION OF WATER AND ELECTROLYTE BALANCE

REGULATION OF THE EFFECTIVE CIRCULATING VOLUME

Adequate tissue perfusion is essential for normal cellular metabolism by providing nutrients and removing waste products. It is not surprising, therefore, that multiple sensors and multiple effectors are involved in this process. As will be seen, the effectors act by influencing the cardiac output, systemic vascular resistance, and renal Na^+ and water excretion.

DEFINITION

The *effective circulating volume* refers to that part of the extracellular fluid (ECF) that is in the vascular space and effectively perfusing the tissues. In general, the effective circulating volume varies directly with the ECF volume. Both of these parameters also vary with total body Na^+ stores since Na^+ salts are the primary

Table 9-1 Potential independence of effective circulating volume from other measurable hemodynamic parameters

Clinical condition	Effective circulating volume	ECF volume	Plasma volume	Cardiac output
Na⁺-depleted normal subjects	↓	↓	↓	↓
Hypoalbuminemia	↓	↑	↓	↓
Heart failure	↓	↑	↑	↓
Hepatic cirrhosis	↓	↑	↑	N/↑

extracellular solutes which act to hold water within the extracellular space (see page 26). As a result, *the regulation of Na⁺ balance (by alterations in urinary Na⁺ excretion) and the maintenance of the effective circulating volume are closely related functions.* Na⁺ loading will tend to produce volume expansion; Na⁺ loss will lead to volume depletion.

It is important to note that the effective circulating volume is not a measurable entity but refers to the rate of perfusion of the capillary circulation. Although it varies with the ECF volume in normal subjects, *it may be independent of the ECF volume, the plasma volume, or even the cardiac output in a variety of disease states* (Table 9-1). In patients with hypoalbuminemia, for example, the fall in the plasma oncotic pressure results in fluid movement out of the vascular space into the interstitium (see Fig. 1-16). This extravascular sequestration of fluid lowers the plasma volume and therefore the cardiac output and effective circulating volume. As the kidney then retains NaCl and water in an attempt to restore normovolemia (see below), most of the retained fluid accumulates in the interstitium, where it may become clinically apparent as edema. The net result may be persistent plasma and effective volume depletion even though the total extracellular volume (which includes the interstitial fluid) is markedly increased (see Chap. 16) [1].

In congestive heart failure, on the other hand, the effective circulating volume is reduced because of a primary decrease in cardiac output. As a result, there is compensatory fluid retention by the kidney. This fluid initially expands the venous volume, thereby raising the venous and intracapillary hydrostatic pressures. The latter change, as with hypoalbuminemia, promotes fluid movement into the interstitium. The net effect is effective volume depletion in association with increases in both the plasma and total ECF volumes.

Finally, the effective volume may be independent of the cardiac output. In the presence of an arteriovenous fistula, the cardiac output is elevated by an amount equal to the flow through the fistula. However, this fluid can be considered to be circulating ineffectively since it bypasses the capillary circulation. In patients with severe hepatic cirrhosis, for example, the cardiac output is frequently elevated, due in part to multiple arteriovenous fistulas throughout the body such as the spider angiomas on the skin [2]. Despite the high cardiac output, these patients frequently behave as if they are hypovolemic [3], probably because the cirrhosis-induced elevation in hepatic sinusoidal pressure promotes the movement of fluid out of the vascular space into the peritoneum (see Chap. 16). The ensuing renal Na^+ and water retention expands both the ECF and plasma volumes [4]. The latter is mostly due to fluid accumulation in the dilated but slowly circulating portal venous system, the former to continued intraperitoneal fluid accumulation (called ascites). Thus, hepatic cirrhosis may be associated with effective circulating volume depletion despite elevations in the plasma and ECF volumes and cardiac output.

In summary, the effective circulating volume is an unmeasured entity that reflects tissue perfusion and may be independent of other hemodynamic parameters (Table 9-1). The presence of effective volume depletion is usually made by demonstrating renal Na^+ retention as evidenced by a urine Na^+ concentration less than 10 to 15 meq/L.

EFFECTIVE CIRCULATING VOLUME AND RENAL SODIUM EXCRETION

The kidney is the primary regulator of Na^+ and volume balance as renal Na^+ excretion responds in an appropriate manner to changes in the effective circulating volume. When there is an increase in volume, as after a Na^+ load [5] or closure of an arteriovenous fistula [6], Na^+ excretion rises in an attempt to lower the volume toward normal. Conversely, the kidney retains Na^+ in the presence of effective volume depletion.

The time course of the response to variations in Na^+ intake is illustrated in Fig. 9-1. If dietary intake is abruptly increased in a patient on a low-sodium diet, only one-half of the excess intake is excreted on the first day. The remainder is retained, augmenting body Na^+ stores. This elevates the plasma osmolality which stimulates both thirst and the secretion of ADH (see page 135). The increments in water intake and renal water reabsorption produce water retention, resulting in increases in the effective circulating volume and body weight and the return of the plasma osmolality to normal. (This process of osmolal regulation is discussed in detail in Chap. 10.)

On subsequent days, a progressively greater fraction of the excess intake is excreted (and less retained) until, by 4 to 5 days, a new steady state is achieved in which renal Na^+ excretion matches intake. This new steady state is characterized by a mild increase in the effective circulating volume due to the Na^+ and water retained on the first 4 days [6a]. The same sequence occurs, in reverse, if Na^+ intake is now reduced. Negative Na^+ balance occurs until there has been enough loss of volume to lower Na^+ excretion to the reduced level of intake. It should be noted that, with lesser variations in dietary intake, the full renal response may be achieved within 1 to 2 days rather than 4 to 5 days.

Thus, a high-sodium diet is characterized by increases in volume and Na^+ excretion and a low-sodium diet by decreases in volume and Na^+ excretion. These findings suggest that *changes in volume constitute the signal that allows urinary Na^+ excretion to vary appropriately with fluctuations in Na^+ intake.*

To maintain Na^+ and volume balance, this regulatory system must be extremely sensitive. A normal adult with a GFR of 150 L/day and a Na^+ concentration in the

Pattern of sodium balance in a normal human

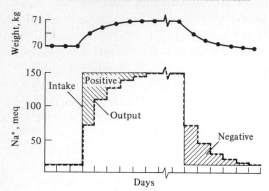

Figure 9-1 Effect of abrupt changes in Na$^+$ intake on body weight and renal Na$^+$ excretion in a normal human. The shaded areas refer to changes in total body Na$^+$ stores due to the difference between intake and excretion. See text for details. *(From L. E. Earley, Clinical Disorders of Fluid and Electrolyte Metabolism, M. H. Maxwell and C. R. Kleeman (eds.), McGraw-Hill, New York, 1972.)*

plasma water of 150 meq/L (see Table 1-6) filters 22,500 meq of Na$^+$ per day. Thus, enhancing Na$^+$ intake and subsequent urinary excretion from 10 to 150 meq/day (as in Fig. 9-1) involves an increase in the fraction of the filtered Na$^+$ that is excreted from 0.04 percent to only 0.67 percent.

REGULATION OF THE EFFECTIVE CIRCULATING VOLUME

The body responds to variations in the effective circulating volume in two steps: (1) the change is sensed by the volume receptors; and (2) these receptors then activate a series of effectors which restore normovolemia by varying vascular resistance, cardiac output, and renal Na$^+$ and water excretion.

Volume Receptors

The primary volume receptors are in the cardiopulmonary circulation, the carotid sinuses and aortic arch, and the kidney [7–10]. Although it is volume that is being regulated, it is difficult to conceive of receptors that sense total extracellular, plasma, or capillary volume. What is actually being sensed at most of the renal and extrarenal volume receptors is pressure (or stretch).† This allows effective volume control since pressure and volume are directly related. For example, volume depletion induced by vomiting is sequentially associated with reductions in venous return to the heart, intracardiac filling pressures, cardiac output, and systemic blood pressure.

In the kidney, the major volume receptors are the stretch receptors in the afferent arteriole and perhaps the macula densa cells in the early distal tubule. These receptors affect volume balance by influencing the activity of the renin-angiotensin-aldosterone system (see Chaps. 3 and 8).

In contrast, the extrarenal receptors primarily govern the activity of the sympathetic nervous system, although the release of other hormones may also be affected [11]. The role of cardiopulmonary receptors has been demonstrated in humans by

---

†The mechanism by which pressure receptors function as volume receptors is discussed on p. 136.

the response to immersion to the neck in warm water [9]. In this setting, the hydrostatic pressure of the water on the lower extremities results in a redistribution of intravascular fluid from the legs to the chest. The ensuing increase in central blood volume (and subsequently in cardiac output) is associated with a marked increase in Na^+ and water excretion in an attempt to restore normovolemia (Fig. 9-2a). Decreased secretion of aldosterone induced by the volume expansion is only partly responsible for the enhanced Na^+ excretion, as increased secretion of a natriuretic hormone may also contribute [9,11].

A similar natriuretic response to neck immersion can be demonstrated in many patients with hepatic cirrhosis and ascites who excrete little Na^+ in the basal state (Fig. 9-2b) [12]. This observation supports the view that the reduced Na^+ excretion in cirrhosis is due to effective volume depletion even though these patients have, as described above, elevations in the plasma volume and cardiac output.

Although multiple receptors are involved in the regulation of the effective circulating volume, no single receptor appears to be of primary importance. As an example, Na^+ balance is well maintained in the presence of cardiac denervation [13], indicating that although cardiac receptors may be important in normal subjects, other receptors (including the carotid sinus and afferent glomerular arteriole) are capable of maintaining volume homeostasis. It may be that the net volume status is normally determined in the brain by integration of the individual signals from the different peripheral receptors. In this regard, there is some experimental evidence supporting a central role in Na^+ excretion, perhaps mediated by a humoral mechanism [14]. Alternatively, volume changes have direct mechanical effects on cardiac output and blood pressure, and the latter can influence NaCl and water excretion when other regulatory systems have failed (see "Pressure Natriuresis" below).

Effectors

Multiple effectors are involved in volume control, influencing both systemic hemodynamics and urinary Na^+ excretion (Table 9-2). Since these regulatory systems (except for the sympathetic nervous system) have been discussed previously in Chaps. 3, 5, and 8, their effects will only be briefly reviewed in this section.

Sympathetic nervous system Sympathetic neural tone and the secretion of catecholamines (norepinephrine and epinephrine) from the adrenal medulla are reduced by volume expansion and enhanced by volume depletion [8,15–17]. When volume depletion occurs, e.g., because of vomiting, venous return to the heart is reduced, resulting in a fall in cardiac output. From the formula relating pressure, output, and resistance,†

$$\text{Mean arterial pressure} = \text{cardiac output} \times \text{systemic vascular resistance}$$

---

†The product of the cardiac output and the systemic vascular resistance actually equals the change in pressure across the circulation, i.e., mean arterial pressure minus mean venous pressure. However, since the mean venous pressure (normal equals 1 to 7 mmHg) is normally so much lower than the mean arterial pressure, only a slight error results from ignoring the venous pressure.

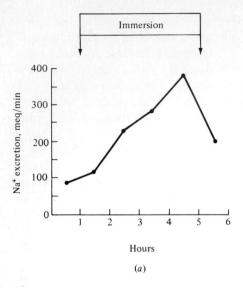

(a)

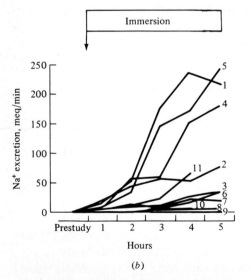

(b)

Figure 9-2 Effect of water immersion to the neck on sodium excretion in (*a*) six normal subjects ingesting 150 meq of Na$^+$ per day and (*b*) eleven patients with hepatic cirrhosis who excrete little Na$^+$ in the basal state. Sodium excretion tends to be increased in both settings, mediated at least in part by cardiopulmonary receptors sensing the associated expansion of the central blood volume. (*Adapted from M. Epstein, Circ. Res., 39:619, 1976 and M. Epstein, D. S. Pins, N. Schneider, and R. Levinson, J. Lab. Clin. Med., 87:822, 1976. By permission of the American Heart Association, Inc.*)

the reduction in cardiac output will lower the blood pressure. This decrease in pressure is sensed by the arterial baroreceptors, resulting in a change in baroreceptor afferent discharge to the vasomotor centers in the brain stem [8]. These centers then induce an increase in peripheral sympathetic tone, initiating a series of events that act to restore normal tissue perfusion (Fig. 9-3). First, there is venous constriction. Normally, 70 percent of the vascular volume is contained within the venous system

Table 9-2 Principle effectors involved in volume regulation

Systemic hemodynamics
  Sympathetic nervous system
  Angiotensin II
Renal Na^+ excretion
  Glomerular filtration rate
  Aldosterone
  Angiotensin II
  Intrarenal hemodynamics
  Sympathetic nervous system
  Natriuretic hormone
  Pressure natriuresis
  Medullary interstitial pressure
  Plasma Na^+ concentration
  Other

[18]. By decreasing the volume of this reservoir, blood delivery to the heart is augmented [16]. Second, myocardial contractility and heart rate are increased, and this, in combination with the enhanced venous return, increases the cardiac output. Third, there is arterial vasoconstriction, increasing systemic vascular resistance, which raises the blood pressure toward normal. Fourth, renin secretion is enhanced by hypotension, mostly because of the increase in sympathetic tone [19,20]. This results in the generation of the vasoconstrictor, angiotensin II. Fifth, renal tubular Na^+ reabsorption is enhanced because of the increases in angiotensin II and aldosterone secretion and, possibly, because of a direct adrenergic effect on renal function (see below). This renal Na^+ retention acts to restore normovolemia. These cardiovascular changes are reversed by volume expansion as sympathetic activity is reduced, minimizing the increases that occur in cardiac output and blood pressure [15].

The importance of this regulatory system is illustrated in Fig. 9-4. Although normal subjects easily tolerate the removal of 500 mL of blood (the equivalent of donating one unit of blood), patients with idiopathic autonomic insufficiency can develop severe hypotension. These patients also have postural hypotension as they cannot compensate for the pooling of blood in the legs that occurs when one assumes the erect position.

It should be noted that these alterations in sympathetic tone with changes in the effective circulating volume are mostly compensatory and that appropriate changes in renal Na^+ excretion are required for the restoration of normal volume balance. Thus, an increase in volume can be corrected only by the renal excretion of the excess Na^+ and water, and a decrease in volume due to fluid loss only by the ingestion and renal retention of exogenous Na^+ and water. However, when effective volume depletion is due to heart failure or to hepatic cirrhosis with ascites, the associated renal Na^+ retention is only compensatory as complete correction cannot occur without reversal of the underlying process.

Angiotensin II The physiology of the renin-angiotensin-aldosterone system is discussed in detail in Chap. 3. Reviewed briefly, this system has little effect in the basal state. However, renin secretion is enhanced in hypovolemic disorders, resulting in the generation of angiotensin II, which has two effects: It raises the blood pressure by arterial vasoconstriction, and it induces renal Na^+ retention both directly and by increasing the secretion of aldosterone. As with sympathetic blockade, inhibition of the renin-angiotensin system in effectively hypovolemic subjects can lead to marked hypotension [21,22].

Regulation of renal Na^+ excretion Renal Na^+ excretion varies directly with the effective circulating volume. When the effective volume is expanded, the urine Na^+ concentration can exceed 100 meq/L. In contrast, the urine can be rendered virtually

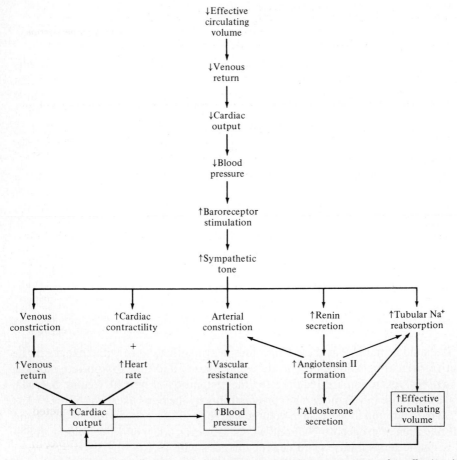

Figure 9-3 Hemodynamic responses induced by the sympathetic nervous system after effective circulating volume depletion.

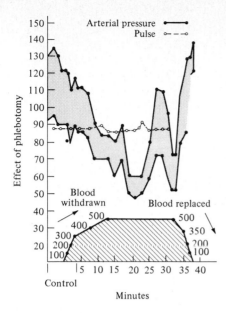

Figure 9-4 Effect of rapid removal and replacement of whole blood on the arterial blood pressure during recumbency in a patient with idiopathic autonomic insufficiency. (*Reproduced from H. N. Wagner, Jr., J. Clin. Invest., 36:1319, 1957, by copyright permission of the American Society for Clinical Investigation.*)

Na$^+$-free (urine Na$^+$ concentration as low as 1 meq/L) in the presence of volume depletion and normal renal function. These homeostatic changes in Na$^+$ excretion can result from alterations both in the filtered load, i.e., the GFR, and in tubular reabsorption, which is affected by multiple factors. As will be seen, *an abnormality in any one factor does not preclude the maintenance of Na$^+$ balance,* a finding indicative of the substantial overlap involved in volume regulation.

Glomerular filtration rate The GFR tends to increase with volume expansion and decrease with volume depletion, both of which can contribute to the associated changes in Na$^+$ excretion [23]. However, alterations in the GFR are not required for the maintenance of Na$^+$ balance. For example, most patients with less than end-stage renal failure, in whom the GFR is reduced, are still able to adjust Na$^+$ excretion to match intake by varying the rate of tubular reabsorption [24]. The importance of tubular reabsorption is also illustrated by the phenomenon of *glomerulotubular balance* in which a primary alteration in GFR without change in volume does not lead to a substantial change in urinary excretion since tubular reabsorption varies in the same direction (see page 97) [23].

Thus, it appears that *variations in tubular reabsorption constitute the main adaptive response to fluctuations in the effective circulating volume.* The approximate contribution of the different nephron segments to Na$^+$ reabsorption is depicted in Table 9-3. In conditions associated with variations in the effective circulating volume, changes in reabsorption may occur at several sites in the nephron. For example, mild volume depletion in the rat results in a decrease in Na$^+$ excretion that is mostly due to enhanced collecting tubule reabsorption [25], an effect that is mediated at least

Table 9-3 Anatomic distribution of NaCl reabsorption in the monkey†

Tubule segment	Percent filtered NaCl reabsorbed
Proximal tubule	70
Loop of Henle	17
Distal tubule and connecting segment	5
Collecting tubules	7
Total	99

†Data from C. M. Bennett, B. M. Brenner, and R. Berliner, J. Clin. Invest., 47:203, 1968, by copyright permission of The American Society for Clinical Investigation.

in part by aldosterone. With more marked hypovolemia, proximal reabsorption is also enhanced [25,26]. This response may be due to altered peritubular capillary hemodynamics, perhaps induced by angiotensin II and norepinephrine (see page 95). Reabsorption in the loop of Henle may also be increased in this setting, an effect that may be mediated by a decrease in medullary interstitial pressure (see below) [27]. The reverse changes occur with volume expansion as collecting tubule and then proximal and loop NaCl reabsorption may be reduced [28–31].

Aldosterone Aldosterone is a mineralocorticoid hormone that promotes volume expansion by enhancing Na^+ reabsorption in the cortical collecting tubule (see Chap. 8). It plays an important role in Na^+ balance since the secretion of aldosterone (via activation of the renin-angiotensin system) is inversely related to variations in Na^+ intake. For example, an increase in Na^+ intake, which tends to expand the effective circulating volume, suppresses the release of renin and aldosterone, thereby favoring the excretion of the excess Na^+ in the urine. It seems likely that, in the normal subject, aldosterone may be the fine modulator of Na^+ excretion, allowing the kidney to respond to daily fluctuations in Na^+ intake. As described above, a seemingly large change in dietary Na^+ intake from 10 to 150 meq/day requires less than a 1 percent change in the fraction of the filtered Na^+ that has to be excreted. Since aldosterone affects the reabsorption of only 1 to 2 percent of the filtered Na^+, it is well suited to meet this need.

Despite this role in volume regulation, abnormalities in aldosterone secretion are not necessarily associated with disturbances in Na^+ balance. For example, adrenalectomized patients treated with replacement doses of a mineralocorticoid are still able to maintain Na^+ balance even though they are unable to vary the level of mineralocorticoid secretion [32]. Similarly, subjects given aldosterone or patients with an autonomous, aldosterone-secreting adrenal adenoma retain fluid for only a few days and then undergo a spontaneous diuresis that returns the volume status toward normal. This phenomenon of *aldosterone escape* (see Chap. 8) is due to decreased

Na^+ reabsorption at some other site in the nephron, perhaps in the loop of Henle [30]. These observations illustrate the importance of other factors in the control of tubular Na^+ reabsorption, particularly when there is an abnormality in aldosterone release.

Angiotensin II In addition to stimulating the secretion of aldosterone, angiotensin II also enhances Na^+ reabsorption [33,34]. Thus, when volume depletion increases renin secretion, angiotensin II as well as aldosterone may contribute to the fall in Na^+ excretion. These changes are reversed with volume expansion. It is not known if the effect of angiotensin II is direct or is mediated by the alterations induced in renal hemodynamics.

Intrarenal hemodynamics The hydrostatic and oncotic pressures in the peritubular capillary are an important determinant of Na^+ reabsorption in the proximal tubule† (see Chap. 5). These forces in turn are influenced by the resistances at the afferent and efferent glomerular arterioles, which are in part under neurohormonal control. In hypovolemic states, the associated increases in angiotensin II production and sympathetic activity can lead to arteriolar constriction which alters peritubular capillary hemodynamics in a direction that promotes enhanced proximal Na^+ reabsorption. The opposite changes occur with volume expansion.

However, these changes in proximal reabsorption are not necessarily reflected in the final urine [36]. For example, a reduction in the plasma protein concentration should lower the peritubular capillary oncotic pressure, resulting in diminished proximal reabsorption. In humans with the nephrotic syndrome, urinary protein loss results in hypoalbuminemia and the predicted reduction in proximal Na^+ reabsorption [37]. Nevertheless, these patients may avidly retain Na^+ as enhanced loop and distal nephron reabsorption compensate for the decrease in proximal reabsorption [37]. This is an appropriate response since hypoalbuminemia generally leads to effective circulating volume depletion due to movement of fluid out of the vascular space into the interstitium [1].

Variations in the *distribution of renal blood flow* also may participate in the regulation of Na^+ excretion (see page 69). In hypovolemic states, there is a reduction in blood flow to the outer cortex with a preferential increase in perfusion to the inner cortex, the site of the juxtamedullary nephrons. Theoretically, this could promote Na^+ retention since the juxtamedullary nephrons have long loops of Henle (see Fig. 2-3) and therefore a greater reabsorptive surface area. Conversely, inner cortical perfusion is decreased and outer cortical perfusion enhanced in volume expansion. However, the physiologic significance of these changes is uncertain since a cause-and-effect relationship has not been proved and, in some conditions, Na^+ excretion can vary without change in the pattern of renal perfusion [38]. Furthermore, redistribution of blood flow is not necessarily associated with redistribution of glomerular

---

†In contrast, peritubular capillary hemodynamics do not appear to play a role in the regulation of tubular reabsorption in the more distal segments. [35].

filtration, making it difficult to determine the physiologic significance of these changes [39].

Sympathetic nervous system As described above, sympathetic neural tone is affected by alterations in the effective circulating volume, increasing with volume depletion and decreasing with volume expansion. These responses maintain cardiac output and blood pressure and may play a role in the regulation of renal Na^+ excretion [40]. Sympathetic neural tone and circulating catecholamines appear to increase tubular Na^+ reabsorption, and this may be important in the compensatory renal Na^+ retention seen with volume depletion [17,41]. This response may result from the interplay of three factors: a direct effect on proximal tubular reabsorption [40,42]; altered proximal peritubular capillary hemodynamics resulting from renal vasoconstriction (see page 95) [43]; and activation of the renin-angiotensin-aldosterone system [19,20].

Attempts to evaluate the significance of these adrenergic effects on Na^+ excretion by the use of sympathetic blocking agents have produced conflicting results. This appears to be due to the counteracting renal and extrarenal actions of the sympathetic nervous system. Although there is a direct adrenergic enhancement of Na^+ reabsorption, the associated increases in cardiac output and blood pressure, i.e., in the effective circulating volume, tend to reduce Na^+ reabsorption. This difference can be illustrated by the response to the sympatholytic agent, guanethidine. If it is given in doses that do not reduce the systemic pressure, the adrenergic effects on the kidney are blocked, resulting in an increase in Na^+ excretion [44]. However, when it is used to lower the blood pressure in patients with hypertension, the hypotensive response predominates and Na^+ retention occurs [45,46]. This constitutes the rationale for the use of diuretics in combination with sympathetic blockers in the treatment of hypertension since the increase in volume produced by the latter tends to raise the blood pressure [45]. The mechanism of the antinatriuretic effect of hypotension is unclear but may be due to a direct effect of systemic blood pressure on net Na^+ reabsorption (see below).

Also complicating the evaluation of the effect of the sympathetic nervous system on Na^+ excretion is the demonstration that renal sympathetic tone may occasionally be regulated independent of that in other organs. This effect appears to be mediated by volume receptors in the left side of the heart as an increase in left-sided pressure results in a *reduction* in renal sympathetic tone [47,48] even if total sympathetic activity is enhanced [49].

The potential importance of this cardiorenal reflex is illustrated by the following experiment [49]. The induction of hypotensive hemorrhage (which lowers left-sided pressures) resulted in a decrease in renal blood flow of as much as 90 percent and a cessation of urine output, both of which were due in large part to a hypotension-induced increase in sympathetic activity and subsequent marked renal vasoconstriction. In contrast, a similar reduction in systemic blood pressure due to an acute myocardial infarction (which raises left-sided pressures) was associated with only a 25 percent fall in renal blood flow and an adequate urine output. This relative maintenance of renal perfusion and urine flow presumably was due to reduced renal sympathetic tone resulting from the elevation in left-sided heart pressure. Although it is uncertain how important the left-sided heart receptors are in humans [50], a cardiorenal reflex can explain the clinical observation that acute renal failure due to postischemic acute tubular necrosis frequently follows hypotension due to sepsis, surgery, or hemorrhage but is much less common after a myocardial infarction [51].

In summary, the net effect of the sympathetic nervous system on renal Na^+ reabsorption is determined by the balance between its renal and extrarenal actions. However, the adrenergic effect does not appear to be essential since Na^+ balance is maintained following renal transplantation in which the kidney is denervated [52].

In contrast to the Na^+-retaining effects of norepinephrine and epinephrine, dopamine decreases Na^+ reabsorption in the straight segment of the proximal tubule [53]. Since dopamine production is increased by volume expansion [54], it may contribute to the ensuing natriuresis. However, the physiologic importance of dopamine in this setting remains to be proven.

Natriuretic hormone The evidence supporting the existence of a natriuretic hormone released (perhaps from the brain or the cardiac atria) in response to volume expansion has been presented in Chap. 8 [55,55a]. In addition to facilitating Na^+ excretion in hypervolemic subjects [11], this hormone may play an important role in maintaining Na^+ balance in chronic renal failure [24,56]. In this setting in which the GFR and therefore the filtered load of Na^+ are reduced, tubular reabsorption must be decreased to keep Na^+ excretion at a constant level, and there is evidence that this is achieved at least in part by a natriuretic hormone [56]. A deficiency of a natriuretic hormone, on the other hand, may contribute to the Na^+ retention that occurs in hepatic cirrhosis [57,58]. However, the exact role of an as yet unidentified natriuretic hormone (or hormones) in the maintenance of Na^+ balance cannot be determined until its physiologic characteristics are more completely understood.

Pressure natriuresis In normal subjects, a small elevation in blood pressure results in a relatively large increase in the urinary excretion of Na^+ and water (Fig. 9-5) [59]. Although the mechanism by which this occurs is incompletely understood, this relationship may be important in the maintenance of volume balance. For example, an increase in Na^+ intake initially results in sequential increases in plasma volume, venous return to the heart, cardiac output, and blood pressure. The last effect may then promote the excretion of the excess Na^+, returning circulatory hemodynamics to normal.

In contrast to the other effectors, pressure natriuresis does not require neurally or humorally mediated sensor mechanisms since changes in volume directly affect the blood pressure [59]. Furthermore, the influence of other factors on Na^+ excretion will modify but not basically alter this relationship. For example, the chronic administration of a *subpressor* amount of angiotensin II enhances Na^+ reabsorption. As this induces Na^+ retention, the ensuing volume expansion gradually raises the blood pressure until Na^+ excretion is increased back to a level equal to intake. The new

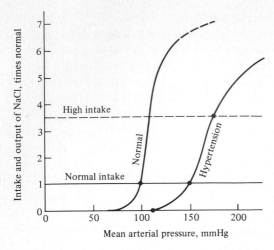

Figure 9-5 Relationship between arterial pressure and urinary NaCl and water excretion in normal subjects and in patients with hypertension. *(From A. C. Guyton, T. G. Coleman, A. W. Cowley, Jr., K. W. Scheel, R. D. Manning, Jr., and R. A. Norman, Jr., Am. J. Med., 52:584, 1972.)*

steady state is characterized by mild volume expansion (due to the Na^+ retained when excretion was less than intake) and therefore persistent hypertension [60]. In terms of Fig. 9-5, the pressure natriuresis curve has been shifted to the right; that is, Na^+ balance is maintained but at a higher than normal blood pressure to overcome the sodium-retaining effect of angiotensin II. In comparison, augmenting Na^+ intake does not produce a large increase in blood pressure in normal subjects since renin secretion and therefore angiotensin II and aldosterone levels are reduced, thereby promoting excretion of the excess Na^+ [60].

The mechanism by which changes in blood pressure affect Na^+ excretion is uncertain. There is some evidence that alterations in reabsorption in the loop of Henle are of primary importance [61,62], an effect that could result from changes in medullary interstitial pressure which varies directly with the systemic blood pressure.

Medullary interstitial pressure An increase in medullary interstitial pressure (induced, for example, by hypertension or ureteral obstruction) can reduce net loop reabsorption [63]. To understand how this might occur, it is necessary to briefly review how *passive* reabsorption occurs in the thin descending and ascending limbs (see Chap. 6) [64,65]. As fluid from the proximal tubule enters the *descending* limb of Henle's loop, it equilibrates osmotically with the hypertonic medullary interstitium, mostly by water movement out of the tubule. If, for example, the osmolality increases from 285 mosmol/kg (isosmotic to plasma with a Na^+ concentration of 150 meq/L) at the corticomedullary junction to 1200 mosmol/kg at the papillary tip (see Fig. 6-3), then more than three-quarters of the water in the tubule must move into the interstitium. This marked loss of water raises the tubular fluid Na^+ concentration from 150 to 600 meq/L, a level well above that in the interstitium. As a result, Na^+ can passively diffuse out of the thin *ascending* limb since this segment, in contrast to the descending limb, is relatively permeable to Na^+ [64,65].

When the hydrostatic pressure in the interstitium is increased, it will tend to counteract the osmotic gradient favoring water reabsorption in the descending limb. The ensuing reduction in descending limb water reabsorption will lower the Na^+ concentration in the fluid entering the thin ascending limb, decreasing the gradient for passive Na^+ reabsorption in that segment [63]. These changes are reversed with a fall in blood pressure as loop reabsorption is enhanced [62,63].

Plasma Na^+ concentration Urinary Na^+ excretion also may be affected by the plasma Na^+ concentration, increasing with hypernatremia and decreasing with hyponatremia [66,67]. This effect may be mediated both by changes in the filtered Na^+ load (GFR times plasma Na^+ concentration) and in tubular reabsorption [66]. However, since the plasma Na^+ concentration is normally maintained within narrow limits by ADH and thirst (see Chap. 10), it cannot play an important role in the daily regulation of Na^+ excretion. Even in patients who are hyponatremic or hypernatremic, variations in the effective circulating volume can override the effects of the plasma Na^+ concentration.† Thus, hyponatremia due to volume expansion, e.g., due to water retention in the syndrome of inappropriate ADH secretion, is associated with enhanced Na^+ excretion (see Chap. 23) although the increase is less than that seen with an equivalent degree of volume expansion in a patient with a normal plasma Na^+ concentration [67]. Similarly, urinary Na^+ excretion is reduced in hypernatremic patients who are volume-depleted due, for example, to lack of replacement of insensible water losses from the skin and respiratory tract (see Chap. 24).

Other Other factors may also influence the rate of Na^+ excretion, although their physiologic importance is uncertain. Included in this group are prostaglandins and kinins, the renal production of which may respond to changes in volume (see Chap. 8).

Summary It is clear that multiple factors affect renal Na^+ excretion and therefore the regulation of the effective circulating volume. It may be that aldosterone and possibly a natriuretic hormone are responsible for day-to-day variations in Na^+ excretion with other factors contributing only if these hormones are ineffective in maintaining Na^+ balance. As Na^+ intake is reduced, for example, the ensuing decrease in volume (Fig. 9-1) results in increased secretion of aldosterone and perhaps reduced secretion of a natriuretic hormone. The net effect is enhanced Na^+ reabsorption in the collecting tubules, which seems to account for the appropriate fall in Na^+ excretion in this setting [25]. With more marked hypovolemia, a decrease in GFR and increase in *proximal* Na^+ reabsorption also contribute to Na^+ retention [25]. The latter change may be mediated by increases in angiotensin II production and sympathetic activity, which can promote proximal Na^+ reabsorption directly or

---

†The plasma Na^+ concentration and effective circulating volume are independent variables; the former being proportional to the *ratio* of Na^+ and water, the latter to the *absolute amounts* of Na^+ and water present (see Chap. 1).

by vasoconstriction-induced alterations in peritubular capillary hemodynamics. A clinical example of this interplay occurs in congestive heart failure in which a low cardiac output is associated with a reduced GFR, increased activity of the renin-angiotensin-aldosterone and sympathetic nervous systems, an elevated filtration fraction (resulting in altered peritubular capillary hemodynamics), and enhanced proximal reabsorption [17,68–70]. To the degree that medullary interstitial pressure is reduced, loop reabsorption may also be increased [63]. A low rate of Na^+ excretion will persist in this setting unless the cardiac output can be normalized (see Chap. 16).

This sequence is reversed with volume expansion as decreased secretion of aldosterone and perhaps increased secretion of natriuretic hormone [11] initially allow excretion of the excess Na^+ by reducing collecting tubule Na^+ reabsorption. With more pronounced hypervolemia, reabsorption may also fall in the proximal tubule [29] and loop of Henle [30,31].

The pressure natriuresis phenomenon may be *the final defense against changes in the effective circulating volume* [59]. In normal subjects, it probably plays a relatively minor role since the other regulatory systems are sufficiently sensitive to maintain Na^+ balance without a large change in extracellular volume or blood pressure [60]. However, if there is an abnormality in one or more of these factors (as with excess angiotensin II), the degree of Na^+ retention that will occur is ultimately limited since the ensuing volume expansion raises the blood pressure, which then enhances Na^+ excretion [60]. Eventually, a new steady state is reestablished in which intake and excretion are equal and the blood pressure is greater than normal.

It must be emphasized that this chapter has dealt with the factors influencing Na^+ balance and the effective circulating volume. These systems are *not* involved in the regulation of the plasma Na^+ concentration and plasma osmolality in which ADH release and thirst are of primary importance (see Chap. 10).

PROBLEMS

9-1 In what direction would the plasma volume, the total extracellular volume, the effective circulating volume, and urinary Na^+ excretion change in the following conditions?

    (*a*) An acute myocardial infarction producing a decrease in the cardiac output

    (*b*) A high-sodium diet

    (*c*) The retention of ingested water due to inappropriate secretion of ADH

    (*d*) The development of hypoalbuminemia due to protein loss in the urine

9-2 What effect would you expect the administration of diuretics (which increase urinary NaCl and water loss) to have on the secretion of renin? Diuretics are used in the treatment of hypertension. Would this effect on renin secretion have any effect on the degree to which the blood pressure would be lowered by the diuretics?

9-3 Assuming that renal function is normal and that there is no obstruction to renal blood flow, which of the following offers the most accurate assessment of the effective circulating volume: cardiac output, plasma volume, systemic blood pressure, or urinary Na^+ excretion?

9-4 In a stable patient chronically taking diuretics for hypertension, what will be the relationship between Na^+ intake and urinary Na^+ excretion?

REFERENCES

1. Meltzer, J. I., H. J. Keim, J. H. Laragh, J. E. Sealey, K-M. Jan, and S. Chien: Nephrotic syndrome: vasoconstriction and hypervolemic types indicated by renin-sodium profiling, Ann. Intern. Med., 91:688, 1979.
2. Kowalski, H. J., and W. H. Abelmann: The cardiac output at rest in Laennec's cirrhosis, J. Clin. Invest., 32:1025, 1953.
3. Bichet, D. G., M. J. Van Putten, and R. W. Schrier: Potential role of increased sympathetic activity in impaired sodium and water excretion in cirrhosis, N. Engl. J. Med., 307:1552, 1982.
4. Lieberman, F. L., and T. B. Reynolds: Plasma volume in cirrhosis of the liver: its relation to portal hypertension, ascites, and renal failure, J. Clin. Invest., 46:1297, 1967.
5. Strauss, M. B., E. Lamdin, W. P. Smith, and D. J. Bleifer: Surfeit and deficit of sodium: a kinetic concept of sodium excretion, Arch. Intern. Med., 102:527, 1958.
6. Epstein, F. H., R. S. Post, and M. McDowell: The effect of an arteriovenous fistula on renal hemodynamics and electrolyte excretion, J. Clin. Invest., 32:233, 1953.
6a. Bonventre, J. V., and A. Leaf: Sodium homeostasis: steady states without a set point, Kidney Int., 21:880, 1982.
7. Skorecki, K. L., and B. M. Brenner: Body fluid homeostasis in man, Am. J. Med., 70:77, 1981.
8. Scher, A.: Control of arterial blood pressure, in *Physiology and Biophysics,* vol. II, T. C. Ruch and H. D. Patton (eds.), Saunders, Philadelphia, 1974.
9. Epstein, M.: Cardiovascular and renal effects of headout water immersion in man, Circ. Res., 39:619, 1976.
10. Migdal, S., E. A. Alexander, and N. G. Levinsky: Evidence that decreased cardiac output is not the stimulus to sodium retention during acute constriction of the vena cava. J. Lab. Clin. Med., 89:809, 1977.
11. Epstein, M., N. S. Bricker, and J. J. Bourgoignie: Presence of a natriuretic factor in urine of normal men undergoing water immersion, Kidney Int., 13:152, 1978.
12. Epstein, M., D. S. Pins, N. Schneider, and R. Levinson: Determinants of deranged sodium and water homeostasis in decompensated cirrhosis, J. Lab. Clin. Med., 87:822, 1976.
13. Gilmore, J. P., and J. M. Daggetti: Response of the chronic cardiac denervated dog to acute volume expansion, Am. J. Physiol., 210:509, 1966.
14. Silva-Netto, C. R., M. de Mello Aires, and G. Malnic: Hypothalamic stimulation and electrolyte excretion: a micropuncture study, Am. J. Physiol., 239:F206, 1980.
15. Frye, R. L., and E. Braunwald: Studies on Starling's law of the heart. I. The circulatory response to acute hypervolemia and its modification by ganglionic blockade, J. Clin. Invest., 39:1043, 1960.
16. Freis, E. D., J. R. Stanton, F. A. Finnerty, Jr., H. W. Schnaper, R. L. Johnson, C. E. Rath, and R. W. Wilkins: The collapse produced by venous congestion of the extremities or by venesection following certain hypotensive agents, J. Clin. Invest., 30:435, 1951.
17. Schrier, R. W., M. H. Humphreys, and R. C. Ufferman: Role of cardiac output and the autonomic nervous system in the antinatriuretic response to acute constriction of the thoracic superior vena cava, Circ. Res., 29:490, 1971.
18. Braunwald, E.: Regulation of the circulation, N. Engl. J. Med., 290:1420, 1974.
19. Haber, E.: The renin-angiotensin system and hypertension, Kidney Int., 15:427, 1979.
20. Gordon, R. D., O. Kuckel, G. W. Liddle, and D. P. Island: Role of the sympathetic nervous system in regulating renin and aldosterone production in man, J. Clin. Invest., 46:599, 1967.

21. Watkins, L., Jr., J. A. Burton, E. Haber, J. R. Cant, F. W. Smith, and A. C. Barger: The renin-angiotensin-aldosterone system in congestive failure in conscious dogs, J. Clin. Invest., 57:1606, 1976.
22. Schroeder, E. T., G. H. Anderson, S. H. Goldman, and D. H. P. Streeten: Effect of blockade of angiotensin II on blood pressure, renin and aldosterone in cirrhosis, Kidney Int., 9:511, 1976.
23. Lindheimer, M. D., R. C. Lalone, and N. G. Levinsky: Evidence that an acute increase in glomerular filtration has little effect on sodium excretion in the dog unless extracellular volume is expanded, J. Clin. Invest., 46:256, 1974.
24. Rose, B. D., *Pathophysiology of Renal Disease,* McGraw-Hill, New York, 1981, chap. 9.
25. Stein, J. H., R. W. Osgood, S. Boonjarern, J. W. Cox, and T. F. Ferris: Segmental sodium reabsorption in rats with mild and severe volume depletion, Am. J. Physiol., 227:351, 1974.
26. Weiner, M. W., E. J. Weinman, M. Kashgarian, and J. P. Hayslett: Accelerated reabsorption in the proximal tubule produced by volume depletion, J. Clin. Invest., 50:1379, 1971.
27. Faubert, P. F., S-Y. Chou, J. G. Porush, I. J. Belizon, and S. Spitalewitz: Papillary plasma flow and tissue osmolality in chronic caval dogs, Am. J. Physiol., 242:F370, 1982.
28. Sonnenberg, H.: Proximal and distal tubular function in salt-deprived and in salt-loaded deoxycorticosterone acetate-escaped rats, J. Clin. Invest., 52:263, 1973.
29. Dirks, J. H., W. J. Cirksena, and R. W. Berliner: The effect of saline infusion on sodium reabsorption by the proximal tubule of the dog, J. Clin. Invest., 44:1160, 1965.
30. Knox, F. G., J. C. Burnett, Jr., D. E. Kohan, W. S. Spielman and J. C. Strand: Escape from the sodium-retaining effects of mineralocorticoids, Kidney Int., 17:263, 1980.
31. Osgood, R. W., H. J. Reineck, and J. H. Stein: Further studies on segmental sodium transport in the rat kidney during expansion of the extracellular fluid volume, J. Clin. Invest., 62:311, 1978.
32. Rosenbaum, J. D., S. Papper, and M. M. Ashley: Variations in renal excretion of sodium independent of change in adrenocortical hormone dosage in patients with Addison's disease, J. Clin. Endocrinol. Metab., 15:1549, 1955.
33. Hall, J. E., A. C. Guyton, N. C. Trippodo, T. E. Lohmeier, R. E. McCaa, and A. W. Cowley, Jr.: Intrarenal control of electrolyte excretion by angiotensin II, Am. J. Physiol., 232:F538, 1977.
34. Hall, J. E., A. C. Guyton, M. J. Smith, Jr., and T. G. Coleman: Chronic blockade of angiotensin II formation during sodium deprivation, Am. J. Physiol., 237:F424, 1979.
35. Diezi, J., M. Nenninger, and G. Giebisch: Distal tubular function in superficial rat tubules during volume expansion, Am. J. Physiol., 239:F228, 1980.
36. Howards, S. S., B. B. Davis, F. G. Knox, F. S. Wright, and R. W. Berliner: Depression of fractional sodium reabsorption by the proximal tubule of the dog without sodium diuresis, J. Clin. Invest, 47:1561, 1968.
37. Grausz, H., R. Lieberman, and L. E. Earley: Effect of plasma albumin on sodium reabsorption in patients with nephrotic syndrome, Kidney Int., 1:147, 1972.
38. Westenfelder, C., J. A. L. Arruda, R. Lockwood, S. Boonjarern, L. Nascimento, and N. A. Kurtzman: Distribution of renal blood flow in dogs with congestive heart failure, Am. J. Physiol., 230:537, 1976.
39. Bruns, F. J., E. A. Alexander, A. L. Riley, and N. G. Levinsky: Superficial and juxtamedullary nephron function during saline loading in the dog, J. Clin. Invest., 53:971, 1974.
40. Moss, N. G.: Renal function and renal afferent and efferent nerve activity, Am. J. Physiol., 243:F425, 1982.
41. Gill, J. R., Jr., and A. G. T. Casper: Role of the sympathetic nervous system in the renal response to hemorrhage, J. Clin. Invest., 48:915, 1969.
42. Bello-Reuss, E.: Effect of catecholamines on fluid reabsorption by the isolated proximal convoluted tubule, Am. J. Physiol., 238:F347, 1980.
43. Schrier, R. W., and T. Berl: Mechanism of effect of alpha adrenergic stimulation with nonrepinephrine on renal water excretion, J. Clin. Invest, 52:502, 1973.
44. Gill, J. R., Jr., D. T. Mason, and F. C. Bartter: Adrenergic nervous system in sodium metabolism: effects of guanethidine and sodium-retaining steroids in normal man, J. Clin. Invest., 43:177, 1964.

45. Dustan, H., R. C. Tarazi, and E. L. Bravo: Dependence of arterial pressure on intravascular volume in treated hypertensive patients, N. Engl. J. Med., 286:861, 1972.

46. Smith, A. J.: Fluid retention produced by guanethidine: changes in body exchangeable sodium, blood volume, and creatinine clearance, Circulation, 31:490, 1965.

47. Thames, M. D., and F. M. Abboud: Reflex inhibition of renal sympathetic nerve activity during myocardial ischemia mediated by left ventricular receptors with vagal afferents in dogs, J. Clin. Invest., 63:395, 1979.

48. Thames, M. D., B. D. Miller, and F. M. Abboud: Baroreflex regulation of renal nerve activity during volume expansion, Am. J. Physiol., 243:H810, 1982.

49. Gorfinkel, H. J., J. P. Szidon, L. J. Hirsch, and A. P. Fishman: Renal performance in experimental cardiogenic shock, Am. J. Physiol., 222:1260, 1972.

50. Cornish, K. G., and J. P. Gilmore: Increased left atrial pressure does not alter renal function in the conscious primate, Am. J. Physiol., 243:R119, 1982.

51. Schrier, R. W.: Acute renal failure, Kidney Int., 15:205, 1979.

52. Blaufox, M. D., E. J. Lewis, P. Jagger, D. Lauler, R. Hickler, and J. P. Merrill: Physiologic responses of the transplanted human kidney, N. Engl. J. Med., 280:62, 1969.

53. Bello-Reuss, E., Y. Higashi, and Y. Kaneda: Dopamine decreases fluid reabsorption in straight portions of rabbit proximal tubule, Am. J. Physiol., 242:F634, 1982.

54. Alexander, R. W., J. R. Gill, Jr., H. Yamabe, W. Lovenberg, and H. R. Keiser: Effects of dietary sodium and of acute saline infusion on the interrelationship between dopamine excretion and adrenergic activity in man, J. Clin. Invest., 54:194, 1974.

55. deWardener, H. E., and E. M. Clarkson: The natriuretic hormone: recent developments, Clin. Sci., 63:415, 1982.

55a. Currie, M. G., D. M. Geller, B. R. Cole, J. G. Boylan, W. YuSheng, S. W. Holmberg, and P. Needleman: Bioactive cardiac substances: potent vasorelaxant activity in mammalian atria, Science, 221:71, 1983.

56. Schmidt, R. W., J. J. Bourgoignie, and N. S. Bricker: On the adaptation in sodium excretion in chronic uremia: the effects of "proportional reduction" of sodium intake, J. Clin. Invest., 53:1376, 1974.

57. Naccarato, R., P. Messa, A. D'Angelo, A. Fabris, M. Messa, M. Chiaramonte, C. Gregolin, and G. Zanon: Renal handling of sodium and water in early chronic liver disease, Gastroenterology, 81:205, 1981.

58. Epstein, M.: Natriuretic hormone and the sodium retention of cirrhosis, Gastroenterology, 81:395, 1981.

59. Guyton, A. C., T. G. Coleman, A. W. Cowley, Jr., K. W. Scheel, R. D. Manning, Jr., and R. A. Norman, Jr.: Arterial pressure regulation: overriding dominance of the kidneys in long-term regulation and in hypertension, Am. J. Med., 52:584, 1972.

60. Hall, J. E., A. C. Guyton, M. J. Smith, Jr., and T. G. Coleman: Blood pressure and renal function during chronic changes in sodium intake: role of angiotensin, Am. J. Physiol., 239:F271, 1980.

61. Stumpe, K. O., H. D. Lowitz, and B. Ochwadt: Fluid reabsorption in Henle's loop and urinary excretion of sodium and water in normal rats and rats with chronic hypertension, J. Clin. Invest., 49:1200, 1970.

62. Levy, M.: Effects of acute volume expansion and altered hemodynamics on renal tubular function in chronic caval dogs, J. Clin. Invest., 51:922, 1972.

63. Knox, F. G., J. I. Mertz, J. C. Burnett, Jr., and A. Haramati: Role of hydrostatic and oncotic pressures in renal sodium reabsorption, Circ. Res., 52:491, 1983.

64. Jamison, R. L., and R. H. Maffly: The urinary concentrating mechanism, N. Engl. J. Med., 295:1059, 1976.

65. Bonventre, J. V., and C. Lechene: Renal medullary concentrating process: an integrative hypothesis, Am. J. Physiol., 239:F578, 1980.

66. Schrier, R. W., and H. E. deWardener: Tubular reabsorption of sodium ion: influence of factors other than aldosterone and glomerular filtration rate, N. Engl. J. Med., 285:1231, 1971.

67. Schrier R. W., R. L. Fein, J. S. McNeil, and W. J. Cirksena: Influence of interstitial fluid volume expansion and plasma sodium concentration on the natriuretic response to volume expansion in dogs, Clin. Sci., 36:371, 1969.
68. Wolff, H. P., R. Koczorek, and E. Buchborn: Hyperaldosteronism in heart disease, Lancet, 2:63, 1957.
69. Heller, B. J., and W. E. Jacobson: Renal hemodynamics in heart disease, Am. Heart J., 39:188, 1950.
70. Bell, N. H., H. P. Schedl, and F. C. Bartter: An explanation for abnormal water retention and hypoosmolality in congestive heart failure, Am. J. Med., 36:351, 1964.

REGULATION OF PLASMA OSMOLALITY

Hypoosmolality and hyperosmolality can produce serious neurologic symptoms and death (see Chaps. 23 to 25) [1,2]. To prevent this, the plasma osmolality (P_{osm}), which is a function of the ratio of body solute to body water (see Chap. 1), is normally maintained within narrow limits by appropriate variations in water intake and water excretion. In this chapter, we will describe the sources of water intake, the sites of water loss from the body, and the roles of antidiuretic hormone (ADH), thirst, and renal water excretion in the regulation of the P_{osm}.

WATER BALANCE

In the steady state, water intake (including that generated from endogenous metabolism) equals water output (Table 10-1). Much of the water output involves obligatory losses in the urine, stool, and, by evaporation, from the moist surfaces of the skin and respiratory tract. The evaporative losses play an important role in thermoregulation since the heat required for evaporation, 0.58 kcal/1.0 mL of water, nor-

Table 10-1 Typical daily water balance in a normal human, assuming a low rate of sweat production†

	Water intake, mL/day			Water output, mL/day	
	Obligatory	Elective		Obligatory	Elective
Source			Source		
Ingested water	400	1000	Urine	500	1000
Water content of food	850		Skin	500	
Water of oxidation	350		Respiratory tract	400	
			Stool	200	
Total	1600	1000	Total	1600	1000

†Under conditions of increased sweat production, the water losses from the skin can increase markedly, occasionally exceeding 5 L/day. When this occurs, thirst is stimulated, resulting in an appropriate increase in the volume of ingested water.

mally accounts for 20 to 25 percent of the heat lost from the body, the remainder occurring by radiation and convection [3]. The net effect is the elimination of the heat produced by body metabolism, thereby preventing the development of hyperthermia.

In contrast to these "insensible" losses, sweat can be called a "sensible" loss. Sweat is a hypotonic fluid (Na^+ concentration equals 30 to 50 meq/L) secreted by the sweat glands in the skin. It also contributes to thermoregulation as the secretion and subsequent evaporation of sweat result in the loss of heat from the body. In the basal state, sweat production is low, but it can increase markedly in the presence of high external temperatures or when endogenous heat production is enhanced, as with exercise, fever, or hyperthyroidism [3]. For example, a subject exercising in a hot, dry climate can lose as much as 1500 mL/h as sweat.

The obligatory renal loss is directly related to solute excretion. If a subject has to excrete 600 mosmol of solute per day (mostly Na^+ and K^+ salts and urea) to remain in the steady state, and the maximum urine osmolality (U_{osm}) is 1200 mosmol/kg, then the excretion of the 600 mosmol will require a minimum volume of 500 mL/day.

Only small amounts of water are normally lost in the stool, averaging 100 to 200 mL/day. However, gastrointestinal losses are increased to a variable degree in patients with vomiting or diarrhea.

To maintain water balance, water must be taken in (or generated) to replace these losses (Table 10-1). Net water intake is derived from three sources: (1) ingested water; (2) water contained in foods, e.g., meat is roughly 70 percent water and certain fruits and vegetables are almost 100 percent water; and (3) water produced from the oxidation of carbohydrates, proteins, and fats. If the latter two sources account for 1200 mL/day and the obligatory water loss is 1600 mL/day, then 400 mL must be ingested by drinking. Humans drink more than this minimum requirement for social and cultural reasons, and the extra water is excreted in the urine.

REGULATION OF PLASMA OSMOLALITY

The normal plasma osmolality (P_{osm}) is 275 to 290 mosmol/kg. It usually is held within narrow limits as variations of only 1 to 2 percent initiate mechanisms to return the P_{osm} to normal. These alterations in osmolality are sensed by receptor cells in the hypothalamus which affect water intake (via thirst) and water excretion (via ADH which increases water reabsorption in the collecting tubules; see Chap. 8).

In terms of water balance, water retention, e.g., after a water load, decreases the P_{osm}, and a water deficit, e.g., due to hypotonic sweat loss after exercise on a hot day, increases the P_{osm}. These changes in water balance must be differentiated from conditions of isosmotic fluid loss, e.g., diarrhea, in which solute and water are lost proportionately, producing no direct change in P_{osm}.

The body responds to a water load by suppressing ADH secretion and thirst, resulting in decreased collecting tubule water reabsorption and the excretion of the excess water. The peak diuresis is delayed for 90 to 120 min, the time necessary for the metabolism of previously circulating ADH. As will be seen, the kidney can excrete up to 10 to 20 liters of water per day. Therefore, *persistent water retention resulting in hypoosmolality occurs, with rare exceptions, only in patients with reduced renal water excretion* (see Chap. 23).

The correction of a water deficit (hyperosmolality) requires the intake and retention of exogenous water. This is achieved by increases in thirst and ADH secretion which are induced by the elevation in the P_{osm}. In contrast to the response to hypoosmolality, in which renal water excretion is of primary importance, *increased thirst is the major defense against hyperosmolality*. Although the kidney can minimize water excretion via the effect of ADH, a water deficit can be corrected only by increased dietary intake. An example of the efficiency of the thirst mechanism occurs in patients with complete central diabetes insipidus who, because they secrete no ADH, may excrete more than 10 liters of water per day. Despite this, the P_{osm} remains near normal because the thirst mechanism augments water intake to match output. Thus, *symptomatic hyperosmolality will not occur in a patient with a normal thirst mechanism and access to water* (see Chap. 24).

When the P_{osm} is increased by an increase in body solute, e.g., after a Na^+ load, both the volume and osmoregulatory systems come into play. The increment in Na^+ stores expands the effective circulating volume, promoting the renal excretion of the excess Na^+. Thirst also is stimulated, and the increase in water intake both lowers the P_{osm} toward normal and further expands the volume, thereby enhancing the stimulus to renal Na^+ excretion.

RENAL WATER EXCRETION AND REABSORPTION

In the kidney, the bulk of the filtered water is reabsorbed passively in the proximal tubule and descending limb of Henle's loop (see Chaps. 5 and 6) down an osmotic gradient created by NaCl transport. This serves to maintain the volume of the extracellular fluid. In addition, the kidney contributes to the stability of the P_{osm} by excret-

Table 10-2 Effect of ADH on urine volume in a subject excreting 700 mosmol of solute per day

ADH	U_{osm}, mosmol/kg	Urine volume, L/day
0	70	10
++	350	2
+++	1400	0.5

ing or reabsorbing water without solute. This function is primarily mediated by the presence (water conservation, high U_{osm}) or absence (water excretion, low U_{osm}) of ADH. In normal adults, the U_{osm} can vary from a minimum of 40 to 100 mosmol/kg [4] to a maximum of 900 to 1400 mosmol/kg† [5].

The quantitative importance of ADH on water excretion is depicted in Table 10-2. In a subject excreting 700 mosmol of solute per day, the urine volume can vary 20-fold, depending upon the availability of ADH. For example, in the absence of ADH, the minimum U_{osm} may be 70 mosmol/kg, resulting in the excretion of the 700 mosmol of solute in 10 liters of water. In a normal subject, this degree of polyuria is rarely seen and occurs only after a massive water load. More commonly, there is a moderate amount of ADH present, and the U_{osm} is somewhere between the extremes of 70 mosmol/kg (no ADH) and 1400 mosmol/kg (maximum ADH). If this subject had to excrete 2000 mL of water to remain in water balance, the average U_{osm} would be 350 mosmol/kg, that is, 700 mosmol of solute in 2000 mL. This would require a submaximal ADH effect.

The *urine output also is affected by solute excretion,* which is equal to net solute intake in the steady state. This is particularly important in disorders in which the rate of ADH secretion is relatively constant (Fig. 10-1). In the absence of ADH, for example, the daily urine volume is 10 liters if 700 mosmol of solute is excreted (Table 10-2) but only 5 liters if 350 mosmol of solute is excreted, that is, 350 mosmol of solute at 70 mosmol/kg.

Measurement of Renal Water Excretion

This simple example of the effect of solute excretion demonstrates that water excretion can vary widely without changes in the U_{osm}. Thus, *the U_{osm}, which reflects the kidney's ability to dilute or concentrate the urine, is not an accurate estimate of its quantitative ability to excrete or retain water.*

To measure the *amount of solute-free water* that the kidney can excrete per unit time, one can calculate the *free-water clearance* (C_{H_2O}). If the urine is hypoosmotic to plasma, the total urine volume (V, in mL/min or L/day) can be viewed as having two components: one that contains all the urinary solute in a solution that is isosmotic

---

†Concentrating ability tends to fall with age, probably due to a concomitant reduction in GFR. As a result, the maximum U_{osm} may only be 700 mosmol/kg in an elderly patient [5,6].

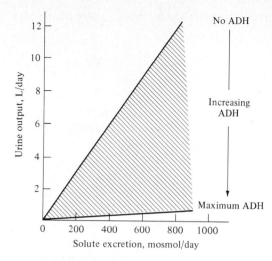

Figure 10-1 Effects of ADH and solute excretion on water excretion. It is assumed that the U_{osm} is 70 mosmol/kg in the absence of ADH and 1400 mosmol/kg with maximum ADH effect.

to plasma (the osmolal clearance, or C_{osm}); and one that contains the solute-free water that makes the urine dilute (the free-water clearance, or C_{H_2O})

$$V = C_{osm} + C_{H_2O} \qquad (10\text{-}1)$$

The C_{osm} can be calculated from the general formula for clearance, $C = UV/P$ (see page 71):

$$C_{osm} = \frac{U_{osm} \times V}{P_{osm}} \qquad (10\text{-}2)$$

If Eq. (10-1) is solved for C_{H_2O}, then

$$C_{H_2O} = V - C_{osm} \qquad (10\text{-}3)$$

The manner in which this formula is used can be illustrated by the following example. If

$$P_{osm} = 280 \text{ mosmol/kg}$$

$$U_{osm} = 70 \text{ mosmol/kg}$$

$$V = 10 \text{ L/day}$$

then

$$C_{H_2O} = 10 - (70 \times 10/280)$$

$$= 7.5 \text{ L/day}$$

Thus, of the 10 liters of urine being excreted, 7.5 liters exist as free water (C_{H_2O}) and 2.5 liters as an isosmotic solution (the C_{osm}).

In the clinical setting, the excretion of large volumes of dilute urine may be *appropriate* if it follows a water load or *inappropriate* if it is due to a primary deficiency of ADII or renal resistance to its effects. In either case, the loss of solute-free water tends to raise the P_{osm} unless unaccompanied by an equivalent increase in water intake.

Physiologic factors affecting C_{H_2O} The excretion of water by the kidney occurs in two basic steps (see Chap. 6):

1. Solute-free water is generated by NaCl reabsorption without water in the medullary and cortical aspects of the ascending limb of the loop of Henle.
2. This water is then excreted by keeping the collecting tubules impermeable to water.

In normal subjects, the volume of free water generated in the loop of Henle (step 1) is primarily dependent upon the volume of water presented to that segment. Collecting tubular impermeability to water (step 2), on the other hand, requires the absence of ADH.

An understanding of the factors that influence C_{H_2O} has important clinical implications in patients with hyponatremia and hypoosmolality. Since the capacity for water excretion is normally so great (as much as 10 to 20 L/day), water retention leading to hyponatremia will occur only if there is a defect in water excretion or rarely if the amount of water ingested exceeds excretory capacity (see Chap. 23). Water excretion will be impaired only if one or both of the steps described above are impaired. This can occur in one of three settings: (1) if less free water is generated because the rate of fluid delivery to the loop of Henle is reduced, as with oliguric renal failure (where less water is filtered) or volume depletion (where less water may be filtered and more is reabsorbed in the proximal tubule); (2) if less free water is generated because loop NaCl reabsorption is inhibited by diuretics; or (3) if ADH is present, as with volume depletion, hypothyroidism, adrenal insufficiency, or the syndrome of inappropriate ADH secretion. These disorders compose most of the differential diagnosis of hyponatremia (see Chap. 23)

Measurement of Renal Water Reabsorption

In addition to the formation of a dilute urine, the kidney is also able to excrete urine with an osmolality exceeding that of the plasma. If the urine is hyperosmotic, the urine volume (V) can again be viewed as having two components: one containing all the urinary solute in an isosmotic solution (the C_{osm}) and one containing the amount of free water that must have been removed from the urine by tubular reabsorption to raise the U_{osm} to the observed hyperosmotic value (the free-water reabsorption; $T^c_{H_2O}$). In this setting,

$$V = C_{osm} - T^c_{H_2O} \qquad (10\text{-}4)$$

$$T^c_{H_2O} = C_{osm} - V \qquad (10\text{-}5)$$

In contrast to the C_{H_2O}, which is equal to the volume of free water *excreted* per unit time, the $T^c_{H_2O}$ is equal to the volume of free water *reabsorbed* per unit time.

For example, if a subject who has developed hyperosmolality due to a water deficit has the following values,

$$P_{osm} = 295 \text{ mosmol/kg}$$

$$U_{osm} = 885 \text{ mosmol/kg}$$

$$V = 1 \text{ L/day}$$

then

$$T^c_{H2O} = \left(\frac{885 \times 1}{295}\right) - 1$$

$$T^c_{H2O} = 2 \text{ L/day}†$$

Thus, 2 L/day of free water is being added to the plasma. This tends to lower the P_{osm} back toward normal, an appropriate response to a water deficit.

The net effect, however, is different if the elevation in ADH release responsible for the high U_{osm} is due to the syndrome of inappropriate ADH secretion. In this setting, the retention of 2 liters of water *that would normally have been excreted* will lead to hypoosmolality and hyponatremia. This illustrates the importance of thinking in terms of T^c_{H2O} rather than of only the U_{osm}. The latter merely indicates the presence of a concentrated urine; the former tells exactly how much water is being retained by the kidney.

Physiologic factors affecting T^c_{H2O} Renal water conservation is dependent upon the following (see Chap. 6):

1. The formation and maintenance of the medullary osmotic gradient
2. Equilibration of the urine in the collecting tubules with the hyperosmotic medullary interstitium

ADH plays an important role in both steps by promoting the medullary accumulation of urea and by increasing the water permeability of the collecting tubules. Medullary hyperosmolality is also dependent upon NaCl reabsorption without water in the ascending limb of Henle's loop, a process that is independent of ADH.

T^c_{H2O} can be impaired by a defect in ADH release, decreased responsiveness of

---

†These values also can be used to calculate the C_{H2O}:

$$C_{H2O} = V - C_{osm}$$

$$= 1 - \frac{885 \times 1}{295} = -2 \text{ L/day}$$

Thus, -2 L/day of free water is being excreted, another way of stating that 2 L/day is being reabsorbed. This illustrates the inverse relationship between the C_{H2O}, which measures water excretion, and the T^c_{H2O}, which represents water reabsorption:

$$C_{H2O} = -T^c_{H2O}$$

the collecting tubule epithelium to ADH, or a primary abnormality in countercurrent function, preventing the maintenance of the hyperosmotic interstitium. When one of these disturbances is present, water excretion increases and the patient may complain of polyuria. If these water losses are not replaced, hyperosmolality will ensue.

In humans on a regular diet, the maximum $T^c_{H_2O}$ is 2 to 2.5 L/day. Although this is much less than the maximum C_{H_2O} of 10 to 20 L/day, it must be emphasized that *the volume of water retained during the correction of a water deficit is dependent upon water intake, mediated by thirst, as well as on renal water conservation.* For example, suppose a normal subject develops a 1000-mL water deficit due to sweat loss after exercise on a hot day. The increase in the P_{osm} induced by the water loss stimulates both ADH release and thirst. If this subject excretes 600 mosmol of solute per day and the urine can be concentrated to 1200 mosmol/kg, the urine volume will be 500 mL. If water intake also is 500 mL, there will be no water retention and no replacement of the water deficit even though the kidney is conserving water maximally. To restore water balance, water intake must exceed output (urine plus insensible losses) by 1000 mL; this is achieved by the stimulation of thirst.

Summary and Clinical Implications

The ability to excrete urine with an osmolality different from that of the plasma plays a central role in the maintenance of water balance and the P_{osm}. If the P_{osm} is decreased, e.g., after a water load, ADH secretion is inhibited. This results in the excretion of a dilute urine, which increases the P_{osm} to normal. If the P_{osm} is elevated, e.g., because of sweat loss, both ADH release and thirst are stimulated. The combination of renal water conservation (by excreting a concentrated urine) and an increase in water intake results in water retention and a decrease in P_{osm} to normal.

In addition to the urine osmolality, a solute excretion also determines how much water can be excreted (Fig. 10-1). Thus, the quantity of free water excreted or reabsorbed is best measured directly as C_{H_2O} or $T^c_{H_2O}$, rather than being inferred from the urine osmolality. Although the role of solute excretion may at first glance appear to be of interest only to the physiologist, it becomes clinically important in several situations. This is particularly true when ADH release is abnormal as in the syndrome of inappropriate ADH secretion or diabetes insipidus. In the former disorder, water retention and hyponatremia generally occur because there is persistent ADH release, resulting in an inappropriately high urine osmolality (see Chap. 23). If for example, solute excretion is normal at 750 mosmol but the minimum U_{osm} is 375 mosmol/kg (versus <100 mosmol/kg in normal subjects), maximum water excretion is only 2 L/day. Water intake above this level will not be excreted. One way to treat this disorder is to increase the rate of solute excretion, by using a high-salt, high-protein diet or by administering urea [7]. If solute excretion is increased by 50 percent to 1125 mosmol/day, water excretion will rise to 3 L/day (1125 mosmol/day $\div$ 375 mosmol/kg = 3 L/day), thereby making water retention less likely.

On the other hand, diabetes insipidus is characterized by partial or complete ADH-lack, resulting in enhanced water excretion (see Chap. 24). Patients with this disorder present with polyuria and secondarily increased thirst. (Hypernatremia usu-

Table 10-3 Relation of urine osmolality and Na$^+$ excretion to plasma osmolality and the effective circulating volume

	Urine		ADH release
	Na$^+$	Osmolality	
Plasma osmolality			
Increased	0†	↑	↑
Decreased	0	↓	↓
Effective circulating volume			
Increased	↑	↓	↓
Decreased	↓	↑	↑

†The urine osmolality is determined primarily by the availability of ADH. Since ADH release is controlled by osmolality and volume, the urine osmolality is affected by both of these parameters. Urinary Na$^+$ excretion, on the other hand, is mostly a function of the effective circulating volume with a much lesser contribution from the plasma osmolality or the plasma Na$^+$ concentration.

ally doesn't occur since the thirst mechanism allows water intake to match excretion.) If solute excretion is 750 mosmol/day but the urine is persistently dilute at 75 mosmol/kg, daily urine output will be 10 L. One way to treat this disorder is to reduce solute excretion by limiting dietary salt and protein, thereby reducing water excretion and alleviating the patient's main complaints.†

In summary, water intake is the major determinant of the urine volume in normal subjects by its effect on ADH secretion (Table 10-2). However, when ADH secretion is relatively fixed (as in the syndrome of inappropriate ADH secretion or diabetes insipidus), *water intake no longer affects the urine volume* and the rate of solute excretion assumes primary importance.

VOLUME REGULATION AND OSMOREGULATION

It is important to understand the differences between volume regulation and osmoregulation. The latter is governed by osmoreceptors influencing ADH and thirst; the former is sensed by multiple volume receptors which activate effectors such as aldosterone (see Chap. 9). ADH increases water reabsorption (and therefore the U$_{osm}$) but does not affect Na$^+$ transport; aldosterone enhances Na$^+$ reabsorption but not directly that of water. Thus, *osmoregulation is achieved by changes in water balance, volume regulation primarily by changes in Na$^+$ balance* (Table 10-3). For example, a water load results in the suppression of ADH release and the excretion of a dilute urine without a significant alteration in Na$^+$ excretion. In contrast, volume depletion due to vomiting is associated with a fall in Na$^+$ excretion and, since ADH secretion

†This is clearly not the optimal treatment, which consists of the administration of some form of ADH. However, dietary manipulation is a central part of the therapy of *nephrogenic* diabetes insipidus in which ADH secretion is normal but a dilute urine is produced because of end-organ resistance (see Chap. 24) [8].

is also stimulated by hypovolemia (see Fig. 8-6), a rise in the U_{osm}. These responses are appropriate since the ensuing Na^+ and water retention tend to restore normovolemia.

PROBLEMS

10-1 Diarrheal fluid usually is isosmotic to plasma even though its ionic composition differs from that of the plasma. What effect will the loss of 2 liters of diarrheal fluid have on (*a*) the effective circulating volume; (*b*) urinary Na^+ excretion; (*c*) the plasma osmolality; (*d*) the plasma Na^+ concentration; (*e*) ADH secretion; (*f*) the urine osmolality; and (*g*) thirst?

   If such a patient ingested water, what would happen to the plasma osmolality?

10-2 What is the relation between the plasma Na^+ concentration and urinary Na^+ excretion?

10-3 A patient with volume depletion due to vomiting has hyponatremia and hypoosmolality. What are the factors that may contribute to the inability to restore normal osmolality by excreting the excess water in the urine?

10-4 The following data were obtained from a normal subject who had taken no fluids for 12 h. At the end of each 30-min collection period, the bladder was emptied and a blood specimen drawn.

Period	P_{osm}, mosmol/kg	Volume, mL/min	U_{osm}, mosmol/kg
1	286	0.5	1040

Calculate the free-water reabsorption (T^c_{H2O}). After this control collection, the subject drinks 1200 mL of water, and six additional 30-min collections are made.

2	286	0.8	650
3	284	5.6	92
4	282	10.4	50
5	281	10.0	51
6	282	4.0	128
7	284	1.3	398

Calculate the free-water clearance (C_{H2O}) from the collection in period 4. Explain the sequence of events that led to the changes in the U_{osm} and urine volume.

10-5 A subject switches from a regular diet to one consisting primarily of beer, and is then given a water load. What will the effect of the dietary change be on the minimum U_{osm} that can be achieved and the maximum C_{H2O}?

REFERENCES

1. Pollock, A. S., and A. I. Arieff: Abnormalities of cell volume regulation and their functional consequences, Am. J. Physiol., 239:F195, 1980.
2. Arieff, A. I., F. Llach, and S. G. Massry: Neurological manifestations and morbidity of hyponatremia: correlation with brain water and electrolytes, Medicine, 55:121, 1976.
3. Guyton, A. C.: *Textbook of Medical Physiology*, 6th ed. Saunders, Philadelphia, 1981, chap. 72.

4. Schoen, E. J.: Minimum urine total solute concentration in response to water loading in normal men, J. Appl. Physiol., 10:267, 1957.
5. Lindeman, R. D., H. C. van Buren, and L. G. Raisz: Osmolar renal concentrating ability in healthy young men and hospitalized patients without renal disease, N. Engl. J. Med., 262:1306, 1960.
6. Sporn, I. N., R. G. Lancestremere, and S. Papper: Differential diagnosis of oliguria in aged patients, N. Engl. J. Med., 267:130, 1962.
7. Decaux, G., and F. Genette: Urea for long-term treatment of syndrome of inappropriate secretion of antidiuretic hormone, Br. Med. J., 2:1081, 1981.
8. Earley, L. E., and J. Orloff: The mechanisms of antidiuresis associated with the administration of hydrochlorothiazide to patients with vasopressin-resistant diabetes insipidus, J. Clin. Invest., 41:1988, 1962.

ELEVEN

ACID-BASE PHYSIOLOGY

INTRODUCTION

As with the other components of the extracellular fluid, the H^+ concentration is maintained within narrow limits. The normal extracellular H^+ concentration is approximately 40 nanomol/L (nanomol/L equals 10^{-6} mmol/L), roughly one-millionth the millimole per liter concentrations of Na^+, K^+, Cl^-, and HCO_3^-. The reg-

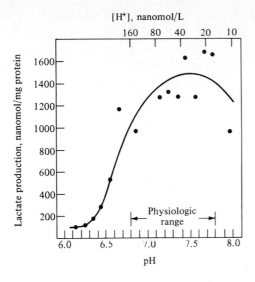

Figure 11-1 Influence of H^+ concentration and pH on lactate production by leukocytes. *(From M. L. Halperin, H. P. Connors, A. S. Relman, and M. L. Karnovsky, J. Biol. Chem., 244:384, 1969.)*

ulation of the H^+ concentration at this level is essential for normal cellular function because of the high reactivity of H^+ ions, particularly with proteins [1]. This property is related to the relatively small size of hydronium ions, the hydrated form of H^+,† in comparison with that of sodium and potassium ions. As a result, H^+ ions are more strongly attracted to negatively charged portions of molecules and are more tightly bound than Na^+ or K^+. When there is a change in the H^+ concentration, proteins gain or lose H^+ ions, resulting in alterations in charge distribution, molecular configuration, and consequently in protein function. For example, the rate of glycolysis (as measured by the rate of lactate production) varies inversely with the H^+ concentration, increasing as the H^+ concentration is reduced (Fig. 11-1). This change in cellular metabolism is mediated by a similar inverse relationship between the H^+ concentration and the activity of several glycolytic enzymes, particularly phosphofructokinase [1].

Under normal conditions, the H^+ concentration varies little from the normal value of 40 nanomol/L. This occurs even though acids and bases are continually being added to the extracellular fluid. The process of H^+ regulation involves three basic steps: (1) chemical buffering by the extracellular and intracellular buffers, (2) control of the carbon dioxide level in the blood by alterations in the rate of alveolar ventilation, and (3) control of the bicarbonate concentration in the blood by changes in renal H^+ excretion. This chapter will review the basic principles of acid-base physiology, including the role of buffers in preventing large changes in the H^+ concentration. The roles of ventilation and renal H^+ excretion in acid-base homeostasis will be presented in Chap. 12.

†In the aqueous environment in the body, H^+ combines with H_2O and exists primarily as the hydronium ion, H_3O^+. For simplicity, H^+ will be used in place of H_3O^+ for the remainder of this discussion.

ACIDS AND BASES

Using the definitions proposed by Brønsted, an acid is defined as a substance which can donate H^+ and a base as a substance which can accept H^+ [2]. These properties are independent of charge. For example, H_2CO_3, HCl, NH_4^+, and $H_2PO_4^-$ all can act as acids:

$$H_2CO_3 \rightleftharpoons H^+ + HCO_3^-$$

$$HCl \rightleftharpoons H^+ + Cl^-$$

$$NH_4^+ \rightleftharpoons H^+ + NH_3$$

$$H_2PO_4^- \rightleftharpoons H^+ + HPO_4^{2-}$$

Acid   Base

There are two classes of acids that are physiologically important: carbonic acid (H_2CO_3) and noncarbonic acids. This distinction is important because of the different rates of production and routes of excretion of these acids. Each day, the metabolism of carbohydrates and fats results in the generation of approximately 15,000 mmol of CO_2. Although CO_2 is not an acid, it combines with H_2O to form H_2CO_3 (see below). Thus, if the endogenously produced CO_2 were not excreted, there would be a progressive accumulation of acid. This is prevented by the excretion of CO_2 by the lungs. Noncarbonic acids also are produced, primarily from the metabolism of proteins. For example, the oxidation of sulfur-containing amino acids results in the formation of H_2SO_4 [3]. In comparison with CO_2, only 50 to 100 meq of H^+ is produced from these sources per day; these H^+ ions are excreted by the kidney.

Law of Mass Action

The law of mass action states that the velocity of a reaction is proportional to the product of the concentrations of the reactants. For example, water can dissociate into hydrogen and hydroxyl ions:†

$$H_2O \rightleftharpoons H^+ + OH^-$$

The velocity with which this reaction moves to the right is equal to

$$v_1 = k_1[H_2O]$$

where k_1 is the rate constant for this reaction. Similarly, the velocity with which the reaction moves to the left can be expressed by

$$v_2 = k_2[H^+][OH^-]$$

---

†This reaction actually should be written

$$H_2O + H_2O \rightleftharpoons H_3O^+ + OH^-$$

At equilibrium, $v_1 = v_2$. Therefore,

$$k_1[H_2O] = k_2[H^+][OH^-]$$

$$K' = \frac{k_1}{k_2} = \frac{[H^+][OH^-]}{[H_2O]} \tag{11-1}$$

Since the H_2O concentration is relatively constant in the body fluids,

$$K_w = [H^+][OH^-] \tag{11-2}$$

where K_w is equal to the product of the two constants, K' and $[H_2O]$. At body temperature, $K_w = 2.4 \times 10^{-14}$. Thus for distilled water,

$$[H^+][OH^-] = 2.4 \times 10^{-14}$$

$$[H^+] = 1.55 \times 10^{-7} \text{ mol/L}$$

$$[OH^-] = 1.55 \times 10^{-7} \text{ mol/L}$$

Since the normal H^+ concentration in the extracellular fluid is 40 nanomol/L, we can see that the extracellular fluid is slightly less acid than water (H^+ concentration equals 155 nanomol/L).

The law of mass action can be written for the dissociation of all the acids and bases in the body. For example, for the dissociation of an acid HA into $H^+ + A^-$,

$$K_a = \frac{[H^+][A^-]}{[HA]} \tag{11-3}$$

where K_a is the apparent *ionization* or *dissociation constant* for this acid. In the body, K_a has a single value for the dissociation of each acid. Although the K_a can vary slightly with changes in temperature, solute concentration, and H^+ concentration [4,5], these parameters are held relatively constant under normal conditions. Since the same principles can be applied to the dissociation of a base, $BOH \rightleftharpoons B^+ + OH^-$, the behavior of bases will not be discussed separately [6].

Acids and bases may be strong or weak. Strong acids are those which are essentially completely ionized in the body. Since most of the acid exists as H^+ and A^-, a strong acid has a relatively high K_a. HCl and NaOH are examples of a strong acid and a strong base. In comparison, $H_2PO_4^-$ is only 80 percent dissociated at the normal extracellular H^+ concentration and is considered a weak acid. As we will see, weak acids are the principal buffers in the body.

pH

The pH of a solution can be defined by the following relationship:

$$pH = -\log [H^+] \tag{11-4}$$

In the laboratory, the H^+ concentration of the blood is measured with a glass membrane electrode which is permeable only to H^+. The diffusion of H^+ ions between the blood and the fluid in the electrode results in the generation of a measurable electrical potential (E_m) across the membrane [7]. The magnitude of this potential is proportional to the logarithm of the ratio of the H^+ concentration in the two compartments according to the Nernst equation:

$$E_m = 61 \log \frac{[H^+]_e}{[H^+]_b}$$

where the subscripts e and b refer to the fluid within the electrode and the blood, respectively. Since $[H^+]_e$ is a known value,

$$E_m \propto \log \frac{1}{[H^+]_b}$$

The log $(1/a)$ is equal to $-\log a$. Therefore,

$$E_m \propto -\log [H^+]_b$$

Since pH $= -\log [H^+]$,

$$E_m \propto pH\dagger$$

It is important to note that the pH varies inversely with the H^+ concentration. Thus, an increase in the H^+ concentration reduces the pH, and a decrease in the H^+ concentration elevates the pH. The relationship between the H^+ concentration and pH within the physiologic range is depicted in Table 11-1. In general, the range of H^+ concentration that is compatible with life is 16 to 160 nanomol/L (pH equals 7.80 to 6.80). Since the normal pH is approximately 7.40, the normal H^+ concentration can be calculated from

$$pH = -\log [H^+]$$

$$\log [H^+] = -7.40$$

Taking the antilogarithm of both sides,

$$[H^+] = \text{antilog} (-7.40)$$

$$= \text{antilog} (0.60 - 8)$$

The antilogarithm of 0.60 is 4, that of -8 is 10^{-8}. Therefore,

$$[H^+] = 4 \times 10^{-8} \text{ mol/L}$$

$$= 40 \text{ nanomol/L}$$

†Actually, the membrane potential and the pH are proportional to the activity of H^+, that is, to the random movement of H^+ across the membrane, not to its molar concentration. Although the activity of H^+ (a_{H+}) is directly proportional to the H^+ concentration,

$$a_{H+} = c[H^+]$$

the value of c is dependent upon the ionic strength of the solution. In concentrated ionic solutions, ionic interaction, i.e., the attraction of H^+ to anions, can retard the random movement of H^+ so that the a_{H+} is significantly less than the H^+ concentration. However, the body fluids are relatively dilute, and it can be assumed without much error that the a_{H+} is equal to the H^+ concentration, that is, $c = 1$.

Table 11-1 Relationship between the pH and the H^+ concentration in the physiologic range

pH†	$[H^+]$, nanomol/L
7.80	16
7.70	20
7.60	26
7.50	32
7.40	40
7.30	50
7.20	63
7.10	80
7.00	100
6.90	125
6.80	160

†To convert from pH to $[H^+]$ see Chap. 17.

The relative merits of measuring the acidity of a solution in terms of pH or H^+ concentration have been the subject of much debate [7,8]. Although this debate is beyond the scope of this discussion, both pH and H^+ concentration will be used to familiarize the reader with these concepts.

Henderson-Hasselbalch Equation

Equation (11-3) can be rearranged in the following manner:

$$[H^+] = K_a \frac{[HA]}{[A^-]} \tag{11-5}$$

If we take the negative logarithm of both sides,

$$-\log [H^+] = -\log K_a - \log \frac{[HA]}{[A^-]}$$

Substituting pH for $-\log [H^+]$, $+\log ([A^-]/[HA])$ for $-\log ([HA]/[A^-])$, and defining pK_a as $-\log K_a$ (the H^+ concentration and K_a being expressed in units of moles per liter),

$$pH = pK_a + \log \frac{[A^-]}{[HA]} \tag{11-6}$$

This is the Henderson-Hasselbalch equation, which can be written for the dissociation of any weak acid. Using the Brønsted definition, in which A^- acts as a base and HA as an acid, this equation becomes

$$pH = pK_a + \log \frac{base}{acid} \tag{11-7}$$

For example, for the reaction

$$H_2PO_4^- \rightleftharpoons H^+ + HPO_4^{2-}$$

the relationship between the concentrations of the reactants can be expressed either by the law of mass action or the Henderson-Hasselbalch equation:

$$[H^+] = K_a \frac{[H_2PO_4^-]}{[HPO_4^{2-}]} \tag{11-8}$$

$$pH = pK_a + \log \frac{[HPO_4^{2-}]}{[H_2PO_4^-]} \tag{11-9}$$

For this reaction, $K_a = 1.6 \times 10^{-7}$ mol/L (or 160 nanomol/L) and $pK_a = 6.80$. To show how these equations can be used, let us calculate the HPO_4^{2-} and $H_2PO_4^-$ concentrations in the extracellular fluid if the total phosphate concentration is 1 mmol/L and the H^+ concentration equals 40 nanomol/L (pH is 7.40). From the law of mass action,

$$40 = 160 \frac{[H_2PO_4^-]}{[HPO_4^{2-}]}$$

or

$$\frac{[HPO_4^{2-}]}{[H_2PO_4^-]} = 4$$

Since the total phosphate concentration is 1 mmol/L,

$$[HPO_4^{2-}] = 0.8 \text{ mmol/L}$$

$$[H_2PO_4^-] = 0.2 \text{ mmol/L}$$

The same results can be obtained from the Henderson-Hasselbalch equation:

$$7.40 = 6.80 + \log \frac{[HPO_4^{2-}]}{[H_2PO_4^-]}$$

$$\log \frac{[HPO_4^{2-}]}{[H_2PO_4^-]} = 0.60$$

Since the antilogarithm of $0.60 = 4$,

$$\frac{[HPO_4^{2-}]}{[H_2PO_4^-]} = 4$$

It should be noted that phosphate also can exist as PO_4^{3-} and H_3PO_4:

$$PO_4^{3-} + H^+ \overset{1}{\rightleftharpoons} HPO_4^{2-} + H^+ \overset{2}{\rightleftharpoons} H_2PO_4^- + H^+ \overset{3}{\rightleftharpoons} H_3PO_4$$

However, only trace amounts of PO_4^{3-} and H_3PO_4 are present in the body since the pK_a of reaction 1 (pK_{a1} equals 12.4) is much higher and that of reaction 3 (pK_{a3} equals 2.0) is much lower than the extracellular pH of 7.40. For example, for reaction 1,

$$7.40 = 12.40 + \log \frac{[PO_4^{3-}]}{[HPO_4^{2-}]}$$

$$\frac{[PO_4^{3-}]}{[HPO_4^{2-}]} = \text{antilog} \ (-5) = 10^{-5}$$

Thus, at pH of 7.40, there is only one molecule of PO_4^{3-} present for every 10^5 molecules of HPO_4^{2-}.

BUFFERS

One of the major ways in which large changes in H^+ concentration are prevented is by *buffering*. The body buffers, which are primarily weak acids, are able to take up or release H^+ so that changes in the H^+ concentration are minimized. For example, phosphate is an effective buffer:

$$H_2PO_4^- \rightleftharpoons H^+ + HPO_4^{2-}$$

If H^+ ions are added to the extracellular fluid, they will combine with HPO_4^{2-} to form $H_2PO_4^-$. Conversely, if H^+ ions are lost from the extracellular fluid, the reaction will move to the right as H^+ ions are released from $H_2PO_4^-$. In contrast, strong acids, such as HCl, are poor buffers at the body pH since they are almost completely ionized and cannot bind H^+ ions.

The efficiency of phosphate buffering can be appreciated from the following example. Let us assume that in 1 liter of solution there are 10 mmol each of HPO_4^{2-} and $H_2PO_4^-$ as the Na^+ salts. From Eq. (11-8),

$$[H^+] = K_a \frac{[H_2PO_4^-]}{[HPO_4^{2-}]}$$

$$= 160 \times \frac{10}{10}$$

$$= 160 \text{ nanomol/L} \qquad (\text{pH} = 6.80)$$

[Note that when the concentrations of acid ($H_2PO_4^-$) and base (HPO_4^{2-}) are equal, $[H^+] = K_a$ or pH $= pK_a$.] If 2 mmol of HCl is added to this solution, these H^+ ions can combine with HPO_4^{2-}:

$$HCl + Na_2HPO_4 \rightleftharpoons NaCl + NaH_2PO_4$$

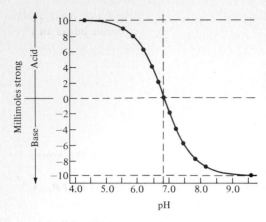

Figure 11-2 Titration curve of 1 liter of a 20 mmol/L NaH_2PO_4, Na_2HPO_4 solution. Initially, HPO_4^{2-} concentration equals $H_2PO_4^-$ concentration equals 10 mmol/L and H^+ concentration equals 160 nanomol/L (pH equals 6.80). The different points represent the effects on the pH of the solution of the addition of a strong acid or base. *(From J. W. Woodbury, Physiology and Biophysics, 20th ed., T. C. Ruch and H. C. Patton (eds.), Saunders, Philadelphia, 1974.)*

If we assume that virtually all the added H^+ is taken up by HPO_4^{2-}, then the HPO_4^{2-} concentration will fall to 8 mmol/L and the $H_2PO_4^-$ concentration will increase to 12 mmol/L. The new H^+ concentration will be

$$[H^+] = 160 \times \frac{12}{8}$$

$$= 240 \text{ nanomol/L} \qquad (\text{pH} = 6.62)$$

Thus, there has been an increase of 80 nanomol/L in the H^+ concentration even though 2 mmol/L or 2 million nanomol/L of H^+ has been added to the solution. As a result, more than 99.99 percent of the excess H^+ ions has been taken up or buffered by HPO_4^{2-}. If no buffers had been present, the H^+ concentration would have been 2 million nanomol/L (pH of 2.70).

If more H^+ ions are added or if H^+ ions are removed by adding NaOH,

$$NaOH + NaH_2PO_4 \rightleftharpoons Na_2HPO_4 + H_2O$$

the change in pH (or H^+ concentration) can be calculated in a similar manner. If the new pH is plotted against the amount of acid or base added, the result is the buffer curve in Fig. 11-2. Although the shape of the curve is sigmoidal, there is a linear midregion (pH equals 5.80 to 7.80) in which relatively large amounts of acid or base can be added without much change in pH. Thus, *a buffer is most efficient when the pH of the solution is within ± 1.0 pH unit of its pK_a.* If the pH is outside these limits, then a small amount of acid or base will produce a relatively large change in pH.

Bicarbonate/Carbon Dioxide Buffer System

Carbonic acid can dissociate into a hydrogen ion and a bicarbonate ion:

$$H_2CO_3 \rightleftharpoons H^+ + HCO_3^-$$

The pK_a of this reaction is 3.57 (K_a equals 2.72×10^{-4}) [9]. Since this is far from the normal pH of 7.40, it seems as if HCO_3^- would be an ineffective buffer in the body. However, H_2CO_3 is formed from the hydration of CO_2, and this buffer system can be more accurately described by the following series of reactions†:

$$CO_2 \rightleftharpoons CO_2 + H_2O \rightleftharpoons H_2CO_3 \rightleftharpoons H^+ + HCO_3^- \qquad (11\text{-}10)$$
$$\begin{pmatrix} \text{gas} \\ \text{phase} \end{pmatrix} \begin{pmatrix} \text{aqueous} \\ \text{phase} \end{pmatrix}$$

Dissolved carbon dioxide All gases dissolve in water (that is, enter the aqueous phase) to some extent. The degree to which this occurs is proportional to the partial pressure of the gas in the solution. In humans, the partial pressure of CO_2 (P_{CO_2}) in the arterial blood is in equilibrium with that in the alveolar air and normally is approximately 40 mmHg. At 37°C (normal body temperature), the amount of CO_2 dissolved in the plasma is

$$[CO_2]_{dis} = 0.03 P_{CO_2} \qquad (11\text{-}11)$$
$$= 0.03 \times 40 = 1.2 \text{ mmol/L}$$

where 0.03 is the solubility constant for CO_2 in the plasma.

Hydration of carbon dioxide The equilibrium of the reaction

$$[CO_2]_{dis} + H_2O \rightleftharpoons H_2CO_3$$

normally is far to the left so that there is approximately 340 molecules of CO_2 in the solution for each molecule of H_2CO_3 [9]. Nevertheless, an increase in the P_{CO_2} increases the $[CO_2]_{dis}$ and, therefore, the H_2CO_3 concentration. Thus, CO_2, which is not an acid, increases the acidity of the solution through the formation of H_2CO_3.

In certain tissues, e.g., red blood cells and ion-secreting tissues such as the renal tubular epithelium, the rate of the hydration and dehydration reactions is enhanced by the enzyme carbonic anhydrase. The importance of this enzyme in renal H^+ secretion will be discussed in Chap. 12.

Dissociation of carbonic acid The degree to which H_2CO_3 dissociates into $H^+ + HCO_3^-$ [Eq. (11-10)] can be appreciated from the law of mass action for this reaction:

$$K_a = \frac{[H^+][HCO_3^-]}{[H_2CO_3]}$$

---

†An additional reaction can occur, as HCO_3^- can dissociate into hydrogen and carbonate ions:

$$HCO_3^- \rightleftharpoons H^+ + CO_3^{2-}$$

However, the pK_a of this reaction is 9.8 so that only trace amounts of carbonate are present in the physiologic pH range.

Since the $K_a = 2.72 \times 10^{-4}$ and the normal $[H^+] = 40 \times 10^{-9}$ mol/L,

$$2.72 \times 10^{-4} = \frac{40 \times 10^{-9}[HCO_3^-]}{[H_2CO_3]}$$

$$\frac{[HCO_3^-]}{[H_2CO_3]} = 6.8 \times 10^3$$

Thus, there are approximately 6800 molecules of HCO_3^- for each molecule of H_2CO_3.

Law of mass action for bicarbonate/carbon dioxide buffer system Since the concentration of H_2CO_3 is so low in relation to the $[CO_2]_{dis}$ (1:340) and the HCO_3^- concentration (1:6800), the reactions

$$[CO_2]_{dis} + H_2O \rightleftharpoons H_2CO_3 \rightleftharpoons H^+ + HCO_3^-$$

can be simplified to

$$[CO_2]_{dis} + H_2O \rightleftharpoons H^+ + HCO_3 \tag{11-12}$$

The law of mass action for this reaction is

$$K_a = \frac{[H^+][HCO_3^-]}{[CO_2]_{dis}[H_2O]}$$

or if $K_a' = K_a[H_2O]$ since the concentration of water is constant

$$K_a' = \frac{[H^+][HCO_3^-]}{[CO_2]_{dis}} \tag{11-13}$$

In plasma at 37° C, K_a' is equal to 800 nanomol/L (800×10^{-9} mol/L, $pK_a' = 6.10$). If we now solve this equation for $[H^+]$,

$$[H^+] = 800 \times \frac{[CO_2]_{dis}}{[HCO_3^-]} \tag{11-14}$$

Substituting $0.03 P_{CO_2}$ for $[CO_s]_{dis}$,

$$[H^+] = 24 \times \frac{P_{CO_2}}{[HCO_3^-]} \quad \text{(nanomol/L)} \tag{11-15}$$

Since the normal H^+ concentration is 40 nanomol/L and P_{CO_2} is 40 mmHg, the normal HCO_3^- concentration can be calculated from Eq. (11-15):

$$40 = 24 \times \frac{40}{[HCO_3^-]}$$

$$[HCO_3^-] = 24 \text{ mmol/L}$$

These relationships also can be expressed by the Henderson-Hasselbalch equation:

$$pH = 6.10 + \log \frac{[HCO_3^-]}{0.03 P_{CO_2}} \qquad (11\text{-}16)$$

where 6.10 is the pK_a'.

The HCO_3^- concentration usually is measured in the laboratory in one of two ways. The first way is indirect as the pH and P_{CO_2} are measured and the HCO_3^- concentration calculated from the Henderson-Hasselbalch equation. The second way involves adding a strong acid to the solution and measuring the amount of CO_2 generated (e.g., as the added H^+ combines with the plasma, HCO_3^-, H_2CO_3, and then CO_2 are formed). This method measures the total CO_2 content, which is equal to all the forms by which CO_2 is carried in the blood:

$$\text{Total } CO_2 \text{ content} = [HCO_3^-] + [CO_2]_{dis} + [H_2CO_3]$$

Since the H_2CO_3 concentration is very low, it can be omitted. If $0.03 P_{CO_2}$ is substituted for $[CO_2]_{dis}$, then

$$[HCO_3^-] = \text{total } CO_2 - 0.03 P_{CO_2} \qquad (11\text{-}17)$$

If the normal HCO_3^- concentration is 24 mmol/L and the normal P_{CO_2} equals 40 mmHg, the total CO_2 content will be $24 + (0.03 \times 40)$ or 25.2 mmol/L. When the total CO_2 content is measured, Eq. (11-16) must be modified in the following way:

$$pH = 6.10 + \log \frac{\text{total } CO_2 - 0.03 P_{CO_2}}{0.03 P_{CO_2}} \qquad (11\text{-}18)$$

For the sake of simplicity, only the HCO_3^- concentration will be used in this discussion.

Buffering by bicarbonate As described above, the most efficient buffering occurs within 1.0 pH unit of the pK. Although the pK_a' for the HCO_3^-/CO_2 system is 1.30 pH units less than the normal extracellular pH of 7.40, *this system buffers extremely well because the P_{CO_2} can be regulated by changes in alveolar ventilation* (see Chap. 12). An increase in ventilation augments CO_2 excretion and lowers the P_{CO_2}; a reduction in ventilation decreases CO_2 excretion, resulting in an increase in the P_{CO_2}. Thus, as H^+ ions are buffered by HCO_3^-, an increase in the P_{CO_2} [as Eq. (11-10) is driven to the left] can be prevented by an increase in alveolar ventilation, thereby enhancing the effectiveness of HCO_3^- buffering.

The importance of this ability to regulate ventilation can be illustrated by the following example. Let us assume that 1 liter of plasma, in which HCO_3^- is the only buffer, has the following composition:

$$[H^+] = 40 \text{ nanomol/L} \qquad (pH = 7.40)$$

$$[HCO_3^-] = 24 \text{ mmol/L}$$

$$P_{CO_2} = 40 \text{ mmHg}$$

$$[CO_2]_{dis} = 0.03 P_{CO_2} = 1.2 \text{ mmol/L}$$

How many millimoles of HCl would have to be added to this solution to raise the H^+ concentration to 80 nanomol/L (pH equals 7.10)? As each millimole of H^+ combines with HCO_3^-, there will be an equimolar *decrease* in the HCO_3^- concentration and *increase* in the $[CO_2]_{dis}$ [Eq. (11-12)]. Thus, the new HCO_3^- concentration will be $24 - x$ and the new $[CO_2]_{dis}$ will be $1.2 + x$. From Eq. (11-14),

$$[H^+] = 800 \times \frac{[CO_2]_{dis}}{[HCO_3^-]}$$

$$80 = 800 \times \frac{1.2 + x}{24 - x}$$

$$x = 1.1 \text{ mmol/L}$$

This represents substantial buffering in that the H^+ concentration has increased only from 40 nanomol/L to 80 nanomol/L even though 1.1 mmol/L (or 1.1 million nanomol/L) has been added to the solution. However, the H^+ concentration has risen to a potentially dangerous level after adding only 1.1 mmol/L of H^+ per liter. In this sense, the degree of buffering is relatively ineffective. The increase in the $[CO_2]_{dis}$ to 2.3 mmol/L in this setting is equivalent to a P_{CO_2} of 77 mmHg ($0.03 \times 77 = 2.3$).

If, however, ventilation could be increased so that the P_{CO_2} remained constant at 40 mmHg (and therefore the $[CO_2]_{dis}$ at 1.2 mmol/L), then

$$80 = 800 \times \frac{1.2}{24 - x}$$

$$x = 12 \text{ mmol/L}$$

Thus, the ability to maintain the P_{CO_2} at a constant level increases the efficiency of HCO_3^- buffering 11-fold. Furthermore, if ventilation could be sufficiently enhanced to reduce the P_{CO_2} below 40 mmHg, there would be an additional increase in the buffering capacity of HCO_3^-. For example, if the P_{CO_2} were lowered to 20 mmHg ($[CO_2]_{dis} = 0.03 \times 20 = 0.6$), then 18 mmol of H^+ could be added to each liter of plasma before the H^+ concentration increased to 80 nanomol/L:

$$80 = 800 \times \frac{0.6}{24 - x}$$

$$x = 18 \text{ mmol/L}$$

These changes in ventilation, which make the HCO_3^-/CO_2 buffering system so effective, occur in humans since the chemoreceptors controlling ventilation are sensitive to alterations in the H^+ concentration (see Chap. 12). If the H^+ concentration is increased by the addition of HCl to the extracellular fluid, there will be an increase in ventilation resulting in a reduction in the P_{CO_2}. From Eq. (11-15), this is an appropriate response since the decrease in P_{CO_2} will lower the H^+ concentration toward normal. Conversely, a decrease in the H^+ concentration (or increase in the pH) will reduce ventilation.

Isohydric Principle

From the law of mass action [Eq. (11-5)], the acid/base ratio of any weak acid is determined by its K_a and the H^+ concentration of the solution. Since the H^+ concentration affects each buffer, the following relationship is present:

$$[H^+] = K_{a1} \frac{0.03 P_{CO_2}}{[HCO_3^-]} = K_{a2} \frac{[H_2PO_4^-]}{[HPO_4^{2-}]} = K_{a3} \frac{[HA]}{[A^-]} \qquad (11-19)$$

This is called the *isohydric principle*. If the H^+ concentration is altered, the acid/base ratio of all the buffers in the solution is affected. This means that studying the behavior of any one buffer is adequate to predict the behavior of the other buffers in the solution. Clinically, the acid-base status of a patient is expressed in terms of the principal extracellular buffer, the HCO_3^-/CO_2 system:

$$[H^+] = 24 \frac{P_{CO_2}}{[HCO_3^-]}$$

Extracellular Buffers

The body buffers are located in the extracellular and intracellular fluids and in bone. As described above, the ability of a particular buffer to protect the pH is proportional to its concentration and its pK_a in relation to the body pH. In the extracellular fluid, HCO_3^- is the most important buffer due to both its relatively high concentration and the ability to vary the P_{CO_2} through changes in alveolar ventilation. For example, if H_2SO_4 is added to the extracellular fluid from the metabolism of the sulfur-containing amino acid methionine, the excess H^+ will be buffered primarily by HCO_3^-,

$$H_2SO_4 + 2NaHCO_3 \rightarrow Na_2SO_4 + 2H_2CO_3 \rightarrow 2CO_2 + 2H_2O \qquad (11-20)$$

The CO_2 produced by this reaction is excreted by the lungs.

Although HCO_3^- is an effective buffer for noncarbonic acids, it cannot buffer H_2CO_3 because the combination of H^+ with HCO_3^- results in the regeneration of H_2CO_3:

$$H_2CO_3 + HCO_3^- \rightarrow HCO_3^- + H_2CO_3 \qquad (11-21)$$

Consequently, H_2CO_3 is buffered primarily by the intracellular buffers (see below).

There are other, quantitatively less important buffers in the extracellular fluid, including inorganic phosphates (plasma phosphate concentration of 1 mmol/L versus 24 mmol/L of HCO_3^-) and the plasma proteins (Pr^-):

$$H^+ + Pr^- \rightleftharpoons HPr \qquad (11-22)$$

Intracellular and Bone Buffers

The primary intracellular buffers are proteins, organic and inorganic phosphates, and, in the erythrocyte, hemoglobin (Hb^-):

$$H^+ + Hb^- \rightleftharpoons HHb \qquad (11-23)$$

In addition, bone carbonate (CO_3^{2-}) represents a large store of buffer (approximately 35,000 meq [10]) which contributes to the buffering of acid and base loads [10–14]. For example, after an acid load, bone carbonate is released into extracellular fluid. This is accompanied by the uptake of extracellular phosphate by bone or by the release of Ca^{2+} or Na^+ from bone [13–16]:

$$-Ca-O-CO_2-Na + HCl \rightarrow -Ca-Cl + NaHCO_3 \qquad (11\text{-}24)$$

Although it is difficult to measure the exact contribution of bone carbonate, it has been estimated that as much as 40 percent of the buffering of an acute acid load takes place in bone [12,16]. The role of the bone buffers may be even greater in the presence of a chronic acid load such as that seen in patients with renal failure [14]. Parathyroid hormone plays at least a permissive role in bone buffering since this process can be largely abolished by parathyroidectomy [12,17].

Bone and intracellular buffers also participate in protecting the pH in the presence of base loads. For example, increased deposition of carbonate in bone has been demonstrated after the administration of $NaHCO_3$ [16]. In addition, some of the excess HCO_3^- combines with free H^+ to form H_2CO_3 and then CO_2 and H_2O which are excreted. This results in a fall in the H^+ concentration, which drives Eqs. (11-22) and (11-23) to the left. The ensuing release of H^+ from proteins and hemoglobin raises the H^+ concentration toward normal.

Chemical Buffering of Acids and Bases

Acidosis and alkalosis In a variety of clinical conditions (see Chaps. 17 to 21), the arterial H^+ concentration may be abnormal. An increase in the H^+ concentration (or decrease in the pH) is called *acidemia;* a decrease in the H^+ concentration (or increase in the pH) is called *alkalemia.* Processes which tend to raise or lower the H^+ concentration are called acidosis and alkalosis, respectively.

In general, acidosis induces acidemia and alkalosis induces alkalemia. However, the difference between these concepts becomes important in those patients who have mixed acid-base disturbances in which both acidotic and alkalotic processes may coexist. In this setting, the net pH may be acidemic even though a disorder which induces an alkalosis is also present.

From Eq. (11-15),

$$[H^+] = 24 \frac{P_{CO_2}}{[HCO_3^-]}$$

a primary elevation in the P_{CO_2} causes an acidemia whereas a decrease in the P_{CO_2} causes alkalemia. Since the P_{CO_2} is regulated by the rate of alveolar ventilation, these disturbances are referred to as *respiratory acidosis* and *respiratory alkalosis.*

The H^+ concentration also varies inversely with the plasma HCO_3^- concentration. Processes which primarily lower or raise the plasma HCO_3^- concentration are called *metabolic acidosis* and *metabolic alkalosis,* respectively.

Table 11-2 Summary of data from the infusion of HCl into five nephrectomized dogs†

Weight (kg)	18.9
HCl infused (mmol)	180
Final plasma pH	7.07
Change in total extracellular (mmol)	
Na^+	+ 65
K^+	+ 28
HCO_3^-	− 78
Cl^-	+170
Percent HCl neutralized by:	
Extracellular HCO_3^-	43
Intracellular buffers	57
Na^+ exchange	36
K^+ exchange	15
Cl^- entry	6

†Data adapted from R. C. Swan and R. F. Pitts, J. Clin. Invest., 34:205, 1955, by copyright permission of The American Society for Clinical Investigation.

Buffer responses to acid and base loads The importance of the body buffers in protecting the pH can be appreciated from the data in Table 11-2. In these experiments, *metabolic acidosis* (with acidemia) was induced in dogs by the infusion of HCl. The dogs were nephrectomized to eliminate the effect of changes in renal H^+ excretion. The total extracellular amounts of Na^+, K^+, HCO_3^-, and Cl^- were calculated from the product of the extracellular fluid volume (estimated from the volume of distribution of SO_4^{2-} which is limited to the extracellular fluid) and the plasma electrolyte concentrations, correcting for the Gibbs-Donnan equilibrium across the capillary membrane (see Fig. 1-10).

An average of 180 mmol of HCl was administered to each dog (average weight was 18.9 kg). Let us assume that the total body water was 60 percent of the body weight, or 11.3 liters. If 180 mmol of H^+ were distributed through 11.3 liters of distilled water, the H^+ concentration would be 16 mmol/L (pH of 1.80), a concentration incompatible with life. In comparison, the arterial pH of the dogs fell only from 7.40 to 7.07 (H^+ concentration of 86 nanomol/L). This was associated with a reduction in the plasma HCO_3^- concentration from 24 to 7 mmol/L (by the combination of extracellular HCO_3^- with the excess H^+) and a compensatory increase in alveolar ventilation lowering the P_{CO_2} from 40 to 25 mmHg. Even if the P_{CO_2} had remained at 40 mmHg, the pH would have been 6.87 [from pH = 6.10 + log 7/ (0.03 × 40)]. Thus, the body buffers were extremely effective in minimizing the reduction in the arterial pH.

The relative contributions of the intracellular and extracellular buffers to this process can be estimated from the changes in the quantities of Na^+, K^+, HCO_3^-, and Cl^- in the extracellular fluid. The administered H^+ ions either remain in the extra-

Cell   Extracellular fluid

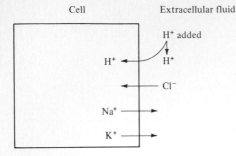

Figure 11-3 Effect of an HCl load on extracellular Cl^-, Na^+, and K^+. As H^+ enters the cells to be buffered, either Cl^- follows H^+ into the cells or intracellular Na^+ and K^+ leave the cells and move into the extracellular fluid. These ion shifts are reversed when H^+ ions are removed from the extracellular fluid.

cellular fluid or enter the cells (Fig. 11-3). The H^+ ions that stay in the extracellular fluid are buffered by HCO_3^- (and, to a much lesser degree, the plasma proteins), resulting in a decrease in the amount of extracellular HCO_3^-. If H^+ ions enter the cells, then, to maintain electroneutrality, either Cl^- will follow H^+ into the cells (primarily into the red blood cells where H^+ will be buffered by Hb^-) or Na^+ ions and K^+ ions will leave the cells (and bone) and enter the extracellular fluid. From Table 11-2, of the 180 mmol of H^+ infused, 78 mmol has been buffered by HCO_3^- and 103 mmol has entered the cells: 65 mmol in exchange for Na^+, 28 mmol in exchange for K^+, and 10 mmol followed by Cl^- (180 mmol of Cl^- was infused but only 170 mmol remained in the extracellular fluid).† These results are depicted schematically in Fig. 11-4.

Buffering by the extracellular and intracellular buffers follows a characteristic time course that is dependent upon the rapidity with which the administered H^+ ions move into the different fluid compartments. Buffering by plasma HCO_3^- occurs almost immediately, whereas approximately 15 min is required for H^+ to diffuse into the interstitial space to be buffered by interstitial HCO_3^-. H^+ entry into the cell occurs more slowly as buffering by cell buffers is not complete until 2 to 4 h have elapsed [18].

It should be noted that the transcellular exchange of H^+ for K^+ that follows a H^+ load may result in a potentially serious elevation in the plasma K^+ concentration, e.g., from the normal of 4 meq/L to as high as 7 to 8 meq/L in severe metabolic acidemia (see Chap. 13). A similar increase may occur in the plasma Na^+ concentration because Na^+ also leaves the cells. However, variations of several milliequivalents per liter are not physiologically important since the normal plasma Na^+ concentration is approximately 140 meq/L.

The response to *respiratory acidosis* (high P_{CO_2}) differs from that to metabolic acidosis in that there is virtually no extracellular buffering since HCO_3^- is not an effective buffer for H_2CO_3 [Eq. (11-21)]. As the P_{CO_2} increases, the elevation in H^+ concentration is initially minimized by a buffer-induced rise in the plasma HCO_3^- concentration [Eq. (11-15)]. This HCO_3^- is derived from two major sources: (1)

---

†An alternative explanation for the intracellular movement of Cl^- is that Cl^- enters the red blood cell in exchange for intracellular HCO_3^-. This HCO_3^- enters the extracellular fluid and buffers the excess H^+. The net effect is the same as the movement of HCl into the cell.

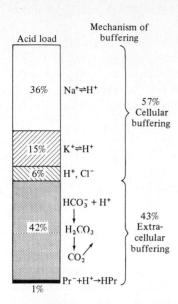

Metabolic acidosis

Figure 11-4 Mechanisms of buffering of strong acid infused intravenously in the dog. *(From R. F. Pitts, Physiology of the Kidney and Body Fluids, 3d ed. Copyright © 1974 by Year Book Medical Publishers, Inc., Chicago. Used by permission. Adapted from R. C. Swan, and R. F. Pitts, J. Clin. Invest., 34:205, 1955, by copyright permission of The American Society for Clinical Investigation.)*

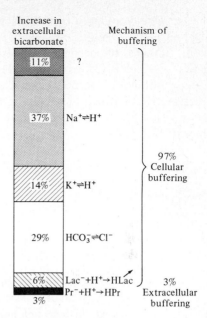

Respiratory acidosis

Figure 11-5 Mechanisms of buffering of CO_2 in respiratory acidosis in the dog. The source of approximately 11 percent of the increase in the extracellular HCO_3^- has not been identified. *(From R. F. Pitts, Physiology of the Kidney and Body Fluids, 3d ed. Copyright © 1974 by Year Book Medical Publishers, Inc., Chicago. Used by permission. Adapted from G. Giebisch, L. Berger, and R. F. Pitts, J. Clin. Invest., 34:231, 1955, by copyright permission of The American Society for Clinical Investigation.)*

Extracellular H_2CO_3 dissociates into HCO_3^- ions and H^+ ions, the latter moving into the cells (and bone) in exchange for intracellular Na^+ and K^+, and (2) HCO_3^- is released from erythrocytes in exchange for extracellular Cl^- (Fig. 11-5). The latter process occurs in the following manner. CO_2 diffuses into the erythrocyte where it combines with H_2O to form H_2CO_3. This reaction is catalyzed by the enzyme carbonic anhydrase. H_2CO_3 is then buffered by Hb^-:

$$H_2CO_3 + Hb^- \rightarrow HHb + HCO_3^-$$

It is this HCO_3^- that moves into extracellular fluid.

Of lesser importance is the uptake of H^+ by the plasma proteins and by extracellular lactate:

$$Na\ lactate + H_2CO_3 \rightarrow lactic\ acid + NaHCO_3^-$$

The lactic acid produced by this reaction is metabolized within the cells either into CO_2 and H_2O or via gluconeogenesis into glucose.

In humans, these buffers increase the plasma HCO_3^- concentration approximately 1 mmol/L for each 10 mmHg elevation in the P_{CO_2} (see Chap. 20). The degree to which this response protects the H^+ concentration can be appreciated if we calculate the effects of increasing the P_{CO_2} from 40 to 80 mmHg. If there is no buffering and the plasma HCO_3^- concentration remains constant, then the new H^+ concentration will be

$$[H^+] = 24 \times \frac{80}{24} = 80 \text{ nanomol/L} \qquad (pH = 7.10)$$

However, a 40-mmHg elevation in the P_{CO_2} normally will induce roughly a 4 mmol/L increase in the plasma HCO_3^- concentration. In this situation,

$$[H^+] = 24 \times \frac{80}{28} = 69 \text{ nanomol/L} \qquad (pH = 7.17)$$

As illustrated by this example, the buffer-induced elevation in the plasma HCO_3^- concentration is not particularly effective in protecting the H^+ concentration in respiratory acidosis. As will be seen in Chap. 12, the most effective defense against respiratory acidosis is a further increase in the plasma HCO_3^- concentration produced by enhanced renal H^+ excretion.

The intracellular and extracellular buffers also protect the pH in metabolic and respiratory alkalosis, as the buffer reactions move in the opposite direction from that observed in the acidemic conditions.† Thus, H^+ ions are released, not taken up, by the buffers, e.g.,

$$HPr \rightarrow H^+ + Pr^-$$

$$H_2PO_4^- \rightarrow H^+ + HPO_4^{2-}$$

These H^+ ions then react with HCO_3^-, resulting in an appropriate reduction in the plasma HCO_3^- concentration, which tends to lower the elevated pH toward normal. To the degree that these H^+ ions are derived from cell buffers, their movement into the extracellular fluid occurs in exchange for extracellular Na^+ and K^+, which enter the cells. Thus, the plasma concentrations of Na^+ and K^+, which tend to rise with acidemia, may fall with alkalemia.

INTRACELLULAR pH

The intracellular pH may be measured by comparing the distribution of a weak acid or base between the extracellular and intracellular fluids (see Appendix) [19] or by the technique of nuclear magnetic resonance [20]. In a wide variety of tissues, the intracellular pH has been noted to be lower than that in the extracellular fluid

---

†Although the buffers involved are similar, the percentage contributions of the individual intracellular and extracellular buffers are somewhat different in alkalemia than in acidemia [6].

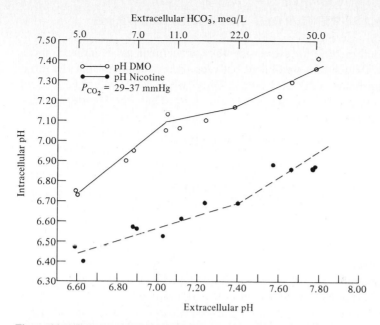

Figure 11-6 Relationship between skeletal muscle cell pH and the extracellular pH in metabolic acidosis and alkalosis. A similar relationship is present in respiratory acidosis and alkalosis. The cell pH can be seen to be heterogeneous as evidenced by the difference between measuring the pH with a weak acid (DMO) or a weak base (nicotine). *(From S. Adler, J. Clin. Invest., 51:256, 1972, by copyright permission of The American Society for Clinical Investigation.)*

although it varies from organ to organ. For example, at a normal extracellular pH of 7.40, the mean pH of skeletal muscle is 7.02 [20] and that of the renal tubular cell is 7.32 [21]. The factors responsible for these differences are poorly understood. However, it is clear that the extracellular pH is an important determinant of the intracellular pH. As illustrated in Fig. 11-6, the intracellular pH varies in a parallel direction with changes in the extracellular pH. This relationship is extremely important in the clinical setting. The principal physiologic effect of changes in pH is on protein function. Since the cells are the functioning units in the body, it is the intracellular pH that is of primary importance, yet it is only the extracellular (plasma) pH that can be easily measured in patients. Fortunately, this still allows an accurate assessment of acid-base status because of the direct relationship between these two parameters.

In contrast to the extracellular pH, *the pH within the cell is not uniform.* When the cell pH is measured with a weak acid, for example 5,5-dimethyl-2,4-oxazolidinedione (DMO), which is preferentially bound to the alkaline regions in the cell, the pH of the skeletal muscle cell is 7.17. However, when a weak base, e.g., nicotine, is used, which is preferentially bound to the more acidic areas in the cell, the muscle cell pH is 6.69 (Fig. 11-6) [22]. This heterogeneity is not surprising in view of the number of different functioning organelles within the cell, e.g., the mitochondria, nucleus, and endoplasmic reticulum.

APPENDIX: MEASUREMENT OF INTRACELLULAR pH

The most widely used method for estimating the intracellular pH is indirect and is based upon the distribution of a weak acid between the extracellular and the intracellular fluids. The primary weak acid used is DMO which has a pK_a of 6.13 at the concentration and temperature of the body fluids [19]. Thus, the Henderson-Hasselbalch equation for the reaction

$$HDMO \rightleftharpoons H^+ + DMO^-$$

can be written

$$pH = 6.13 + \log \frac{[DMO^-]}{[HDMO]} \tag{11-25}$$

With DMO, two assumptions are made: (1) that the pK_a in the cell is the same as that in the extracellular fluid and (2) that the undissociated acid (HDMO), being lipid-soluble, equilibrates across the cell membrane whereas the polar compound DMO^- crosses the membrane very slowly, if at all (Fig. 11-7). Using these assumptions, the intracellular pH can be estimated in the following way:

1. The extracellular pH is measured and, from Eq. (11-25), the $[DMO^-]/[HDMO]$ ratio is calculated. At the normal pH of 7.40, this ratio is approximately 20:1.
2. The extracellular DMO concentration, that is, $[DMO^-] + [HDMO]$, is measured and, since the $[DMO^-]/[HDMO]$ is known, the HDMO concentration in the extracellular fluid can be calculated. It is assumed that the HDMO concentration in the extracellular fluid is the same as that in the cell.
3. The extracellular and intracellular volumes are measured by using markers limited to these compartments (see Chap. 1).
4. The total quantity of DMO in the extracellular fluid is calculated from the product of the extracellular volume and the extracellular DMO concentration.
5. The total DMO in the cell is calculated from the known amount of DMO administered minus the quantity in the extracellular fluid.
6. The cell DMO concentration is then calculated from the total DMO in the cells divided by the intracellular volume.

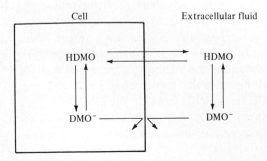

Figure 11-7 Distribution of HDMO and DMO^- between the cell and the extracellular fluid. Since HDMO is lipid-soluble, it is able to equilibrate across the cell membrane, reaching equal concentrations in both compartments. Once in the cell, HDMO dissociates into $H^+ + DMO^-$ (the latter is polar and cannot cross the cell membrane). The extent of this reaction is dependent upon the cell pH. See text for details.

7. Since the DMO$^-$ concentration in the cell equals the total DMO concentration in the cell minus the HDMO concentration in the cell (both of which are known), the intracellular pH can be calculated from

$$pH = 6.13 + \log \frac{[DMO]_{cell} - [HDMO]_{cell}}{[HDMO]_{cell}}$$

PROBLEMS

11-1 How do buffers minimize change in the H$^+$ concentration? What factors determine how effective a buffer will be?

11-2 The sequential changes in the plasma HCO$_3^-$ concentration and arterial pH produced by the rapid intravenous administration of 90 meq of HCO$_3^-$ to a 70-kg man are depicted in the accompanying table. What accounts for the progressive fall in the plasma HCO$_3$ concentration between 10 and 180 min?

Time, min	[HCO$_3^-$], meq/L	Arterial pH
0	24	7.40
10	32	7.51
20	29	7.48
180	27	7.45

11-3 If a patient has a P_{CO_2} that is fixed at 40 mmHg, what factors will determine how much the extracellular pH will fall after an acid load?

REFERENCES

1. Relman, A. S.: Metabolic consequences of acid-base disorders, Kidney Int., 1:347, 1972.
2. Relman, A. S.: What are "acids" and "bases"?, Am. J. Med., 17:435, 1954.
3. Lennon, E. J., J. Lemann, Jr., and J. R. Litzow: The effects of diet and stool composition on the net external acid balance of normal subjects, J. Clin. Invest., 45:1601, 1966.
4. Trenchard, D., M. I. M. Noble, and A. Guz: Serum carbonic acid pK_1' abnormalities in patients with acid-base disturbances, Clin. Sci., 32:189, 1967.
5. Hood, I., and E. J. M. Campbell: Is pK OK?, N. Engl. J. Med., 306:864, 1982.
6. Pitts, R. F.: *Physiology of the Kidney and Body Fluids,* Yearbook, Chicago, 1974, chap. 11.
7. Huckabee, W. E.: Henderson vs. Hasselbalch, Clin. Res., 9:116, 1961.
8. Hills, A. G.: pH and the Henderson-Hasselbalch equation, Am. J. Med., 55:131, 1973.
9. Malnic, G., and G. Giebisch: Mechanism of renal hydrogen ion secretion, Kidney Int., 1:280, 1972.
10. Goodman, A. D., J. Lemann, Jr., E. J. Lennon, and A. S. Relman: Production, excretion and net balance of fixed acid in patients with renal acidosis, J. Clin. Invest., 44:495, 1965.
11. Lemann, J., Jr., J. R. Litzow, and E. J. Lennon: The effects of chronic acid-base loads in normal man: further evidence for the participation of bone mineral in the defense against chronic metabolic acidosis, J. Clin. Invest., 45:1608, 1966.
12. Fraley, D. S., and S. Adler: An extrarenal role for parathyroid hormone in the disposal of acute acid loads in rats and dogs, J. Clin. Invest., 63:985, 1979.

13. Lemann, J., Jr., and E. J. Lennon: Role of diet, gastrointestinal tract and bone in acid-base homeostasis, Kidney Int., 1:275, 1972.
14. Litzow, J. R., J. Lemann, Jr., and E. J. Lennon: The effect of treatment of acidosis on calcium balance in patients with chronic azotemic renal disease, J. Clin. Invest., 46:280, 1967.
15. Kaye, M., A. J. Frueh, and M. Silverman: A study of vertebral bone powder from patients with chronic renal failure, J. Clin. Invest., 49:442, 1970.
16. Burnell, J. M.: Changes in bone sodium and carbonate in metabolic acidosis and alkalosis in the dog, J. Clin. Invest., 50:327, 1971.
17. Arruda, J. A. L., V. Alla, H. Rubinstein, M. Cruz-Soto, S. Sabatini, D. C. Batlle, and N. A. Kurtzman: Parathyroid hormone and extrarenal acid buffering, Am. J. Physiol., 239:F533, 1980.
18. Schwartz, W. B., K. J. Orming, and R. Porter: The internal distribution of hydrogen ions with varying degrees of metabolic acidosis, J. Clin. Invest., 36:373, 1957.
19. Waddell, W. J., and T. C. Butler: Calculation of intracellular pH from the distribution of 5,5-dimethyl-2,4-oxazolidinedione (DMO): application to skeletal muscle of the dog, J. Clin. Invest., 38:720, 1959.
20. Ross, B. D., G. K. Radda, D. G. Gadian, G. Rocker, M. Esiri, and J. Falconer-Smith: Examination of a case of suspected McArdle's syndrome by ^{31}P nuclear magnetic resonance, N. Engl. J. Med., 304:1338, 1980.
21. Struyvenberg, A., R. B. Morrison, and A. S. Relman: Acid-base behavior of separated canine renal tubule cells, Am. J. Physiol., 214:1155, 1968.
22. Adler, S.: The simultaneous determination of muscle cell pH using a weak acid and weak base, J. Clin. Invest., 51:256, 1972.

REGULATION OF ACID-BASE BALANCE

INTRODUCTION

Acid-base homeostasis can be easily understood if viewed in terms of the HCO_3^-/CO_2 buffering system:

$$H^+ + HCO_3^- \rightleftharpoons H_2CO_3 \rightleftharpoons H_2O + CO_2 \qquad (12\text{-}1)$$

At equilibrium, the relationship between the reactants can be expressed by the law of mass action (see Chap. 11)

$$[H^+] = 24 \times \frac{P_{CO_2}}{[HCO_3^-]} \tag{12-2}$$

or by the Henderson-Hasselbalch equation

$$pH = 6.10 + \log \frac{[HCO_3^-]}{0.03 P_{CO_2}} \tag{12-3}$$

This system plays a central role in the maintenance of acid-base balance because the HCO_3^- concentration and P_{CO_2} can be regulated independently, the former by changes in renal H^+ excretion and the latter by changes in the rate of alveolar ventilation.

These processes are extremely important because acids and to a lesser degree bases are continually being added to the body from endogenous metabolic processes. For example, the metabolism of carbohydrates and fats (primarily derived from the diet) results in the production of approximately 15,000 mmol of CO_2 per day. Since CO_2 combines with H_2O to form H_2CO_3, severe acidemia would ensue if this CO_2 were not excreted by the lungs. In addition, the metabolism of proteins and other substances results in the generation of noncarbonic acids and bases [1]. These acids are derived mostly from the oxidation of sulfur-containing amino acids and the hydrolysis of phosphoester acids:

$$Methionine + O_2 \rightarrow urea + CO_2 + SO_4^{2-} + 2H^+$$

$$R\text{-}H_2PO_4 + H_2O \rightarrow ROH + 0.8HPO_4^{2-}/0.2H_2PO_4^- + 1.8\ H^+$$

The major source of alkali is the oxidation of organic anions such as citrate:

$$Citrate^- + 4.5O_2 \rightarrow 5CO_2 + 3H_2O + HCO_3^-$$

On a normal diet, the net effect is the production of 50 to 100 meq of H^+ per day in adults.

The response of the body to these acid and base loads occurs in three stages: (1) chemical buffering by the extracellular and intracellular buffers (see Chap. 11); (2) changes in alveolar ventilation to control the P_{CO_2}; and (3) alterations in renal H^+ excretion to regulate the plasma HCO_3^- concentration. For example, the H_2SO_4 produced from the oxidation of sulfur-containing amino acids is initially buffered in the extracellular fluid by HCO_3^-:

$$H_2SO_4 + 2NaHCO_3 \rightarrow Na_2SO_4 + 2H_2CO_3 \rightarrow 2H_2O + 2CO_2 \tag{12-4}$$

The CO_2 produced by this reaction is excreted by the lungs. Although HCO_3^- minimizes the increase in the H^+ concentration, the excess H^+ ions must be excreted by the kidney to prevent progressive depletion of HCO_3^- and the other body buffers and severe metabolic acidemia.

Under normal conditions, the steady state is preserved as the rates of excretion of H^+ and CO_2 equal their rates of production. As a result, the H^+ concentration and pH are maintained within narrow limits. The normal values for the parameters are†

	pH	[H$^+$], nanomol/L	P_{CO_2}, mmHg	[HCO$_3^-$], mmol/L
Arterial	7.37–7.43	37–43	36–44	22–26
Venous	7.32–7.38	42–48	42–50	23–27

The decrease in the pH (and increase in the H$^+$ concentration) in venous blood is due to the uptake of metabolically produced CO$_2$ in the capillary circulation. However, since it is the arterial blood that perfuses the tissues and affects cell function, it is the regulation of the arterial pH that is of primary importance. The remainder of this chapter will describe the manner in which H$^+$ ions are excreted by the kidney and the factors controlling both this process and the elimination of CO$_2$ by ventilation.

RENAL HYDROGEN EXCRETION

The kidney contributes to acid-base balance by regulating H$^+$ excretion so that the plasma HCO$_3$ concentration remains within appropriate limits. This involves two basic steps: (1) reabsorption of the filtered HCO$_3^-$; and (2) excretion of the 50 to 100 meq of H$^+$ produced per day.

Virtually all of the filtered HCO$_3^-$ must be reabsorbed before the daily H$^+$ load can be excreted *since HCO$_3^-$ loss in the urine is equivalent to the addition of H$^+$ to the body* (both H$^+$ and HCO$_3^-$ being derived from the dissociation of H$_2$CO$_3$). The quantitative importance of this process should not be underestimated. A normal subject with a GFR of 180 L/day and a plasma HCO$_3^-$ concentration of 24 meq/L filters and then must reabsorb approximately 4300 meq of HCO$_3^-$ each day.

The second step in renal acid-base regulation, excretion of the 50- to 100-meq daily H$^+$ load, is accomplished by the combination of H$^+$ ions either with urinary buffers (referred to as titratable acidity) or with ammonia to form ammonium— NH$_3$ + H$^+$ → NH$_4^+$. Excretion of free H$^+$ ions essentially does not occur since the minimum urine pH that can be achieved in humans is 4.5. This represents a maximum free H$^+$ concentration of less than 0.04 meq/L.

The reabsorption of HCO$_3^-$ and the formation of titratable acidity and NH$_4^+$ all occur by H$^+$ secretion from the tubular cell into the lumen (Figs. 12-1 to 12-3). This process involves the following sequence:

†Since the valence of H$^+$ and HCO$_3^-$ is 1, the normal concentrations of these ions also can be expressed in terms of nanoequivalents per liter (nanoeq/L) or milliequivalents per liter (meq/L), respectively (see Chap. 1). Thus the normal arterial H$^+$ concentration is 37 to 43 nanoeq/L, and the normal HCO$_3^-$ concentration is 22 to 26 meq/L.

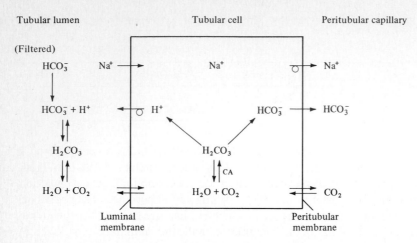

Figure 12-1 Major cellular and luminal events in HCO_3^- reabsorption. Intracellular H_2CO_3 is formed from $CO_2 + H_2O$ in a reaction catalyzed by carbonic anhydrase (CA) and then rapidly dissociates into H^+ and HCO_3^-. The HCO_3^- is returned to the peritubular capillary across the peritubular membrane whereas the H^+ is secreted into the lumen across the luminal membrane by a process that is at least in part linked to Na^+ reabsorption. The secreted H^+ combines with filtered HCO_3^- to form H_2CO_3 and the $CO_2 + H_2O$, which are reabsorbed. The net effect is HCO_3^- reabsorption even though the HCO_3^- returned to the systemic circulation is not the same as the filtered HCO_3^-.

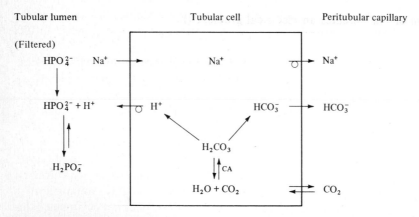

Figure 12-2 Formation of titratable acidity which is primarily due to buffering by urinary HPO_4^{2-}. Note that a new HCO_3^- is generated in the peritubular capillary by H^+ secretion.

1. The secreted H^+ ions are generated within the tubular cell from CO_2 and H_2O in a reaction catalyzed by the enzyme carbonic anhydrase. The H_2CO_3 formed from the hydration of CO_2 instantaneously dissociates into H^+ and HCO_3^-.†

†Carbonic anhydrase also may catalyze the reaction $OH^- + CO_2 \rightleftharpoons HCO_3^-$. The OH^- is derived from the dissociation of water—$HOH \rightleftharpoons OH^- + H^+$—and therefore is associated with the production of H^+. The net effect is the same in that the hydration of CO_2 results in the formation of H^+ and HCO_3^-.

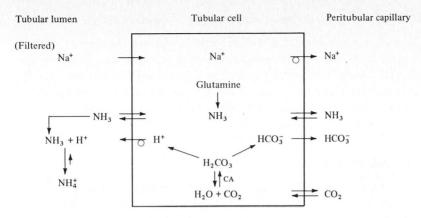

Tubular lumen   Tubular cell   Peritubular capillary

Figure 12-3 Formation of NH_4^+ in the tubule. Note that each H^+ secreted results in the generation of a new HCO_3^- in the peritubular capillary.

2. This H^+ is then secreted into the lumen in a process that is in part coupled to the passive movement of luminal Na^+ into the cell down its concentration gradient (see page 92). In the proximal tubule, this coupling is mediated by a carrier protein in the brush border of the luminal membrane which recognizes both Na^+ and H^+, transporting luminal Na^+ into the cell and cellular H^+ into the lumen [2,3]. The linkage is somewhat different in the distal nephron where the reabsorption of Na^+ creates an electrical gradient (lumen negative; see Fig. 13-9) which then promotes the secretion of II^+ into the lumen [4,5]. In addition to these passive mechanisms, active H^+ secretion independent of Na^+ transport probably also occurs, particularly in the distal nephron [5a].
3. The HCO_3^- generated from the dissociation of H_2CO_3 moves passively across the peritubular membrane into the peritubular capillary and is then returned to the systemic circulation. This process occurs by facilitated diffusion as a carrier is present in the peritubular membrane that specifically promotes HCO_3^- transport [2,6].

The net effect is that the secretion of each H^+ ion is associated with the generation of one HCO_3^- ion in the plasma. If the secreted H^+ combines with filtered HCO_3^-, the result is HCO_3^- reabsorption (Fig. 12-1). This maintains the plasma HCO_3^- concentration by preventing HCO_3^- excretion. However, if the secreted H^+ combines with the urinary buffers, for example, HPO_4^{2-} or NH_3, a new HCO_3^- is added to the peritubular capillary (Figs. 12-2 and 12-3). This results in an increase in the plasma HCO_3^- concentration to replace the HCO_3^- lost in buffering the daily H^+ load [Eq. (12-4)].

Net Hydrogen Excretion

Since the urinary concentration of free H^+ is extremely low, the net quantity of H^+ excreted in the urine is equal to the amount of H^+ excreted as titratable acidity and NH_4^+ minus any H^+ added to the body because of urinary HCO_3^- loss:

$$\text{Net } H^+ \text{ excretion} = \text{titratable acidity} + NH_4^+ - \text{urinary } HCO_3^- \qquad (12\text{-}5)$$

In the steady state, the net amount of H^+ excreted is equal to the normal H^+ load of 50 to 100 meq/day. However, this value can exceed 300 meq/day (primarily due to enhanced NH_4^+ excretion) if acid production is increased (see below). Net H^+ excretion also can have a negative value if a large amount of HCO_3^- is lost in the urine. This may appropriately occur after the ingestion of citrate-containing fruit juices since the citrate is metabolized to HCO_3^-.

Bicarbonate Reabsorption

Filtered HCO_3^- is reabsorbed by H^+ secretion, most of which (90 percent)† occurs in the proximal tubule. In the tubular fluid, the secreted H^+ combines with filtered HCO_3^- to form H_2CO_3 (Fig. 12-1). The H_2CO_3 dissociates into $CO_2 + H_2O$, which are then reabsorbed. Thus, HCO_3^- reabsorption is indirect. The HCO_3^- added to the peritubular capillary is derived from the intracellular dissociation of H_2CO_3 whereas the filtered HCO_3^- is removed from the tubular fluid as $CO_2 + H_2O$.

It should be noted that two separate reactions occur in the tubular lumen: (1) the combination of H^+ with HCO_3^- to form H_2CO_3, and (2) the dehydration of H_2CO_3:

$$\overset{1}{\text{}} \qquad \overset{2}{\text{}}$$
$$H^+ + HCO_3^- \rightleftharpoons H_2CO_3 \rightleftharpoons CO_2 + H_2O \qquad (12\text{-}6)$$

The dehydration of H_2CO_3 into $CO_2 + H_2O$ normally proceeds relatively slowly. However, the brush border of the proximal tubular cells contains carbonic anhydrase which accelerates this reaction [7–9]. Consequently, there is no accumulation of H_2CO_3 in the proximal tubular fluid. From the law of mass action, the maintenance of a low H_2CO_3 concentration drives the reaction

$$H^+ + HCO_3^- \rightleftharpoons H_2CO_3$$

to the right, thereby keeping the H^+ concentration at a relatively low level (Fig. 12-4a). By minimizing this increase in the tubular fluid H^+ concentration, carbonic anhydrase minimizes the gradient against which H^+ is secreted, thus facilitating the secretory process (and possibly minimizing passive back-diffusion of H^+ out of the lumen). The result is the proximal reabsorption of 90 percent of the filtered HCO_3^- with a reduction in the tubular fluid pH of only 0.6 units (Fig. 12-5).

The importance of this system can be appreciated from the response to the administration of a carbonic anhydrase inhibitor such as benzolamide which enters the cells to a limited degree and therefore inhibits the luminal but not the intracellular enzyme [9]. In this setting, the dehydration of H_2CO_3 in the lumen is slowed, resulting in increases in the H_2CO_3 and H^+ concentrations‡ (Fig. 12-4b). Since H^+

---

†In addition to its role in acid-base balance, the preferential reabsorption of HCO_3^- in the proximal tubule creates gradients that permit about one-third of net proximal reabsorption to occur by passive mechanisms (see p. 93).

‡It is also possible at this time to demonstrate a *disequilibrium* pH in the proximal tubule [9]. If the proximal tubular fluid P_{CO_2} and HCO_3^- concentration are measured after benzolamide, the *calculated* pH from the Henderson-Hasselbalch equation [Eq. (12-3)] is 7.18. However, the *measured* pH

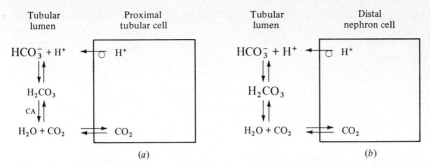

Figure 12-4 Effect of luminal carbonic anhydrase (CA) on HCO_3^- reabsorption. (*a*) When CA is present, as in the brush border of the proximal tubule, the luminal H_2CO_3 that is formed from the combination of secreted H^+ and filtered HCO_3^- rapidly dissociates into CO_2 and H_2O. As a result, the secreted H^+ is removed from the tubule as CO_2, preventing a large increase in the luminal H^+ concentration which could retard further H^+ secretion. (*b*) If there is no luminal carbonic anhydrase activity, as in the distal nephron or in the proximal tubule after the administration of a carbonic anhydrase inhibitor, the dehydration of luminal H_2CO_3 proceeds slowly. Consequently, the luminal H^+ and H_2CO_3 concentrations increase after a relatively small amount of H^+ secretion. The exaggerated elevation in the H^+ concentration tends to diminish total H^+ secretion, thereby decreasing the capacity for HCO_3^- reabsorption.

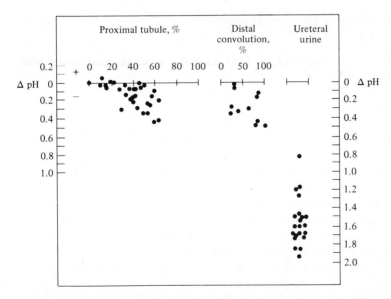

Figure 12-5 Change in pH (ΔpH) of the tubular fluid along the nephron of the rat. (*From C. W. Gottschalk, W. Lassiter, E. William, and M. Mylle, Am. J. Physiol., 198:581, 1960.*)

is only 6.68. This error in the calculated pH results from the fact that the pK_a' of 6.10 is applicable only when the H_2CO_3 concentration is very low in comparison to the dissolved CO_2 and HCO_3^- concentrations (see p. 212). The 0.50 unit pH difference (or disequilibrium pH) after benzolamide is presumably due to the accumulation of excess acid as H_2CO_3. Normally, the calculated and measured pH values are the same, i.e., there is no disequilibrium, since luminal carbonic anhydrase promotes the rapid dissociation of H_2CO_3.

secretion must now proceed against a steeper cell-to-tubular fluid gradient, further H$^+$ secretion is impaired. This results in a marked decrease in proximal HCO$_3^-$ reabsorption and HCO$_3^-$ loss in the urine [9,10]. This ability to induce a HCO$_3^-$ diuresis makes a carbonic anhydrase inhibitor useful in the treatment of patients with metabolic alkalosis (see Chap. 18).

The 10 percent of the filtered HCO$_3^-$ that leaves the proximal tubule is reabsorbed in the more distal segments, again by H$^+$ secretion (Fig. 12-4b) [4]. In contrast to the proximal tubule, carbonic anhydrase does not seem to be present in the lumen of the distal nephron [7,11]. Thus, H$_2$CO$_3$ accumulates in the tubular fluid, resulting in a greater increase in the tubular fluid H$^+$ concentration than occurs with H$^+$ secretion and HCO$_3^-$ reabsorption in the proximal tubule (Fig. 12-5). However, the distal nephron is able to generate and maintain a very steep plasma-to-tubular H$^+$ gradient; e.g., a urine pH of 4.5 represents a H$^+$ concentration almost 1000 times that of the plasma. As a result, the absence of luminal carbonic anhydrase does not prevent the distal reabsorption of essentially all of the relatively small amount of HCO$_3^-$ delivered from the proximal tubule.

In summary, the filtered HCO$_3^-$ is reabsorbed in a process involving H$^+$ secretion from the tubular cell into the lumen. The presence of carbonic anhydrase in the proximal tubular lumen allows the bulk of the filtered HCO$_3^-$ to be reabsorbed in this segment with only a minimal change in the tubular fluid pH.

Titratable Acidity

Several weak acids are filtered at the glomerulus and may act as buffers in the urine. Their ability to do so is proportional to the quantity of the buffer present and to its pK_a. The latter is important since maximum buffering occurs at ± 1.0 pH unit from the pK_a (see Fig. 11-2). Because of its favorable pK_a (6.8) and relatively high concentration, the bulk of the urinary buffering is performed by HPO$_4^{2-}$ (H$^+$ + HPO$_4^{2-}$ → H$_2$PO$_4^-$) with lesser contributions from other weak acids such as uric acid (pK_a of 5.75) and creatinine (pK_a of 4.97). This process is referred to as *titratable acidity* (Fig. 12-2) since it is measured by the amount of NaOH that must be added to a 24-h urine collection to titrate the urine pH back to a pH of 7.40, the same pH as that in the plasma. Under normal conditions, 10 to 40 meq/day of H$^+$ is buffered by these weak acids.

The ability of phosphate to buffer H$^+$ can be illustrated by the following example (Table 12-1). From the Henderson-Hasselbalch equation for the

Table 12-1 Effect of tubular fluid pH on buffering by HPO$_4^{2-}$ if 50 mmol of phosphate is excreted

Segment	pH	Quantity (in mmol) of HPO$_4^{2-}$	Quantity (in mmol) of H$_2$PO$_4^-$	Amount buffered by HPO$_4^{2-}$ (in mmol)
Filtrate	7.40	40	10	0
Proximal tubule	6.80	25	25	15
Final urine	4.80	0.5	49.5	39.5

$HPO_4^{2-}/H_2PO_4^-$ system

$$pH = 6.8 + \log \frac{[HPO_4^{2-}]}{[H_2PO_4^-]} \qquad (12\text{-}7)$$

the ratio of HPO_4^{2-} to $H_2PO_4^-$ is 4:1 at the normal arterial pH of 7.40. If 50 mmol of phosphate is excreted in the urine (the remainder of the filtered phosphate being reabsorbed), then 40 mmol exists as HPO_4^{2-} and 10 mmol as $H_2PO_4^-$ in the glomerular filtrate. (The quantities of H_3PO_4 and PO_4^{3-} in the physiologic pH range are negligible.) If the tubular fluid pH is lowered to 6.8 in the proximal tubule by H^+ secretion, then, from Eq. (12-7), the ratio of HPO_4^{2-} to $H_2PO_4^-$ will fall to 1:1. As a result, there will now be 25 mmol of both HPO_4^{2-} and $H_2PO_4^-$ in the tubule. This represents the buffering of 15 mmol (or 15 million nanomol) of H^+ by HPO_4^{2-} with an increase in the free H^+ concentration from 40 nanomol/L (pH of 7.40) to only 160 nanomol/L (pH of 6.80). Thus, over 99.99 percent of the secreted H^+ has been buffered. If the tubular fluid pH is lowered further to 4.8 (H^+ concentration of 0.016 mmol/L) in the collecting tubules, essentially all the HPO_4^{2-} will have been converted to $H_2PO_4^-$, as a total of 39.5 mmol of H^+ will have been buffered by HPO_4^{2-} (Table 12-1).

In summary, *the amount of H^+ buffered by HPO_4^{2-} increases as the tubular fluid pH is reduced.* When the urine pH reaches its minimum of 4.5 to 5.0, further buffering by HPO_4^{2-} cannot occur unless there is an increase in phosphate excretion. Since phosphate excretion does not increase markedly in the presence of an acid load, the ability to enhance net H^+ excretion by increased formation of titratable acidity is limited. An exception occurs in diabetic ketoacidosis where large amounts of β-hydroxybutyrate (pK_a equals 4.8) are excreted in the urine (see Chap. 25). These ketoacids can act as urinary buffers, augmenting titratable acid excretion by as much as 150 meq/day [12]. This effect is due both to the high concentration of ketoacids present and to the proximity of the pK_a of β-hydroxybutyrate to the acid urine pH.

Ammonium

Ammonia (NH_3) is a base which can combine with H^+ to form ammonium (NH_4^+):

$$NH_3 + H^+ \rightarrow NH_4^+ \qquad (12\text{-}8)$$

The pK_a of the NH_3/NH_4^+ system is 9.3. Therefore, at the normal arterial pH of 7.40, the ratio of NH_3 to NH_4^+ is almost 1:100. In the urine, the ratio is even lower. If the urine pH is 5.3, only 1 in 10^4 molecules exists as NH_3. In either fluid, the very low concentration of NH_3 makes NH_3 an ineffective buffer. Instead, NH_3 acts by *passive, nonionic diffusion* to allow H^+ to be bound with a minimal change in the free H^+ concentration, i.e., in the urine pH. This process is *dependent upon the solubility characteristics of NH_3 and NH_4^+.* NH_3 is lipid-soluble and is able to cross cell membranes passively in a manner similar to CO_2. In contrast, NH_4^+, a polar compound, is water-soluble and crosses membranes poorly. Once NH_4^+ is formed in the lumen, it cannot diffuse out. Thus, as NH_3 combines with secreted H^+, the H^+ becomes trapped in the lumen as NH_4^+ (Fig. 12-3).

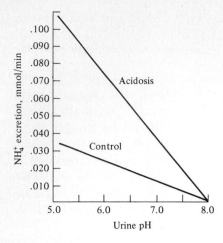

Figure 12-6 Effect of urinary and arterial pH on NH_3 excretion. Lowering the arterial pH (that is, acidemia) increases cellular NH_3 production from glutamine. Lowering the urine pH enhances the trapping of NH_3 as NH_4^+. *(Redrawn from R. F. Pitts, Fed. Proc., 7:418, 1948.)*

The initial step in NH_4^+ excretion is the generation of NH_3 within the tubular cells from the metabolism of amino acids, particularly but not solely glutamine [13,14].

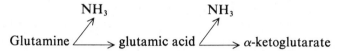

As the NH_3 concentration within the cell increases, NH_3 diffuses out of the cell down concentration gradients into the tubular lumen and the peritubular capillary (Fig. 12-3). The NH_3 preferentially moves into that compartment with the lower NH_3 concentration; this is dependent upon two factors: the relative rates of flow and the urine pH. As flow increases, the NH_3 that has diffused out of the cell is washed away, thereby maintaining a favorable gradient for further diffusion. Since capillary blood flow is much greater than that in the tubule, this promotes NH_3 movement into the capillary.

The main determinant favoring NH_3 diffusion into the tubule and, consequently, NH_4^+ excretion is the urine pH (Fig. 12-6). Remember that virtually all the NH_3 in the urine exists as NH_4^+. Thus, as the urine is acidified by H^+ secretion, NH_3 combines with the secreted H^+ to form NH_4^+, thereby keeping the urinary NH_3 concentration at low levels and promoting the diffusion of more NH_3 into the lumen to bind H^+. *It is this continued diffusion of NH_3 out of the cells that allows the NH_3/NH_4^+ system to act as an effective "buffer"* even though its pK_a is so far from the urine pH. There is so little NH_3 initially present in the lumen (only one-hundredth that of NH_4^+ at the plasma and initial filtrate pH of 7.4) that it would be rapidly depleted by combination with H^+ if not for continuous replenishment from the cells.

In addition to the urine pH, NH_4^+ excretion also is dependent upon the rate of NH_3 production by the cells. The metabolism of glutamine, which results in the generation of NH_3, is pH-dependent, appropriately increasing with acidemia and decreasing with alkalemia [13]. *The ability to augment NH_3 production and NH_4^+ excretion is the main adaptive response to the kidney to an acid load* as the ensuing increase in net H^+ excretion eliminates some of the excess acid. Although only 30 to

50 meq of H^+ normally is excreted as NH_4^+ per day, NH_4^+ excretion can increase to more than 250 meq/day in the presence of metabolic acidosis [12]. This is in contrast to titratable acidity, which can be only mildly increased in most acidemic patients.

The mechanism by which glutamine metabolism is enhanced by acidemia is incompletely understood [15]. This response takes 4 to 5 days to reach its maximum and may be due to one or both of two factors: (1) increased glutamine entry into the mitochondria, the site of glutamine degradation; and (2) enhanced activity of an enzyme (or enzymes) involved in glutamine metabolism [16–18]. Since the renal tubular cell pH tends to vary directly with the arterial pH (Fig. 12-7), it has been suggested that the pH dependence of glutamine metabolism may be mediated by changes in intracellular pH [19,20]. However, the direct effect of pH alone does not appear to account for all of the increase in NH_3 production [15,21], and it is possible that other factors, including corticosteroid hormones, may play a contributory role [22,23].

Urine pH

As depicted in Fig. 12-5, the tubular fluid pH falls progressively because of H^+ secretion throughout the nephron. In humans, the pH can be lowered to 4.5 to 5.0 in the collecting tubules, representing a plasma-to-tubular fluid H^+ gradient of almost 1:1000 (arterial pH of 7.40 versus urine pH of 4.50). The inability to make the urine more acid may reflect a limit on the strength of the H^+ pump or on the impermeability of the tubular epithelium, which prevents the backflux of H^+ ions out of the lumen down this concentration gradient.

This ability to lower the urine pH is important because the formation of titratable acidity and NH_4^+ are pH-dependent, both increasing as the urine is made more acid (Table 12-1, Fig. 12-6). If the minimum urine pH were higher at 5.5 to 6.0 (which is still less than that of the plasma), titratable acid and NH_4^+ excretion would fall and excretion of the daily H^+ load might be prevented. This appears to be the mechanism responsible for the acidemia in patients with type 1 (distal) renal tubular acidosis (see Chap. 19).

The pH-dependence of titratable acid and NH_4^+ formation also means that these processes (as well as HCO_3^- reabsorption) occur throughout the nephron as the urine is made more acid (Fig. 12-5). The sites at which they are most likely to occur can be appreciated from the isohydric principle which states that all three buffer systems must be in equilibrium:

$$pH = 6.1 + \log \frac{[HCO_3^-]}{0.03 P_{CO_2}} = 6.8 + \log \frac{[HPO_4^{2-}]}{[H_2PO_4^-]} = 9.3 + \log \frac{[NH_3]}{[NH_4^+]}$$

Thus, a secreted H^+ ion will preferentially be buffered by that system with the highest concentration and/or pK_a closest to the pH [24]. In the proximal tubule, most secreted H^+ ions are utilized for HCO_3^- reabsorption because of the high concentration of HCO_3^- and the ability to minimize the reduction in pH by the action of luminal carbonic anhydrase. In contrast, most H^+ ions secreted in the medullary collecting tubule (where the urine pH is reduced to its lowest value) combine with

NH_3 [24] since virtually all the HCO_3^- has been reabsorbed and most of the HPO_4^{2-} has already been buffered (which occurs when the pH is below 5.8, that is, more than 1 pH unit from the pK_a of 6.8).

REGULATION OF RENAL HYDROGEN EXCRETION

The preceding section discussed how the kidney excretes H^+. In this section, we will review the factors that determine exactly how much H^+ is excreted. The *arterial pH* is the major physiologic regulator of this process as it allows acid excretion to vary with day-to-day changes in the dietary H^+ load [25]. In addition, the rate of H^+ secretion can be influenced by the effective circulating volume, the plasma K^+ concentration, the arterial P_{CO_2}, and aldosterone.

Arterial pH

The effect of the arterial pH on H^+ excretion is mediated at least in part by parallel changes in the renal tubular cell pH (Fig. 12-7) [10,25a]. For example, the handling of the daily H^+ load may occur in the following sequence. The excess H^+ ions are initially buffered by HCO_3^- and the cell buffers. The ensuing small reduction in the arterial pH then lowers the renal cell pH, probably by H^+ entry into the cell. This increased availability of H^+ ions promotes H^+ secretion into the lumen. The intra-

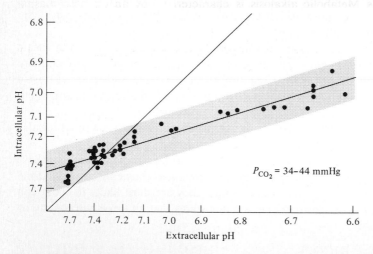

Figure 12-7 Relationship between extracellular and renal tubular cell pH when the former is varied by changing the extracellular HCO_3^- concentration while keeping the P_{CO_2} constant. The tubular cell pH (shaded area) changes directly with, although not to the same degree as, the extracellular pH. (The thin solid line represents the line of equivalence of extracellular and intracellular pH.) A similar relationship exists when the extracellular pH is changed by varying the P_{CO_2} and holding the extracellular HCO_3^- concentration constant. These alterations in cell pH, that is, in cell H^+ concentration, may be responsible for the changes in H^+ secretion which occur with changes in extracellular pH. (*Adapted from A. Struyvenberg, R. B. Morrison, and A. S. Relman, Am. J. Physiol., 214:1155, 1968.*)

cellular acidosis also stimulates NH_3 formation [15,19], permitting some of the secreted H^+ to be trapped in the lumen as NH_4^+. It should be remembered that each H^+ ion secreted results in the generation of a HCO_3^- ion in the plasma (Figs. 12-1 to 12-3), replacing that lost during buffering and restoring the pH to normal.

The role of the arterial pH is also important during acid-base disturbances. For example, acidemia can be produced by a decrease in the plasma HCO_3^- concentration (metabolic acidosis) or an elevation in the P_{CO_2} (respiratory acidosis) [see Eq. (12-3)]. In either disorder, H^+ excretion is augmented (primarily as NH_4^+) which tends to raise the plasma HCO_3^- concentration and return the pH toward normal. Conversely, net H^+ excretion normally is diminished in alkalemic states. These pH-induced changes in H^+ excretion begin within 24 h but are not completed for 4 to 5 days [14,26,27]. The reasons for this delay are not known but may include alterations in the H^+ secretory pump or NH_3 production.

Metabolic acidosis Net acid excretion may be dramatically increased in metabolic acidosis. This is mostly due to enhanced NH_4^+ excretion since titratable acidity is generally limited by the amount of phosphate in the urine. In diabetic ketoacidosis, however, urinary ketone anions can act as titratable acids. In this setting, net H^+ excretion can exceed 500 meq/day [12], resulting in the generation of an equivalent amount of HCO_3^- in the extracellular fluid and an elevation of the pH toward normal.

Metabolic alkalosis Metabolic alkalosis is characterized by an elevated plasma HCO_3^- concentration and an alkaline pH. The normal renal response to a HCO_3^- load is to excrete the excess HCO_3^- in the urine.† However, this adaptive response frequently is blunted. When metabolic alkalosis is associated with volume depletion, e.g., due to vomiting, the stimulation of H^+ secretion and HCO_3^- reabsorption by hypovolemia prevents the excretion of the excess HCO_3^- and correction of the alkalemia (see below).

Respiratory acidosis and alkalosis Disturbances in alveolar ventilation induce changes in CO_2 elimination and, consequently, in the P_{CO_2}. Hyperventilation augments CO_2 loss, resulting in a fall in the P_{CO_2} (hypocapnia) and alkalemia; hypoventilation, on the other hand, impairs CO_2 elimination, producing an increase in the P_{CO_2} (hypercapnia) and acidemia. Although correction of these acid-base disorders requires the restoration of normal alveolar ventilation, the kidney can minimize the changes in arterial pH by varying H^+ excretion and HCO_3^- reabsorption.

From Eq. (12-3), the arterial pH is a function of the HCO_3^-/P_{CO_2} ratio. To maintain this ratio and, therefore, the arterial pH at a near normal level, an elevation in the plasma HCO_3^- concentration is an appropriate response to hypercapnia, and a reduction in the plasma HCO_3^- concentration is an appropriate response to hypocapnia. These changes occur because the P_{CO_2} is an important determinant of H^+ secre-

---

†In addition to decreased reabsorption, HCO_3^- *secretion* in the cortical collecting tubule may also contribute to the bicarbonaturia in this setting [27a]. How this occurs is not known.

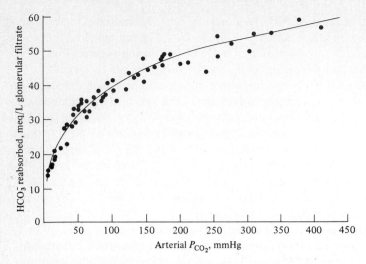

Figure 12-8 Relationship between arterial P_{CO_2} and HCO_3^- reabsorption. Note the curve is steepest in the physiologic range (P_{CO_2} of 15 to 90 mmHg). *(Reproduced from F. C. Rector, Jr., D. W. Seldin, A. D. Roberts, Jr., and J. S. Smith, J. Clin. Invest., 39:1706, 1960, by copyright permission of The American Society for Clinical Investigation.)*

tion and HCO_3^- reabsorption (Fig. 12-8). When the P_{CO_2} is low, H^+ secretion is diminished, resulting in HCO_3^- loss in the urine and decreased NH_4^+ excretion, both of which lower the plasma HCO_3^- concentration [28,29]. Conversely, an elevation in the P_{CO_2} enhances H^+ excretion (as NH_4^+), resulting in HCO_3^- generation in the plasma and an increase in the plasma HCO_3^- concentration [27]. It should also be noted that the renal compensation to respiratory acid-base disorders is incomplete, returning the arterial pH toward, but not all the way to, normal (see Chaps. 20 and 21).

The mechanism by which the P_{CO_2} affects H^+ secretion and HCO_3^- reabsorption is incompletely understood. It had been thought that changes in the intracellular pH induced by those in the extracellular fluid were of primary importance. However, recent studies suggest that the P_{CO_2} has a *direct* effect on H^+ secretion that is independent of the extracellular pH [30,31]. How this might occur is not known, although alterations in the intracellular pH may still be involved.

Effective Circulating Volume

Effective circulating volume depletion may have important effects on net renal H^+ excretion, independent of the state of acid-base balance. The mechanism by which this occurs is as follows. The renal response to hypovolemia is to conserve Na^+ (see Chap. 9). In this situation, the urine Na^+ concentration can be reduced to less than 5 meq/L. If the filtrate Na^+ concentration is 145 meq/L and the filtrate Cl^- concentration is 115 meq/L, then only 115 meq/L of Na^+ can be reabsorbed with Cl^-. Since Cl^- is the only quantitatively important reabsorbable anion in the filtrate, further Na^+ reabsorption must be accompanied by H^+ or K^+ secretion to maintain

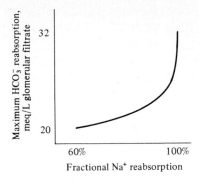

Figure 12-9 Relationship between HCO_3^- reabsorption and fractional Na^+ reabsorption in patients with renal failure. A similar relationship exists in normal subjects. *(Adapted from E. Slatopolsky, P. Hoffsten, M. Purkerson, and N. S. Bricker, J. Clin. Invest., 49:988, 1970, by copyright permission of The American Society for Clinical Investigation).*

electroneutrality. Thus, when Na^+ is conserved maximally, as in volume depletion, H^+ secretion is increased. This is illustrated by the direct relationship between Na^+ reabsorption and HCO_3^- reabsorption (Fig. 12-9) (see page 99). When the fraction of the filtered Na^+ that is reabsorbed is very high, there is no intrinsic limit to renal H^+ secretory capacity and, therefore, HCO_3^- reabsorption [32,33]. Conversely, HCO_3^- reabsorption is reduced by volume expansion, which diminishes fractional Na^+ reabsorption [33,34]. These changes occur primarily in the proximal tubule, the major site of HCO_3^- reabsorption.

The availability of Cl^- as a reabsorbable anion and the state of the effective circulating volume can also affect H^+ secretion in the distal nephron. If, for example, Na_2SO_4 (SO_4^{2-} being a relatively nonreabsorbable anion) is administered to a normovolemic (non-sodium-avid) subject, it will be excreted in the urine. However, if given to a volume-depleted subject, the Na^+ will be retained, and since SO_4^{2-} cannot be reabsorbed, H^+ (as NH_4^+) and K^+ secretion must be increased (Fig. 12-10) [35]. In contrast, if NaCl is administered, Na^+ can be retained with Cl^- without affecting H^+ and K^+ secretion. These changes probably occur in the cortical collecting tubule since they require the presence of aldosterone [36], which acts at this site and the secretion of which is increased by volume depletion (see Chap. 8).

The effects of Na^+ reabsorption and anion reabsorbability on H^+ secretion may have important clinical implications since *the attempt to maintain volume may occur at the expense of the systemic pH.* For example, vomiting induces both metabolic alkalosis (in which the plasma HCO_3^- concentration is elevated) and hypovolemia

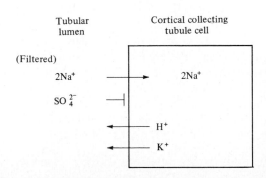

Figure 12-10 Events occurring after Na^+ reabsorption across the luminal membrane of the cortical collecting tubule cell. In a sodium-avid state, the presentation of Na^+ with a nonreabsorbable anion to the cortical collecting tubule enhances H^+ (as NH_4^+) and K^+ secretion. In contrast, if NaCl is presented to this segment, Na^+ will be reabsorbed with Cl^-, with little effect on H^+ and K^+ secretion.

due to the loss of HCl and water (see Chap. 18). Although the alkalemia can be simply corrected by the urinary excretion of the excess HCO_3^-, this does not occur because the associated Na^+ avidity enhances HCO_3^- reabsorption.

Furthermore, there is a paradoxical situation in that *the administration of acid will not necessarily correct the alkalemia.* If HNO_3 is given (NO_3^- being relatively nonreabsorbable), it will be buffered by extracellular HCO_3^-:

$$HNO_3 + NaHCO_3 \rightarrow NaNO_3 + H_2CO_3 \rightarrow CO_2 + H_2O$$

As the $NaNO_3$ is presented to the cortical collecting tubule, Na^+ will be retained and H^+ excretion enhanced. This is similar to the fate of Na_2SO_4 as shown in Fig. 12-10. The net effect is the excretion of the administered HNO_3 as NH_4NO_3 [37]. Thus, the arterial pH will be unchanged as an acid urine is excreted despite the presence of systemic alkalemia. In contrast, if acid is given as HCl, NaCl will be formed from the extracellular buffering by $NaHCO_3$. When this reaches the cortical collecting tubule, Na^+ will be reabsorbed with Cl^- and not exchanged for H^+. Therefore, the administered H^+ will be retained and the alkalemia will be corrected.

Rather than giving HCl, the alkalemia can be reversed more easily by promoting HCO_3^- excretion in the urine. From Fig. 12-9, HCO_3^- reabsorption can be decreased and urinary excretion increased by diminishing the stimulus to Na^+ retention. This can be achieved by reexpanding the effective circulating volume with NaCl. In comparison, the administration of Na^+ with another anion, such as SO_4^{2-}, will be ineffective since it will increase renal H^+ secretion and aggravate the alkalemia.

In summary, the state of the effective circulating volume and the reabsorbability of the anion accompanying Na^+ to the cortical collecting tubule are important determinants of renal H^+ secretion. For the reasons outlined above, *the correction of metabolic alkalosis in a volume-depleted (sodium-avid) subject requires the administration of the only reabsorbable anion, Cl^-, either as NaCl, HCl, or if hypokalemia is present, KCl* (see Chap. 18).

Plasma Potassium Concentration

Potassium is another potential influence on renal H^+ secretion as a reciprocal relationship has been demonstrated between the plasma K^+ concentration and HCO_3^- reabsorption (Fig. 12-11) [38,39]. This effect may be related to changes in the cell H^+ concentration induced by cation shifts between the cell and the extracellular fluid (Fig. 12-12). For example, as hypokalemia develops due to gastrointestinal or urinary losses, K^+ leaves the cell down a concentration gradient. To maintain electroneutrality, H^+ (and Na^+) enter the cell, resulting in an intracellular acidosis [40–42]. This increase in H^+ concentration in the renal tubular cells may account for the enhanced H^+ secretion and HCO_3^- reabsorption observed with K^+ depletion. This effect is not limited to HCO_3^- reabsorption since the intracellular acidosis also stimulates NH_3 production and NH_4^+ excretion. These changes are reversed with hyperkalemia as K^+ moves into and H^+ out of cells. The ensuing intracellular alkalosis may then account for the associated reductions in HCO_3^- reabsorption and NH_4^+ excretion [38,43].

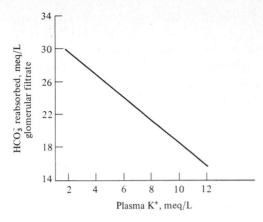

Figure 12-11 Renal tubular reabsorption of HCO_3^- as a function of the plasma K^+ concentration. *(Adapted from G. R. Fuller, M. B. MacLeod, and R. F. Pitts, Am. J. Physiol., 182:111, 1956.)*

To summarize, hypokalemia tends to increase net acid excretion, which promotes the development of metabolic alkalosis. In contrast, hyperkalemia reduces net acid excretion which, by causing H^+ retention, favors the development of metabolic acidosis. In some patients with hyperkalemia due to hypoaldosteronism, for example, the associated metabolic acidosis can be corrected solely by lowering the plasma K^+ concentration [43].

Aldosterone

Aldosterone, acting in the cortical collecting tubule, promotes H^+ secretion in three ways: by directly stimulating the secretory pump; by the need to maintain electroneutrality as Na^+ reabsorption is enhanced; and by maintenance of a normal plasma K^+ concentration (see Chap. 8) [44]. Although aldosterone secretion does not appear to be directly influenced by changes in pH, this hormone plays at least a permissive role in renal H^+ excretion as evidenced by the reductions in H^+ and NH_4^+ excretion and the metabolic acidosis that usually accompany aldosterone deficiency (see Chap. 19) [43–45]. Hyperaldosteronism, on the other hand, is commonly associated with increased net H^+ excretion and metabolic alkalosis (see Chap. 18) [46].

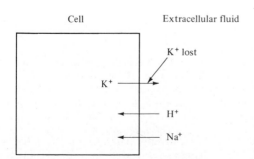

Figure 12-12 Reciprocal cation shifts of K^+, H^+, and Na^+ between the cells, including renal tubular cells, and the extracellular fluid. In the presence of hypokalemia, K^+ moves out of the cells down a concentration gradient. Since the cell anions (primarily proteins and organic phosphates) are unable to cross the cell membrane, electroneutrality is maintained by the entry of Na^+ and H^+ into the cell. The increase in cell H^+ concentration may be responsible for the increased H^+ secretion and HCO_3^- reabsorption seen with hypokalemia. On the other hand, hyperkalemia causes H^+ and Na^+ to leave the cells, resulting in a fall in H^+ secretion and HCO_3^- reabsorption.

EFFECT OF ARTERIAL pH ON VENTILATION

Alveolar ventilation provides the oxygen necessary for oxidative metabolism and eliminates the CO_2 produced by these metabolic processes. Therefore, it is appropriate that the main physiologic stimuli to respiration are a reduction in the P_{O_2} (hypoxemia) and an elevation in the P_{CO_2} [47,48]. The CO_2 stimulus to ventilation primarily occurs in chemosensitive areas in the respiratory center in the brain stem, which appear to respond to CO_2-induced changes in the cerebral interstitial pH [49]. This effect is extremely important in the maintenance of the arterial pH since roughly 15,000 mmol of CO_2 is produced daily from endogenous metabolism, added to the capillary blood, and then eliminated via the lungs. In contrast, hypoxemia is primarily sensed by peripheral chemoreceptors in the carotid bodies which are located near the bifurcation of the carotid arteries [48,50].

Alveolar ventilation also is affected by metabolic acid-base disorders, a response that is mediated by both the peripheral and, more importantly, the central chemoreceptors [47,51]. In metabolic acidosis, minute ventilation can increase from the normal 5.0 L/min to greater than 30 L/min as the arterial pH falls from 7.40 to 7.00 (Fig. 12-13). This is an appropriate compensatory response since the concomitant fall in the P_{CO_2} *raises* the pH toward normal [see Eq. (12-3)] [52,53]. Conversely, hypoventilation with a consequent elevation in the P_{CO_2} *lowers* the pH toward normal when the plasma HCO_3^- concentration is increased in a patient with metabolic alkalosis [54,55].

The potential importance of these respiratory compensations to metabolic acidosis and alkalosis can be appreciated from the following hypothetical example. In

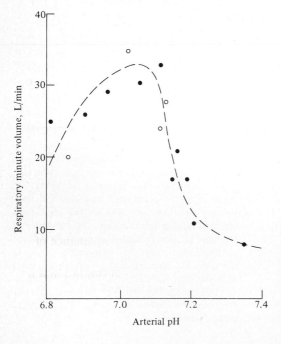

Figure 12-13 Relationship between respiratory minute volume and arterial pH in patients with diabetic ketoacidosis. *(Reproduced from S. S. Kety, B. D. Polis, C. S. Nadler, and C. F. Schmidt, J. Clin. Invest., 27:500, 1948, by copyright permission of the American Society for Clinical Investigation.)*

diabetic ketoacidosis (see Chap. 25), the increased production of ketoacids is buffered in part in the extracellular fluid, resulting in a reduction in the plasma HCO_3^- concentration. If the plasma HCO_3^- concentration were reduced to 6 meq/L and the P_{CO_2} remained at the normal 40 mmHg, then

$$pH = 6.1 + \log \frac{6}{0.03 \times 40} = 6.80$$

However, if ventilation were stimulated by the acidemia and the P_{CO_2} decreased to 15 mmHg, then

$$pH = 6.1 + \log \frac{6}{0.03 \times 15} = 7.22$$

Thus, the respiratory compensation has turned a life-threatening reduction in pH into one that is much less dangerous. In contrast to the renal compensation to respiratory acid-base disorders, which begins on the first day and is not complete for 3 to 5 days, the respiratory response to metabolic acidosis or alkalosis is relatively rapid, beginning within minutes and reaching its maximum within 12 to 24 h [51,56].

Despite the effectiveness of the respiratory compensation, the pH is protected for only a few days since the initially beneficial change in P_{CO_2} then alters renal HCO_3^- reabsorption. In metabolic acidosis, for example, the compensatory fall in P_{CO_2} decreases HCO_3^- reabsorption (see Fig. 12-8) and, therefore, the plasma HCO_3^- concentration and arterial pH. The net effect is that after several days *the arterial pH is the same as it would have been if no respiratory compensation had occurred*† [30]. Fortunately, most forms of severe metabolic acidosis are acute (ketoacidosis, lactic acidosis, ingestions; see Chap. 18) so that the associated hypocapnia does protect the pH. Similar considerations apply to the hypercapnia seen with metabolic alkalosis as the rise in P_{CO_2} leads to a further elevation in the plasma HCO_3^- concentration [31].

These results indicate that the P_{CO_2} has a direct effect on renal H^+ secretion that is independent of the arterial pH and may have deleterious effects on acid-base balance. Although reductions in HCO_3^- reabsorption and the plasma HCO_3^- concentration are *appropriate* when there is a primary decrease in P_{CO_2} in respiratory alkalosis, they are *inappropriate* when the P_{CO_2} is secondarily diminished in metabolic acidosis since the acidemia will be enhanced.

SUMMARY

From the Henderson-Hasselbalch equation, the arterial pH is a function of the $[HCO_3^-]/0.03P_{CO_2}$ ratio. Three processes are involved in the maintenance of the arterial pH: (1) the extracellular and intracellular buffers act to minimize changes in pH induced by an acid or base load; (2) the plasma HCO_3^- concentration is held within narrow limits by the regulation of renal H^+ excretion; and (3) the P_{CO_2} is

†An example of this phenomenon is depicted in Table 19-2.

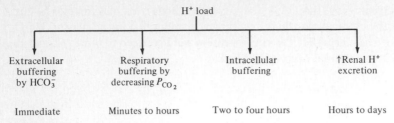

Figure 12-14 Sequential response to a H^+ load, culminating in the restoration of acid-base balance by the renal excretion of the excess H^+.

controlled by variations in alveolar ventilation. How these processes interact to protect the pH can be appreciated from the response to a HCl load (Fig. 12-14). Extracellular buffering of the excess H^+ by HCO_3^- occurs almost immediately. Within several minutes, the respiratory compensation begins, resulting in hyperventilation, a decrease in the P_{CO_2}, and an increase in the pH toward normal. Within 2 to 4 h, the intracellular buffers (primarily proteins and organic phosphates) provide further buffering as H^+ ions enter the cells in exchange for intracellular K^+ and Na^+. These responses act to prevent wide swings in the arterial pH until acid-base homeostasis can be restored by the renal excretion of the H^+ load as NH_4^+ and titratable acidity. This corrective response begins on the first day and is complete within 3 to 5 days [14,26]. This sequence is reversed in the presence of metabolic alkalosis.

The response to changes in pH induced by changes in the P_{CO_2} is somewhat different. There is virtually no extracellular buffering since HCO_3^- cannot effectively buffer H_2CO_3 (see page 215). Similarly, there is no compensatory change in alveolar ventilation since the primary disturbance is one of abnormal respiration. Thus the intracellular buffers (including hemoglobin) and changes in renal H^+ excretion constitute the only protective mechanisms against respiratory acidosis or alkalosis. For example, if the P_{CO_2} is increased, the intracellular buffers act to increase the plasma HCO_3^- concentration, thereby minimizing the degree of acidemia (Fig. 12-15). This process is complete within 10 to 30 min [57]. Since these buffers increase the plasma HCO_3^- concentration by only 1 meq/L for each 10 mmHg increase in the P_{CO_2},[†] they are relatively ineffective in protecting the pH. If the hypercapnia persists, however, there will be an appropriate increase in renal H^+ excretion, resulting in a further elevation in the plasma HCO_3^- concentration. It is this renal compensation, which begins in hours but is not complete for several days [27], that constitutes the main defense against respiratory acidosis. Even if the P_{CO_2} is chronically elevated at 80 mmHg, the pH usually is not much lower than 7.30[†] because of the effectiveness of the renal compensation. This sequence is reversed with respiratory alkalosis as there is an appropriate reduction in the plasma HCO_3^- concentration primarily because of HCO_3^- loss in the urine. It should be noted that the renal responses to alterations in the P_{CO_2} are compensatory but not corrective. Acid-base homeostasis will not be restored until alveolar ventilation is normalized.

[†] The changes in the plasma HCO_3 concentration seen with acute and chronic respiratory acidosis and alkalosis are presented in detail in Chaps. 20 and 21.

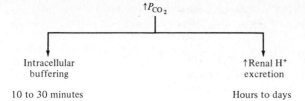

Figure 12-15 Response to an increase in the P_{CO_2}. Although these changes increase the pH toward normal, acid-base homeostasis will not be restored until ventilation is normalized.

PROBLEMS

12-1 What is the primary adaptive response by the kidney to an increased H^+ load? What effect will this H^+ load have on alveolar ventilation and subsequently on the P_{CO_2}?

12-2 The daily H^+ load is excreted in the urine as titratable acidity and NH_4^+. Would H^+ retention leading to metabolic acidosis occur if there were (a) a marked reduction in titratable acid excretion, e.g., due to a decrease in the plasma phosphate concentration; and (b) a marked reduction in NH_3 formation?

12-3 Equal amounts of H^+, as HCl or H_2SO_4, are given over several days to a volume-depleted subject. Which acid will produce the greater increase in the arterial H^+ concentration?

12-4 The following values are obtained on a 24-h urine collection:

$$\text{Phosphate} = 60 \text{ mmol}$$
$$\text{pH} = 5.8$$

If the arterial pH is 7.40 and the pK_a for phosphate is 6.80, how many millimoles of H^+ are excreted as titratable acidity using HPO_4^{2-} as a buffer? Is NH_4^+ excretion included in the measurement of titratable acidity?

12-5 What effect will a low P_{CO_2} have on the following?
    (a) The arterial pH
    (b) HCO_3^- reabsorption
    (c) The plasma HCO_3^- concentration

12-6 A patient with persistent vomiting develops metabolic alkalosis due to the loss of HCl in gastric juice. Why isn't the condition corrected spontaneously by the excretion of the excess HCO_3^- in the urine?

REFERENCES

1. Lennon, E. J., J. Lemann, Jr., and J. R. Litzow: The effects of diet and stool composition on the net external acid balance of normal subjects, J. Clin. Invest., 45:1601, 1966.
2. Kinsella, J. L., and P. S. Aronson: Properties of the Na^+–H^+ exchanger in renal microvillus membrane vesicles, Am. J. Physiol., 238:F461, 1980.
3. Chan, Y. L., and G. Giebisch: Relationship between sodium and bicarbonate transport in the rat proximal convoluted tubule, Am. J. Physiol., 240:F222, 1981.
4. Arruda, J. A. L., and N. A. Kurtzman: Mechanisms and classification of deranged distal urinary acidification, Am. J. Physiol., 239:F515, 1980.
5. Tam, S-C., M. B. Goldstein, B. J. Stinebaugh, C. B. Chen, A. Gougoux, and M. L. Halperin: Studies on the regulation of hydrogen ion secretion in the collecting duct in vivo: evaluation of factors that influence the urine minus blood PCO₂ difference, Kidney, Int., 20:636, 1981.
5a. Berry, C. A., and D. G. Warnock: Acidification in the in vitro perfused tubule, Kidney Int., 22:507, 1982.
6. Cohen, L. H., A. Mueller, and P. R. Steinmetz: Inhibition of the bicarbonate exit step in urinary acidification by a disulfonic stilbene, J. Clin. Invest., 61:981, 1978.

7. Dobyan, D. C., and R. E. Bulger: Renal carbonic anhydrase, Am. J. Physiol., 243:F311, 1982.
8. Lucci, M. S., D. G. Warnock, and F. C. Rector, Jr.: Carbonic anhydrase-dependent bicarbonate reabsorption in the rat proximal tubule, Am. J. Physiol., 236:F58, 1979.
9. Lucci, M. S., L. R. Pucacco, T. D. DuBose, Jr., J. P. Kokko, and N. W. Carter: Direct evaluation of acidification by rat proximal tubule: role of carbonic anhydrase, Am. J. Physiol., 238:F372, 1980.
10. Rector, F. C., Jr.: Sodium, bicarbonate, and chloride absorption by the proximal tubule, Am. J. Physiol., 244:F461, 1983.
11. Vieira, F. J., and G. Malnic: Hydrogen ion secretion by rat renal cortical tubules as studied by an antimony microelectrode, Am. J. Physiol., 214:710, 1968.
12. Clarke, E., B. M. Evans, and I. M. MacIntyre: Acidosis in experimental electrolyte depletion, Clin. Sci., 14:421, 1955.
13. Pitts, R. F.: *Physiology of the Kidney and Body Fluids,* Year Book, Chicago, 1974.
14. Tizianello, A., G. Deferrari, G. Garibotto, C. Robaudo, N. Acquarone, and G. M. Ghiggeri: Renal ammoniagenesis in an early stage of metabolic acidosis in man, J. Clin. Invest., 69:240, 1982.
15. Tannen, R. L.: Ammonia metabolism, Am. J. Physiol., 235:F265, 1978.
16. Simpson, D. P.: Mitochondrial transport functions and renal metabolism, Kidney Int., 23:785, 1983.
17. Welbourne, T. C., and G. T. Bazer: Mitochondrial glutamine permeability and renal ammonia production in metabolic acidosis, Am. J. Physiol., 239:E51, 1980.
18. Goodman, D., R. E. Fuisz, and G. F. Cahill, Jr.: Renal gluconeogenesis in acidosis, alkalosis, and potassium deficiency: its possible role in regulation of renal ammonia production, J. Clin. Invest., 45:612, 1966.
19. Kamm, D. E., and G. L. Strope: Glutamine and glutamate metabolism in renal cortex from potassium-depleted rats, Am. J. Physiol., 224:1241, 1973.
20. Adler, S., B. Anderson, and B. Zett: Regulation of citrate metabolism by cell pH in potassium-depleted rat diaphragm, Kidney Int.. 6:92, 1974.
21. Tannen, R. L., and A. S. Kunin: Effect of pH on ammonia production by renal mitochondria, Am. J. Physiol., 231:1631, 1976.
22. Smythe, G. A., and L. Lazarus: Regulation of renal cortex ammoniagenesis. I. Stimulation of renal cortex ammoniagenesis in vitro by plasma isolated from acutely acidotic rats, J. Clin. Invest., 53:117, 1974.
23. Welbourne, T. C.: Acidosis activation of the pituitary-adrenal-renal glutaminase I axis, Endocrinology, 99:1071, 1976.
24. Buerkert, J., D. Martin, and D. Trigg: Segmental analysis of the renal tubule in buffer production and net acid formation, Am. J. Physiol., 244:F442, 1983.
25. Malnic, G., and G. Giebisch: Mechanism of renal hydrogen ion secretion, Kidney Int., 1:280, 1972.
25a. Alpern, R. J., M. G. Cogan, and F. C. Rector, Jr.: Effects of extracellular fluid volume and plasma bicarbonate concentration on proximal acidification in the rat, J. Clin. Invest., 71:736, 1983.
26. Kraut, J. A., J. B. Wish, S. J. Sweet, S. S. Weinstein, and J. J. Cohen: Failure of increased sodium avidity to facilitate renal acid excretion in dogs fed sulfuric acid, Kidney Int., 20:50, 1981.
27. Polak, A., G. D. Haynie, R. M. Hays, and W. B. Schwartz: Effects of chronic hypercapnia on electrolyte and acid-base equilibrium. I. Adaptation, J. Clin. Invest., 40:1223, 1961.
27a. Lombard, W. E., J. P. Kokko, and H. R. Jacobson: Bicarbonate transport in cortical and outer medullary collecting tubules, Am. J. Physiol., 244:F289, 1983.
28. Gennari, J. F., M. B. Goldstein, and W. B. Schwartz: The nature of the renal adaptation to chronic hypocapnia, J. Clin. Invest., 51:1722, 1972.
29. Gougoux, A., W. D. Kaehny, and J. J. Cohen: Renal adaptation to chronic hypocapnia: dietary constraints in achieving H^+ retention, Am. J. Physiol., 229:1330, 1975.
30. Madias, N., W. B. Schwartz, and J. J. Cohen: The maladaptive renal response to secondary hypocapnia during chronic HCl acidosis in the dog, J. Clin. Invest., 60:1393, 1977.
31. Madias, N. E., H. H. Adrogué, and J. J. Cohen: Maladaptive response to secondary hypercapnia in chronic metabolic alkalosis, Am. J. Physiol., 238:F283, 1980.
32. Kurtzman, N. A.: Regulation of renal bicarbonate reabsorption by extracellular volume, J. Clin. Invest., 49:586, 1970.
33. Slatopolsky, E., P. Hoffsten, M. Purkerson, and N. S. Bricker: On the influence of extracellular

fluid volume expansion and of uremia on bicarbonate reabsorption in man, J. Clin. Invest., 49:988, 1970.

34. Purkerson, M., H. Lubowitz, R. White, and N. S. Bricker: On the influence of extracellular fluid volume expansion on bicarbonate reabsorption in the rat, J. Clin. Invest., 48:1754, 1969.

35. Schwartz, W. B., R. L. Jenson, and A. S. Relman: Acidification of the urine and increased ammonium excretion without change in acid-base equilibrium: sodium reabsorption as a stimulus to the acidifying process, J. Clin. Invest., 34:673, 1955.

36. Kurtzman, N. A., M. G. White, and P. W. Rogers: Aldosterone deficiency and renal bicarbonate reabsorption, J. Lab. Clin. Med., 77:931, 1971.

37. Tannen, R. L., H. L. Bleich, and W. B. Schwartz: The renal response to acid loads in metabolic alkalosis: an assessment of the mechanisms regulating acid excretion, J. Clin. Invest., 45:562, 1966.

38. Fuller, G. R., M. B. MacLeod, and R. F. Pitts: Influence of administration of potassium salts on the renal tubular reabsorption of bicarbonate, Am. J. Physiol., 182:111, 1955.

39. Chan, Y. L., B. Biagi, and G. Giebisch: Control mechanisms of bicarbonate transport across the rat proximal convoluted tubule, Am. J. Physiol., 242:F532, 1982.

40. Cooke, R. E., W. Segar, D. B. Cheek, F. Coville, and D. C. Darrow: The extrarenal correction of alkalosis associated with potassium deficiency, J. Clin. Invest., 31:798, 1952.

41. Adler, S., B. Zett, and B. Anderson: The effect of acute potassium depletion on muscle cell pH in vitro, Kidney Int., 2:159, 1972.

42. Wilson, A. F., and D. H. Simmons: Relationships between potassium, chloride, intracellular and extracellular pH in dogs, Clin. Sci. 39:731, 1970.

43. Szylman, P., O. S. Better, C. Chaimowitz, and A. Rosler: Role of hyperkalemia in the metabolic acidosis of isolated hypoaldosteronism, N. Engl. J. Med., 294:361, 1976.

44. Hulter, H. N., L. P. Ilnicki, J. A. Harbottle, and A. Sebastian: Impaired renal H^+ secretion and NH_3 production in mineralocorticoid-deficient glucocorticoid-replete dogs, Am. J. Physiol., 232:F136, 1977.

45. DeFronzo, R. A.: Hyperkalemia and hyporeninemic hypoaldosteronism, Kidney Int., 17:118, 1980.

46. Kassirer, J. P., A. M. Landon, D. M. Goldman, and W. B. Schwartz: On the pathogenesis of metabolic alkalosis in hyperaldosteronism, Am. J. Med., 49:306, 1970.

47. Lambertsen, C. J.: Chemical control of respiration at rest, in *Medical Physiology,* V. B. Mountcastle (ed.), 14th ed., Mosby, St. Louis, 1980.

48. Berger, A. J., R. A. Mitchell, and J. W. Severinghaus: Regulation of respiration, N. Engl. J. Med., 297:92, 138, 194, 1977.

49. Fencl, V., T. B. Miller, and J. R. Pappenheimer: Studies on the respiratory response to disturbances of acid-base balance, with deductions concerning the ionic composition of cerebral interstitial fluid, Am. J. Physiol., 210:459, 1966.

50. Lugliani, R., B. J. Whipp, C. Seard, and K. Wasserman: Effect of bilateral carotid-body resection on ventilatory control at rest and during exercise in man, N. Engl. J. Med., 285:1105, 1971.

51. Mitchell, R. A., and M. M. Singer: Respiration and cerebrospinal fluid pH in metabolic acidosis and alkalosis, J. Appl. Physiol., 20:905, 1965.

52. Albert, M. S., R. B. Dell, and R. W. Winters: Quantitative displacement of acid-base equilibrium in metabolic acidosis, Ann. Intern. Med., 66:312, 1967.

53. Bushinsky, D. A., F. L. Coe, C. Katzenberg, J. P. Szidon, and J. H. Parks: Arterial P_{CO_2} in chronic metabolic acidosis, Kidney Int., 22:311, 1982.

54. van Ypersele de Strihou, C., and A. Frans: The respiratory response to chronic metabolic alkalosis and acidosis in disease, Clin. Sci. Mol. Med., 54:439, 1973.

55. Javaheri, S., N. S. Shore, B. D. Rose, and H. Kazemi: Compensatory hypoventilation in metabolic alkalosis, Chest, 81:296, 1982.

56. Pierce, N. F., D. S. Fedson, K. L. Brigham, R. C. Mitra, R. B. Sack, and A. Mondol: The ventilatory response to acute base deficit in humans, Ann. Intern. Med., 72:633, 1970.

57. Brackett, N. C., Jr., J. J. Cohen, and W. B. Schwartz: Carbon dioxide titration curve of normal man. Effect of increasing degrees of acute hypercapnia on acid-base equilibrium, N. Engl. J. Med., 272:6, 1965.

POTASSIUM HOMEOSTASIS

INTRODUCTION

The total body K^+ stores in a normal adult are approximately 3000 to 4000 meq (50 to 55 meq/kg body weight). In contrast to Na^+, which is restricted primarily to the extracellular fluid (ECF), K^+ is basically an intracellular cation, with 98 percent of body K^+ being located in the cells. This can be appreciated from the disparity between the K^+ concentrations in the two compartments: cell K^+ concentration of 150 meq/L versus extracellular (and plasma) K^+ concentration of only 4 to 5 meq/L (see Table 1-6).The location of Na^+ and K^+ in the different fluid compartments is maintained by the active Na^+-K^+-ATPase pump in the cell membrane which pumps Na^+ out of and K^+ into the cell in a 3:2 ratio (see page 11) [1,2].

Potassium has two major physiologic functions. First, it plays an important role in cell metabolism, participating in the regulation of such processes as protein and glycogen synthesis [3,4]. As a result, a variety of cell functions may become impaired in conditions of K^+ imbalance. For example, patients with K^+ depletion often complain of polyuria (increased urine output). This is due in part to a reduced ability to concentrate the urine, resulting from a decreased responsiveness to antidiuretic hormone (ADH) (Fig. 13-1). This resistance to ADH appears to be due both to interference with the generation and action of cyclic AMP, the intracellular mediator of the effects of ADH (see Fig. 8-2), and to impairment of the countercurrent mechanism [5–7].

Second, the *ratio* of the K^+ concentrations in the cell and the ECF is the major determinant of the resting membrane potential (E_m) across the cell membrane. This relationship can be expressed by the following formula

$$E_m = -61 \log \frac{r[K^+]_{cell} + 0.01\,[Na^+]_{cell}}{r[K^+]_{ecf} + 0.01\,[Na^+]_{ecf}} \qquad (13\text{-}1)$$

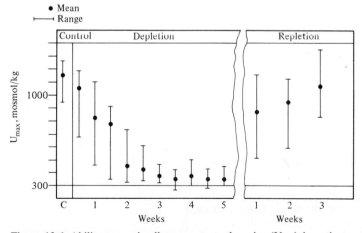

Figure 13-1 Ability to maximally concentrate the urine (U_{max}) in patients with progressive potassium depletion. The average K^+ deficit was 350 meq or about 10 percent of the total body K^+. *(From M. Rubini, J. Clin. Invest., 40:2215, 1961, by copyright permission of The American Society for Clinical Investigation.)*

where r is the Na^+/K^+ active transport ratio of 3:2 and 0.01 is the relative membrane permeability of Na^+ to K^+ [2]. It is the resting potential that sets the stage for the generation of the action potential that is essential for normal neural and muscular function. Thus, both hypokalemia (low plasma K^+ concentration) and hyperkalemia (high plasma K^+ concentration) can result in potentially fatal muscle paralysis and cardiac arrhythmias by altering the resting potential of skeletal and cardiac muscle.

The pathophysiologic effects of K^+ imbalance will be discussed in detail in Chaps. 26 to 28. The remainder of this chapter will deal with the two functions responsible for the maintenance of a normal plasma K^+ concentration: (1) the distribution of K^+ between the cells and the extracellular fluid; and (2) the urinary excretion of the K^+ added to the extracellular fluid from the diet and endogenous cellular breakdown.

DISTRIBUTION OF POTASSIUM BETWEEN THE CELLS AND THE EXTRACELLULAR FLUID

Regulation of the internal distribution of K^+ must be extremely sensitive since the movement of as little as 1.5 to 2 percent of the cell K^+ into the ECF can result in a potentially fatal increase in the plasma K^+ concentration to as high as 8 meq/L or more. A variety of factors, both physiologic and pathologic, can influence this process (Table 13-1) [8,9]. The most important in the normal regulation of K^+ balance are Na^+-K^+-ATPase, catecholamines, insulin, and the plasma K^+ concentration.

Sodium-Potassium-ATPase

The Na^+-K^+-ATPase pump is the major factor responsible for normal K^+ distribution. An example of its importance in humans can be seen when Na^+-K^+-ATPase is partially inhibited by a massive overdose of digitalis, a drug useful in the treatment of heart disease. In this setting, marked hyperkalemia (plasma K^+ concentration up

Table 13-1 Factors influencing the distribution of K^+ between the cells and the extracellular fluid

Physiologic
Na^+-K^+-ATPase
Catecholamines
Insulin
Plasma K^+ concentration
Exercise
Pathologic
Chronic diseases
Arterial pH
Rate of cell breakdown
Hyperosmolality

to 13.5 meq/L) may occur because of the relative inability of K^+ to enter the cells [10].

The result is somewhat different when the pump is impaired chronically. In a variety of chronic diseases such as heart failure and renal failure, Na^+-K^+-ATPase activity may be reduced, presumably due to some defect in cell metabolism [11–13]. As a result, K^+ leaves and Na^+ enters the cells down passive gradients. The net effect is as much as a 10 to 15 percent reduction in total body K^+ stores in association with a high cell Na^+ concentration, a low cell K^+ concentration, but no change in the plasma K^+ concentration because the excess extracellular K^+ has time to be excreted in the urine [12–15].

In addition to the basal distribution of K^+, there is a frequent exchange of K^+ between the ECF and the cells because of variations in dietary intake. For example, three large glasses of orange juice contain approximately 40 meq of K^+. If after ingestion, this K^+ remained entirely in the extracellular fluid (the normal extracellular volume being roughly 17 liters in a 70-kg man), there would be a potentially dangerous 2.4 meq/L increase in the plasma K^+ concentration. This is prevented by the rapid entry of most of the K^+ load into the cells [16,17]. Within 6 to 8 h, K^+ balance is then restored by the urinary excretion of the ingested K^+ [18,19]. Although the initial elevation in the plasma K^+ concentration may directly promote the intracellular movement of K^+, catecholamines and insulin also play an important role in this process.

Catecholamines

It has been known for some time that catecholamines can affect K^+ distribution, with α-receptors impairing and β_2-receptors promoting the cellular uptake of K^+ [20,20a]. The physiologic importance of these effects in humans is suggested by the observation that the increment in the plasma K^+ concentration after a K^+ load is greater and more prolonged if the subject has been pretreated with a β-adrenergic blocker such as propranolol (Fig. 13-2) [21,22]. This difference is due to as much as a 50 percent reduction in the cellular uptake of the excess K^+, most of which occurs in skeletal muscle and the liver [22]. Conversely, the administration of epinephrine, a β-adrenergic agonist, promotes the intracellular movement of K^+, thereby decreasing the elevation in the plasma K^+ concentration [21,22].

The mechanism by which catecholamines enhance cell K^+ uptake is incompletely understood. Since plasma epinephrine and norepinephrine levels do not rise after a K^+ load [22], a positive feedback loop—K^+ load $\rightarrow \uparrow$ catecholamines $\rightarrow \uparrow$ uptake—does not appear to be present. It may be, therefore, that the facilitation of K^+ uptake results from a *permissive* effect of basal catecholamine levels. Alternatively, some other unidentified β-adrenergic agonist may be involved [23].

Insulin

Insulin promotes the entry of K^+ into skeletal muscle and the liver [24,25], a property which has made the administration of insulin (with glucose to prevent hypoglycemia) an effective form of therapy for hyperkalemia (see Chap. 28). It has also

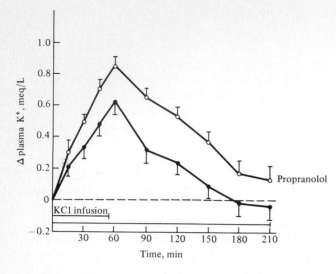

Figure 13-2 Changes in the plasma K^+ concentration after a K^+ load in the absence (solid circles) or presence (open circles) of the β-adrenergic blocker propranolol. *(From R. N. Rosa, P. Silva, J. B. Young, L. Landsberg, R. S. Brown, J. W. Rowe, and F. H. Epstein, N. Engl. J. Med., 302:431, 1980. Reprinted by permission from the New England Journal of Medicine.)*

become clear that endogenous insulin plays a physiologic role in the regulation of the plasma K^+ concentration [8,17,26]. This is illustrated by studies in subjects with very low basal insulin levels due to an infusion of somatostatin which impairs pancreatic insulin secretion. In this setting, the baseline plasma K^+ concentration rises (by 0.4 to 0.5 meq/L) and a K^+ load induces a greater than normal increase in the plasma K^+ concentration (Fig. 13-3). These changes are reversed by an infusion of insulin [17].

As with catecholamines, it appears that *basal* insulin levels *permissively* allow K^+ entry into cells since a K^+ load does not promote insulin release in humans [26]. However, if insulin availability is increased, either by giving glucose or insulin, there will be a further tendency for K^+ to move into the cells. This effect lasts only several hours since other factors (perhaps the plasma K^+ concentration itself) then cause K^+ to move back into the extracellular fluid [26a].

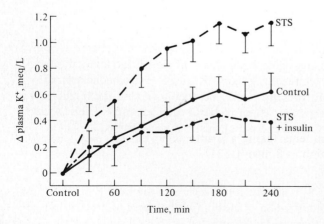

Figure 13-3 The effect of a potassium chloride infusion on the plasma K^+ concentration in the control state, with somatostatin (STS), or with somatostatin and insulin. *(Adapted from R. A. DeFronzo, R. S. Sherwin, M. Dillingham, R. Hendler, W. Tamborlane, and P. Felig, J. Clin. Invest., 61:472, 1978, by copyright permission of The American Society for Clinical Investigation.)*

It should be noted that the primary physiologic effect of insulin and catecholamines is probably in facilitating the disposition of a K^+ load. Although a deficiency of these hormones may initially elevate the baseline plasma K^+ concentration, this is usually transient since the excess K^+ is eventually excreted in the urine. Thus, the fasting plasma K^+ concentration is typically normal in patients treated with β-adrenergic blockers and in patients with type 1 (insulin-dependent) diabetes mellitus given enough insulin to prevent marked hyperglycemia [17,21,22].

Plasma Potassium Concentration

The combination of insulin deficiency and sympathetic blockade impairs but does not prevent the intracellular movement of K^+ after a K^+ load, indicating that other factors must also be involved [16,27]. One of these is probably the plasma K^+ concentration itself. After a K^+ load, for example, the initial elevation in the plasma K^+ concentration promotes K^+ entry into the cells, perhaps by passive mechanisms. Conversely, the loss of K^+ from the ECF due to gastrointestinal or renal losses results first in a fall in the plasma K^+ concentration and secondarily in the movement of K^+ from the cells into the ECF to minimize the degree of hypokalemia.

The net effect is that, in most situations, the plasma K^+ concentration varies directly with body K^+ stores, decreasing with K^+ depletion and increasing with K^+ retention [9]. There are, however, some exceptions to this rule, including chronic diseases as described above (in which Na^+-K^+-ATPase activity appears to be reduced), exercise, changes in the arterial pH or rate of cell breakdown, and an increase in the effective plasma osmolality. In these disorders, clinically significant hyperkalemia or hypokalemia may result from redistribution of K^+ between the cells and the ECF without change in body K^+ stores.

Exercise

Potassium is released from cells during exercise. This is an appropriate response since the local increase in the plasma K^+ concentration has a vasodilatory effect which contributes to the increase in blood flow (and therefore energy delivery) to the exercising muscle [28]. The elevation in the systemic plasma K^+ concentration is less pronounced and related to the degree of exercise: 0.3 to 0.4 meq/L with slow walking [29], 0.7 meq/L with moderate exercise [30], and as much as 2.0 meq/L with severe exercise to exhaustion [31,32]. These changes are reversed after several minutes of rest as K^+ moves back into the cells [32].

The hyperkalemia associated with exercise is generally mild and produces no symptoms. However, it can lead to a potentially dangerous elevation in the plasma K^+ concentration in the presence of some other abnormality in K^+ handling. For example, severe exercise in patients taking a β-adrenergic blocker may lead to a plasma K^+ concentration as high as 8 meq/L [31].

The effect of exercise may also be relevant in the venipuncture procedure used to obtain blood for measurement of the plasma K^+ concentration. After a tourniquet is applied, the patient is frequently instructed to repeatedly clench and unclench his or her fist in an attempt to increase local blood flow and make the veins more apparent. This can result in as much as a 2 meq/L elevation in the plasma K^+ concentration, leading to erroneous evaluation of the state of K^+ balance [33].

Arterial pH

Changes in acid-base balance may have important effects on the plasma K^+ concentration, particularly in those forms of metabolic acidosis due to an excess of mineral acid (as occurs with renal failure or diarrhea). In this setting, almost 60 percent of the excess H^+ ions is buffered in the cells (see Fig. 11-4). Since the major extracellular anion Cl^- enters the cells only to a limited degree, electroneutrality is maintained by K^+ and Na^+ movement into the ECF (Fig. 13-4). The result is a variable increase in the plasma K^+ concentration of 0.2 to 1.7 meq/L for every 0.1-unit fall in the arterial pH [34]. (The plasma Na^+ concentration may also rise, but an elevation of a few milliequivalents per liter is not physiologically important since the normal value is 140 meq/L, not 4 to 5 meq/L as for potassium.)

This relationship, however, does not apply to metabolic acidosis due to excessive production of organic acids (lactic acidosis, ketoacidosis, and certain ingestions). In these disorders, acidemia does not lead to a significant rise in the plasma K^+ concentration (Fig. 13-5) [34–37]. The mechanism by which this occurs is not known but may be related to the ability of the organic anion (such as β-hydroxybutyrate in ketoacidosis) to follow H^+ into the cell, thereby removing the necessity for K^+ to leave the cell [34,38].

The change in the plasma K^+ concentration is also much less prominent in metabolic alkalosis [34]. Although H^+ tends to move out of and K^+ into the cells in this disorder, there is only a small reduction in the plasma K^+ concentration induced by the rise in pH. This may be due in part to the observation that there is less intracellular buffering (and therefore less transcellular H^+ movement) in metabolic alkalosis than in metabolic acidosis (33 percent versus 57 percent) [39]. Large changes in the plasma K^+ concentration also do not occur in respiratory acidosis and alkalosis [34]. However, the mechanism responsible for this relative lack of K^+ movement is not well understood [34].

In summary, substantial pH-induced changes in the plasma K^+ concentration occur only in metabolic acidosis due to an excess of mineral acid. Although hyper-

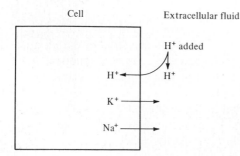

Figure 13-4 Reciprocal cation shifts of H^+, K^+, and Na^+ between the cells and the extracellular fluid. In the presence of a mineral acid load, H^+ moves into the cells where it is buffered. To maintain electroneutrality, K^+ and Na^+ leave the cells, resulting in an increase in the plasma K^+ concentration.

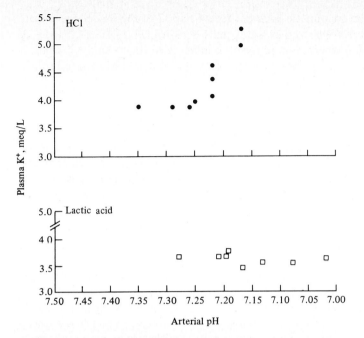

Figure 13-5 Change in the plasma K^+ concentration in relation to the arterial pH in experimentally induced hydrochloric acidosis (HCl is a mineral acid) and lactic acidosis in dogs. *(From G. O. Perez, J. R. Oster, and C. A. Vaamonde, Nephron, 27:233, 1981.)*

kalemia can occur in the organic acidoses, it results from factors other than the pH. In diabetic ketoacidosis, for example, hyperkalemia is commonly present due to insulin deficiency, renal failure, and the hyperglycemia-induced elevation in the plasma osmolality (see below) [35].

Rate of Cell Breakdown

Any condition which enhances cell breakdown (such as severe trauma) results in the release of K^+ into the ECF. The degree to which this will elevate the plasma K^+ concentration is related to the ability of other cells to take up the excess K^+ and of the kidney to augment K^+ excretion. On the other hand, conditions associated with a rapid *increase in cell production* can result in K^+ movement into the cells and hypokalemia. This sequence may occur after the administration of folic acid or vitamin B_{12} to patients with megaloblastic anemia who respond with a rapid and marked increase in the production of red cells and platelets [40].

Hyperosmolality

The plasma K^+ concentration may rise by as much as 0.4 to 0.8 meq/L for every 10 mosmol/kg elevation in the effective plasma osmolality (due primarily to hypernatremia or hyperglycemia) [41]. Since hyperosmolality results in the movement of water out of cells down an osmotic gradient (see Chap. 24), the associated hyper-kalemia may be due to *solvent drag* as water pulls K^+ with it into the ECF [41]. A clinical example of this phenomenon may be the increase in the plasma K^+ concentration that commonly accompanies hyperglycemia [42,43], although insulin deficiency also may contribute to this response.

Summary

In the basal state, the distribution of K^+ between the cells and the ECF is primarily governed by the Na^+-K^+-ATPase pump in the cell membrane. Although they are not so important in the basal state, catecholamines and insulin play a major role in promoting the cellular uptake of K^+ after a dietary load. This prevents a potentially serious elevation in the plasma K^+ concentration until the kidney can restore K^+ balance by excreting the excess K^+. These hormones appear to act permissively since their rate of secretion is not enhanced by K^+.

The plasma K^+ concentration may also directly influence K^+ distribution as K^+ moves into the cells with hyperkalemia and out of the cells with hypokalemia. As a result, the plasma K^+ concentration generally reflects the state of total body K^+ stores. This relationship may be disturbed, however, in a variety of conditions (such as exercise, certain forms of metabolic acidosis, or hyperosmolality) in which internal redistribution can lead to changes in the plasma K^+ concentration without change in K^+ stores.

RENAL POTASSIUM EXCRETION

Although small amounts of K^+ are lost each day in stool (5 to 10 meq) and sweat (0 to 10 meq), the kidney plays the major role in the maintenance of K^+ balance, appropriately varying K^+ secretion with changes in dietary intake (normal is 40 to 120 meq/day). *The primary event in urinary K^+ excretion is the secretion of K^+ from the tubular cell into the lumen in the distal nephron, particularly in the cortical collecting tubule*† [44,45]. Although a substantial amount of K^+ is filtered, almost all of this is reabsorbed (in part actively) in the proximal tubule and passively in the loop of Henle. Changes in K^+ excretion largely reflect changes in cortical collecting tubule K^+ secretion. The medullary collecting tubule normally reabsorbs K^+ [44,46],†† but K^+ secretion may occur with K^+ loading [47].

A typical cortical collecting tubule cell is depicted in Fig. 13-6. The final step, K^+ movement from the cell into the lumen, appears to be passive [44,48] and, therefore, is governed by those factors affecting passive transport (see Chap. 1):

---

†The anatomy and physiologic functions of the cortical distal nephron have recently been clarified (see Chap. 7). One of the findings has been that the cortical collecting tubule, *not* the distal tubule, is the primary site of K^+ secretion and the site at which aldosterone acts [45,46].

††Some of the K^+ reabsorbed in the medulla appears to reenter the tubule in the descending limb of the loop of Henle or perhaps the end of the proximal straight tubule. The physiologic role of this K^+ recycling is uncertain [46a].

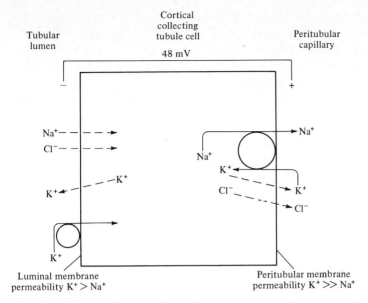

Figure 13-6 Schematic representation of some of the properties of a cortical collecting tubule cell. The high cell K^+ concentration, maintained by the active transport of K^+ into the cell across the peritubular membrane, the low tubular fluid K^+ concentration, the transepithelial potential difference (-48 mV, lumen negative), and the relatively high luminal membrane permeability to K^+ all favor the passive secretion of K^+ from the cell into the lumen. In addition to K^+ secretion, some K^+ reabsorption may occur by an active pump in the luminal membrane. *(Redrawn from G. Giebisch, Nephron, 6:260, 1969.)*

1. The concentration gradient across the luminal membrane, which is equal to the cell K^+ concentration minus the tubular fluid K^+ concentration
2. The electrical gradient, which is determined by the transepithelial potential difference
3. The K^+ permeability of the luminal membrane

As will be seen, *aldosterone* and the *plasma K^+ concentration* are the major physiologic regulators of K^+ secretion. The *flow rate to the distal nephron and* the *potential difference* generated by Na^+ reabsorption also are important, but they play a permissive rather than a regulatory role.

Cell Potassium Concentration

K^+ secretion varies directly with the cell K^+ concentration. Although the total cell K^+ concentration is similar to that in other cells at 140 to 150 meq/L, only about 25 to 35 percent (35 to 55 meq/L) is contained within the rapidly exchangeable transport pool that is available for K^+ secretion [49,50]. However, the size of this pool is not constant, appropriately increasing after a K^+ load and decreasing with K^+ depletion [45,49]. Aldosterone and the plasma K^+ concentration appear to be the prime determinants of this response. In addition, alterations in the arterial pH may also affect the cell K^+ concentration and therefore K^+ excretion.

Aldosterone Aldosterone plays a major role in K^+ homeostasis by augmenting K^+ secretion in the cortical collecting tubule (see Chap. 8) [51,52]. After a K^+ load, aldosterone secretion is directly enhanced, thereby promoting the excretion of the excess K^+ in the urine. Conversely, the release of this hormone is reduced with K^+ depletion, which tends to preserve K^+ balance by minimizing further urinary losses.

Aldosterone appears to act both by stimulating the Na^+-K^+-ATPase pump at the peritubular membrane, leading to an elevation in the cell K^+ concentration, and by increasing the permeability of the luminal membrane to K^+ [50,53]. In addition, the associated enhancement of Na^+ reabsorption may indirectly promote K^+ secretion by increasing the transepithelial potential difference (see below).

Plasma potassium concentration The plasma K^+ concentration can directly affect K^+ excretion by influencing the size of the K^+ transport pool; this effect is in part mediated by changes in Na^+-K^+-ATPase activity [54–56]. Thus, the mild elevation in the plasma K^+ concentration that follows a K^+ load promotes the excretion of the excess K^+ both by stimulating the secretion of aldosterone and by increasing the cell K^+ concentration.

The importance of the plasma K^+ concentration on urinary K^+ excretion is illustrated in Fig. 13-7 [56]. Dogs were adrenalectomized, given aldosterone replacement at different doses, and then studied at different levels of K^+ intake. As intake was increased, there was a gradual elevation in the plasma K^+ concentration. When aldosterone replacement was at a normal level (50 μg/day, middle curve), urinary K^+ excretion showed little change until the plasma K^+ concentration exceeded 4.2 meq/ L. At this point, K^+ excretion increased markedly in an attempt to maintain a normal plasma K^+ concentration. This presumably reflected a direct effect of the plasma

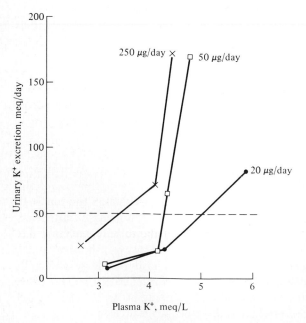

Figure 13-7 Mean values for plasma potassium concentration and steady-state urinary potassium excretion in adrenalectomized dogs given different levels of aldosterone replacement and studied at increasing levels of K^+ intake. The dashed line represents the effects seen when K^+ excretion is 50 meq/day. See text for details. *(From D. B. Young, and A. W. Paulsen, Am. J. Physiol., 244:F28, 1983.)*

K^+ concentration since aldosterone levels, Na^+ intake, and the urine output were relatively constant. In intact animals, aldosterone secretion will rise after a K^+ load, resulting in even more efficient excretion of K^+ (left curve, Fig. 13-7).

The experiments in Fig. 13-7 also demonstrate the effect of chronic changes in aldosterone secretion on the plasma K^+ concentration. If, for example, K^+ intake and excretion are 50 meq/day, the plasma K^+ concentration will be approximately 4.3 meq/L in normal dogs (50 μg/day), 3.4 meq/L with hyperaldosteronism (250 μg/day), and 5.0 meq/L with hypoaldosteronism (20 μg/day). When less aldosterone is available, urinary K^+ excretion becomes less efficient and a higher plasma K^+ concentration is required to establish a new steady state in which intake again equals excretion. Thus, *hypo*aldosteronism is associated with hyperkalemia whereas primary *hyper*aldosteronism tends to induce hypokalemia (see Chaps. 27 and 28).

Arterial pH As described above, changes in the arterial pH produce reciprocal H^+ and K^+ shifts between the cells and the ECF. As a result, K^+ tends to move into cells with alkalemia and out of cells with acidemia. These changes in the cell K^+ concentration in the kidney may be reflected in the rate of K^+ excretion which increases in alkalemia and falls in acidemia [57–59]. However, these effects are transient and are frequently overridden by concurrent variations in distal nephron flow rate and transepithelial potential difference (see below) [59,60].

Tubular Fluid Potassium Concentration and Distal Flow Rate

In addition to being affected by changes in the cell K^+ concentration, the concentration gradient favoring secretion varies inversely with the tubular fluid K^+ concentration. Since almost all the filtered K^+ is reabsorbed in the proximal tubule and loop of Henle, the K^+ concentration in the fluid entering the distal nephron is very low and may be less than 1 meq/L.† As K^+ is secreted into the lumen in the cortical collecting tubule, the increase in the tubular fluid K^+ concentration will reduce the concentration gradient across the cell membrane, which tends to retard further secretion. The degree to which this occurs is related to the distal flow rate (Fig. 13-8) [61,62]. As distal flow increases, the increment in the tubular fluid K^+ concentration produced by K^+ secretion is minimized because the secreted K^+ is washed away. As a result, the K^+ concentration in the lumen is kept at a relatively low level, maintaining a favorable gradient for continued K^+ secretion.

Conversely, there is an exaggerated rise in the tubular fluid K^+ concentration and a reduction in net K^+ secretion when distal flow is diminished due, for example, to volume depletion. In this setting, the increase in K^+ secretion following aldosterone is impaired in comparison to that seen in subjects in normal fluid balance [63].

The relationship between K^+ secretion and distal flow rate plays an important role in allowing aldosterone to regulate Na^+ and K^+ balance independently (see

---

†This can be appreciated from the micropuncture data in Fig. 4-3. The ratio of the tubular fluid K^+ concentration to the plasma K^+ concentration in the early distal tubule is 0.2. If the normal plasma K^+ concentration is 4.0 meq/L, the tubular fluid K^+ concentration at this site will be approximately 0.8 meq/L.

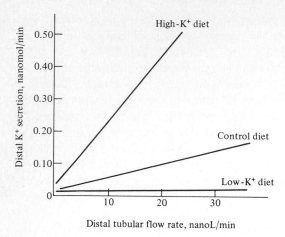

Figure 13-8 Combined effects of dietary intake and distal tubular flow rate on distal K^+ secretion. *(From R. M. Khuri, M. Wiederholt, N. Strieder, and G. Giebisch, Am. J. Physiol., 228:1249, 1975.)*

Table 8-2). For example, a Na^+ load expands the extracellular volume, resulting in a reduction in renin secretion and therefore that of aldosterone. Although the latter change promotes the excretion of the excess Na^+ (by decreasing cortical collecting tubule Na^+ reabsorption), it should also lead to K^+ retention and hyperkalemia. This does not happen, however, since volume expansion tends to increase the GFR and diminish proximal Na^+ reabsorption (see Chap. 9), both of which augment distal flow rate. The enhanced flow counteracts the effect of the fall in aldosterone release, resulting in no change in K^+ excretion [64]. Similarly, the combination of increased aldosterone secretion and reduced distal flow in hypovolemic states allows Na^+ to be conserved without substantially affecting K^+ balance.

Changes in distal flow rate can also explain the effect of some diuretics, such as furosemide and the thiazides, on K^+ balance. These agents, which inhibit Na^+ and water reabsorption proximal to the K^+ secretory site, frequently cause K^+ wasting and hypokalemia (see Chap. 16). The increase in urinary K^+ loss is at least in part due to the augmented distal flow [65]. Increased secretion of aldosterone (due to the diuretic-induced volume depletion) and decreased passive K^+ reabsorption at the site at which the diuretic acts may also contribute to this response.

The combined effects of dietary intake and tubular flow rate on K^+ secretion are illustrated in Fig. 13-8 [61]. Remember that passive K^+ secretion is proportional to the concentration gradient across the luminal membrane: the cell K^+ concentration (actually the K^+ concentration in the transport pool) minus the tubular fluid K^+ concentration. Since a high-potassium diet increases the cell K^+ concentration and a high urine flow rate reduces the tubular fluid K^+ concentration, it is not surprising that the maximum rate of K^+ secretion is seen when both are present. The flow rate is less of a factor when the cell K^+ concentration is reduced by a low-potassium diet.

Transepithelial Potential Difference

Since K^+ is a charged particle, its secretion also is affected by the transepithelial potential difference across the tubular cell. The normal potential difference in the

K^+-secreting cells is approximately -48 mV (lumen negative). This potential is generated by the transport of Na^+ from the lumen into the peritubular capillary (Fig. 13-9). Since Na^+ is positively charged, its reabsorption makes the lumen relatively electronegative. Cl^- follows down this electrical gradient after a finite time lag, and it is this delay that is responsible for the observed potential difference. The importance of Na^+ in the generation of this potential can be seen if the Na^+ in the lumen is replaced by a nonreabsorbable cation such as choline$^+$ [66]. In this setting, the potential difference falls to zero (Fig. 13-9). On the other hand, replacing the Cl^- in the lumen with a poorly reabsorbed anion, such as SO_4^{2-}, increases the anion delay and augments the potential difference. The enhanced luminal electronegativity favors the transfer of K^+, a positively charged ion, from the cell into the lumen [48,49].

The importance of the Na^+-generated potential difference on K^+ secretion can be illustrated by the response to the diuretic, amiloride [49,65]. This agent impairs the entry of luminal Na^+ into the cells of the distal nephron, apparently by decreasing the Na^+ permeability of the luminal membrane [67]. The net effect is diminished Na^+ reabsorption, a reduction in the transepithelial potential difference, and a

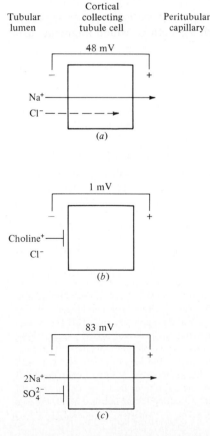

(a)

(b)

(c)

Figure 13-9 Representation of the electrical events at a typical cortical collecting tubule cell. (*a*) Na^+ is actively transported from the lumen into the capillary. Cl^- follows after a finite time lag; this delay is responsible for the transepithelial potential difference of -48 mV, lumen negative. (*b*) Replacing Na^+ in the lumen with the nonreabsorbable cation choline essentially eliminates the potential difference. Thus, it is the active transport of Na^+ that generates the potential difference. (*c*) Replacing Cl^- in the lumen with the nonreabsorbable anion SO_4^{2-} increases the anion delay and enhances the potential difference. This change in the potential difference favors the secretion of the cations K^+ and H^+ into the lumen. (*Data from G. Giebisch, G. Malnic, R. M. Klose, and E. E. Windhager, Am. J. Physiol., 211:560, 1966.*)

marked fall in K^+ secretion. Since amiloride has no known direct effect on K^+ handling, it is likely that the decrease in the potential difference is responsible for the change in K^+ secretion† [49].

Changing the potential difference by administering Na^+ with a nonreabsorbable anion also can affect K^+ excretion. For example, a volume-depleted subject has a strong stimulus to Na^+ reabsorption in the cortical collecting tubule that is mediated by aldosterone. In this situation, the administration of Na_2SO_4 results in Na^+ reabsorption without SO_4^{2-} and, consequently, increases in the potential difference (Fig. 13-9) and K^+ secretion [68]. In contrast, if Na^+ balance is normal, there is no stimulus to retain the excess Na^+ and Na_2SO_4 is excreted with only a small elevation in K^+ secretion [68]. An interesting clinical example of this phenomenon has been reported with the intravenous administration of the penicillin derivative carbenicillin. Carbenicillin is given as sodium carbenicillin and can act as a nonreabsorbable anion, resulting in urinary K^+ loss and hypokalemia [69,70].

Potassium Permeability of the Luminal Membrane

Changes in the luminal membrane permeability of K^+ can affect K^+ secretion. As described above, part of the kaliuretic effect of aldosterone appears to be due to an increase in the K^+ permeability of the luminal membrane [50]. Whether other factors affect membrane permeability in physiologic states is not known.

Potassium Wasting in Metabolic Acidosis

In certain conditions, several of the factors that affect K^+ secretion may be operative at one time. For example, metabolic acidosis should reduce urinary K^+ excretion by decreasing the cell K^+ concentration. However, in type 2 (proximal) renal tubular acidosis, proximal HCO_3^- reabsorption is impaired (see Chap. 19). This results in the increased delivery of Na^+, the relatively poorly reabsorbed anion HCO_3^-, and water to the distal secretory site. These effects overcome the direct effect of acidosis, and K^+ loss ensues [71]. A similar effect is seen in diabetic ketoacidosis where Na^+ is delivered to the distal tubule with the ketoacid anions, β-hydroxybutyrate and acetoacetate.

A different mechanism governs K^+ wasting in type 1 (distal) renal tubular acidosis. In this disorder, there is an inability to reduce the urine pH below 5.5 to 6.0

†These results indicate that Na^+ reabsorption plays an important, though indirect, role in K^+ secretion. Its effect is *permissive* in that an adequate potential difference must be present but alterations in K^+ secretion do not require changes in Na^+ reabsorption. This is in contrast to earlier suggestions that there was a tight 1:1 exchange of luminal Na^+ for cellular K^+ at the secretory site. This theory was based in part upon the apparent direct relationship between distal Na^+ delivery and K^+ secretion, as K^+ secretion falls when Na^+ delivery is reduced and rises when Na^+ delivery is enhanced. However, since Na^+ and water reabsorption are linked in the proximal tubule and loop of Henle (see Chaps. 5 and 6), changes in Na^+ delivery are associated with parallel changes in water delivery, i.e., in the distal flow rate, and it is the latter that is primarily responsible for the alterations in K^+ secretion [62]. In hypovolemic states, for example, Na^+ delivery falls but is not a limiting factor because it still exceeds the amount of K^+ secreted by more than 10-fold [48].

in the distal nephron, resulting in a decrease in net H^+ secretion. Na^+ that cannot be reabsorbed with Cl^- (about 25 meq/L) must be reabsorbed in exchange for H^+ or K^+. When H^+ secretion is diminished, K^+ secretion must increase to preserve Na^+ balance, and hypokalemia commonly ensues [72]. This effect probably is mediated by an increase in the potential difference as Na^+ is reabsorbed without an anion. If anions which can accept H^+ with little change in the urine pH, for example, buffers such as HCO_3^- or HPO_4^{2-}, are presented to the distal nephron, H^+ secretion can increase without reaching the limiting pH of 5.5 to 6.0. This eliminates the need for excessive K^+ secretion. Thus, the administration of $NaHCO_3$ or Na_2HPO_4 will correct both the acidosis and, at least in part, K^+ wasting by augmenting the secretion of H^+ and reducing that of K^+ [72,73].

Renal Response to Potassium Depletion and Potassium Loading

The regulation of K^+ excretion can be summarized by reviewing the renal responses to changes in K^+ balance. K^+ excretion appropriately falls with K^+ depletion. Initially, this response is mediated by reduced aldosterone secretion [74], the latter representing a direct effect of K^+ on the adrenal zona glomerulosa cells [75]. Within several days, however, a decrease in the cell K^+ concentration in the distal nephron assumes primary importance. At this time, neither the administration of aldosterone [74] nor increasing distal flow rate (Fig. 13-8) substantially enhances urinary K^+ loss. The fall in K^+ excretion appears to be due to both reduced secretion and increased active reabsorption [44,48], the latter occurring primarily in the medullary collecting tubule [46,76]. The net effect is that K^+ excretion can be lowered to 15 to 25 meq/day with a total K^+ deficit of 50 to 150 meq, and to 5 to 15 meq/day with more severe K^+ depletion [77].

On the other hand, K^+ excretion increases after a K^+ load due, as described above, to enhanced aldosterone secretion and the direct effect of the plasma K^+ concentration on that in the cell. As a result, normal subjects can maintain K^+ balance even if K^+ intake is slowly increased from the normal of about 80 meq/day to 500 meq/day or more [78]. This ability to handle what might be a lethal K^+ load if given acutely is called *K^+ adaptation* and is due both to enhanced K^+ entry into the extrarenal cells and, more importantly, to more rapid K^+ excretion in the urine [47,79,80]. Increased secretion of aldosterone appears to be responsible, by an unknown mechanism, for the extrarenal adaptation [79] and may contribute to, but does not explain all of, the renal response [55,81].

The increase in urinary K^+ excretion during adaptation is due to enhanced K^+ secretion throughout the late distal nephron, including the short connecting segment (see Fig. 2-3), and the cortical and medullary collecting tubules [46,47,82]. Although the medullary collecting tubule does not secrete K^+ in the basal state, it may account for as much as one-half the increase in K^+ excretion that occurs with K^+ loading [47]. Na^+-K^+-ATPase activity in these distal segments, but not in other nephron segments, is enhanced during adaptation [54,55], probably due both to aldosterone (which acts primarily in the cortical collecting tubule) [51,52] and to a small elevation in the plasma K^+ concentration. This elevation in Na^+-K^+-ATPase can facilitate K^+ excretion by increasing K^+ entry into the cell across the peritubular membrane, thereby augmenting the size of the K^+ transport pool.†

The major clinical example of K^+ adaptation is chronic renal failure in which the combination of a constant K^+ intake and fewer functioning nephrons requires an increase in K^+ secretion per nephron [83,84]. This allows K^+ balance to be maintained even in advanced renal failure (where the GFR is very low) as long as the urine output and therefore the distal flow rate are adequate [85]. In this setting, Na^+-K^+-ATPase activity in the distal nephron is increased when K^+ intake is normal but not when intake is restricted in proportion to the fall in GFR, a situation in which enhanced K^+ excretion per nephron is not required [86]. This finding suggests that the increase in Na^+-K^+-ATPase activity is appropriate and specific, not incidentally induced by renal insufficiency.

In addition, K^+ adaptation is also associated with an aldosterone-induced increase in Na^+-K^+-ATPase activity and K^+ secretion in the colon [87]. Although this is not likely to be quantitatively important in patients with relatively normal renal function, enhanced stool losses may account for the excretion of as much as 30 to 50 percent of dietary K^+ intake in patients with end-stage renal failure on chronic dialysis [88].

SUMMARY

The maintenance of a normal plasma K^+ concentration is dependent upon the ability of K^+ to enter the cells, where it achieves high concentrations, and upon the urinary excretion of the net dietary intake. After a K^+ load, most of the K^+ is initially taken up by the cells, a response that is facilitated by basal levels of catecholamines and insulin. This cell uptake minimizes the increase in the plasma K^+ concentration, pending the excretion of the excess K^+ in the urine.

Urinary K^+ excretion is largely a function of secretion in the distal nephron, particularly the cortical collecting tubule. The main factors modulating this process are (1) the cell K^+ concentration, which is affected by the plasma K^+ concentration and aldosterone, both of which increase after K^+ load and decrease with K^+ depletion; (2) the tubular fluid K^+ concentration at the secretory site, which is affected by the distal flow rate; and (3) the transepithelial potential difference, which is generated by Na^+ reabsorption. Aldosterone and the plasma K^+ concentration are the major physiologic regulators of K^+ secretion in that they vary appropriately with changes in intake. Distal flow rate and the potential difference are more *permissive*

---

†These reabsorptive and secretory changes appear to be mediated by different cell types, not by alterations in function in a uniform population of cells. The collecting tubules are composed of principal cells (60 to 70 percent) and intercalated cells (30 to 40 percent) [46]. The increase in K^+ reabsorption seen with K^+ depletion occurs in the intercalated cells in the medullary, but not the cortical collecting tubule [46,76]. K^+ adaptation, on the other hand, is associated with enhanced K^+ secretion in the principal cells in the cortex and medulla with no change in intercalated cell function [46,82]. Other than aldosterone, which presumably acts on the principal cells in the cortical collecting tubule, it is not known what factors are responsible for the difference between principal and intercalated cell function.

in that they do not change directly with K^+ balance but relatively normal values are required for adequate K^+ secretion.

PROBLEMS

13-1 How do changes in distal tubular flow rate affect K^+ excretion?

13-2 In a patient with primary hyperaldosteronism due to an adrenal adenoma, what effect would increased Na^+ intake have on urinary K^+ excretion? How does this differ from the response in normal subjects?

13-3 K^+ depletion is most often due to urinary or gastrointestinal losses of K^+. What test would be helpful in differentiating between these disorders?

13-4 Untreated patients with effective circulating volume depletion due to heart failure or hepatic cirrhosis (see Chap. 9) are generally normokalemic even though the activity of the renin-angiotensin-aldosterone system is frequently increased. Why doesn't aldosterone promote excess urinary K^+ loss in this setting? What would happen to K^+ excretion if a diuretic such as furosemide were then given to increase Na^+ and water excretion?

13-5 ADH increases water reabsorption in the collecting tubules. In patients with central diabetes insipidus, the urine output can exceed 10 L/day, because of decreased collecting tubule water reabsorption. What effect would this high-output state have on K^+ excretion?

REFERENCES

1. Sweadner, K. J., and S. M. Goldin: Active transport of sodium and potassium ions: mechanism, function, and regulation, N. Engl. J. Med., 302:777, 1980.
2. DeVoe, R. D., and P. C. Maloney: Principles of cell homeostasis, in *Medical Physiology*, V. B. Mountcastle (ed.), 14th ed., Mosby, St. Louis, 1980.
3. Cannon, P. L., L. E. Frazier, and R. H. Hughes: Sodium as a toxic ion in potassium deficiency, Metabolism, 2:297, 1953.
4. Lubin, M.: Intracellular potassium and control of protein synthesis, Fed. Proc., 23:994, 1964.
5. Beck, N., and S. K. Webster: Impaired urinary concentrating ability and cyclic AMP in K^+-depleted rat kidney, Am. J. Physiol., 231:1204, 1976.
6. Manitius, A., H. Levitin, D. Beck, and F. H. Epstein: On the mechanism of impairment of renal concentrating ability in potassium deficiency, J. Clin. Invest., 39:684, 1960.
7. Peterson, L. N.: On the mechanism of the concentrating defect in potassium depletion: the role of papillary plasma flow, Kidney Int., 19:252, 1981.
8. Bia, M. J., and R. A. DeFronzo: Extrarenal potassium homeostasis, Am. J. Physiol., 240:F257, 1981.
9. Sterns, R. H., M. Cox, P. U. Feig, and I. Singer: Internal potassium balance and the control of the plasma potassium concentration, Medicine, 60:339, 1981.
10. Reza, M. J., R. B. Kovick, K. I. Shine, and M. L. Pearce: Massive intravenous digoxin overdosage, N. Engl. J. Med., 291:777, 1974.
11. Minkoff, L., G. Gaertner, M. Darab, C. Mercier, and M. L. Levin: Inhibition of brain sodium-potassium ATPase in uremic rats, J. Lab. Clin. Med., 80:71, 1972.
12. Bilbrey, G. L., N. W. Carter, M. G. White, J. F. Schilling, and J. P. Knochel: Potassium deficiency in chronic renal failure, Kidney Int., 4:423, 1973.
13. Edmondson, R. P. S., R. D. Thomas, R. J. Hilton, J. Patrick, and N. F. Jones: Leucocyte electrolytes in cardiac and non-cardiac patients receiving diuretics, Lancet, 1:12, 1974.
14. Flear, C. T. G., W. T. Cooke, and A. Quinton: Serum-potassium levels as an index of body content, Lancet, 1:458, 1957.

15. Casey, T. H., W. H. J. Summerskill, and A. L. Orvis: Body and serum potassium in liver disease. I. Relationship to hepatic function and associated factors, Gastroenterology, 48:198, 1965.

16. Silva, P., and K. Spokes: Sympathetic system in potassium homeostasis, Am. J. Physiol., 241:F151, 1981.

17. DeFronzo, R. A., R. S. Sherwin, M. Dillingham, R. Hendler, W. V. Tamborlane, and P. Felig: Influence of basal insulin and glucagon secretion on potassium and sodium metabolism: studies with somatostatin in normal dogs and in normal and diabetic human beings, J. Clin. Invest., 61:472, 1978.

18. Winkler, A. W., H. E. Hoff, and P. K. Smith: The toxicity of orally administered potassium salts in renal insufficiency, J. Clin. Invest., 20:119, 1941.

19. DeFronzo, R. A., P. A. Taufield, H. Black, P. McPhedran, and C. R. Cooke: Impaired renal tubular potassium secretion in sickle cell disease, Ann. Intern. Med., 90:310, 1979.

20. Knochel, J. P.: Role of glucoregulatory hormones in potassium homeostasis, Kidney Int., 11:443, 1977.

20a. Brown, M. J., D. C. Brown, and M. B. Murphy: Hypokalemia from beta$_2$-receptor stimulation by circulating epinephrine, N. Engl. J. Med., 309:1414, 1983.

21. Rosa, R. M., P. Silva, J. B. Young, L. Landsberg, R. S. Brown, J. W. Rowe, and F. H. Epstein: Adrenergic modulation of extrarenal potassium disposal, N. Engl. J. Med., 302:431, 1980.

22. DeFronzo, R. A., M. Bia, and G. Birkhead: Epinephrine and potassium homeostasis, Kidney Int., 20:83, 1981.

23. Hiatt, N., L. W. Chapman, M. B. Davidson, J. A. Sheinkopf, and H. Mack: β-receptor mediated transfer in potassium loaded nephrectomized dogs, Proc. Soc. Exp. Biol. Med., 167:525, 1981.

24. Zierler, K. L., and D. Rabinowitz: Effect of very small concentrations of insulin on forearm metabolism. Persistence of its action on potassium and free fatty acids without its effects on glucose, J. Clin. Invest., 43:950, 1964.

25. Fenn, W. O.: The deposition of potassium and phosphate with glycogen in rat livers, J. Biol. Chem., 128:297, 1939.

26. Cox, M., R. H. Sterns, and I. Singer: The defense against hyperkalemia: the roles of insulin and aldosterone, N. Engl. J. Med., 299:525, 1978.

26a. Minaker, K. L., and J. W. Rowe: Potassium homeostasis during hyperinsulinemia: effect of insulin level, β-blockade, and age, Am. J. Physiol., 242:E373, 1982.

27. DeFronzo, R. A., R. Lee, A. Jones, and M. Bia: Effect of insulinopenia and adrenal hormone deficiency on acute potassium tolerance, Kidney Int., 17:586, 1980.

28. Knochel, J. P., and E. M. Schlein: On the mechanism of rhabdomyolysis in potassium depletion, J. Clin. Invest., 51:1750, 1972.

29. Sessard, J., M. Vincent, G. Annat, and C. A. Bizollon: A kinetic study of plasma renin and aldosterone during changes of posture in man, J. Clin. Endocrinol. Metab., 42:20, 1976.

30. Carlsson, E., E. Fellenius, P. Lundborg, and L. Svensson: β-adrenoceptor blockers, plasma potassium, and exercise (letter), Lancet, 2:424, 1978.

31. Lim, M., R. A. F. Linton, C. B. Wolff, and D. M. Band: Propranolol, exercise and arterial plasma potassium, Lancet, 2:591, 1981.

32. Coester, N., J. C. Elliott, and U. C. Luft: Plasma electrolytes, pH, and ECG during and after exhaustive exercise, J. Appl. Physiol., 34:677, 1973.

33. Brown, J. J., R. H. Chinn, D. L. Davies, R. Fraser, A. F. Lever, R. J. Rae, and J. I. S. Robertson: Falsely high plasma potassium values in patients with hyperaldosteronism, Br. Med. J., 2:18, 1970.

34. Adrogué, H. J., and N. E. Madias: Changes in plasma potassium concentration during acute acid-base disturbances, Am. J. Med., 71:456, 1981.

35. Fulop, M.: Serum potassium in lactic acidosis and ketoacidosis, N. Engl. J. Med., 300:1087, 1979.

36. Orringer, C. E., J. C. Eustace, C. D. Wunsch, and L. B. Gardner: Natural history of lactic acidosis after grand-mal seizures: a model for the study of an anion-gap acidosis not associated with hyperkalemia, N. Engl. J. Med., 297:796, 1977.

37. Perez, G. O., J. R. Oster, and C. A. Vaamonde: Serum potassium concentration in acidemic states, Nephron, 27:233, 1981.

38. Owen, O. E., H. Markus, S. Sarshik, and M. Mozzoli: Relationship between plasma and muscle concentrations of ketone bodies and free fatty acids in fed, starved, and alloxan-diabetic states, Biochem. J., 134:499, 1973.
39. Pitts, R. F.: *Physiology of the Kidney and Body Fluids*, Year Book, Chicago, 1974, chap. 11.
40. Lawson, D. H., R. M. Murray, and J. L. W. Parker: Early mortality in the megaloblastic anaemias, Q. J. Med., 41:1, 1972.
41. Moreno, M., C. Murphy, and C. Goldsmith: Increase in serum potassium resulting from the administration of hypertonic mannitol and other solutions, J. Lab. Clin. Med., 73:291, 1969.
42. Nicolis, G. L., T. Kahn, A. Sanchez, and J. L. Gabrilove: Glucose-induced hyperkalemia in diabetic subjects, Arch. Intern. Med., 141:48, 1981.
43. Viberti, G. C.: Glucose-induced hyperkalaemia: a hazard for diabetics?, Lancet, 1:690, 1978.
44. Wright, F. S.: Sites and mechanisms of potassium transport along the renal tubule, Kidney Int., 11:415, 1977.
45. Wright, F. S.: Potassium transport by successive segments of the mammalian nephron, Fed. Proc., 40:2398, 1981.
46. Stanton, B. A., D. Biemesderfer, J. B. Wade, and G. Giebisch: Structural and functional study of the rat distal nephron: effects of potassium adaptation and depletion, Kidney Int., 19:36, 1981.
46a. Strokes, J. B.: Consequences of K^+ recycling in the renal medulla. Effects on Ion transport by the medullary thick ascending limb of Henle's loop, J. Clin. Invest., 70:219, 1982.
47. Schon, D. A., K. A. Backman, and J. P. Hayslett: Role of the medullary collecting duct in potassium excretion in potassium-adapted animals, Kidney Int., 20:655, 1981.
48. Malnic, G., R. M. Klose, and G. Giebisch: Micropuncture study of distal tubular potassium and sodium transport in rat nephron, Am. J. Physiol., 211:529, 1966.
49. Garcia-Filho, E., G. Malnic, and G. Giebisch: Effects of changes in electrical potential difference on tubular potassium transport, Am. J. Physiol., 238:F235, 1980.
50. Hierholzer, K., and M. Wiederholt: Some aspects of distal tubular solute and water transport, Kidney Int., 9:198, 1976.
51. Garg, L. C., M. A. Knepper, and M. B. Burg: Mineralocorticoid effects on Na-K-ATPase in individual nephron segments, Am. J. Physiol., 240:F536, 1981.
52. Stokes, J. B., M. J. Ingram, A. D. Williams, and D. Ingram: Heterogeneity of the rabbit collecting tubule: localization of mineralocorticoid hormone action to the cortical portion, Kidney Int., 20:340, 1981.
53. Petty, K. J., J. P. Kokko, and D. Marver: Secondary effect of aldosterone on Na-K-ATPase activity in the rabbit cortical collecting tubule, J. Clin. Invest., 68:1514, 1981.
54. Silva, P., B. D. Ross, A. N. Charney, A. Besarab, and F. H. Epstein: Potassium transport by the isolated perfused kidney, J. Clin. Invest., 56:862, 1975.
55. Doucet, A., and A. I. Katz: Renal potassium adaptation: Na-K-ATPase activity along the nephron after chronic potassium loading, Am. J. Physiol., 238:F380, 1980.
56. Young, D. B., and A. W. Paulsen: Interrelated effects of aldosterone and plasma potassium on potassium excretion, Am. J. Physiol., 244:F28, 1983.
57. Malnic, G., M. deMello Aires, and G. Giebisch: Potassium transport across renal distal tubules during acid-base disturbances, Am. J. Physiol., 221:1192, 1971.
58. Barker, E. S., R. B. Singer, J. R. Elkinton, and J. K. Clark: The renal response in man to acute experimental respiratory alkalosis and acidosis, J. Clin. Invest., 36:515, 1957.
59. Stanton, B. A., and G. Giebisch: Effects of pH on potassium transport by renal distal tubule, Am. J. Physiol., 242:F544, 1982.
60. Gennari, F. J., and J. J. Cohen: Role of the kidney in potassium homeostasis: lesson from acid-base disturbances, Kidney Int., 8:1, 1975.
61. Khuri, R. M., M. Wiederholt, N. Strieder, and G. Giebisch: Effects of flow rate and potassium intake on distal tubular potassium transfer, Am. J. Physiol., 228:1249, 1975.
62. Good, D. W., and F. S. Wright: Luminal influences on potassium secretion: sodium concentration and fluid flow rate, Am. J. Physiol., 236:F192, 1979.
63. Seldin, D., L. Welt, and J. Cort: The role of sodium salts and adrenal steroids in the production of hypokalemic alkalosis, Yale J. Biol. Med., 29:229, 1956.

64. Young, D. B., and R. E. McCaa: Role of the renin-angiotensin system in potassium control, Am. J. Physiol., 238:R359, 1980.

65. Duarte, C. G., F. Chomety, and G. Giebisch: Effect of amiloride, ouabain, and furosemide on distal tubular function in the rat, Am. J. Physiol., 221:632, 1971.

66. Giebisch, G., G. Malnic, R. M. Klose, and E. E. Windhager: Effect of ionic substitutions on distal potential differences in rat kidney, Am. J. Physiol., 211:560, 1966.

67. Benos, D. J.: Amiloride: a molecular probe of sodium transport in tissues and cells, Am. J. Physiol., 242:C131, 1982.

68. Schwartz, W. B., R. L. Jenson, and A. S. Relman: Acidification of the urine and increased ammonium excretion without change in acid-base equilibrium: sodium reabsorption as a stimulus to the acidifying process, J. Clin. Invest., 34:673, 1955.

69. Lipner, H. T., F. Ruzany, M. Dasgupta, P. D. Lief, and N. Bank: The behavior of carbenicillin as a nonreabsorbable anion, J. Lab. Clin. Med., 86:183, 1975.

70. Klastersky, J., B. Vanderkelen, D. Daneua, and M. Mathieu: Carbenicillin and hypokalemia (letter), Ann. Intern. Med., 78:774, 1973.

71. Sebastian, A., E. McSherry, and R. C. Morris, Jr.: On the mechanism of renal potassium wasting in renal tubular acidosis with the Fanconi syndrone (type 2 RTA), J. Clin. Invest., 50:231, 1971.

72. Sebastian, A., E. McSherry, and R. C. Morris, Jr.: Renal potassium wasting in renal tubular acidosis (RTA): its occurrence in types 1 and 2 RTA despite sustained correction of systemic acidosis, J. Clin. Invest., 50:667, 1971.

73. Gill, J. R., Jr., N. H. Bell, and F. C. Bartter: Impaired conservation of sodium and potassium in renal tubular acidosis and its correction by buffer anions, Clin. Sci. 33:577, 1967.

74. Linas, S. L., L. N. Peterson, R. J. Anderson, G. A. Aisenbrey, F. R. Simon, and T. Berl: Mechanism of renal potassium conservation in the rat, Kidney Int., 15:601, 1979.

75. Aguitera, G., and K. J. Catt: Loci of action of regulators of aldosterone biosynthesis in isolated glomerulosa cells, Endocrinology, 104:1046, 1978.

76. Stetson, D. L., J. B. Wade, and G. Giebisch: Morphologic alterations in the rat medullary collecting duct following potassium depletion, Kidney Int., 17:45, 1980.

77. Squires, R. D., and E. J. Huth: Experimental potassium depletion in normal human subjects. I. Relation of ionic intakes to the renal conservation of potassium, J. Clin. Invest., 38:1134, 1959.

78. Talbott, J. H., and R. S. Schwab: Recent advances in the biochemistry and therapeusis of potassium salts, N. Engl. J. Med., 222:585, 1940.

79. Alexander, E. A., and N. G. Levinsky: An extrarenal mechanism of potassium adaptation, J. Clin. Invest., 47:740, 1968.

80. Hayslett, J. P., and H. J. Binder: Mechanism of potassium adaptation, Am. J. Physiol., 243:F103, 1982.

81. Silva, P., J. P. Hayslett, and F. H. Epstein: The role of Na-K-activated adenosine triphosphatase in potassium adaptation. Stimulation of enzymatic activity by potassium loading, J. Clin. Invest., 52:2665, 1973.

82. Rastegar, A. R., D. Biemesderfer, M. Kashgarian, and J. P. Hayslett: Changes in membrane surfaces of collecting duct cells in potassium adaptation, Kidney Int., 18:293, 1980.

83. Schultze, R. G., D. D. Taggart, H. Shapiro, J. P. Pennell, S. Caglar, and N. S. Bricker: On the adaptation in potassium excretion associated with nephron reduction in the dog, J. Clin. Invest., 50:1061, 1971.

84. Bourgoignie, J. J., M. Kaplan, J. Pincus, G. Gavellas, and A. Rabinovitch: Renal handling of potassium in dogs with chronic renal insufficiency, Kidney Int., 20:482, 1981.

85. Gonick, H. C., C. R. Kleeman, M. E. Rubini, and M. H. Maxwell: Functional impairment in chronic renal disease. III. Studies of potassium excretion, Am. J. Med. Sci., 261:281, 1971.

86. Schon, D. A., P. Silva, and J. P. Hayslett: Mechanism of potassium excretion in renal insufficiency, Am. J. Physiol., 227:1323, 1974.

87. Bastl, C., J. P. Hayslett, and H. J. Binder: Increased large intestinal secretion of potassium in renal insufficiency, Kidney Int., 12:9, 1977.

88. Hayes, C. P., Jr., and R. R. Robinson: Fecal potassium excretion in patients on chronic intermittent hemodialysis, Trans. Am. Soc. Artif. Intern. Organs, 11:242, 1965.

PHYSIOLOGIC APPROACH TO ACID-BASE AND ELECTROLYTE DISORDERS

FOURTEEN

MEANING AND APPLICATION OF URINE CHEMISTRIES

SODIUM EXCRETION
  Fractional Excretion of Sodium
CHLORIDE EXCRETION
POTASSIUM EXCRETION
URINE OSMOLALITY
URINE pH

As will be discussed in the ensuing chapters, measurement of the urinary electrolyte concentrations, osmolality, and pH play an important role in the diagnosis and management of a variety of disorders. This chapter will briefly review the meaning of these parameters and the settings in which they may be helpful (Table 14-1). It is important to emphasize that there are *no fixed normal values* since the kidney varies the rate of excretion to match net dietary intake and endogenous production. Thus, interpretation of a given test requires knowledge of the patient's clinical state. For example, the urinary excretion of 125 meq of Na^+ per day may be appropriate for a subject on a regular diet but represents inappropriate renal Na^+ wasting in a patient who is volume-depleted.

In addition to their clinical usefulness, these tests are simple to perform and are widely available. In most circumstances, a random urine specimen is sufficient, although a 24-h collection to determine the daily rate of solute excretion is occasionally indicated. When K^+ depletion is due to extrarenal losses, for example, the urinary K^+ excretion should fall below 25 meq/day. A random measurement, however,

Table 14-1 Clinical application of urine chemistries

Parameter	Uses
Na^+ excretion	Assessment of volume status
	Differential diagnosis of hyponatremia and acute renal failure
	Dietary compliance in patients with hypertension
	Evaluation of calcium and uric acid excretion in stone-formers
Cl^- excretion	Similar to that for Na^+ excretion
	Differential diagnosis of metabolic alkalosis
K^+ excretion	Differential diagnosis of hypokalemia
Osmolality	Differential diagnosis of hyponatremia, hypernatremia, and acute renal failure
pH	Diagnosis of renal tubular acidosis
	Efficacy of treatment of metabolic alkalosis

may be confusing in some patients. If the urine output is only 500 mL/day because of the associated volume depletion, then the appropriate excretion of only 20 meq of K^+ per day will be associated with an apparently high urine K^+ concentration of 40 meq/L (20 meq/day ÷ 0.5 L/day = 40 meq/L).

SODIUM EXCRETION

The kidney varies the rate of Na^+ excretion to maintain the effective circulating volume, a response that is mediated by a variety of factors including the renin-angiotensin-aldosterone system.† As a result, the urine Na^+ concentration can be used as an estimate of the patient's volume status. In particular, a urine Na^+ concentration below 10 to 15 meq/L is generally indicative of hypovolemia. This finding is especially useful in the differential diagnosis of both *hyponatremia* and *acute renal failure*. The two major causes of hyponatremia are effective volume depletion and the syndrome of inappropriate antidiuretic hormone secretion (SIADH). The urine Na^+ concentration should be low in the former but greater than 20 meq/L in the SIADH, which is characterized by water retention and volume expansion (see Chap. 23).

Similar considerations apply to acute renal failure which is most often due to volume depletion or acute tubular necrosis [1]. The urine Na^+ concentration usually exceeds 20 meq/L in the latter in part because of the associated tubular damage and

---

†The concept of the effective circulating volume and how it is regulated is discussed in detail in Chap. 9. In particular, effective volume depletion includes true volume depletion (due to gastrointestinal or renal losses) as well as edematous states such as heart failure and hepatic cirrhosis in which tissue perfusion is respectively reduced because of a primary decrease in cardiac output or sequestration of fluid in the splanchnic circulation and peritoneum (as ascites).

a consequent inability to maximally reabsorb Na^+ [1,2]. Measuring the fractional excretion of Na^+ and the urine osmolality also can help to differentiate between these conditions (see below).

In normal subjects, urinary Na^+ excretion roughly equals average dietary intake. Thus, measurement of urinary Na^+ excretion (by obtaining a 24-h collection) can be used to check dietary compliance in patients with essential hypertension. Restriction of Na^+ intake is frequently an important component of the therapeutic regimen [3], and adequate adherence should result in the excretion of less than 100 meq/day. The concurrent use of diuretics does not interfere with the utility of this test. For example, a thiazide diuretic initially increases Na^+ and water excretion by reducing Na^+ reabsorption in the distal tubule. However, the ensuing volume depletion enhances Na^+ reabsorption both in the distal nephron (via aldosterone) and the proximal tubule to prevent progressive fluid loss [4]. The net effect is the establishment within 1 week of a new steady state in which the plasma volume is somewhat diminished but Na^+ excretion is again equal to intake [5].

Measurement of urinary Na^+ excretion is also important when evaluating patients with recurrent kidney stones. A 24-h urine collection is typically obtained in this setting to determine if calcium or uric acid excretion is increased, either of which can predispose to stone formation [5a]. However, the tubular reabsorption of both calcium and uric acid is linked to that of Na^+ (see Chap. 5). Thus, hypovolemia will increase their reabsorption and may mask the presence of underlying hypercalciuria or hyperuricosuria. Concomitant measurement of the rate of Na^+ excretion can indicate the degree to which Na^+ balance may be contributing; a value above 100 meq/day suggests Na^+ reabsorption is not an important limiting factor for calcium or uric acid excretion.

Despite its usefulness, there are some pitfalls in relying upon the measurement of Na^+ excretion as an index of volume status. A low urine Na^+ concentration, for example, may be seen in *normovolemic patients* who have selective renal or glomerular ischemia due to bilateral renal artery stenosis or acute glomerulonephritis [2,6]. On the other hand, the urine Na^+ concentration may be inappropriately high in the presence of volume depletion if there is a defect in tubular Na^+ reabsorption. This may occur with the use of diuretics,† in aldosterone deficiency, or in advanced renal failure [7,8].

The urine Na^+ concentration can also be influenced by the rate of water reabsorption. If there is a selective decrease in water reabsorption due to the absence of ADH (called diabetes insipidus), the urine output can exceed 10 L/day. In this setting, the daily excretion of 100 meq of Na^+ will be associated with a urine Na^+ concentration of 10 meq/L or less, incorrectly suggesting the presence of volume depletion. Conversely, a high rate of water reabsorption can raise the urine Na^+ concentration and mask the presence of hypovolemia. To correct for the effect of water reabsorption, the renal handling of Na^+ can be evaluated directly by calculating the fractional excretion of Na^+ (FE_{Na+}).

---

†Although chronic diuretic use does not prevent the attainment of a new steady state, urinary Na^+ excretion that is equal to intake is still inappropriately high in a hypovolemic patient.

Fractional Excretion of Sodium

The FE_{Na+} can be calculated from a random urine specimen [2]:

$$FE_{Na+}(\%) = \frac{\text{quantity of Na}^+ \text{ excreted}}{\text{quantity of Na}^+ \text{ filtered}} \times 100$$

The quantity of Na^+ excreted is equal to the product of the urine Na^+ concentration (U_{Na+}) and the urine volume (V); the quantity of Na^+ filtered is equal to the product of the plasma Na^+ concentration (P_{Na+}) and the glomerular filtration rate (or creatinine clearance $= U_{cr} \times V/P_{cr}$). Therefore,

$$
\begin{aligned}
FE_{Na+}(\%) &= \frac{U_{Na+} \times V}{P_{Na+} \times (U_{cr} \times V/P_{cr})} \times 100 \\
&= \frac{U_{Na+} \times P_{cr}}{P_{Na+} \times U_{cr}} \times 100
\end{aligned}
$$

The primary use of the FE_{Na+} is in patients with acute renal failure. As described above, a low urine Na^+ concentration favors the diagnosis of volume depletion, whereas a high value points toward acute tubular necrosis. However, a level between 20 and 40 meq/L may be seen with either disorder [2]. This overlap, which is due in part to variations in the rate of water reabsorption, can be minimized by calculating the FE_{Na+} [2,9]. Na^+ reabsorption is appropriately enhanced in hypovolemic states, and the FE_{Na+} is less than 1 percent; i.e., more than 99 percent of the filtered Na^+ has been reabsorbed. In contrast, tubular damage leads to an FE_{Na+} in excess of 2 to 3 percent in acute tubular necrosis. One major exception occurs when acute tubular necrosis is superimposed upon chronic effective volume depletion (see Chap. 9) due to hepatic cirrhosis, congestive heart failure, or severe burns. In these settings, the FE_{Na+} may remain under 1 percent because of the intense stimulus to Na^+ retention [10–12]. A similar finding may be seen when acute renal failure follows the administration of radiocontrast media [13]. The mechanism responsible for the low FE_{Na+} in this disorder is uncertain.

The FE_{Na+} is much less useful in patients with normal renal function. If the GFR is 180 L/day and the plasma Na^+ concentration is 150 meq/L, then 27,000 meq of Na^+ will be filtered each day. Thus, the FE_{Na+} will be under 1 percent if daily Na^+ intake is in the normal range of 125 to 250 meq even if the patient is normovolemic. As a result, the urine Na^+ concentration is more helpful in diagnosing volume depletion unless a very low FE_{Na+} is obtained, e.g., less than 0.2 percent.

CHLORIDE EXCRETION

Chloride is reabsorbed with sodium throughout the nephron (see Chaps. 5 to 7). As a result, the rate of excretion of these ions is usually similar, and measurement of the urine Cl^- concentration generally adds little to the information obtained from the more routinely measured urine Na^+ concentration. However, experimental studies employing a high-Na^+, low-Cl^- diet indicate that Cl^- can be conserved independent of Na^+, a response that appears to be mediated in part by active Cl^- transport in the inner medullary collecting tubule [14].

Similar findings may be seen in humans. As many as 30 percent of hypovolemic patients will have more than a 15-meq/L difference between the urine Na^+ and Cl^- concentrations [15]. This is due to the excretion of Na^+ with another anion (HCO_3^-, carbenicillin) or Cl^- with another cation (ammonium, calcium). Thus, it may be helpful to measure the urine Cl^- concentration in a patient who seems to be volume-depleted but has a somewhat elevated urine Na^+ concentration. This most often occurs in metabolic alkalosis in which the urinary excretion of some of the excess HCO_3^- as $NaHCO_3$ can elevate the urine Na^+ concentration (in some cases to over 100 meq/L). This can mask the presence of underlying hypovolemia, which, in the patient with metabolic alkalosis, is usually due to vomiting or prior diuretic use. The urine Cl^- concentration, however, will remain appropriately low (unless some diuretic effect persists; see Chap. 18).

POTASSIUM EXCRETION

Potassium excretion varies appropriately with intake, a response that is mediated primarily by aldosterone and the direct effect of changes in the plasma K^+ concentration (see Chap. 13). If K^+ depletion occurs, urinary K^+ excretion can fall to a minimum of 5 to 25 meq/day [16]. As a result, measurement of K^+ excretion may aid in the diagnosis of unexplained hypokalemia. An appropriately low value suggests either extrarenal losses (usually from the gastrointestinal tract) or the use of diuretics (if the collection has been obtained after the diuretic effect has worn off). Values greater than 25 meq/day indicate at least a component of renal K^+ wasting (see Chap. 27).

Measurement of K^+ excretion is less helpful in patients with hyperkalemia. If K^+ intake is increased *slowly,* normal subjects can take in and excrete more than 500 meq of K^+ per day (normal daily intake is 40 to 120 meq) without a substantial elevation in the plasma K^+ concentration [17]. Thus, if one excludes an *acute* K^+ load (due to excess intake or release from cells), hyperkalemia almost always results from an inability of the kidneys to excrete K^+ adequately (see Chap. 28).

URINE OSMOLALITY

Variations in the urine osmolality (U_{osm}) play a central role in the regulation of the plasma osmolality (P_{osm}) and Na^+ concentration, a response that is primarily mediated by osmoreceptors in the hypothalamus that influence the secretion of ADH (see Chap. 10). After a water load, there is a transient reduction in the P_{osm}, which lowers ADH release. This diminishes water reabsorption in the collecting tubules, resulting in the excretion of the excess water in a dilute urine ($U_{osm} < P_{osm}$). Water restriction, on the other hand, sequentially raises the P_{osm}, ADH secretion, and renal water reabsorption, resulting in water retention and the excretion of a concentrated urine ($U_{osm} > P_{osm}$).

These relationships allow the U_{osm}† to be helpful in the differential diagnosis of both *hyponatremia* and *hypernatremia*. Hyponatremia with hypoosmolality should shut off ADH release. As a result, a maximally dilute urine should be excreted with the U_{osm} falling below 100 mosmol/kg. If this is found, then the hyponatremia is probably due to excess water intake at a rate that exceeds normal excretory capacity. More commonly, the U_{osm} is inappropriately high and the hyponatremia results from an inability of the kidneys to excrete free water normally (see Chap. 23). In contrast, hypernatremia with hyperosmolality should stimulate ADH secretion, and the U_{osm} should exceed 600 to 800 mosmol/kg. If a concentrated urine is found, then extrarenal water loss (from the respiratory tract or skin) or the administration of Na^+ in excess of water is responsible for the elevation in the plasma Na^+ concentration. A U_{osm} below that of the plasma represents primary renal water loss due to ADH lack or resistance (see Chap. 24).

The U_{osm} (in addition to the FE_{Na^+}) also may be helpful in distinguishing volume depletion from acute tubular necrosis as the cause of acute renal failure. Tubular dysfunction in the latter condition results in the excretion of urine with an osmolality between 1.0 and 1.2 times that of the plasma [18,19]. In contrast, hypovolemia is a potent stimulus to ADH secretion (see Fig. 8-6), which should result in a U_{osm} exceeding 450 mosmol/kg if there is no underlying renal disease [18,19]. Thus, a high U_{osm} *excludes* the diagnosis of acute tubular necrosis. The finding of an isosmotic urine, however, is less useful diagnostically. It is consistent with acute tubular necrosis but does not rule out volume depletion since there may be a concomitant impairment in concentrating ability, a common finding in the elderly or in patients with severe reductions in the GFR [20,21].

URINE pH

The urine pH generally reflects the degree of acidification of the urine and normally varies with systemic acid-base balance. The major clinical use of the urine pH occurs in patients with metabolic acidosis. The appropriate response to this disturbance is to increase urinary acid excretion, with the urine pH falling below 5.3 and usually below 5.0. Values above 5.3‡ in adults and 5.6 in children indicate abnormal urinary

---

†The specific gravity can also be used as an estimate of urinary concentration. It is, however, less accurate than the urine osmolality, being influenced by the size as well as the number of particles (see Fig. 1-8).

‡The diagnostic use of the urine pH requires the demonstration of a sterile urine. Infection with any of the urinary pathogens which produce urease results in the metabolism of urinary urea into ammonia. The excess ammonia (NH_3) directly elevates the urine pH according to the Henderson-Hasselbalch equation (see Chap. 11):

$$pH = 9.3 + \log \frac{[NH_3]}{[NH_4^+]}$$

nary acidification and the presence of renal tubular acidosis. Distinction between the various types of renal tubular acidosis can then be made by measurement of the urine pH and fractional excretion of HCO_3^- (using a formula similar to that for the FE_{Na^+} described above) at different plasma HCO_3^- concentrations (see Chap. 19).

Monitoring the urine pH may also be helpful in assessing the efficacy of treatment of metabolic alkalosis associated with true volume depletion due to vomiting or diuretics. This disorder is characterized by elevations in the plasma HCO_3^- concentration and pH but an inappropriately acid urine (pH < 6.0) because the volume depletion prevents the excretion of the excess HCO_3^- as $NaHCO_3$. As fluids are given to restore normovolemia, $NaHCO_3$ excretion will ultimately increase and the urine pH will exceed 7.0. A persistently low urine pH usually indicates inadequate volume repletion.

REFERENCES

1. Rose, B. D., *Pathophysiology of Renal Disease,* McGraw-Hill, New York, 1981, chap. 3.
2. Miller, T. R., R. J. Anderson, S. L. Linas, W. L. Henrich, A. S. Berns, P. A. Gabow, and R. W. Schrier: Urinary diagnostic indices in acute renal failure: a prospective study, Ann. Intern. Med., 89:47, 1978.
3. MacGregor, G. A., N. D. Markandu, F. E. Best, D. M. Elder, J. M. Cam, G. A. Sagnella, and M. Squires: Double-blind randomised crossover trial of moderate sodium restriction in essential hypertension, Lancet, 1:351, 1982.
4. Martinez-Maldonado, M., G. Eknoyan, and W. N. Suki: Diuretics in nonedematous states: physiological basis for the clinical use, Arch. Intern. Med., 131:797, 1973.
5. Maronde, R. F., M. Milgrom, N. D. Vlachakis, and L. Chan: Response of thiazide-induced hypokalemia to amiloride, J. Am. Med. Assoc., 249:237, 1983.
5a. Rose, B. D.: *Pathophysiology of Renal Disease,* McGraw-Hill, New York, 1981, pp. 674, 677.
6. Besarab, A., R. S. Brown, N. T. Rubin, E. Salzman, L. Wirthlin, T. Steinman, R. R. Atlia, and J. J. Skillman: Reversible renal failure following bilateral renal artery occlusive disease: clinical features, pathology, and the role of surgical revascularization, J. Am. Med. Assoc., 235:2838, 1976.
7. Coleman, A. J., M. Arias, N. W. Carter, F. C. Rector, Jr., and D. W. Seldin: The mechanism of salt-wasting in chronic renal disease, J. Clin. Invest., 45:1116, 1966.
8. Danovitch, G. M., J. J. Bourgoignie, and N. S. Bricker: Reversibility of the "salt-losing" tendency of chronic renal failure, N. Engl. J. Med., 296:14, 1977.
9. Oken, D. E.: On the differential diagnosis of acute renal failure, Am. J. Med., 71:916, 1981.
10. Diamond, J. R., and D. C. Yoburn: Nonoliguric acute renal failure associated with a low fractional excretion of sodium, Ann. Intern. Med., 96:597, 1982.
11. Planas, M., T. Wachtel, H. Frank, and L. W. Henderson: Characterization of acute renal failure in the burned patient, Arch. Intern. Med., 142:2087, 1982.
12. Hilberman, M., B. D. Myers, B. J. Carrie, G. Derby, R. L. Jamison, and E. B. Stinson: Acute renal failure following cardiac surgery, J. Thorac. Cardiovasc. Surg., 77:880, 1979.
13. Fang, L. S. T., R. A. Sirota, T. H. Ebert, and N. S. Lichenstein: Low fractional excretion of sodium with contrast media-induced acute renal failure, Arch. Intern. Med., 140:531, 1980.
14. Luke, R. G.: Effect of adrenalectomy on the renal response to chloride depletion in the rat, J. Clin. Invest., 54:1329, 1974.
15. Sherman, R. A., and R. P. Eisinger: The use (and misuse) of urinary sodium and chloride measurements, J. Am. Med. Assoc., 247:3121, 1982.
16. Squires, R. D., and E. J. Huth: Experimental potassium depletion in normal human subjects. I. Relation of ionic intakes to the renal conservation of potassium, J. Clin. Invest., 38:1134, 1959.

17. Talbott, J. H., and R. S. Schwab: Recent advances in the biochemistry and therapeusis of potassium salts, N. Engl. J. Med., 222:585, 1940.
18. Luke, R. G., J. D. Briggs, M. E. M. Allison, and A. C. Kennedy: Factors determining response to mannitol in acute renal failure, Am. J. Med. Sci., 259:168, 1970.
19. Jones, L. W., and M. H. Weil: Water, creatinine and sodium excretion following circulatory shock with renal failure, Am. J. Med., 51:314, 1971.
20. Sporn, I. N., R. G. Lancestremere, and S. Papper: Differential diagnosis of oliguria in aged patients, N. Engl. J. Med., 267:130, 1962.
21. Levinsky, N. G., D. G. Davidson, and R. W. Berliner: Effects of reduced glomerular filtration on urine concentration in presence of antidiuretic hormone, J. Clin. Invest., 38:730, 1959.

HYPOVOLEMIC STATES

The extracellular fluid is contained in two major compartments: the interstitium and the vascular space (see Chap. 1). In a variety of clinical disorders, depletion of the extracellular fluid usually develops because of fluid losses from the body. As a result, *there are reductions in the plasma and interstitial volumes* which, if severe, cause a potentially fatal decrease in tissue perfusion. Fortunately, early diagnosis and treatment can restore normovolemia in almost all cases.

Table 15-1 Etiology of true volume depletion

A. Gastrointestinal losses
   1. Gastric: vomiting or nasogastric suction
   2. Intestinal, pancreatic, or biliary: diarrhea, fistulas, ostomies, or tube drainage
   3. Bleeding
B. Renal losses
   1. Salt and water: diuretics, osmotic diuresis, adrenal insufficiency, or salt-wasting nephropathies
   2. Water: central or nephrogenic diabetes insipidus
C. Skin and respiratory losses
   1. Insensible losses from skin and respiratory tract
   2. Sweat
   3. Burns
   4. Skin lesions
   5. Marked bronchorrhea
   6. Drainage of a large pleural effusion
D. Sequestration into a third space
   1. Intestinal obstruction or peritonitis
   2. Crush injury or skeletal fractures
   3. Acute pancreatitis
   4. Bleeding
   5. Obstruction of a major venous system

ETIOLOGY

True volume depletion occurs when fluid is lost from the extracellular fluid at a rate exceeding net intake. These losses may occur from the gastrointestinal tract, skin, or lungs; in the urine; or by acute sequestration in the body in a "third space" that is not in equilibrium with the extracellular fluid (Table 15-1).

When these losses occur, two factors tend to protect against the development of hypovolemia. First, dietary Na^+ and water intake are generally far above basal needs. Therefore, relatively large losses must occur unless intake is concomitantly reduced, e.g., due to vomiting. Second, the kidney normally minimizes further urinary losses by enhancing Na^+ and water reabsorption. This renal response explains, for example, why patients given a diuretic for hypertension usually do not become very volume-depleted. Although a thiazide diuretic inhibits NaCl reabsorption in the distal tubule, the initial volume loss stimulates the renin-angiotensin-aldosterone system (and possibly other compensatory mechanisms), resulting in increased proximal and collecting tubule Na^+ reabsorption (see Chap. 9) [1,2]. This balances the diuretic effect, preventing progressive volume depletion.

Gastrointestinal Losses

Each day approximately 3 to 6 liters of fluid is secreted by the stomach, pancreas, gallbladder, and intestines into the lumen of the gastrointestinal tract. Almost all this fluid is reabsorbed, with only 100 to 200 mL being lost in the stool. However, if reabsorption is decreased (as with external drainage) or secretion is increased (as with diarrhea), volume depletion may ensue.

Acid-base disturbances frequently occur with gastrointestinal losses, depending upon the site from which the fluid is lost (see Table 1-6). Secretions from the stomach contain high concentrations of H^+ and Cl^-. As a result, vomiting and nasogastric suction are generally associated with metabolic alkalosis. In contrast, intestinal, pancreatic, and biliary secretions are relatively alkaline with high concentrations of HCO_3^-. Thus, the loss of these fluids due to diarrhea, laxative abuse, fistulas, ostomies, or tube drainage tends to cause metabolic acidosis. Hypokalemia is also commonly associated with these disorders since K^+ is present in all the gastrointestinal secretions.

Acute bleeding from any site in the gastrointestinal tract is another common cause of volume depletion. Electrolyte disturbances usually do not occur in this setting (except for shock-induced lactic acidosis) since it is plasma not gastrointestinal secretions that is lost. The clinical picture is somewhat different with chronic blood loss, e.g., due to a colonic carcinoma, in that hypovolemia is typically absent since there is time for compensatory renal Na^+ and water retention.

Renal Losses

Under normal conditions, renal Na^+ and water excretion is adjusted to match intake. In a 70-kg adult male, approximately 180 liters is filtered across the glomerular capillaries each day. More than 98 to 99 percent of the filtrate is then reabsorbed by the tubules, resulting in a urine output averaging 1 to 2 L/day. Thus, a small (1 to 2 percent) reduction in tubular reabsorption can lead to a 2- to 4-liter increase in Na^+ and water excretion, which, if not replaced, can result in severe volume depletion.

Salt and water loss A variety of conditions can lead to excessive urinary excretion of NaCl and water (Table 15-1). *Diuretics* inhibit active Na^+ transport at different sites in the nephron, resulting in an increased rate of excretion (see Chap. 16). Although they are frequently given to remove fluid in edematous patients, they can produce true hypovolemia if used in excess.

The presence of large amounts of nonreabsorbed solutes in the tubule also can inhibit Na^+ and water reabsorption, resulting in an *osmotic diuresis* (see page 119). Common clinical examples occur in uncontrolled diabetes mellitus and in renal failure in which glucose and urea, respectively, act as the osmotic diuretics. With severe hyperglycemia, urinary losses can induce a net fluid deficit of as much as 8 to 10 liters (see Chap. 25).

Adrenal insufficiency is frequently associated with hypovolemia since aldosterone and, to a lesser degree, cortisol stimulate Na^+ reabsorption and H^+ and K^+ secretion in the cortical collecting tubule. When the secretion of these hormones is reduced, there is a tendency toward Na^+ loss and K^+ and H^+ retention, the last two resulting in hyperkalemia and metabolic acidosis.

Variable degrees of salt wasting are also present in many renal diseases. Most patients with renal insufficiency (GFR less than 25 mL/min) are unable to maximally conserve Na^+ if acutely placed on a low-sodium diet. These patients may have an *obligatory* Na^+ loss of 10 to 40 meq/day, in contrast to normal subjects who can lower Na^+ excretion to less than 5 meq/day [3]. This degree of Na^+ wasting is usually not important since normal Na^+ balance is maintained as long as the patient is on a regular diet.

In rare cases, a more severe degree of Na^+ wasting is present in which obligatory urinary losses may exceed 100 meq of Na^+ and 2 liters of water per day. In this setting, hypovolemia will ensue unless the patient maintains a high Na^+ intake. This picture of a *salt-wasting nephropathy* is most often seen in tubular and interstitial diseases such as medullary cystic disease, chronic pyelonephritis, polycystic renal disease, and, infrequently, after the release of bilateral urinary tract obstruction (referred to as a postobstructive diuresis)† [4–7].

Three factors are thought to contribute to this variable salt-wasting: the osmotic diuresis produced by increased urea excretion in the remaining functioning nephrons; direct damage to the tubular epithelium which, in severe cases, can impair the response to aldosterone; and, perhaps most important, an inability to acutely shut off natriuretic forces [3,5,10–12]. Patients with renal insufficiency tend to have a decreased number of functioning nephrons. Therefore, if Na^+ intake remains normal, they must be able to augment Na^+ excretion per functioning nephron to maintain Na^+ balance [10]. This requires a fall in tubular Na^+ reabsorption that may be mediated at least in part by a natriuretic hormone (see Chap. 8). Thus, the salt wasting that occurs when Na^+ intake is abruptly lowered could represent persistent activation of these natriuretic forces. Consistent with this hypothesis is the observation that apparent salt-wasters (with acute obligatory losses of as much as 300 meq/day) can maintain Na^+ balance on an intake of only 5 meq/day if intake is reduced over a period of weeks rather than rapidly [12].

Therapy of renal salt wasting must be directed toward establishing the level of Na^+ intake required to remain in Na^+ balance. Enough Na^+ must be given to replace obligatory losses. This can be determined by measuring 24-h urinary Na^+ excretion when the patient is on a sodium-restricted diet (30 to 40 meq/day). Sodium intake can then be set just above the level of Na^+ excretion. It should not be assumed, however, that a patient with salt wasting has a normal ability to excrete a salt load. Some patients with renal insufficiency who become hypovolemic with Na^+ restriction may retain Na^+ and develop edema and hypertension if placed on a high-sodium diet. In these patients, the range of Na^+ intake compatible with the maintenance of Na^+ balance is relatively narrow and must be determined empirically.

Salt-wasting nephropathies may also be complicated by hypokalemia or hyperkalemia. If the impairment in tubular reabosrption occurs in the proximal tubule or loop of Henle, there will be an increase in Na^+ and water delivery to the K^+ secretory site in the cortical collecting tubule, resulting in enhanced K^+ secretion and possibly K^+ depletion and hypokalemia. In severe cases, daily K^+ losses in the urine can exceed 200 meq [7]. However, *hyper*kalemia is more common and may be due to aldosterone resistance [5] or, if unreplaced Na^+ losses lead to volume depletion, to reduced renal perfusion which limits K^+ excretion [13].

---

†Although a postobstructive diuresis is common, it is generally *appropriate* in that the diuresis consists of fluid retained during the period of obstruction [8,9]. In this sense, it is not a salt-wasting condition and complete replacement of urinary losses is unnecessary.

Water loss In some patients, volume depletion results from a selective increase in urinary water excretion. This is due to decreased water reabsorption in the collecting tubules, where antidiuretic hormone (ADH) promotes the reabsorption of water but not Na^+. Therefore, an impairment in either ADH secretion (central diabetes insipidus) or the renal response to ADH (nephrogenic diabetes insipidus) may be associated with the excretion of relatively large volumes (up to 10 to 20 L/day in severe cases) of dilute urine (see Chap. 24). As water is lost, the initial elevation in the plasma osmolality and Na^+ concentration stimulates thirst (see Chap. 8). As a result, water balance is usually maintained since intake is increased to match the excess losses. However, water loss, hypovolemia, and persistent hypernatremia will ensue in infants, comatose patients (neither of whom have ready access to water), or those with a defective thirst mechanism.

Skin and Respiratory Losses

Each day approximately 700 to 1000 mL of water is lost by evaporation from the skin and respiratory tract (see Chap. 10). Since heat is required for the evaporation of water, these insensible losses play an important role in thermoregulation, allowing the dissipation of some of the heat generated from body metabolism. When external temperatures are high or metabolic heat production is increased (as with fever or exercise), further heat can be lost by the evaporation of sweat (a "sensible" loss) from the skin. Although sweat (Na^+ concentration equals 30 to 50 meq/L) production is low in the basal state, it can exceed 1000 mL/h in a subject exercising in a hot, dry climate. As with the water loss in patients with diabetes insipidus, negative water balance due to these insensible and sensible losses is usually prevented by the thirst mechanism. However, the cumulative sweat Na^+ losses can lead to hypovolemia. This is particularly true in cystic fibrosis, where the sweat Na^+ concentration may be much greater than normal [14].

In addition to its role in thermoregulation, the skin acts as a barrier which prevents the loss of interstitial fluid to the external environment. When this barrier is interrupted by burns or skin lesions, e.g., exfoliative dermatitis or pemphigus, a large volume of fluid can be lost [15]. In contrast to the insensible and sweat losses, which are primarily water, the fluid lost in these disorders has an electrolyte composition similar to that of the plasma and contains a variable amount of protein. Thus, the replacement therapy in a burn patient may differ from that in a patient with increased insensible or sweat losses.

Although rare, pulmonary losses other than those by evaporation can lead to volume depletion. This most often occurs in patients with alveolar cell carcinoma who may have a marked increase in bronchial secretions (bronchorrhea) [16] or in patients that have continuous drainage of an active, usually malignant pleural effusion.

Sequestration into a Third Space

Volume depletion can be produced by the loss of interstitial and intravascular fluid into a third space which is not in equilibrium with the extracellular fluid. For example, a patient with a fractured hip may lose 1500 to 2000 mL of blood into the tissues adjacent to the fracture. Although this fluid will be resorbed back into the extracellular fluid over a period of days to weeks, the acute reduction in the blood volume can lead to severe volume depletion if this fluid is not replaced. Other examples of this phenomenon can occur with intestinal obstruction, severe pancreatitis, crush injuries, bleeding, e.g., due to trauma or a ruptured abdominal aortic aneurysm, peritonitis, and obstruction of a major venous system, e.g., inferior vena caval ligation for recurrent pulmonary emboli.

The main difference between these disorders and, for example, the sequestration of fluid that occurs in the peritoneal cavity (as ascites) in patients with hepatic cirrhosis is the rate of fluid accumulation. In cirrhosis, this occurs relatively slowly, allowing time for renal Na^+ and water retention to replenish the effective circulating volume. As a result, cirrhotic patients usually have symptoms of edema rather than those of hypovolemia. However, if ascites accumulates rapidly, e.g., after the acute removal of a large volume of ascitic fluid, hypotension and shock can occur (see below).

HEMODYNAMIC RESPONSES TO VOLUME DEPLETION

Volume depletion induces a characteristic sequence of compensatory hemodynamic responses. The initial volume deficit results in decreases in the plasma volume, venous return to the heart, and cardiac output. From the relationship between mean arterial pressure, cardiac output, and systemic vascular resistance,†

$$\text{Mean arterial pressure} = \text{cardiac output} \times \text{resistance}$$

the fall in cardiac output lowers the systemic blood pressure. This is sensed by the carotid sinus baroreceptors, which induce an increase in sympathetic activity that tends to restore the blood pressure toward normal (see Fig. 9-3). This is achieved by enhanced venous return, cardiac contractility, and heart rate (all of which act to elevate the cardiac output) and by increased vascular resistance due both to direct sympathetic effects and to enhanced secretion of renin from the kidney, resulting in the generation of the potent vasoconstrictor angiotensin II.

If the volume deficit is small (about 10 percent of the blood volume, which is equivalent to donating 500 mL of blood), these sympathetic effects return the cardiac output and blood pressure to normal or near normal although the heart rate is likely to be increased. In contrast, a marked fall in blood pressure will ensue if the sympathetic response does not occur, as a result of autonomic insufficiency, for example (see Fig. 9-4) [17,18].

With more severe hypovolemia (15 to 25 percent of the blood volume), there is

---

†The product of the cardiac output and the systemic vascular resistance actually equals the change in pressure across the circulation, i.e., mean arterial pressure minus mean venous pressure. However, since the mean venous pressure (normal equals 1 to 7 mmHg) is normally so much lower than the mean arterial pressure, only a slight error results from ignoring the venous pressure.

increased sympathetic and angiotensin II–mediated vasoconstriction. Although this may maintain the blood pressure when the patient is recumbent, hypotension may occur when the upright position is assumed, leading to postural dizziness. The intense vasoconstriction also results in the shunting of blood away from the mesenteric, renal, and musculocutaneous circulations with the relative preservation of cerebral and coronary blood flow. At this point, the compensatory sympathetic responses are maximal, and any further fluid loss will induce marked hypotension, even in recumbency, and eventually shock (see below).

SYMPTOMS

Three sets of symptoms can occur in hypovolemic patients: (1) those related to the manner in which fluid loss occurs, e.g., vomiting, diarrhea, or, in a patient with renal losses, polyuria; (2) those due to volume depletion; and (3) those due to the electrolyte and acid-base disorders which can accompany volume depletion.

The symptoms induced by hypovolemia are primarily related to the decrease in tissue perfusion. The earliest complaints may include lassitude, easy fatigability, thirst, muscle cramps, and postural dizziness. With more severe degrees of volume depletion, abdominal pain, chest pain, or lethargy and confusion may develop as a result of mesenteric, coronary, or cerebral ischemia. These symptoms usually are reversible, although tissue necrosis may develop if the low-flow state is allowed to persist.

Symptomatic hypovolemia most often occurs in patients with isosmotic Na^+ and water depletion in whom most of the fluid deficit comes from the extracellular fluid. In contrast, in patients with pure water loss due to insensible losses or diabetes insipidus, the elevation in the plasma osmolality (and Na^+ concentration) causes water to move down an osmotic gradient from the cells into the extracellular fluid. The net result is that about two-thirds of the water lost comes from the intracellular fluid. Consequently, these patients are likely to exhibit the symptoms of hypernatremia (produced by the water deficit) before those of extracellular fluid depletion.

A variety of electrolyte and acid-base disorders may occur, depending upon the composition of the fluid that is lost (see below). These include hypo- and hypernatremia, hypo- and hyperkalemia, and metabolic alkalosis and acidosis. The more serious symptoms produced by these disturbances include muscle weakness (hypo- and hyperkalemia), polyuria and polydipsia (hypokalemia and hyperglycemia), and lethargy, confusion, seizures, and coma (hypo- and hypernatremia and hyperglycemia).†

An additional symptom that appears to occur only in primary adrenal insufficiency is extreme salt craving. Patients with this disease frequently give a history of heavily salting all foods (including foods not usually salted) and even of eating salt that they have sprinkled on their hands. The mechanism responsible for this appropriate increase in salt intake is not known.

---

†The symptoms produced by these disorders are discussed in detail in the following chapters.

EVALUATION OF THE HYPOVOLEMIC PATIENT

Physical Examination

The physical examination may suggest the presence of volume depletion even in a patient thought to be normovolemic. A decrease in the interstitial volume can be detected by the examination of the skin and mucous membranes; a decrease in the plasma volume is indicated by reductions in the systemic blood pressure and the venous pressure in the jugular veins.

Skin and mucous membranes If the skin on the thigh, calf, or forearm is pinched in normal subjects, it will immediately return to its normally flat state when the pinch is released. This elastic property, called *turgor,* is partially dependent upon the interstitial volume of the skin and subcutaneous tissue. When the interstitial volume is reduced, the skin turgor is decreased and the skin flattens more slowly after the pinch is released. In younger patients, decreased skin turgor is a reliable clue to the presence of volume depletion. In older patients (more than 55 to 60 years old), however, reduced skin turgor is a common finding because of a primary decrease in the elasticity of the skin and therefore is a less valid indicator of hypovolemia. In these patients, the skin elasticity is usually best preserved on the inner aspect of the thighs and the skin overlying the sternum. Decreased turgor at these sites is suggestive of volume depletion.

Although reduced skin turgor is an important clinical finding, normal turgor *does not exclude* the presence of hypovolemia. This is particularly true with mild volume deficits in young patients, whose skin is very elastic, and in obese patients since fat deposits under the skin prevent the changes in subcutaneous turgor from being appreciated.

In addition to having reduced turgor, the skin is usually dry. The tongue and oral mucosa may also be dry in the hypovolemic patient since salivary secretions are commonly decreased in this setting.

Examination of the skin may also be helpful in the diagnosis of primary adrenal insufficiency. Because of impaired secretion of cortisol (which inhibits the release of ACTH), there is hypersecretion of ACTH, which can result in increased pigmentation of the skin, especially in the palmar creases and buccal mucosa.

Arterial blood pressure As described above, the arterial blood pressure changes from near normal with mild hypovolemia to low in the upright position and then, with progressive volume depletion, to persistently low regardless of posture. Postural hypotension leading to dizziness may be the patient's major complaint and is strongly suggestive of hypovolemia in the absence of an autonomic neuropathy or the use of sympatholytic drugs for hypertension.

It should be noted that the definition of normal blood pressure in this setting is dependent upon the patient's basal value. Although 120/80 is considered "normal," it is actually low in a hypertensive patient whose usual blood pressure is 180/100.

Venous pressure The reduction in the vascular volume seen with hypovolemia is primarily in the venous circulation (which normally contains 70 percent of the blood volume), resulting in a decrease in venous pressure. Thus, measurement of the venous pressure is useful both in the diagnosis of hypovolemia and in assessing the adequacy of volume replacement.

In most patients, the venous pressure can be estimated accurately from the examination of the external jugular vein which runs across the sternocleidomastoid muscle. The patient should initially be recumbent, with the head turned slightly away from the side to be examined. The external jugular vein can be identified by placing the forefinger just above the clavicle and pressing lightly. This will occlude the external jugular vein, which will distend as blood from the head enters the vein. The vein usually can be seen more easily by shining a beam of light obliquely across the neck. When the occlusion is released, the venous pressure will be approximately equal to the *vertical distance* between the upper level of the fluid column within the vein and the level of the right atrium (estimated as being 5 to 6 cm posterior to the sternal angle of Louis). To prevent distention of the vein by blood flowing back toward the heart from the head, the venous pressure should be measured with the vein occluded superiorly by the examiner's finger. If the vein is distended throughout its length, the patient's trunk should be elevated to 15 to 45° until an upper level can be seen. In a patient with a markedly increased venous pressure due to right ventricular failure, the external jugular vein may remain distended even when the patient is upright. The normal venous pressure is 1 to 8 cmH$_2$O or 1 to 6 mmHg (1.36 cmH$_2$O is equal to 1.0 mmHg).

Rarely, kinking or obstruction of the external jugular vein at the base of the neck causes an increase in the external jugular venous pressure which does not reflect a similar change in right atrial pressure. If this is suspected, e.g., if an elevated venous pressure is found in a patient with no evidence or history of cardiac or pulmonary disease, or if the jugular veins cannot be seen in a patient with a fat neck, the central venous, i.e., right atrial, pressure can be measured directly by the insertion of a catheter [19].

Relationship between right atrial and left atrial pressures The filling pressures in the heart are important determinants of cardiac output since the contractility of cardiac muscle and therefore the stroke volume increase as the filling pressure is increased (Fig. 15-1). If there is no obstruction to flow across the mitral valve, the left atrial pressure will be equal to the left ventricular end-diastolic pressure (LVEDP), that is, to the filling pressure in the left ventricle. The left atrial pressure can be estimated clinically by the measurement of the pulmonary capillary wedge pressure [20]. A flow-directed balloon catheter (such as a Swan-Ganz catheter) is passed through the right ventricle into the pulmonary artery. When the balloon is inflated, the catheter will occlude a small branch of the pulmonary artery. Since there is no forward flow past the catheter, the pressure that is measured reflects the pressure transmitted back from the left atrium. In general, there is a predictable relationship between the right and left atrial pressures, the latter being greater by

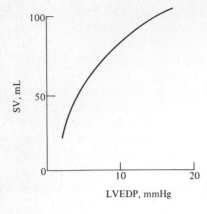

Figure 15-1 Frank-Starling curve relating stroke volume (SV) to left ventricular end-diastolic pressure (LVEDP). *(Adapted from J. N. Cohn, Am. J. Med., 55:351, 1973.)*

approximately 5 mmHg (Fig. 15-2) [21]. When the right atrial (or central venous) pressure is reduced, the LVEDP also is decreased, and this tends to lower the cardiac output. Conversely, a high central venous pressure is associated with a high left atrial pressure, which predisposes toward the development of pulmonary edema.

Although it is the LVEDP (not the right atrial pressure) that is an important determinant of left ventricular output and therefore tissue perfusion, measurement of the central venous pressure is useful because of its direct relationship to the LVEDP. However, there are two clinical settings in which the central venous pressure is not an accurate estimate of the LVEDP (Fig. 15-2). In patients with pure

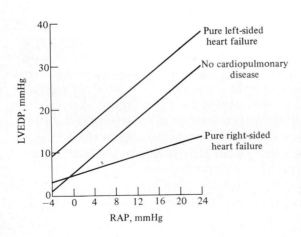

Figure 15-2 Relationship between left ventricular end-diastolic pressure (LVEDP) and mean right artrial pressure (RAP) in three groups of patients. In subjects without cardiopulmonary disease, the LVEDP exceeds the RAP by about 5 mmHg and varies directly with the RAP. In patients with pure right-sided heart failure, e.g., due to chronic pulmonary disease, relatively large changes in the RAP can occur with little change in the LVEDP. In contrast, the LVEDP is much greater than the RAP (average equals 13.6 mmHg) in patients with pure left-sided heart failure, e.g., due to an acute myocardial infarction. This graph is somewhat simplified since the standard deviations within each group have been omitted. *(Adapted from J. N. Cohn, F. E. Tristani, and I. M. Khatri, J. Clin. Invest., 48:2008, 1969, by copyright permission of the American Society for Clinical Investigation.)*

Table 15-2 Laboratory changes in hypovolemic states

Urine Na^+ concentration less than 10 to 15 meq/L
Urine osmolality greater than 450 mosmol/kg
BUN/plasma creatinine ratio greater than 20:1 with a normal urinalysis
Variable effects on plasma Na^+, K^+, and HCO_3^- concentrations
Occasional elevations in the hematocrit and plasma albumin concentration

left-sided heart failure, e.g., due to an acute myocardial infarction, the wedge pressure is increased but, since right ventricular function is normal, the central venous pressure tends to be normal. In this setting, treating a low central venous pressure with volume expanders can precipitate pulmonary edema. On the other hand, in patients with pure right-sided heart failure, e.g., due to cor pulmonale, the central venous pressure tends to exceed the LVEDP. Since these patients may have high central venous pressures even in the presence of volume depletion, the central venous pressure cannot be used as a guide to therapy.

Patients in shock The symptoms and physical findings that have been described apply to patients with mild to moderate volume depletion who are still able to maintain an adequate level of tissue perfusion. However, as the degree of hypovolemia becomes more severe, e.g., from the loss of 30 percent of the blood volume from a bleeding duodenal ulcer, there is a marked reduction in tissue perfusion, resulting in a clinical syndrome referred to as *hypovolemic shock*.† This syndrome is associated with a marked increase in sympathetic activity and is characterized by tachycardia, cold, clammy extremities, cyanosis, a low urine output (less than 15 mL/h), and agitation and confusion due to reduced cerebral blood flow. Although hypotension is generally present, it is not required for the diagnosis of shock since some patients vasoconstrict enough to maintain a relatively normal blood pressure. Therapy to restore tissue perfusion must be begun immediately to prevent both ischemic tissue damage and irreversible shock (see below).

Laboratory Data

Hypovolemia can produce a variety of changes in the composition of the urine and blood (Table 15-2). In addition to confirming the presence of volume depletion, these changes can give important clues to the pathogenesis of the fluid loss and to the appropriate replacement therapy.

Urine sodium concentration The response of the kidney to volume depletion is to conserve Na^+ and water in an attempt to expand the extracellular volume. With the exception of those disorders in which Na^+ reabsorption is impaired, the urine Na^+

---

†In addition to volume depletion, shock can be caused by several other conditions, including cardiac failure (cardiogenic shock) and sepsis. A discussion of these disorders is beyond the scope of this chapter and has been reviewed elsewhere [22].

Table 15-3 Urine Na$^+$ concentration in volume depletion

Less than 10 to 15 meq/L	Greater than 20 meq/L
Gastrointestinal losses	Renal disease
Skin losses	Diuretics (while the drug is acting)
Third-space losses	Osmotic diuresis
Diuretics (late)	Adrenal insufficiency
	Some patients with metabolic alkalosis

concentration will usually be less than 10 to 15 meq/L and may be as low as 1 meq/L (Table 15-3). This increase in tubular Na$^+$ reabsorption is mediated by several factors, including increased aldosterone secretion (via the renin-angiotensin system), enhanced sympathetic tone, and altered intrarenal hemodynamics (see Chap. 9).

The urine Cl$^-$ concentration is usually similar to that of Na$^+$ in hypovolemic states since Na$^+$ and Cl$^-$ are generally reabsorbed together. An exception occurs when Na$^+$ is excreted with another anion [22a]. This is most often seen in metabolic alkalosis, where the need to excrete the excess HCO$_3^-$ (as NaHCO$_3$) may raise the urine Na$^+$ concentration despite the presence of volume depletion. Thus, the urine Cl$^-$ concentration frequently is a better index of volume status in this setting (see Chap. 18) and should be measured when an apparently hypovolemic patient has what appears to be an inappropriately high urine Na$^+$ concentration.

Even if the physical examination is not diagnostic of hypovolemia, the finding of a low urine Na$^+$ concentration is virtually pathognomonic of reduced tissue perfusion.† However, a low urine Na$^+$ concentration does not necessarily mean that the patient has true volume depletion, since edematous patients with congestive heart failure, hepatic cirrhosis with ascites, and the nephrotic syndrome also avidly conserve Na$^+$. As will be described in Chap. 16, these patients have a depleted effective circulating volume due to a primary reduction in cardiac output (heart failure) or to the movement of fluid out of the vascular space into the peritoneal cavity (cirrhosis) or into the interstitium (nephrotic syndrome). The differentiation between edematous states and true volume depletion usually is made easily from the physical examination.

Urine osmolality The renal retention of water in hypovolemic states is mediated in part by ADH which is secreted in response to the decrease in tissue perfusion (see Chap. 8). As a result, the urine is relatively concentrated with an osmolality usually exceeding 450 mosmol/kg. This response may not be seen, however, if concentrating ability is impaired by renal disease, an osmotic diuresis, the administration of diuretics, or central or nephrogenic diabetes insipidus (see Chap. 24). For example, both severe volume depletion (which impairs urea accumulation in the renal medulla) [25] and hypokalemia (which induces ADH resistance; see Fig. 13-1) can limit the increase in the urine osmolality in some patients.

---

†There are some conditions in which the kidneys avidly retain Na$^+$ independent of systemic fluid balance. This is most likely to occur when there is selective renal or glomerular hypoperfusion, as with bilateral renal artery stenosis or acute glomerulonephritis [23,24].

Urinary concentration can also be assessed by measuring the specific gravity. This test, however, is less accurate than the osmolality since it is dependent upon the size as well as the number of solute particles in the urine (see Fig. 1-8). As a result, it should be used only if the osmolality cannot be measured; a value above 1.015 is suggestive of a concentrated urine, as is usually seen with hypovolemia.

BUN and plasma creatinine concentration In most circumstances the BUN (blood urea nitrogen) and plasma creatinine concentration vary inversely with the glomerular filtration rate (GFR), increasing as the GFR falls (see Fig. 3-11). Thus, serial measurements of these parameters can be used to look for changes in the GFR. However, an elevation in the BUN can also be produced by an increase in the rate of urea production or tubular reabsorption. As a result, the plasma creatinine concentration is a more reliable estimate of the GFR since it is produced at a relatively constant rate by skeletal muscle and is not reabsorbed by the renal tubules.

In normal subjects and those with uncomplicated renal disease, the BUN/ plasma creatinine ratio is approximately 10:1. However, this value may be substantially elevated in hypovolemic states because of the associated changes in tubular reabsorption [26]. In general, about 40 to 50 percent of filtered urea is reabsorbed, much of this occurring in the proximal tubule, where it is passively linked to the reabsorption of Na^+ and water (see Chap. 5). Thus, the increase in proximal Na^+ reabsorption in volume depletion produces a parallel increase in urea reabsorption. The net effect is a fall in urea excretion and elevations in the BUN and the BUN/plasma creatinine ratio, often to greater than 20:1. This selective rise in the BUN is called *prerenal azotemia*. The plasma creatinine concentration will increase in this setting only if the degree of hypovolemia is severe enough to lower the GFR.

Although the BUN/plasma creatinine ratio is helpful in the evaluation of hypovolemic patients, it is subject to misinterpretation since it is also affected by the rate of urea production. A high ratio may be due solely to increased urea production (as with gastrointestinal bleeding), whereas a normal ratio may occur in some patients with hypovolemia if urea production is reduced. This can be illustrated by the following example:

Case history 15-1 A 40-year-old man with a history of peptic ulcer disease was seen after 2 weeks of persistent vomiting. On physical examination, the patient's blood pressure was normal but his neck veins were flat and skin turgor was reduced. The laboratory data included:

$$BUN = 62 \text{ mg/dL}$$
$$\text{Plasma [creatinine]} = 6.6 \text{ mg/dL}$$
$$\text{Urine } [Na^+] = 7 \text{ meq/L}$$
$$\text{Urine osmolality} = 602 \text{ mosmol/kg}$$

COMMENT. In this patient, the low urine Na^+ concentration, the high urine osmolality, and the physical examination all pointed to hypovolemia. This diagnosis was confirmed by the return of the BUN and plasma creatinine concentration to normal levels with volume repletion. The failure of the BUN to increase out of proportion to the plasma creatinine concentration probably was due in part to reduced protein intake in a patient with vomiting.

Urinalysis In patients with elevations in the BUN and plasma creatinine concentration, examination of the urine is an important diagnostic tool. The urinalysis is generally normal in hypovolemic states since the kidney is not diseased. This is in contrast to most, but not all, of the other causes of renal insufficiency in which the urinalysis reveals protein, cells, and/or casts [27].

Hypovolemia and renal disease The laboratory diagnosis of hypovolemia may be difficult to establish in patients with underlying renal disease. In this setting, the urine Na^+ concentration may exceed 10 meq/L and the urine osmolality may be less than 350 mosmol/kg since renal insufficiency impairs the ability to maximally conserve Na^+ [3,10] and to concentrate the urine [28]. In addition, the urinalysis may be abnormal due to the primary disease.

Despite these difficulties, making the correct diagnosis is important since volume depletion is a *reversible* cause of worsening renal function in contrast to progression of the underlying disease. The history and physical examination (?vomiting, diarrhea, use of diuretics, or decreased skin turgor) may be helpful in some patients, but the diagnosis can be confirmed only by a positive response to a trial of fluid repletion.

A somewhat different problem may arise when it is not known whether acute renal failure is due to volume depletion (as in Case history 15-1) or to some intrinsic renal disease such as acute tubular necrosis. A random urine measurement may not be diagnostic since a high rate of water reabsorption can occasionally raise the urine Na^+ concentration to greater than 20 meq/L in hypovolemia and a relatively low rate of water reabsorption can reduce the urine Na^+ concentration by dilution to below 20 meq/L in some patients with acute tubular necrosis [24]. In this setting, calculation of the fraction of the filtered Na^+ that is excreted may be very helpful since a value under 1 percent is highly suggestive of volume depletion (see Chap. 14) [24].

Plasma sodium concentration In most hypovolemic states (such as vomiting, diarrhea, or bleeding), the fluid that is lost is isosmotic to plasma and will not directly change the plasma osmolality or the plasma Na^+ concentration (see page 652). However, volume depletion is a stimulus to both thirst and ADH secretion, and the increases in water intake and renal water reabsorption can result in water retention and *hypo*natremia (Table 15-4). Alternatively, *hyper*natremia can occur when water is lost in excess of solute. This can be seen with increased insensible or sweat losses and with central or nephrogenic diabetes insipidus. Even with these disorders, symptomatic hypernatremia is unusual in an alert patient since the increase in the plasma osmolality stimulates thirst and the ingested water returns the plasma Na^+ concentration toward normal.

Table 15-4 Plasma Na^+ concentration in volume depletion

May be greater than 150 meq/L	May be less than 135 meq/L
Insensible and sweat losses Central or nephrogenic diabetes insipidus	All other forms of volume depletion

Table 15-5 Acid-base disorders that may occur in volume depletion

Metabolic acidosis	Metabolic alkalosis
Diarrhea	Vomiting or nasogastric suction
Intestinal fistulas, ostomies, or tube drainage	Diuretics (excluding the K^+-sparing diuretics)
Renal failure	
Adrenal insufficiency	
Shock (lactic acidosis)	
Ketoacidosis (uncontrolled diabetes mellitus)	

A clinical illustration of the effects of volume depletion on the plasma Na^+ concentration can be seen in a febrile patient with diarrhea. The loss of diarrheal fluid, which is isosmotic to plasma, does not affect the plasma Na^+ concentration. However, fever increases water loss as sweat, and volume depletion stimulates thirst, resulting in enhanced water intake. Most commonly, the increases in water loss and water intake are of the same magnitude and there is little change in the plasma Na^+ concentration. If water intake is not increased, as in infants who do not have access to water, the insensible and sweat losses can lead to negative water balance and hypernatremia. Conversely, positive water balance and hyponatremia can occur in hypovolemic adults who are able to satisfy their thirst.

Plasma potassium concentration Either hypokalemia or hyperkalemia can occur in hypovolemic patients. The former is much more common, because of K^+ loss from the gastrointestinal tract or in the urine. Hyperkalemia may be seen in several settings. First, the plasma K^+ concentration may be elevated in some forms of metabolic acidosis. As the excess H^+ ions enter the cells to be buffered, intracellular K^+ moves into the extracellular fluid to maintain electroneutrality (see Chap. 13). Thus, a patient may have an elevated plasma K^+ concentration even if total body K^+ stores are reduced. Second, there may be an inability to excrete the dietary K^+ load in the urine because of renal failure, adrenal insufficiency (due to decreased secretion of aldosterone), or volume depletion itself since the delivery of Na^+ and water to the K^+ secretory site in the cortical collecting tubule will be reduced [13].

Acid-base balance The effect of fluid loss on acid-base balance is variable. Although most patients remain in normal acid-base balance, either metabolic alkalosis or metabolic acidosis can occur (Table 15-5). Patients with vomiting or nasogastric suction or those given diuretics tend to develop metabolic alkalosis because of H^+ loss and volume contraction (see Chap. 18). On the other hand, HCO_3^- loss (due to diarrhea or intestinal fistulas) or reduced renal H^+ excretion (due to renal failure or adrenal insufficiency) can lead to metabolic acidosis (see Chap. 19). In addition, lactic acidosis can occur in shock and ketoacidosis in uncontrolled diabetes mellitus.

Hematocrit and plasma albumin concentration Since the red blood cells and albumin are essentially limited to the vascular space, a reduction in the plasma volume due to volume depletion tends to elevate both the hematocrit and plasma albumin concentration. If these changes are evident, they are suggestive of hemoconcentration due to volume depletion. However, they are frequently absent because of underlying anemia and/or hypoalbuminemia, due, for example, to bleeding or renal disease.

Summary An accurate history and physical examination can help to determine both the presence and etiology of volume depletion. In the patient in whom the diagnosis cannot be made from the history, laboratory data can provide important clues to the correct diagnosis. This can be demonstrated by the following example.

> **Case history 15-2** A 38-year-old woman was admitted with a 2-day history of weakness and postural dizziness. She denied any history of vomiting, diarrhea, melena, or drugs. On physical examination, the blood pressure was 110/60 recumbent and fell to 80/50 erect. The pulse was 100 and regular. The neck veins were flat with the patient recumbent. The skin turgor was poor, and the mucous membranes were dry. The remainder of the examination was normal. The laboratory data were:
>
> $$\text{Plasma } [Na^+] = 140 \text{ meq/L} \qquad \text{Arterial pH} = 7.25$$
> $$[K^+] = 3.2 \text{ meq/L} \qquad P_{CO_2} = 28 \text{ mmHg}$$
> $$[Cl^-] = 116 \text{ meq/L} \qquad \text{Urine } [Na^+] = 9 \text{ meq/L}$$
> $$[HCO_3^-] = 12 \text{ meq/L} \qquad \text{osmolality} = 584 \text{ mosmol/kg}$$
> $$\text{BUN} = 40 \text{ mg/dL}$$
> $$[\text{Creatinine}] = 1.3 \text{ mg/dL}$$
>
> COMMENT Although the etiology was not apparent, the physical examination was consistent with moderately severe volume depletion. The low urine Na^+ concentration suggested both that renal function was normal and that renal salt wasting and adrenal insufficiency were not responsible for the hypovolemia. The presence of metabolic acidosis and hypokalemia raised the possibility of diarrhea as the cause of the volume depletion. Upon closer questioning, a history of laxative abuse with multiple bowel movements each day was obtained.

TREATMENT

Oral Therapy

The aims of therapy are to restore normovolemia and to correct any associated acid-base or electrolyte disorders that may be present. In patients with mild volume depletion, increasing dietary Na^+ and water intake either by altering the diet or using NaCl tablets may be sufficient to correct the volume deficit. Oral solutions containing glucose and electrolytes can also be used to treat persistent or severe diarrhea [29]. The addition of glucose both provides extra calories and promotes small intestinal Na^+ reabsorption since there is coupled transport of Na^+ and glucose at this site, similar to that in the proximal tubule (see page 92).

With more severe hypovolemia or in patients unable to take oral fluids (due to nausea or vomiting), volume repletion requires the administration of intravenous fluids.

Intravenous Solutions

A wide variety of intravenous solutions are available. The compositions of the most commonly used solutions are listed in Table 15-6. The content of each solution determines the clinical situation in which it will be most useful.

Dextrose solutions Since glucose is rapidly metabolized to $CO_2 + H_2O$, the administration of dextrose solutions is physiologically equivalent to administering distilled water.† Thus, the main indication for the use of dextrose in water is to provide free water to replace insensible losses or to correct hypernatremia due to a water deficit. More concentrated dextrose solutions (20 and 50%) are available and are used to provide calories (1 g of glucose equals 4 kcal). When these solutions are given, the patient should be watched carefully for the development of significant hyperglycemia.

Saline solutions Most hypovolemic patients are both Na^+- and water-depleted, e.g., by diarrhea or vomiting. In this situation, isotonic, hypotonic, or hypertonic saline solutions can be used to correct both deficits. Isotonic saline (0.9%) has a Na^+ concentration of 154 meq/L, similar to that in the plasma water (see page 22). Half-isotonic saline (0.45%, Na^+ concentration of 77 meq/L) is more dilute than the plasma, and each liter can be viewed as being composed of 500 mL of isotonic saline and 500 mL of free water. On the other hand, hypertonic saline (3%, Na^+ concentration of 513 meq/L) is more concentrated than the plasma, and each liter can be viewed as containing 1000 mL of isotonic saline plus 359 meq of Na^+.

The plasma Na^+ concentration can be used to help determine which solution should be used. Because it contains free water, half-isotonic saline (or dextrose in quarter-isotonic saline) should be used in patients with hypernatremia who have a greater deficit of water than solute. On the other hand, hypovolemic patients with hyponatremia have a greater deficit of solute than water and should be treated with isotonic or hypertonic saline (see Chap. 23). If the plasma Na^+ concentration is normal, either half-isotonic or isotonic saline can be given. The former has the advantage of containing free water which can replace continued insensible water losses.

Dextrose in saline solutions The indications for the use of these solutions are the same as those for the saline solutions. The addition of glucose provides a small amount of calories (5% dextrose equals 50 g of glucose per liter or 200 kcal per liter).

Alkalinizing solutions The primary uses of $NaHCO_3$ are in the treatment of metabolic acidosis or severe hyperkalemia. $NaHCO_3$ is most commonly administered as a 7.5% solution in 50-mL ampuls containing 44 meq of Na^+ and 44 meq of HCO_3^-. This can be given intravenously over 5 min or added to other intravenous solutions.

---

†Distilled water cannot be given intravenously since it produces hemolysis (see page 19). This problem can be prevented by the osmotic effect of adding dextrose.

Table 15-6 Composition of commonly used intravenous solutions†

Solution	Solute	Concentrations, g/100 mL	Ionic concentration, meq/L					Total mosmol/L
			$[Na^+]$	$[K^+]$	$[Ca^{2+}]$	$[Cl^-]$	$[HCO_3^-]$	
Dextrose in water								
5.0%	Glucose	5.0	—	—	—	—	—	278
10.0%	Glucose	10.0	—	—	—	—	—	556
Saline								
Hypotonic (0.45%, half-normal)	NaCl	0.45	77	—	—	77	—	154
Isotonic (0.9%, normal)	NaCl	0.90	154	—	—	154	—	308
Hypertonic	NaCl	3.0	513	—	—	513	—	1026
		5.0	855	—	—	855	—	1710
Dextrose in saline								
5% in 0.22%	Glucose	5.0	—	—	—	—	—	355
	NaCl	0.22	38.5	—	—	38.5	—	—
5% in 0.45%	Glucose	5.0	—	—	—	—	—	432
	NaCl	0.90	77	—	—	77	—	—
5% in 0.9%	Glucose	5.0	—	—	—	—	—	586
	NaCl	0.90	154	—	—	154	—	—

Solution	Salt							
Alkalinizing solutions								
Hypertonic sodium bicarbonate (0.6 M)	NaHCO$_3$	5.0	595	—	—	—	595	1190
Hypertonic sodium bicarbonate (0.9 M)‡	NaHCO$_3$	7.5	893	—	—	—	893	1786
Polyionic solutions								
Ringer's	NaCl	0.86	147	—	—	156	—	309
	KCl	0.03	—	4	—	—	—	—
	CaCl$_2$	0.03	—	—	5	—	—	—
Lactated Ringer's	NaCl	0.60	130	—	—	109	—	272
	KCl	0.03	—	4	—	—	—	—
	CaCl$_2$	0.02	—	—	3	—	—	—
	Na lactate	0.31	—	—	—	—	28§	—
Potassium chloride††	KCl	14.85	—	2††	—	—	—	—

†Adapted from A. Arieff, *Clinical Disorders of Fluid and Electrolyte Metabolism*, 2d ed., M. H. Maxwell and C. R. Kleeman (eds.). McGraw-Hill, New York, 1972.

‡The 0.9 M solution of NaHCO$_3$ usually is available in the clinical setting in 50-mL ampuls containing 44 meq of Na$^+$ and 44 meq of HCO$_3^-$. This solution can be infused intravenously or added to other solutions.

§Lactated Ringer's solution contains 28 meq/L of lactate, which is converted in the body to HCO$_3^-$.

††The KCl solution is available in 20- to 50-mL ampuls which can be added to other solutions to provide K$^+$. The K$^+$ concentration in this solution is 2 meq/mL.

$NaHCO_3$ should not be added to solutions containing calcium, e.g., Ringer's lactate, since Ca^{2+} and HCO_3^- can combine to form the insoluble salt $CaCO_3$.

Polyionic solutions Ringer's solution contains physiologic concentrations of K^+ and Ca^{2+} in addition to NaCl. Lactated Ringer's solution has a composition even closer to that of the extracellular fluid, containing 28 meq of lactate per liter, which is rapidly metabolized into HCO_3^- in the body. Although they may seem more physiologic, there is no evidence that these solutions offer any advantages when compared with isotonic saline. Lactated Ringer's solution probably should not be used in lactic acidosis since the ability to convert lactate into HCO_3^- is reduced.

Potassium chloride KCl is available in a highly concentrated solution containing 2 meq of K^+ per milliliter. When used to repair a K^+ deficit, 10 to 60 meq of K^+ (5 to 30 mL) can be added to 1 liter of any of the above solutions (see Chap. 27). K^+ should never be given as an intravenous bolus since it can produce a potentially fatal increase in the plasma K^+ concentration.

Plasma volume expanders Since Na^+ salts freely cross the capillary wall, the administration of saline solutions expands both the intravascular and interstitial volumes. When free water is provided, as with dextrose or hypotonic saline solutions, there is also an increase in the intracellular volume as two-thirds of the free water enters the cells. Thus, dextrose in water expands the extracellular volume only one-third as much as an equivalent volume of isotonic saline which is limited to the extracellular fluid (see Fig. 1-12). In contrast, albumin and dextran remain in the vascular space and selectively expand the plasma volume.

Albumin is available as pooled human albumin which has been treated with heating and filtration to eliminate the risk of hepatitis. (Since it may cause hepatitis, pooled human plasma is rarely used as a volume expander.) When given as a 25% solution (25 g/dL) which is markedly hyperoncotic (normal plasma albumin concentration is 4 to 5 g/dL), albumin increases the plasma oncotic pressure, thereby drawing several times its volume of fluid into the vascular space from the interstitium. Albumin also can be given as a 5% solution in isotonic saline, which is similar to administering plasma.

Dextran is a nonmetabolizable sucrose polysaccharide that can be given as low-molecular-weight dextran (average molecular weight is 40,000) or as clinical dextran (average molecular weight is 75,000). In addition to expanding the plasma volume, these solutions also decrease blood viscosity, lower red blood cell and platelet aggregation, and have anticoagulant properties. Since there is no evidence that dextran is more effective than albumin and it has the added side effects of anticoagulation, which can increase bleeding, allergic reactions, and renal failure due to the sludging of dextran in the tubules, albumin is the preferred plasma volume expander [30].

Blood In patients with anemia, particularly those who are actively bleeding, the administration of blood may be necessary to maintain oxygen transport to the tissues. When blood is given, packed red blood cells are used in most situations since saline or albumin can be administered in place of the plasma, whose components can be used for other purposes, e.g., platelets and antihemophilic globulin.

Which fluid should be used? The composition of the appropriate replacement fluid varies from patient to patient. The type of fluid lost, the plasma K^+ concentration, the plasma osmolality, and acid-base balance all must be taken into account. For example, relatively hypotonic solutions should be used in hyperosmolal patients with hypernatremia or hyperglycemia, and isotonic or hypertonic solutions should be used in hypoosmolal patients with hyponatremia. All the solutes in an intravenous solution must be included when calculating its effective osmolality since K^+, the primary intracellular solute, is as osmotically active as Na^+. Thus, 1 liter of isotonic saline is osmotically equivalent to 1 liter of half-isotonic saline (Na^+ concentration of 77 meq/L) to which 77 meq of K^+ has been added. The major exception is glucose, which is rapidly metabolized in the body to CO_2 and H_2O and therefore is only transiently osmotically active.

A patient with diabetes insipidus who develops hypernatremia due to water loss can be treated with dextrose solutions alone. In contrast, a patient who has lost both solutes and water may require more complex replacement therapy. This can be illustrated by the following example:

Case history 15-3 A 37-year-old woman is seen after several days of severe diarrhea and poor oral intake. Findings of physical examination were consistent with moderately severe volume depletion. The laboratory data included:

Plasma $[Na^+]$ = 142 meq/L   Plasma osmolality = 285 mosmol/kg
$[K^+]$ = 3.9 meq/L   Arterial pH = 7.22
$[Cl^-]$ = 114 meq/L P_{CO_2} = 20 mmHg
$[HCO_3^-]$ = 8 meq/L   Urine $[Na^+]$ = 4 meq/L

COMMENT In addition to volume depletion, this patient had metabolic acidosis and probably K^+ depletion since the plasma K^+ concentration was normal in the presence of acidemia. In view of the normal plasma Na^+ concentration and osmolality, replacement fluid should be mildly hypotonic to provide free water that will replace insensible water losses. An appropriate intravenous solution for this patient would be 1 liter of dextrose in quarter-isotonic saline (Na^+ concentration equal to 38.5 meq/L) to which 44 meq of Na^+ (as $NaHCO_3$) and 40 meq of K^+ (as KCl) have been added. This solution contains HCO_3^- and K^+ to correct the acidemia and K^+ depletion and is slightly hypotonic to plasma, having a $Na^+ + K^+$ concentration of 122 meq/L.

In addition to the infusion of dextrose and electrolyte solutions, blood may be required if the patient is actively bleeding or has marked anemia. Volume repletion with solutions other than blood expands the plasma volume and will lower the hematocrit by dilution. Thus, a severe degree of anemia may be masked on admission and become apparent only with volume replacement.

The primary indication for the use of albumin-containing solutions is in protein-losing states such as burns or occasionally the nephrotic syndrome. Although these solutions have also been used in the treatment of shock or severe hypovolemia, they now appear to offer no substantial advantages over the pure electrolyte solutions (see below).

Volume Deficit

It is usually difficult to estimate the volume deficit in a hypovolemic patient. Although knowledge of the patient's normal weight is helpful, this is frequently not obtainable. If hyponatremia or hypernatremia is present, the respective Na^+ and water deficits can be estimated from the following formulas:[†]

$$Na^+ \text{ deficit (in meq)} = 0.6 \times \text{body weight (in kg)} \times (140 - \text{plasma } [Na^+])$$

$$\text{Water deficit (in liters)} = 0.5 \times \text{body weight} \times \left(\frac{\text{plasma } [Na^+]}{140} - 1 \right)$$

However, these formulas only estimate the amount of Na^+ in a hyponatremic patient and the volume of water in a hypernatremic patient that would have to be retained to return the plasma Na^+ concentration to the normal value of 140 meq/L. This ignores any isosmotic fluid deficit which may also be present. For example, the formula for the water deficit is relatively accurate for a patient with diabetes insipidus who has lost only water, but it underestimates the deficit in a hypernatremic patient with diarrhea and increased insensible losses who has lost both Na^+ and water.

Since the loss of extracellular fluid, normally 20 percent of the body weight, results in hemoconcentration and an increase in the hematocrit, the extracellular deficit can be calculated from the change in the hematocrit (Hct) according to a formula similar to that for the water deficit:

$$\text{Extracellular fluid deficit} = 0.2 \times \text{body weight} \times \left(\frac{\text{Hct}}{\text{normal Hct}} - 1 \right)$$

However, since there is a relatively wide range, from 38 to 45 percent, in the normal hematocrit and the patient may be anemic, e.g., because of gastrointestinal bleeding, this formula is useful only if the patient's normal hematocrit is known and if bleeding has not occurred.

In summary, the fluid deficit in a hypovolemic patient usually cannot be calculated precisely. As a result, the adequacy of volume repletion must be evaluated from the findings on physical examination and laboratory data. As volume expansion occurs, the skin turgor should improve and there should be increases in the body weight, arterial pressure (if there has been a fall in blood pressure), venous pressure, urine output, and urine Na^+ concentration. For patients who start with a low urine Na^+ concentration, serial measurements of the urine Na^+ concentration can be used as an index of the degree to which normovolemia has been restored. If the urine Na^+ concentration remains under 15 meq/L, the kidney is sensing persistent volume depletion and more fluids should be given.[‡]

[†]These formulas are derived in Chaps. 23 and 24. The formula for the Na^+ deficit assumes that the patient has true hyponatremia, not pseudohyponatremia due to hyperglycemia or hyperlipidemia (see page 494).

[‡]This excludes edematous patients, e.g., those with heart failure or hepatic cirrhosis with ascites, in whom the low urine Na^+ concentration is an indication of effective circulating volume depletion but not of the need for more fluids.

In addition to the search for evidence of reversal of volume depletion, the patient should be carefully observed for the onset of pulmonary edema due to the administration of an excessive volume of fluid. This can be done by listening to the chest for rales and to the heart for gallop rhythms and by following a chest x-ray. In elderly patients requiring large amounts of fluid or with a history of heart disease, measurement of the central venous pressure or, if indicated, the pulmonary capillary wedge pressure can be used as a guide to therapy.

Rate of Volume Replacement

As with other water and electrolyte disorders, the immediate aim of therapy is to get the patient out of danger. With the exception of patients with hypotension, shock, or severe associated electrolyte disturbances, e.g., hypokalemia or hypernatremia, *gradual* intravenous replacement will restore normovolemia while minimizing the risk of volume overload and pulmonary edema. Since there are no set rules that can be used, the rate of fluid replacement is somewhat arbitrary. A regimen that has been successful is the infusion of the appropriate replacement fluids at the rate of 50 to 100 mL/h *in excess of the sum* of the urine output, estimated insensible losses (approximately 30 to 40 mL/h), and any other losses that may be present (such as diarrhea or tube drainage).

It must be emphasized that the *aim of therapy is not to administer fluids but to induce positive fluid balance.* Suppose a patient with severe diarrhea has losses averaging 75 mL/h. If fluid is administered at the rate of 75 mL/h plus estimated insensible losses, there will be no positive fluid balance and no correction of the hypovolemic state. A similar problem with continuing losses can occur in central diabetes insipidus where the urine volume can exceed 500 mL/h. In these patients, the administration of ADH will reduce the urine output and make volume repletion easier to achieve (see Chap. 24).

Treatment of Hypovolemic Shock

Hypovolemic shock is most often due to bleeding or third-space sequestration, although a similar picture may be produced by any of the causes of true volume depletion. Before discussing the therapy of this disorder, it is important to first review its pathophysiology [22,31]. As described above, progressive volume depletion is associated with increasing degrees of sympathetic and angiotensin II–mediated vasoconstriction. This response initially maintains the blood pressure and cerebral and coronary perfusion. However, the combination of a hypovolemia-induced decrease in cardiac output and intense vasoconstriction results in a marked reduction in splanchnic, renal, and musculocutaneous blood flow that can ultimately lead to lactic acidosis, ischemic tissue necrosis, organ failure, and the release of intracellular contents (such as lysosomal enzymes) into the systemic circulation. Intestinal ischemia may also result in the development of endotoxinemia.

Early therapy is important to prevent hypovolemic shock from becoming *irreversible.* As depicted in Fig. 15-3*a*, experimentally induced hemorrhagic shock in a

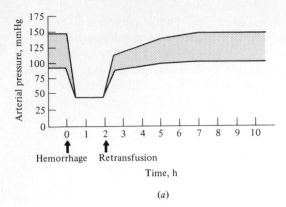

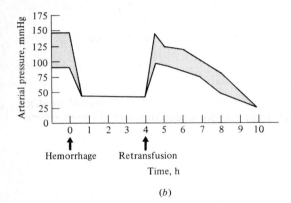

Figure 15-3 Reversibility of experimental hemorrhagic shock in the dog. (*a*) If the mean arterial pressure is reduced to 35 to 40 mmHg for less than 2 h. the infusion of the shed blood will restore a normal blood pressure. (*b*) If the period of hypotension is extended to 4 h before the shed blood is returned, most of the dogs die within 24 h despite retransfusion. (*From R. C. Lillihei and R. H. Dietzman, Principles of Surgery, S. I. Schwartz, R. C. Lillihei, G. T. Shires, F. C. Spencer, and E. H. Storer (eds.), McGraw-Hill, New York, 1974.)*

dog can be reversed if the blood that has been removed is reinfused within 2 h. However, if return of the shed blood is delayed for 4 h or longer, there is only a transient increase in blood pressure and the animal dies within 24 h (Fig. 15-3*b*). A similar phenomenon appears to occur in humans, although substantially more than 4 h may be required before volume repletion becomes ineffective [31].

Irreversible shock seems to be associated with an inability to expand the plasma volume persistently and raise the blood pressure. One factor that contributes to this problem is relative relaxation of vasoconstriction at the arterial end of the capillary with little change at the venous end [31]. It is not known how this occurs, although circulating endorphins and locally produced vasodilator prostaglandins may be important [32]. The net effect is that the administered fluid is initially pooled in the capillaries (particularly in the splanchnic circulation) since exit from the capillaries is delayed by venous constriction. The ensuing elevation in the capillary hydrostatic pressure (plus a possible increase in capillary permeability that may be induced by lysosomal enzymes released from damaged tissues or the local accumulation of neutrophils)† favors the movement of fluid out of the vascular space into the interstitium

---

†Tissue breakdown in shock can lead to activation of the complement pathway. C_{5a} is a potent chemotactic factor, resulting in the aggregation of neutrophils which, by releasing proteolytic enzymes, can damage the vascular wall and promote fluid accumulation in the interstitium [32a].

[31,32,32a]. This produces plasma volume depletion, hemoconcentration, increased viscosity, and red blood cell aggregation, all of which further impair the capillary circulation. In addition, persistent hypotension (which is maintained in part by the arteriolar dilation) can lead to tissue necrosis, e.g., mesenteric or myocardial infarction, which makes recovery less likely.

With these potential hazards in mind, a rational therapeutic program can be begun. Patients with shock should have careful monitoring of their arterial pressure, central venous pressure (or, in case of cardiac or pulmonary disease, the pulmonary capillary wedge pressure), arterial pH, hematocrit, urine output, and mental status. In addition, therapy must be directed toward the underlying disease; for example, surgery in a patient with a ruptured abdominal aortic aneurysm.

The immediate aim of therapy in hypovolemic shock must be to *restore tissue perfusion by the administration of fluids.* The use of vasopressors such as dopamine or norepinephrine will not correct the underlying volume deficit and may intensify the problem in the capillary circulation, further reducing tissue perfusion and predisposing toward ischemic damage.

Which fluids should be given? The choice of replacement fluids depends upon the type of fluid lost. Patients who are bleeding may require the administration of large amounts of blood. This can be given most rapidly under pressure through several intravenous catheters. In general, the hematocrit should not be raised over 35 percent. Increasing the hematocrit above this level is not necessary for oxygen transport and may produce an increase in blood viscosity which can lead to stasis in the already impaired capillary circulation.

The optimal form of fluid replacement other than blood in the treatment of hypovolemic shock is controversial [33]. The main problem has centered on the relative merits of albumin-containing solutions versus pure electrolyte solutions (such as isotonic saline or Ringer's lactate). Those physicians favoring the use of albumin have claimed that it has two advantages: (1) it is a more effective plasma volume expander since it remains in the vascular space (in contrast to saline, two-thirds of which enters the interstitium); and (2) it is less likely to induce pulmonary edema since the increase in plasma oncotic pressure favors the movement of fluid out of the pulmonary interstitium into the vascular space.

However, recent controlled studies have *failed to confirm* either of these potential advantages [34–37]. Albumin and electrolyte solutions have been shown to be *equally effective* in producing volume repletion, although 2.5 to 3 times as much saline must be given because of its extravascular distribution [34]. This is not a deleterious effect since the interstitial volume also becomes depleted in severe hypovolemia as a result of both fluid losses and the movement of interstitial fluid into the cells. The latter problem may be due to an ischemia-induced impairment in the activity of the Na^+-K^+-ATPase pump in the cell membrane [22]. Extracellular Na^+ normally diffuses passively into the cells since the cell Na^+ concentration of about 12 meq/L is much less than that in the extracellular fluid. This Na^+ is then extruded from the cells by the Na^+-K^+-ATPase pump. Consequently, a reduction in the activity of this pump allows interstitial Na^+ and water to accumulate in the cells (see Fig. 1-11), thereby reducing the interstitial volume.

Table 15-7 Comparison of approximate Starling's forces in the pulmonary and skeletal muscle capillaries[†]

	Pulmonary	Skeletal muscle
Hydrostatic pressure		
Capillary (mean)	7	17
Interstitium	−3.3	−4
Mean gradient	10.3	21
Oncotic pressure		
Capillary	26	26
Interstitium	16	5.3
Mean gradient	10	20.7
Net gradient favoring filtration (hydrostatic minus oncotic gradients)	+0.3	+0.3

†Units are mmHg.

Albumin is also not more effective in improving pulmonary function and may, in some cases, actually promote the development of pulmonary edema [34,35]. Two factors appear to contribute to this problem. First, the limitation of albumin to the vascular space can lead to an excessive elevation in the pulmonary capillary wedge pressure. Second, patients with hypovolemic shock frequently develop the adult respiratory distress syndrome which is associated with an increase in pulmonary capillary permeability [32a]. In this setting, administrated albumin can cross the capillary membrane and enter the pulmonary interstitium, thereby enhancing interstitial oncotic pressure and the tendency to interstitial fluid accumulation [35,38].

One concern with the use of saline alone has been that the development of dilutional hypoalbuminemia could lead to pulmonary edema [39,40]. However, although hypoalbuminemia can produce peripheral edema in this setting [33,41], pulmonary edema is unusual [34–36,41]. The reason for this difference appears to be that capillary hemodynamics in the pulmonary circulation differs from that in skeletal muscle (Table 15-7) [42,43]. In the latter, a high hydrostatic pressure gradient is balanced by an oncotic pressure gradient of roughly equal magnitude (see page 32). In contrast, the hydrostatic pressure is lower in the lung (see Fig. 15-2), as is the oncotic pressure gradient, since the pulmonary capillary is more permeable to albumin [42]. When hypoalbuminemia develops, the marked drop in the oncotic pressure gradient leads to peripheral edema. Pulmonary edema is less likely to occur since hypoalbuminemia also lowers the previously high pulmonary interstitial oncotic pressure, resulting in a smaller change in the oncotic pressure gradient [41].

In summary, electrolyte solutions are preferable in the treatment of severe hypovolemia in the absence of significant underlying hypoalbuminemia. In normoalbuminemic patients, albumin has no proven advantage, may exacerbate pulmonary problems, and is about 50 times more expensive [34,37].

Rate of fluid replacement As much as 1 to 2 liters of fluid should be given in the first hour in an attempt to restore adequate tissue perfusion as rapidly as possible. It is impossible to predict in a given patient what the total fluid deficit will be, particularly if bleeding or third-space sequestration continues. Consequently, further fluids should be administered while monitoring the central venous or preferably the pulmonary capillary wedge pressure. Fluids should be given at the initial rapid rate as long as the cardiac filling pressures and the systemic blood pressure remain low. The development of peripheral edema does not necessarily indicate that fluids should be stopped since, as described above, it may be due to dilutional hypoalbuminemia even though plasma volume depletion persists [33].

Other aspects of therapy Rarely, a patient does not show an adequate response to volume repletion, as occurs when the filling pressures in the heart are increased but signs of poor peripheral perfusion, including oliguria, decreased pulse pressure, and cool extremities, persist. In this setting, which is probably the equivalent of irreversible shock, a pressor agent such as dopamine can be tried. If this is ineffective, vasodilators such as phentolamine (an α-adrenergic blocker) or chlorpromazine (a potent vasodilator when given intravenously) have been used, but their efficacy has not been proved [22].

In addition to the restoration of volume and blood pressure, other problems may require treatment, including hypoxemia, lactic acidosis, and oliguria. In the presence of circulatory failure, hypoxemia will intensify the defect in cell metabolism. Thus, the correction of hypoxemia by the administration of oxygen, and endotracheal intubation and mechanical ventilation if necessary, is an essential part of management. Patients who develop the adult respiratory distress syndrome may also require the use of positive end-expiratory pressure to maintain adequate oxygenation [44].

Since oxidative metabolism is impaired, increased lactic acid production and lactic acidosis commonly occur. The therapy of lactic acidosis will be discussed in Chap. 19. Stated briefly, lactate is metabolized to HCO_3^- so that the restoration of tissue perfusion will spontaneously correct the acidemia. Therefore for a patient in shock, $NaHCO_3$ should be administered only in doses sufficient to maintain the arterial pH at about 7.20. Not only is raising the pH above this level unnecessary, it may actually reduce oxygen delivery to the tissues by increasing the affinity of hemoglobin for oxygen [45].

Oliguria is an almost invariable accompaniment of shock and is due primarily to reduced renal perfusion. The glomerular filtration rate begins to fall when the mean arterial pressure is less than 80 to 90 mmHg and ceases entirely when the mean arterial pressure is less than 45 to 50 mmHg [46]. The restoration of tissue perfusion will initially normalize renal function, but prolonged ischemia can lead to renal failure due to acute tubular necrosis [47,48]. Increasing the urine output during the hypotensive period with intravenous mannitol (12.5 to 25 g) and/or furosemide (80 to 320 mg) has been recommended as a means of protecting against the development of renal failure [48–52]. In particular, mannitol increases the GFR and renal blood flow by inducing renal vasodilation [46,53] and minimizes intratubular obstruction in part by minimizing ischemic tubular damage and therefore sloughing of the cells into the lumen [52]. Furosemide is somewhat less effective [52] but can be given with the mannitol. With either drug, urinary losses should be replaced since the aim is to induce a diuresis, not further volume depletion. Repeated doses should not be administered if a diuresis does not occur since the incidence of toxic side effects is increased [47]. Furthermore, neither agent appears to improve renal function once acute tubular necrosis has become established [54,55].

PROBLEMS

15-1 A 75-year-old woman is admitted to the hospital with the acute onset of severe abdominal pain. When examined, the patient is agitated, her extremities are cold and clammy, and her blood pressure is 60/30. Her abdomen is distended and tympanitic with diffuse tenderness. The results of the laboratory workup include a hematocrit of 53 percent. An arteriogram shows complete occlusion of one of the branches of the superior mesenteric artery. What is the etiology of the shock state in this patient? What fluids would you administer?

Prior to surgery, a total of 7 liters of fluid is administered to the patient to maintain her blood pressure. Throughout this period she is virtually anuric. At surgery, 40 cm of infarcted ileum is removed. Six hours after surgery, the patient is doing well when a marked increase in the urine output to nearly 1000 mL/h is noted. Her urine osmolality is 350 mosmol/kg; her urine Na^+ concentration is 95 meq/L. What might be responsible for this increase in output? How would you treat the patient at this time?

15-2 Compare the effects of the loss of water (due to increased insensible losses or diabetes insipidus) to the loss of an equal volume of an isotonic Na^+ solution (due to diuretics or diarrhea) on the extracellular volume and the arterial blood pressure.

15-3 What is the role of pure dextrose solutions in the treatment of hypovolemic shock?

15-4 A 75-year-old woman develops volume depletion due to the excessive administration of diuretics. Prior to the administration of diuretics, the patient had a normal BUN and plasma creatinine concentration. After a 6-kg weight loss over 10 days, poor skin turgor is present, and the central venous pressure is 1 cmH$_2$O. The following laboratory data are obtained:

$$BUN = 208 \text{ mg/dL}$$
$$\text{Plasma [creatinine]} = 5.7 \text{ mg/dL}$$
$$\text{Urine } [Na^+] = 5 \text{ meq/L}$$
$$\text{Urine output} = 25 \text{ mL/h}$$
$$\text{Urinalysis} = \text{normal}$$

After the administration of 5 liters of half-isotonic saline over 18 h, the central venous pressure is 3 cmH$_2$O, the skin turgor has improved, and the results of repeat laboratory studies are:

$$BUN = 160 \text{ mg/dL}$$
$$\text{Urine } [Na^+] = 45 \text{ meq/L}$$
$$\text{Urine output} = 80 \text{ mL/h}$$

   (a) Why have the urine Na^+ concentration and urine output increased?

   (b) Does the repeat central venous pressure indicate persistent volume depletion?

   (c) Why is the repeat BUN still elevated despite volume repletion?

15-5 A 74-year-old man is admitted from a nursing home with a 3-day history of recurrent vomiting and diarrhea. The results of the physical examination are consistent with volume depletion. The laboratory data reveal:

$$\text{Plasma } [Na^+] = 155 \text{ meq/L}$$
$$[K^+] = 3.0 \text{ meq/L}$$
$$[Cl^-] = 117 \text{ meq/L}$$
$$[HCO_3^-] = 25 \text{ meq/L}$$

What intravenous solution would you use for replacement therapy? How rapidly should it be administered?

15-6 A 72-year-old woman is found confused on the floor of her apartment. No history is obtainable except that she has a history of hypertension. The physical examination reveals a blood pressure of 110/70, reduced skin turgor, and flat neck veins. The following laboratory data are obtained:

$$BUN = 62 \text{ mg/dL}$$
$$\text{Plasma [creatinine]} = 1.8 \text{ mg/dL}$$
$$[Na^+] = 138 \text{ meq/L}$$
$$[K^+] = 3.1 \text{ meq/L}$$
$$[Cl^-] = 100 \text{ meq/L}$$
$$[HCO_3^-] = 29 \text{ meq/L}$$

(*a*) Is the blood pressure normal? (*b*) Could this patient's volume depletion be due to the lack of replacement of insensible losses?

REFERENCES

1. Dirks, J. H., W. J. Cirksena, and R. W. Berliner: Micropuncture study of the effect of various diuretics on sodium reabsorption by the proximal tubules of the dog, J. Clin. Invest., 45:1875, 1966.
2. Martinez-Maldonado, M., G. Eknoyan, and W. N. Suki: Diuretics in nonedematous states: physiological basis for the clinical use, Arch. Intern. Med., 131:797, 1973.
3. Coleman, A. J., M. Arias, N. W. Carter, F. C. Rector, Jr., ard D. W. Seldin: The mechanism of salt-wasting in chronic renal disease, J. Clin. Invest., 45:1116, 1966.
4. Thorn, G. W., G. F. Koepf, and M. Clinton, Jr.: Renal failure simulating adrenocortical insufficiency, N. Engl. J. Med., 231:76, 1944.
5. Uribarri, J., M. S. Oh, and H. J. Carroll: Salt-losing nephropathy. Clinical presentation and mechanisms, Am. J. Nephrol., 3:193, 1983.
6. Strauss, M. B.: Clinical and pathological aspects of cystic disease of the renal medulla: an analysis of eighteen cases, Ann. Intern. Med., 57:373, 1962.
7. Bricker, N. S., E. I. Shwayri, J. B. Reardan, D. Kellogg, J. P. Merrill, and J. H. Holmes: An abnormality in renal function resulting from urinary tract obstruction, Am. J. Med., 23:554, 1957.
8. Howards, S. S.: Post-obstructive diuresis: a misunderstood phenomenon, J. Urol., 110:537, 1973.
9. Rose, B. D., *Pathophysiology of Renal Disease,* McGraw-Hill, New York, 1981, chap. 7.
10. Rose, B. D., *Pathophysiology of Renal Disease,* McGraw-Hill, New York, 1981, p. 423.
11. Yeh, B. P. Y., D. J. Tomko, W. K. Stacy, E. S. Bear, H. T. Haden, and W. F. Falls, Jr.: Factors influencing sodium and water excretion in uremic man, Kidney Int., 7:103, 1975.
12. Danovitch, G. M., J. J. Bourgoignie, and N. S. Bricker: Reversibility of the "salt-losing" tendency of chronic renal failure, N. Engl. J. Med., 296:14, 1977.
13. Popovtzer, M. M., F. H. Katz, W. F. Pinggera, J. Robinette, C. C. Halgrimson, and D. E. Butkus: Hyperkalemia in salt-wasting nephropathy: study of the mechanism, Arch. Intern. Med., 132:203, 1973.
14. di Sant'Agnese, P. A., R. C. Darling, G. A. Perera, and E. Shea: Abnormal electrolyte composition of sweat in cystic fibrosis of the pancreas: clinical singificance and relationship to the disease, Pediatrics, 12:549, 1953.
15. Moyer, C. A., H. W. Margraf, and W. W. Monafo: Burn shock and extravascular sodium deficiency: treatment with Ringer's solution with lactate, Arch. Surg., 90:799, 1965.
16. Dwek, J. H., C. Charytan, I. Stachura, and A. Kaganowicz: Salt-wasting bronchorrhea and its mechanisms, Arch. Intern. Med., 137:791, 1977.
17. Freis, E. D., J. R. Stanton, F. A. Finnerty, Jr., H. W. Schnaper, R. L. Johnson, C. E. Rath, and R. W. Wilkins: The collapse produced by venous congestion of the extremities or by venesection following certain hypotensive agents, J. Clin. Invest., 30:435, 1951.

18. Wagner, H. N., Jr.: The influence of autonomic vasoregulatory reflexes on the rate of sodium and water excretion in man, J. Clin. Invest., 36:1319, 1957.

19. Franch, R. H.: Examination of the blood, urine, and extravascular fluids, including circulation time and venous pressure, in *The Heart,* 3d ed., J. W. Hurst, R. B. Logue, R. C. Schlant, and N. K. Wenger (eds.), McGraw-Hill, New York, 1974.

20. Rackley, C. E.: The use of the Swan-Ganz catheter, in *The Heart: Arteries and Veins,* 4th ed., J. W. Hurst, R. B. Logue, R. C. Schlant, and N. K. Wenger (eds.), McGraw-Hill, New York, 1978.

21. Cohn, J. N., F. E. Tristani, and I. M. Khatri: Studies in clinical shock and hypotension. VI. Relationship between left and right ventricular function, J. Clin. Invest., 48:2008, 1969.

22. Sobel, B. E.: Cardiac and noncardiac forms of acute circulatory collapse (shock), in *Heart Disease: A Textbook of Cardiovascular Medicine,* E. Braunwald (ed.), Saunders, Philadelphia, 1980.

22a. Sherman, R. A., and R. P. Eisinger: The use (and misuse) of urinary sodium and chloride measurements, J. Am. Med. Assoc., 247:3121, 1982.

23. Besarab, A., R. S. Brown, N. T. Rubin, E. Salzman, L. Wirthlin, T. Steinman, R. R. Atlia, and J. J. Skillman: Reversible renal failure following bilateral renal artery occlusive disease: clinical features, pathology, and the role of surgical revascularization, J. Am. Med. Assoc., 235:2838, 1976.

24. Miller, T. R., R. J. Anderson, S. L. Linas, W. L. Henrich, A. S. Berns, P. A. Gabow, and R. W. Schrier: Urinary diagnostic indices in acute renal failure: a prospective study, Ann. Intern. Med., 89:47, 1978.

25. Levinsky, N. G., D. G. Davidson, and R. W. Berliner: Effects of reduced glomerular filtration and urine concentration in presence of antidiuretic hormone, J. Clin. Invest., 38:730, 1959.

26. Dossetor, J. B.: Diagnosis and treatment, creatininemia versus uremia: the relative significance of blood urea nitrogen and serum creatinine concentrations in azotemia, Ann. Intern. Med., 65:1287, 1966.

27. Rose, B. D.: *Pathophysiology of Renal Disease,* McGraw-Hill, New York, 1981, chap. 2.

28. Dorhout-Mees, E. J.: Relation between maximal urine concentration, maximal water reabsorption capacity, and mannitol clearance in patients with renal disease, Br. Med. J., 1:1159, 1959.

29. Santosham, M., R. S. Daum, L. Dillman, J. L. Rodriguez, S. Luque, R. Russell, M. Kourany, R. W. Ryder, A. V. Bartlett, A. Rosenberg, A. S. Benenson, and R. B. Sack: Oral rehydration therapy of infantile diarrhea, N. Engl. J. Med., 306:1070, 1982.

30. Data, J. L., and A. S. Nies: Dextran 40, Ann. Intern. Med., 81:500, 1974.

31. Zweifach, B. W., and A. Fronek: The interplay of central and peripheral factors in irreversible hemorrhagic shock, Prog. Cardiovasc. Dis., 18:147, 1975.

32. Curtis, M. T., and A. M. Leter: Protective actions of naloxone in hemorrhagic shock, Am. J. Physiol., 239:H416, 1980.

32a. Jacob, H. S.: The role of activated complement and granulocytes in shock states and myocardial infarction, J. Lab. Clin. Med., 98:645, 1981.

33. Shine, K. I., M. Kuhn, L. S. Young, and J. H. Tillisch: Aspects of the management of shock, Ann. Intern. Med., 93:723, 1980.

34. Virgilio, R. W., C. L. Rice, D. E. Smith, D. R. James, C. K. Zarins, C. F. Hobelmann, and R. M. Peters: Crystalloid vs. colloid resuscitation: is one better?, Surgery, 85:129, 1979.

35. Weaver, D. M., A. M. Ledgerwood, C. E. Lucas, R. Higgins, D. L. Bouwman, and S. D. Johnson: Pulmonary effects of albumin resuscitation for severe hypovolemic shock, Arch. Surg., 113:387, 1978.

36. Moss, G. S., R. J. Lowe, J. Jilek, and H. D. Levine: Colloid or crystalloid in the resuscitation of hemorrhagic shock: a controlled clinical trial, Surgery, 89:434, 1981.

37. Monafo, W.: Expensive salt water, Surgery, 80:525, 1981.

38. Holcroft, J. W., and D. D. Trunkey: Extravascular lung water following hemorrhagic shock in the baboon: comparison between resuscitation with Ringer's lactate and Plasmanate, Ann. Surg., 180:408, 1974.

39. Stein, L., J-J. Beraud, J. Cavanilles, P. da Luz, M. H. Weil, and H. Shubin: Pulmonary edema during fluid infusion in the absence of heart failure, J. Am. Med. Assoc., 229:65, 1975.

40. Levine, O. R., R. B. Millins, R. M. Senior, and A. P. Fishman: The application of Starling's law of capillary exchange to the lungs, J. Clin. Invest., 46:934, 1967.

41. Zarins, C. K., C. L. Rice, R. M. Peters and R. W. Virgilio: Lymph and pulmonary response to isobaric reduction in plasma oncotic pressure in baboons, Circ. Res. 43:925, 1978.
42. Taylor, A. E.: Capillary fluid filtration: Starling forces and lymph flow, Circ. Res., 49:557, 1981.
43. Brace, R. A.: Progress toward resolving the controversy of positive vs. negative interstitial fluid pressure, Circ. Res., 49:281, 1981.
44. Rinaldo, J. E., and R. M. Rogers: Adult respiratory-distress syndrome, N. Engl. J. Med., 306:900, 1982.
45. Bellingham, A. J., J. C. Detter, and C. Lenfant: Regulatory mechanisms of hemoglobin oxygen affinity in acidosis and alkalosis, J. Clin. Invest., 50:700, 1971.
46. Johnston, P. A., D. B. Bernard, N. S. Perrin, and N. G. Levinsky: Prostaglandins mediate the vasodilatory effect of mannitol in the hypoperfused rat kidney, J. Clin. Invest., 68:127, 1981.
47. Rose, B. D., *Pathophysiology of Renal Disease,* McGraw-Hill, New York, 1981, chap. 3.
48. Schrier, R. W.: Acute renal failure, Kidney Int., 15:209, 1979.
49. Warren, S. E., and R. C. Blantz: Mannitol, Arch. Intern. Med., 141:493, 1981.
50. Patak, R. V., S. Z. Fadem, M. D. Lifschitz, and J. H. Stein: Study of factors which modify the development of norepinephrine-induced acute renal failure in the dog, Kidney Int., 15:227, 1979.
51. Burke, T. J., R. E. Cronin, K. L. Duchin, L. N. Peterson, and R. W. Schrier: Ischemia and tubule obstruction during acute renal failure in dogs: mannitol in protection, Am. J. Physiol., 238:F305, 1980.
52. Hanley, M. J., and K. Davidson: Prior mannitol and furosemide infusion in a model of ischemic acute renal failure, Am. J. Physiol., 241:F556, 1981.
53. Braun, W. E., and L. S. Lilienfield: Renal hemodynamic effects of hypertonic mannitol infusions, Proc. Soc. Exp. Biol. Med., 114:1, 1963.
54. Luke, R. G., J. D. Briggs, M. E. M. Allison, and A. C. Kennedy: Factors determining response to mannitol in acute renal failure, Am. J. Med. Sci., 259:168, 1970.
55. Brown, C. B., C. S. Ogg, and J. S. Cameron: High dose frusemide in acute renal failure: a controlled trial, Clin. Nephrol., 15:90, 1981.

SIXTEEN

EDEMATOUS STATES AND THE USE OF DIURETICS

TREATMENT
 General Principles
 Clinical Use of Diuretics
 Congestive Heart Failure
 Hepatic Cirrhosis
 Nephrotic Syndrome
 Primary Renal Sodium Retention
 Idiopathic Edema
 Other Causes of Edema

Edema is defined as a palpable swelling produced by expansion of the interstitial fluid volume. A variety of clinical conditions are associated with the development of edema, including heart failure, cirrhosis of the liver, and the nephrotic syndrome. In this chapter, we will discuss the pathophysiology and etiology of edema and the use of diuretic agents to remove the excess fluid.

PATHOPHYSIOLOGY

There are two basic steps involved in the interstitial accumulation of fluid that leads to edema: (1) an alteration in capillary hemodynamics is present that favors the movement of fluid from the vascular space into the interstitium; and (2) dietary Na^+ and water are retained by the kidney. The importance of the latter in the development of edema should not be underestimated. Edema does not become clinically apparent until the interstitial volume has increased by at least 2.5 to 3 liters. Since the normal plasma volume is only about 3 liters, it is clear that patients would die from hemoconcentration and shock if the edema fluid were derived only from the plasma. This does not occur because of the sequence depicted in Fig. 16-1. The initial movement of fluid from the vascular space into the interstitium reduces the plasma volume and consequently tissue perfusion. In response to these changes, the kidney retains Na^+ and water. Some of this fluid stays in the vascular space, returning the plasma volume toward normal. However, because of the alteration in capillary hemodynamics, most of the retained fluid enters the interstitium and eventually becomes apparent as edema. The net effect is a marked expansion of the total extracellular volume (as edema) in the presence of a plasma volume that may be reduced. This illustrates an important point to which we will return in the section on treatment. The renal Na^+ and water retention is this setting is an *appropriate* compensation in that it restores the plasma volume, even though it also augments the degree of edema. Thus, removing the edema fluid, e.g., by using diuretics, may reduce the plasma volume and tissue perfusion, occasionally to clinically significant levels.

  The hemodynamic effects are somewhat different when the primary abnormality is inappropriate renal fluid retention. In this setting, both the plasma and interstitial

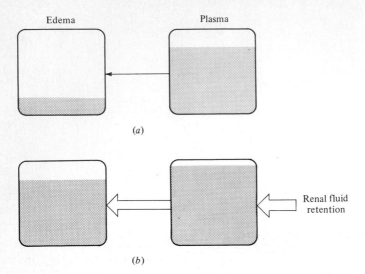

Edema   Plasma

(a)

Renal fluid
retention

(b)

Figure 16-1 Pathophysiology of edema formation when there is an alteration in capillary hemodynamics favoring the movement of fluid out of the plasma into the interstitium, e.g., because of hypoalbuminemia in the nephrotic syndrome. The normal plasma volume is depicted as the full size of the plasma square. The shaded area in the edema square refers to the increase in the interstitial fluid volume as edema. The initial reduction in the plasma volume produced by the loss of fluid into the interstitium (a) stimulates renal Na$^+$ and water retention (b). This appropriately restores the plasma volume toward normal but, because of the altered capillary hemodynamics, much of the retained fluid enters the interstitium and becomes apparent as edema.

volumes are expanded and there are no deleterious effects from removal of the excess fluid. This is an example of overfilling of the vascular tree in contrast to the underfilling described above. It most often occurs with renal insufficiency but may also be seen in hepatic cirrhosis and with the use of certain drugs.

Capillary Hemodynamics

The exchange of fluid between the plasma and the interstitium is determined by the hydrostatic and oncotic pressures in each compartment. The relationship between these parameters can be expressed by Starling's law (see Chap. 1):

Net filtration = K_f (hydrostatic pressure gradient − oncotic pressure gradient)

$$= K_f ([P_{cap} - P_{if}] - [\pi_P - \pi_{if}]) \tag{16-1}$$

where K_f is the net permeability of the capillary wall, P_{cap} and P_{if} are the mean capillary and interstitial hydrostatic pressures, and π_P and π_{if} are the oncotic pressures of the proteins in the plasma and interstitium.

The approximate normal values for these parameters in skeletal muscle are depicted in Fig. 16-2 [1]. As can be seen, the capillary hydrostatic pressure (17 mmHg) which pushes fluid out of the capillary and the plasma oncotic pressure (26

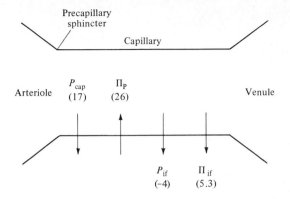

Figure 16-2 Schematic representation of the hemodynamic factors controlling fluid movement across the capillary wall in skeletal muscle. The numbers in parentheses represent the approximate normal values, in mmHg, for each of the factors. The *negative* value for P_{if} is probably generated by the removal of interstitial fluid by the lymphatic vessels [1]. The net effect is a small gradient favoring filtration of 0.3 mmHg.

mmHg) which pulls fluid into the vascular space are quantitatively the most important. There is normally a small gradient of about 0.3 mmHg favoring filtration out of the vascular space; the fluid that is filtered is then returned to the systemic circulation by the lymphatics.

These values for skeletal muscle do not apply to all visceral capillaries (see Tables 3-1 and 15-7) [2]. In the liver, for example, the hepatic sinusoids are highly permeable to proteins, thereby tending to abolish the oncotic pressure gradient; that is, π_{if} almost equals π_P. To some degree, this is balanced by a lower capillary hydrostatic pressure since a substantial part of hepatic blood flow is derived from the portal vein, a low-pressure system. Nevertheless, there is a larger gradient favoring filtration than in skeletal muscle, although the filtered fluid is again removed by the lymphatics.

The development of edema requires an alteration in one or more of these forces in a direction that favors an increase in net filtration. This can be produced by an increase in capillary hydrostatic pressure, interstitial oncotic pressure, or capillary permeability, or by a reduction in the plasma oncotic pressure (Table 16-1).

Increased capillary hydrostatic pressure Capillary hydrostatic pressure, although generated by cardiac contraction, is relatively insensitive to alterations in arterial pressure. This stability is due to autoregulatory changes in resistance at the precapillary sphincter (Fig. 16-2) which determine the extent to which the arterial pressure is transmitted to the capillary (see page 35). If the arterial pressure is increased, the sphincter constricts, minimizing the elevation in capillary hydrostatic pressure. This explains why patients with hypertension do not routinely develop edema. Conversely, the sphincter dilates when the arterial pressure is reduced. This decreases the pressure drop across the sphincter, allowing the capillary pressure (as well as blood flow) to be maintained.

In contrast, the resistance at the venous end of the capillary is not well regulated. Consequently, changes in venous pressure result in parallel changes in capillary hydrostatic pressure. The venous pressure is increased in two settings: (1) when the blood volume is expanded, augmenting the volume in the venous system, and (2) when there is venous obstruction. Examples of edema due to volume expansion occur

Table 16-1 Etiology of edematous states

Increased capillary hydrostatic pressure
  A. Increased blood volume due to renal Na^+ retention
    1. Heart failure
    2. Primary renal Na^+ retention
      a. Renal failure
      b. Drugs: minoxidil, diazoxide, calcium-channel blockers (especially nifedipine), nonsteroidal anti-inflammatory drugs, fludrocortisone, estrogens
      c. Refeeding edema
      d. ?Hepatic cirrhosis
      e. ?Idiopathic edema
    3. Pregnancy and premenstrual edema
  B. Venous obstruction
    1. Hepatic cirrhosis or hepatic venous obstruction
    2. Acute pulmonary edema
    3. Local venous obstruction

Decreased plasma oncotic pressure
  A. Protein loss
    1. Nephrotic syndrome
    2. Protein-losing enteropathy
  B. Reduced albumin synthesis
    1. Liver disease
    2. Malnutrition

Increased capillary permeability
  A. Burns
  B. Trauma
  C. Inflammation
  D. Allergic reactions
  E. Adult respiratory distress syndrome
  F. ?Idiopathic edema
  G. ?Malignant ascites

Increased interstitial oncotic pressure
  A. Lymphatic obstruction
  B. Increased capillary permeability
  C. Hypothyroidism

in heart failure, renal failure, and with the use of certain drugs. Edema due to venous obstruction is commonly seen with cirrhosis of the liver in which there is a marked increase in hepatic sinusoidal pressure or with deep venous thrombosis in the lower extremities.

Decreased plasma oncotic pressure The capillary hydrostatic pressure is effectively balanced by the oncotic pressure of the plasma proteins. The plasma proteins, primarily albumin, exert this osmotic pressure because they are essentially limited to the vascular space, crossing the capillary wall to only a slight degree (see Chap. 1). As a result, a reduction in the plasma albumin concentration (hypoalbuminemia) can lead to edema. This is most often due to albumin loss in the urine in the nephrotic syndrome or to decreased hepatic albumin synthesis due to liver disease or malnutrition.

Increased capillary permeability An increase in capillary permeability promotes the development of edema both directly and by permitting albumin to move into the interstitium, thereby diminishing the oncotic pressure gradient. This cause of edema may be operative in patients with burns, allergic reactions, or any of the conditions associated with the adult respiratory distress syndrome. Capillary permeability is also moderately increased in patients with diabetes mellitus [2a]. This change may enhance the severity of edema which, in these patients, is usually due to heart failure or the nephrotic syndrome.

Increased interstitial oncotic pressure Under normal conditions, small amounts of protein are filtered across the capillary and then returned to the circulation by the lymphatics. If the lymphatics are obstructed, these proteins accumulate in the interstitium, increasing the interstitial oncotic pressure.

Safety factors against the development of edema Since there is normally a small gradient favoring filtration, it might be expected that only a minor change in these hemodynamic forces would lead to edema. However, experimental and clinical observations indicate that there must be a 10- to 15-mmHg increase in the gradient favoring filtration before edema can be detected [2–4]. Two factors appear to be responsible for this. First, as fluid moves into the interstitium, there is an increase in interstitial hydrostatic pressure and, by dilution, a decrease in interstitial oncotic pressure, both of which tend to retard further filtration. Second, the elevation in interstitial hydrostatic pressure can enhance lymphatic flow severalfold so that much of the excess fluid can be removed from the interstitium. Thus, edema will not occur until these factors are overcome. With hypoalbuminemia, for example, this usually requires a fall in the plasma albumin concentration to below 2.5 to 3.0 g/dL (normal equals 4 to 5 g/dL) [5].

Renal Sodium Retention

The retention of fluid by the kidney in edematous states results from one of two basic mechanisms. In some patients, the primary problem is an inability to excrete the Na^+ and water that have been ingested. This most often occurs in patients with renal disease. More commonly, fluid retention is an appropriate compensatory response to effective circulating volume depletion (Fig. 16-1).

As described in Chap. 9, the *effective circulating volume* is an unmeasurable entity that refers to the fluid in the vascular space that is effectively perfusing the tissues. In most instances, this is directly proportional to the cardiac output. Thus, when the cardiac output is reduced because of primary heart disease or a decrease in venous return (as with hypoalbuminemia, which lowers the plasma volume by promoting fluid movement into the interstitium), the kidney attempts to restore the effective circulating volume by retaining Na^+ and water.

However, the effective circulating volume is not always related to the cardiac output. If an arteriovenous fistula is created, there will be no initial change in cardiac output, yet tissue perfusion will be reduced since the blood flowing through the fistula is bypassing the capillary circulation. In response to this, the kidney retains Na^+ and water, thereby increasing the blood volume and the cardiac output [6]. The new steady state is characterized by a cardiac output that exceeds the baseline level by an amount equal to the flow rate through the fistula. A clinical correlate of this experiment occurs in patients with cirrhosis of the liver and ascites (the accumulation of fluid within the peritoneal cavity), who frequently have an increased cardiac output [7]. Despite this, they may avidly retain Na^+ [8]. This disparity between the high cardiac output and the renal response suggesting effective circulating volume depletion is due in part to the presence of multiple arteriovenous fistulas throughout the body, such as spider angiomata in the skin [7]. Thus, much of the cardiac output is circulating ineffectively.

It should be noted that the two mechanisms of renal Na^+ retention—primary renal dysfunction and the response to effective circulating volume depletion—are not mutually exclusive. This appears to be particularly true in hepatic cirrhosis. In some patients with this disorder, fluid accumulation in the dilated splanchnic venous system and in the peritoneum results in effective circulating volume depletion. However, cirrhosis may also directly promote Na^+ reabsorption independent of changes in volume (see "Etiology: Hepatic cirrhosis and ascites" below) [9,10].

In summary, *heart failure, the nephrotic syndrome, and hepatic cirrhosis frequently represent conditions of effective circulating volume depletion even though the total extracellular volume is markedly expanded,* primarily because of the edema fluid in the interstitium. The renal response to these disorders is the same as it is to true volume depletion, as occurs with vomiting: to try to restore tissue perfusion by retaining Na^+ and water. Thus, the urine Na^+ concentration frequently is less than 10 meq/L in these conditions.

This renal response may be due both to a hypovolemia-induced fall in GFR and, more importantly, an increase in tubular reabsorption. The latter may occur throughout the nephron, as enhanced proximal, loop, and collecting tubular reabsorption have been demonstrated [11–17]. A variety of factors may contribute to these changes in tubular function. Effective volume depletion is sensed by pressure receptors in the heart, aorta, and afferent glomerular arteriole, leading to activation of the renin-angiotensin-aldosterone and sympathetic nervous systems and perhaps to reduced secretion of a natriuretic hormone (see Chap. 9) [18–22]. However, no single factor has been identified as being of primary importance. For example, although many edematous patients have hyperaldosteronism, Na^+ retention can occur with normal or even low levels of aldosterone secretion [9,18,23]. Furthermore, aldosterone, which stimulates Na^+ reabsorption in the cortical collecting tubule (see Chap. 8), cannot explain the increases in proximal and loop reabsorption that are frequently present [11–16].

Fluid retention in edematous states is usually isosmotic. Increases in proximal and loop Na^+ reabsorption create osmotic gradients that result in enhanced water reabsorption (in the loop, occurring in the descending limb; see Chap. 6). In contrast, Na^+ is initially reabsorbed without water in the water-impermeable segments of the distal nephron (see Chap. 7). The addition of this Na^+ to the extracellular fluid raises the plasma osmolality, resulting in the stimulation of ADH secretion and thirst. The ensuing increases in collecting tubule water reabsorption and water intake then return the plasma osmolality to normal.

Summary

The development of edema requires both an alteration in capillary hemodynamics favoring fluid movement into the interstitium and renal Na^+ and water retention. When the former predominates (as with hypoalbuminemia or major venous obstruction), there is an initial fall in the plasma volume. Edema then occurs because the compensatory retention of Na^+ and water by the kidney permits the plasma volume to be maintained at near normal levels while much of the excess fluid accumulates in the interstitium (Fig. 16-1). However, *generalized edema will not occur if Na^+ retention is prevented by eliminating Na^+ from the diet.* In this setting, the initial movement of fluid into the interstitium will significantly reduce the plasma volume. This will decrease both the arterial and venous pressures and consequently the capillary hydrostatic pressure. The reduction in the capillary hydrostatic pressure counteracts the effect of hypoalbuminemia, and fluid entry into the interstitium will cease.

Similar considerations apply to conditions in which renal Na^+ retention is primary, as in renal failure. If dietary Na^+ intake is limited, there will be no fluid retention, increase in capillary hydrostatic pressure, or edema.

ETIOLOGY

Congestive Heart Failure

Congestive heart failure can be produced by a variety of cardiac disorders, including coronary artery disease, hypertension, valvular disease, cor pulmonale, and cardiomyopathies. The edema in heart failure is due to an increase in venous pressure which results in an increase in capillary hydrostatic pressure. The site of edema accumulation is variable and is dependent upon the nature of the cardiac disease. Patients with only left ventricular dysfunction, e.g., due to aortic stenosis or a myocardial infarction, initially accumulate fluid in the lungs and develop pulmonary edema. In contrast, patients with pure right ventricular failure, e.g., due to cor pulmonale, develop dependent edema in the lower extremities and perhaps ascites. Patients with biventricular failure, e.g., due to a cardiomyopathy, have both peripheral and pulmonary edema.

In acute pulmonary edema due to a myocardial infarction, the left ventricular disease results in elevated left ventricular end-diastolic and left atrial pressures which are transmitted *back* through the pulmonary veins to the pulmonary capillaries. In general, the pulmonary capillary pressure must exceed 18 to 20 mmHg (normal is 5 to 12 mmHg) before pulmonary edema occurs [24,25].

More commonly, in patients with chronic congestive heart failure, the increase in venous pressure is produced by increased blood volume secondary to renal Na^+ and water retention, not by the obstructive effect of a diseased heart. This is called

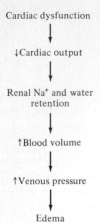

Cardiac dysfunction

↓Cardiac output

Renal Na⁺ and water retention

↑Blood volume

↑Venous pressure

Edema

Figure 16-3 Forward hypothesis of heart failure in which a reduction in cardiac output is the primary event.

the *forward hypothesis of heart failure* (Fig. 16-3) [26,27]. By this theory, the primary event is a reduction in cardiac output.† This is sensed by the kidney, which responds by increasing Na⁺ and water reabsorption. As this process continues, blood volume expands beyond the capacity of the venous system to distend, resulting in increased venous pressure and edema. One of the principal pieces of evidence supporting this theory comes from the studies of patients who are edema-free and have normal venous pressures after their heart failure has been treated with dietary Na⁺ restriction, digitalis, and diuretics [27]. If their Na⁺ intake is increased, they will retain Na⁺ *before* there is an elevation in venous pressure. Only after sufficient amounts of Na⁺ and water have been retained will the venous pressure rise and edema occur.

The cardiac response to this fluid retention is depicted in Fig. 16-4. The upper curve represents the normal Frank-Starling relationship between stroke volume and the left ventricular end-diastolic pressure (LVEDP) [28]. If the LVEDP rises, there will be an increase in cardiac contractility and stroke volume due to stretching of the cardiac muscle. In the presence of mild cardiac failure (middle curve), there will be an initial reduction in stroke volume and cardiac output (line AB). As renal Na⁺ and water retention increases the plasma volume, there will be an elevation in the cardiac filling pressure which will tend to normalize the stroke volume (line BC). (In addition, other factors such as an increase in sympathetic tone and heart rate will act to restore a normal cardiac output.) At this point, the patient will be in a new steady state of compensatory heart failure in which the stroke volume and cardiac output are normal, Na⁺ excretion matches Na⁺ intake, and the activity of the renin-angiotensin-aldosterone system has returned to normal [18]. However, the restora-

---

†This hypothesis is also applicable to patients with high-output heart failure due, for example, to arteriovenous fistulas or hyperthyroidism. In this setting, the patients still behave as if they are effectively hypovolemic because the cardiac output is inappropriately low in relation to tissue requirements [6].

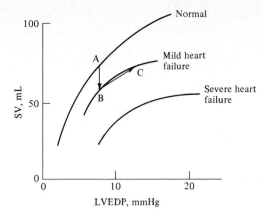

Figure 16-4 Frank-Starling curves relating stroke volume (SV) to left ventricular end-diastolic pressure (LVEDP) in normal subjects and patients with heart failure. See text for details. *(Adapted from J. N. Cohn, Am. J. Med., 55:131, 1973.)*

tion of tissue perfusion has occurred only after there has been an increase in the LVEDP, perhaps to a level sufficient to produce pulmonary edema.

There are several points that deserve emphasis in this simple example of mild to moderate heart failure.

1. It demonstrates again the dual effects of fluid retention in edematous states: a beneficial increment in cardiac output and a potentially harmful elevation in venous pressure. The major increase in cardiac output occurs as the LVEDP rises to 12 to 15 mmHg. There is little further effect on cardiac output above this level, but pulmonary edema becomes more likely [25]. These relationships have important implications for therapy (see "Treatment: Congestive heart failure" below).
2. It illustrates that vascular congestion (that is, an elevated LVEDP) and a low cardiac output do not have to occur together in patients with heart disease. At point B, the patient is in a low-output state but there is no congestion; at point C, the patient is congested but has a normal output.
3. The Frank-Starling relationship in Fig. 16-4 varies with exercise. Patients with moderate heart disease may have a normal cardiac output at rest but may be unable to increase it adequately with exertion. This relative decrease in the effective circulating volume can lead sequentially to activation of the sympathetic nervous system, renal vasoconstriction and ischemia, sodium retention, and ultimately edema [29,30]. In this setting, limiting physical activity may produce substantial improvement.
4. Patients with mild to moderate heart disease may have no edema with dietary Na^+ restriction but may retain Na^+ and possibly become edematous if given a Na^+ load [31]. Suppose points A and C in Fig. 16-4 reflect the hemodynamic state on a low-Na^+ diet. An increase in Na^+ intake will initially expand the intravascular volume and raise the LVEDP. In the normal subject (point A) who is still on the ascending limb of the Frank-Starling curve, this will produce an increase in stroke volume and cardiac output, which will then promote the excretion of the excess Na^+. In contrast, a similar elevation in the LVEDP in the

patient with heart failure (point C), who is on a flatter part of the curve, will produce less of an elevation in cardiac output and consequently in Na^+ excretion. Edema in this setting will be associated with a relatively high rate of Na^+ excretion that is still inappropriately low in relation to intake. Limiting dietary Na^+ may be sufficient to alleviate the edema.

The situation is somewhat different with severe heart failure (lower curve, Fig. 16-4). In this setting, the stroke volume can never be normalized by increasing the LVEDP as the plateau in stroke volume occurs earlier and at a lower level than in mild heart failure. Two factors appear to account for this plateau. First, the heart may simply have reached its maximum capacity to contract. Second, the Frank-Starling relationship really applies to left ventricular end-diastolic *volume* since it is the stretching of the muscle that is responsible for the enhanced contractility [28]. The more easily measured LVEDP is used since, in relatively normal hearts, pressure and volume vary in parallel. However, cardiac compliance may be greatly reduced with severe heart disease [32]. As a result, a small increase in volume produces a large elevation in LVEDP but no substantial stretching of the cardiac muscle and therefore little change in cardiac output.

Cor Pulmonale The pathogenesis of edema in cor pulmonale due to chronic obstructive pulmonary disease is different from that in other forms of heart failure. In this disorder, the cardiac output and GFR are usually normal or near normal both in the resting state and with exercise [33–35]. Edema seems to occur almost exclusively in patients with hypercapnia, suggesting that the high P_{CO_2} rather than primary cardiac dysfunction may be responsible for the Na^+ retention. Hypercapnia is associated with an appropriate increase in HCO_3^- reabsorption, which serves to minimize the change in arterial pH (see Chap. 20). Na^+ is retained in this setting because electroneutrality must be maintained and because proximal HCO_3^- reabsorption creates gradients that promote the passive reabsorption of NaCl and water (see page 93). In addition, hypercapnic patients with edema have a low rate of renal blood flow (mechanism unknown) but a relatively normal GFR [34,35]. The associated elevation in the filtration fraction (GFR/renal plasma flow) can contribute to Na^+ retention by further increasing net proximal reabsorption (see page 95).

Hepatic Cirrhosis and Ascites

Ascites is the accumulation of fluid in the peritoneal cavity. It can be seen in a variety of conditions, including severe acute or chronic hepatic disease (particularly hepatic cirrhosis), heart failure, and the nephrotic syndrome, and with tumor implants on the peritoneum. In the last condition, both increased capillary permeability and lymphatic obstruction may contribute to the development of ascites.

In patients with liver disease, the ascitic fluid is derived from the hepatic sinusoids and enters the peritoneum by moving across the hepatic capsule. The principal factor in the development of hepatic ascites is postsinusoidal obstruction (due to fibrosis or hepatic venous obstruction) leading to an increase in the hydrostatic pres-

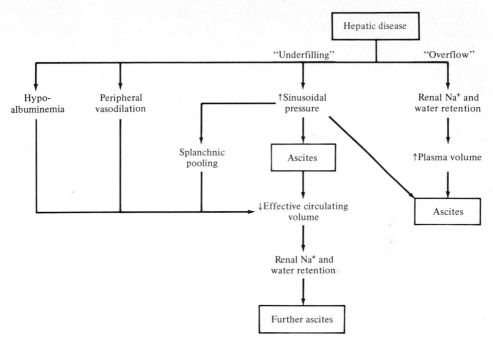

Figure 16-5 Underfilling and overflow theories of the pathogenesis of fluid retention and ascites in hepatic disease. In some patients, both mechanisms may be operative. In addition, the combination of increased femoral venous pressure (resulting from an ascites-induced elevation in intraperitoneal pressure) and hypoalbuminemia can lead to peripheral edema.

sure in the sinusoids† [36,37]. Hypoalbuminemia due to decreased hepatic synthesis may also be present and can contribute to the development of peripheral edema, particularly in the lower extremities. It does not play a major role in hepatic ascites, however, since the sinusoids are normally freely permeable to albumin. As a result, there is essentially no oncotic pressure gradient acting to hold fluid in the vascular space [2,38].

Two safety factors act to limit the degree of ascites: increased lymph flow and elevation in the intraperitoneal pressure. Enhanced hepatic lymph flow can initially remove the edema fluid. On the other hand, abnormalities in the thoracic duct, into which the hepatic lymphatics drain, may contribute to the development of ascites [37]. Once the rate of edema formation is sufficient to overcome the protective lymphatic effect, the ensuing elevation in intraperitoneal pressure is the only factor that retards further fluid accumulation [38,39].

Underfilling versus overflow The traditional underfilling theory of fluid retention and ascites formation is illustrated in Fig. 16-5. Hepatic disease produces a variety

---

†Although portal venous hypertension is also present, this alone does not lead to ascites formation [36,38].

of changes that contribute to the development of effective hypovolemia including: increased hepatic sinusoidal pressure which leads to fluid accumulation both in the peritoneum and in the dilated splanchnic venous system; hypoalbuminemia; and peripheral vasodilation (due in part to both the formation of arteriovenous fistulas and enhanced production of prostaglandins) which augments vascular capacity [40]. As a result, the kidney appropriately retains Na^+ and water. This tends to restore normovolemia but also promotes further ascites formation since the retained fluid preferentially accumulates in the peritoneum because of the high hepatic sinusoidal pressure.

A variety of observations are consistent with the underfilling hypothesis. In particular, many cirrhotic patients with ascites have increased plasma levels of those hormones whose secretion is enhanced by hypovolemia: renin, norepinephrine, and antidiuretic hormone [41]. Furthermore, the removal of some of the edema fluid with diuretics can cause renal insufficiency (presumably due to hypoperfusion) [42], whereas volume expansion produced by ascites reinfusion or immersion to the neck in warm water (which redistributes fluid to the cardiopulmonary circulation, thereby increasing the cardiac output) can improve renal function and lower renin secretion (see Fig. 9-2) [43,44,44a].

The following observations, however, have led to an alternate hypothesis that liver disease directly promotes Na^+ retention and that the excess fluid (or overflow) then leads to ascites formation [9,10,40].

1. Patients with cirrhosis *without* ascites or in whom the ascites has been reinfused by a peritoneovenous shunt (see below) are still unable to excrete a Na^+ load *normally* even though the plasma volume is increased, renal function is normal, and renin secretion is suppressed [21,22,45,46]. Similar findings are present in cirrhosis in experimental animals in whom portal hypertension is prevented by performing a portacaval shunt prior to the induction of cirrhosis. In this setting, renal Na^+ and water retention occurs *before* the development of ascites [10,47].

2. Sera from sodium-loaded patients in one study demonstrated less in vivo natriuretic activity than seen in similarly treated normal subjects [21]. This raises the possibility that cirrhosis may lead to decreased secretion of a natriuretic hormone (see Chap. 8). Consistent with this theory is the suggestion that the abnormal Na^+ retention occurs in the distal nephron [45,47a], the presumed site of action of natriuretic hormone.

3. Some patients with cirrhosis and ascites do not have the increased activity of the renin-angiotensin-aldosterone system that would be expected if they were underfilled [8,9,48]. This finding, however, is not necessarily indicative of overflow since these patients could have been underfilled and hyperreninemic initially. The ensuing fluid retention could then have restored normovolemia, thereby lowering renin secretion.

4. The plasma volume [49] and cardiac output [7] are frequently elevated in hepatic cirrhosis. Although these findings can be explained, respectively, by sequestration of slowly circulating blood in the dilated splanchnic system and by arteriovenous

fistulas (such as spider angiomas of the skin), they are also compatible with primary overfilling.

In summary, the relative roles of underfilling and overflow in hepatic ascites in humans are uncertain [40]. It may be that *overflow is primary early in the disease and that underfilling becomes more important with the development of severe hepatic fibrosis* which intensifies the stimulus to ascites formation. In experimental cirrhosis, for example, a small degree of Na^+ retention precedes any obvious hemodynamic changes and is consistent with overflow. However, the major part of fluid retention occurs after peripheral vasodilation and possible underfilling are present [10]. Similarly, patients with high plasma renin activities and probable underfilling tend to have more advanced disease, as evidenced by more ascites, a lower GFR and rate of Na^+ excretion, and shorter survival, than those in whom renin secretion is normal [8,48].

Nephrotic Syndrome

The nephrotic syndrome can be produced by a variety of renal disorders and is characterized by an increase in the permeability of the glomerular capillary wall to proteins [50]. Consequently, more albumin and globulins are filtered and then lost in the urine, resulting in a reduction in the plasma oncotic pressure. In addition, catabolism of some of the filtered albumin by the proximal tubular cells and decreased hepatic albumin synthesis may contribute to the decrease in the plasma albumin concentration [51,52].

Both underfilling and overflow may contribute to the development of edema in the nephrotic syndrome (Fig. 16-6) [53]. Hypoalbuminemia favors the movement of fluid out of the vascular space into the interstitium, leading to effective volume depletion and renal Na^+ and water retention. Since there is a 10- to 15-mmHg safety factor protecting against edema, the plasma albumin concentration usually has to fall below 2.5 to 3.0 g/dL for edema to occur [5]. This sequence is most often seen when the GFR is relatively normal. In other patients, there is primary fluid retention by the kidney due either to a low GFR or perhaps some direct effect of the kidney disease itself [16a,53,54]. Edema is now due primarily to overfilling and is associated with low renin and aldosterone levels [53,54]. However, hypoalbuminemia will also contribute to the development of edema in these patients.

Primary Renal Sodium Retention

Patients with normal cardiac and hepatic function may develop edema if there is a primary renal abnormality preventing the excretion of Na^+ and water (Fig. 16-7). This is most frequently seen with acute or chronic renal failure in which the low GFR favors Na^+ retention. Patients with acute glomerulonephritis may be particularly prone to develop edema. In this disorder, the glomeruli are diseased but tubular function is initially normal. As a result, these patients behave as if they had reduced renal perfusion and avidly reabsorb Na^+ in addition to filtering less [55,56].

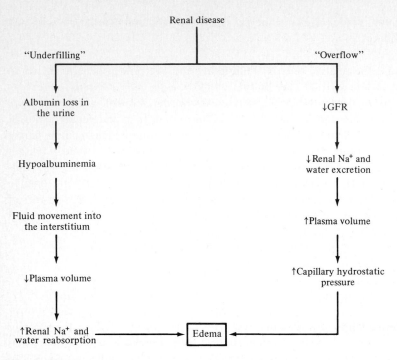

Figure 16-6 Underfilling and overflow mechanisms of edema in the nephrotic syndrome and renal insufficiency.

Certain drugs can also enhance renal Na^+ reabsorption (Table 16-1). Many antihypertensive agents induce Na^+ retention [57], particularly peripheral vasodilators such as minoxidil, diazoxide, and calcium-channel blockers (especially nifedipine) [58–60]. The mechanism by which this occurs is uncertain. Since hypertensives require a high blood pressure to maintain Na^+ excretion (see Fig. 9-5), the fall in pressure induced by these agents may directly impair Na^+ excretion.

Nonsteroidal anti-inflammatory drugs are widely used in the treatment of rheumatologic disorders and appear to act by inhibiting renal prostaglandin synthesis. Since prostaglandins maintain renal perfusion and may promote Na^+ excretion (see Chap. 8), decreasing their production can lead to Na^+ retention and edema [61,62].

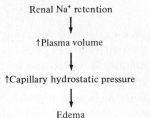

Figure 16-7 Pathogenesis of edema in primary renal Na^+ retention.

Edema is particularly likely to occur in a patient with underlying heart failure, hepatic cirrhosis, or nephrotic syndrome, conditions of effective hypovolemia in which basal prostaglandin synthesis is increased.

Fludrocortisone is a synthetic mineralocorticoid used in the treatment of hypo-aldosteronism. Although this drug initially causes fluid retention, edema is unusual because of the phenomenon of aldosterone escape (see page 144). Estrogens (alone or in oral contraceptives) also may promote fluid retention, primarily in patients with impaired estrogen metabolism due to hepatic disease [63,64].

Another example of primary renal Na$^+$ retention occurs in refeeding edema. Patients who have fasted for as little as three days display marked Na$^+$ retention and possibly edema after refeeding with carbohydrates [65]. Although it is unclear how this occurs, both the increase in insulin secretion and reduction in glucagon secretion induced by the carbohydrate meal may play a role [66,67]. A similar phenomenon may occur during the treatment of diabetic ketoacidosis [68].

Finally, the edema of hepatic cirrhosis (see above) and idiopathic edema may be due in part to primary retention of fluid by the kidney.

Idiopathic Edema

Idiopathic edema refers to a disorder occurring in young, menstruating women in the absence of cardiac, hepatic, or renal disease [69]. Fluid retention may initially occur premenstrually but eventually becomes persistent. Emotional problems and obesity commonly are part of this syndrome.

The etiology of idiopathic edema is uncertain. Fluid retention is most prominent when the patient is upright, during which time the patient has an exaggerated fall in the plasma volume, GFR, and urinary Na$^+$ excretion [69,70]. These observations have led to the suggestion that the primary defect is either an increase in capillary permeability or an elevation in capillary hydrostatic pressure due to dilation of the precapillary sphincter [69,71]. How these changes might occur is not known.

Diuretic-induced edema Another theory postulates that idiopathic edema may be paradoxically induced by the chronic administration of diuretics [72,73]. According to this hypothesis, patients are initially begun on a diuretic for a minor degree of fluid retention. As therapy is continued, persistent diuretic-induced hypovolemia results in the activation of sodium-retaining mechanisms, particularly the renin-angiotensin-aldosterone system. If the diuretic is then stopped, the patient may be unable to *acutely* shut off these sodium-retaining mechanisms, resulting in edema formation and the *mistaken* assumption that chronic diuretic therapy is indicated. If, however, the patient is maintained without diuretics for 1 to 3 weeks, a spontaneous diuresis will frequently ensue with resolution of the edema (Fig. 16-8).

Idiopathic edema should be differentiated from premenstrual edema, which occurs in many women [74]. This disorder is mild and self-limited with a diuresis beginning with or shortly after the onset of menses. The fluid retention is thought to be humorally mediated, as estrogens or possibly prolactin may play a role [74].

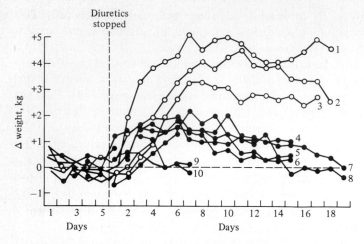

Figure 16-8 Changes in weight before and after stopping diuretics in 10 patients with idiopathic edema. Patients 4 to 10 returned to their baseline weight within 20 days. Patients 1 to 3 maintained their weight gain but eventually lost the extra weight after the institution of a low-sodium diet. *(From G. A. MacGregor, J. E. Roulston, N. D. Markandu, J. C. Jones, and H. E. deWardener, Lancet, 1:397, 1979.)*

Other Causes of Edema

Many other conditions can affect capillary hemodynamics and produce edema (Table 16-1). Of greatest clinical importance are burns, laryngeal edema secondary to an allergic reaction, and the adult respiratory distress syndrome. The latter disorder is characterized by pulmonary edema due primarily to an increase in pulmonary capillary permeability [75,76]. A variety of conditions can cause the adult respiratory distress syndrome, including sepsis, shock, trauma, oxygen toxicity, and severe renal failure [75].

Mild edema of the lower extremities occurs in up to 75 percent of pregnant women. Pregnancy is normally associated with the retention of 500 to 900 meq of Na^+ and 12 to 13 L of water, much of which is used to meet the needs of the growing fetus [77]. The degree of Na^+ retention is independent of alterations in intake, suggesting that this response is under strict regulation. The mechanism by which it occurs is uncertain as several factors may contribute, including hormones (aldosterone and estrogens) and a fall in blood pressure (to a mean of 105/60 at the end of the second trimester) due to smooth muscle relaxation [74,77]. Since the fluid retention is primarily physiologic, diuretic therapy for edema is usually not indicated [77].

Hypothyroidism (myxedema) may be associated with mild edema and, less commonly, serous effusions in the pericardium, pleura, or peritoneum [78,79]. These patients have a marked increase in interstitial albumin and other proteins due to movement out of the capillary [78]. Although this is indicative of an increase in capillary permeability, the excess interstitial protein and fluid would normally be returned to the systemic circulation by the lymphatics. However, lymphatic flow is low or normal in myxedema [78], not increased as in other edema states [80]. This may be due to binding of the filtered proteins to excess interstitial mucopolysaccharides, thereby preventing their removal by the lymphatics [78]. These abnormalities are corrected by the administration of thyroid hormone.

SYMPTOMS AND DIAGNOSIS

A complete discussion of the many diseases that can produce heart failure, cirrhosis, or the nephrotic syndrome and the methods used in their diagnosis is beyond the scope of this chapter. In this section, we will attempt to demonstrate how an understanding of the pathophysiology of edematous states can be used to arrive at the proper diagnosis. Of particular importance are (1) the pattern of distribution of edema, which reflects those capillaries with altered hemodynamic forces, and (2) the venous pressures (or the filling pressures) in the heart (Table 16-2).

Pulmonary Edema

Patients with pulmonary edema complain primarily of shortness of breath and orthopnea. When pulmonary edema is due to an acute myocardial infarction, chest pain also may be a prominent symptom. Physical examination usually reveals a tachypneic, diaphoretic patient with wet rales on auscultation of the chest and possibly gallop rhythms and heart murmurs. The diagnosis should be confirmed by a chest x-ray since other disorders which require different therapy may produce similar findings, e.g., a pulmonary embolus.

 Although cardiac disease is the most common cause of pulmonary edema, it can also be produced by those disorders associated with primary renal Na^+ retention or the adult respiratory distress syndrome. If the correct diagnosis cannot be established from the history, physical examination, and laboratory data, measurement of the pulmonary capillary wedge pressure can be extremely helpful. The wedge pressure exceeds 18 to 20 mmHg when pulmonary edema is due to heart disease or primary

Table 16-2 Physical and laboratory findings in the major edematous states

Disorder	Pulmonary edema	Central venous pressure	Ascites and/or pedal edema	Proteinuria
Left-sided heart failure	+	Variable	−	0
Right-sided heart failure	−	↑	+	0 − + + + +
Hepatic cirrhosis	−	↓ −N	+	0 − + +
Nephrotic syndrome	−	↓ −N	+	+ + + +
Primary renal Na^+ retention	+	↑	+	0 − + + + +
Idiopathic edema	−	↓ −N	+	0

renal Na^+ retention [24] but is relatively normal in the setting of increased capillary permeability in the adult respiratory distress syndrome [75].

Pulmonary edema does not occur in uncomplicated hepatic cirrhosis or the nephrotic syndrome because the primary movement of fluid out of the vascular space into the peritoneum or interstitium prevents the increase in venous return necessary to elevate the pulmonary capillary pressure [81]. Even with overfilling in cirrhosis, the elevation in hepatic sinusoidal pressure promotes the accumulation of most of the excess fluid in the peritoneum. Hypoalbuminemia alone does not lead to pulmonary edema [82] since the pulmonary capillary is relatively permeable to albumin (see Table 15-7) [2]. As a result, the pulmonary circulation does not normally depend upon a favorable oncotic pressure gradient to hold fluid in the vascular space. This is similar to the lack of effect of hypoalbuminemia on hepatic ascites formation (see above).

Peripheral Edema and Ascites

In comparison to the potentially life-threatening nature of pulmonary edema, peripheral edema and ascites are cosmetically undesirable but produce less serious symptoms. These include swollen legs, difficulty in walking, increased abdominal girth, and shortness of breath due to pressure on the diaphragm in patients with tense ascites. Peripheral edema can be detected by the presence of pitting after pressure is applied to the edematous area. Since peripheral edema locates preferentially in the dependent areas, it is primarily found in the lower extremities in ambulatory patients and over the sacrum in patients at bed rest. Ascites is typically associated with abdominal distention, and shifting dullness and a fluid wave on percussion of the abdomen. Patients with the nephrotic syndrome may also have prominent periorbital edema due to the low tissue pressure in this area.

The distribution of edema and the measurement of the filling pressure on the right side of the heart can aid in the differential diagnosis of heart failure, cirrhosis, the nephrotic syndrome, and primary renal Na^+ retention (Table 16-2). This is particularly important since chronic right-sided heart failure can produce cirrhosis due to chronic passive congestion of the liver and proteinuria, which on rare occasions can approach the nephrotic range [83,84].

Heart failure Patients with right-sided heart failure have peripheral edema and, in severe cases, ascites and edema of the abdominal wall. Shortness of breath is commonly present and may be due to underlying pulmonary disease or coexistent left ventricular failure. The edema in these disorders is due to an increase in venous pressure behind the right side of the heart. Thus, the pressures in the right atrium and subclavian vein are elevated. This can be detected as distention of the jugular neck veins or measured directly with a central venous pressure catheter.†

---

†The technique for evaluating the jugular neck veins is presented on page 287.

Hepatic cirrhosis Cirrhotic patients develop ascites and edema in the lower extremities, because of an increase in the venous pressure below the hepatic vein. As a result, the venous pressure above the hepatic vein, i.e., in the vena cava and right atrium, usually is reduced or normal [81] but not elevated as in right-sided heart failure.† The presence of other signs of portal hypertension such as distended abdominal wall veins and splenomegaly also is suggestive of primary hepatic disease. However, since chronic right-sided heart failure can produce cirrhosis, these findings are not sufficient to establish the correct diagnosis. For example:

> **Case history 16-1** A 56-year-old man had a 3-year history of ascites which had become so severe as to require removal of 3 to 4 liters of ascitic fluid by paracentesis every 3 weeks. The patient had been told he had cirrhosis, although a liver biopsy was not performed. He had no history of hepatic disease and was only a social drinker. The physical examination revealed no acute distress, a soft abdomen with marked ascites, and moderate pedal edema. The heartbeat was irregularly irregular. The heart sounds were distant, and no murmurs were heard. Several spider angiomata were present. The jugular neck veins were distended to the angle of the jaw at 45°. The electrocardiogram showed atrial fibrillation and low voltage. There was 3+ proteinuria; the plasma albumin concentration was 2.9 g/dL; and the liver function tests were mildly abnormal.

> COMMENT Despite the features suggestive of cirrhosis, the distended neck veins pointed toward the presence of right-sided heart failure. The central venous pressure was measured at 18 mmHg (normal is 1 to 6 mmHg). Further evaluation confirmed the diagnosis of constrictive pericarditis. After pericardiectomy, the patient had a complete recovery.

Nephrotic syndrome Patients with the nephrotic syndrome have periorbital and peripheral edema and occasionally ascites. Since the primary abnormality is hypoalbuminemia leading to the movement of fluid out of the vascular space, the venous pressures in the heart are normal or slightly reduced. Consequently, pulmonary edema is not present, and the central venous pressure is not elevated in the absence of heart failure, e.g., due to hypertension, or renal failure. The diagnosis can be established by the demonstration of heavy proteinuria, greater than 3.5 g/24 h, and hypoalbuminemia, usually less than 2.5 g/dL. Hyperlipidemia also is frequently present.

Renal sodium retention The physical findings associated with primary renal Na^+ retention are similar to those seen with biventricular heart failure. Since these patients are *volume-expanded*, both pulmonary and peripheral edema may be present and the neck veins should be distended because of a high central venous pressure. These disorders can be differentiated from primary cardiac disease by the presence of severe renal failure (as evidenced by a marked elevation in the BUN and plasma creatinine concentration), by a history of taking a Na^+-retaining drug, and, most importantly, by the presence of normal cardiac function when the excess fluid is removed.

---

†One exception to this general rule occurs with tense ascites in which upward pressure on the diaphragm increases the intrathoracic pressures. The central venous pressure, however, rapidly falls to normal with removal of a small amount of ascites to reduce the intraperitoneal pressure [81].

Other Patients with idiopathic edema behave as if they are volume-depleted because of the exaggerated fall in the plasma volume in the erect position and the concomitant effect of diuretics [69–73]. As a result, they have peripheral edema, but the central venous pressure is normal and pulmonary edema does not occur.

In addition to the above disorders, edema may occur as a result of local changes in capillary hemodynamics. For example, a patient with a postphlebitic syndrome after an episode of thrombophlebitis may develop *unilateral* pedal edema because of an increase in the venous pressure limited to that extremity. This is different from the generalized edematous states in which bilateral edema should be present.

PHYSIOLOGY OF DIURETICS

Introduction

Diuretics are useful in the treatment of edema because they impair the active transport of Na^+, resulting in an increase in the urinary excretion of Na^+ and water. To understand the effects of diuretics, it is important to review how Na^+ and water are handled by the kidney. In a normal adult, the glomerular filtration rate is approximately 135 to 180 L/day or 95 to 125 mL/min. About 99 percent of the filtrate is then reabsorbed by the tubules, resulting in a urine output of 1 to 2 L/day, which in the steady state is equal to water intake. Each of the nephron segments contributes in a varying degree to this process (see Table 9-3). In the proximal tubule, roughly 70 percent of the filtrate is reabsorbed in an isosmotic manner, primarily by active Na^+ transport. In the medullary loop of Henle, approximately 15 to 20 percent of the filtered NaCl and 5 to 10 percent of the filtered water are reabsorbed. This reabsorption of NaCl in excess of water is the essential step in both urinary concentration and dilution (see below). The remainder of tubular reabsorption occurs in the distal nephron segments, including the cortical collecting tubule where Na^+ reabsorption and K^+ and H^+ secretion are enhanced by aldosterone.

The effect of a diuretic is determined not only by the amount of Na^+ and water reabsorption which has been inhibited but also by the ability of the nephron segments distal to the drug's site of action to increase the amount of Na^+ and water they reabsorb. This is particularly true of the proximal tubule and loop of Henle. For example, if the glomerular filtration rate is increased by 5 L/day, there may be relatively little change in the urine output, because of a proportionate increase in tubular reabsorption. This is referred to as glomerulotubular balance (see page 97). Similarly, decreasing proximal reabsorption in an edematous state does not result in a large increase in Na^+ and water excretion because of the ability of the more distal segments, primarily the loop of Henle, to reabsorb the increased load [15,85]. As a result, agents which inhibit proximal reabsorption, such as acetazolamide, are not particularly effective diuretics when used alone [86]. In contrast, diuretics which act in the loop of Henle, such as furosemide, are extremely potent since the distal nephron is unable to reabsorb all of the increased load.

The sites at which diuretics act have been determined from experiments using micropuncture and isolated tubular segments (see Chap. 4) and from the effects of these drugs on urinary concentration, dilution, and K^+ excretion. As described in Chap. 6 (see Fig. 6-9), the excretion of a concentrated urine is dependent upon production of medullary interstitial hypertonicity by NaCl reabsorption without water in the medullary ascending limb of Henle's loop; the excretion of a dilute urine, on the other hand, is dependent upon lowering the tubular fluid osmolality by NaCl reabsorption without water in the *medullary and cortical* aspects of the ascending limb and, to a lesser degree, the more distal nephron segments. Thus, any diuretic which acts in the medullary ascending limb (furosemide or ethacrynic acid) will impair both urinary concentration and dilution, whereas a diuretic which acts distal to the medullary ascending limb (such as a thiazide) will impair dilution but not concentration.

The effect of a diuretic on K^+ excretion can also be used to localize its site of action. K^+ *excretion* is primarily a function of K^+ *secretion* from the cortical collecting tubule cell into the lumen by a process that varies directly with the urinary flow rate to the distal nephron (see Chap. 13). Thus, any diuretic which acts proximal to the secretory site will increase distal flow rate and K^+ secretion. On the other hand, a diuretic which inhibits Na^+ reabsorption at the distal secretory site will reduce K^+ secretion.

The sites of action and pharmacologic properties of the most commonly used diuretics are depicted in Fig. 16-9 and Table 16-3 [87,88]. Diuretics which act in the proximal tubule, loop of Henle, and distal nephron are available. As will be seen, knowledge of the site of action provides the rationale for the use of diuretic combinations in patients with resistant edema. Drugs which act at different sites are likely to have additive or even potentiating effects on electrolyte and water excretion.

Acetazolamide

Acetazolamide inhibits the enzyme carbonic anhydrease, which plays an important role in proximal tubular HCO_3^- reabsorption (see page 230). As a result, acetazolamide produces a $NaHCO_3^-$ diuresis which is associated with enhanced K^+ excretion due to the increase in flow to the distal nephron [89]. NaCl is also lost, even though acetazolamide has no known direct action on NaCl transport [90]. This effect is presumably due to diminished *passive* proximal NaCl reabsorption, the gradient for which is primarily created by HCO_3^- reabsorption (see page 93).

Acetazolamide is a relatively weak diuretic because of the ability of the more distal segments to reclaim much of the excess solute and water coming out of the proximal tubule [86]. In addition, its diuretic action is self-limited since it becomes relatively ineffective in the presence of the metabolic acidosis that results from the loss of HCO_3^- in the urine. Nevertheless, acetazolamide can be useful in two clinical settings: (1) in patients with metabolic alkalosis since the loss of excess HCO_3^- tends to restore acid-base balance (see Chap. 18); and (2) in combination with other diuretics in refractory edema (see below).

Acetazolamide can be given orally or intravenously in a dose of 250 to 375 mg once a day. It begins to act within 1 h, and the diuresis is complete within 6 to 8 h.

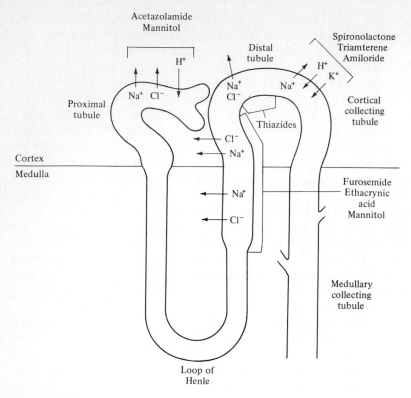

Figure 16-9 Schematization of the probable major sites of action of the most commonly used diuretics.

Side effects are infrequent unless electrolyte disturbances develop as a result of the diuresis.

Mannitol

Mannitol is a nonreabsorbable polysaccharide which acts as an osmotic diuretic, inhibiting Na^+ and water reabsorption in the proximal tubule and loop of Henle [91–93]. A similar effect can be produced by the administration of glycerol or urea, or with glucosuria in uncontrolled diabetes mellitus. In contrast to other diuretics, the osmotic diuretics produce a relative water diuresis since water is lost in excess of NaCl [91,92].

The main clinical use of mannitol as a diuretic is in the early stages of acute oliguria in an attempt to prevent progression to acute renal failure (see page 305) [94,95]. It is not commonly used as a diuretic in generalized edema states except in combination with other diuretics for refractory edema and in certain patients with cirrhosis and hyponatremia since it enhances water excretion more than Na^+ excretion [14].

Table 16-3 Pharmacologic properties of diuretics

Drug	Site of action	Characteristics of diuresis
Acetazolamide	Proximal tubule	Loss of Na^+, HCO_3^-, Cl^-, and K^+ Ineffective in metabolic acidosis
Mannitol	Proximal tubule Loop of Henle	Loss of water in excess of NaCl
Furosemide Ethacrynic acid	Medullary and cortical ascending limbs of the loop of Henle, ± inner medullary collecting tubule	Loss of Na^+, Cl^-, and K^+ May lose up to 35–40% of filtered Na^+ Effective in acidosis and alkalosis Furosemide may have minor action in proximal tubule Inhibit concentration and dilution
Thiazides	Distal tubule, ± proximal tubule, inner medullary collecting tubule	Loss of Na^+, Cl^-, and K^+ May lose up to 5% of filtered Na^+ Inhibit dilution, not concentration Mild carbonic anhydrase inhibitor
Spironolactone Amiloride Triamterene	Collecting tubules	Loss of Na^+ and Cl^- K^+ sparing

Mannitol is administered intravenously over several minutes in doses of 12.5 to 25 g (50 to 100 mL of a 25% solution). In patients with relatively normal renal function, a diuresis begins within 15 to 30 min and lasts for 3 to 4 h.

The major side effects of mannitol are circulatory overload and hyperosmolality. Since it is given as a hypertonic solution, mannitol causes water to move out of the cells into the interstitium and vascular space. (This constitutes the rationale for its use in cerebral edema [96].) This increase in the plasma volume can precipitate pulmonary edema in patients with borderline cardiac function. Thus, the use of mannitol is contraindicated in patients with heart failure.

Mannitol can also produce a clinically important increase in the plasma osmolality in one of two ways. First, by producing a water diuresis, the repeated administration of mannitol can lead to a water deficit, hypernatremia, and hyperosmolality [97]. Second, the hypertonic mannitol may be retained in patients with renal failure, directly increasing the plasma osmolality. In this setting, there is a paradoxical reduction in the plasma Na^+ concentration because of dilution by water movement out of the cells [98].

Loop Diuretics

Furosemide and ethacrynic acid act primarily in the thick segment of the medullary and cortical ascending limbs of Henle's loop[†] but also appear to inhibit NaCl reabsorption in the inner medullary collecting tubule [87,88,99]. The major loop effect of furosemide, acting in the tubular lumen, is to inhibit the coupled transport of Na^+ and Cl^- from the lumen into the cell [100,101]; it is not known if ethacrynic acid shares the same mechanism, as it may act by interfering with oxidative metabolism [88]. Since reabsorption in the loop of Henle is essential for concentration and dilution, loop diuretics tend to result in the production of large volumes of relatively isosmotic urine [102,103]. K^+ excretion is also enhanced because of the increase in flow to the distal nephron.

In addition, both drugs produce renal vasodilation, resulting in an increase in renal blood flow. This effect may be mediated by vasodilator prostaglandins, although the mechanism responsible for their release is uncertain [104,105]. Thiazides, which act more distally, and acetazolamide do not induce this response and may actually reduce renal blood flow [104,106].

Furosemide and ethacrynic acid[‡] are the most powerful diuretics available, potentially leading to the excretion of as much as 30 to 40 percent of the filtered Na^+ and water when given in high doses [15,88,102]. Consequently, the excessive use of these drugs can produce severe volume depletion and electrolyte disturbances.

Furosemide and ethacrynic acid also increase venous capacitance acutely by causing venodilation [107,108]. This effect appears to be mediated at least in part by the release of prostaglandins from the kidney [108a]. The combination of venodilation and their diuretic properties makes these drugs an important adjunct to therapy in acute pulmonary edema since the decrease in venous return will lower the pulmonary capillary hydrostatic pressure (see below).

Furosemide Furosemide can be administered orally or intravenously. When given orally, it is rapidly absorbed and produces a diuresis that begins within 1 h and lasts 6 to 8 h. It acts more quickly when given intravenously, beginning within minutes and lasting 2 to 3 h. The usual starting dose is 20 to 40 mg, which can be increased gradually until the desired effect is achieved. In patients with advanced renal failure, higher doses (which may exceed 500 mg) are usually required [109,110]. Furosemide, like ethacrynic acid and the thiazides, is highly protein-bound in the circulation. As a result, it enters the urine primarily by secretion by the organic acid secretory pathway in the proximal tubule, not by glomerular filtration [109]. The retention of other organic acid anions in renal failure (such as urate and hippurate) results in less efficient furosemide secretion. Consequently, a higher dose is required to achieve a urinary level adequate to produce a diuresis.

†Passive NaCl reabsorption is also reduced in the thin ascending limb since active transport in the thick segment indirectly generates the gradient for passive thin limb reabsorption (see page 113).

‡A new loop diuretic, bumetanide, has recently become available. Its efficacy appears similar to that of furosemide and ethacrynic acid [106a].

Relatively few side effects, other than those induced by the diuresis, are associated with furosemide. Deafness, which is usually reversible, may occur and is thought to be due to changes in endolymph formation. This complication is most likely to occur when high doses are given intravenously to patients with renal failure [110,111]. In this setting, the rate of drug administration should probably not exceed 4 mg/min [111].

Ethacrynic acid Although chemically different from furosemide, ethacrynic acid shares most of its pharmacologic properties. It can be given orally or intravenously and has a similar time course of action. The usual starting dose is 25 to 50 mg, which can be increased gradually to a maximum dose of 200 to 300 mg. Side effects, other than those due to the diuresis, are uncommon. Gastrointestinal symptoms of anorexia, nausea, and vomiting can be minimized by giving the drug with meals. Deafness may occur with somewhat greater frequency than with furosemide.

Thiazides

The thiazide diuretics inhibit Na^+ reabsorption in the distal tubule and, to a lesser degree, the inner medullary collecting tubule [112,113]. Since they do not affect reabsorption in the medullary ascending limb, concentrating ability is not impaired. However, the distal tubule does contribute to urinary dilution, so the thiazides reduce diluting ability [102]. As with furosemide and ethacrynic acid, the delivery of fluid to the K^+ secretory site is enhanced, resulting in increased K^+ excretion. The thiazides also have a moderate degree of carbonic anhydrase inhibitory activity, thereby leading to reduced $NaHCO_3$ and $NaCl$ reabsorption in the proximal tubule [112]. In most subjects, this does not contribute to the diuresis since the excess fluid leaving the proximal tubule is reclaimed in the loop of Henle [112]. When given in maximum dosage, the thiazides can lead to the output of as much as 3 to 5 percent of the filtered Na^+ and water. The lesser potency of these drugs in comparison with the loop diuretics is due to the fact that less Na^+ is reabsorbed in the thiazide-sensitive segments than in the ascending limb.

Many thiazide derivatives are available which differ only in their duration of action. In addition, chlorthalidone and metolazone are chemically different from the thiazides but have a similar site of action. Hydrochlorothiazide is one of the most widely used thiazides. It begins to act within 1 h and lasts for 8 to 12 h. The usual dose is 25 to 50 mg, once or twice daily. Higher doses are not likely to produce a greater diuresis. The thiazides are usually ineffective when the GFR is below 25 mL/ min, although metolazone may continue to act in this setting.

The frequency of side effects unrelated to diuresis is higher with the thiazides than with the loop diuretics. Skin rashes, including exfoliative dermatitis, leukopenia, thrombocytopenia, and acute pancreatitis all have been reported in association with the thiazides. Hyperglycemia may develop in patients with no history of diabetes, and diabetics may have to increase their dose of insulin or oral hypoglycemics. The hyperglycemic effect is at least in part related to hypokalemia, which impairs insulin secretion [114].

Potassium-Sparing Diuretics

Since active Na^+ reabsorption in the cortical collecting tubule creates an electrical gradient that promotes K^+ secretion (see Chap. 13), diuretics which inhibit Na^+ reabsorption at this site also *decrease* K^+ excretion. These agents (spironolactone, amiloride, and triamterene) are used in edematous patients in three settings: (1) in hepatic cirrhosis, in which they may be more effective than loop diuretics (see "Treatment: Hepatic cirrhosis" below); (2) in patients who have become hypokalemic due to a thiazide or loop diuretic; and (3) in combination with other diuretics in refractory edema. These diuretics are not usually given alone (except in hepatic cirrhosis) since they induce a relatively small increment in urinary Na^+ excretion.

A potential side effect associated with the use of these drugs is hyperkalemia [115–117]. This is most likely to develop in patients with underlying renal disease, particularly if they are also receiving K^+ supplements [115].

Spironolactone Spironolactone is an antagonist of aldosterone, competing with it for binding to the aldosterone receptor in the cytoplasm of the cortical collecting tubule cell [118]. The usual dose of spironolactone is 25 to 150 mg twice a day. It has a relatively slow onset of action, taking from 12 to 72 h (and occasionally longer) before a diuresis is seen. Side effects other than hyperkalemia are not uncommon and include gastrointestinal complaints, gynecomastia and impotence in men, and menstrual irregularities in women [115,119].

Amiloride Amiloride has a different mechanism of action, as it blocks the Na^+ channels in the luminal membrane that permit passive Na^+ entry into the cell [120]. The usual dose is 5 to 10 mg given once a day. Side effects are less common than with spironolactone and include gastrointestinal disturbances and skin rashes [121]. The relatively low incidence of complications and once-a-day dosage may make amiloride the preferred potassium-sparing diuretic.

Triamterene The mechanism by which triamterene acts is uncertain. In contrast to spironolactone, it is not an aldosterone antagonist and does not require the presence of aldosterone to be effective. The usual dose is 100 mg, once or twice daily. Side effects are infrequent and include headache, gastrointestinal disturbances, renal calculi composed of triamterene [122], and reversible renal failure by an unknown mechanism when given with indomethacin [123].

Other Diuretics

Other agents are occasionally used to increase the urine output in edematous patients. *Organomercurials* are potent diuretics that appear to act primarily in the thick ascending limb of the loop of Henle [88]. However, they have several major disadvantages when compared with the other loop diuretics. These include the necessity for parenteral administration, which makes them unsuited for outpatient use; an increased frequency of side effects such as hypersensitivity to mercury and renal insufficiency; and a loss of effect in the presence of the metabolic alkalosis that frequently follows their use (see below). As a result of these problems, the organomercurials are now rarely given.

When *glucocorticoids* are given in high doses, e.g., 40 mg of prednisone, they augment the delivery of Na^+ and water out of the proximal tubule. This effect may be due to an increase in GFR that results from renal vasodilation [124]. Glucocorticoids also stimulate Na^+ reabsorption in the more distal segments so that the net effect is a decrease in Na^+ excretion [125]. However, when administered with diuretics that block loop and distal reabsorption, the proximal effect of glucocorticoids can lead to an enhanced diuresis [126].

Side Effects of Diuretic Therapy

In addition to the side effects of the individual drugs, diuresis itself can produce a variety of potentially serious disturbances (Table 16-4).

Volume depletion Not uncommonly, an edematous patient is started on daily diuretics, which are then continued even after the edema has disappeared. The result may be the induction of volume depletion with symptoms of weakness, malaise, muscle cramps, and postural dizziness. Effective circulating volume depletion also can follow the appropriate use of diuretics. How this occurs will be discussed in the section on treatment.

Azotemia If the effective circulating volume is decreased by diuretics, renal perfusion may be reduced as well. This can result in a fall in urea and creatinine excretion, producing increases in BUN and plasma creatinine concentration. Since the problem is one of renal perfusion and not one of renal function, this is referred to as prerenal azotemia (see page 103).

Metabolic alkalosis Those diuretics causing NaCl loss (furosemide, ethacrynic acid, and the thiazides) can produce metabolic alkalosis in two ways (see Chap. 18) [128,129]: (1) by the contraction of the extracellular volume around a constant amount of extracellular HCO_3^- (called a contraction alkalosis); and (2) by increasing renal H^+ loss since more Na^+ is presented to the distal nephron, where Na^+ is

Table 16-4 Complications of diuretic therapy

Volume depletion
Azotemia
Metabolic alkalosis
Hypokalemia
Hyperkalemia and metabolic acidosis
Hyperuricemia
Hyponatremia
Hypomagnesemia

exchanged for H^+ as well as K^+. Although NaCl will reverse the alkalosis, the administration of NaCl is not desirable in patients with edema. In this setting, acetazolamide can be used to correct the alkalosis by promoting HCO_3^- loss in the urine.

Hypokalemia As described above, the non-K^+-sparing diuretics all augment K^+ excretion. The subsequent development of hypokalemia is potentially serious in two groups of edematous patients: (1) those taking digitalis, who are made more susceptible to digitalis-toxic arrhythmias, and (2) those with severe liver disease in whom K^+ depletion may precipitate hepatic coma (see below). Hypokalemia can usually be prevented by the concurrent administration of K^+-sparing diuretics or K^+ supplementation. Since K^+ is lost with Cl^- in the urine, KCl should be given. K^+ preparations containing potassium bicarbonate or potassium citrate should not be used since they are not likely to be effective in the alkalemic patient. A dose of 30 to 60 meq of K^+ per day is usually sufficient to maintain a normal plasma K^+ concentration, although higher doses may be needed initially when the patient is K^+-depleted.

Hyperkalemia and metabolic acidosis Since spironolactone, amiloride, and triamterene reduce distal H^+ and K^+ secretion, their use can result in hyperkalemia and metabolic acidosis [115–117,127]. Prevention is the best therapy, and these drugs should be used with great caution, if at all, in patients with renal failure and those receiving K^+ supplements.

Hyperuricemia Hyperuricemia is a frequent finding in patients on diuretic therapy. This appears to be secondary to the effects of the diuresis on Na^+ reabsorption rather than reflecting a direct effect on uric acid handling by the kidney [130]. Uric acid is both reabsorbed and secreted in the proximal tubule with little change occurring in the more distal segments. As with other solutes, uric acid reabsorption in the proximal tubule varies directly with that of Na^+ (see Chap. 5). Thus, as proximal Na^+ reabsorption is enhanced in response to the reduction in the effective circulating volume induced by diuretics [131], there is a concomitant increase in uric acid reabsorption [130,132]. The net effect is a fall in uric acid excretion and hyperuricemia.

Treatment of this complication in asymptomatic patients is *unnecessary* [130,133]. Gouty arthritis is uncommon in this setting, occurring mostly in patients with a personal or family history of gout. Similarly, acute renal failure due to uric acid precipitation in the distal nephron is not a problem, since the hyperuricemia is due to a primary decrease in the distal delivery and subsequent excretion of uric acid.

Hyponatremia Hyponatremia is a common occurrence in edematous states both before and after therapy with diuretics. The mechanism by which this occurs is discussed in detail in Chap. 23 and is related primarily to the effective circulating volume depletion present in edematous states and frequently exacerbated by diuretics (see below). In many patients, this leads to an increase in ADH secretion which, by impairing water excretion, promotes water retention and the development of hyponatremia [134]. The appropriate treatment is water restriction since the administration of Na^+ will increase the severity of the edema. If diuretics are to be continued, furosemide or ethacrynic acid should be used. Since the loop diuretics interfere with concentrating ability (in contrast to the thiazides, which act in the cortex), they will limit renal water retention, thereby minimizing the tendency to hyponatremia (see page 488) [135].

Hypomagnesemia Magnesium depletion, which is usually mild, may occur with the use of loop diuretics or the thiazides but not the K^+-sparing diuretics [135a]. This is primarily due to inhibition of magnesium reabsorption at the site of diuretic action since most of the filtered magnesium is reabsorbed in the loop of Henle (see Chap. 5). The potential consequences of this complication include lethargy, further hypokalemia, and an increased incidence of arrhythmias [135a]. Magnesium supplementation is usually able to restore normal magnesium balance.

TREATMENT

General Principles

Before discussing the therapy of the specific disorders, it is important to consider the following questions, which apply to all edematous states:

1. When must edema be treated?
2. What are the consequences of the removal of edema fluid?
3. How rapidly should edema fluid be removed?

When must edema be treated? It is important to recognize that *pulmonary edema is the only form of edema that is life-threatening and demands immediate treatment.*† *In all other edematous states, the removal of the excess fluid can proceed more slowly since it is of no danger to the patient.* This is particularly true of cirrhosis where hypokalemia, alkalosis, and rapid fluid shifts induced by diuretics can precipitate hepatic coma or the hepatorenal syndrome (see below).

What are the consequences of the removal of edema fluid? As described above, the retention of Na^+ and water by the kidney in most edematous states is compensatory in that it acts to increase the effective circulating volume. The major exception occurs in conditions associated with primary renal Na^+ retention where the effective circulating volume as well as the total extracellular volume is expanded.

If the retention of edema fluid is compensatory, then the removal of edema fluid, e.g., by diuretics, should diminish the effective circulating volume. To the degree that the fluid lost by diuresis comes from the plasma volume, there will be a decrease in

---

†Laryngeal edema, e.g., due to an anaphylactic reaction, also is potentially fatal. However, this is a special form of edema requiring epinephrine, corticosteroids, and, if needed, tracheostomy, not fluid removal.

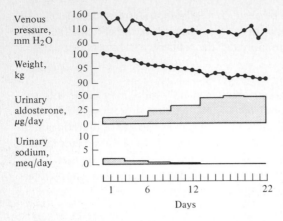

Venous pressure, mm H_2O

Weight, kg

Urinary aldosterone, $\mu g/day$

Urinary sodium, meq/day

Days

Figure 16-10 Effect of edema removal using a sodium-exchange resin in a patient with congestive heart failure. The loss of sodium and edema was accompanied by a gradual decrease in urinary sodium excretion and a marked increase in the excretion of aldosterone. *(Reproduced from L. E. Duncan, Jr., G. W. Liddle, and F. C. Bartter, J. Clin. Invest., 35:1299, 1956, by copyright permission of The American Society for Clinical Investigation.)*

venous return to the heart and therefore in the cardiac filling pressures. From the Frank-Starling relationship (Fig. 16-4), this reduction in the LVEDP should lower the stroke volume in both normal and failing hearts, resulting in a fall in the cardiac output and consequently in tissue perfusion.†

There is a great deal of evidence that this sequence occurs commonly in edematous patients. First, there frequently is a reduction in the cardiac output after the administration of diuretics to patients with either acute or chronic heart failure [136–138]. Second, many patients with heart failure, hepatic cirrhosis, and the nephrotic syndrome demonstrate an increase in the secretion of renin, aldosterone, and ADH after edema fluid has been removed (Fig. 16-10) [8,139–141]. Since these hormones are secreted in response to a decrease in the effective circulating volume (see Chap. 8), the changes in hormone secretion suggest that treatment of the edema frequently impairs tissue perfusion.

Despite the reduction that may occur in the effective circulating volume, *most patients benefit from the appropriate use of diuretics.* For example, in one study, the exercise tolerance of patients with heart failure and pulmonary congestion was significantly improved by diuretics even though the cardiac output fell by an average of 20 percent [137]. This suggests that in patients with mild to moderate heart failure it is primarily pulmonary congestion that limits the patients' activity and that moderate reductions in the cardiac output can be well tolerated. A similar positive response to diuretics typically occurs in patients with peripheral edema and ascites with the relief of symptoms of fatigue and feeling bloated, and of the cosmetic problems associated with the edema.

However in some patients the decrease in the effective circulating volume is sufficient to significantly impair tissue perfusion. This occurs most commonly in two settings: (1) in patients with a low baseline effective circulating volume, e.g., in severe cardiac failure, and (2) after the excessive use of diuretics. Since a reduction in the effective circulating volume decreases renal plasma flow and then glomerular filtra-

---

†This assumes that increased sympathetic neural tone, which can increase both myocardial contractility and heart rate, is unable to completely compensate for the reduction in the stroke volume.

tion rate, increases in BUN and plasma creatinine concentration can be used as an index of impaired tissue perfusion. Other changes which can occur are weakness, fatigue, postural hypotension, and lethargy and confusion due to decreased cerebral blood flow. These principles can be illustrated by the following example:

Case history 16-2 A previously well 46-year-old man was admitted to the hospital with acute pulmonary edema due to an acute myocardial infarction. As part of his initial therapy, he was given intravenous furosemide and then continued on an oral dose of 40 mg of furosemide per day. The pulmonary edema rapidly cleared, and the patient was having an uneventful recovery when it was noted on the tenth hospital day that his BUN had risen from 10 mg/dL on admission to 110 mg/dL and his plasma creatinine concentration increased from 1 to 4.5 mg/dL. There had been a 6-kg weight loss since admission. It was felt that the patient had acute tubular necrosis secondary to his pulmonary edema.

The physical examination revealed that the patient was in no acute distress. The vital signs showed a small reduction in blood pressure since admission. The results of the cardiac examination were normal. The chest was clear to percussion and auscultation. The neck veins were flat, and the skin turgor was poor.

The examination of the urine revealed a normal sediment with a urine Na^+ concentration of 2 meq/L and urine osmolality of 550 mosmol/kg.

COMMENT There were many signs in this patient pointing toward volume depletion secondary to excessive diuresis as the cause of the increased BUN and creatinine concentration, including weight loss, poor skin turgor, flat neck veins, a low urine Na^+ concentration and high urine osmolality, and an increase in BUN out of proportion to the elevation in the plasma creatinine concentration. As a result, the diuretics were discontinued and the patient given a high-sodium diet while being carefully observed for the recurrence of heart failure. After 6 days of this regimen, his skin turgor, BUN, and creatinine concentration had returned to normal.

However, a reduction in the effective circulating volume in response to diuretic therapy is not always due to overdiuresis. This can be appreciated from the following case history:

Case history 16-3 A 64-year-old woman with chronic congestive heart failure due to arteriosclerotic heart disease was admitted to the hospital. In addition to pulmonary edema, she also had signs of right-sided heart failure, including distended neck veins and pedal edema. After 3 days of diuretic therapy, there had been a 5-kg weight loss with marked clinical improvement. However, a mild degree of pulmonary edema persisted. During this period, the BUN had risen from 20 to 90 mg/dL with an increase in the plasma creatinine concentration from 1.2 to 3.0 mg/dL. The urine findings were similar to those for the patient in Case history 16-2.

COMMENT This represents another example of reduced tissue perfusion due to diuretics. However, in this case, edema persisted. Thus, this patient had such severe heart disease that she could not be both edema-free and have a normal plasma creatinine concentration. As shown in Fig. 16-4, the stroke volume can vary directly with the left ventricular end-diastolic pressure even in severe heart failure. In this patient, the cardiac output was relatively well maintained only at filling pressures so high that they caused both pulmonary and peripheral edema. When the filling pressures were reduced to a degree sufficient to eliminate the edema, the cardiac output and tissue perfusion were sacrificed. In this setting, therapy also must be aimed at increasing the cardiac output by using digitalis or vasodilators (see below).

These responses to diuretics can also occur in hepatic cirrhosis and the nephrotic syndrome (except for those patients in whom overfilling is of primary importance)

[42,53,142]. Regardless of the cause, it is important to recognize that increases in the BUN and plasma creatinine concentration may be due to decreased renal perfusion and not necessarily to intrinsic renal disease. Conversely, if the BUN and plasma creatinine concentration remain constant during diuresis, it can be assumed that the effective circulating volume is being reasonably well maintained.

In contrast to the hemodynamic changes that may be seen in heart failure or cirrhosis, impaired renal perfusion should not occur after the appropriate use of diuretics in patients with primary renal Na^+ retention. In these conditions, the effective circulating volume is increased by fluid retention. Although diuretics reduce the effective circulating volume, it will be from an initially high level back toward normal.

How rapidly should edema fluid be removed? When diuretics are administered, the fluid that is lost initially comes from the plasma. This results in a reduction in the venous pressure and consequently in capillary hydrostatic pressure, thereby favoring the restoration of the plasma volume by the movement of the edema fluid back into the vascular space. The rapidity with which this occurs is variable. In patients with generalized edema due to heart failure, the nephrotic syndrome, or primary Na^+ retention, the edema fluid can be mobilized rapidly. Thus, in patients with severe degrees of edema, the removal of 2 to 3 liters of edema fluid in 24 h can be accomplished without much reduction in the plasma volume. In contrast, hepatic ascites equilibrates much more slowly with the plasma since only one capillary bed, the hepatic sinusoid, is involved. Direct measurements have indicated that 930 mL/day is the maximum amount of ascites that can be absorbed into the vascular space and that 200 to 400 mL/day is the level that can be achieved by most patients [141]. If diuresis proceeds more rapidly in a patient with ascites but without peripheral edema, the ascitic fluid will be unable to completely replenish the plasma volume (Fig. 16-11). This can result in hypotension or the precipitation of hepatic coma or the hepatorenal syndrome [141,143].

Dietary sodium restriction Restricting sodium intake is beneficial in edematous patients since it limits the quantity of fluid that can be retained. In a patient whose edema is inadequately controlled with diuretics, dietary compliance can sometimes be assessed by measuring the rate of Na^+ excretion with a 24-h urine collection (while diuretic therapy is continued). A value above 100 meq/day indicates that the patient can excrete Na^+ and that persistent edema is due in part to excessive Na^+ intake. A low value (below 40 meq/day), on the other hand, indicates very avid Na^+ retention (independent of intake) and the need for more diuretics.

Avoidance of nonsteroidal anti-inflammatory drugs Nonsteroidal anti-inflammatory drugs (including high doses of aspirin) inhibit prostaglandin synthesis and should be avoided if possible in edematous patients with effective circulating volume depletion. As described in Chap. 8, the renal secretion of prostaglandins is low in the basal state but is stimulated by vasoconstrictors such as angiotensin II and norepinephrine, the production of which is increased by hypovolemia. In this setting, the

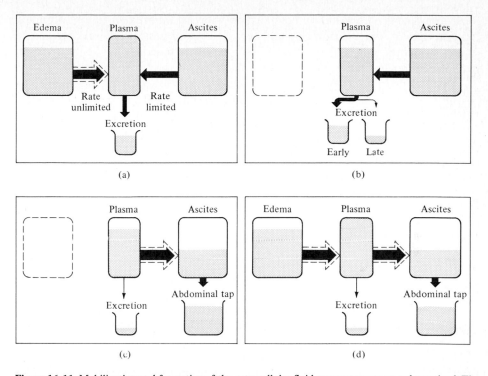

Figure 16-11 Mobilization and formation of the extracellular fluid compartments are schematized. The key factors are that the mobilization rate of ascites is limited, that of edema relatively unlimited. With spontaneous or induced diuresis, optimal therapy results (*a*) when the rates of mobilization of edema and ascites equal the rate of diuresis; the edema reservoir then protects the plasma volume. Sustained diuresis in the absence of edema (*b*) may be at the expense of plasma volume, with eventual oliguria, azotemia, and electrolyte abnormalities. The same may follow paracentesis (*c*), since ascites tends to restore itself at the plasma's expense. But if edema is present (*d*), it is mobilized, protects the plasma volume, and shifts to the ascites compartment. Diuresis and paracentesis thus entail less risk when edema is present. *(From G. J. Gabuzda, Hosp. Pract., 8 (No. 8):67, 1973.)*

vasodilator prostaglandins minimize the degree of renal ischemia and may promote Na$^+$ excretion. If a nonsteroidal anti-inflammatory drug is given, reversible reductions in renal blood flow, GFR, and Na$^+$ excretion frequently ensue [144–147] and the patient may show a diminished response to diuretics [144,148,149]. The one exception may be sulindac (Clinoril), which does not appear to interfere with renal prostaglandin synthesis, perhaps because it is the only NSAID that has no active drug excreted in the urine [150,150a].

Clinical Use of Diuretics

To minimize the volume and electrolyte disturbances associated with diuretic therapy, it is important to use the least amount of diuretic that is required to control the edema. Not uncommonly, patients who have been made edema-free can be well controlled on a low-sodium diet in combination with the *intermittent* use of diuretics, e.g., alternate day or twice weekly regimens. Whenever possible, diuretics should be given during the day and not before bedtime. This schedule markedly improves patient comfort by allowing a full night's sleep.

In mild degrees of edema, either a thiazide or a low dose of furosemide, for example, 20 mg, is usually sufficient to control the edema. If this regimen is ineffective, then furosemide in higher doses or ethacrynic acid should be used. Since the efficacy of loop diuretics is dose-dependent, it is important to establish for each patient what dose is required to achieve the desired effect. For example, if 40 mg of furosemide is ineffective, then multiple 40-mg doses also will be ineffective. The *individual* dose should be increased in increments of 20 to 40 mg until a diuresis is obtained. If daily diuretics are required, a K^+-sparing diuretic or K^+ supplementation may have to be added to prevent hypokalemia. The former has the extra advantage of augmenting the diuresis.

In some patients, the diuretic dose has to be increased with time. This is usually due to more severe hypovolemia that can result from either progression of the underlying disease or the effect of the diuretic. In this setting, the kidney augments Na^+ reabsorption in the proximal tubule and the collecting tubules [131,151], a response that is mediated at least in part by increased activity of the renin-angiotensin-aldosterone and sympathetic nervous systems (see Chap. 9) [8,139–141]. In addition, less of the diuretic may enter the urine if renal perfusion becomes impaired [152,153]. These effects counterbalance the direct diuretic action, resulting in a new steady state in which further fluid removal requires more diuretic therapy. This sequence also explains why nonedematous hypertensive patients treated with a thiazide lose fluid only during the first few days of therapy and do not become progressively volume-depleted [154,154a].

Refractory edema Infrequently, a patient does not respond to conventional oral diuretic therapy, even when high doses of furosemide, for example, 480 mg, or ethacrynic acid are used. Although they act at the same site and are not additive when used together, occasionally a patient who is resistant to furosemide will respond to ethacrynic acid and vice versa. If both loop diuretics are unable to produce a diuresis, there are at least four possible explanations: (1) The diuretic is not well absorbed; (2) the diuretic is effective but reabsorption is increased distally, thereby preventing the excretion of the Na^+ and water not reabsorbed in the loop; (3) the diuretic would be effective but the volume of fluid delivered to the diuretic-sensitive site in the loop is limited by the responses of the kidney to effective circulating volume depletion: a reduction in the glomerular filtration rate and an increase in proximal reabsorption; and (4) the diuretic never reaches the tubules, due to decreased filtration and secretion.

An initial step in the management of refractory edema (Table 16-5) can be to switch to intravenous diuretics since intestinal absorption may be impaired in severely edematous patients [155]. This effect is presumably due to interstitial edema in the intestinal wall and is reversible with removal of the edema. Some patients who are initially unresponsive to 200 to 240 mg of furosemide orally may respond to as little as 40 mg intravenously [155].

Table 16-5 Treatment of refractory edema

Increase to maximum oral dose of furosemide or ethacrynic acid
Switch to intravenous loop diuretic
Add distal diuretics
  Thiazide or metolazone
  K^+-sparing diuretic
Add proximal diuretic
  Acetazolamide
  Mannitol
  Albumin
  Glucocorticoids
Peritoneal dialysis or hemodialysis

If high doses of an intravenous loop diuretic (such as 320 mg of furosemide given slowly to minimize ototoxicity) are ineffective, a thiazide (or metolazone) should be added to inhibit distal reabsorption. This combination can frequently lead to a marked diuresis of as much as 5 L/day (Fig. 16-12) [156,157]. The potency of this combination may be related to both the *proximal and distal* actions of the thiazide. The former is normally not important because of increased loop reabsorption [112]; this is now blocked by furosemide. It is prudent to begin with low doses, for example, 25 mg of hydrochlorothiazide or 1.25 to 2.5 mg of metolazone, to prevent an excessive diuresis and potentially severe electrolyte abnormalities [157,157a]. A K^+-sparing diuretic can also be added to this regimen to minimize urinary K^+ losses.

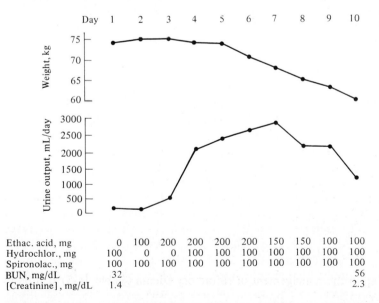

	Day 1	2	3	4	5	6	7	8	9	10
Ethac. acid, mg	0	100	200	200	200	200	150	150	100	100
Hydrochlor., mg	100	0	0	100	100	100	100	100	100	100
Spironolac., mg	100	100	100	100	100	100	100	100	100	100
BUN, mg/dL	32									56
[Creatinine], mg/dL	1.4									2.3

Figure 16-12 Use of combination diuretic therapy in refractory congestive heart failure. Ethacrynic acid or hydrochlorothiazide in combination with spironolactone was ineffective but the use of all three drugs produced a significant diuresis. After the loss of 13 kg and marked clinical improvement, there was a moderate rise in the BUN and plasma creatinine concentration, suggesting a decrease in renal perfusion.

In certain patients with advanced disease, blockade of loop and distal reabsorption is ineffective because of a marked reduction in the delivery of fluid out of the proximal tubule. In this setting, increasing loop delivery with acetazolamide, mannitol,† or glucocorticoids may substantially enhance the urine output [14,126,158]. In addition, the infusion of salt-poor albumin in patients with severe hypoalbuminemia due to the nephrotic syndrome can transiently increase the effective circulating volume, resulting in an elevation in the GFR and enhanced fluid delivery to the diuretic-sensitive sites [159]. Another measure that can be used in severely nephrotic patients is tight wrapping of the legs, resulting in decreased peripheral edema, increased venous return to the heart, and expansion of the effective circulating volume [160].

Finally, patients with severe renal insufficiency of any cause may become resistant to diuretic therapy because of an inability of the diuretics to enter the tubules by filtration and tubular secretion [109]. Extremely high-dose furosemide therapy (up to 2400 mg/day) may be effective in this situation, but irreversible deafness can occur [110,111]. If this regimen does not promote a diuresis or if side effects occur, peritoneal dialysis or hemodialysis can be used to correct the fluid overload.

Use of diuretics in nonedematous states Diuretics are also useful in a variety of nonedematous conditions [151]. These include hypertension [154,161], hypercalcemia or hypercalciuria (see page 104), metabolic alkalosis (see page 388), diabetes insipidus (see page 539), and type 2 (or proximal) renal tubular acidosis (see page 423). In addition, spironolactone, triamterene, or amiloride may be effective in K^+-wasting conditions such as primary hyperaldosteronism or Bartter's syndrome (see page 591).

Congestive Heart Failure

Acute Acute pulmonary edema is a medical emergency requiring immediate treatment to restore tissue oxygenation and perfusion. Tissue oxygenation can be increased by the administration of oxygen, the reversal of bronchoconstriction (if present) with aminophylline, and the improvement of pulmonary gas exchange by correcting the pulmonary edema [162].

Pulmonary edema is treated by decreasing venous return to the heart, thereby reducing the elevated pulmonary capillary pressure and allowing the edema fluid to reenter the vascular space. This can be accomplished in two ways: by lowering the plasma volume with furosemide or a phlebotomy, and by increasing the capacitance of the venous system with rotating tourniquets or by causing venodilation with parenteral morphine or a loop diuretic [107,108,162]. Venodilation can also be induced by the intravenous administration of nitroprusside or nitroglycerin [162]. These drugs, which can cause hypotension, are most safely given after a Swan-Ganz cath-

---

†Mannitol should not be used in patients with left-sided heart failure since the ensuing volume expansion can precipitate pulmonary edema.

eter has been inserted to monitor intracardiac filling pressures. The aim is to reduce the pulmonary capillary wedge pressure to 15 to 18 mmHg: low enough to alleviate pulmonary edema but not so low as to substantially reduce the cardiac output (see Fig. 16-4) [24,163].

Inotropic therapy to enhance cardiac contractility is indicated in patients who are hypotensive or show evidence of a low-output state. Digitalis derivatives are relatively ineffective in this acute setting, and intravenous agents such as dopamine or dobutamine should be used [164].

Patients with acute pulmonary edema may also have severe acid-base disturbances such as respiratory and metabolic acidosis (primarily lactic acidosis) [165]. Specific therapy may not be necessary since reversal of the pulmonary edema is usually sufficient to restore acid-base balance by improving gas exchange and by allowing metabolism of the excess lactate into bicarbonate (see page 428) [165,166]. In severe cases, tracheal intubation to control ventilation and $NaHCO_3$ to buffer the excess acid may be required to prevent life-threatening reductions in pH. There is some danger involved in the use of $NaHCO_3$ since it tends to expand the extracellular volume, increasing the severity of the edema. If this becomes a problem because of a continuing need for $NaHCO_3$, dialysis can be used to treat both the acidemia and the volume overload.

Chronic The therapeutic regimen is different for patients with chronic heart failure. In mild cases, diuretics plus dietary Na^+ restriction (1 to 2 g of sodium per day) may be sufficient to control the edema. Digitalis derivatives such as digoxin can be used as an alternate initial therapy and in patients who do not respond to diuretics alone. Digitalis acts to improve the cardiac output by increasing cardiac contractility† [168,169]. It is particularly effective in patients with rapid atrial fibrillation. By prolonging the refractory period of the atrioventricular node, digitalis slows the ventricular rate, allowing better ventricular filling and augmenting the cardiac output.

Although the many side effects produced by digitalis toxicity are beyond the scope of this discussion, it is important to note that an increased incidence of arrhythmias occurs with concomitant hypokalemia, a common complication of diuretic therapy [170]. Therefore, the plasma K^+ concentration must be followed carefully in patients taking both digitalis and diuretics. K^+ supplements or a K^+-sparing diuretic should be added if the plasma K^+ concentration falls below 3.5 meq/L.

Periods of rest may be a useful adjunct to therapy. Some patients have a cardiac output that is relatively normal at rest but does not increase adequately with exercise [29]. It is during this latter period that renal Na^+ retention is most intense [29,30]. Thus, limiting physical activity may produce substantial clinical improvement since maximizing tissue perfusion will maximize Na^+ excretion.

---

†Digitalis is most effective when heart failure is due to myocardial dysfunction, as with hypertensive or ischemic heart disease. When there is obstruction to flow, as in mitral stenosis, digitalis is not very effective (unless used to control a rapid arrhythmia) and therapy should consist of diuretics and relieving the obstruction if possible [167].

Patients with *severe* heart failure may not respond to this regimen or may develop signs of reduced tissue perfusion (due to a diuretic-induced fall in LVEDP) such as an elevation in the BUN and plasma creatinine concentration. In this situation, vasodilator therapy should be tried, preferably with initial monitoring with a Swan-Ganz catheter [163,164,171,172]. The principle underlying the use of these agents is that, by reducing systemic vascular resistance, the work required by the left ventricle to pump blood into the aorta is diminished, resulting in an increase in cardiac output. This applies only to severe heart failure since vascular resistance does not substantially limit the cardiac output in mild to moderate cardiac disease [172].

Those agents that are primarily arterial dilators (hydralazine and minoxidil) increase the cardiac output with little change in the pulmonary capillary pressure [173]. In contrast, those drugs that are both arterial and venous dilators (nitroprusside, prazosin, and captopril) will enhance the cardiac output and lower the pulmonary capillary pressure, thereby also relieving pulmonary congestion [172–175]. Oral nitrates, which are primarily venodilators, are not usually used alone to augment cardiac output but can be given in combination with an arterial dilator such as hydralazine.

The relative efficacy of the oral vasodilators is variable. However, hydralazine and captopril may be preferable in patients who have developed renal insufficiency since they augment renal blood flow, in contrast to prazosin, which may not [176,177]. The reason for this difference is not known.

Inotropic agents can increase the effect of vasodilators. Thus, the combination of dopamine or dobutamine with nitroprusside can substantially enhance the cardiac output and lower the pulmonary capillary pressure [164]. However, these drugs must be given intravenously and are not useful for chronic management. Oral inotropic agents other than digitalis are currently undergoing experimental evaluation but are not yet widely used clinically. These include the sympathetic agonist ephedrine, the β-adrenergic agonist pirbuterol, and amrinone [178].

In summary, the effect of inotropic agents and vasodilators can be viewed schematically in terms of the Frank-Starling curves in Fig. 16-4. As a result of improved cardiac performance, the relationship between LVEDP and stroke volume can be shifted upward from the lower curve of severe heart failure to the middle curve of mild to moderate heart failure. This allows an increased cardiac output to be maintained at a lower filling pressure and therefore less tendency toward pulmonary edema.

Hepatic Cirrhosis

Diuretics must be used with caution in hepatic disease. Rapid diuresis can result in plasma volume depletion (Fig. 16-11) and is not necessary since ascites, although disfiguring, is rarely of any danger to the patient. Therefore, the aim of therapy should be gradual removal of the ascitic fluid.

In addition, cirrhotic patients are prone to develop the electrolyte problems frequently seen with diuretics, such as hyponatremia, hypokalemia, and metabolic alkalosis [42]. This is particularly important because there have been reports of patients who went into hepatic coma coincident with the onset of hypokalemia and metabolic alkalosis and who awakened following no therapy other than KCl replacement [179,180]. Hypokalemia may promote the development of hepatic coma by increasing ammonia (NH_3) production by the renal tubular cells† [181,182]. This can add to the already high blood ammonia levels that may be present in severe liver disease. Concomitant metabolic alkalosis may also contribute to this problem. From the Henderson-Hasselbalch equation for the ammonia/ammonium buffer system:

$$pH = 9.3 + \log \frac{[NH_3]}{[NH_4^+]}$$

an elevation in arterial pH will increase the arterial concentration of lipid-soluble NH_3, which can then diffuse down a concentration gradient into the brain cells.

These considerations suggest that diuretic therapy in severe hepatic disease is *safer* with a K^+-sparing drug (such as spironolactone) than with a loop diuretic or a thiazide. In nonazotemic patients, spironolactone (in doses of 150 to 300 mg/day) is also likely to be as or *more effective* than furosemide [182a]. This response, which is the opposite of that generally seen, is probably due to the very high aldosterone levels that tend to occur in this setting. As a result, the aldosterone-sensitive site in the cortical collecting tubule is able to reclaim much of the furosemide-induced increase in fluid delivered out of the loop of Henle, thereby diminishing the diuretic effect.

Removal of ascitic fluid by paracentesis also may have deleterious hemodynamic effects. As the intraabdominal pressure is reduced, fluid rapidly moves from the vascular space into the peritoneum. If peripheral edema is present, this fluid can be mobilized and the plasma volume will be well maintained (Fig. 16-11). However, if the patient has no peripheral edema, the reaccumulation of ascites partially occurs at the expense of the plasma volume. In most cases, paracentesis is well tolerated with little change in the cardiac output, particularly if less than 1000 mL is removed [183,184]. The main indications for paracentesis are for diagnosis, e.g., to rule out peritonitis, and to reduce the intraabdominal pressure when tense ascites is responsible for shortness of breath or impending rupture of an umbilical hernia [183].

With these caveats in mind, a rational therapeutic program for the patient with cirrhosis and ascites should include the following: (1) discontinuation of hepatotoxic agents, particularly alcohol [185]; (2) a high-calorie diet, starting with a low-protein content (since the NH_3 produced from the metabolism of amino acids may precipitate hepatic coma), which is gradually increased if tolerated; (3) a water intake of 1000 mL/day, which can be increased if hyponatremia does not develop; and (4)

---

†As described in Chap. 12, renal NH_3 production is stimulated by acidemia, which probably is sensed as a fall in intracellular pH. As hypokalemia develops, K^+ moves out of the cells to replete extracellular K^+ stores. To maintain electroneutrality, extracellular H^+ and Na^+ move into the cells. This will cause an intracellular acidosis that may be responsible for the increase in NH_3 production by the renal tubular cells [182].

extreme dietary sodium restriction to 250 mg/day. In patients who are unable to tolerate such severe Na^+ restriction or in whom a spontaneous diuresis does not occur, cautious diuretic therapy can be begun. In view of the frequently elevated aldosterone levels and the potential danger of hypokalemia and metabolic alkalosis, a potassium-sparing diuretic such as spironolactone should be used and a thiazide or furosemide added as necessary. If the BUN and plasma creatinine concentration begin to rise after diuretic therapy, the diuretics should be discontinued unless some other cause for renal impairment can be identified.

The success of this regimen is dependent in part upon the degree to which the liver can recover. If recovery occurs, then a spontaneous diuresis may ensue along with a general improvement in the nutritional status of the patient. Diuretic-resistant ascites, baseline urinary Na^+ excretion below 10 meq/day, and persistent hypoalbuminemia and malnutrition are suggestive of advanced hepatic disease and are associated with poor long-term survival [8,185].

Peritoneovenous shunt Patients with refractory ascites or those who develop renal insufficiency after the appropriate use of diuretics generally have a good response to the insertion of a peritoneovenous shunt [43,186]. This shunt, which drains into the internal jugular vein, chronically reinfuses the ascites into the vascular space. The ensuing effective volume expansion results in increased Na^+ and water excretion, a marked reduction in the degree of ascites, and improved renal function [43].

Despite the dramatic successes that can occur with this modality, *it should be used only for refractory, symptomatic ascites* because of a relatively high incidence of potentially fatal complications. These include infection of the shunt, which can lead to bacteremia, disseminated intravascular coagulation due to entry into the bloodstream of endotoxin or procoagulant material in the ascitic fluid, variceal bleeding from a volume expansion–induced elevation in portal venous pressure, and small bowel obstruction [43,187,188]. In addition, tumor emboli to the lungs can occur when the shunt is used to treat severe malignant ascites [43].

Nephrotic Syndrome

The main aim of therapy in the nephrotic syndrome is, if possible, to reverse the glomerular disease, e.g., with glucocorticoids in minimal change disease [50]. Pending this potential response, diuretics and dietary Na^+ restriction can be used to control the edma. However, diuretics must be used with care in those patients in whom underfilling due to hypoalbuminemia is of primary importance (Fig. 16-6). In this setting, fluid loss can lead to further effective volume depletion and a decline in renal function [53,142,189]. This may be less of a problem when there is underlying renal insufficiency since the baseline plasma volume may be increased due to primary renal Na^+ retention [53].

Edema refractory to diuretics is unusual in the nephrotic syndrome in the absence of severe renal insufficiency. When it does occur, intermittent albumin infusions and tight wrapping of the legs may be beneficial [159,160].

Primary Renal Sodium Retention

Primary renal sodium retention of any cause can be safely treated with diuretics. However, high doses of furosemide with or without a thiazide may be required in patients with renal failure [109,110,156,157] or those taking minoxidil or diazoxide for severe hypertension [58,59]. If diuretics are ineffective in renal failure, either peritoneal dialysis or hemodialysis can be used to remove the excess fluid.

Idiopathic Edema

Since diuretic-induced edema may be operative in many of these patients, initial therapy should consist of a low-Na^+ diet and *stopping the diuretics* for 3 to 4 weeks (see Fig. 16-8). The patient should be advised that this will initially lead to weight gain and reassured that diuretics can be always be reinstituted if a spontaneous diuresis does not ensue.

 If it becomes evident that a diuretic is required, the lowest effective dose should be used and given in the early evening since the edema primarily accumulates during the daytime when the patient is erect [69]. The patient should remain recumbent as much as possible after the diuretic is taken.

 In rare cases, diuretics may be relatively ineffective even in high doses. In this setting, the administration of amphetamines (which have sympathomimetic activity) or the sympathetic agonist ephedrine may induce a diuresis and aid in controlling the edema [69,190]. These drugs may act by constricting the precapillary sphincter, thereby lowering the capillary hydrostatic pressure and retarding fluid movement out of the capillary. Tight wrapping of the legs may also be beneficial in this setting by minimizing the postural hypovolemia.

 Preliminary reports have suggested that captopril (a converting enzyme inhibitor that impairs the generation of angiotensin II; see Fig. 3-5) may also be effective in this disorder [191]. The mechanism by which it might act is uncertain but decreased angiotensin II production will reduce the secretion of aldosterone. Captopril must be used with caution since the fall in angiotensin II levels can induce hypotension in these previously normotensive patients.

Other Causes of Edema

Pulmonary edema due to increased capillary permeability in the adult respiratory distress syndrome usually occurs in severely ill patients, e.g., with sepsis or shock, and has a high mortality rate. Treatment consists of attempting to correct the underlying abnormality, lowering the pulmonary capillary pressure to the lowest level compatible with an adequate cardiac output, the possible use of high doses of corticosteroids (2 g of methylprednisolone), and maintaining tissue oxygenation [75,192]. Mechanical ventilation with positive end-expiratory pressure is frequently necessary since the prolonged use of high partial pressures of oxygen in the inspired air will lead to oxygen toxicity and eventual worsening of the hypoxemia.

Patients with mild forms of localized edema due to venous or lymphatic obstruction usually respond to thiazides if therapy is required. Myxedema, on the other hand, should be treated with thyroid hormone, not diuretics.

PROBLEMS

16-1 The appropriate use of diuretics may lower the effective circulating volume in which of the following edematous states?

(a) Congestive heart failure

(b) Nephrotic syndrome

(c) Renal failure

(d) Hepatic cirrhosis and ascites due to overflow

(e) Adult respiratory distress syndrome

How would you detect this change?

16-2 A previously well 45-year-old man is admitted with the acute onset of crushing chest pain and dyspnea. Medical evaluation confirms the diagnosis of acute myocardial infarction with pulmonary edema. After treatment with oxygen, intermittent positive pressure breathing, and diuretics, he becomes edema-free. Because of the diuresis, his weight has fallen by 3 kg within 24 h after admission. At this time, he is noted to be oliguric with a urine Na^+ concentration of 4 meq/L. His BUN has increased from 10 to 28 mg/dL. What is probably responsible for the oliguria and increase in BUN? Is his total extracellular volume greater than, equal to, or less than normal? What modes of therapy might return his BUN and urine output to normal?

16-3 A patient with chronic congestive heart failure is treated with digitalis, furosemide, and triamterene. He is clinically improved, but a moderate degree of pulmonary and peripheral edema persists. His blood tests prior to therapy were normal. Repeat blood tests revealed:

$$\text{Plasma } [Na^+] = 140 \text{ meq/L}$$

$$[K^+] = 4.5 \text{ meq/L}$$

$$[Cl^-] = 83 \text{ meq/L}$$

$$[HCO_3^-] = 40 \text{ meq/L}$$

$$\text{Arterial pH} = 7.57$$

$$\text{Plasma [uric acid]} = 12 \text{ mg/dL} \quad (\text{normal} = 2.5 \text{ to } 8.0 \text{ mg/dL})$$

What are the mechanisms responsible for the development of metabolic alkalosis (high pH, high HCO_3^- concentration) and hyperuricemia? How would you treat these disturbances?

REFERENCES

1. Brace, R. A.: Progress toward resolving the controversy of positive vs. negative interstitial fluid pressure, Circ. Res., 49:281, 1981.

2. Taylor, A. E.: Capillary fluid filtration: Starling forces and lymph flow, Circ. Res., 49:557, 1981.

2a. Bollinger, A., J. Frey, K. Jäger, J. Furrer, J. Seglias, and W. Siegenthaler: Patterns of diffusion through skin capillaries in patients with long-term diabetes, N. Engl. J. Med., 307:1305, 1982.

3. Wiederhielm, C. A.: Dynamics of transcapillary fluid exchange, J. Gen. Physiol., 52:29s, 1968.

4. Landis, E., and J. R. Pappenheimer: Exchange of substances through the capillary walls, in

Handbook of Physiology, sec. 2, Circulation, vol. II, W. F. Hamilton and P. Dow (eds.), American Physiological Society, Washington, D.C., 1963.

5. Darrow, D. C., E. B. Hopper, and M. K. Cory: Plasmapheresis edema. I. The relation of reduction of serum proteins to edema and the pathological anatomy accompanying plasmapheresis, J. Clin. Invest., 11:683, 1932.

6. Epstein, F. H., and T. B. Ferguson: The effect of the formation of an arteriovenous fistula upon blood volume, J. Clin. Invest., 34:434, 1955.

7. Kowalski, H. J., and W. H. Abelmann: The cardiac output at rest in Laennec's cirrhosis, J. Clin. Invest., 32:1025, 1953.

8. Arroyo, V., J. Bosch, J. Gaya-Beltran, D. Kravetz, L. Estrada, F. Rivera, and J. Rodés: Plasma renin activity and urinary sodium excretion as prognostic indicators in nonazotemic cirrhosis with ascites, Ann. Intern. Med., 94:198, 1981.

9. Wilkinson, S. P., and R. Williams: Renin-angiotensin-aldosterone system in cirrhosis, Gut, 21:545, 1980.

10. Levy, M., and J. B. Allotey: Temporal relationships between urinary salt retention and altered systemic hemodynamics in dogs with experimental cirrhosis, J. Lab. Clin. Med., 92:560, 1978.

11. Stein, J. H., R. W. Osgood, S. Boonjarern, J. W. Cox, and T. F. Ferris: Segmental sodium reabsorption in rats with mild and severe volume depletion, Am. J. Physiol., 227:351, 1974.

12. Weiner, M. W., E. J. Weinman, M. Kashgarian, and J. P. Hayslett: Accelerated reabsorption in the proximal tubule produced by volume depletion, J. Clin. Invest., 50:1379, 1971.

13. Bell, N. H., H. P. Schedl, and F. C. Bartter: An explanation for abnormal water retention and hypoosmolality in congestive heart failure, Am. J. Med., 36:351, 1964.

14. Schedl, H. P., and F. C. Bartter: An explanation for and experimental correction of the abnormal water diuresis in cirrhosis, J. Clin. Invest., 39:248, 1960.

15. Grausz, H., R. Lieberman, and L. E. Earley: Effect of plasma albumin on sodium reabsorption in patients with nephrotic syndrome, Kidney Int., 1:47, 1972.

16. Levy, M.: Effects of acute volume expansion and altered hemodynamics on renal tubular function in chronic caval dogs, J. Clin. Invest., 51:922, 1972.

16a. Ichikawa, I., H. G. Rennke, J. R. Hoyer, K. F. Badr, N. Schor, J. L. Troy, C. P. Lechene, and B. M. Brenner: Role for intrarenal mechanisms in the impaired salt excretion of experimental nephrotic syndrome, J. Clin. Invest., 71:91, 1983.

17. Knox, F. G., J. I. Mertz, J. C. Burnett, Jr., and A. Haramati: Role of hydrostatic and oncotic pressures in renal sodium reabsorption, Circ. Res., 52:491, 1983.

18. Watkins, L., Jr., J. A. Burton, E. Haber, J. R. Cant, F. W. Smith, and A. C. Barger: The renin-angiotensin-aldosterone system in congestive failure in conscious dogs, J. Clin. Invest., 57:1606, 1976.

19. Schrier, R. W., M. H. Humphreys, and R. C. Ufferman: Role of cardiac output and the autonomic nervous system in the antinatriuretic response to acute constriction of the thoracic superior vena cava, Circ. Res., 29:490, 1971.

20. Bourgoignie, J. J., K. H. Hwang, E. Ipakchi, and N. S. Bricker: The presence of a natriuretic factor in urine of patients with chronic uremia: the absence of the factor in nephrotic uremic patients, J. Clin. Invest., 53:1559, 1974.

21. Naccarato, R., P. Messa, A. D'Angelo, A. Fabris, M. Messa, M. Chiaramonte, C. Gregolin, and G. Zanon: Renal handling of sodium and water in early chronic liver disease, Gastroenterology, 81:205, 1981.

22. Epstein, M.: Natriuretic hormone and the sodium retention of cirrhosis, Gastroenterology, 81:395, 1981.

23. Chonko, A. M., W. H. Bay, J. H. Stein, and T. F. Ferris: The role of renin and aldosterone in the salt retention of edema, Am. J. Med., 63:881, 1977.

24. McHugh, T. J., J. Forrester, L. Adler, D. Zion, and H. J. C. Swan: Pulmonary vascular congestion in acute myocardial infarction: hemodynamic and radiologic correlations, Ann. Intern. Med., 76:29, 1972.

25. Forrester, J. S., G. Diamond, K. Chatterjee, and H. J. C. Swan: Medical therapy of acute myo-

cardial infarction by application of hemodynamic subsets, N. Engl. J. Med., 295:1356, 1404, 1976.

26. Starr, I., Jr.: The role of "static blood pressure" in abnormal increments of venous pressure, especially in heart failure: clinical and experimental studies, Am. J. Med. Sci., 199:40, 1940.

27. Warren, J. V., and E. A. Stead: Fluid dynamics in chronic congestive heart failure: an interpretation of the mechanisms producing the edema, increased plasma volume, and elevated venous pressure in certain patients with prolonged congestive failure, Arch. Intern. Med., 73:138, 1944.

28. Noble, M. I. M.: The Frank-Starling curve, Clin. Sci., 54:1, 1978.

29. Higgins, C. B., S. F. Vatner, D. Franklin, and E. Braunwald: Effects of experimentally produced heart failure on the peripheral vascular response to severe exercise in conscious dogs, Circ. Res., 31:186, 1972.

30. Millard, R. W., C. B. Higgins, D. Franklin, and S. F. Vatner: Regulation of the renal circulation during severe exercise in normal dogs and dogs with experimental heart failure, Circ. Res., 31:881, 1972.

31. Braunwald, E., W. H. Plauth, and A. G. Morrow: A method for the detection and quantification of impaired sodium excretion, Circulation, 32:223, 1965.

32. Gault, J. H., J. W. Covell, E. Braunwald, and J. Ross, Jr.: Left ventricular performance following correction of free aortic regurgitation, Circulation, 42:773, 1970.

33. Campbell, E. J. M., and D. S. Short: The cause of oedema in cor pulmonale, Lancet, 1:1184, 1960.

34. Richens, J. M., and P. Howard: Oedema in cor pulmonale, Clin. Sci., 62:255, 1982.

35. Farber, M. O., L. R. Roberts, M. R. Weinberger, G. L. Robertson, N. S. Fineberg, and F. Manfredi: Abnormalities of sodium and H_2O handling in chronic obstructive lung disease, Arch, Intern. Med., 142:1326, 1982.

36. Skorecki, K. L., and B. M. Brenner: Body fluid homeostasis in congestive heart failure and cirrhosis with ascites, Am. J. Med., 72:323, 1982.

37. Witte, M. H., C. L. Witte, and A. E. Dumont: Progress in liver disease: physiological factors involved in the causation of cirrhotic ascites, Gastroenterology, 61:742, 1971.

38. Zink, J., and C. V. Greenway: Intraperitoneal pressure in formation and reabsorption of ascites in cats, Am. J. Physiol., 233:H185, 1977.

39. Henriksen, J. H., J. G. Stage, P. Schlichting, and K. Winkler: Intraperitoneal pressure: ascitic fluid and splanchnic vascular pressures, and their role in prevention and formation of ascites, Scand. J. Clin. Lab. Invest., 40:493, 1980.

40. Better, O. S., and R. W. Schrier: Disturbed volume homeostasis in patients with cirrhosis of the liver, Kidney Int., 23:303, 1983.

41. Bichet, D. G., M. J. VanPutten, and R. W. Schrier: Potential role of increased sympathetic activity in impaired sodium and water excretion in cirrhosis, N. Engl. J. Med., 307:1552, 1982.

42. Sherlock, S., B. Senewiratne, A. Scott, and J. G. Walker: Complications of diuretic therapy in hepatic cirrhosis, Lancet, 1:1049, 1966.

43. Epstein, M.: Peritoneovenous shunt in the management of ascites and the hepatorenal syndrome, Gastroenterology, 82:790, 1982.

44. Schroeder, E. T., G. H. Anderson, and H. Smulyan: Effects of a portacaval or peritoneovenous shunt on renin in the hepatorenal syndrome, Kidney Int., 15:54, 1979.

44a. Epstein, M., R. Levinson, J. Sancho, E. Haber, and R. Re: Characterization of the renin-aldosterone system in decompensated cirrhosis, Circ. Res., 41:818, 1977.

45. Wilkinson, S. P., I. K. Smith, H. Moodie, L. Poston, and R. Williams: Studies on mineralocorticoid 'escape' in cirrhosis, Clin. Sci., 56:401, 1979.

46. Greig, P. D., L. M. Blendis, B. Langer, B. R. Taylor, and R. F. Colapinto: Renal and hemodynamic effects of the peritoneovenous shunt. II. Long-term effects, Gastroenterology, 80:119, 1981.

47. Levy, M.: Sodium retention and ascites formation in dogs with experimental portal cirrhosis, Am. J. Physiol., 233:F572, 1977.

47a. Levy, M.: Sodium retention in dogs with cirrhosis and ascites: efferent mechanisms, Am. J. Physiol., 233:F586, 1977.

48. Wilkinson, S. P., I. K. Smith, and R. Williams: Changes in plasma renin activity in cirrhosis: a reappraisal based on studies in 67 patients and "low renin" cirrhosis, Hypertension, 1:125, 1979.
49. Lieberman, F. L., and T. B. Reynolds: Plasma volume in cirrhosis of the liver: its relation to portal hypertension, ascites, and renal failure, J. Clin. Invest., 46:1297, 1967.
50. Rose, B. D.: *Pathophysiology of Renal Disease,* McGraw-Hill, New York, 1981, chap. 4.
51. Rothschild, M. A., M. Oratiz, and S. S. Schreiber: Albumin synthesis, N. Engl. J. Med., 286:816, 1972.
52. Katz, J., G. Bonorris, and A. L. Sellers: Effect of nephrectomy on plasma albumin catabolism in experimental nephrosis, J. Lab. Clin. Med., 63:680, 1964.
53. Meltzer, J. I., H. J. Keim, J. H. Laragh, J. E. Sealey, K-M. Jan, and S. Chien: Nephrotic syndrome: vasoconstriction and hypervolemic types indicated by renin-sodium profiling, Ann. Intern. Med., 91:688, 1979.
54. Brown, E. A., N. D. Markandu, J. E. Roulston, B. E. Jones, M. Squires, and G. A. MacGregor: Is the renin-angiotensin-aldosterone system involved in the sodium retention in the nephrotic syndrome?, Nephron, 32:102, 1982.
55. Wagnild, J. P., and F. D. Gutmann: Functional adaptation of nephrons in dogs with acute progressing to chronic experimental glomerulonephritis, J. Clin. Invest., 57:1575, 1976.
56. Miller, T. R., R. J. Anderson, S. L. Linas, W. L. Henrich, A. S. Berns, P. A. Gabow, and R. W. Schrier: Urinary diagnostic indices in acute renal failure: a prospective study, Ann. Intern. Med., 89:47, 1978.
57. Finnerty, F. A., Jr.: Relationship of extracellular fluid volume to the development of drug resistance in the hypertensive patient, Am. Heart J., 81:563, 1971.
58. Pettinger, W. A.: Minoxidil and the treatment of severe hypertension, N. Engl. J. Med., 303:922, 1980.
58a. Terry, R. W.: Nifedipine therapy in angina pectoris: evaluation of safety and side effects, Am. Heart J., 104:681, 1982.
59. Mroczek, W. J., and W. R. Lee: Diazoxide therapy: use and risks, Ann. Intern. Med., 85:529, 1976.
60. Pohl, J. E. F., H. Thurston, and J. D. Swales: The antidiuretic action of diazoxide, Clin. Sci., 42:145, 1972.
61. Dunn, M. J., and E. J. Zambraski: Renal effects of drugs that inhibit prostaglandin synthesis, Kidney Int., 18:609, 1980.
62. Feldman, D., D. S. Loose, and S. Y. Tan: Nonsteroidal anti-inflammatory drugs cause sodium and water retention in the rat, Am. J. Physiol., 234:F490, 1978.
63. Christy, N. P., and J. C. Shaver: Estrogens and the kidney, Kidney Int., 6:366, 1974.
64. Preedy, J. R. K., and E. H. Aitken: The effect of estrogen on water and electrolyte metabolism. II. Hepatic disease, J. Clin. Invest., 35:430, 1956.
65. Veverbrants, E., and R. A. Arky: Effects of fasting and refeeding. I. Studies on sodium, potassium and water excretion on a constant electrolyte and fluid intake, J. Clin. Endocrinol. Metab., 29:55, 1969.
66. DeFronzo, R. A., C. R. Cooke, R. Andres, G. R. Faloona, and P. J. Davis: The effect of insulin on renal handling of sodium, potassium, calcium and phosphate in man, J. Clin. Invest., 55:845, 1975.
67. Spark, R. F., R. A. Arky, P. R. Boulter, C. D. Saudek, and J. T. O'Brian: Renin, aldosterone, and glucagon in the natriuresis of fasting, N. Engl. J. Med., 292:1335, 1975.
68. Saudek, C. D., P. R. Boulter, R. H. Knopp, and R. A. Arky: Sodium retention accompanying insulin treatment of diabetes mellitus, Diabetes, 23:240, 1974.
69. Streeten, D. H. P.: Idiopathic edema: pathogenesis, clinical features, and treatment, Metabolism, 25:353, 1978.
70. Edwards, O. M., and R. I. S. Bayliss: Idiopathic oedema of women, Q. J. Med., 45:125, 1976.
71. Coleman, M., M. Horwith, and J. L. Brown: Idiopathic edema: studies demonstrating protein-leaking angiopathy, Am. J. Med., 49:106, 1970.
72. MacGregor, G. A., P. R. W. Tasker, and H. E. deWardener: Diuretic-induced oedema, Lancet 1:489, 1975.

73. MacGregor, G. A., N. D. Markandu, J. E. Roulston, J. C. Jones, and H. E. deWardener: Is "idiopathic" oedema idiopathic?, Lancet, 1:397, 1979.

74. Levy, M., and J. F. Seely, Pathophysiology of edema formation, in *The Kidney*, 2d ed., B. M. Brenner and F. C. Rector, Jr. (eds.), W. B. Saunders, Philadelphia, 1981, pp. 760–764.

75. Rinaldo, J. E., and R. M. Rogers: Adult respiratory-distress syndrome: changing concepts of lung injury and repair, N. Engl. J. Med., 306:900, 1982.

76. Fein, A., R. F. Grossman, J. G. Jones, E. Overland, L. Pitts, J. F. Murray, and N. C. Staub: The value of edema fluid protein measurement in patients with pulmonary edema, Am. J. Med., 67:32, 1979.

77. Lindheimer, M. D., and A. I. Katz: Sodium and diuretics in pregnancy, N. Engl. J. Med., 288:891, 1973.

78. Parving, H. H., J. M. Hansen, S. L. Nielsen, N. Rossing, O. Munck, and N. Lassen: Mechanisms of edema formation in myxedema: increased protein extravasation and relatively slow lymphatic drainage, N. Engl. J. Med., 301:460, 1979.

79. Sachdev, Y., and R. Hall: Effusions into body cavities in hypothyroidism, Lancet, 1:564, 1975.

80. Hollander, W., P. Reilly, and B. A. Burrows: Lymphatic flow in human subjects as indicated by the disappearance of I^{131}-labeled albumin from the subcutaneous tissue, J. Clin. Invest., 40:222, 1961.

81. Guazzi, M., A. Polese, F. Magrini, C. Fiorentini, and M. T. Olivar: Negative influences of ascites on the cardiac function of cirrhotic patients, Am. J. Med., 59:165, 1975.

82. Zarins, C. K., C. L. Rice, R. M. Peters, and R. W. Virgilio: Lymph and pulmonary response to isobaric reduction in plasma oncotic pressure in baboons, Circ. Res., 43:925, 1978.

83. Race, G. A., C. H. Scheifley, and J. E. Edwards: Albuminuria in congestive heart failure, Circulation, 13:329, 1956.

84. Bohrer, M. P., W. M. Deen, C. R. Robertson, and B. M. Brenner: Mechanisms of angiotensin II-induced proteinuria in the rat, Am. J. Physiol., 233:F13, 1977.

85. Auld, R. B., E. A. Alexander, and N. S. Levinsky: Proximal tubular function in dogs with thoracic caval construction, J. Clin. Invest., 50:2150, 1971.

86. Relman, A. S., A. Leaf, and W. B. Schwartz: Oral administration of a potent carbonic anhydrase inhibitor ("Diamox"). II. Its use as a diuretic in patients with severe congestive heart failure, N. Engl. J. Med., 250:800, 1954.

87. Seely, J. F., and J. H. Dirks: Site of action of diuretic drugs, Kidney Int., 11:1, 1977.

88. Burg, M. B.: Tubular chloride transport and the mode of action of some diuretics, Kidney Int., 9:189, 1976.

89. Leaf, A., W. B. Schwartz, and A. S. Relman: Oral administration of a potent carbonic anhydrase inhibitor ("Diamox"). I. Changes in electrolyte and acid-base balance, N. Engl. J. Med., 250:759, 1954.

90. Mathisen, O., T. Monclair, H. Holdaas, and F. Kiil: Bicarbonate as mediator of proximal tubular NaCl reabsorption and glomerulotubular balance, Scand. J. Lab. Clin. Invest., 38:7, 1978.

91. Seely, J. F., and J. H. Dirks: Micropuncture study of hypertonic mannitol diuresis in the proximal and distal tubule of the dog kidney, J. Clin. Invest., 48:2330, 1969.

92. Gennari, F. J., and J. P. Kassirer: Osmotic diuresis, N. Engl. J. Med., 291:714, 1974.

93. Mathisen, Ø., M. Raeder, and F. Kiil: Mechanism of osmotic diuresis, Kidney Int., 19:431, 1981.

94. Rose, B. D., *Pathophysiology of Renal Disease*, McGraw-Hill, New York, 1981, chap. 3.

95. Warren, S. E., and R. C. Blantz: Mannitol, Arch. Intern. Med., 141:493, 1981.

96. Fishman, R. A.: Brain edema, N. Engl. J. Med., 293:706, 1975.

97. Gipstein, R. M., and J. D. Boyle: Hypernatremia complicating prolonged mannitol diuresis, N. Engl. J. Med., 272:1116, 1965.

98. Aviram, A., A. Pfau, J. W. Czaczkes, and T. D. Ullman: Hyperosmolality with hyponatremia caused by inappropriate administration of mannitol, Am. J. Med., 42:648, 1967.

99. Wilson, D. R., U. Honrath, and H. Sonnenberg: Furosemide action on collecting ducts: effect of prostaglandin synthesis inhibition, Am. J. Physiol., 244:F666, 1983.

100. Warnock, D. G., and J. Eveloff: NaCl entry mechanisms in the luminal membrane of the renal tubule, Am. J. Physiol., 242:F561, 1982.

101. Oberleithner, H., W. Guggino, and G. Giebisch: Mechanism of distal tubular chloride transport in *Amphiuma* kidney, Am. J. Physiol., 242:F331, 1982.

102. Suki, W. N., F. C. Rector, Jr., and D. W. Seldin: The site of action of furosemide and other sulfonamide diuretics in the dog, J. Clin. Invest., 44:1458, 1965.

103. Earley, L. E., and R. M. Friedler: Renal tubular effects of ethacrynic acid, J. Clin. Invest., 43:1495, 1964.

104. Patak, R. V., S. Z. Fadem, S. G. Rosenblatt, M. D. Lifschitz, and J. H. Stein: Diuretic-induced changes in renal blood flow and prostaglandin E excretion in the dog, Am. J. Physiol., 236:F494, 1979.

105. Gerber, J. G., and A. S. Nies: Furosemide-induced vasodilation: importance of the state of hydration and filtration, Kidney Int., 18:454, 1980.

106. Tucker, B. J., R. W. Steiner, L. C. Gushwa, and R. C. Blantz: Studies on the tubuloglomerular feedback system in the rat: the mechanism of reduction in filtration rate with benzolamide, J. Clin. Invest., 62:993, 1978.

106a. Flamenbaum, W., and R. Friedman: Pharmacology, therapeutic efficacy, and adverse effects of bumetanide, a new "loop" diuretic, Pharmacotherapy, 2:213, 1982.

107. Dikshit, K., J. K. Vyden, J. S. Forrester, K. Chatterjee, R. Prakash, and H. J. C. Swan: Renal and extrarenal hemodynamic effects of furosemide in congestive heart failure after acute myocardial infarction, N. Engl. J. Med., 288:1087, 1973.

108. Austin, S. M., B. F. Schreiner, D. H. Kramer, P. M. Shah, and P. N. Yu: The acute hemodynamic effects of ethacrynic acid and furosemide in patients with chronic postcapillary pulmonary hypertension, Circulation, 53:364, 1976.

108a. Bourland, W. A., D. K. Day, and H. E. Williams: The role of the kidney in the early nondiuretic action of furosemide to reduce elevated left atrial pressure in the hypervolemic dog, J. Pharmacol. Expt. Therap., 202:221, 1977.

109. Rane, A., J. P. Villeneuve, W. J. Stone, A. S. Nies, G. R. Wilkinson, and R. A. Branch: Plasma binding and disposition of furosemide in the nephrotic syndrome and in uremia, Clin. Pharmacol. Therap., 24:199, 1978.

110. Brown, C. B., C. S. Ogg, and J. S. Cameron: High dose frusemide in acute renal failure: a controlled trial. Clin. Nephrol., 15:90, 1981.

111. Gallagher, K. L., and J. K. Jones: Furosemide-induced ototoxicity, Ann. Intern. Med., 91:744, 1979.

112. Kunau, R. T., Jr., D. R. Weller, and H. L. Webb: Clarification of the site of action of chlorothiazide in the rat nephron, J. Clin. Invest., 56:401, 1975.

113. Wilson, D. R., U. Honrath, and H. Sonnenberg: Thiazide diuretic effect on medullary collecting duct in the rat, Kidney Int., 23:711, 1983.

114. Helderman, J. H., D. Elahi, D. K. Andersen, G. S. Raizes, J. D. Tobin, D. Shocken, and R. Andres: Prevention of the glucose intolerance of thiazide diuretics by maintenance of body potassium, Diabetes, 32:106, 1983.

115. Greenblatt, D. J., and J. Koch-Weser: Adverse reactions to spironolactone, J. Am. Med. Assoc., 225:40, 1973.

116. Jaffey, L., and A. Martin: Malignant hyperkalemia after amiloride/hydrochlorothiazide treatment, Lancet, 1:1272, 1981.

117. Cohen, A. B.: Hyperkalemic effects of triamterene, Ann. Intern. Med., 65:521, 1966.

118. Corvol, P., M. Claire, M. E. Oblin, K. Geering, and B. Rossier: Mechanism of the antimineralocorticoid effects of spironolactones, Kidney Int., 20:1, 1981.

119. Loriaux, D. L., R. Menard, A. Taylor, J. C. Pita, and R. Santen: Spironolactone and endocrine dysfunction, Ann. Intern. Med., 85:630, 1976.

120. Benos, D. J.: Amiloride: a molecular probe of sodium transport in tissues and cells, Am. J. Physiol., 242:C131, 1982.

121. Anonymous: Amiloride: a potassium-sparing diuretic, Med. Lett. Drugs Therap., 23(26):109, 1981.

122. Werness, P. G., J. H. Bergert, and L. H. Smith: Triamterene urolithiasis: solubility, pK, effect on crystal formation, and matrix binding of triamterene and its metabolites, J. Lab. Clin. Med., 99:254, 1982.

123. Favre, L., P. Glasson, and M. B. Vallotton: Reversible acute renal failure from combined triamterene and indomethacin, Ann. Intern. Med., 96:317, 1982.

124. Baylis, C., and B. M. Brenner: Mechanism of the glucocorticoid-induced increase in glomerular filtration rate, Am. J. Physiol., 234:F166, 1978.

125. Yunis, S. L., D. D. Bercovitch, R. M. Stein, M. F. Levitt, and M. H. Goldstein: Renal tubular effects of hydrocortisone and aldosterone in normal hydropenic man: comment on site of action, J. Clin. Invest., 43:1668, 1964.

126. Jick, H.: The use of glucocorticoids for diuresis in patients with fluid retention not resulting from renal disease, Ann. N.Y. Acad. Sci., 139:512, 1966.

127. Gabow, P. A., S. Moore, and R. W. Schrier: Spironolactone-induced hyperchloremic acidosis in cirrhosis, Ann. Intern. Med., 90:338, 1979.

128. Garella, S., B. S. Chang, and S. I. Kahn: Dilution acidosis and contraction alkalosis: review of a concept, Kidney Int., 8:279, 1975.

129. Cannon, P. J., H. O. Heinemann, M. S. Albert, J. H. Laragh, and R. W. Winters: "Contraction" alkalosis after diuresis of edematous patients with ethacrynic acid, Ann. Intern. Med., 62:979, 1965.

130. Steele, T. H., and S. Oppenheimer: Factors affecting urate excretion following diuretic administration in man, Am. J. Med., 47:564, 1969.

131. Dirks, J. H., W. J. Cirksena, and R. W. Berliner: Micropuncture study of the effect of various diuretics on sodium reabsorption by the proximal tubules of the dog, J. Clin. Invest., 45:1875, 1966.

132. Weinman, E. J., G. Eknoyan, and W. N. Suki: The influence of the extracellular fluid volume on the tubular reabsorption of uric acid, J. Clin. Invest., 45:1875, 1966.

133. Liang, M. H., and J. F. Fries: Asymptomatic hyperuricemia: the case for conservative management, Ann. Intern. Med., 88:666, 1978.

134. Schrier, R. W., and D. G. Bichet: Osmotic and non-osmotic control of vasopressin release and the pathogenesis of impaired water excretion in adrenal, thyroid, and edematous disorders, J. Lab. Clin. Med., 98:1, 1981.

135. Szatalowicz, V. L., P. D. Miller, J. W. Lacher, J. A. Gordon, and R. W. Schrier: Comparative effect of diuretics on renal water excretion in hyponatraemic oedematous disorders, Clin. Sci., 62:235, 1982.

135a. Swales, J. D.: Magnesium deficiency and diuretics, Br. Med. J., 2:1377, 1982.

136. Cohn, J. N.: Blood pressure and cardiac performance, Am. J. Med., 55:351, 1973.

137. Stampfer, M., S. E. Epstein, G. D. Beiser, and E. Braunwald: Hemodynamic effects of diuresis at rest and during intense exercise in patients with impaired cardiac function, Circulation, 37:900, 1968.

138. Lal, S., J. G. Murtagh, A. M. Pollock, E. Fletcher, and P. F. Binnion: Acute hemodynamic effects of furosemide in patients with normal and raised left atrial pressures, Br. Heart J., 31:711, 1969.

139. Duncan, L. E., Jr., G. W. Liddle, and F. C. Bartter: The effect of changes in body sodium on extracellular fluid volume and aldosterone and sodium excretion by normal and edematous men, J. Clin. Invest., 35:1299, 1956.

140. Wolff, H. P., Kh. R. Koczorek, and E. Buchborn: Aldosterone and antidiuretic hormone (adiuretin) in liver disease, Acta Endocrinol., 27:45, 1958.

141. Shear, L., S. Ching, and G. J. Gabuzda: Compartmentalization of ascites and edema in patients with hepatic cirrhosis, N. Engl. J. Med., 282:1391, 1970.

142. Connolly, M. E., O. M. Wrong, and N. F. Jones: Reversible renal failure in idiopathic nephrotic syndrome with minimal glomerular changes, Lancet, 1:665, 1968.

143. Conn, H. O.: A rational approach to the hepatorenal syndrome, Gastroenterology, 65:321, 1973.

144. Dunn, M. J., and E. J. Zambraski: Renal effects of drugs that inhibit prostaglandin synthesis, Kidney Int., 18:609, 1980.

145. Zipser, R. D., J. C. Hoefs, P. F. Speckhart, P. K. Zia, and R. Horton: Prostaglandins: modulators of renal function and pressor resistance in chronic liver disease, J. Clin. Endocrinol. Metab., 48:895, 1979.

146. Arisz, L., A. J. M. Donker, J. R. H. Brentjens and G. K. van der Hem: The effect of indomethacin on proteinuria and kidney function in the nephrotic syndrome, Acta Med. Scand., 199:121, 1976.

147. Walsh, J. J., and R. C. Venuto: Acute oliguric renal failure induced by indomethacin: possible mechanism, Ann. Intern. Med., 91:47, 1979.

148. Brater, D. C.: Analysis of the effect of indomethacin on the response to furosemide in man: effect of dose of furosemide, J. Pharmacol. Exp. Therap., 210:386, 1979.

149. Laiwah, A. C. Y., and R. A. Mactier: Antagonistic effect of non-steroidal anti-inflammatory drugs on frusemide-induced diuresis in cardiac failure, Br. Med. J., 2:714, 1981.

150. Ciabattoni, G., F. Pugliese, G. A. Cinotti, and C. Patrono: Renal effects of anti-inflammatory drugs, Eur. J. Rheumatol. Inflam., 3:210, 1980.

150a. Bunning, R. D., and W. F. Barth: Sulindac: a potentially renal-sparing nonsteroidal anti-inflammatory drug, J. Am. Med. Assoc., 248:2864, 1982.

151. Martinez-Maldonado, M., G. Eknoyan, and W. N. Suki: Diuretics in nonedematous states: physiological basis for the clinical use, Arch. Intern. Med., 131:797, 1973.

152. Fuller, R., C. Hoppel, and S. T. Ingalls: Furosemide kinetics in patients with hepatic cirrhosis, Clin. Pharmacol. Therap., 30:461, 1981.

153. Nomura, A., H. Yasuda, M. Minami, T. Akimoto, K. Miyazaki, and T. Arita: Effect of furosemide in congestive heart failure, Clin. Pharmacol. Therap., 30:177, 1981.

154. Leth, A.: Changes in plasma and extracellular fluid volumes in patients with essential hypertension during long-term treatment with hydrochlorothiazide, Circulation, 42:479, 1970.

154a. Maronde, R. F., M. Milgrom, N. D. Vlachakis, and L. Chan: Response of thiazide-induced hypokalemia to amiloride, J. Am. Med. Assoc., 249:237, 1983.

155. Odlund, B. O. G., and B. Freeman: Diuretic resistance: reduced bioavailability and effect of oral frusemide, Br. Med. J., 1:1577, 1980.

156. Wollam, G. L., R. C. Tarazi, E. L. Bravo, and H. P. Dustan: Diuretic potency of combined hydrochlorothiazide and furosemide therapy in patients with azotemia, Am. J. Med., 72:929, 1982.

157. Brown, E., and G. MacGregor: Synergistic action of metolazone and frusemide (letter), Br. Med. J., 2:1611, 1981.

157a. Oster, J. R., M. Epstein, and S. Smoller: Combination therapy with thiazide-type and loop diuretic agents for resistant sodium retention, Ann. Intern. Med., 99:405, 1983.

158. Brest, A. N., R. Seller, G. Onesti, O. Ramirez, C. Swartz, and J. H. Moyer: Clinical selection of diuretic drugs in the management of cardiac edema, Am. J. Cardiol., 22:168, 1968.

159. Eder, H. A., H. D. Lauson, F. P. Chinard, R. L. Greif, G. C. Cotzias, and D. D. van Slyke: A study of the mechanisms of edema formation in patients with the nephrotic syndrome, J. Clin. Invest., 33:636, 1954.

160. Bank, N.: External compression for treatment of resistant edema (letter), N. Engl. J. Med., 302:969, 1980.

161. Rose, B. D.: *Pathophysiology of Renal Disease,* McGraw-Hill, New York, 1981, chap. 11.

162. Forrester, J. S., G. Diamond, K. Chatterjee, and H. J. C. Swan: Medical therapy of acute myocardial infarction by application of hemodynamic subsets, N. Engl. J. Med., 295:1356, 1404, 1976.

163. Crexells, C., K. Chatterjee, J. S. Forrester, K. Dikshit, and H. J. C. Swan: Optimal level of filling pressure in the left side of the heart in acute myocardial infarction, N. Engl. J. Med., 289:1263, 1973.

164. Cohn, J. N., and J. A. Franciosa: Selection of vasodilator, inotropic or combined therapy for the management of heart failure, Am. J. Med., 65:181, 1978.

165. Aberman, A., and M. Fulop: The metabolic and respiratory acidosis of acute pulmonary edema, Ann. Intern. Med., 76:173, 1972.

166. Fulop, M., M. Horowitz, A. Aberman, and E. Jaffee: Lactic acidosis in pulmonary edema due to left ventricular failure, Ann. Intern. Med., 79:180, 1973.

167. Beiser, G. D., S. E. Epstein, M. Stampfer, B. Robinson, and E. Braunwald: Studies on digitalis. XVII. Effects of ouabain on the hemodynamic response to exercise in patients with mitral stenosis in normal sinus rhythm, N. Engl. J. Med., 278:131, 1968.

168. Arnold, S. B., R. C. Byrd, W. Meister, K. Melman, M. D. Cheitlin, J. D. Bristow, W. W. Parmley, and K. Chatterjee: Long-term digitalis therapy improves left ventricular function in heart failure, N. Engl. J. Med., 303:1443, 1980.

169. Lee, D. C-S., R. A. Johnson, J. B. Bingham: M. Leahy, R. E. Dinsmore, A. H. Goroll, J. B. Newell, H. W. Strauss, and E. Haber: Heart failure in outpatients: a randomized trial of digoxin versus placebo, N. Engl. J. Med., 306:699, 1982.

170. Shapiro, W., and K. Taubert: Hypokalemia and digoxin-induced arrhythmias, Lancet, 2:604, 1975.

171. Mason, D. T.: Symposium on vasodilator and inotropic therapy, Am. J. Med., 65:101, 1978.

172. Cohn, J. N., and J. A. Franciosa: Therapy of cardiac failure, N. Engl. J. Med., 297:27, 254, 1977.

173. Franciosa, J. A., G. Pierpont, and J. N. Cohn: Hemodynamic improvement after oral hydralazine in left ventricular failure: a comparison with nitroprusside infusion in 16 patients, Ann. Intern. Med., 86:388, 1977.

174. Colucci, W. S., J. Wynne, B. L. Holman and E. Braunwald: Long-term therapy of heart failure with prazosin: a randomized double blind trial, Am. J. Cardiol., 45:337, 1980.

175. Dzau, V. J., W. S. Colucci, G. H. Williams, G. Curfman, L. Meggs, and N. K. Hollenberg: Sustained effectiveness of converting-enzyme inhibition in patients with severe congestive heart failure, N. Engl. J. Med., 302:1373, 1980.

176. Magorien, R. D., D. W. Triffon, C. E. Desch, W. H. Bay, D. V. Unverferth, and C. V. Leier: Prazosin and hydralazine in congestive heart failure, Ann. Intern. Med., 95:5, 1981.

177. Creager, M. A., J. L. Halperin, D. B. Bernard, D. P. Faxon, C. D. Melidossian, H. Gavras, and T. J. Ryan: Acute regional circulatory and renal hemodynamic effects of converting-enzyme inhibition in patients with congestive heart failure, Circulation, 64:483, 1981.

178. Weber, K. T.: New hope for the failing heart, Am. J. Med., 72:665, 1982.

179. Gabuzda, G. J., and P. W. Hall, III: Relation of potassium depletion to renal ammonium metabolism and hepatic coma, Medicine, 45:481, 1966.

180. Artz, S. A., I. C. Paes, and W. W. Faloon: Hypokalemia-induced hepatic coma in cirrhosis: occurrence despite neomycin therapy, Gastroenterology, 51:1046, 1966.

181. Baertl, J. M., S. M. Sancelta, and G. J. Gabuzda: Relation of acute potassium depletion to renal ammonium metabolism in patients with cirrhosis, J. Clin. Invest., 42:696, 1963.

182. Kamm, D. E., and G. L. Strope: Glutamine and glutamate metabolism in renal cortex from potassium-depleted rats, Am. J. Physiol., 224:1241, 1973.

182a. Pérez-Ayuso, R. M., V. Arroyo, R. Planas, J. Gaya, F. Bory, A. Rimola, F. Riversa, and J. Rodés: Random comparative study of efficacy of furosemide vs. spironolactone in nonazotemic cirrhosis with ascites: relationship between the diuretic response and the activity of the renin-aldosterone system, Gastroenterology, 84:961, 1983.

183. Gabuzda, G. J.: Cirrhosis, ascites, and edema: clinical course related to management, Gastroenterology, 58:546, 1970.

184. Knauer, C. M., and H. M. Lower: Hemodynamics in the cirrhotic patient during paracentesis, N. Engl. J. Med., 276:491, 1967.

185. Capone, R. R., I. Buhac, R. C. Kohberger, and J. A. Balint: Resistant ascites in alcoholic liver cirrhosis, Am. J. Dig. Dis., 23:867, 1978.

186. Wapnick, S., S. Grossberg, M. Kinney, and H. LeVeen: LeVeen continuous peritoneal-jugular shunt, J. Am. Med. Assoc., 237:131, 1977.

187. Greig, P. D., B. Langer, L. M. Blendis, B. R. Taylor, and M. F. X. Glynn: Complications after peritoneovenous shunting for ascites, Am. J. Surg., 139:125, 1980.

188. Harmon, D. C., Z. Demerjian, L. Ellman, and J. E. Fischer: Disseminated intravascular coagulation with the peritoneovenous shunt, Ann. Intern. Med., 90:774, 1979.

189. Garnet, E. S., and C. E. Webber: Changes in blood volume produced by treatment in the nephrotic syndrome, Lancet, 2:798, 1967.

190. Speller, P. J., and D. H. P. Streeten: Mechanism of the diuretic action of D-amphetamine, Metabolism, 13:453, 1964.

191. Docci, D., and F. Turci: Captopril in idiopathic edema (letter), N. Engl. J. Med., 308:1102, 1983.

192. Prewitt, R. M., J. McCarthy, and L. D. H. Wood: Treatment of acute low pressure pulmonary edema in dogs, J. Clin. Invest., 67:409, 1981.

INTRODUCTION TO SIMPLE AND MIXED ACID-BASE DISORDERS

Disturbances of acid-base homeostasis are common clinical problems. The individual disorders are discussed in Chaps. 18 to 21. This chapter will review the basic principles of acid-base physiology, the general mechanisms by which abnormalities can occur, and an approach to evaluating patients with acid-base disorders.

ACID-BASE PHYSIOLOGY

Free H^+ ions are present in the body fluids in extremely low concentrations. The normal H^+ concentration in the extracellular fluid is roughly 40 nanoeq/L, approximately one-millionth the milliequivalent per liter concentrations of Na^+, K^+, Cl^-, and HCO_3^-.† Despite this low concentration, the maintenance of a stable H^+ concentration is required for normal cellular function, as small fluctuations in the H^+ concentration have important effects on the activity of cellular enzymes. For example, the rate of glycolysis varies inversely with the H^+ concentration (see Fig. 11-1). This is mediated by a similar H^+ dependence of several glycolytic enzymes, particularly phosphofructokinase. Because of these effects on enzyme function, there is a relatively narrow range of H^+ concentration that is compatible with life, from 16 to 160 nanoeq/L (pH equals 7.80 to 6.80).

Under normal conditions, the H^+ concentration varies little from the normal value of 40 nanoeq/L. The body buffers play an important role in this regulatory process as they are able to take up or release H^+ ions to prevent large changes in the H^+ concentration. There are a variety of buffers in the extracellular and intracellular fluids, most of which are weak acids (which can release H^+ ions) and their ionized salts (which can take up H^+ ions). The most important extracellular buffer is HCO_3^-, which combines with H^+ according to the following reaction:

$$H^+ + \underset{\text{Salt}}{HCO_3^-} \rightleftharpoons \underset{\text{Weak acid}}{H_2CO_3} \rightleftharpoons H_2O + CO_2 \tag{17-1}$$

The relationship between the concentrations of H^+, HCO_3^-, and CO_2 can be expressed by the law of mass action (see page 212):

$$[H^+] = K_a' \times \frac{0.03 P_{CO_2}}{[HCO_3^-]} \tag{17-2}$$

where K_a' is the dissociation constant for this reaction and $0.03 P_{CO_2}$ represents the solubility of CO_2 in the plasma. If the H^+ concentration is measured in nanomoles per liter (nanomol/L), the value for K_a' is approximately 800 nanomol/L. If this is substituted in Eq. (17-2), then

$$[H^+] = 24 \times \frac{P_{CO_2}}{[HCO_3^-]} \tag{17-3}$$

Eq. (17-2) can also be expressed in logarithmic terms as the Henderson-Hasselbalch equation:

$$pH = 6.10 + \log \frac{[HCO_3^-]}{0.03 P_{CO_2}} \tag{17-4}$$

---

†These concentrations can also be expressed in terms of molarity. Since the valence of H^+ is 1+, 40 nanoeq/L equals 40 nanomol/L (see Chap. 1).

where *pH equals* $-log\ [H^+]$ (the H^+ concentration being measured in moles per liter) and $6.10 = -log\ K_a'$ (or $-log\ 800 \times 10^{-9}$ mol/L). At the normal H^+ concentration of 40 nanomol/L (or 40×10^{-9} mol/L),

$$pH = -log\ (40 \times 10^{-9})$$

$$= -(log\ 40 + log\ 10^{-9})$$

Since $log\ 40 = 1.6$ and $log\ 10^{-9} = -9$,

$$pH = -(1.6 - 9) = 7.40$$

Although the acidity of the extracellular fluid is measured as the pH, it is frequently easier to think in terms of the H^+ concentration. Therefore, the following chapters will use both the pH and H^+ concentration to permit the reader to become familiar with both concepts. It is important to recognize the inverse relationship between the pH and the H^+ concentration. An increase in the H^+ concentration reduces the pH, and a decrease in the H^+ concentration increases the pH (Table 17-1).

Measurement of pH

The pH and P_{CO_2} are measured on blood drawn anaerobically (to prevent the loss of CO_2 from the blood into the air) in a heparinized syringe [1]. The pH is measured by an electrode permeable only to H^+ ions (see page 205) and the P_{CO_2} by a CO_2 electrode. The HCO_3^- concentration can then be calculated from the Henderson-Hasselbalch equation or measured directly. Although the calculated and measured values for the plasma HCO_3^- concentration are usually similar, they may occasionally differ by as much as 7 to 8 meq/L. In this setting, it is most likely that the measured level is correct. Calculation of the plasma HCO_3^- concentration from the

Table 17-1 Relationship between the pH and the H^+ concentration in the physiologic range

pH	$[H^+]$, nanomol/L
7.80	16
7.70	20
7.60	26
7.50	32
7.40	40
7.30	50
7.20	63
7.10	80
7.00	100
6.90	125
6.80	160

Henderson-Hasselbalch equation assumes that $-\log K'_a$ is 6.10 and that 0.03 is the solubility constant for CO_2 in the plasma. Although these values are generally correct, they may vary in patients with acute acid-base disorders, and if used in such cases they may lead to an erroneous calculated HCO_3^- concentration [2,3].

The normal range of arterial and venous values for the pH, H^+ concentration, P_{CO_2}, and HCO_3^- concentration are:

	pH	$[H^+]$	P_{CO_2}	$[HCO_3^-]$
Arterial	7.37–7.43	37–43 nanoeq/L	36–44 mmHg	22–26 meq/L
Venous	7.32–7.38	42–48 nanoeq/L	42–50 mmHg	23–27 meq/L

The decrease in the pH in venous blood is due to the uptake of metabolically produced CO_2 in the capillary circulation. Although arterial blood is generally used to measure the pH (in part because it also allows measurement of arterial oxygenation), venous blood is easier to obtain and just as accurate if drawn without a tourniquet from a well-perfused area.

It should be noted that most laboratories measure the total CO_2 content and not the HCO_3^- concentration. This is equal to the sum of the various chemical forms by which CO_2 is carried in the blood:

$$\text{Total } CO_2 \text{ content } = HCO_3^- + \text{dissolved } CO_2 + H_2CO_3$$

The H_2CO_3 concentration is negligible, and the dissolved CO_2 is equal to $0.03 P_{CO_2}$. Therefore, the total CO_2 content normally exceeds the HCO_3^- concentration by 1.2 mmol/L (0.03×40). For the sake of simplicity, the following discussion will refer only to the HCO_3^- concentration since it is the plasma HCO_3^- concentration that is directly affected by changes in renal H^+ secretion (see below) and by the addition of acid or alkaline loads to the extracellular fluid.

Regulation of Hydrogen Concentration

Bicarbonate is the principal buffer in the extracellular fluid because of both its high concentration and the ability of the body to control independently the plasma HCO_3^- concentration and the P_{CO_2} (see Chap. 12). *The plasma HCO_3^- concentration is regulated by the changes in the rate of H^+ secretion from the renal tubular cell into the tubular lumen.* Each H^+ ion that is secreted results in the generation of one HCO_3^- ion in the extracellular fluid (see Figs. 12-1 to 12-3). If H^+ secretion is reduced, less acid is excreted in the urine. The net effect of this retained H^+ is a decrease in the plasma HCO_3^- concentration. If, on the other hand, H^+ secretion is increased, all the filtered HCO_3^- is reabsorbed plus extra H^+ ions are excreted in combination with urinary buffers and NH_3. The loss of these H^+ ions results in the addition of HCO_3^- ions to the extracellular fluid and an elevation in the plasma HCO_3^- concentration.

In comparison with HCO_3^-, CO_2 is eliminated by the lungs. Thus, *the P_{CO_2} is* *regulated by the rate of alveolar ventilation.* Hyperventilation enhances CO_2 excretion and lowers the P_{CO_2}; hypoventilation reduces CO_2 excretion and raises the P_{CO_2}. Although CO_2 is not an acid since it contains no H^+ ions, it acts as an acid in the body by combining with water to form carbonic acid (H_2CO_3) [Eq. (17-1)].

The kidneys and lungs play a central role in the maintenance of acid-base balance since they continuously excrete acid. Each day approximately 15,000 mmol of CO_2 is produced by endogenous metabolism and then excreted by the lungs. Similarly, a normal diet generates 50 to 100 meq of H^+ per day, mostly from the production of H_2SO_4 from metabolism of sulfur-containing amino acids [4]. These excess H^+ ions are initially buffered by HCO_3^- and the cellular and bone buffers to minimize the fall in pH (see page 217). Acid-base balance is then restored by increased urinary H^+ excretion.

When acid-base disturbances do occur, renal and respiratory functions change in an attempt to normalize the pH. From the law of mass action:

$$[H^+] = 24 \times \frac{P_{CO_2}}{[HCO_3^-]}$$

it can be seen that *the H^+ concentration is related to the $P_{CO_2}/[HCO_3^-]$ ratio, not their absolute values.* If the H^+ concentration is increased, regardless of cause, it can be reduced toward normal by a decrease in the P_{CO_2} and/or an increase in the plasma HCO_3^- concentration. Both of these changes occur as alveolar ventilation and renal H^+ secretion are enhanced by an elevation in the H^+ concentration. Conversely, alveolar ventilation and H^+ secretion are diminished when the H^+ concentration is reduced. The resultant increase in the P_{CO_2} and decrease in the plasma HCO_3^- concentration increase the H^+ concentration toward normal. These changes in renal and respiratory function appear to be due primarily to the effects of the extracellular H^+ concentration on the renal tubular cell H^+ concentration and on the peripheral and central chemoreceptors controlling ventilation (see Chap. 12).

ACID-BASE DISORDERS

Definitions

A change in the extracellular pH may be seen when renal or respiratory function is abnormal or when an acid or base load overwhelms excretory capacity. *Acidemia* is defined as a decrease in the blood pH (or an increase in the H^+ concentration), and *alkalemia* as an elevation in the blood pH (or a reduction in the H^+ concentration). On the other hand, *acidosis* and *alkalosis* refer to processes that tend to lower and raise the pH, respectively. In most conditions, an acidotic process leads to acidemia and an alkalotic process to alkalemia. However, this may not be true in patients with mixed acid-base disturbances in whom the final pH depends upon the balance between the different disorders that are present (see below).

Changes in the plasma pH and H^+ concentration can be induced by alterations

Table 17-2 Characteristics of the primary acid-base disorders

Disorders	pH	$[H^+]$	Primary disturbance	Compensatory response
Metabolic acidosis	↓	↑	↓ $[HCO_3^-]$	↓ P_{CO_2}
Metabolic alkalosis	↑	↓	↑ $[HCO_3^-]$	↑ P_{CO_2}
Respiratory acidosis	↓	↑	↑ P_{CO_2}	↑ $[HCO_3^-]$
Respiratory alkalosis	↑	↓	↓ P_{CO_2}	↓ $[HCO_3^-]$

in the P_{CO_2} or plasma HCO_3^- concentration [Eqs. (17-3) and (17-4)]. Since the P_{CO_2} is regulated by respiration, primary abnormalities in the P_{CO_2} are called *respiratory acidosis* (high P_{CO_2}) and *respiratory alkalosis* (low P_{CO_2}). In contrast, primary changes in the plasma HCO_3^- concentration are referred to as *metabolic acidosis* (low HCO_3^-) and *metabolic alkalosis* (high HCO_3^-). In each of these disorders, compensatory renal or respiratory responses act to minimize the change in H^+ concentration by minimizing the change in the $P_{CO_2}/[HCO_3^-]$ ratio (Table 17-2). To achieve this, *the compensatory response always changes in the same direction as the primary disturbance.* Thus, a high P_{CO_2} in respiratory acidosis results in enhanced renal H^+ excretion and an elevation in the plasma HCO_3^- concentration.

Table 17-2 also illustrates that the *diagnosis of an acid-base disorder requires measurement of the pH*. Simply looking at the plasma HCO_3^- concentration (which is routinely measured with the plasma Na^+, K^+, and Cl^- concentrations) is not sufficient. A high value, for example, can be seen both in metabolic alkalosis (where it is the primary problem) and in respiratory acidosis (where it represents the appropriate renal compensation). These disorders can be differentiated by the plasma pH.

Metabolic Acidosis

Metabolic acidosis is characterized by a fall in the plasma HCO_3^- concentration and a low pH (or high H^+ concentration). It can be due either to HCO_3^- loss (as with diarrhea) or to the buffering of a noncarbonic acid such as lactic acid or retained diet-generated sulfuric acid, as occurs in renal failure:

$$H_2SO_4 + 2NaHCO_3 \rightarrow Na_2SO_4 + 2H_2CO_3 \rightarrow 2CO_2 + 2H_2O$$

The reduction in pH stimulates ventilation, resulting in a compensatory decrease in the P_{CO_2}. Ultimate restoration of the pH usually depends upon renal excretion of the excess acid, a process that takes several days.

Metabolic Alkalosis

Metabolic alkalosis results from an elevation in the plasma HCO_3^- concentration and is associated with a high pH (or low H^+ concentration). This disorder can be produced by HCO_3^- administration or, more commonly, by H^+ loss, as with vomiting or the use of diuretics. The respiratory compensation consists of hypoventilation and an increase in the P_{CO_2}.

Renal excretion of the excess HCO_3^- (as $NaHCO_3$) can rapidly correct the pH. However, this does not occur in many patients because HCO_3^- reabsorptive capacity is enhanced, usually due to concomitant volume depletion. In this setting, the kidney avidly conserves Na^+ and therefore cannot excrete $NaHCO_3$ (see Chap. 18).

Respiratory Acidosis

Respiratory acidosis is due to decreased effective alveolar ventilation, resulting in reduced pulmonary excretion of CO_2 and, consequently, an increase in the plasma P_{CO_2} (hypercapnia). The renal compensation consists of enhanced H^+ excretion, which raises the plasma HCO_3^- concentration. This response, however, takes 3 to 5 days to reach completion [5]. As a result, two different acid-base disorders may occur: *acute* respiratory acidosis in which there may be a dramatic fall in pH, and *chronic* respiratory acidosis in which the pH is relatively well protected due to the renal compensation (see Chap. 20). Similar considerations apply to respiratory alkalosis but not to metabolic acidosis or alkalosis since the respiratory compensation to these disorders is rapid, beginning within minutes and being complete within 12 to 24 h [6].

Respiratory Alkalosis

The primary disturbance in respiratory alkalosis is hyperventilation, resulting in a fall in the plasma P_{CO_2} (hypocapnia) and an increase in the pH (or reduction in the H^+ concentration). The compensatory response consists of diminished renal H^+ secretion, producing HCO_3^- loss in the urine and an appropriate decrease in the plasma HCO_3^- concentration.

Mixed Acid-Base Disorders

It is not uncommon for more than one of the above primary disorders to be present. Suppose a patient has a low arterial pH and is therefore acidemic. In this setting, a low plasma HCO_3^- concentration indicates metabolic acidosis and a high P_{CO_2} indicates respiratory acidosis. If both are present, then the patient has a combined metabolic and respiratory acidosis. Similar reasoning can lead to the diagnosis of a combined metabolic and respiratory alkalosis in a patient with an elevated pH, a high plasma HCO_3^- concentration, and a low P_{CO_2}.

Knowledge of the extent of the renal and respiratory compensations allows more complex disturbances to be diagnosed. The responses listed in Table 17-3 have been empirically derived from observations in humans with the different acid-base disorders [5,7–10]. A simple example can illustrate how this information can be utilized. A patient with a salicylate overdose is found to have the following arterial blood values:

Table 17-3 Renal and respiratory compensations to primary acid-base disturbances in humans

Disorder	Primary change	Compensatory response
Metabolic acidosis	$\downarrow[HCO_3^-]$	1.2-mmHg decrease in P_{CO_2} for every 1-meq/L fall in $[HCO_3^-]$
Metabolic alkalosis	$\uparrow[HCO_3^-]$	0.7-mmHg increase in P_{CO_2} for every 1-meq/L elevation in $[HCO_3^-]$
Respiratory acidosis	$\uparrow P_{CO_2}$	
Acute		1.0-meq/L increase in $[HCO_3^-]$ for every 10-mmHg rise in P_{CO_2}
Chronic		3.5-meq/L elevation in $[HCO_3^-]$ for every 10-mmHg rise in P_{CO_2}
Respiratory alkalosis	$\downarrow P_{CO_2}$	
Acute		2.0-meq/L fall in $[HCO_3^-]$ for every 10-mmHg reduction in P_{CO_2}
Chronic		5.0-meq/L decrease in $[HCO_3^-]$ for every 10-mmHg reduction in P_{CO_2}

$$pH = 7.45$$

$$P_{CO_2} = 20 \text{ mmHg}$$

$$[HCO_3^-] = 13 \text{ meq/L}$$

As described above, *evaluation of acid-base status begins with the pH*. The slightly high pH indicates that the patient is alkalemic. This can be due to a high $[HCO_3^-]$ or a low P_{CO_2}. Since only the latter is present, the primary diagnosis is respiratory alkalosis, most likely acute given the history. In this disorder, the body buffers will reduce the plasma HCO_3^- concentration by 2 meq/L for every 10-mmHg decrease in the P_{CO_2} (Table 17-3). Thus, the $[HCO_3^-]$ should fall from 24 to 20 meq/L as the P_{CO_2} drops acutely from 40 to 20 mmHg. The actual $[HCO_3^-]$ of 13 meq/L is lower than expected, suggesting that the patient has a combined respiratory alkalosis and metabolic acidosis. This disturbance is frequently seen with a salicylate overdose [11].

It is also useful to note that the renal and respiratory compensations return the pH toward but rarely to normal. Thus, a normal pH in the presence of changes in the P_{CO_2} and plasma HCO_3^- concentration immediately suggests a mixed disorder. For example, the following arterial blood values:

$$pH = 7.40$$

$$P_{CO_2} = 60 \text{ mmHg}$$

$$[HCO_3^-] = 36 \text{ meq/L}$$

are due to a combination of respiratory acidosis (elevated P_{CO_2}) and metabolic alkalosis (high plasma HCO_3^- concentration). This disorder is most often due to diuretic therapy in a patient with severe chronic lung disease.

Finally, an arterial P_{CO_2} of 40 mmHg or a plasma HCO_3^- concentration of 24 meq/L is not always normal. A patient with metabolic acidosis should hyperventilate to minimize the reduction in pH. On the average, the P_{CO_2} falls 1.2 mmHg for every 1-meq/L fall in the plasma HCO_3^- concentration [7]. Thus, a 16-meq/L reduction in the plasma HCO_3^- concentration from 24 to 8 meq/L should lower the P_{CO_2} by about 19 mmHg (16 $\times$ 1.2) from 40 to 21 mmHg. In this setting, the new pH will be 7.20. If, however, the P_{CO_2} remains at 40 mmHg, then the degree of acidemia will be more severe,

$$pH = 6.10 + \log \frac{8}{0.03 \times 40} = 6.92$$

Since the P_{CO_2} of 40 mmHg is *inappropriately high* by 19 mmHg, this patient has a combined metabolic and respiratory acidosis.

Acid-Base Map

If the relationship between the arterial pH (or H^+ concentration), P_{CO_2}, and HCO_3^- concentration in the different acid-base disorders is plotted, the result is the *acid-base map* in Fig. 17-1. The stippled areas represent the responses of otherwise normal subjects to metabolic and respiratory acidosis and alkalosis, including the appropriate compensatory responses which should be present. Thus, a given increase in the P_{CO_2} is associated with a lesser decrease in pH in chronic respiratory acidosis than in acute respiratory acidosis. This difference is due to the compensatory increase in the plasma HCO_3^- concentration seen with chronic respiratory acidosis.

Values between the stippled areas, on the other hand, represent mixed acid-base disturbances. This can be appreciated by plotting the three mixed disorders described above: point A lies between respiratory alkalosis and metabolic acidosis, point B between respiratory acidosis and metabolic alkalosis (even though the pH is normal), and point C between metabolic and respiratory acidosis.

As mentioned above, the diagnostic approach used in this and the following four chapters is based upon the observed in vivo compensatory responses of patients with the different acid-base disorders [5,7–10]. In vitro measurements such as whole blood buffer base, standard bicarbonate, and base deficit offer no advantages and frequently are confusing [12]. Consequently, they will not be used in this text.

CLINICAL USE OF HYDROGEN CONCENTRATION

Although the acidity of the blood is measured in terms of pH, it is somewhat difficult to use logarithms at the bedside. In contrast, the calculation of the H^+ concentration is much easier. As stated in Eq. (17-3),

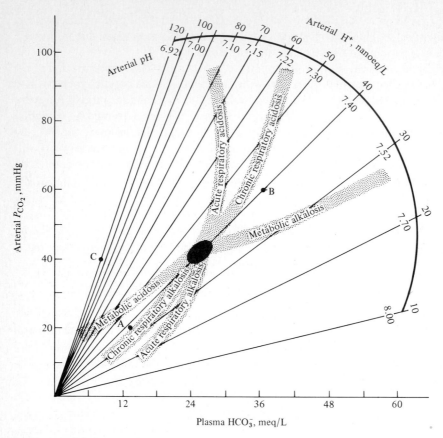

Figure 17-1 Acid-base map describing the relationships between the arterial pH, H^+ concentration, P_{CO_2}, and HCO_3^- concentration. The dark area in the center represents the range of normal values for these parameters; the stippled areas represent the different simple acid-base disturbances. Points A, B, and C indicate the three mixed acid-base disorders discussed in the text. (From J. T. Harrington, J. J. Cohen, and J. P. Kassirer: Mixed acid-base disturbances, in *Acid/Base,* J. J. Cohen and J. P. Kassirer (eds.), Little, Brown, Boston, 1982.)

$$[H^+] = 24 \times \frac{P_{CO_2}}{[HCO_3^-]}$$

If the normal P_{CO_2} is 40 mmHg and HCO_3^- concentration is 24 meq/L, then the normal H^+ concentration is 40 nanoeq/L. To use this formula, one only has to know how to convert the measured pH into H^+ concentration, a process involving simple calculations (Table 17-1) [13]. If one begins at a pH of 7.40 and H^+ concentration of 40 nanoeq/L, then for every 0.10 increase in pH, one must multiply the H^+ concentration by 0.8; for every 0.10 decrease in pH, one must multiply the H^+ concentration by 1.25. For example,

$$pH = 7.30 \quad [H^+] = 40 \times 1.25 = 50 \text{ nanoeq/L}$$

$$pH = 7.20 \quad [H^+] = 40 \times 1.25 \times 1.25 = 63 \text{ nanoeq/L}$$

$$pH = 7.50 \quad [H^+] = 40 \times 0.8 = 32 \text{ nanoeq/L}$$

Values at less than 0.10-unit steps can be estimated from interpolation. A pH of 7.27 is three-tenths of the way between 7.30 and 7.20. Since the H^+ concentration increases by 13 nanoeq/L (from 50 to 63 nanoeq/L) as the pH falls from 7.30 to 7.20, the H^+ concentration at a pH of 7.27 can be calculated from

$$[H^+] = 50 + 0.3 \times 13 = 54 \text{ nanoeq/L}$$

The following example illustrates how this equation can be used in the clinical setting. Suppose a patient with salicylate intoxication is found to have the following arterial values, which are consistent with a mild metabolic acidosis:

$$pII = 7.32$$

$$P_{CO_2} = 30 \text{ mmHg}$$

$$[HCO_3^-] = 15 \text{ meq/L}$$

An important facet of therapy in this disorder is to alkalinize the blood, which will decrease the concentration of salicylate in the tissues (see Chap. 19). If the aim is to raise the arterial pH to 7.45 (H^+ concentration equal to 36 nanoeq/L), to what level does the plasma HCO_3^- concentration have to be increased to achieve this goal? If the P_{CO_2} remains constant, then

$$[H^+] = 24 \times \frac{P_{CO_2}}{[HCO_3^-]}$$

$$36 = 24 \times \frac{30}{[HCO_3^-]}$$

$$[HCO_3^-] = 20 \text{ meq/L}$$

POTASSIUM BALANCE IN ACID-BASE DISORDERS

The internal distribution of K^+ can be influenced by changes in the arterial pH. In acidemic states, for example, some of the excess H^+ is buffered intracellularly. To maintain electroneutrality, K^+ moves into the extracellular fluid, raising the plasma K^+ concentration. These cation shifts are reversed in alkalemia as the plasma K^+ concentration tends to fall. These changes, however, are generally of minor clinical importance except in some forms of metabolic acidosis (see page 254) [14]. In this setting, monitoring the plasma K^+ concentration during correction of the acidemia may be extremely important since potentially serious degrees of hypokalemia can occur.

PROBLEMS

17-1 Convert the following values for arterial pH to H^+ concentration:
(a) 7.60
(b) 7.15
(c) 7.24

17-2 What acid-base disorders are represented by the following sets of arterial blood tests:

	pH	P_{CO_2}, mmHg	$[HCO_3^-]$, meq/L
(a)	7.32	28	14
(b)	7.47	20	14
(c)	7.08	49	14
(d)	7.51	49	38

17-3 A patient with severe diarrhea has the following results from laboratory tests:

$$\text{Arterial pH} = 6.98$$

$$P_{CO_2} = 13 \text{ mmHg}$$

$$[HCO_3^-] = 3 \text{ meq/L}$$

What is the acid-base disorder? To get the patient out of danger, the initial aim of therapy is to increase the pH to 7.20. Assuming the P_{CO_2} remains constant, to what level must the plasma HCO_3^- concentration be increased to reach a pH of 7.20? If the P_{CO_2} increased to 18 mmHg with therapy, then what HCO_3^- concentration would be required for the pH to reach 7.20?

REFERENCES

1. Biswas, C. K., J. M. Ramos, B. Agroyannis, and D. N. S. Kerr: Blood gas analysis: effect of air bubbles in syringe and delay in estimation, Br. Med. J., 1:923, 1982.
2. Hood, I., and E. J. M. Campbell: Is pK OK?, N. Engl. J. Med., 306:864, 1982.
3. Trenchard, D., M. I. M. Noble, and A. Guz: Serum carbonic acid pK_1' abnormalities in patients with acid-base disturbances, Clin. Sci., 32:189, 1967.
4. Lennon, E. J., J. Lemann, Jr., and J. R. Litzow: The effects of diet and stool composition on the net external acid balance of normal subjects, J. Clin. Invest., 45:1601, 1966.
5. Polak, A., G. D. Haynie, R. M. Hays, and W. B. Schwartz: Effects of chronic hypercapnia on electrolyte and acid-base equilibrium. I. Adaptation, J. Clin. Invest., 40:1223, 1961.
6. Pierce, N. F., D. S. Fedson, K. L. Brigham, R. C. Mitra, R. B. Sack, and A. Mondal: The ventilatory response to acute base deficit in humans, Ann. Intern. Med., 72:633, 1970.
7. Bushinsky, D. A., F. L. Coe, C. Katzenberg, J. P. Szidon, and J. H. Parks: Arterial P_{CO_2} in chronic metabolic acidosis, Kidney Int., 22:311, 1982.
8. Javaheri, S., N. S. Shore, B. D. Rose, and H. Kazemi: Compensatory hypoventilation in metabolic alkalosis, Chest, 81:296, 1982.
9. Arbus, G. S., L. A. Hebert, P. R. Levesque, B. E. Etsen, and W. B. Schwartz: Characterization and clinical application of "the significance band" for acute respiratory alkalosis, N. Engl. J. Med., 280:117, 1969.
10. Gennari, J. F., M. B. Goldstein, and W. B. Schwartz: The nature of the renal adaptation to chronic hypocapnia, J. Clin. Invest., 51:1722, 1972.
11. Gabow, P. A., R. J. Anderson, D. E. Potts, and R. W. Schrier: Acid-base disturbances in the salicylate-intoxicated adult, Arch. Intern. Med., 138:1481, 1978.

12. Schwartz, W. B., and A. S. Relman: A critique of the parameters used in the evaluation of acid-base disorders: "whole blood buffer base" and "standard bicarbonate" compared with blood pH and plasma bicarbonate concentration, N. Engl. J. Med., 268:1382, 1963.
13. Fagan, T. J.: Estimation of hydrogen ion concentration (letter), N. Engl. J. Med., 288:915, 1973.
14. Adrogué, H. J., and N. E. Madias: Changes in plasma potassium concentration during acute acid-base disturbances, Am. J. Med., 71:456, 1981.

EIGHTEEN

METABOLIC ALKALOSIS

The introduction to acid-base disorders presented in Chap. 17 should be read before proceeding with this discussion. Primary metabolic alkalosis is characterized by an elevation in the arterial pH (or a reduction in the H^+ concentration), an increase in the plasma HCO_3^- concentration, and compensatory hypoventilation, resulting in a rise in the P_{CO_2}. A high HCO_3^- concentration, however, is not diagnostic of metabolic alkalosis since it can also represent the renal compensation to chronic respiratory acidosis. These disorders can be differentiated by measurement of the systemic pH which is reduced in chronic respiratory acidosis. In addition, a plasma HCO_3^- concentration of 40 meq/L or more is indicative of metabolic alkalosis since this level is greater than that achieved by the renal compensation to chronic hypercapnia (see Fig. 20-3).

PATHOPHYSIOLOGY

The pathophysiology of metabolic alkalosis is most easily understood if we ask two separate questions:

1. How do patients become alkalotic?
2. Why do they stay alkalotic?

Generation of Metabolic Alkalosis

A primary elevation in the plasma HCO_3^- concentration most often results from enhanced H^+ loss from the gastrointestinal tract (primarily from gastric secretions) or in the urine (Table 18-1). These H^+ ions are derived from the intracellular dissociation of H_2CO_3 (see page 227):

$$CO_2 + H_2O \rightleftharpoons H_2CO_3 \rightleftharpoons H^+ + HCO_3^-$$

Table 18-1 Causes of metabolic alkalosis

H^+ loss
  A. Gastrointestinal loss
      1. Removal of gastric secretions (vomiting or nasogastric suction)†
      2. Congenital chloridorrhea
  B. Renal loss
      1. Mineralocorticoid excess†
      2. Diuretics†
      3. Postchronic hypercapnia
      4. Low chloride intake
      5. High dose carbenicillin or other penicillin derivative
      6. Hypercalcemia and hypoparathyroidism (including the milk-alkali syndrome)
  C. H^+ movement into the cells
      1. Hypokalemia†
      2. ?Refeeding

HCO_3^- retention
  A. Massive blood transfusion
  B. Administration of $NaHCO_3$
  C. Milk-alkali syndrome

Contraction alkalosis†

†Most common causes.

Extracellular fluid

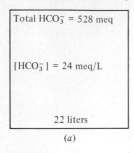

(a)

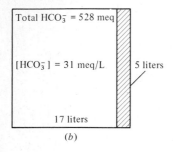

(b)

Figure 18-1 Mechanism of contraction alkalosis. (*a*) The volume and HCO_3^- concentration of the extracellular fluid in a 70-kg man whose normal extracellular volume is increased from 17 to 22 liters because of congestive heart failure. (*b*) If diuretics induce the loss of the excess 5 liters of isotonic NaCl, there will be a selective reduction in the extracellular volume. Since the quantity of extracellular HCO_3^- is initially unchanged, the HCO_3^- concentration in the extracellular fluid will increase from 24 to 31 meq/L.

Thus, for each milliequivalent of H^+ lost, there will be an equimolar generation of HCO_3^-.

Hyperbicarbonatemia can also be produced by the administration of HCO_3^-, H^+ movement into the cells, and certain forms of volume contraction. A transcellular H^+ shift typically occurs with hypokalemia. As the plasma K^+ concentration falls, K^+ moves out of the cells into the extracellular fluid. To maintain electroneutrality, there is a reciprocal shift of H^+ (and Na^+) into the cells [1–3]. The net effect is an extracellular alkalosis with a paradoxical intracellular acidosis [1]. K^+ repletion can reverse the H^+ shift and at least partially restore the extracellular pH toward normal [2,3].

A *contraction alkalosis* occurs when NaCl and water are lost but not HCO_3^-. In this setting, which is most commonly due to diuretics, the extracellular volume is diminished but the quantity of extracellular HCO_3^- is initially unchanged. As a result, the plasma HCO_3^- concentration rises (Fig. 18-1) [4].

Maintenance of Metabolic Alkalosis

The kidney possesses the ability to correct a metabolic alkalosis by excreting the excess HCO_3^- in the urine. The capacity for HCO_3^- excretion is quite high. Normal subjects given 1000 meq of $NaHCO_3$ per day for 2 weeks develop only a minor increase in the plasma HCO_3^- concentration as almost all the HCO_3^- is excreted in the urine [5]. Since a much smaller HCO_3^- load is presented to the body by the disorders which cause metabolic alkalosis, *the perpetuation of metabolic alkalosis requires an impairment in renal HCO_3^- excretion* (Table 18-2). This can result from

Table 18-2 Primary factors responsible for the reduction in HCO$_3^-$ excretion necessary for the perpetuation of metabolic alkalosis

Decreased HCO$_3^-$ excretion
  A. Decreased filtration
      1. Renal failure
  B. Increased reabsorption
      1. Effective circulating volume depletion
      2. K$^+$ depletion

decreased HCO$_3^-$ filtration, as in renal failure, or, as is most commonly the case, from enhanced HCO$_3^-$ reabsorption.

HCO$_3^-$ is reabsorbed by H$^+$ secretion from the tubular cell into the lumen. H$^+$ secretion and HCO$_3^-$ reabsorption are enhanced in three situations: acidemia (either metabolic or respiratory), volume depletion, and hypokalemia (see Chap. 12). In acidemic states, the increase in H$^+$ secretion represents an attempt to excrete the excess H$^+$ and return the pH toward normal. In contrast, the augmentation of H$^+$ secretion and HCO$_3^-$ reabsorption in the presence of volume and K$^+$ depletion acts to prevent the restoration of acid-base balance in patients with metabolic alkalosis.

Volume depletion The mechanism by which volume depletion augments H$^+$ secretion and consequently HCO$_3^-$ reabsorption in metabolic alkalosis is illustrated in Fig. 18-2. In the presence of hypovolemia, Na$^+$ reabsorption is enhanced in an attempt to restore normovolemia (see Chap. 9). To maintain electroneutrality, Na$^+$ reabsorption must be accompanied by the reabsorption of Cl$^-$, or by the secretion of H$^+$ or, to a much lesser degree, K$^+$. Of the 140 meq of Na$^+$ in each liter of glomerular filtrate, only 115 meq can be reabsorbed with Cl$^-$, and the remaining 25 meq must be reabsorbed in exchange for H$^+$ or K$^+$. In metabolic alkalosis, the increase in the plasma HCO$_3^-$ concentration is associated with a decrease in the plasma Cl$^-$ con-

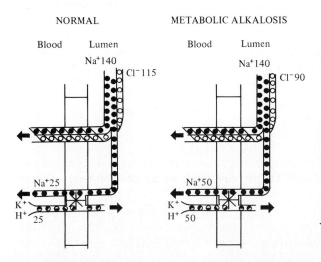

NORMAL METABOLIC ALKALOSIS

Blood Lumen   Blood Lumen

Na$^+$140   Na$^+$140

Cl$^-$115   Cl$^-$90

Na$^+$25   Na$^+$50

K$^+$ K$^+$
H$^+$ 25   H$^+$ 50

Figure 18-2 Effect of hypochloremia on H$^+$ and K$^+$ secretion in metabolic alkalosis. When the kidney is sodium-avid, virtually all the Cl$^-$ will be reabsorbed. To maintain electroneutrality, the remaining Na$^+$ must be reabsorbed in exchange for H$^+$ and K$^+$. When the Cl$^-$ concentration is reduced, as in metabolic alkalosis, H$^-$ secretion increases, resulting in enhanced HCO$_3^-$ reabsorption. *(Adapted from J. P. Kassirer and W. B. Schwartz, Am. J. Med., 40:10, 1966.)*

centration, e.g., due to Cl^- loss secondary to vomiting or diuretic therapy. Thus, less Cl^- is available for reabsorption with Na^+. As a result, Na^+ conservation requires an increase in H^+ secretion and therefore in HCO_3^- reabsorption. The net effect is that the body chooses to maintain volume at the expense of the extracellular pH, the correction of which would require the loss of $NaHCO_3$ in the urine, resulting in further volume depletion [6,7]. The almost total reabsorption of filtered HCO_3^- that may occur in this setting is manifested by the paradoxical finding of an *acid urine pH despite the presence of an extracellular alkalemia.*

It is important to remember what we mean by volume depletion. We are referring to the *effective circulating volume,* i.e., that volume that is effectively perfusing the tissues (see Chap. 9). Effective circulating volume depletion can be produced not only by actual volume loss (as with vomiting or diuretics) but also by a primary reduction in cardiac output in heart failure or by the sequestration of the plasma water in extravascular sites, e.g., as ascites in hepatic cirrhosis. In each of these conditions, the kidney will attempt to restore normal tissue perfusion by retaining Na^+. This occurs even though the total extracellular fluid volume is increased in patients with congestive heart failure or hepatic cirrhosis.

Hypokalemia As described above, hypokalemia is associated with an intracellular acidosis; it is this increase in the renal tubular cell H^+ concentration that appears to be at least in part responsible for the enhanced H^+ secretion and HCO_3^- reabsorption. In addition, severe hypokalemia may cause, by an unknown mechanism, a reduction in chloride reabsorption in the distal nephron [8,9]. As a result, Na^+ reabsorption at this site is preferentially associated with H^+ secretion and, therefore, an increased tendency to the development of metabolic alkalosis [9].

Summary The combined effects of fractional Na^+ reabsorption, which is enhanced by volume depletion and reduced by volume expansion, and K^+ balance on HCO_3^- reabsorption are depicted in Fig. 18-3. These relationships have important therapeutic implications since the removal of the stimulus to increased HCO_3^- reabsorption (volume depletion or hypokalemia) allows HCO_3^- to be lost in the urine, resulting in the correction of the alkalosis.

It should be emphasized that hypovolemia has *two separate and independent* effects in metabolic alkalosis. To the degree that renal HCO_3^- reabsorption is

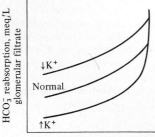

Figure 18-3 HCO_3^- reabsorption as a function of the fraction of the filtered Na^+ that is reabsorbed and K^+ balance. HCO_3^- reabsorption is enhanced in the presence of volume depletion (high fractional Na^+ reabsorption) and hypokalemia. *(Redrawn from N. A. Kurtzman, M. G. White, and P. W. Rogers, Arch. Intern. Med., 131:702, 1973. By permission of the American Medical Association, copyright 1973.)*

increased, volume depletion will tend to perpetuate the alkalosis regardless of the underlying cause. However, hypovolemia will *produce* an alkalosis by contraction only when the fluid lost contains more chloride than bicarbonate, as with the administration of diuretics. In comparison, bleeding will not lead to metabolic alkalosis since the concentrations of chloride and bicarbonate that are lost are similar to those in the plasma.

Respiratory Compensation

The development of alkalemia is sensed by the chemoreceptors controlling ventilation, resulting in hypoventilation and an appropriate increase in the P_{CO_2}. On the average, the P_{CO_2} increases 0.7 mmHg for every 1.0-meq/L increment in the plasma HCO_3^- concentration [10]. Thus, if the plasma HCO_3^- concentration is 34 meq/L (or 10 meq/L greater than normal), there should be a 7-mmHg increase in the P_{CO_2}—from the normal 40 mmHg to 47 mmHg. Values significantly different from the predicted P_{CO_2} represent superimposed respiratory acidosis or alkalosis.

The respiratory compensation may be partially or completely impaired in the presence of hypoxemia or underlying respiratory alkalosis. For example, patients with heart failure and cirrhosis of the liver frequently develop metabolic alkalosis due to diuretic therapy. However, both of these disorders may be associated with a primary respiratory alkalosis (see Chap. 21), which can prevent the appropriate compensatory hypoventilation.

Hypoxemia is a potent stimulus to ventilation. Thus, the respiratory compensation to metabolic alkalosis is frequently self-limiting since hypoventilation tends to lower the P_{O_2}. The level at which this is likely to occur can be estimated from the relationship between the arterial P_{O_2} and P_{CO_2} (see page 454):

$$\text{Arterial } P_{O_2} = P_{iO_2} - 1.25 P_{CO_2} - 10 \qquad (18\text{-}1)$$

where the P_{iO_2} refers to the partial pressure of oxygen in the inspired air (normally 150 mmHg at sea level) and 10 refers to the normal alveolar-arterial oxygen gradient (in mmHg) [11]. From this relationship we can see that every 1.0-mmHg increase in the arterial P_{CO_2} induced by hypoventilation will reduce the arterial P_{O_2} by 1.25 mmHg.

In humans, hypoxemia begins to stimulate ventilation when the P_{O_2} falls below 70 to 80 mmHg (see Fig. 21-2) [12]. From Eq. (18-1), we can calculate how high the P_{CO_2} can rise in metabolic alkalosis before the P_{O_2} reaches 70 mmHg and prevents further hypoventilation:

$$70 = 150 - 1.25 P_{CO_2} - 10$$

$$P_{CO_2} = 56 \text{ mmHg}$$

Thus, it is unusual for the P_{CO_2} to exceed 55 mmHg in a patient with metabolic alkalosis who is breathing room air. This limitation occurs even earlier in a patient with underlying pulmonary disease and resting hypoxemia. Whenever hypoxemia limits the hypoventilatory response to metabolic alkalosis, the administration of oxygen will increase the P_{O_2}, thereby removing the hypoxemic stimulus to ventilation and allowing further compensatory CO_2 retention.

ETIOLOGY

Metabolic alkalosis can be produced by a variety of disorders (Table 18-1), most of which are characterized by enhanced HCO_3^- reabsorption due to volume and/or K^+ depletion.

Gastrointestinal Hydrogen Loss

Removal of gastric secretions Gastric juice contains high concentrations of HCl and lesser concentrations of KCl. Each milliequivalent of H^+ secreted generates 1 meq of HCO_3^-. Under normal conditions, the increase in the plasma HCO_3^- concentration is only transient since the entry of the H^+ into the duodenum stimulates an equal amount of pancreatic HCO_3^- secretion. However, if the gastric juice is removed, either by vomiting or nasogastric suction, there is no stimulus to HCO_3^- secretion. The result is an increase in the plasma HCO_3^- concentration and metabolic alkalosis [6]. The tendency toward alkalosis is enhanced by the concomitant volume and K^+ depletion which frequently are present.

A similar sequence can occur when patients with advanced renal failure are treated with antacids containing magnesium hydroxide and, for hyperkalemia, a cation-exchange resin [12a]. Antacids do not normally cause a metabolic alkalosis since the hydroxide component buffers gastric H^+ whereas the magnesium combines with HCO_3^- secreted from the pancreas to form insoluble $MgCO_3$, which cannot be absorbed. Thus, both H^+ and HCO_3^- are lost and acid-base balance is unaffected. In the presence of the cation-exchange resin, however, a Mg^{2+}-resin complex is formed, leaving HCO_3^- soluble in the intestinal lumen and able to be absorbed. The renal failure is important in perpetuating the alkalosis since it prevents the excretion of the excess HCO_3^-.

Congenital chloridorrhea Since the enteric fluids below the stomach are alkaline, diarrhea usually is associated with metabolic acidosis. In a rare condition, *congenital chloridorrhea,* an acid diarrhea is formed and alkalosis is produced [13]. This appears to be due to a specific intestinal defect in Cl^- reabsorption and HCO_3^- secretion. The fecal Cl^- concentration is very high, about 140 meq/L, and the fecal pH is low.

Renal Hydrogen Loss

Mineralocorticoid excess and hypokalemia Aldosterone stimulates the secretion of H^+ and K^+, as well as the reabsorption of Na^+, in the cortical collecting tubule (see Chap. 8). Therefore, the excessive secretion of aldosterone (or any other mineralocorticoid) can lead to H^+ loss and metabolic alkalosis. Hypokalemia due to urinary K^+ loss plays an important contributory role in this process [7,9,14]. If K^+ depletion is prevented, there is a lesser increment in net H^+ excretion and only a minor elevation in the plasma HCO_3^- concentration [14]. The potentiating effect of hypokalemia may be due both to the associated intracellular acidosis [1] and to the decrease in distal chloride reabsorption [9].

For these effects to occur, there must be adequate delivery of Na^+ and water to the distal secretory site (see page 259). When distal delivery is reduced due to effective circulating volume depletion, K^+ and H^+ secretion may be relatively unchanged despite the presence of secondary hyperaldosteronism [15]. Thus, uncomplicated patients with heart failure or hepatic cirrhosis typically have a normal plasma K^+ concentration and are not alkalemic. However, hypokalemia and metabolic alkalosis may rapidly ensue if distal delivery is enhanced by the administration of diuretics.

Primary mineralocorticoid excess occurs in a variety of uncommon conditions, including primary hyperaldosteronism, Cushing's syndrome, congenital adrenal hyperplasia, Bartter's syndrome, chronic licorice ingestion, and renin-secreting tumors. These disorders are discussed in Chap. 27 since hypokalemia is typically more prominent than metabolic alkalosis.

Diuretics Most diuretics, other than those that are K^+-sparing, tend to induce metabolic alkalosis. Both volume contraction (Fig. 18-1) and, more importantly, increased urinary H^+ loss contribute to this problem [4,16,17]. The latter is primarily due to enhanced distal H^+ secretion that results from the interplay of three factors: hypersecretion of aldosterone due to the associated hypovolemia; increased distal flow since furosemide, ethacrynic acid, and the thiazides inhibit NaCl and water reabsorption proximal to the secretory site (see Chap. 16); and the concomitant development of hypokalemia.

Posthypercapnic alkalosis Chronic respiratory acidosis is associated with a compensatory increase in H^+ secretion and therefore in renal HCO_3^- reabsorption (see Fig. 12-8) [18]. If a patient with this disorder is treated with mechanical ventilation, the P_{CO_2} can be rapidly returned toward normal. However, the plasma HCO_3^- concentration will remain elevated, resulting in metabolic alkalosis. Since this acute increase in arterial pH can produce serious neurologic abnormalities and death, the P_{CO_2} should be lowered slowly and carefully in patients with chronic hypercapnia [19].

In addition to the increase in renal HCO_3^- reabsorption, chronic respiratory acidosis tends to produce, in an unknown manner, Cl^- loss in the urine, resulting in hypochloremia and volume depletion [20]. If the P_{CO_2} is returned toward normal, the kidney will be unable to excrete all the excess HCO_3^- in the plasma and the alkalosis will persist until Cl^- balance is restored [20].

Low chloride intake Metabolic alkalosis has resulted from the inadvertent administration to infants of formula containing Na^+ but almost no Cl^- [21]. The ensuing Cl^- deficiency diminishes the amount of Cl^- in the renal tubule that is available for reabsorption. Thus, more tubular Na^+ reabsorption must occur in exchange for secreted H^+ to maintain electroneutrality, leading to increased urinary H^+ loss and metabolic alkalosis.

High-dose carbenicillin or penicillin A similar problem may occur with the Na^+ load associated with the intravenous administration of high doses of Na^+ carbenicillin or other penicillin derivative [22,23]. Carbenicillin, for example, contains 4.7 meq of Na^+ per gram or 141 meq if 30 grams are given. As the Na^+ carbenicillin is filtered, carbenicillin acts as a nonreabsorbable anion. Consequently, some distal Na^+ reabsorption must occur in exchange for K^+ and H^+, resulting in hypokalemia and metabolic alkalosis [24].

Hypercalcemia and hypoparathyroidism Renal H^+ secretion and HCO_3^- reabsorption are increased by Ca^{2+} and reduced by parathyroid hormone (PTH) [25,26]. These effects tend to balance out, and the arterial pH is relatively normal when both parameters are either elevated (PTH-induced hypercalcemia) or reduced (hypocalcemia due to hypoparathyroidism) [27,28]. However, a mild metabolic alkalosis may occur with non-PTH-induced hypercalcemia since both the elevation in the plasma Ca^{2+} concentration and low PTH levels (hypercalcemia suppresses PTH secretion) promote net HCO_3^- reabsorption [29].

Similar factors may be important in the *milk-alkali syndrome* in which the chronic ingestion of milk and antacids containing calcium carbonate leads to hypercalcemia and metabolic alkalosis [30,31]. In normal subjects, the HCO_3^- generated from the ingested carbonate (CO_3^{2-}) would be rapidly excreted in the urine. However, hypercalcemia, reduced PTH secretion, and renal insufficiency (which is commonly present and probably due to the hypercalcemia) all contribute to the maintenance of the metabolic alkalosis in this disorder [31].

Intracellular Shift of Hydrogen

Hypokalemia Hypokalemia is a frequent finding in patients with metabolic alkalosis. This association is due to several factors. As described above, hypokalemia contributes to the production and maintenance of metabolic alkalosis by causing H^+ to move into cells [1–3] and by increasing renal H^+ secretion and HCO_3^- reabsorption [7–9]. In addition, the common causes of metabolic alkalosis (diuretics, vomiting, mineralocorticoid excess) directly induce K^+ as well as H^+ loss.

Refeeding Patients who are refed carbohydrate after a prolonged fast can develop metabolic alkalosis [32]. Since there is neither volume contraction nor a demonstrable increase in urinary acid excretion, it has been proposed that an intracellular shift of H^+ may be responsible. The mechanism by which this might occur is unknown.

Refeeding is also associated with Na^+ retention, which is responsible for perpetuation of the alkalosis [32]. Increased secretion of insulin, resulting from the carbohydrate ingestion, may contribute to this response.

Retention of Bicarbonate

Because of the ability of the kidney to excrete HCO_3^-, it is difficult to produce more than a small increment in the plasma HCO_3^- concentration by the chronic administration of as much as 1000 meq of HCO_3^- per day [5]. However, a significant alkalemia can be produced by the acute infusion of base or by the chronic administration of alkali in a patient in whom renal HCO_3^- excretion is impaired (as in the milk-alkali syndrome described above).

Massive blood transfusion Organic anions are rapidly metabolized in the body to HCO_3^- [33–35]. For example,

$$CH_3CHOHCOO^- \text{ (lactate)} + 3O_2 \rightarrow 2CO_2 + 2H_2O + HCO_3^-$$

The same is true for acetate, citrate, and, in the presence of insulin, the anions of the ketoacids. Therefore, the administration of these anions is the same as giving HCO_3^-. Most bank blood is anticoagulated with acid-citrate-dextran. Each unit (500 mL) of blood contains 16.8 meq of citrate which can be metabolized into HCO_3^-. Although citric acid also is present, it has only a transient effect on the systemic pH since it is rapidly converted into CO_2 and H_2O. As a rule, more than eight units of blood must be given acutely to produce a significant elevation in the arterial pH [36].

A similar problem may occur after the administration of some human plasma protein fractions (Protenate, Plasmatein) which are used as volume expanders [37]. The solutions contain acetate (as a source of HCO_3^-) and citrate (as a preservative) in a total concentration of 40 to 50 meq/L. The metabolism of these anions to HCO_3^- can lead to metabolic alkalosis.

Administration of sodium bicarbonate The most common indication for $NaHCO_3$ is in the treatment of metabolic acidosis. However, $NaHCO_3$ can result in a metabolic alkalosis if given in excess amounts. This is particularly true in lactic acidosis [38] and ketoacidosis [35], in which the metabolism of the associated anion, lactate or β-hydroxybutyrate, results in the regeneration of the HCO_3^- initially lost in buffering the H^+ load. Since HCO_3^- is replaced by an anion which is metabolized back to HCO_3^-, there is no loss of *potential* HCO_3^- from the body (excluding those anions excreted in the urine) [35]. Therefore, correction of the underlying abnormality— by the restoration of tissue perfusion or the administration of insulin—results in the return of acid-base balance toward normal. The excessive use of exogenous HCO_3^- in these disorders can create an excess of potential HCO_3^- and lead to the development of metabolic alkalosis. In extreme cases, the systemic pH has reached 7.90 with the plasma HCO_3^- concentration exceeding 60 to 70 meq/L after the indiscriminate use of $NaHCO_3$ during cardiopulmonary resuscitation [38].

Contraction Alkalosis

NaCl and water loss without HCO_3^- can result in a contraction alkalosis with the ensuing volume depletion then perpetuating the alkalosis. This is most commonly seen after the administration of diuretics, such as furosemide or ethacrynic acid [4]. It may also occur with vomiting (even in patients with achlorhydria in whom NaCl replaces HCl in the gastric secretions) or with cystic fibrosis, where the sweat Cl^- concentration can exceed 70–100 meq/L [38a,38b]. As illustrated in the hypothetical example in Fig. 18-1, a 5-liter decrease in the extracellular volume produces an initial 7-meq/L elevation in the plasma HCO_3^- concentration. This direct effect of contraction, however, is partially counteracted by the release of H^+ from the cell buffers, resulting in a lesser increment in the plasma HCO_3^- concentration and pH [16]:

$$HCO_3^- + HBuf \rightarrow Buf^- + H_2CO_3 \rightarrow CO_2 + H_2O$$

SYMPTOMS

Patients with metabolic alkalosis may be asymptomatic or complain of symptoms related either to volume depletion, e.g., weakness, muscle cramps, and postural dizziness, or to hypokalemia, e.g., polyuria, polydipsia, and muscle weakness. The neurologic symptoms of paresthesias, carpopedal spasm, and light-headedness seen with respiratory alkalosis occur less frequently in patients with metabolic alkalosis. This may be due to the fact that HCO_3^-, a polar compound, crosses the blood-brain barrier much more slowly than the lipid-soluble CO_2, producing a lesser increase in the cerebrospinal fluid pH [39].

DIAGNOSIS

Urine Chloride Concentration

The etiology of metabolic alkalosis almost always is obtainable from the history. If there is no pertinent history, then the most likely diagnoses are surreptitious vomiting or diuretic ingestion or one of the causes of mineralocorticoid excess. The urine Cl^- concentration can be helpful in differentiating between these conditions (Table 18-3). The combination of hypovolemia and hypochloremia in patients with vomiting or those taking diuretics should induce maximum renal Cl^- conservation, usually lowering the urine Cl^- concentration to less than 15 meq/L. (This excludes the period during which the diuretic is acting, when Cl^- excretion is elevated.) These patients may also show the physical findings of volume depletion such as reduced skin turgor, flat neck veins, and postural hypotension. In contrast, the physical signs of hypovolemia are absent and the urine Cl^- concentration exceeds 20 meq/L in patients with mineralocorticoid excess or HCO_3^- retention who are generally volume-expanded.

Metabolic alkalosis is the major clinical setting in which the urine Cl^- concentration is a more accurate estimate of volume status than is the urine Na^+ concentration. Although hypovolemia leads to Na^+ retention, this may be counteracted by the necessity for Na^+ to be excreted with the excess HCO_3^-. As depicted in Fig. 18-3, HCO_3^- reabsorption increases with enhanced Na^+ reabsorption. In a hypovolemic

Table 18-3 Urine Cl⁻ concentration in patients with metabolic alkalosis

Urine Cl⁻ concentration	
Less than 15 meq/L	Greater than 20 meq/L
Vomiting	Mineralocorticoid excess
Nasogastric suction	Diuretics (early)
Diuretics (late)	HCO_3^- load
Posthypercapnia	Severe hypokalemia (plasma $[K^+]$
Low chloride intake	<2.0 meq/L)
Refeeding	

patient, the maximum reabsorptive capacity for HCO_3^- may rise from the normal of 26 to 28 meq/L up to, for example, 36 meq/L of glomerular filtrate (see Fig. 5-9). If, however, the plasma HCO_3^- concentration is 42 meq/L, $NaHCO_3$ will be lost in the urine despite the presence of volume depletion. In this setting, the urine Na^+ concentration will be elevated but the urine Cl^- concentration will remain appropriately low.

The urine Cl^- concentration may not be useful in patients who are unable to maximally conserve Cl^- because of a defect in tubular reabsorption. This abnormality may occur with renal insufficiency (see page 281) or severe hypokalemia (plasma K^+ concentration below 2.0 meq/L) in which distal Cl reabsorption is impaired [8,9,40]. In these settings, the urine Cl^- concentration may be elevated despite the presence of volume depletion.

Metabolic Alkalosis versus Respiratory Acidosis

An elevated plasma HCO_3^- concentration, hypercapnia, hypoxemia, and an increase in the P_{CO_2} after oxygen may be found in chronic respiratory acidosis as well as in metabolic alkalosis (see Chap. 20). If uncomplicated, these disorders can be easily differentiated by measuring the arterial pH. However, this differential becomes more difficult when the patient with possible chronic hypercapnia develops a superimposed metabolic alkalosis. For example, consider the following case history:

Case history 18-1 A 45-year-old man with a long smoking history reports 1 week of recurrent vomiting and has the following arterial blood values on room air:

$$pH = 7.49$$

$$P_{CO_2} = 55 \text{ mmHg}$$

$$[HCO_3^-] = 40 \text{ meq/L}$$

$$P_{O_2} = 68 \text{ mmHg}$$

COMMENT The high P_{CO_2} is compatible with either an appropriate respiratory compensation to metabolic alkalosis or underlying lung disease in this chronic smoker. The simplest way to establish the correct diagnosis is to treat the metabolic alkalosis and follow the P_{CO_2}. If the P_{CO_2} remains elevated, then the patient has an underlying respiratory acidosis. It also may be helpful to calculate the alveolar-arterial (A-a) oxygen gradient (normal is 5 to 15 mmHg) [11]. If we solve Eq. (18-1) for the (A-a) gradient, we have

$$(\text{A-a}) \ O_2 \ \text{gradient} = P_{iO_2} - 1.25 P_{CO_2} - P_{O_2}$$

$$= 150 - (1.25 \times 55) - 68$$

$$= 13 \ \text{mmHg}$$

Since the (A-a) O_2 gradient is almost always elevated in patients with chronic respiratory acidosis, especially those with chronic obstructive pulmonary disease due to smoking, the finding of a normal (A-a) O_2 gradient suggests that this patient has a primary metabolic alkalosis and no significant underlying lung disease. However, the converse is not necessarily true. If the gradient had been increased, this would not have been diagnostic of chronic respiratory acidosis since an increased gradient can be seen in many acute and chronic pulmonary diseases not associated with CO_2 retention.

The degree of hypercapnia may also be helpful in deciding whether the patient has underlying chronic respiratory acidosis. Values above 55 to 60 mmHg are unusual in uncomplicated metabolic alkalosis because this is the level at which hypoxemia limits further hypoventilation. Rarely, patients with severe alkalemia will display more marked hypercapnia [41]. It may be that these patients have a diminished hypoxic drive to ventilation [42] so that the fall in P_{O_2} does not limit the compensatory response.

TREATMENT

Metabolic alkalosis can be corrected most easily by the urinary excretion of the excess HCO_3^-. This does not occur spontaneously because, in the patient with relatively normal renal function, HCO_3^- reabsorption is increased because of volume and/or K^+ depletion (Fig. 18-3). Therefore, *the aim of therapy is to restore volume and K^+ balance, which will reduce HCO_3^- reabsorption and allow the excess HCO_3^- to be excreted in the urine.* As will be seen, *this requires the administration of Cl^- as NaCl, KCl, or HCl* [43–45]. In addition, correction of K^+ depletion directly lowers the plasma HCO_3^- concentration because of the reciprocal shift of K^+ into and H^+ out of the cells [2,3].

Saline-Responsive Alkalosis

The most common causes of metabolic alkalosis are vomiting, nasogastric suction, and diuretics. In these disorders (and with posthypercapnia and a low chloride intake), the perpetuation of the alkalosis is due to the associated volume depletion. The mainstay of therapy is reexpansion of the effective circulating volume by the oral or intravenous administration of NaCl and water, e.g., as half-isotonic or isotonic saline. This acts to correct the alkalosis in two ways: by reversal of the contraction component and by removing the stimulus to renal Na^+ retention, thereby permitting $NaHCO_3$ excretion in the urine. The therapeutic effectiveness of this regimen can be followed at the bedside by measuring the urine pH. When $NaHCO_3$ is appropriately lost in the urine, the urine pH, frequently initially acid due to the enhanced H^+ secretion, exceeds 7.0 and may reach 8.0. If the urine Cl^- concentration is measured during this $NaHCO_3$ diuresis, it will initially remain low as the Cl^- deficit is corrected by the retention of the administered Cl^-.

The efficacy of this therapy is dependent upon the administration of Na^+ with the only reabsorable anion, Cl^- [7,43–45]. As this Na^+ enters the glomerular filtrate, it is reabsorbed with Cl^-, resulting in volume expansion. However, if Na^+ is given with another, nonreabsorbable anion, as Na_2SO_4, for example, then the reabsorption of this Na^+ in the distal nephron must be accompanied by H^+ (or K^+) secretion (see Fig. 12-10) [46]. Since H^+ secretion results in the generation of HCO_3^- in the plasma, the net effect is the addition of $NaHCO_3$ to the extracellular fluid and aggravation of the alkalemia.

Although $NaCl$ will correct the alkalosis, it will not correct the K^+ depletion which may be present. As with Na^+, the administration of K^+ with any anion other than Cl^- results in an increase in H^+ secretion, preventing the correction of the alkalosis [45,47]. This is important clinically since many of the commercial K^+ supplements contain HCO_3^-, acetate, or citrate. Only KCl will be effective.

Saline-Resistant Alkalosis

The administration of saline is occasionally ineffective in correcting the alkalosis (Table 18-4). This typically occurs in those disorders in which K^+ depletion not hypovolemia is responsible for perpetuation of the alkalosis.

Mineralocorticoid excess States of primary mineralocorticoid excess are characterized by mild volume expansion and high urinary Na^+ excretion (due to aldosterone escape; see page 144). Since renal sodium avidity is absent and volume is not the limiting factor in HCO_3^- reabsorption, it is not surprising that the alkalosis in these patients is resistant to $NaCl$ [7]. In these disorders, the combination of hypokalemia and hypersecretion of aldosterone is responsible for the increases in H^+ secretion and HCO_3^- reabsorption that are necessary for the perpetuation of the alkalosis. The correction of the hypokalemia with KCl tends to correct the alkalosis in two ways:

Table 18-4 Causes of metabolic alkalosis according to their ability to be corrected by the administration of NaCl

Saline-responsive	Saline-resistant
Vomiting	Mineralocorticoid excess
Nasogastric suction	Severe hypokalemia
Diuretics	Edematous states
Posthypercapnia	Renal failure
Low chloride intake	

by allowing increased HCO_3^- excretion, and by causing H^+ ions to move out of the cells into the extracellular fluid [2,3].

Successful treatment can also be achieved by interfering with the excessive mineralocorticoid activity [7]. This can be done by surgery, e.g., by removal of an adrenal adenoma, or by the use of a K^+-sparing diuretic such as amiloride or the aldosterone antagonist spironolactone [48].

Severe hypokalemia Patients with metabolic alkalosis and volume depletion of any cause may be resistant to saline therapy in the presence of severe K^+ depletion [40]. The deficit in total body exchangeable K^+ usually is greater than 800 to 1000 meq, and the plasma K^+ concentration generally is less than 2.0 meq/L. As described above, the urine Cl^- concentration is greater than 15 meq/L in this setting despite the presence of hypovolemia [40]. This defect in Cl^- conservation may explain why marked K^+ depletion interferes with the corrective effect of NaCl. If Cl^- reabsorption is impaired [8] and the availability of K^+ for exchange with Na^+ is limited, then Na^+ reabsorption must be accompanied by increased H^+ secretion and HCO_3^- reabsorption [9], thereby preventing a HCO_3^- diuresis.

These effects of severe hypokalemia are readily reversible. The replacement of only one-half of the K^+ deficit will normalize Cl^- reabsorption and restore saline responsiveness as the administration of saline will now be effective in correcting the alkalosis [40].

Edematous states Patients with congestive heart failure, cirrhosis with ascites, or the nephrotic syndrome may develop metabolic alkalosis, most commonly due to diuretic therapy. Since these disorders are usually associated with a reduction in the effective circulating volume, there is compensatory renal Na^+ retention and therefore an increase in HCO_3^- reabsorption. However, the administration of saline is not indicated since it may increase the degree of edema, perhaps precipitating pulmonary edema in the presence of heart failure. In this setting, acetazolamide, HCl, or dialysis may be used to correct the alkalosis.

Acetazolamide (250 to 375 mg, once or twice a day, given orally or intravenously) is a carbonic anhydrase inhibitor which increases the renal excretion of $NaHCO_3$ (see page 331). This serves the dual purpose of treating both the edema and the alkalosis. As with the use of saline in saline-responsive states, the efficacy of acetazolamide can be measured by following the urine pH. If the drug substantially increases HCO_3^- excretion, the urine pH should exceed 7.0. K^+ balance must be carefully followed since acetazolamide, as with most other diuretics, increases urinary K^+ excretion [49].

Acetazolamide can also be used in edematous patients with cor pulmonale and chronic hypercapnia. However, there are some potential problems in this setting as acute, transient elevations in the P_{CO_2} may be seen and an excess fall in the plasma HCO_3^- concentration can lead to marked *acidemia* (see page 457).

If acetazolamide is ineffective and the alkalemia is moderately severe, HCl can be used to lower the plasma HCO_3^- concentration [50]:

$$HCl + NaHCO_3 \rightarrow NaCl + H_2CO_3 \rightarrow CO_2 + H_2O$$

As with the administration of Na^+ and K^+, the infusion of H^+ is effective only if given with the reabsorbable anion Cl^-. If H^+ is given as HNO_3, $NaNO_3$ is formed from the extracellular buffering of the HNO_3:

$$HNO_3 + NaHCO_3 \rightarrow NaNO_3 + H_2CO_3 \rightarrow CO_2 + H_2O$$

The delivery of this Na^+ to the distal nephron with the nonreabsorbable anion NO_3^- results in the reabsorption of Na^+ in exchange for H^+. The net effect is the excretion of the administered H^+ and persistence of the alkalosis [51].

The amount of HCl required to normalize the plasma HCO_3^- concentration is equal to the HCO_3^- excess, which can be estimated from

$$HCO_3^-\ excess = HCO_3^-\ space \times HCO_3^-\ excess\ per\ liter$$

In metabolic alkalosis, the HCO_3^- space is approximately 50 percent of the lean body weight [52]. If the normal plasma HCO_3^- concentration is 24 meq/L, then,

$$HCO_3^-\ excess = 0.5 \times lean\ body\ weight\ (kg) \times (plasma\ [HCO_3^-] - 24)$$

Thus, in a 60-kg patient with a plasma HCO_3^- concentration of 40 meq/L,

$$HCO_3^-\ excess = 0.5 \times 60 \times (40 - 24) = 480\ meq$$

It should be noted that this formula underestimates the acid requirement of a patient in a nonsteady state. For example, if nasogastric suction is removing large amounts these losses must be added on to the initial estimate of the HCO_3^- excess.

HCl is usually given as an isotonic solution (150 meq each of H^+ and Cl^- in 1 liter of distilled water) over 8 to 24 h [50]. Since HCl is very corrosive, it is generally infused into a major vein, such as the subclavian or femoral vein. However, a peripheral vein can be safely used if the HCl is buffered in an amino acid solution and infused with a fat emulsion [53].

Ammonium chloride and arginine hydrochloride, which result in the formation of HCl, have *little clinical use* since they may lead to appreciable toxicity. Ammonium chloride is converted into HCl and ammonia in the liver. This makes this drug contraindicated in patients with advanced liver disease. In addition, an ammonia-related metabolic encephalopathy characterized by lethargy and coma may occur even in patients with normal hepatic and renal function [54]. Arginine hydrochloride, on the other hand, can induce potentially life-threatening hyperkalemia [55,56]. This effect is thought to result from the movement of cellular potassium into the extracellular fluid as the cationic arginine enters the cells.

As an alternative to HCl, peritoneal dialysis or hemodialysis can be used to remove the excess HCO_3^-. However, a low-acetate solution must be used since normal dialysis solutions contain 35 to 40 meq/L of acetate or lactate (which are metabolized in the body into HCO_3^-) [57].

Renal failure Rarely, a patient with oliguric renal failure develops metabolic alkalosis, usually due to nasogastric suction. In this setting, either HCl or dialysis can be used if the alkalemia is severe [57].

PROBLEMS

18-1 A patient with cirrhosis and ascites is admitted to the hospital with acute gastrointestinal bleeding due to ruptured esophageal varices. He is taken to surgery where a portacaval shunt is performed. He is given a total of 19 units of blood before and during the surgery. Although the ascites was removed during the surgery, it begins to reaccumulate postoperatively. His laboratory tests were normal preoperatively, but the following values are obtained 12 h after surgery:

$$\text{Arterial pH} = 7.53$$

$$P_{CO_2} = 50 \text{ mmHg}$$

$$[HCO_3^-] = 40 \text{ meq/L}$$

What is responsible for the development of the metabolic alkalosis? What would you expect the urine pH and Na^+ concentration to be? How would you correct the alkalosis?

18-2 A 45-year-old woman with peptic ulcer disease reports 6 days of persistent vomiting. On physical examination, the blood pressure is found to be 100/60 without postural change, the skin turgor is decreased, and the jugular neck veins are flat. The initial laboratory data are:

Plasma $[Na^+] = 140$ meq/L	BUN $= 80$ meq/dL
$[K^+] = 2.2$ meq/L	[Creatinine] $= 1.9$ mg/dL
$[Cl^-] = 86$ meq/L	Urine pH $= 5.0$
$[HCO_3^-] = 42$ meq/L	$[Na^+] = 2$ meq/L
Arterial pH $= 7.53$	$[K^+] = 21$ meq/L
$P_{CO_2} = 53$ mmHg	$[Cl^-] = 3$ meq/L

How would you treat this patient?

Twenty-four hours after appropriate therapy has been started, the plasma HCO_3^- concentration is 30 meq/L. The following urinary values are obtained:

$$\text{Urine } [Na^+] = 100 \text{ meq/L}$$

$$[K^+] = 20 \text{ meq/L}$$

$$[Cl^-] = 3 \text{ meq/L}$$

How do you account for the discrepancy between the high urine Na^+ concentration and the low urine Cl^- concentration?

REFERENCES

1. Adler, S., B. Anderson, and B. Zett: The effect of acute potassium depletion on muscle cell pH in vitro, Kidney Int., 2:159, 1972.
2. Cooke, R. E., W. Segar, D. B. Cheek, F. Coville, and D. Darrow: The extrarenal correction of alkalosis associated with potassium deficiency, J. Clin. Invest., 31:798, 1952.
3. Orloff, J., T. Kennedy, Jr., and R. W. Berliner: The effect of potassium in nephrectomized rats with hypokalemic alkalosis. J. Clin. Invest., 32:538, 1953.
4. Cannon, P. J., H. O. Heinemann, M. S. Albert, J. H. Laragh, and R. W. Winters: "Contraction" alkalosis after diuresis of edematous patients with ethacrynic acid, Ann. Intern. Med., 62:979, 1965.
5. Van Goidsenhoven, G., O. V. Gray, A. V. Price, and P. H. Sanderson: The effect of prolonged administration of large doses of sodium bicarbonate in man, Clin. Sci., 13:383, 1954.

6. Kassirer, J. P., and W. B. Schwartz: The response of normal man to selective depletion of hydrochloric acid, Am. J. Med., 40:10, 1966.

7. Kurtzman, N. A., M. G. White, and P. W. Rogers: Pathophysiology of metabolic alkalosis, Arch. Intern. Med., 131:702, 1973.

8. Luke, R. G., F. S. Wright, N. Fowler, M. Kashgarian, and G. H. Giebisch: Effects of potassium depletion on renal tubular chloride transport in the rat, Kidney Int., 14:414, 1978.

9. Hulter, H. N., J. F. Sigala, and A. Sebastian: K^+ deprivation potentiates the renal alkalosis-producing effect of mineralocortocoid, Am. J. Physiol., 235:F298, 1978.

10. Javaheri, S., N. S. Shore, B. D. Rose, and H. Kazemi: Compensatory hypoventilation in metabolic alkalosis, Chest, 81:296, 1982.

11. Snider, G. L.: Interpretation of the arterial oxygen and carbon dioxide partial pressure, Chest, 63:801, 1973.

12. Weil, J. V., E. Byrne-Quinn, E. Sadal, W. O. Friesen, B. Underhill, G. F. Filley, and R. F. Grover: Hypoxic ventilatory drive in normal man, J: Clin. Invest., 49:1061, 1970.

12a. Madias, N. E., and A. S. Levey: Metabolic alkalosis due to absorption of "nonabsorbable" antacids, Am. J. Med., 74:155, 1983.

13. Gorden, P., and H. Levitin: Congenital alkalosis with diarrhea: a sequel to Darrow's original description, Ann. Intern. Med., 78:876, 1973.

14. Kassirer, J. P., A. M. London, D. M. Goldman, and W. B. Schwartz: On the pathogenesis of metabolic alkalosis in hyperaldosteronism, Am. J. Med., 49:306, 1970.

15. Seldin, D., L. G. Welt, and J. Cort: The role of sodium salts and adrenal steroids in the production of hypokalemic alkalosis, Yale J. Biol. Med., 29:229, 1956.

16. Garella, S., B. S. Chang, and S. I. Kahn: Dilution acidosis and contraction alkalosis: review of a concept, Kidney Int., 8:279, 1975.

17. Bosch, J. P., M. H. Goldstein, M. F. Levitt, and T. Kahn: Effect of chronic furosemide administration on hydrogen and sodium excretion in the dog, Am. J. Physiol. 232:F397, 1977.

18. Polak, A., G. D. Haynie, R. M. Hays, and W. B. Schwartz: Effects of chronic hypercapnia on electrolyte and acid-base equilibrium. I. Adaptation, J. Clin. Invest., 40:1223, 1961.

19. Rotheram, E. B., Jr., P. Safar, and E. D. Robin: CNS disorder during mechanical ventilation in chronic pulmonary disease, J. Am. Med. Assoc. 189:993, 1964.

20. Schwartz, W. B., R. M. Hays, A. Polak, and G. Haynie: Effects of chronic hypercapnia on electrolyte and acid-base equilibrium. II. Recovery with special reference to the influence of chloride intake, J. Clin. Invest., 40:1238, 1961.

21. Linshaw, M. A., H. L. Harrison, A. B. Gruskin, J. Prebis, J. Harris, R. Stein, M. R. Jayaram, D. Preston, J. DiLiberti, J. H. Baluarte, A. Elzouki, and N. Carroll: Hypochloremic alkalosis in infants associated with soy protein formula, J. Pediatr., 96:635, 1980.

22. Klastersky, J., B. Vanderkelen, D. Daneua, and M. Mathieu: Carbenicillin and hypokalemia (letter), Ann. Intern. Med., 78:774, 1973.

23. Brunner, F. P., and P. G. Frick: Hypokalemia, metabolic alkalosis, and hypernatremia due to "massive" sodium penicillin therapy, Br. Med. J., 4:550, 1968.

24. Lipner, H. I., F. Ruzany, M. Dasgupta, P. D. Lief, and N. Bank: The behavior of carbenincillin as a nonreabsorbable anion, J. Lab. Clin. Med., 86:183, 1975.

25. McKinney, T. D., and P. Myers: Effect of calcium and phosphate on bicarbonate and fluid transport by proximal tubules in vitro, Kidney Int., 21:433, 1982.

26. Crumb, C. K., M. Martinez-Maldonado, G. Eknoyan, and W. N. Suki: Effects of volume expansion, purified parathyroid extract, and calcium on renal bicarbonate absorption in the dog, J. Clin. Invest., 54:1287, 1974.

27. Coe, F. L.: Magnitude of metabolic acidosis in primary hyperparathyroidism, Arch. Intern. Med., 134:262, 1974.

28. Hulter, H. N., A. Sebastian, R. D. Toto, E. L. Bonner, and L. P. Ilnicki: Renal and systemic acid-base effects of the chronic administration of hypercalcemia-producing agents: calcitrol, PTH, and intravenous calcium, Kidney Int., 21:445, 1982.

29. Heinemann, H. O.: Metabolic alkalosis in patients with hypercalcemia, Metabolism, 14:1137, 1965.

30. McMillan, D. E., and R. B. Freeman: The milk-alkali syndrome: a study of the acute disorder with comments on the development of the chronic condition, Medicine, 44:485, 1965.

31. Orwoll, E. S.: The milk-alkali syndrome: current concepts, Ann. Intern. Med., 97:241, 1982.
32. Stinebaugh, B. J., and F. X. Schloeder: Glucose-induced alkalosis in fasting subjects: relationship to renal bicarbonate reabsorption during fasting and refeeding, J. Clin. Invest., 51:1326, 1972.
33. Cohen, R. D., R. A. Iles, D. Barnett, M. E. O. Howell, and J. Strunin: The effect of changes in lactate uptake on the intracellular pH of the perfused rat liver, Clin. Sci., 41:159, 1971.
34. Fulop, M., M. Horowitz, A. Aberman, and E. Jaffé: Lactic acidosis in pulmonary edema due to left ventricular failure, Ann. Intern. Med., 79:180, 1973.
35. Seldin, D. W., and R. Tarail: The metabolism of glucose and electrolytes in diabetic acidosis, J. Clin. Invest., 29:552, 1950.
36. Litwin, M., L. Smith, and F. D. Moore: Metabolic alkalosis following massive transfusion, Surgery, 45:805, 1959.
37. Rahilly, G. T., and T. Berl: Severe metabolic alkalosis caused by administration of plasma protein fraction in end-stage renal failure, N. Engl. J. Med., 301:824, 1979.
38. Mattar, J. A., M. H. Weil, H. Shubin, and L. Stein: Cardiac arrest in the critically ill. II. Hyperosmolal states following cardiac arrest, Am. J. Med., 56:162, 1974.
38a. Hunt, J. N., and B. Wan: Electrolytes of mammalian gastric juice, in *Handbook of Physiology*, sec. 6, Alimentary canal, vol. II, Secretion, C. F. Code (ed.), American Physiological Society, Washington, D.C., 1967.
38b. Beckerman, R. C., and L. M. Taussig: Hypoelectrolytemia and metabolic alkalosis in infants with cystic fibrosis, Pediatrics, 63:580, 1979.
39. Mitchell, R. A., C. T. Carman, J. W. Severinghaus, B. W. Richardson, M. M. Singer, and S. Shnider: Stability of cerebrospinal fluid pH in chronic acid-base disturbances in blood, J. Appl. Physiol., 20:443, 1965.
40. Garella, S., J. A. Chazan, and J. J. Cohen: Saline-resistant metabolic alkalosis or "chloride-wasting nephropathy," Ann. Intern. Med., 73:31, 1970.
41. Javaheri, S., and E. A. Nardell: Severe metabolic alkalosis: a case report, Br. Med. J., 2:1016, 1981.
42. Collins, D. D., C. H. Scoggin, C. W. Zwillich, and J. V. Weil: Hereditary aspects of decreased hypoxic reponse, J. Clin. Invest., 62:105, 1978.
43. Cohen, J. J.: Correction of metabolic alkalosis by the kidney after isometric expansion of extracellular fluid, J. Clin. Invest., 47:1181, 1968.
44. Kassirer, J. P., and W. B. Schwartz: Correction of metabolic alkalosis in man without repair of potassium deficiency, Am. J. Med., 40:19, 1966.
45. Schwartz, W. B., C. E. van Ypersele de Strihou, and J. P. Kassirer: Role of anions in metabolic alkalosis and potassium deficiency, N. Engl. J. Med., 279:630, 1968.
46. Schwartz, W. B., R. L. Jenson, and A. S. Relman: Acidification of the urine and increased ammonium excretion without change in acid-base equilibrium: sodium reabsorption as a stimulus to the acidifying process, J. Clin. Invest., 34:673, 1955.
47. Bleich, H. L., R. L. Tannen, and W. B. Schwartz: The induction of metabolic alkalosis by correction of potassium deficiency, J. Clin. Invest., 45:573, 1966.
48. Griffing, G. T., A. G. Cole, S. A. Aurecchia, B. H. Sindler, P. Komanicky, and J. C. Melby: Amiloride in primary hyperaldosteronism, Clin. Pharmacol. Therap., 31:56, 1982.
49. Leaf, A., W. B. Schwartz, and A. S. Relman: Oral administration of a potent carbonic anhydrase inhibitor ("Diamox"). I. Changes in electrolyte and acid-base balance, N. Engl. J. Med., 250:759, 1954.
50. Abouna, G., P. Veazey, and D. Terry, Jr.: Intravenous infusion of hydrochloric acid for treatment of severe metabolic alkalosis. Surgery, 75:194, 1974.
51. Tannen, R. L., H. L. Bleich, and W. B. Schwartz: The renal response to acid loads in metabolic alkalosis: an assessment of the mechanism regulating acid excretion, J. Clin. Invest., 45:562, 1966.
52. Adrogué, J. H., J. Brensilver, J. J. Cohen, and N. E. Madias: Influence of steady-state alterations in acid-base equilibrium on the fate of administered bicarbonate in the dog, J. Clin. Invest., 71:867, 1983.
53. Knutsen, O. H.: New method for administration of hydrochloric acid in metabolic alkalosis, Lancet, 1:953, 1983.

54. Warren, S. E., A. R. H. Swerdlin, and S. M. Steinberg: Treatment of alkalosis with ammonium chloride: a case report, Clin. Pharmacol. Therap., 25:624, 1979.
55. Bushinsky, D. A., and F. J. Gennari: Life-threatening hyperkalemia induced by arginine, Ann. Intern. Med., 89:632, 1978.
56. Hertz, P., and J. A. Richardson: Arginine-induced hyperkalemia in renal failure patients, Arch. Intern. Med., 130:778, 1972.
57. Swartz, R. D., J. E. Rubin, R. S. Brown, H. M. Yager, T. I. Steinman, and H. S. Frazier: Correction of postoperative metabolic alkalosis and renal failure by hemodialysis, Ann. Intern. Med., 86:52, 1977.

METABOLIC ACIDOSIS

The introduction to acid-base disorders in Chap. 17 should be read before proceeding with this discussion. Metabolic acidosis is a clinical disturbance characterized by a low arterial pH (or an increased H^+ concentration), a reduced plasma HCO_3^- concentration, and compensatory hyperventilation resulting in a decrease in the P_{CO_2}. A low plasma HCO_3^- concentration, however, is not diagnostic of metabolic acidosis since it also results from the renal compensation to chronic respiratory alkalosis. This disorder can be differentiated from metabolic acidosis by measurement of the arterial pH. In addition, a plasma HCO_3^- concentration of 10 meq/L or less is indicative of metabolic acidosis since the renal compensation to chronic hypocapnia does not produce this degree of hypobicarbonatemia (see Chap. 21).

PATHOPHYSIOLOGY

From the reaction of H^+ with the primary extracellular buffer, HCO_3^-,

$$H^+ + HCO_3^- \rightleftharpoons H_2CO_3 \rightleftharpoons CO_2 + H_2O \qquad (19\text{-}1)$$

we can see that metabolic acidosis can be produced in two ways: *by the addition of H^+ or the loss of HCO_3^-*. The latter increases the H^+ concentration by driving the buffering reaction to the left.

The response of the body to an increase in the arterial H^+ concentration involves four processes (see Chaps. 11 and 12): extracellular buffering, intracellular buffering, respiratory buffering, and the renal excretion of the H^+ load. The first three act to minimize the increase in H^+ concentration until the kidneys restore acid-base balance by eliminating the excess H^+ in the urine. Since each of these processes has important clinical implications, they will be considered separately. It should be noted that these responses are reversed in the presence of metabolic alkalosis.

Response to an Acid Load

Extracellular buffering Due to its high concentration, HCO_3^- is the most important buffer in the extracellular fluid [Eq. (19-1)]. The ability of HCO_3^- to prevent large changes in the arterial pH (or H^+ concentration) can be appreciated if we use the law of mass action to express the relationship between H^+, HCO_3^-, and P_{CO_2} (see page 212):

$$[H^+] = 24 \times \frac{P_{CO_2}}{[HCO_3^-]} \qquad (19\text{-}2)$$

If the normal P_{CO_2} is 40 mmHg and plasma HCO_3^- concentration is 24 meq/L (equal to 24 mmol/L), then

$$[H^+] = 24 \times \frac{40}{24}$$

$$= 40 \text{ nanoeq/L} \qquad (pH = 7.40)$$

Let us assume that 12 meq of H^+ is added to each liter of the extracellular fluid. As this H^+ is buffered by HCO_3^-, the plasma HCO_3^- concentration will fall from 24 to 12 meq/L. If the P_{CO_2} remains constant,

$$[H^+] = 24 \times \frac{40}{12}$$

$$= 80 \text{ nanoeq/L} \quad (pH = 7.10)$$

Even though 12 meq (or 12 million nanoeq) of H^+ has been added to each liter, the free H^+ concentration has increased by only 40 nanoeq/L or 40×10^{-6} meq/L. Thus, more than 99.99 percent of the extra H^+ has been taken up by HCO_3^-, thereby preventing the H^+ concentration from exceeding 160 nanoeq/L (pH of 6.80), the highest H^+ concentration generally compatible with life.

Intracellular buffering and the plasma potassium concentration H^+ also is able to enter the cells and be taken up by the cell buffers, particularly proteins and phosphates:

$$H^+ + Pr^- \rightleftharpoons HPr$$

On the average, 55 to 60 percent of an acid load will eventually be buffered by the cells (or by bone carbonate), although higher values may occur with severe acidemia [1,2]. As a result, the addition of 12 meq of H^+ to each liter of extracellular fluid will lower the plasma HCO_3^- concentration by 5 meq/L or less, not by 12 meq/L. If the new plasma HCO_3 concentration is 19 meq/L and the P_{CO_2} is unchanged, then

$$[H^+] = 24 \times \frac{40}{19}$$

$$= 51 \text{ nanoeq/L} \quad (pH = 7.29)$$

a value which is in contrast to the H^+ concentration of 80 nanoeq/L if there were no cellular or bone buffering.

The intracellular entry of H^+ in metabolic acidosis has been considered to be accompanied in part by the movement of K^+ out of the cell to maintain electroneutrality [3]. This cation shift can result in hyperkalemia even though total body K^+ stores frequently are depleted because of gastrointestinal and/or renal losses of K^+. For reasons that are unclear, however, the extracellular movement of K^+ does *not* occur in organic acidoses such as lactic acidosis, ketoacidosis, or that following certain ingestions (see page 254) [4]. Thus, an acidemia-induced elevation in the plasma K^+ concentration would be expected only in a nonorganic metabolic acidosis, as might occur with diarrhea or renal failure. With diarrhea, however, the plasma K^+ concentration may still be below normal since the effect of acidemia is counteracted by the intestinal K^+ loss.

It should be noted, however, that hyperkalemia may occur with lactic acidosis or ketoacidosis, but by mechanisms other than acidemia. In diabetic ketoacidosis, for example, the combination of insulin deficiency (which retards K^+ entry into cells) and hyperglycemia (which pulls water and, by solvent drag, K^+ out of the cells) frequently leads to an elevation in the plasma K^+ concentration (see Chap. 25). Similarly, hypoperfusion-induced tissue breakdown and renal failure in lactic acidosis can also result in hyperkalemia.

Respiratory buffering Acidemia stimulates the chemoreceptors controlling respiration, resulting in an increase in alveolar ventilation (see Chap. 12). As a result, the P_{CO_2} will fall, which will initially return the pH toward normal in a patient with metabolic acidosis. The increase in ventilation, characterized more by an increase in tidal volume than in respiratory rate, may be quite striking, reaching a maximum of approximately 30 L/min (normal is 5 to 6 L/min) if the arterial pH falls to 7.00 [5]. This degree of hyperventilation (called Kussmaul's respiration) is usually apparent on physical examination and should alert the physician to the possible presence of metabolic acidosis.

Studies in otherwise normal patients with metabolic acidosis have revealed that, on the average, the P_{CO_2} will fall 1.2 mmHg for every 1.0 meq/L reduction in the plasma HCO_3^- concentration down to a minimum P_{CO_2} of 10 to 15 mmHg [6]. For example, if an acid load lowers the plasma HCO_3^- concentration from the normal 24 meq/L to 9 meq/L, this decrease of 15 meq/L should be associated with an 18-mmHg (15 × 1.2) fall in the P_{CO_2} to 22 mmHg (pH of 7.23) (Table 19-1). Thus, *in pure metabolic acidosis* with a *plasma HCO_3^- concentration of 9 meq/L, the normal P_{CO_2} is roughly 22 mmHg, not 40 mmHg.* Values different from this P_{CO_2} represent mixed acid-base disorders. If the P_{CO_2} remained at 40 mmHg (pH of 6.98), this would represent a *combined metabolic and respiratory acidosis,* as might occur in a patient with chronic lung disease. If, on the other hand, the P_{CO_2} were 15 mmHg (pH of 7.40), this would reflect a *combined metabolic acidosis and respiratory alkalosis,* as might be seen with salicylate intoxication (see below).

Although the respiratory compensation initially minimizes the fall in pH, this protective effect appears to last for only a few days. This limitation occurs because the fall in the P_{CO_2} directly lowers renal HCO_3 reabsorption, resulting in inappropriate HCO_3^- loss in the urine and a further reduction in the plasma HCO_3^- concentration [7]. The net effect is that *the arterial pH in chronic metabolic acidosis is the same whether or not compensation has occurred.* [7]. As shown in the example in Table 19-2, the arterial pH is 7.29 in uncompensated metabolic acidosis. The compensatory 6-mmHg decrease in the P_{CO_2} then lowers the plasma HCO_3^- concentration from 19 to 16 meq/L, returning the arterial pH to 7.29. Fortunately, severe metabolic acidosis is usually acute (lactic acidosis, ketoacidosis, ingestions) and the fall in P_{CO_2} is protective in this setting.

Table 19-1 Arterial measurements in hypothetical acid-base disorders

Acid-base status	Plasma $[HCO_3^-]$, meq/L	P_{CO_2}, mmHg	Arterial pH
Normal	24	40	7.40
Pure metabolic acidosis	9	22	7.23
Combined metabolic and respiratory acidosis	9	40	6.98
Combined metabolic acidosis and respiratory alkalosis	9	15	7.40

Table 19-2 Arterial pH in chronic metabolic acidosis with and without respiratory compensation

Clinical state	Arterial		
	pH	HCO_3^-, meq/L	P_{CO_2}, mmHg
Baseline	7.40	24	40
Metabolic acidosis			
No compensation	7.29	19	40
Compensation			
Acute	7.37	19	34
Chronic	7.29	16	34

Renal hydrogen excretion The metabolism of a normal diet results in the generation of 50 to 100 meq of H^+ per day,† which must then be excreted in the urine if acid-base balance is to be maintained (see Chap. 12). This process involves two basic steps. First, the filtered HCO_3^- must be reabsorbed since urinary HCO_3^- loss would directly lower the plasma HCO_3^- concentration. Ninety percent of HCO_3^- reabsorption occurs in the proximal tubule, and the remainder in the distal nephron.

Second, the kidney is able to excrete H^+ by the secretion of H^+ from the tubular cell into the lumen. The amount of H^+ that can be excreted as free H^+ is limited by the minimum urine pH of 4.50 that can be achieved in humans. Since this is equal to a free H^+ concentration of less than 0.04 meq/L, the excretion of significant quantities of H^+ requires the buffering of H^+ in the urine. This is accomplished by the combination of the secreted H^+ with the urinary buffers (particularly HPO_4^{2-}) or with NH_3,

$$H^+ + HPO_4^{2-} \rightarrow H_2PO_4^-$$

$$H^+ + NH_3 \rightarrow NH_4^+$$

(The former process is referred to as titratable acidity.) In general, 10 to 40 meq of H^+ is excreted each day as titratable acidity and 30 to 60 meq as NH_4^+.

In the absence of therapy with $NaHCO_3$, the correction of a metabolic acidosis usually requires the urinary excretion of the excess H^+. The kidney responds to an increased H^+ load by augmenting cellular NH_3 production and consequently NH_4^+ excretion. In patients with metabolic acidosis, NH_4^+ excretion can exceed 250 meq/day [8]. In contrast, there generally is only a limited ability to enhance titratable acid excretion since phosphate excretion remains relatively constant.

Generation of Metabolic Acidosis

From this discussion, it can be seen that metabolic acidosis can result from an inability of the kidney to excrete the dietary H^+ load or from an increase in the generation of H^+, because of either the addition of H^+ or the loss of HCO_3^- (Table 19-3).

---

†In children, the daily H^+ load is 2 to 3 meq/kg.

Table 19-3 Causes of metabolic acidosis

Inability to excrete the dietary H^+ load
  A. Diminished NH_3 production
     1. Renal failure†
     2. Hypoaldosteronism (type 4 renal tubular acidosis)†
  B. Diminished H^+ secretion
     1. Type 1 (distal) renal tubular acidosis

Increased H^+ load or HCO_3^- loss
  A. Lactic acidosis†
  B. Ketoacidosis†
  C. Ingestions
     1. Salicylates
     2. Methanol or formaldehyde
     3. Ethylene glycol
     4. Paraldehyde
     5. Sulfur
     6. Toluene
     7. Ammonium chloride
     8. Hyperalimentation fluids
  D. Massive rhabdomyolysis
  E. Gastrointestinal HCO_3^- loss
     1. Diarrhea and fistulas†
     2. Ureterosigmoidostomy
     3. Cholestyramine
  F. Renal HCO_3^- loss
     1. Type 2 (proximal) renal tubular acidosis
     2. Renal failure

†Most common causes.

Decreased H^+ excretion produces a slowly developing acidemia since only that fraction of the 50- to 100-meq daily H^+ load that is not excreted will be retained. By comparison, an acute increase in the H^+ load (as with lactic acidosis) can overwhelm renal excretory capacity, leading to the rapid onset of metabolic acidosis.

Anion Gap

The anion gap refers to the difference between the plasma concentrations of the major cation (Na^+) and the major anions ($Cl^- + HCO_3^-$). The approximate normal values for these ions are 140, 105, and 24 meq/L, respectively, leading to an anion gap of 9 to 14 meq/L. The negative charges on the plasma proteins account for most of the missing anions, as the charges of the other cations (K^+, Ca^{2+}, and Mg^{2+}) and anions (phosphate, sulfate, and organic anions) tend to balance out [9,10]. Determining the anion gap is very helpful in the differential diagnosis of metabolic acidosis since the causes of this disorder can be divided into those which elevate the anion gap and those which do not (Table 19-4).

   As acid accumulates in the body, there is rapid extracellular buffering by HCO_3^-. If the acid is HCl, then

$$HCl + NaHCO_3 \rightarrow NaCl + H_2CO_3 \rightarrow H_2O + CO_2 \qquad (19\text{-}3)$$

Table 19-4 Anion gap in metabolic acidosis

Normal anion gap (hyperchloremic acidosis)
  A. Gastrointestinal loss of HCO_3^-
    1. Diarrhea and fistulas
    2. Ureterosigmoidostomy
    3. Cholestyramine
  B. Renal HCO_3^- loss
    1. Type 2 (proximal) renal tubular acidosis
    2. Renal insufficiency
  C. Ingestions
    1. Ammonium chloride
    2. Hyperalimentation fluids
  D. Renal dysfunction
    1. Tubulointerstitial diseases
    2. Hypoaldosteronism (type 4 renal tubular acidosis)
    3. Type 1 (distal) renal tubular acidosis
  E. Some cases of ketoacidosis, particularly during treatment

High anion gap†
  A. Ketoacidosis: β-hydroxybutyrate
  B. Lactic acidosis: lactate
  C. Renal insufficiency: SO_4^{2-}, HPO_4^{2-}
  D. Ingestions
    1. Salicylate: organic anions, especially lactate
    2. Ethylene glycol: glycolate, oxalate
    3. Methanol or formaldehyde: formate
    4. Paraldehyde: organic anions
    5. Sulfur: SO_4^{2-}
    6. Toluene: hippurate
  E. Massive rhabdomyolysis: organic anions

†The substances after the colon represent the major retained anions in the high anion gap acidoses.

The net effect is the meq-for-meq replacement of extracellular HCO_3^- by Cl^-. Since the sum of Cl^- and HCO_3^- concentrations remains constant, the anion gap is unchanged. Because of the increase in the plasma Cl^- concentration, this is often referred to as a *hyperchloremic acidosis*. Gastrointestinal or renal loss of $NaHCO_3$ produces the same result, i.e., the exchange of HCO_3^- for Cl^-, since the kidney retains NaCl in an effort to preserve the volume of the extracellular fluid.

Conversely, if H^+ accumulates with any anion other than Cl^-, extracellular HCO_3^- will be replaced by an unmeasured anion (A^-):

$$HA + NaHCO_3 \rightarrow NaA + H_2CO_3 \rightarrow CO_2 + H_2O \qquad (19\text{-}4)$$

As a result, there will be a decrease in the sum of Cl^- and HCO_3^- concentrations and an increase in the anion gap. In these disorders (Table 19-4), identification of the specific disease process usually can be obtained by measuring the serum concentrations of BUN, creatinine, glucose, and lactate, and by checking the serum for the presence of ketones and intoxicants (salicylates, methanol, and ethylene glycol).

Case history 19-1 A 27-year-old man with insulin-dependent diabetes mellitus has not been taking his insulin and is admitted to the hospital in a semicomatose condition. The following laboratory data are obtained:

Arterial pH $= 7.10$	$[Na^+] = 140$ meq/L
$P_{CO_2} = 20$ mmHg	$[K^+] = 7.0$ meq/L
[Glucose] $= 800$ mg/dL	$[Cl^-] = 105$ meq/L
Serum ketones $= 4+$	$[HCO_3^-] = 6$ meq/L
	Anion gap $= 29$ meq/L

COMMENT The high anion gap, hyperglycemia, and ketonemia all point to the diagnosis of diabetic ketoacidosis. Note that the increase in the anion gap of approximately 18 meq/L (from 11 to 29) is the same as the fall in the plasma HCO_3^- concentration (from 24 to 6 meq/L).

Δ Anion gap/Δ plasma HCO_3^- concentration In addition to the value of finding an elevated anion gap, the relationship between the increase in the anion gap and the fall in the plasma HCO_3^- concentration may also be helpful diagnostically. Although Eq. (19-4) seems to imply that there should be a 1:1 relationship, this is frequently not the case. As described above, more than 50 percent of the excess H^+ is buffered by the cells, *not by HCO_3^-*. In contrast, most of the excess anions remain in the extracellular fluid since they are charged particles that do not easily cross the cell membrane. Therefore, the increase in anion gap usually exceeds the fall in the plasma HCO_3^- concentration; in lactic acidosis, for example, the ratio averages about 1.6:1 [11]. Although the same principles apply to ketoacidosis, the ratio usually is close to 1:1 in this disorder because the loss of ketoacid anions in the urine (which lowers the anion gap) tends to balance the effect of intracellular buffering of H^+ [11,11a]. Anion loss in the urine is much less prominent in lactic acidosis because the tubular reabsorptive capacity for lactate exceeds that for the ketoacid anions [12]. Thus, calculating the Δ anion gap/Δ plasma HCO_3^- concentration may help differentiate ketoacidosis from other organic acidoses, although some overlap clearly exists.

The ketonuria also accounts for the observation that a normal anion gap acidosis may occur during the treatment phase of ketoacidosis [11]. In the above case history, there is an 18-meq/L elevation in the anion gap. After the administration of insulin, these ketoacid anions will be metabolized to HCO_3^- (see below). Thus, the anion gap will return to normal but the plasma HCO_3^- concentration will increase by about 8 meq/L, not 18 meq/L, since most of the generated HCO_3^- will enter the cells to replenish the cell buffers. At this time, the plasma HCO_3^- concentration will be 14 meq/L, the pH will still be acid, but there will be no excess anion gap (i.e., the patient will have a normal anion gap acidosis). The acidemia in this setting is due to two factors: the previous production of ketoacids, and the excretion of the organic anions, which, if retained, could have been converted into HCO_3^-.

In addition to this sequence during treatment, some patients with ketoacidosis excrete ketones in the urine so efficiently that the anion gap is relatively normal *before* any therapy has been instituted [11a]. These findings illustrate a very important concept: Whatever the size of the anion gap, *it is the retention of H^+, not of the particular anion, that is responsible for the acidemia.*

The relationship between the changes in the anion gap and plasma bicarbonate concentration may also suggest the presence of mixed acid-base disturbances. Consider the following example:

Case history 19-2 A previously well 55-year-old woman is admitted with a complaint of severe vomiting for 5 days. Physical examination reveals postural hypotension, tachycardia, and diminished skin turgor. The laboratory findings include

$$\text{Arterial pH} = 7.23$$

$$P_{CO_2} = 22 \text{ mmHg}$$

$$\text{Plasma } [Na^+] = 140 \text{ meq/L}$$

$$[K^+] = 3.4 \text{ meq/L}$$

$$[Cl^-] = 77 \text{ meq/L}$$

$$[HCO_3^-] = 9 \text{ meq/L}$$

$$\text{Anion gap} = 54 \text{ meq/L}$$

$$\text{Ketones} = \text{trace}$$

$$[\text{Creatinine}] = 2.1 \text{ mg/dL}$$

COMMENT This patient has a high anion gap metabolic acidosis. Lactic acidosis is most likely in view of the physical findings and lack of significant ketonemia, renal failure, or a history of an ingestion. However, the anion gap of 54 meq/L is markedly increased (42 meq/L above normal) with a much smaller 15-meq/L drop in the plasma HCO_3^- concentration to 9 meq/L. This disparity can be explained by a concomitant *metabolic alkalosis* due to vomiting, which raised the plasma HCO_3^- concentration without affecting the anion gap. Proof of this diagnosis came from evaluating the response to fluid repletion. As tissue perfusion was restored and metabolism of the excess lactate generated HCO_3^- (see below), the plasma HCO_3^- concentration rose from 9 to 37 meq/L and the pH became alkalemic. Thus, the 42-meq/L elevation in the anion gap was actually associated with a 28-meq/L fall in the plasma HCO_3^- concentration, a 1.5:1 ratio that is typical of lactic acidosis [11].

Anion gap in renal failure To understand what happens to the anion gap in renal failure, we must first review the normal handling of acids. The dietary acid load is primarily due to the generation of H_2SO_4 from the metabolism of sulfur-containing amino acids [13]. To maintain the steady state, both the $2H^+$ and the SO_4^{2-} are excreted in the urine as $(NH_4)_2SO_4$. The renal excretion of SO_4^{2-}, which is poorly reabsorbed by the tubules, is mostly determined by the GFR. Although the excretion of H^+ (primarily as NH_4^+) occurs by tubular secretion, the quantitative ability to excrete NH_4^+ usually varies directly with the GFR [14], which is an index of the functioning renal mass. Thus, the reduction in the GFR in renal failure usually diminishes both NH_4^+ and SO_4^{2-} excretion. As a result, SO_4^{2-} retention producing an increase in the anion gap occurs at the same time as H^+ retention and the development of metabolic acidosis. (Other retained anions in renal failure include phosphate, urate, and hippurate.)

The situation may be different in tubulointerstitial diseases involving the renal medulla predominantly, such as pyelonephritis or obstructive uropathy. In these conditions, tubular function is likely to be affected to a greater degree than glomerular filtration since the glomeruli are located in the cortex. Consequently, NH_4^+ excretion may be diminished before there is anion retention. The filtered SO_4^{2-} will now be excreted as Na_2SO_4, not $(NH_4)_2SO_4$. To preserve the extracellular volume, NaCl will be retained, resulting in a hyperchloremic acidosis (due to the retention of H^+ and Cl^-) with a normal anion gap. Thus, the anion gap may be normal or elevated in patients with renal failure and metabolic acidosis.

Anion gap in other conditions Small changes in the anion gap occur in a variety of disorders other than metabolic acidosis [9,10,15]. A low anion gap, for example, may be seen when there are fewer net negative charges on the plasma proteins. This occurs with hypoalbuminemia or in some patients with multiple myeloma who have a cationic paraprotein [16]. In contrast, patients with metabolic alkalosis may have an elevated anion gap. This is due both to a hypovolemia-induced elevation in the plasma albumin concentration and an alkalemia-induced increase in the anionic charge per albumin molecule (since the pH is now further away from the isoelectric point of 5.4) [17].

ETIOLOGY AND DIAGNOSIS

This section will review the pathogenesis, etiology, and diagnosis of the different disorders that can cause metabolic acidosis. It will also include some specific aspects of therapy, although the general principles involved in the treatment of metabolic acidosis will be discussed separately later in the chapter.

Lactic Acidosis

Metabolism Lactic acid is derived from the metabolism of pyruvic acid; this reaction is catalyzed by lactate dehydrogenase and involves the conversion of NADH into NAD^+ (reduced and oxidized nicotine adenine dinucleotide, respectively) (Fig. 19-1). Normal subjects produce 15 to 20 mmol/kg of lactic acid per day, most of

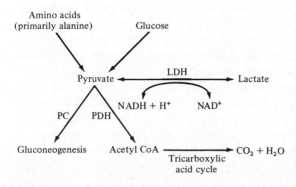

Figure 19-1 Major pathways of pyruvate and lactate metabolism. LDH = lactate dehydrogenase, PDH = pyruvate dehydrogenase, PC = pyruvate carboxylase, and NADH and NAD^+ = reduced and oxidized nicotine adenine dinucleotide, respectively.

which is generated by the metabolism of glucose and alanine [18]. This lactic acid is rapidly buffered, in large part by extracellular HCO_3^-, resulting in the generation of lactate:

$$CH_3-CHOH-COOH + NaHCO_3 \rightarrow Na\ lactate + H_2CO_3 \rightarrow CO_2 + H_2O$$

In the liver and, to a lesser degree, the kidney, lactate is converted into $CO_2 + H_2O$ (80 percent) or glucose (20 percent). Either of these reactions results in the regeneration of the HCO_3^- initially lost in buffering the lactic acid [19]:

$$Lactate^- + 3O_2 \rightarrow HCO_3^- + 2CO_2 + 2H_2O \qquad (19\text{-}5)$$

$$2\ lactate^- + 2H_2O + 2CO_2 \rightarrow 2HCO_3^- + glucose \qquad (19\text{-}6)$$

The normal plasma lactate concentration is 0.5 to 1.5 meq/L. Lactic acidosis is considered to be present if the plasma lactate level exceeds 4 to 5 meq/L in an acidemic patient.

Pathogenesis and etiology Excess lactate can accumulate when there is increased lactate production and/or diminished lactate utilization. The former can occur by three mechanisms: enhanced pyruvate production, reduced pyruvate utilization, or, most commonly, an altered redox state within the cell in which pyruvate is preferentially converted into lactate (Table 19-5) [18]. During glycolysis, NADH is generated and then reoxidized to NAD^+ in the mitochondria. If oxidation is impaired, as with decreased oxygen delivery in shock, NADH will accumulate. This promotes the conversion of pyruvate to lactate, a reaction which also regenerates NAD^+ (Fig. 19-1).

In certain disorders, the primary role of overproduction is clear. For example, plasma lactate levels may transiently be as high as 15 meq/L during a grand mal seizure [26], and 20 to 25 meq/L with maximal exercise, with the systemic pH falling to as low as 6.80 [27]. Studies in these patients have demonstrated that the maximum rate of lactate metabolism can reach 320 meq/h [18].

This high rate of lactate metabolism suggests that *there must be some component of decreased utilization* in those disorders in which lactate overproduction occurs more slowly [32]. In shock, for example, the reduction in perfusion to the liver, as well as enhanced lactate synthesis, contributes to the development of lactic acidosis. Severe acidemia (pH < 7.10) may also play a role in perpetuating this process by directly inhibiting hepatic and renal lactate utilization [18,36].

The most common cause of lactic acidosis is an altered redox state due to *shock*. This may be seen with hemorrhage, cardiac failure, or sepsis and generally indicates a poor prognosis unless tissue perfusion can be rapidly restored.

The association of lactic acidosis with *diabetes mellitus* is less certain since many cases in the past were associated with the use of phenformin. Nevertheless, a moderate degree of lactic acidosis may occur in some patients with diabetic ketoacidosis [24]. This may be due to a combination of hypovolemia-induced lactate overproduction and reduced lactate utilization, perhaps resulting from decreased pyruvate dehydrogenase activity (Fig. 19-1) [18,23].

Table 19-5 Etiology of lactic acidosis

Increased lactate production
  A. Increased pyruvate production
    1. Enzymatic defects in glycogenolysis or gluconeogenesis (as with glycogen storage disease, type 1) [20]
    2. Respiratory alkalosis (including salicylate intoxication) [18,21]
  B. Impaired pyruvate utilization
    1. Decreased activity of pyruvate dehydrogenase or pyruvate carboxylase
      a. Congenital [22]
      b. ?Role in diabetes mellitus, Reye's syndrome [23–25]
  C. Altered redox state favoring pyruvate conversion to lactate
    1. Enhanced metabolic rate
      a. Grand mal seizure [26]
      b. Severe exercise [27]
      c. Hypothermic shivering [28]
    2. Decreased oxygen delivery
      a. Shock
      b. Cardiac arrest
      c. Carbon monoxide poisoning ($\downarrow O_2$ uptake by hemoglobin) [29]
      d. Severe hypoxemia ($P_{O_2} < 25$–30 mmHg) [30]
    3. Reduced oxygen utilization
      a. Cyanide intoxication ($\downarrow$oxidative metabolism) [31]
      b. ?Phenformin [32]
  D. D-Lactic acidosis [33,34]

Primary decrease in lactate utilization
  A. Liver disease [18]
  B. Alcoholism [35]
  C. Severe acidemia (pH < 7.10) [18,36]

Mechanism uncertain
  A. Diabetes mellitus [18]
  B. Malignancy [37–39]
  C. Hypoglycemia [18]
  D. Idiopathic

    The pathogenesis of the lactic acidosis found with *malignancies* is also unclear [37–39]. Anaerobic metabolism due to dense clusters of tumor cells and/or metastatic replacement of the hepatic parenchyma have been proposed, but lactic acidosis has occurred in patients with relatively small tumor loads [37,39]. Direct lactate production by the neoplastic cells has also been suggested, but this would not explain the rarity of tumor-induced lactic acidosis. Regardless of the cause, removal of the tumor (by chemotherapy, irradiation, or surgery) leads to correction of the acidosis [37,39].

    A mild degree of lactic acidosis may be seen in *alcoholism* [35]. In this condition, lactate production is usually normal but lactate utilization is diminished because of impaired hepatic gluconeogenesis [35]. Although lactate levels generally do not exceed 3 meq/L in this setting, alcohol ingestion can potentiate the severity of other disorders associated with the overproduction of lactate.

    Finally, a different form of lactic acidosis can occur in patients with bacterial overgrowth due, for example, to jejunoileal bypass or small-bowel resection. In these patients, intestinal bacteria (?gram-positive anaerobes) can metabolize glucose into D-lactic acid [33,34]. This is then absorbed and produces a metabolic acidosis. Endogenous utilization of D-lactate does not occur since the L-lactate dehydrogenase present in humans recognizes only L-lactate. Oral antimicrobial agents such as neomycin or vancomycin effectively treat this disorder.

Diagnosis Although the diagnosis of lactate acidosis can be made definitively only by the demonstration of an elevated plasma lactate concentration,† there may be many suggestive clues in the history, physical examination, laboratory data, and response to therapy. These include a high anion gap, the presence of one of the disorders that can cause lactic acidosis, and continuing production of acid as evidenced by an inability of exogenous HCO_3^- to raise the plasma HCO_3^- concentration. The presence of an organic acidosis (primarily lactic acidosis or ketoacidosis) should also be suspected if effective treatment of the underlying problem (such as fluid repletion in hypovolemic shock) leads to a spontaneous elevation in the plasma HCO_3^- concentration. This occurs because metabolism of the organic anion, in this case lactate, results in the regeneration of HCO_3^- [Eqs. (19-5) and (19-6)].

The specific treatment of lactic acidosis is discussed below (see "Treatment: Use of bicarbonate in lactic acidosis and ketoacidosis").

Ketoacidosis

The biochemistry of ketoacidosis is discussed in detail in Chap. 25. Stated briefly, free fatty acids are converted in the liver into triglycerides, $CO_2 + H_2O$, or the ketoacids, acetoacetic acid and β-hydroxybutyric acid. Overproduction of ketoacids resulting in metabolic acidosis requires two factors: (1) an increase in free fatty acid delivery to the liver due to enhanced lipolysis, and (2) a resetting of hepatocyte function such that the free fatty acids are converted preferentially into ketoacids and not triglycerides [40]. For these changes to occur, there must be both diminished insulin activity and enhanced secretion of glucagon. The latter is induced by insulin deficiency (which is the major physiologic stimulus to glucagon release) and is primarily responsible for the alteration in hepatic fatty acid metabolism [40,41].

Etiology *Diabetes mellitus* is the most common cause of ketoacidosis. Hyperglycemia is invariably present, with the plasma glucose concentration usually exceeding 400 mg/dL (see Chap. 25). *Fasting* can also result in a ketosis which tends to be self-limited. The lack of carbohydrate intake lowers the blood sugar, removing the stimulus to insulin secretion and stimulating that of glucagon. However, ketoacid levels do not exceed 10 meq/L, in part because ketonemia promotes insulin secretion [41–43], thereby limiting further ketone formation. The combination of *alcohol ingestion* and poor dietary intake is another cause of ketoacidosis [44,45]. The

---

†Detection of D-lactate requires a separate enzymatic assay that uses D-lactate dehydrogenase and measures the generation of NADH as the lactate is converted to pyruvate (see Fig. 19-1) [33,34].

decrease in carbohydrate intake plus the inhibition of gluconeogenesis by alcohol [35] results in the necessary changes in insulin and glucagon secretion. In addition, ethanol directly enhances lipolysis [46]. Finally, a variety of congenital *organic acidemias* (such as methylmalonic or isovaleric acidemia) are associated with ketoacidosis [47,48]. The mechanisms responsible for the increased ketone synthesis in these disorders are not completely understood.

Diagnosis Confirmation of the presence of ketoacidosis requires the demonstration of ketonemia. This is generally done with nitroprusside (Acetest) tablets. A 4+ reaction with serum diluted 1:1 is strongly suggestive of ketoacidosis. However, these tablets react with acetoacetate and acetone (produced by the decarboxylation of acetoacetic acid), but not with β-hydroxybutyrate [49]. The latter ketoacid is formed from the reduction of the β-aldehyde group of acetoacetate in a reaction utilizing NADH. β-Hydroxybutyrate makes up about 75 percent of the circulating ketones in diabetic ketoacidosis, but this value can reach 90 percent when NADH levels are elevated with concurrent lactic acidosis [49] or in alcoholic ketoacidosis (where NADH is generated from the oxidation of ethanol to acetic acid) [44]. In these settings, nitroprusside tablets may underestimate the degree of ketonemia and ketonuria. Clinical awareness of the possibility of ketoacidosis is essential since an assay for β-hydroxybutyrate is not available in most hospitals. An indirect way around this problem is to add a few drops of hydrogen peroxide to a urine specimen. This will nonenzymatically convert β-hydroxybutyrate to acetoacetate, which will then be detectable by a nitroprusside tablet or dipstick [50].

Treatment Although insulin is the keystone to therapy in diabetic ketoacidosis, it may be dangerous in alcoholism or fasting where the baseline plasma glucose concentration may be low. In these conditions, the administration of glucose will augment endogenous insulin secretion, diminish that of glucagon, and normalize fatty acid metabolism.

Renal Failure

Renal insufficiency produces a sequence of changes in renal acid excretion. With the initial reduction in GFR, hydrogen balance is maintained by increased ammonium excretion per functioning nephron [51,52]. However, total ammonium excretion begins to fall when the GFR is less than 40 to 50 mL/min [53,54]. The net effect is a metabolic acidosis resulting from an inability to excrete all of the daily H^+ load. Decreased titratable acidity may also contribute in patients with advanced disease in whom net phosphorus intake (and therefore excretion) is restricted [53]. An additional factor may be HCO_3^- wasting due to reduced tubular HCO_3^- reabsorption [53]. Both hyperparathyroidism (which impairs proximal HCO_3^- reabsorption) and decreased Na^+ reabsorption (necessitated by a constant Na^+ intake in the presence of fewer functioning nephrons) may contribute to this response (see page 239) [55–57].

The mechanism responsible for the fall in total ammonium excretion is uncertain. However, ammonium excretion per total GFR in renal failure is similar to the maximum achieved in normal subjects following an acid load [52]. This suggests that the reduction in total ammonium excretion may simply reflect the fact that ammonium production is already proceeding at a maximal rate but there are fewer functioning nephrons [52].

As the patient approaches end-stage renal failure, the plasma HCO_3^- concentration gradually falls and then stabilizes at 12 to 20 meq/L. Although hydrogen ions continue to be retained, a further reduction in the plasma HCO_3^- concentration is prevented by buffering of the excess H^+ by bone [58]. A more marked degree of acidemia in relation to the GFR is usually due to a superimposed abnormality such as a tubulointerstitial disease which directly impairs H^+ secretion, hypoaldosteronism in which hyperkalemia is a prominent finding (see below), or another cause of metabolic acidosis such as diarrhea. The latter can lead to a severe metabolic acidosis since patients with renal failure cannot compensate by increasing renal acid excretion.

Treatment Correction of the metabolic acidosis is frequently not necessary in adults with renal failure. The limited fall in the plasma HCO_3^- concentration plus the respiratory compensation usually maintain the arterial pH near 7.30, a level that poses no danger to the patient. Furthermore, raising the pH in the presence of hypocalcemia may precipitate tetany. Indications for $NaHCO_3$ therapy include a fall in the plasma HCO_3^- concentration below 12 meq/L, symptoms such as dyspnea or drowsiness, persistent hyperkalemia since raising the pH will drive K^+ into the cells, and children in whom acidemia can impair growth [59]. Although fluid overload and volume-dependent hypertension are common problems in patients with renal failure, $NaHCO_3$ therapy is usually well tolerated [60].

Ingestions

Salicylates Aspirin (acetylsalicylic acid) is rapidly converted into salicylic acid in the body. Although there is no absolute correlation between the plasma salicylate concentration and symptoms, most patients show signs of intoxication when the plasma level exceeds 40 to 50 mg/dL (therapeutic range is 20 to 35 mg/dL) [61]. Early symptoms include tinnitus, vertigo, nausea, vomiting, and diarrhea; more severe intoxication can cause altered mental status, coma, and death. Fatal overdosage can occur after the ingestion of 10 to 30 g by adults and as little as 3 g by children. The diagnosis can be made with certainty only by measurement of the plasma salicylate concentration.

A variety of acid-base disturbances may occur with salicylate intoxication. Salicylates stimulate the respiratory center directly, resulting in a fall in the P_{CO_2} and respiratory alkalosis as the earliest acid-base abnormality [61–63]. Metabolic acidosis may then ensue, due primarily to the accumulation of organic acids, particularly lactate [18,64]. The respiratory alkalosis, which normally promotes lactate production to minimize the rise in pH [21], appears to play an important role in this process. In experimental animals, lactate accumulation does not occur if the initial fall in P_{CO_2} is prevented, but gradually becomes prominent if hypocapnia is allowed to occur [64]. Salicylic acid itself (mol wt 180) plays only a minimal role since a plasma concentration of 50 mg/dL is less than 3 meq/L.

The net effect of these changes is that most adults have either a respiratory alkalosis or a mixed respiratory alkalosis–metabolic acidosis. Pure metabolic acidosis is unusual [63]. In addition, approximately one-third of adults will also ingest one or more other medications, many of which are respiratory depressants and can lead to respiratory acidosis [63].

Treatment The serious neurologic toxicity of salicylates, including death, is related to the cerebral tissue salicylate concentration, and a reduction in this level must be the first goal of therapy. This can in part be achieved by alkalinization of the plasma. To appreciate how this works, it is important to note that salicylic acid (HS) is a weak acid with a pK_a of 3.0. Thus, the Henderson-Hasselbalch equation for the reaction

$$H^+ + S^- \rightleftharpoons HS \tag{19-7}$$

can be expressed as

$$pH = 3.0 + \log \frac{[S^-]}{[HS]}$$

where S^- represents the salicylate anion. At the normal pH of 7.40, the ratio of HS to S^- is about 1:25,000, that is, only 0.004 percent of the total extracellular salicylate exists as HS. HS is nonpolar, lipid-soluble, and able to cross cell membranes; S^- is polar and crosses membranes poorly. As a result, the plasma and central nervous system (CNS) HS concentrations are in diffusion equilibrium, but not the S^- concentrations (Fig. 19-2).

If the systemic pH is increased, Eq. (19-7) will move to the left. As the plasma HS concentration falls, HS will leave the CNS (and other tissues) down a concen-

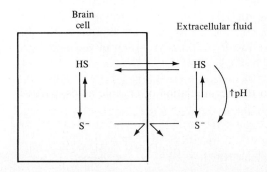

Figure 19-2 Schematic representation of the equilibrium distribution of salicylate (S^-) and salicylic acid (HS) between the extracellular fluid and the brain cell. HS is lipid-soluble and is in diffusion equilibrium; S^- is not since it cannot readily cross the cell membrane. Alkalinization of the extracellular fluid causes the reaction $H^+ + S^- \rightleftharpoons HS$ to move to the left, reducing the extracellular HS concentration. This allows cellular HS to diffuse into the extracellular fluid down a concentration gradient. The decrease in the cellular HS concentration then causes some cellular S^- to be converted to HS, thereby promoting continued HS diffusion out of the cell. Similar considerations explain the increased movement of salicylate from the tubular cell into the lumen when the urine is alkalinized.

tration gradient where it will be trapped in the plasma as S^-. The fall in the CNS HS concentration then causes Eq. (19-7) to move to the right in the brain cell. This maintains the cellular HS concentration, thus promoting further drug movement out of the CNS. For example, increasing the arterial pH from 7.20 to 7.50 will decrease the fractional concentration of HS from 0.006 to 0.003 percent. Although this change appears small, it will promote a significant reduction in the tissue salicylate concentration. Note that there will be an initial increment in the plasma salicylate concentration, but it is the tissue levels that are dangerous to the patient.

A second goal of treatment is rapid drug elimination from the body. Salicylate initially enters the urine by glomerular filtration and through secretion by the organic acid secretory pathway in the proximal tubule. The rate of salicylate excretion can be markedly enhanced by alkalinization of the urine, which, by the same process of nonionic diffusion, converts urinary HS to S^-, thereby promoting the diffusion of more HS from the tubular cells into the urine. Thus, raising the urine pH from 6.5 to 8.1 by the administration of sodium bicarbonate can increase total salicylate excretion more than fivefold [64a].

The efficiency of salicylate removal can also be enhanced by hemodialysis [61]. This procedure should be considered when the plasma salicylate concentration exceeds 70 to 80 mg/dL, especially in a patient with impaired renal function.

Another frequent problem of uncertain etiology is a low cerebrospinal fluid glucose concentration [61], which may contribute to the neurologic abnormalities. Therefore, the administration of glucose should be part of the initial therapy in all patients with salicylate intoxication.

In summary, the administration of alkali in a dose sufficient to raise the systemic pH to between 7.45 and 7.50 is an important component of therapy in the patient with an aspirin overdose and metabolic acidosis. If respiratory alkalosis is the primary disorder, this will be self-protective and further alkalinization is not necessary.

Methanol Methanol (wood alcohol, CH_3OH) is a component of shellac, varnish, sterno, and other commercial preparations. It is metabolized to formaldehyde (in a reaction catalyzed by alcohol dehydrogenase) and then formic acid. Symptoms and the high anion gap metabolic acidosis are usually delayed for 12 to 36 h after ingestion since they are due to accumulation of the metabolites, particularly formic acid [65,66]. Early complaints include weakness, nausea, headache, and decreased vision, which can then progress to blindness, coma, and death [65,67]. Fundoscopic examination may reveal a retinal sheen due to retinal edema. The minimum lethal dose is 50 to 100 mL, although smaller amounts can lead to permanent blindness. Similar clinical and acid-base disturbances may result from the ingestion of formaldehyde [68].

The diagnosis is made by a specific serum assay for methanol. In addition, the presence of methanol intoxication may be suspected indirectly by the demonstration of an *osmolal gap* between the measured and calculated plasma osmolality (see page 29) [69]:

$$\text{Calculated } P_{osm} \cong 2 \times \text{plasma } [Na^+] + \frac{[\text{glucose}]}{18} + \frac{\text{BUN}}{2.8}$$

Methanol is a small molecule (mol wt 32) that can achieve high osmolal concentrations in the plasma. A level of 80 mg/dL, for example, is equivalent to 25 mosmol/kg (see page 15). Thus, the measured plasma osmolality will exceed the calculated value by this amount. Ethanol and isopropanol ingestions or the administration of hypertonic mannitol can produce a similar effect but do not directly lead to metabolic acidosis. Ethylene glycol is larger (mol wt 62) and therefore less likely to produce an osmolal gap, whereas salicylates (mol wt 180) have only a minor effect since their plasma level is almost always less than 5 mosmol/kg [69]. It should be noted that chronic (but not acute) renal failure can produce misleading results because the retention of small solutes can induce an osmolal gap (independent of the drug in question) that may exceed 10 mosmol/kg [69a].

Treatment Prompt treatment is required to prevent death or permanent residua such as blindness. In addition to correcting the acidemia with $NaHCO_3$ and administering oral charcoal to minimize further drug absorption, there are two basic aspects of therapy in the presence of severe poisoning: administration of ethanol and hemodialysis [67,70,71].

The rationale for the administration of intravenous or oral ethanol is based upon the fact that the enzyme alcohol dehydrogenase, which is necessary for the metabolism of methanol into its toxic metabolites, has a much greater affinity for ethanol than for methanol (or ethylene glycol) [70,72]. This effect is most prominent when the plasma ethanol concentration is about 100 to 200 mg/dL, a level that can generally be achieved by the following regimen: a loading dose of 0.6 g/kg plus an hourly maintenance dose of 66 mg/kg in nondrinkers, 154 mg/kg in drinkers, and 240 mg/kg once hemodialysis is started [70,73]. If oral ethanol is given, the dose may have to be doubled if charcoal has been administered [73]. Regardless of the mode of administration, the plasma ethanol concentration should be monitored since adjustments in dosage will be required in some patients.

Hemodialysis is used to remove both the parent compound and metabolites. Sorbent-based hemodialysis systems should be avoided since drug clearance may be impaired because of rapid saturation of the cartridges [74]. Drug removal is also much slower with peritoneal dialysis [71], which should not be used unless hemodialysis is not available.

In general, ethanol is begun at the time of diagnosis and continued until the plasma methanol concentration is below 20 mg/dL [67]. Hemodialysis should also be instituted if the plasma level is greater than 50 mg/dL, more than 30 mL have been ingested, acidemia is present, or visual acuity is decreased [67].

Ethylene glycol Ethylene glycol is a component of antifreeze which is metabolized into a variety of toxic metabolites that are responsible for the clinical symptoms and metabolic acidosis (Fig. 19-3). After ingestion there are three clinical stages of varying severity [75,76]. During the first 12 h, neurologic symptoms predominate, ranging from drunkenness to coma. This is followed by the onset of cardiopulmonary abnormalities, such as tachypnea and pulmonary edema, and then flank pain and renal failure. The latter is primarily due to toxic damage to the tubules, although

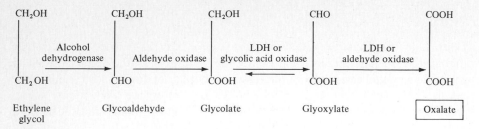

Figure 19-3 Pathway for metabolism of ethylene glycol to oxalate. LDH = lactate dehydrogenase. *(From M. F. Parry and R. Wallach, Am. J. Med., 57:143, 1974.)*

plugging of the tubules by the precipitation of oxalate crystals may also contribute [77]. The lethal dose of ethylene glycol is approximately 100 mL.

The diagnosis, suspected from the history and the possible presence of oxalate and hippurate crystals in the urine, can be confirmed by the demonstration of ethylene glycol in the serum. It should be noted that mannitol, which is frequently used to promote a diuresis after drug ingestions, can give a false-positive result in the standard ethylene glycol assay using sodium periodate and Schiff's aldehyde reagent [78]. The diagnosis may also be suspected indirectly by the presence of an osmolal gap in the plasma (see "Methanol" above). However, this is generally less prominent than with methanol since ethylene glycol is a larger molecule.

Treatment The specific treatment of ethylene glycol intoxication is identical to that for methanol: administration of ethanol and hemodialysis [73,75]. The administration of ethanol is equally important in this setting since alcohol dehydrogenase again promotes the production of toxic metabolites (Fig. 19-3).

Other toxins A variety of other toxic ingestions can rarely lead to metabolic acidosis. *Paraldehyde,* especially that which has been stored for more than 2 months, can result in the accumulation of organic acids such as acetic and chloroacetic acid [79]. It may be, however, that some of these cases actually represent alcoholic ketoacidosis [50]. The diagnosis can be established by the demonstration of acetaldehyde in the urine. The ingestion of *elemental sulfur,* used as a folk remedy, is associated with the generation of sulfuric acid [80]. The anion gap remains normal in this disorder since the excess sulfate is rapidly excreted in the urine. Sniffing of *toluene,* a component of paint thinners, model glues, and transmission fluid, can produce a metabolic acidosis in one of two ways: by inducing type 1 (distal) renal tubular acidosis (see below) and by causing the accumulation of organic acids (probably hippuric acid) [81]. The anion gap will be elevated in the latter condition. Finally, *ammonium chloride* has been used as a mild diuretic. In the liver,

$$NH_4Cl \rightarrow NH_3 + HCl$$

Therefore, the excessive use of this compound can lead to a normal anion gap (hyperchloremic) metabolic acidosis [82].

Hyperalimentation fluids The administration of hyperalimentation fluids can produce a metabolic acidosis by two mechanisms. First, some of these solutions (especially neoaminosol) contain an excess of cationic amino acids such as arginine and lysine. When these amino acids are utilized, H^+ ions are formed [83]:

$$R - NH_3^+ + O_2 \rightarrow urea + CO_2 + H_2O + H^+$$

This is in addition to the H^+ generated by the sulfur-containing amino acids in the solution. Second, starved patients may become hypophosphatemic when fed, resulting in a fall in phosphate and therefore titratable acid excretion. In this setting, the H^+ load associated with the metabolism of the administered protein (as with casein hydrolysates) is not completely excreted and metabolic acidosis ensues [84].

Gastrointestinal Loss of Bicarbonate

Diarrhea and fistulas The intestinal fluids below the stomach, including pancreatic and biliary secretions, are relatively alkaline. The net base in these fluids consists of HCO_3^- as well as organic anions, which if absorbed would be metabolized to HCO_3^- [85]. As a result, diarrhea, a villous adenoma, or the removal of pancreatic, biliary, or intestinal secretions (by tube drainage, fistulas, or vomiting if there is intestinal obstruction) can lead to metabolic acidosis. This is also true of occult laxative abuse, which should be considered in any patient with a hyperchloremic metabolic acidosis of unknown etiology [86].

Ureterosigmoidostomy Implantation of the ureters into the sigmoid colon has been used to treat patients with neurologic bladder abnormalities, obstructive uropathy due to locally invasive tumor, or surgical removal of the bladder for carcinoma. A hyperchloremic metabolic acidosis is a relatively common complication of this procedure and is due to two factors. First, the colon has an anion exchange pump: luminal Cl^- being reabsorbed as HCO_3^- is secreted [87]. Thus, when Cl^- enters the colon (in this case from the urine), Cl^- and HCO_3^- are exchanged and HCO_3^- is lost [88]. Second, the colon reabsorbs NH_4^+ which is derived from both the urine and urea-splitting bacteria in the colon [89]. In the liver, the NH_4^+ is metabolized into NH_3 and H^+.

In recent years, this complication has been minimized by implantation of the ureters into a short loop of ileum which opens at the abdominal wall (ureteroileostomy). With this procedure, contact time between the urine and intestine is normally too short for significant anion exchange to occur.

Cholestyramine Cholestyramine chloride is an orally administered resin used in the treatment of hypercholesterolemia. It is nonreabsorbable and can act as an anion-exchange resin, exchanging its Cl^- for endogenous HCO_3^- and producing a metabolic acidosis [90].

Table 19-6 Characteristics of the different types of renal tubular acidosis†

	Type 1 (distal)	Type 2 (proximal)	Type 4
Basic defect	Decreased acidification in the distal nephron	Diminished proximal HCO_3^- reabsorption	Aldosterone deficiency or resistance
Urine pH during acidemia	>5.3	Variable: >5.3 if above reabsorptive threshold, <5.3 if below	<5.3
Plasma $[HCO_3^-]$ (untreated)	May be below 10 meq/L	Usually 14 to 20 meq/L	Usually above 15 meq/L
Fractional excretion of HCO_3^- at normal plasma $[HCO_3^-]$	<3% in adults; may reach 5–10% in children	>15–20%	
Diagnosis	Response to $NaHCO_3$ or ammonium chloride load	Response to $NaHCO_3$	Measure plasma aldosterone concentration
Plasma $[K^+]$	Usually reduced; may be normal; rarely elevated	Normal or reduced	Elevated
Therapeutic HCO_3^- requirement to normalize plasma $[HCO_3^-]$, meq/kg per day	1–2 in adults, 4–14 in children	10–15	1–3; may require none if hyperkalemia corrected
Nonelectrolyte complications	Nephrocalcinosis and renal stones; rickets or osteomalacia is uncommon	Rickets or osteomalacia	None

†What had been called type 3 RTA is actually a variant of type 1 (see below).

Renal Tubular Acidosis

Renal tubular acidosis refers to those conditions in which metabolic acidosis results from diminished net tubular H^+ secretion [91,92]. There are three major types of RTA, the characteristics of which are summarized in Table 19-6. Although these disorders are relatively unusual in adults (with the exception of type 4 RTA), they provide interesting examples of the different ways in which the renal regulation of acid-base balance can be impaired.

The acidosis associated with renal failure could also be included in this group. However, NH_4^+ excretion per total GFR in this disorder is equal to that achieved in acidemic patients with normal renal function [52]. Thus, the major problem is too few functioning nephrons, not diminished tubular function. In addition, the ability to acidify the urine (urine pH < 5.3) is maintained in renal failure in contrast to type 1 and, in some circumstances, type 2 RTA [53,54].

Type 1 (distal) RTA Type 1 RTA is characterized by a decrease in net H^+ secretion in the collecting tubules such that the urine pH, which can normally be lowered to a minimum of 4.5 to 5.0 in these segments, remains above 5.3 . This defect in acidification diminishes NH_4^+ and titratable acid excretion, thereby preventing excretion of all of the daily acid load. As a result, there is continued H^+ retention, leading to a progressive reduction in the plasma HCO_3^- concentration, which may fall below 10 meq/L.

Three mechanisms have been postulated to account for the inability to acidify the urine in type 1 RTA: a direct impairment in the H^+ secretory pump, an increase in membrane permeability which allows back-diffusion of H^+ (or H_2CO_3) to occur, or a primary defect in Na^+ transport since Na^+ reabsorption normally creates an electrical gradient (lumen negative; see Fig. 13-9) that facilitates net H^+ secretion [93]. It is likely that each of these mechanisms may contribute in different patients [94]. For example, amphotericin B appears to increase membrane permeability, whereas lithium and obstructive uropathy probably impair Na^+ reabsorption [93–95].

Although proximal HCO_3^- reabsorption is intact in this disorder, variable degrees of bicarbonaturia are present due to the high urine pH. If, for example, the urine P_{CO_2} is 46 mmHg (similar to that in venous blood), then, from the Henderson-Hasselbalch equation, the urinary HCO_3^- concentration will vary with the urine pH:†

$$pH = pK'_a + \log \frac{[HCO_3^-]}{0.03 P_{CO_2}}$$

At a urine pH below 6.0, the urinary HCO_3^- concentration is negligible. In adults with type 1 RTA, the urine pH is usually less than 6.5, resulting in a relatively mild degree of urinary HCO_3^- loss with less than 3 percent of the filtered HCO_3^- being excreted.‡ In children, however, the minimum urine pH is generally higher and HCO_3^- losses are greater. When the urine pH exceeds 7.0, the fractional excretion of HCO_3^- can reach 5 to 10 percent, thereby making an important contribution to the development of acidemia. This syndrome, which has been called type 3 RTA, occurs in infants who within a few years have a lower urine pH and follow a course more typical of type 1 RTA [97].

Incomplete type 1 RTA Some patients with defective urinary acidification do not become acidemic, a syndrome that is referred to as incomplete type 1 RTA [54,98]. Although a high urinary pH impairs the efficiency with which urinary NH_3 combines with H^+ to form NH_4^+ (see Fig. 12-6), this can be overcome by enhanced NH_3 production by the tubular cell, which allows net acid excretion to remain normal. It is not known why patients with complete type 1 RTA are not able to similarly augment NH_3 production; it may be that this represents a later stage in the disease in which there is a greater impairment in tubular function [91,99].

---

†The urine pK'_a varies with the total electrolyte concentration and may be somewhat different from the plasma value of 6.10 [96].

‡The fractional excretion of HCO_3^- can be calculated from the following formula (similar to that for the fractional excretion of Na^+; see Chap. 14):

$$FE_{HCO_3^-}(\%) = \frac{\text{urine } [HCO_3^-] \times \text{plasma [creatinine]}}{\text{plasma } [HCO_3^-] \times \text{urine [creatinine]}} \times 100$$

Table 19-7 Clinical spectrum of type 1 (distal) RTA†

Primary (no obvious systemic disease)
  A. Sporadic
  B. Genetically transmitted

Autoimmune disorders
  A. Dysgammaglobulinemia
      1. Hyperglobulinemic purpura
      2. Cryoglobulinemia
  B. Sjögren's syndrome
  C. Chronic active hepatitis
  D. Primary biliary cirrhosis
  E. Thyroiditis
  F. Fibrosing alveolitis

Disorders causing nephrocalcinosis
  A. Idiopathic hypercalciuria
      1. Sporadic
      2. Hereditary
  B. Primary hyperparathyroidism
  C. Hyperthyroidism
  D. Vitamin D intoxication
  E. Medullary sponge kidney
  F. Hereditary fructose intolerance (after chronic fructose ingestion)
  G. Wilson's disease
  H. Fabry's disease

Drug- or toxin-induced nephropathy
  A. Amphotericin B
  B. Toluene
  C. Analgesics
  D. Lithium
  E. Cyclamate

Other renal diseases
  A. Pyelonephritis
  B. Obstructive uropathy
  C. Renal transplantation
  D. Leprosy

Genetically transmitted systemic diseases
  A. Ehlers-Danlos syndrome
  B. Hereditary elliptocytosis
  C. Sickle cell anemia
  D. Marfan's syndrome
  E. Carbonic anhydrase B deficiency

Hepatic cirrhosis

---

†From R. C. Morris, Jr., and A. Sebastian, in *Clinical Disorders of Fluid and Electrolyte Metabolism,* 3d ed., M. H. Maxwell and C. R. Kleeman (eds.), McGraw-Hill, New York, 1980, chap. 18.

Etiology Many different conditions have been associated with type 1 RTA (Table 19-7). Autoimmune disorders such as Sjögren's syndrome and, in recreational drug abusers, the sniffing of toluene are probably the most common identifiable causes of this disorder in adults whereas hereditary RTA is most common in children [91,99a].

Diagnosis The presence of type 1 RTA should be suspected in any patient with metabolic acidosis and a urine pH greater than 5.3 in adults or 5.6 in children [54,100]. In the absence of a high urine pH due to infection with a urea-splitting organism, the only other disorder that can produce this combination is type 2 RTA. These conditions can be differentiated by the response to a HCO_3^- load, as the urine pH and fractional excretion of HCO_3^- will increase (as the plasma HCO_3^- concentration is elevated) only in type 2 RTA (see below). If, however, the plasma HCO_3^- concentration is normal (as in the incomplete form) or only mildly reduced, the diagnosis can be established by giving an acid load as NH_4Cl† in a dose of 0.1 g/kg [54]. This should induce a 4- to 5-meq/L fall in the plasma HCO_3^- concentration within 4 to 6 h. The urine pH will remain above 5.3 in type 1 RTA but will be less than this value and usually below 5.0 in type 2 RTA and in normal subjects since acidemia stimulates maximal urinary acidification.

Nephrocalcinosis Hypercalciuria, hyperphosphaturia, nephrolithiasis (with calcium phosphate or struvite stones), and nephrocalcinosis are frequently associated with type 1 RTA [102,103]. Acidemia can increase calcium phosphate release from bone [58] and reduce the tubular reabsorption of these ions [104,105]. Both of these changes will lead to hypercalciuria and hyperphosphaturia, the severity of which is proportional to the fall in the plasma HCO_3^- concentration [105a]. In addition, hypercalciuria precedes the acidemia in some patients with hereditary RTA, suggesting that hypercalciuria is the primary problem and that calcium-induced tubular damage is then responsible for the RTA [91,106,106a]. Nephrocalcinosis does not generally occur in incomplete type 1 RTA (where acidemia is not present) unless hypercalciuria is the primary defect [91].

Once calcium and phosphate excretion are increased, the persistently elevated urine pH promotes calcium phosphate precipitation and the development of nephrocalcinosis and nephrolithiasis. Decreased urinary excretion of citrate, occurring by an uncertain mechanism, also contributes to this problem, [91,99,103,107].‡ Citrate normally forms a nondissociable but soluble complex with calcium, thereby decreasing the quantity of the calcium available for precipitation.

---

†It is difficult to administer NH_4Cl to young children. Arginine hydrochloride, which is also metabolized to HCl, may be preferable in this setting [101].

‡Citrate excretion is generally reduced in metabolic acidosis. This response appears to be mediated by the associated reduction in proximal tubular cell pH which promotes citrate entry into the mitochondria [108]. As a result, the cytoplasmic citrate concentration falls which promotes citrate reabsorption from the tubular lumen into the cell. However, the observations that hypocitraturia can occur without acidemia in incomplete type 1 RTA and that correction of the acidemia may not normalize citrate excretion [99,102] suggest that the fall in citrate excretion may in part be a manfestation of the underlying metabolic defect that is responsible for the RTA [91].

All of the above changes may respond to early and complete correction of the acidemia: less calcium phosphate release from bone, enhanced tubular reabsorption of these ions, a possible increase in urinary citrate, and prevention of nephrocalcinosis and nephrolithiasis [91,92,102,103]. However, alkali therapy alone is less likely to prevent nephrocalcinosis in those patients with hereditary RTA in whom hypercalciuria is the primary defect [91,106]. In this setting, conventional therapy for calcium stones such as a thiazide diuretic or orthophosphates may be effective [108a].

Type 2 (proximal) RTA A different problem is present in type 2 RTA: decreased proximal HCO_3^- reabsorption (Table 19-6) [91,92]. Normal subjects who are euvolemic reabsorb essentially all the filtered HCO_3^- until the HCO_3^- concentration in the plasma and, therefore, in the glomerular filtrate exceeds 26 meq/L (see Chap. 5). Above this level, the excess HCO_3^- is appropriately excreted in the urine. Approximately 90 percent of this HCO_3^- reabsorption occurs in the proximal tubule.

In type 2 RTA, proximal HCO_3^- reabsorption is reduced as is total HCO_3^- reabsorptive capacity. If, for example, only 17 meq/L of glomerular filtrate can be reabsorbed,† then HCO_3^- will be lost in the urine until the plasma HCO_3^- concentration reaches 17 meq/L. At this point, all the filtered HCO_3^- can be reabsorbed and a new steady state is achieved. Thus, type 2 RTA is a *self-limiting disorder* in which the plasma HCO_3^- concentration is usually between 14 and 20 meq/L [91,92]. This is in contrast to the progressive acidemia in type 1 RTA.

The difference between types 1 and 2 RTA can also be appreciated by examining the relationship between urinary HCO_3^- excretion and the plasma HCO_3^- concentration (Fig. 19-4). In normal subjects, HCO_3^- does not significantly appear in the urine until the plasma HCO_3^- concentration exceeds 26 meq/L. Thus, the urine pH will be very acid (<6.0) below this level. This relationship is shifted to a lower plasma HCO_3^- concentration in type 2 RTA. If maximal reabsorptive capacity is 17 meq/L, then increasing amounts of HCO_3^- will be excreted in the urine if $NaHCO_3$ is administered to raise the plasma HCO_3^- concentration above this level. By the time the plasma HCO_3^- concentration reaches 22 to 24 meq/L, more than 15 to 20 percent of the filtered HCO_3^- will appear in the urine and the urine pH will exceed 7.5. In contrast, the urine pH can be made maximally acid (<5.0) if the plasma HCO_3^- concentration is much below 17 meq/L since all the filtered HCO_3^- can now be absorbed and distal acidification is normal.

The curve in type 1 RTA, on the other hand, is similar to normal subjects except that the elevated urine pH obligates a *fixed* degree of bicarbonaturia. This is generally of minor importance in adults but can be substantial in children. The urine pH and amount of urinary HCO_3^- do not increase further as the plasma HCO_3^- concentration is raised toward normal.

---

†Although this represents a fall in bicarbonate reabsorptive capacity of 9 meq/L (from 26 to 17 meq/L), the decrease in proximal reabsorption may be substantially greater since much of the excess HCO_3^- can be reclaimed in the distal nephron. In experimental animals, for example, carbonic anhydrase inhibition (see Chap. 12) can inhibit up to 80 percent of proximal HCO_3^- reabsorption, yet only 25 to 30 percent of the filtered HCO_3^- appears in the urine due to enhanced distal reabsorption [108b].

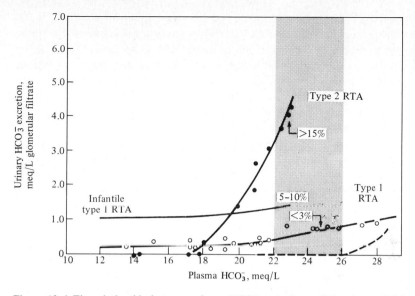

Figure 19-4 The relationship between urinary HCO_3^- excretion and the plasma HCO_3^- concentration in normal subjects (dashed line) and in patients with type 1 and type 2 RTA as $NaHCO_3$ is administered to raise the plasma HCO_3^- concentration. In the latter condition, there is little urinary HCO_3^- and an acid urine pH when the plasma HCO_3^- concentration is below the maximal reabsorptive capacity. Above this level, however, there is a rapid increase in HCO_3^- excretion such that more than 15 percent of the filtered HCO_3^- load is excreted at a normal plasma HCO_3^- concentration (shaded area). Patients with type 1 RTA, on the other hand, are similar to normal subjects except that there is a fixed degree of bicarbonaturia obligated by the high urine pH. In adults, this is generally less than 3 percent of the filtered load but can reach 5 to 10 percent in infantile type 1 RTA in which there is a higher minimal urine pH. (*Adapted from A. Sebastian, E. McSherry, and R. C. Morris, Jr., in The Kidney*, B. M. Brenner and F. C. Rector, Jr. (eds.), Saunders, Philadelphia, 1976, chap. 16.)

The defect in HCO_3^- reabsorption in type 2 RTA may occur alone or as part of the Fanconi syndrome in which a variety of other proximal functions may be impaired, including the reabsorption of phosphate, glucose, amino acids, and urate [109]. In this setting, metabolic acidosis is accompanied by hypophosphatemia, hypouricemia, aminoaciduria, and glucosuria at a normal plasma glucose concentration.

Many theories have been proposed to explain the tubular dysfunction in type 2 RTA, and it is likely that each may contribute in selected patients [91,109]. In conditions associated with chronic hypocalcemia, for example, both the low calcium concentration and the associated secondary hyperparathyroidism can decrease proximal HCO_3^- reabsorption [110,111]. In idiopathic cases, on the other hand, anatomic changes in the proximal tubule [112], decreased activity of carbonic anhydrase (the enzyme that facilitates HCO_3^- reabsorption; see page 230) [113], and inhibition of Na^+-K^+-ATPase (the enzyme regulating active Na^+ transport; see Chap. 5) [114] have all been described.

Etiology A variety of congenital and acquired disorders can cause type 2 RTA (Table 19-8). Idiopathic RTA and cystinosis are probably most commonly responsible in children; chronic hypocalcemia due to some form of vitamin D deficiency and multiple myeloma (which may be latent) are most often responsible in adults [91,115].

Table 19-8 Clinical spectrum of type 2 (proximal) RTA

Associated with multiple dysfunction of proximal tubule (Fanconi syndrome)
A. Primary (no obvious systemic disease)
   1. Sporadic
   2. Genetically transmitted
B. Genetically transmitted systemic diseases
   1. Cystinosis
   2. Lowe's syndrome
   3. Wilson's disease
   4. Tyrosinemia
   5. Hereditary fructose intolerance (during fructose administration or ingestion)
   6. Pyruvate carboxylase deficiency
C. Disorders associated with chronic hypocalcemia and secondary hyperparathyroidism
   1. Vitamin D deficiency
   2. Vitamin D dependency
D. Drug- or toxin-induced nephropathy
   1. Outdated tetracycline
   2. Methyl-5-chromon (diacramone)
   3. Streptozotocin
   4. Lead
E. Other renal diseases
   1. Amyloidosis
   2. Nephrotic syndrome
   3. Renal transplantation
   4. Sjögren's syndrome
   5. Medullary cystic disease
   6. Paroxysmal nocturnal hemoglobinuria
   7. Renal vein thrombosis
F. Multiple myeloma (with monoclonal immunoglobulin light-chainuria)
   1. Fully expressed
   2. Latent

Unassociated with multiple dysfunction of proximal tubule
A. Primary
   1. Sporadic
     a. Transient
     b. Persisting
   2. Genetically transmitted
B. Osteopetrosis
C. Carbonic anhydrase deficiency
   1. Acetazolamide
   2. Sulfanilamide
D. York-Yendt syndrome
E. Cyanotic congenital heart disease

†From R. C. Morris, Jr., and A. Sebastian, in *Clinical Disorders of Fluid and Electrolyte Metabolism,* 3d ed., M. H. Maxwell and C. R. Kleeman (eds.), McGraw-Hill, New York, 1980, chap 18.

Diagnosis The diagnosis of type 2 RTA is established by raising the plasma HCO_3^- concentration toward normal with an infusion of $NaHCO_3$ (at a rate of 0.5 to 1.0 meq/kg per hour). Although the initial urine pH may be appropriately acid, it will increase rapidly once the reabsorptive threshold is exceeded. As a result, the urine pH will be greater than 7.5 and the fractional excretion of HCO_3^- greater than 15 to 20 percent as the plasma HCO_3^- concentration approaches normal (Fig. 19-4). Patients should also be evaluated for other signs of proximal dysfunction. Hypophosphatemia is particularly important since it can contribute to the development of bone disease.

Bone disease Rickets in children and osteomalacia or osteopenia in adults are relatively common in type 2 RTA, occurring in up to 20 percent of cases [103]. Although these skeletal abnormalities may also be present in other acidemic states, their frequency in type 2 RTA may be due in part to phosphate wasting and acquired vitamin D deficiency.

In contrast to type 1 RTA, nephrocalcinosis and nephrolithiasis do not occur in type 2 RTA [103]. Two factors may combine to protect against this complication: the ability to lower the urine pH which increases the solubility of calcium phosphate; and the presence of nonreabsorbed amino acids and organic anions (including citrate) which can form soluble complexes with calcium [103].

Type 4 RTA Type 4 RTA refers to metabolic acidosis resulting from aldosterone deficiency or resistance. Aldosterone normally promotes distal K^+ and H^+ secretion as well as Na^+ reabsorption (see Chap. 8). As a result, hypoaldosteronism leads to hyperkalemia and metabolic acidosis. Since the elevation in the plasma K^+ concentration is generally more prominent, this disorder is discussed in detail in Chap. 28.

Two factors contribute to the impaired H^+ secretion with aldosterone deficiency: decreased activity of the H^+ secretory pump and, more importantly, reduced NH_3 synthesis induced by hyperkalemia [116–118]. The latter effect may result from a transcellular cation exchange. As some of the excess K^+ enters the cells, H^+ (and Na^+) leave to maintain electroneutrality. The ensuing intracellular alkalosis in the renal tubular cells then retards NH_3 formation (see page 240).

In contrast to type 1 RTA, the urine pH is below 5.3 during acidemia in patients with type 4 RTA [91,92,117]. This finding is compatible with the primary role of decreased NH_3 production in that the ability to acidify the urine is relatively normal, but total acid excretion is reduced due to a lack of available buffers.

Potassium balance in RTA Potassium wasting and hypokalemia are common in both type 1 and type 2 RTA. The degree to which this occurs is variable, depending in part upon whether or not the acidemia has been corrected. In type 2 RTA, K^+ excretion increases markedly with elevation of the plasma HCO_3^- concentration and is directly related to the associated bicarbonaturia (Fig. 19-5) [119]. The increase in delivery of water, Na^+, and the nonreabsorbable anion HCO_3^- to the distal nephron all combine to promote K^+ excretion (see Chap. 13).

Persistent hyperaldosteronism is also present in type 2 RTA [119] and plays an

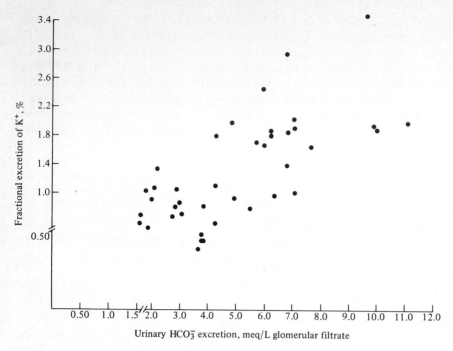

Figure 19-5 Relationship between the fractional excretion of filtered K^+ and urinary HCO_3^- excretion in patients with type 2 RTA in whom the plasma HCO_3^- concentration was maintained at normal levels (22 to 26 meq/L). *(Adapted from A. Sebastian, E. McSherry, and R. C. Morris, Jr., J. Clin. Invest., 50:231, 1971, by copyright permission of The American Society for Clinical Investigation.)*

important role since distal Na^+ reabsorption must be stimulated for $NaHCO_3$ to augment K^+ excretion. Decreased proximal HCO_3^- reabsorption (as occurs in type 2 RTA) will diminish proximal NaCl reabsorption (see page 93), and it is this continued tendency to salt wasting that may be responsible for the hyperaldosteronism in this disorder.

In contrast, K^+ wasting is maximal *before* therapy in type 1 RTA and is partially or completely reversed by correction of the acidemia [120,121]. Since net H^+ secretion is diminished in the distal nephron, there is an impairment in Na^+-H^+ exchange. As a result, more Na^+ must be reabsorbed in exchange for secreted K^+. As the plasma HCO_3^- concentration is elevated toward normal, the urine pH rises because more HCO_3^- is now delivered distally. This allows more distal H^+ secretion to occur without reaching the limiting pH gradient, thereby minimizing the need for excessive K^+ secretion [120,121].

Type 1 RTA may also be associated with *hyper*kalemia. This occurs when the underlying mechanism is a primary decrease in distal Na^+ reabsorption, thereby impairing the generation of the lumen negative potential difference [93,95,122]. In this setting, which most often occurs with obstructive uropathy or sickle cell disease, the diminished potential difference reduces both H^+ and K^+ secretion (see page

260). This can be differentiated from the hyperkalemic acidosis in type 4 RTA by the normally low urine pH in the latter condition [122].

Treatment of RTA Although the general principles involved in the treatment of metabolic acidosis will be discussed later in the chapter, the unique problems presented by the different types of RTA justify their therapy being considered separately. *Complete* correction of the acidemia (plasma HCO_3^- concentration of 22 to 24 meq/L) is usually important in these chronic disorders since it allows normal growth to occur in children [59,91,123], promotes healing of bone disease (with vitamin D and phosphate supplementation if hypophosphatemia is present) [91,103], and, in type 1 RTA, can prevent nephrocalcinosis and nephrolithiasis at least in part by markedly reducing the rate of urinary calcium excretion [91,102,103,105a]. In addition, any underlying cause (drug toxicity, vitamin D deficiency) should be corrected if possible. The adequately treated patient with RTA (but without a complicating systemic disease) should be asymptomatic and able to lead a normal life. Idiopathic type 1 RTA is generally permanent, but type 2 RTA in children may disappear spontaneously after several years [123].

The alkali requirement for correction of type 1 RTA is variable: it must be equal to that fraction of the daily H^+ load that is not excreted *plus* the fixed urinary HCO_3^- losses (Fig. 19-4). The amount given is determined empirically by following the plasma HCO_3^- concentration. In adults, the HCO_3^- losses are realtively small, and usually only 1 to 2 meq/kg per day has to be given. In children, however, the urine pH and fixed HCO_3 losses are higher, and 4 to 14 meq/kg per day (in divided doses) may be necessary [59]. Chronic K^+ replacement may or may not be required since K^+ wasting diminishes but may not disappear as the pH is normalized [120]. However, large doses of KCl may be required initially in patients who present with hypokalemia since the administration of $NaHCO_3$ will drive K^+ into the cells, further reducing the plasma K^+ concentration (see page 635).

Correction of type 2 RTA is more difficult. Raising the plasma HCO_3^- concentration will cause the filtered HCO_3^- load to exceed tubular reabsorptive capacity, resulting in the rapid excretion of much of the administered alkali. To stay ahead of urinary excretion may require 10 to 15 meq/kg per day. In addition, an empirically determined fraction of this must be given as the K^+ salt since there will be a concomitant increase in urinary K^+ excretion (Fig. 19-5). It should be noted that mild metabolic acidosis in adults may not require alkali therapy since complications are unusual and the acidemia is self-limited.

When large doses are necessary in type 2 RTA, adding a thiazide diuretic and a low-salt diet may be helpful [124]. This combination induces mild volume depletion which will increase the proximal reabsorption of Na^+ and, secondarily, that of HCO_3^- (see Chap. 5).

In either type 1 or 2 RTA, alkali therapy is usually administered as $NaHCO_3$ (or $KHCO_3$). To minimize the expense, one 8-oz box of baking soda (a $NaHCO_3$ preparation) can be added to 2.88 L of distilled water [91]. This produces a solution containing 1 meq of HCO_3^- per milliliter, which can last for at least 2 months if kept refrigerated and capped between pourings.

As an alternative, alkali can be given as a 1-meq/mL sodium citrate solution (Bicitra) since citrate is converted in the body into HCO_3^-. Although more expensive, this has the advantage of avoiding the abdominal bloating and eructations frequently produced by large doses of $NaHCO_3$. A modified solution in which one-half the citrate is present as the K^+ salt (Polycitra) can also be used in patients who need K^+ supplementation. This solution contains 2 meq of HCO_3^- per milliliter.

Type 4 RTA is generally treated with mineralocorticoid replacement [117,125]. In edematous or hypertensive patients in whom mineralocorticoid-induced Na^+ retention would be deleterious, a low-K^+ diet and a loop diuretic such as furosemide are usually effective [126].

Massive Rhabdomyolysis

Another rare cause of a high anion gap metabolic acidosis is massive rhabdomyolysis [127]. The presumed mechanism is the release of H^+ and organic anions from the damaged cells. This diagnosis should be suspected if there is a marked elevation in the plasma level of creatine phosphokinase (as well as that of other muscle enzymes) and no other apparent cause for the acidemia.

SYMPTOMS

Metabolic acidosis can result in changes in pulmonary, cardiovascular, neurologic, and bone functions. Since the respiratory compensation results in as much as a four- to eight fold increase in minute ventilation (see Fig. 12-13), the patient may complain of dyspnea on exertion and, with severe acidemia, even at rest. Furthermore, the observation of hyperpnea (affecting the depth more than the rate of ventilation) on physical examination may be the only clue toward the presence of an underlying acidemic state.

A fall in the arterial pH to less than 7.00 to 7.10 can predispose toward potentially fatal ventricular arrhythmias [128] and can reduce both cardiac contractility and the inotropic response to catecholamines [129,130]. This decrease in ventricular function may play an important role in the perpetuation of shock-induced lactic acidosis, and partial correction of the acidemia may be required before tissue perfusion can be restored.

Neurologic symptoms ranging from lethargy to coma have been described in metabolic acidosis. These symptoms appear to be more closely related to the fall in pH in the cerebrospinal fluid (CSF) than that in the plasma [131]. In general, neurologic abnormalities are less prominent in metabolic acidosis than in respiratory acidosis. This may be due to the fact that the lipid-soluble CO_2 crosses the blood-brain barrier much more rapidly than the water-soluble HCO_3^- and, therefore, produces a greater fall in CSF pH [132,133].

Neurologic symptoms may also be induced by factors other than the reduction in pH. These include toxic effects of ingestions and hyperosomalality due to hyperglycemia in diabetic ketoacidosis [134].

Chronic acidemia, as with renal failure or renal tubular acidosis, can lead to a variety of skeletal problems that are probably due in part to negative calcium and phosphate balance (see "Renal tubular acidosis" above). Of particular importance is impaired growth in children [59,135]. Other abnormalities that may occur include osteitis fibrosa (from secondary hyperparathyroidism), rickets in children, and osteomalacia or osteopenia in adults [59,103,136]. Correction of the acidemia may reverse these changes in patients without renal failure [59,103,136]. Therapy is generally less successful with advanced renal disease since other factors also contribute to the bone disease, such as hyperparathyroidism, vitamin D deficiency, and poor nutrition due to anorexia [135,137,138].

In infants and young children, acidemia may also be associated with a variety of nonspecific symptoms such as anorexia, nausea, weight loss, muscle weakness, and listlessness [92]. These changes are reversible with the restoration of acid-base balance.

TREATMENT

General Principles

Correction of the acidemia can usually be achieved by the administration of alkali. In most clinical situations, $NaHCO_3$ is the mainstay of therapy. Sodium citrate, sodium lactate, and tris buffer have also been used but generally offer no particular advantage over $NaHCO_3$ [139,140].

The initial therapeutic goal in patients with severe acidemia is to raise the systemic pH to about 7.20, a level at which arrhythmias become less likely and cardiac contractility and responsiveness to catecholamines will be restored. To achieve this pH usually requires only a small increment in the plasma HCO_3^- concentration. For example, the following arterial blood values were obtained from a patient with chronic diarrhea:

$$pH = 7.10$$

$$P_{CO_2} = 20 \text{ mmHg}$$

$$[HCO_3^-] = 6 \text{ meq/L}$$

The level to which the plasma HCO_3^- concentration must be raised for the pH to reach 7.20 can be calculated from the Henderson-Hasselbalch equation:

$$pH = 6.10 + \log \frac{[HCO_3^-]}{0.03 P_{CO_2}}$$

If we assume that the P_{CO_2} will remain constant, then

$$7.20 = 6.10 + \log \frac{[HCO_3^-]}{0.03 \times 20}$$

Since this equation is difficult to solve at the bedside, it is easier to express the relationship between these parameters in nonlogarithmic terms [Eq. 19-2]:

$$[H^+] = 24 \times \frac{P_{CO_2}}{[HCO_3^-]}$$

Since the H^+ concentration is 63 nanoeq/L at a pH of 7.20 (see Table 17-1),

$$63 = 24 \times \frac{20}{[HCO_3^-]}$$

$$[HCO_3^-] = 8 \text{ meq/L}$$

In fact, this calculation slightly underestimates the initial HCO_3^- requirement since the drive to compensatory hyperventilation will diminish as the acidemia is corrected, resulting in an increase in the P_{CO_2}. If we assume that the P_{CO_2} will rise from 20 to 25 mmHg, then

$$63 = 24 \times \frac{25}{[HCO_3^-]}$$

$$[HCO_3^-] = 10 \text{ meq/L}$$

Thus, *only a small increase in the plasma HCO_3^- concentration is necessary to get the patient out of danger if there is a normal respiratory compensation.*

Rapid administration of HCO_3^- is particularly important in patients with severe metabolic acidosis. In this setting, even a small additional reduction in the plasma HCO_3^- concentration results in a large percentage change and therefore can induce an immediately life-threatening degree of acidemia. For example, lowering the plasma HCO_3^- concentration from 24 to 22 meq/L in a patient with an initial pH of 7.40 and P_{CO_2} of 40 mmHg will have only a minor effect on the pH and H^+ concentration:

$$[H^+] = 24 \times \frac{40}{22} = 44 \text{ nanoeq/L} \qquad (\text{pH} = 7.36)$$

However, a similar 2-meq/L reduction in a patient with an initial pH of 7.11, a plasma HCO_3^- concentration of 4 meq/L, and a P_{CO_2} of 13 mmHg will now decrease the pH to 6.81:

$$[H^+] = 24 \times \frac{13}{2} = 156 \text{ nanoeq/L} \qquad (\text{pH} = 6.81)$$

Regardless of the initial severity, rapid correction of the pH to above 7.20 to 7.25 is not only unnecessary but can induce potentially dangerous reductions in the CSF pH and in tissue oxygen delivery. The administration of $NaHCO_3$ will tend to lower minute ventilation and raise the P_{CO_2}. Since CO_2 crosses the blood-brain barrier much more rapidly than HCO_3^-, the brain will acutely sense only the elevation in the P_{CO_2}. Thus, the CSF pH will become more acid, with possible aggravation of the neurologic symptoms [131]. The increase in the arterial pH will also shift the hemoglobin dissociation curve to the left, increasing the affinity of hemoglobin for oxygen and possibly reducing tissue oxygen delivery.

Bicarbonate Deficit

The amount of HCO_3^- required to correct the acidemia can be calculated from

$$HCO_3^- \text{ deficit} = HCO_3^- \text{ space} \times HCO_3^- \text{ deficit per liter} \qquad (19\text{-}8)$$

In metabolic acidosis, the HCO_3^- space is roughly 70 percent of the lean body weight (in kg)† [2]. If the normal plasma HCO_3^- concentration is 24 meq/L, then

$$HCO_3^- \text{ deficit} = 0.7 \times \text{lean body weight} \times (24 - \text{plasma } [HCO_3^-]) \quad (19\text{-}9)$$

If the above patient with a plasma HCO_3^- concentration of 6 meq/L weighed 70 kg,

$$HCO_3^- \text{ deficit} = 0.7 \times 70 \times (24 - 6) = 882 \text{ meq}$$

Since we want to raise the plasma HCO_3^- concentration only to 10 meq/L, the initial HCO_3^- requirement will be

$$HCO_3^- \text{ deficit} = 0.7 \times 70 \times (10 - 6) = 196 \text{ meq}$$

Thus, 196 meq of HCO_3^- should be given intravenously over the first several hours. If this is effective in raising the pH to a safe level, further HCO_3^- may be unnecessary since renal H^+ excretion will slowly regenerate the lost HCO_3^-. If the initial pH is greater than 7.20 and the underlying process (such as diarrhea) can be controlled, then *exogenous HCO_3^- may not be required* as long as the patient is asymptomatic.

Needless to say, *these are only rough guidelines and cannot replace serial measurements of the plasma pH in the management of the patient.* For example, this calculation will significantly underestimate the HCO_3^- deficit in two situations. First, this formula is dependent upon the presence of a steady state. If the patient is continuously producing acid, as in lactic acidosis, the HCO_3^- requirements will increase with time. Second, the intracellular buffers play a progressively increasing role in severe metabolic acidosis (plasma HCO_3^- concentration less than 10 meq/L). In this setting, the HCO_3^- space can approach 90 percent of the body weight [2].

The degree to which exogenous HCO_3^- will raise the plasma HCO_3^- concentration and pH is also dependent upon when the measurements are made. As described above, excess H^+ ions are buffered first in the extracellular fluid and then in the cells. A similar sequence occurs when HCO_3^- is given to correct a metabolic acidosis. Acutely, the added HCO_3^- is limited to the vascular space, producing a large increase in the plasma HCO_3^- concentration. However, this change is attenuated as the exogeneous HCO_3^- equilibrates through the total extracellular fluid within 15 min and

---

†This is greater than the values of 60 percent in normal subjects and 50 percent in patients with metabolic alkalosis. The possible mechanisms by which the HCO_3^- space varies inversely with the plasma HCO_3^- concentration are discussed in Ref. 2.

then equilibrates with the intracellular buffers in 2 to 4 h.† For example, if we assume that the extracellular volume is 15 liters and the total HCO_3^- space is 49 liters in a 70-kg male, the infusion of 90 meq of HCO_3^- will produce a 6-meq/L increase in the plasma HCO_3^- concentration in 15 min but only a 1.8-meq/L increase in 2 to 4 h. Thus, the plasma pH will be greater if measured at 15 min than after 2 to 4 h, when equilibrium with the intracellular buffers will have occurred. As a result, it should be recognized that measurement of the pH shortly after HCO_3^- has been given may overestimate the effect of the administered HCO_3^-.

Use of Bicarbonate in Lactic Acidosis and Ketoacidosis

The extracellular buffering of H^+ by HCO_3^- results in a decrease in the plasma HCO_3^- concentration [Eq. (19-1)]. Lactic acidosis and ketoacidosis differ from the other causes of metabolic acidosis in that the anion accompanying the H^+ can be metabolized back to HCO_3^- [19,142,143].

$$Lactate^- + 3O_2 \rightarrow 2CO_2 + 2H_2O + HCO_3^-$$

$$\beta\text{-hydroxybutyrate}^- + 4.5O_2 \rightarrow 3CO_2 + 3H_2O + HCO_3^-$$

Thus, there is no loss of "potential HCO_3^-" in these disorders except for the excretion of ketoacid or lactate anions in the urine. In ketoacidosis, insulin alone will raise the plasma HCO_3^- concentration toward normal [143,144]. Improvement in effective tissue perfusion, e.g., by elevating the blood pressure in a patient with shock, will have the same effect in lactic acidosis [142].

The role of HCO_3^- therapy in ketoacidosis is extremely controversial. It has been argued that, since insulin will spontaneously correct the acidemia, the administration of exogenous $NaHCO_3$ may increase the plasma HCO_3^- concentration too rapidly, resulting in clinical deterioration due to the paradoxical fall in the CSF pH and perhaps to a decrease in tissue oxygen delivery [131,145]. Although there is merit to these arguments, it seems prudent in severe metabolic acidosis (pH less than 7.10) to use small amounts of $NaHCO_3$ to raise the pH to a safe level [41,145]. This is particularly true if arrhythmias or hypotension is present since the ability of insulin to increase the plasma HCO_3^- concentration has a latent period of approximately 1 h [144]. $NaHCO_3$ should also be given relatively early to patients with marked acidemia who present with a normal or near normal anion gap. In this setting, the degree of ketonuria has been so large that there are few ketoacid anions remaining that could be converted into HCO_3^- [11a].

The optimal role of HCO_3^- in lactic acidosis is also uncertain. In theory, the arterial pH should be maintained at about 7.15 to 7.20 until the underlying disorder can be corrected. Below this level, lactate utilization may be impaired [18,36] and

---

†This may occur by entry of HCO_3^- into the cells or by the release of intracellular H^+ into the extracellular fluid. Either mechanism will lower the extracellular HCO_3^- concentration.

the possible reduction in cardiac contractility can further reduce tissue perfusion. Above this pH, the enhanced oxygen affinity of hemoglobin can diminish tissue oxygen delivery.

However, clinical and experimental studies have raised questions about the efficacy of any exogenous HCO_3^-. In patients with neoplasia, the administration of HCO_3^- does not substantially increase the plasma HCO_3^- concentration because there is a concomitant elevation in lactate production [38,146,147]. The mechanism by which this occurs is uncertain. It may be that there is a transient increase in pH, which can then promote lactate production by enhancing glycolysis (see Fig. 11-1) [21]. Similar findings have been demonstrated in experimental phenformin-induced lactic acidosis in which the plasma HCO_3^- concentration remained low after administration of HCO_3^- and the mortality rate was not improved when compared with the use of NaCl alone [148].

In addition to its efficacy being unclear, $NaHCO_3$ has potentially serious side effects. The excessive use of $NaHCO_3$ can result in both hypernatremia, since 5 and 7.5% $NaHCO_3$ solutions are very hyperosmolal, and metabolic alkalosis, since lactate will be metabolized back to HCO_3^- when normal cellular metabolism is restored [149]. Large quantities of $NaHCO_3$ can also lead to circulatory overload and pulmonary edema.

There is another problem that may occur in patients with lactic acidosis during a cardiopulmonary arrest or on fixed mechanical ventilation. As the administered $NaHCO_3$ buffers the excess H^+, there will be an increase in CO_2 production:

$$HCO_3^- + H^+ \rightarrow H_2CO_3 \rightarrow CO_2 + H_2O$$

In subjects with adequate ventilation, the CO_2 generated by this reaction will be rapidly excreted by the lungs. However, if ventilation is inadequate or fixed, as frequently occurs during a cardiopulmonary arrest, the administration of $NaHCO_3$ can lead to a significant increase in the P_{CO_2} which will tend to prevent correction of the acidemia [150–152]. In addition, since CO_2 crosses the cell membrane and the blood-brain barrier much more quickly than HCO_3^-, the increase in the P_{CO_2} may exacerbate the intracellular acidosis with possible clinical deterioration [150].

In summary, the efficacy of $NaHCO_3$ in lactic acidosis is not proven. At present, it seems reasonable to use small amounts of $NaHCO_3$ to keep the arterial pH above 7.10 to 7.15. A possible adjunct to intravenous $NaHCO_3$ is peritoneal dialysis with HCO_3^- rather than lactate as a source of buffer in the dialysate [153]. Dialysis can prevent or minimize fluid overload, hypernatremia, and, by removing lactate, the development of alkalemia during the recovery phase.

A new, currently experimental agent, dichloroacetate (DCA) is also being evaluated in the treatment of lactic acidosis. DCA enhances pyruvate dehydrogenase activity, thereby increasing the utilization of pyruvate and subsequently lactate (see Fig. 19-1) [154,155]. It has been effective in some patients [155,156] and dramatically reduces the mortality rate in acute experimental lactic acidosis [157]. Side effects may limit its clinical usefulness [158], although these are most prominent with chronic administration.

Plasma Potassium Concentration

K^+ depletion is common in patients with metabolic acidosis due to gastrointestinal losses (as with diarrhea) and/or renal losses of K^+. Despite this, the initial plasma K^+ concentration may be elevated since metabolic acidemia (except for the organic acidoses) causes K^+ to move out of the cells into the extracellular fluid (see page 254) [4]. A similar effect may occur in diabetic ketoacidosis in which the combination of insulin deficiency and hyperglycemia promote K^+ movement into the extracellular fluid. In these settings, the true state of K^+ balance will become apparent as the pH (or plasma glucose concentration) is normalized, and K^+ supplementation may be necessary. For this reason, *it is extremely important to follow the plasma K^+ concentration as the acidemia is being corrected.*

The danger of hypokalemia is more immediate in the acidemic patient who is initially hypokalemic. Since restoration of normal pH may further reduce the plasma K^+ concentration, $NaHCO_3$ and KCl must be administered concurrently with careful monitoring of the pH, the plasma K^+ concentration, muscle strength, and the electrocardiogram. As much as 40 meq of KCl per hour may be required to prevent life-threatening hypokalemia in some patients (see Chap. 27).

Metabolic Acidosis and Heart Failure

In the presence of left ventricular failure, peritoneal dialysis or hemodialysis can be used to correct both the fluid overload and the acidemia. Acutely, it may be necessary to administer small doses of $NaHCO_3$ (45 to 90 meq) to maintain the pH above 7.15, particularly if hypotension or arrhythmias are present. Since more than half of the HCO_3^- will enter the cells to replenish the intracellular buffers [1,2], there will be much less expansion of the extracellular volume and less exacerbation of the heart failure with $NaHCO_3$ than after an equivalent amount of NaCl, which is distributed almost entirely extracellularly.

PROBLEMS

19-1 The following laboratory tests are obtained from two patients. Would you administer $NaHCO_3$ to either one?

(a)

Plasma $[Na^+]$ = 140 meq/L $[Cl^-]$ = 114 meq/L

$[K^+]$ = 4.2 meq/L $[HCO_3^-]$ = 16 meq/L

(b)

Plasma $[Na^+]$ = 140 meq/L   Arterial pH = 7.32

$[K^+]$ = 4.7 meq/L P_{CO_2} = 14 mmHg

$[Cl^-]$ = 122 meq/L

$[HCO_3^-]$ = 7 meq/L

19-2 A 31-year-old man with a history of epilepsy has a grand mal seizure. Laboratory tests taken immediately after the seizure has stopped reveal:

$$\text{Arterial pH} = 7.14$$

$$P_{CO_2} = 45 \text{ mmHg}$$

$$[Na^-] = 140 \text{ meq/L}$$

$$[K^+] = 4.0 \text{ meq/L}$$

$$[Cl^-] = 98 \text{ meq/L}$$

$$[HCO_3^-] = 17 \text{ meq/L}$$

(a) What is the acid-base disturbance?
(b) Does the patient need $NaHCO_3^-$?
(c) What will happen to his plasma K^+ concentration as the acidemia is corrected?

19-3 If HCO_3^- therapy sufficient to normalize the plasma HCO_3^- concentration is suddenly stopped, match the subsequent course with the type of RTA.

(a) Type 1 RTA in adults (minimum urine pH = 6.5)

(b) Type 1 RTA in infants (minimum urine pH = 7.2)

(c) Type 2 RTA

1. Rapid fall in plasma HCO_3^- concentration, which stabilizes at 16 meq/L
2. Rapid decrease in plasma HCO_3^- concentration, which falls below 10 meq/L
3. Slowly progressive decrease in plasma HCO_3^- concentration to less than 10 meq/L

If the plasma HCO_3^- concentration is raised to 22 meq/L with exogenous $NaHCO_3$, how would you distinguish between these three disorders?

19-4 A 58-year-old man with a history of chronic bronchitis develops severe diarrhea caused by pseudomembranous colitis. It is noted that the volume of diarrheal fluid is approximately 1 L/h. Results of the initial laboratory tests are:

$$\text{Plasma } [Na^+] = 138 \text{ meq/L} \qquad \text{Arterial pH} = 6.97$$

$$[K^+] = 4.8 \text{ meq/L} \qquad P_{CO_2} = 40 \text{ mmHg}$$

$$[Cl^-] = 115 \text{ meq/L}$$

$$[HCO_3^-] = 9 \text{ meq/L}$$

(a) What is the acid-base disorder?
(b) If the patient weighs 80 kg, what is the approximate HCO_3^- deficit?
(c) Assuming the P_{CO_2} remains at 40 mmHg, to what level does the plasma HCO_3^- concentration have to be raised to increase the pH to 7.20? How much exogenous HCO_3^- should this require?
(d) After the administration of this amount of HCO_3^- over 4 h, the plasma HCO_3^- concentration is still 9 meq/L. What is responsible for this inability to raise the plasma HCO_3^- concentration?
(e) What would you estimate the total body K^+ stores in this patient to be?

19-5 A 38-year-old man with hepatic cirrhosis and ascites is admitted to the hospital. The following laboratory data are obtained:

$$\text{Plasma } [Na^+] = 136 \text{ meq/L} \qquad \text{Arterial pH} = 7.45$$

$$[K^+] = 3.7 \text{ meq/L} \qquad P_{CO_2} = 24 \text{ mmHg}$$

$$[Cl^-] = 107 \text{ meq/L} \qquad [\text{Albumin}] = 2.8 \text{ g/dL}$$

$$[HCO_3^-] = 16 \text{ meq/L} \qquad [\text{Globulins}] = 4.8 \text{ g/dL}$$

$$(\text{normal} = 2.5\text{--}3.5 \text{ g/dL})$$

A therapeutic regimen which includes a high-calorie, low-sodium diet and spironolactone is begun. After 2 weeks, the patient's ascites has diminished. Repeat blood tests are obtained:

$$\text{Plasma } [Na^+] = 136 \text{ meq/L} \qquad \text{Arterial pH} = 7.27$$

$$[K^+] = 5.6 \text{ meq/L} \qquad P_{CO_2} = 22 \text{ mmHg}$$

$$[Cl^-] = 113 \text{ meq/L} \qquad \text{Urine pH} = 5.0$$

$$[HCO_3^-] = 10 \text{ meq/L}$$

(a) What was the acid-base disorder on admission?

(b) What was responsible for the fall in the plasma HCO_3^- concentration after the onset of therapy?

(c) Patients with cirrhosis can develop type 1 RTA secondary to hypergammaglobulinemia. What factors are against this being responsible for the development of metabolic acidosis in this patient?

19-6 A 50-year-old woman has chronic renal failure. The following laboratory data are obtained:

$$\text{Plasma } [Na^+] = 137 \text{ meq/L} \qquad \text{BUN} = 108 \text{ mg/dL}$$

$$[K^+] = 5.4 \text{ meq/L} \qquad \text{Plasma } [\text{creatinine}] = 9.2 \text{ mg/dL}$$

$$[Cl^-] = 102 \text{ meq/L} \qquad \text{Arterial pH} = 7.22$$

$$[HCO_3^-] = 10 \text{ meq/L} \qquad P_{CO_2} = 25 \text{ mmHg}$$

Why does metabolic acidosis develop in renal failure? Thirty minutes after the administration of 88 meq of HCO_3^-, repeat blood tests reveal the following:

$$\text{Arterial pH} = 7.38$$

$$P_{CO_2} = 28 \text{ mmHg}$$

$$[HCO_3^-] = 16 \text{ meq/L}$$

In view of the improvement in the pH, no further HCO_3^- is given. However, on the next day, blood tests showed

$$\text{Arterial pH} = 7.28$$

$$P_{CO_2} = 26 \text{ mmHg}$$

$$[HCO_3^-] = 12 \text{ meq/L}$$

What factors might have been responsible for this reduction in the arterial pH?

REFERENCES

1. Schwartz, W. B., K. J. Orning, and R. Porter: The internal distribution of hydrogen ions with varying degrees of metabolic acidosis, J. Clin. Invest., 36:373, 1957.
2. Adrogué, H. J., J. Brensilver, J. J. Cohen, and N. E. Madias: Influence of steady-state alterations in acid-base equilibrium on the fate of administered bicarbonate in the dog, J. Clin. Invest., 71:867, 1983.
3. Burnell, J. M., M. F. Villamil, B. T. Uyeno, and B. H. Scribner: The effect in humans of extra-cellular pH change on the relationship between serum potassium concentration and intracellular potassium, J. Clin. Invest., 35:935, 1956.
4. Adrogué, H. J., and N. E. Madias: Changes in plasma potassium concentration during acute acid-base disturbances, Am. J. Med., 71:456, 1981.

5. Kety, S. S., B. D. Polis, C. S. Nadler, and C. F. Schmidt: The blood flow and oxygen consumption of the human brain in diabetic acidosis and coma, J. Clin. Invest., 27:500, 1948.

6. Bushinsky, D. A., F. L. Coe, C. Katzenberg, J. P. Szidon, and J. H. Parks: Arterial P_{CO_2} in chronic metabolic acidosis, Kidney Int., 22:311, 1982.

7. Madias, N. E., W. B. Schwartz, and J. J. Cohen: The maladaptive renal response to secondary hypocapnia during chronic HCl acidosis in the dog, J. Clin. Invest., 60:1393, 1977.

8. Clarke, E., B. M. Evans, and I. M. MacIntyre: Acidosis in experimental electrolyte depletion, Clin. Sci., 14:421, 1955.

9. Oh, M. S., and J. H. Carroll: The anion gap, N. Engl. J. Med., 297:814, 1977.

10. Emmett, M., and R. G. Narins: Clinical use of the anion gap, Medicine, 56:38, 1977.

11. Oh, M. S., H. J. Carroll, D. A. Goldstein, and I. A. Fein: Hyperchloremic acidosis during the recovery phase of diabetic ketosis, Ann. Intern. Med., 89:925, 1978.

11a. Adrogué, H. J., H. Wilson, A. E. Boyd, W. N. Suki, and G. Eknoyan: Plasma acid-base patterns in diabetic ketoacidosis, N. Engl. J. Med., 307:1603, 1982.

12. Pitts, R. F.: *Physiology of the Kidney and Body Fluids,* 3d ed., Year Book, Chicago, 1974, pp. 86–87.

13. Lennon, E. J., J. Lemann, Jr., and J. R. Litzow: The effects of diet and stool composition on the net external acid balance of normal subjects, J. Clin. Inest., 45:1601, 1966.

14. Lemieux, G., P. Vinay, and P. Cartier: Renal hemodynamics and ammoniagenesis, J. Clin. Invest., 53:884, 1974.

15. Gabow, P. A., W. D. Kaehny, P. V. Fennessey, S. I. Goodman, P. A. Gross, and R. W. Schrier: Diagnostic importance of an increased serum anion gap, N. Engl. J. Med., 303:854, 1980.

16. DeTroyer, A., A. Stolarczyk, D. Zegers de Beyl, and P. Stryckmans: Value of anion-gap determination in multiple myeloma, N. Engl. J. Med. 296:858, 1977.

17. Madias, N. E., J. C. Ayus, and H. J. Adrogué: Increased anion gap in metabolic alkalosis: the role of plasma-protein equivalency, N. Engl. J. Med., 300:1421, 1979.

18. Kreisberg, R. A.: Lactate homeostasis and lactic acidosis, Ann. Intern. Med., 92:227, 1980.

19. Cohen, R. D., R. A. Iles, D. Barnett, M. E. O. Howell, and J. Strunin: The effect of changes in lactate uptake on the intracellular pH of the perfused rat liver, Clin. Sci., 41:159, 1971.

20. Israels, S., J. C. Haworth, H. G. Dunn, and D. A. Applegarth: Lactic acidosis in childhood, Adv. Pediatr., 22:267, 1976.

21. Huckabee, W. E.: Relationships of pyruvate and lactate during anaerobic metabolism. I. Effects of infusion of pyruvate or glucose and of hyperventilation, J. Clin. Invest., 37:244, 1958.

22. Haworth, J. C., T. L. Perry, J. P. Blass, S. Hansen, and N. Urquhart: Lactic acidosis in three sibs due to defects in both pyruvate dehydrogenase and α-ketoglutarate dehydrogenase complexes, Pediatrics, 58:564, 1976.

23. Berger, M., S. A. Hagg, M. N. Goodman, and N. B. Ruderman: Glucose metabolism in perfused skeletal muscle, Biochem. J., 158:191, 1976.

24. Fulop, M., H. D. Hoberman, J. H. Rascoff, N. A. Bonheim, N. P. Dreyer, and H. Tannenbaum: Lactic acidosis in diabetic patients, Arch. Intern. Med., 136:987, 1976.

25. Tonsgard, J. H., P. R. Huttenlocher, and R. A. Thisted: Lactic acidemia in Reye's syndrome, Pediatrics, 69:64, 1982.

26. Orringer, C. E., J. C. Eustace, C. D. Wunsch, and L. B. Gardner: Natural history of lactic acidosis after grand mal seizures: a model for the study of an anion-gap acidosis not associated with hyperkalemia, N. Engl. J. Med., 297:796, 1977.

27. Osnes, J-B., and L. Hermansen: Acid-base balance after maximal exercise of short duration, J. Appl. Physiol., 32:59, 1972.

28. Revler, J. B.: Hypothermia: pathophysiology, clinical settings, and management, Ann. Intern. Med., 89:519, 1978.

29. Buehler, J. H., A. S. Berns, J. R. Webster, W. W. Addington, and D. W. Cugell: Lactic acidosis from carboxyhemoglobinemia after smoke inhalation, Ann. Intern. Med., 82:803, 1975.

30. Eldridge, F.: Blood lactate and pyruvate in pulmonary insufficiency, N. Engl. J. Med., 274:878, 1966.

31. Graham, D. L., D. Laman, J. Theodore, and E. D. Robin: Acute cyanide poisoning complicated by lactic acidosis and pulmonary edema, Arch. Intern. Med., 137:1051, 1977.

32. Arieff, A. I., R. Park, W. J. Leach, and V. C. Lazarowitz: Pathophysiology of experimental lactic acidosis in dogs, Am. J. Physiol., 239:F135, 1980.

33. Barbosa-Saldivar, J. L., C. Boxhill, and H. J. Carroll: D-Lactic acidosis in a man with the short-bowel syndrome, N. Engl. J. Med., 301:249, 1979.

34. Stolberg, L., R. Rolfe, N. Gitlin, J. Merritt, L. Mann, J. Linder, and S. Finegold: D-Lactic acidosis due to abnormal gut flora, N. Engl. J. Med., 306:1344, 1982.

35. Kreisberg, R. A., W. C. Owen, and A. M. Siegal: Ethanol-induced hyperlacticacidemia: inhibition of lactate utilization, J. Clin. Invest., 50:166, 1971.

36. Cohen, R. D., and R. A. Iles: Lactic acidosis: some physiological and clinical considerations, Clin. Sci. Mol. Med., 53:405, 1977.

37. Field, M., J. B. Block, R. Levin, and D. P. Rall: Significance of blood lactate elevations among patients with acute leukemia and other neoplastic proliferative disorders, Am. J. Med., 40:528, 1966.

38. Fraley, D. S., S. Adler, F. J. Bruns, and B. Zett: Stimulation of lactate production by administration of bicarbonate in a patient with a solid neoplasm and lactic acidosis, N. Engl. J. Med., 303:1100, 1980.

39. Nadiminti, Y., J. C. Wang, S-Y. Chou, E. Pineles, and M. S. Tobin: Lactic acidosis associated with Hodgkin's disease, N. Engl. J. Med., 303:15, 1980.

40. Cahill, G. F., Jr.: Ketosis, Kidney Int., 20:416, 1981.

41. Foster, D. W., and J. D. McGarry: The metabolic derangements and treatment of diabetic ketoacidosis, N. Engl. J. Med., 309:159, 1983.

42. Madison, L. L., D. Mebane, R. H. Unger, and A. Lochner: The hypoglycemic action of ketones. II. Evidence for a stimulatory feedback of ketones on the pancreatic beta cells, J. Clin. Invest., 43:408, 1964.

43. Reichard, G. A., Jr., O. E. Owen, A. C. Haff, P. Paul, and W. M. Bortz: Ketone body production and oxidation in fasting obese humans, J. Clin. Invest., 53:508, 1974.

44. Levy, L. J., J. Duga, M. Girgis, and E. E. Gordon: Ketoacidosis associated with alcoholism in nondiabetic subjects, Ann. Intern. Med., 78:213, 1973.

45. Miller, P. D., R. E. Heinig, and C. Waterhouse: Treatment of alcoholic acidosis, Arch. Intern. Med., 138:67, 1978.

46. Lefevre, A. J., H. Adler, and C. Lieber: Effect of ethanol on ketone metabolism, J. Clin. Invest., 49:1775, 1970.

47. Stanbury, J. B., J. B. Wyngaarden, and D. S. Fredrickson (eds.): *The Metabolic Basis of Inherited Disease,* 4th ed., McGraw-Hill, New York, 1978, chaps. 20 and 21.

48. Cohen, J. J.: Methylmalonic acidemia, Kidney Int., 15:311, 1979.

49. Marliss, E. B., J. L. Ohman, Jr., T. T. Aoki, and G. P. Kozak: Altered redox state obscuring ketoacidosis in diabetic patients with lactic acidosis, N. Engl. J. Med., 283:978, 1970.

50. Narins, R. G., E. R. Jones, M. C. Stom, M. R. Rudnick, and C. P. Bastl: Diagnostic strategies in disorders of fluid, electrolyte and acid-base homeostasis, Am. J. Med., 72:496, 1982.

51. Dorhourt-Mees, E. J., M. Machado, E. Slatopolsky, S. Klahr, and N. S. Bricker: The functional adaptation of the diseased kidney. III. Ammonium excretion, J. Clin. Invest., 45:289, 1966.

52. Welbourne, T., M. Weber, and N. Bank: The effect of glutamine administration on urinary ammonium excretion in normal subjects and patients with renal disease, J. Clin. Invest., 51:1852, 1972.

53. Schwartz, W. B., P. W. Hall, R. Hays, and A. S. Relman: On the mechanism of acidosis in chronic renal disease, J. Clin. Invest., 38:39, 1959.

54. Wrong, O., and H. E. F. Davies: The excretion of acid in renal disease, Q. J. Med., 28:259, 1959.

55. Lubowitz, H., M. L. Purkerson, D. B. Rolf, F. Weisser, and N. S. Bricker: Effect of nephron loss on proximal tubular bicarbonate reabsorption in the rat, Am. J. Physiol., 220:457, 1971.

56. Muldowney, F. P., D. V. Carroll, J. F. Donohue, and R. Freaney: Correction of renal bicarbonate wastage by parathyroidectomy, Q. J. Med., 40:487, 1971.

57. Espinel, C. H.: The influence of salt intake on the metabolic acidosis of chronic renal failure, J. Clin. Invest., 56:286, 1975.

58. Litzow, J. R., J. Lemann, Jr., and E. J. Lennon: The effect of treatment of acidosis on calcium balance in patients with chronic azotemic renal disease, J. Clin. Invest., 46:280, 1967.

59. McSherry, E., and R. C. Morris, Jr.: Attainment and maintenance of normal stature with alkali therapy in infants and children with classic renal tubular acidosis, J. Clin. Invest., 61:509, 1978.

60. Husted, F. C., K. D. Nolph, and J. F. Maher: NaHCO$_3$ and NaCl tolerance in chronic renal failure, J. Clin. Invest., 56:414, 1975.

61. Hill, J.: Salicylate intoxication, N. Engl. J. Med., 288:1110, 1973.

61a. Tenny, S. M., and R. M. Miller: The respiratory and circulatory actions of salicylate, Am. J. Med., 19:498, 1955.

62. Winters, R. W., J. S. White, and M. C. Hughes: Disturbances of acid-base equilibrium in salicylate intoxication, Pediatrics, 23:260, 1959.

63. Gabow, P. A., R. J. Anderson, D. E. Potts, and R. W. Schrier: Acid-base disturbances in the salicylate-intoxicated adult, Arch. Intern. Med., 138:1481, 1978.

64. Eichenholz, A., R. O. Mulhausen, and P. S. Redleaf: Nature of acid-base disturbance in salicylate intoxication, Metabolism, 12:164, 1963.

64a. Prescott, L. F., M. Balali-Mood, J. A. J. H. Critchley, A. F. Johnstone, and A. T. Proudfoot: Diuresis or urinary alkalinisation for salicylate poisoning?, Br. Med. J., 2:1383, 1982.

65. Bennett, I. L., Jr., F. H. Cary, G. I. Mitchell, and M. N. Cooper: Acute methyl alcohol poisoning: a review based on experiences in an outbreak of 323 cases, Medicine, 32:431, 1953.

66. McMartin, K. E., J. J. Ambre, and T. R. Tephly: Methanol poisoning in humans subjects: role for formic acid accumulation in the metabolic acidosis, Am. J. Med., 68:414, 1980.

67. Gonda, A., H. Gault, D. Churchill, and D. Hollomby: Hemodialysis for methanol intoxication, Am. J. Med., 64:749, 1978.

68. Eells, J. T., K. E. McMartin, K. Black, V. Virayotha, R. H. Tisdell, and T. R. Tephly: Formaldehyde poisoning, J. Am. Med. Assoc., 246:1237, 1981.

69. Glasser, L., P. D. Sternglanz, J. Combie, and A. Robinson: Serum osmolality and its applicability to drug overdose, Am. J. Clin. Pathol., 60:695, 1973.

69a. Sklar, A. H., and S. L. Linas: The osmolal gap in renal failure, Ann. Intern. Med., 98:480, 1983.

70. McCoy, H. G., R. J. Cipolle, S. M. Ehlers, R. J. Sawchuk, and D. E. Zaske: Severe methanol poisoning: application of a pharmacokinetic model for ethanol therapy and hemodialysis, Am. J. Med., 67:804, 1979.

71. Keyvan-Larijarni, H., and A. M. Tannenberg: Methanol intoxication, Arch. Intern. Med., 134:293, 1974.

72. Freed, C. R., W. H. Bobbitt, R. M. Williams, S. Shoemaker, and A. S. Nies: Ethanol for ethylene glycol poisoning, N. Engl. J. Med., 304:976, 1981.

73. Peterson, C. D., A. J. Collins, J. M. Himes, M. L. Bullock, and W. F. Keane: Ethylene glycol poisoning: pharmacokinetics during therapy with ethanol and hemodialysis, N. Engl. J. Med., 304:21, 1981.

74. Whalen, J. E., C. J. Richards, and J. Ambre: Inadequate removal of methanol and formate using sorbent based regeneration hemodialysis delivery system, Clin. Nephrol., 11:318, 1979.

75. Parry, M. F., and R. Wallach: Ethylene glycol poisoning, Am. J. Med., 57:143, 1974.

76. Case Records of Massachusetts General Hospital, (Case 38-1979), N. Engl. J. Med., 301:650, 1979.

77. Bove, K. E.: Ethylene glycol toxicity, Am. J. Clin. Pathol., 45:46, 1966.

78. Gilmour, I. J., R. J. W. Blanchard, and W. F. Perry: Mannitol gives false-positive biochemical estimations of ethylene glycol, N. Engl. J. Med., 291:51, 1974.

79. Beier, L. S., W. H. Pitts, and H. C. Gonick: Metabolic acidosis occurring during paraldehyde intoxication, Ann. Intern. Med., 58:155, 1963.

80. Blum, J. E., and F. L. Coe: Metabolic acidosis after sulfur ingestion, N. Engl. J. Med., 297:869, 1977.

81. Fischman, C. M., and J. R. Oster: Toxic effects of toluene, J. Am. Med. Assoc., 241:1713, 1979.

82. Relman, A. S., P. F. Shelburne, and A. Talman: Profound acidosis resulting from excessive ammonium chloride in previously healthy subjects, N. Engl. J. Med., 264:848, 1961.

83. Heird, W. C., R.B. Dell, J. M. Driscoll, Jr., B. Grebin, and R. W. Winters: Metabolic acidosis

resulting from intravenous alimentation mixtures containing synthetic amino acids, N. Engl. J. Med., 287:943, 1972.

84. Fraley, D. S., S. Adler, F. Bruns, and D. Segal: Metabolic acidosis after hyperalimentation with casein hydrolysate, Ann. Intern. Med., 88:352, 1978.

85. Teree, T., E. Mirabal-Font, A. Ortiz, and W. Wallace: Stool losses and acidosis in diarrheal disease of infancy, Pediatrics, 36:704, 1965.

86. Schwartz, W. B., and A. S. Relman: Metabolic and renal studies in chronic potassium depletion resulting from overuse of laxatives, J. Clin. Invest., 32:538, 1953.

87. Davis, G. R., S. G. Morawski, C. A. Santa Ana, and J. S. Fordtran: Evaluation of chloride/bicarbonate exchange in the human colon in vivo, J. Clin. Invest., 71:201, 1983.

88. D'Agostino, A., W. F. Leadbetter, and W. B. Schwartz: Alterations in the ionic composition of isotonic saline solution instilled in the colon, J. Clin. Invest., 32:444, 1953.

89. deWardener, H. E.: *The Kidney: An Outline of Normal and Abnormal Structures and Function*, Churchill Livingstone, London, 1973, p. 249.

90. Kleinman, P. K.: Cholestyramine and metabolic acidosis, N. Engl. J. Med., 290:861, 1974.

91. Morris, R. C., Jr., and A. Sebastian: Renal tubular acidosis and the Fanconi Syndrome, in *The Metabolic Basis of Inherited Disease* (5th ed), J. B. Stanbury, J. B. Wyngaarden, D. S. Fredrickson, J. L. Goldstein, and M. S. Brown (eds.), McGraw-Hill, New York, 1983.

92. McSherry, E.: Renal tubular acidosis in childhood, Kidney Int., 20:799, 1981.

93. Arruda, J. A. L., and N. A. Kurtzman: Mechanisms and classification of deranged distal urinary acidification, Am. J. Physiol., 239:F515, 1980.

94. Batlle, D. C., J. T. Sehy, M. K. Roseman, J. A. L. Arruda, and N. A. Kurtzman: Clinical and pathophysiologic spectrum of acquired distal renal tubular acidosis, Kidney Int., 20:389, 1981.

95. Batlle, D. C., J. A. L. Arruda, and N. A. Kurtzman: Hyperkalemic distal renal tubular acidosis associated with obstructive uropathy, N. Engl. J. Med., 304:373, 1981.

96. Portwood, R. M., D. W. Seldin, F. C. Rector, and R. Cade: The relation of urinary CO_2 tension to bicarbonate excretion, J. Clin. Invest., 38:770, 1959.

97. McSherry, E., A. Sebastian, and R. C. Morris, Jr.: Renal tubular acidosis in infants: the several kinds, including bicarbonate-wasting classic renal tubular acidosis, J. Clin. Invest., 51:499, 1972.

98. Buckalew, V. M., Jr., D. K. McCurdy, G. D. Ludwig, L. B. Chaykin, and J. R. Elkington: Incomplete renal tubular acidosis, Am. J. Med., 45:32, 1968.

99. Norman, M. E., N. I. Feldman, R. M. Cohn, K. S. Roth, and D. K. McCurdy: Urinary citrate excretion in the diagnosis of distal renal tubular acidosis, J. Pediatr., 92:394, 1978.

99a. Streicher, H. Z., P. A. Gabow, A. H. Moss, D. Kono, and W. D. Kaehny: Syndromes of toluene sniffing in adults, Ann. Intern. Med., 94:758, 1981.

100. Sebastian, A., E. McSherry, and R. C. Morris, Jr.: Impaired renal conservation of sodium and chloride during sustained correction of systemic acidosis in patients with type 1, classic renal tubular acidosis, J. Clin. Invest., 58:454, 1976.

101. Loney, L. C., L. L. Norling, and A. M. Robson: The use of arginine hydrochloride infusion to assess urinary acidification, J. Pediatr., 100:95, 1982.

102. Coe, F. L., and J. H. Parks: Stone disease in hereditary distal renal tubular acidosis, Ann. Intern. Med., 93:60, 1980.

103. Brenner, R. J., D. B. Spring, A. Sebastian, E. M. McSherry, H. K. Genant, A. J. Palubinskas, and R. C. Morris, Jr.: Incidence of radiographically evident bone disease, nephrocalcinosis, and nephrolithiasis in various types of renal tubular acidosis, N. Engl. J. Med., 307:217, 1982.

104. Marone, C. C., N. L. M. Wong, R. A. L. Sutton, and J. H. Dirks: Effects of metabolic alkalosis on calcium excretion in the conscious dog, J. Lab. Clin. Med., 101:264, 1983.

105. Kempson, S. A.: Effect of metabolic acidosis on renal brush border membrane adaptation to low phosphorus diet, Kidney Int., 22:225, 1982.

105a. Rodriguez-Soriano, J., A. Vallo, G. Castillo, and R. Oliveros: Natural history of primary distal renal tubular acidosis treated since infancy, J. Pediatr., 101:669, 1982.

106. Buckalew, V. M., Jr., M. Purvis, M. Shulman, C. N. Herndon, and D. Rudman: Hereditary renal tubular acidosis, Medicine, 53:229, 1974.

106a. Hamed, I. A., A. W. Czerwinski, B. Coats, C. Kaufman, and D. H. Altmiller: Familial absorptive hypercalciuria and renal tubular acidosis, Am. J. Med., 67:385, 1979.

107. Dedmon, R. E., and O. Wrong: The excretion of organic anion in renal tubular acidosis with particular reference to citrate, Clin. Sci., 22:19, 1962.

108. Simpson, D. P.: Citrate excretion: a window on renal metabolism, Am. J. Physiol., 244:F223, 1983.

108a. Smith, L. H.: Calcium-containing renal stones, Kidney Int., 13:383, 1978.

108b. Dubose, T. D., Jr., and M. S. Lucci: Effect of carbonic anhydrase inhibition on superficial and deep nephron bicarbonate reabsorption in the rat, J. Clin. Invest., 71:55, 1983.

109. Roth, K. S., J. W. Foreman, and S. Segal: The Fanconi syndrome and mechanisms of tubular transport dysfunction, Kidney Int., 20:705, 1981.

110. Crumb, C. K., M. Martinez-Maldonado, G. Eknoyan, and W. N. Suki: Effects of volume expansion, purified parathyroid extract, and calcium on renal bicarbonate absorption in the dog, J. Clin. Invest., 54:1287, 1974.

111. Farrow, S. L., B. J. Stinebaugh, D. Rouse, and W. N. Suki: Effects of hypocalcemia on renal bicarbonate absorption in the dog, Kidney Int., 10:489, 1976.

112. Clay, R. D., E. M. Darmady, and M. Hawkins: The nature of the renal lesion in the Fanconi syndrome, J. Pathol. Bacteriol., 65:551, 1953.

113. Donckerwolcke, R. A., G. J. van Stekelenburg, and H. A. Tiddens: A case of bicarbonate-losing renal tubular acidosis with defective carboanhydrase activity, Arch. Dis. Child., 45:769, 1970.

114. Saldanka, L. F., V. J. Rosen, and H. C. Gonick: Silicon nephropathy, Am. J. Med., 59:95, 1975.

115. Maldonado, J. E., J. A. Velosa, R. A. Kyle, R. D. Wagoner, K. E. Holley, and R. M. Salassa: Fanconi syndrome in adults, Am. J. Med., 58:354, 1975.

116. Hulter, H. N., L. P. Ilnicki, J. A. Harbottle, and A. Sebastian: Impaired renal H^+ secretion and NH_3 production in mineralo-corticoid-deficient, glucocorticoid-replete dogs, Am. J. Physiol., 232:F136, 1977.

117. Sebastian, A., M. Schambelan, S. Lindenfeld, and R. C. Morris, Jr.: Amelioration of metabolic acidosis with fluorocortisone therapy in hyporeninemic hypoaldosteronism, N. Engl. J. Med., 297:576, 1977.

118. Szylman, P., O. S. Better, C. Chaimowitz, and A. Rosler: Role of hyperkalemia in the metabolic acidosis of isolated hypoaldosteronism, N. Engl. J. Med., 294:361, 1976.

119. Sebastian, A., E. McSherry, and R. C. Morris, Jr.: On the mechanism of renal potassium wasting in renal tubular acidosis associated with the Fanconi syndrome (type 2 RTA), J. Clin. Invest., 50:231, 1971.

120. Sebastian, A., E. McSherry, and R. C. Morris, Jr.: Renal potassium wasting in renal tubular acidosis (RTA): its occurrence in types 1 and 2 RTA despite sustained correction of systemic acidosis, J. Clin. Invest., 50:667, 1971.

121. Gill, J. R., Jr., N. H. Bell, and F. C Bartter: Impaired conservation of sodium and potassium in renal tubular acidosis and its correction by buffer anions, Clin. Sci., 33:577, 1967.

122. Batlle, D., K. Itsarayoungyuen, J. A. L. Arruda, and N. A. Kurtzman: Hyperkalemic hyperchloremic metabolic acidosis in sickle cell hemoglobinopathies, Am. J. Med., 72:188, 1982.

123. Nash, M., A. D. Torrado, I. Greifer, A. Spitzer, and C. M. Edelman, Jr.: Renal tubular acidosis in infants and children, J. Pediatr., 80:738, 1972.

124. Donckerwolcke, R. A., G. J. van Stekelenberg, and H. A. Tiddens: Therapy of bicarbonate-losing renal tubular acidosis, Arch. Dis. Child., 45:774, 1970.

125. DeFronzo, R. A.: Hyperkalemia and hyporeninemic hypoaldosteronism, Kidney Int., 17:118, 1980.

126. Sebastian, A., and M. Schambelan: Amelioration of type 4 renal tubular acidosis in chronic renal failure with furosemide, Kidney Int., 12:534, 1977.

127. McCarron, D. A., W. C. Elliott, J. S. Rose, and W. M. Bennett: Severe mixed metabolic acidosis secondary to rhabdomyolysis, Am. J. Med., 67:905, 1979.

128. Gerst, P., W. Fleming, and J. Malm: A quantitative evaluation of the effects of acidosis and alkalosis upon the ventricular fibrillation threshold, Surgery, 59:1050, 1966.

129. Clancy, R. L., H. E. Cingolani, R. A. Taylor, T. P. Graham, Jr., and J. P. Gilmore: Influence of sodium bicarbonate on myocardial performance, Am. J. Physiol., 212:917, 1967.

130. Wildenthal, K., D. S. Mierzwiak, R. W. Myers, and J. H. Mitchell: Effects of acute lactic acidosis on left ventricular performance, Am. J. Physiol., 214:1352, 1968.

131. Posner, J., and F. Plum: Spinal-fluid pH and neurologic symptoms in systemic acidosis, N. Engl. J. Med., 277:605, 1967.

132. Posner, J., A. G. Swanson, and F. Plum: Acid-base balance in cerebrospinal fluid, Arch. Neurol., 12:479, 1965.

133. Mitchell, R. A., C. T. Carman, J. W. Severinghaus, B. W. Richardson, M. M. Singer, and S. Shnider: Stability of cerebrospinal fluid pH in chronic acid-base disturbances in blood, J. Appl. Physiol., 20:443, 1965.

134. Fulop, M., H. Tannenbaum, and N. Dreyer: Ketotic hyperosmolar coma, Lancet, 2:635, 1973.

135. Potter, D. E., and I. Greifer: Statural growth of children with renal disease, Kidney Int., 14:334, 1978.

136. Cunningham, J., L. J. Fraher, T. L. Clemens, P. A., Revell, and S. E. Papapoulos: Chronic acidosis with metabolic bone disease, Am. J. Med., 73:199, 1982.

137. Chesney, R. W., A. V. Moorthy, J. A. Eisman, D. K. Jax, R. B. Mazess, and H. F. DeLuca: Increased growth after oral $1\alpha,25$-vitamin D_3 in childhood osteodystrophy, N. Engl. J. Med., 298:238, 1978.

138. Simmons, J. M., C. J. Wilson, D. E. Potter, and M. A. Holliday: Relation of calorie deficiency to growth failure in children on hemodialysis and the growth response to calorie supplementation, N. Engl. J. Med., 285:653, 1971.

139. Bleich, H. L., and W. B. Schwartz: Tris buffer (Tham): an appraisal of its physiologic effects and clinical usefulness, N. Engl. J. Med., 274:782, 1966.

140. Schwartz, W. B., and W. Waters: Lactate versus bicarbonate: a reconsideration of the therapy of metabolic acidosis, Am. J. Med., 32:831, 1962.

141. Alberti, K. G. M. M., H. Darley, P. M. Emerson, and T. D. R. Hockaday: 2,3-diphosphoglycerate and tissue oxygenation in uncontrolled diabetes mellitus, Lancet, 2:391, 1972.

142. Fulop, M., M. Horowitz, A. Aberman, and E. Jaffé: Lactic acidosis in pulmonary edema due to left ventricular failure, Ann. Intern. Med., 79:180, 1973.

143. Seldin, D. W., and R. Tarail: The metabolism of glucose and electrolytes in diabetic acidosis, J. Clin. Invest., 29:552, 1950.

144. King, A. J., N. J. Cooke, A. McCuish, B. F. Clarke, and B. J. Kirby: Acid-base changes during treatment of diabetic ketoacidosis, Lancet, 1:478, 1974.

145. Kaye, R.: Diabetic ketoacidosis: the bicarbonate controversy, J. Pediatr., 87:156, 1975.

146. Fields, A. L. A., S. L. Wolman, and M. L. Halperin: Chronic lactic acidosis in a patient with cancer: therapy and metabolic consequences, Cancer, 47:2026, 1981.

147. Wainer, R. A., P. H. Wiernik, and W. L. Thompson: Metabolic and therapeutic studies of a patient with acute leukemia and severe lactic acidosis of prolonged duration, Am. J. Med., 55:255, 1973.

148. Arieff, A. I., W. Leach, R. Park, and V. C. Lazarowitz: Systemic effects of $NaHCO_3$ in experimental lactic acidosis in dogs, Am. J. Physiol., 242:F586, 1982.

149. Mattar, J. A., M. H. Weil, H. Shubin, and L. Stein: Cardiac arrest in the critically ill. II. Hyperosmolal states following cardiac arrest, Am. J. Med., 56:162, 1974.

150. Bishop, R. L., and M. L. Weisfeldt: Sodium bicarbonate administration during cardiac arrest. Effect on arterial pH, P_{CO_2}, and osmolality, J. Am. Med. Assoc., 235:506, 1976.

151. Ostrea, E. M., and G. B. Odell: The influence of bicarbonate administration on blood pH in a "closed system": clinical implications, J. Pediatr., 80:671, 1972.

152. Steichen, J. J., and L. I. Kleinman: Studies in acid-base balance. I. Effect of alkali therapy in newborn dogs with mechanically fixed ventilation, J. Pediatr., 91:287, 1977.

153. Vaziri, N. D., R. Ness, L. Wellikson, C. Barton, and N. Greep: Bicarbonate-buffered peritoneal dialysis: an effective adjunct in the treatment of lactic acidosis, Am. J. Med., 67:392, 1979.

154. Stacpoole, P. W., G. W. Moore, and D. M. Kornhauser: Metabolic effects of dichloroacetate in patients with diabetes mellitus and hyperlipoproteinemia, N. Engl. J. Med., 298:526, 1978.

155. Stacpoole, P. W., E. M. Harman, S. H. Curry, T. G. Baumgartner, and R. I. Misbin: Treatment of lactic acidosis with dichloroacetate, N. Engl. J. Med., 309:390, 1983.
156. Coude, F. X., J. M. Saudubray, F. DeMaugre, C. Marsac, J. P. Leroux, and C. Charpentier: Dichloroacetate as treatment for congenital lactic acidosis, N. Engl. J. Med., 299:1365, 1978.
157. Park, R., and A. I. Arieff: Treatment of lactic acidosis with dichloroacetate in dogs, J. Clin. Invest., 70:853, 1982.
158. Stacpoole, P. W., G. W. Moore, and D. M. Kornhauser: Toxicity of chronic dichloroacetate, N. Engl. J. Med., 300:372, 1979.

TWENTY

RESPIRATORY ACIDOSIS

The introduction to acid-base disorders presented in Chap. 17 should be read before proceeding with this discussion. Respiratory acidosis is a clinical disorder characterized by a reduced arterial pH (or increased H^+ concentration), an elevation in the P_{CO_2} (hypercapnia), and a variable increase in the plasma HCO_3^- concentration. Hypercapnia also constitutes the respiratory compensation to metabolic alkalosis. However, in this setting the increase in the P_{CO_2} is appropriate since it lowers the arterial pH toward normal.

PATHOPHYSIOLOGY AND ETIOLOGY

Endogenous metabolism results in the production of approximately 15,000 mmol of CO_2 per day. Although CO_2 is not an acid, it combines with H_2O as it is added to the bloodstream, resulting in the formation of H_2CO_3:

$$CO_2 + H_2O \rightleftharpoons H_2CO_3 \rightleftharpoons H^+ + HCO_3^- \tag{20-1}$$

The ensuing elevation in the H^+ concentration is then minimized because most of the excess H^+ ions combine with the intracellular buffers, particularly hemoglobin (Hb):

$$H_2CO_3 + Hb^- \rightarrow HHb + HCO_3^- \tag{20-2}$$

The HCO_3^- generated by this reaction then leaves the erythrocyte and enters the extracellular fluid in exchange for extracellular Cl^-. The net effect is that CO_2 is primarily carried in the bloodstream as HCO_3^- with little change in the extracellular pH. These processes are reversed in the alveoli. As HHb is oxygenated, H^+ is released. This combines with HCO_3^- to form H_2CO_3 and then CO_2, which is excreted.

Control of Ventilation

Before we discuss how hypercapnia can occur, it is helpful to review briefly the basic aspects of ventilatory regulation. Alveolar ventilation provides the oxygen necessary for oxidative metabolism and eliminates the CO_2 produced by these metabolic porcesses. Therefore, it is appropriate that the main physiologic stimuli to respiration are a reduction in the arterial P_{O_2} (hypoxemia) and an elevation in the arterial P_{CO_2} [1,2]. The CO_2 stimulus to ventilation occurs primarily in chemosensitive areas in the respiratory center in the brainstem which appear to respond to CO_2-induced changes in the cerebral interstitial pH [1,3]. In contrast, the initial hypoxemic enhancement of ventilation is mostly mediated by chemoreceptors in the carotid bodies which are located near the bifurcation of the carotid arteries [2,4]. In normal subjects, these regulatory processes permit adequate oxygenation to be maintained and the arterial P_{CO_2} to be held within narrow limits (40 $\pm$ 4 mmHg) despite the large daily CO_2 load.

Carbon dioxide is normally the major stimulus to ventilation, which is enhanced by even a minor elevation in the arterial P_{CO_2} (Fig. 20-1). In contrast, hypoxemia does not begin to substantially promote ventilation until the arterial P_{O_2} is less than 50 to 55 mmHg. Although lesser degrees of hypoxemia do increase ventilation, this response is initially limited by the ensuing hypocapnia and respiratory alkalosis, which depress respiration. If, however, the arterial P_{CO_2} is held at normal values or is elevated because of intrinsic lung disease, then ventilation begins to be enhanced as the arterial P_{O_2} falls below 70 to 80 mmHg (see Fig. 21-2) [1]. Thus, hypoxemia becomes the major stimulus to ventilation in respiratory acidosis (see below).

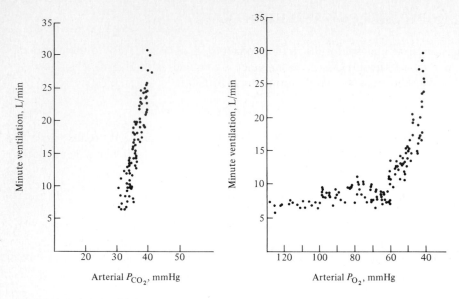

Figure 20-1 Relationship between arterial P_{CO_2} (left panel) and P_{O_2} (right panel) and the respiratory minute volume in normal subjects. *(Adapted from D. W. Hudgel, Control of breathing in asthma, in Bronchial Asthma: Mechanisms and Therapeutics, 2d ed., E. B. Weiss, M. S. Segal, and M. Stein (eds.), Little, Brown, Boston, 1984.)*

Development of Hypercapnia

Since the CO_2 stimulus to ventilation is so strong, hypercapnia and respiratory acidosis are almost always due to a reduction in effective alveolar ventilation, not an increase in CO_2 production. Hypoventilation can occur when there is interference with any step in the ventilatory process (Table 20-1). In patients with reduced respiratory drive or chest wall dysfunction, there tends to be a generalized fall in alveolar ventilation. In contrast, CO_2 retention in intrinsic pulmonary disease is thought to be due primarily to an imbalance between ventilation and perfusion in which there are perfused areas that are underventilated [5].

If ventilatory function is not restored, the decrease in the pH produced by CO_2 retention is minimized by the cell buffers and by increased renal H^+ secretion, both of which result in an elevation in the plasma HCO_3^- concentration. Since the renal response occurs over several days [6], the compensation to acute respiratory acidosis is less effective than that to chronic respiratory acidosis.

Relationship between Hypercapnia and Hypoxemia

All patients with hypercapnia must also be hypoxemic if they are breathing room air since there is a concomitant decrease in oxygen uptake. In most cases, hypoxemia occurs earlier and is more prominent than hypercapnia. Two factors contribute to this difference: (1) CO_2 can diffuse across the alveolar capillary 20 times as quickly

Table 20-1 Causes of acute and chronic respiratory acidosis

Inhibition of the medullary respiratory center
  A. Acute
      1. Drugs: opiates, anesthetics, sedatives
      2. Oxygen in chronic hypercapnia
      3. Cardiac arrest
      4. Central sleep apnea
  B. Chronic
      1. Extreme obesity (Pickwickian syndrome)
      2. Central nervous system lesions (rare)

Disorders of the respiratory muscles and chest wall
  A. Acute
      1. Muscle weakness: crisis in myasthenia gravis, periodic paralysis,
         aminoglycosides, Guillain-Barré syndrome, severe hypokalemia
         or hypophosphatemia
  B. Chronic
      1. Muscle weakness: poliomyelitis, amyotrophic lateral sclerosis,
         multiple sclerosis, myxedema
      2. Kyphoscoliosis
      3. Extreme obesity

Upper airway obstruction
  A. Acute
      1. Aspiration of foreign body or vomitus
      2. Obstructive sleep apnea
      3. Laryngospasm

Disorders affecting gas exchange across the pulmonary capillary
  A. Acute
      1. Exacerbation of underlying lung disease
      2. Adult respiratory distress syndrome
      3. Acute cardiogenic pulmonary edema
      4. Severe asthma or pneumonia
      5. Pneumothorax or hemothorax
  B. Chronic
      1. Chronic obstructive lung disease: bronchitis, emphysema

Mechanical ventilation

as O_2; and (2) as patients attempt to increase ventilation in relatively normal segments of the lung, more CO_2 can be excreted but more O_2 cannot be taken up since the saturation of hemoglobin already approaches 100 percent in these areas.

The early development of hypoxemia stimulates ventilation and helps to delay the onset or minimize the degree of hypercapnia. There is, however, a wide variability (probably inherited) in the sensitivity to the hypoxemic stimulus [7]. Those subjects who are less sensitive to hypoxemia will be more likely to develop respiratory acidosis in the presence of an appropriate cause such as chronic bronchitis or marked obesity (see below).

In addition to this initial protective effect, the hypoxemic drive is particularly important in the presence of chronic respiratory acidosis. In this setting, the respiratory centers appear to become less sensitive to CO_2† [8,9], and hypoxemia constitutes the primary stimulus to ventilation. If oxygen is given to raise the arterial P_{O_2}, alveolar ventilation frequently falls, producing a further elevation in the P_{CO_2}, which can lead to neurologic symptoms [11]. Thus, supplemental oxygen must be given very carefully to patients with chronic hypercapnia.

Understanding the relationship between the arterial P_{O_2} and P_{CO_2} is also helpful in evaluating patients with acute asthma [12]. The combination of mucus plugs and bronchoconstriction initially induces hypoxemia, thereby stimulating ventilation. Thus, a mild to moderate attack is associated with hypocapnia and respiratory *alkalosis*. With increasing severity, the P_{CO_2} rises, eventually to normal and then hypercapnic levels. Thus, the combination of hypoxemia and a "normal" P_{CO_2} of 40 mmHg represents severe disease in the acute asthmatic [12]. This principle can be applied generally to patients with intrinsic lung disease: hypercapnia is a late finding, and even a small elevation in the P_{CO_2} of a few mmHg indicates advanced pulmonary dysfunction.

Acute Respiratory Acidosis

The ability to protect the extracellular pH is different in metabolic and respiratory acidosis. In the former, extracellular and intracellular buffering and the respiratory compensation all minimize the fall in pH (see Chap. 19). In contrast, the body is not so well adapted to handle an acute elevation in the P_{CO_2}. There is virtually no extracellular buffering because HCO_3^- cannot buffer H_2CO_3 [13]:

$$H_2CO_3 + HCO_3^- \rightarrow H_2CO_3 + HCO_3^- \qquad (20\text{-}3)$$

Since the renal response takes time to develop, the cell buffers, particularly hemoglobin and proteins, constitute the only protection against acute hypercapnia:

$$H_2CO_3 + Buf^- \rightarrow HBuf + HCO_3^- \qquad (20\text{-}4)$$

As a result of these buffering reactions, there is an increase in the plasma HCO_3^- concentration, averaging 1 meq/L for every 10-mmHg rise in the P_{CO_2} (Fig.

---

†The acquired decrease in CO_2 sensitivity may in large part be due to the compensatory elevation in the plasma HCO_3^- concentration [8,10]. This returns the pH toward normal, thereby minimizing the acidemic and leaving only the hypoxemic stimulus to ventilation. Furthermore, the reason for the apparent insensitivity to inspired CO_2 can be understood from arithmetic analysis of the law of mass action,

$$[H^+] = 24 \times \frac{P_{CO_2}}{[HCO_3^-]}$$

As the plasma HCO_3^- concentration rises, a given increase in the P_{CO_2} will produce a progressively smaller increment in the arterial H^+ concentration and, therefore, in ventilation. However, the *change in ventilation per change in H^+ concentration* is not diminished (as compared to normal subjects), suggesting that the insensitivity of the respiratory center to CO_2 is more apparent than real [10]. Thus, the compensatory rise in the plasma HCO_3^- concentration has two effects: it protects the pH but, in so doing, it limits the stimulus to maximal alveolar ventilation. If the increase in the plasma HCO_3^- concentration is prevented, the patient will be more acidemic but somewhat less hypercapnic since ventilation will be enhanced [10].

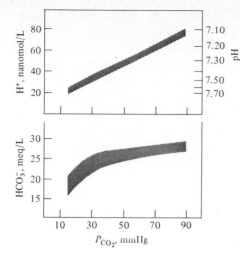

Figure 20-2 Combined significance bands for plasma pH and H^+ and HCO_3^- concentrations in acute hypocapnia and hypercapnia in humans. In uncomplicated acute respiratory acid-base disorders, values for the H^+ and HCO_3^- concentrations will, with an estimated 95 percent probability, fall within the band. Observations lying outside the band would indicate the presence of a complicating metabolic acid-base disturbance. *(From G. S. Arbus, L. A., Herbert, P. R. Levesque, B. E. Etsten, and W. B. Schwartz, N. Engl. J. Med., 280:117, 1969. By permission from the New England Journal of Medicine.)*

20-2) [14,15]. Thus, if the P_{CO_2} is acutely increased to 80 mmHg, there will be approximately a 4-meq/L elevation in the plasma HCO_3^- concentration to 28 meq/L and a potentially serious reduction in the pH to 7.17:

$$pH = 6.10 + \log \frac{28}{0.03(80)} = 7.17$$

This is not very efficient since the pH would have been only slightly lower at 7.10 if there were no buffering and the plasma HCO_3^- concentration had remained at 24 meq/L. A more severe reduction in the pH to below 7.00 can occur when there is a combined respiratory and metabolic acidosis, as with acute pulmonary edema and lactic acidosis due to severe heart failure [16].

Common causes of acute respiratory acidosis include acute exacerbation of underlying lung disease, severe asthma or pneumonia, pulmonary edema, and suppression of the respiratory center following a cardiac arrest, a drug overdose, or the administration of oxygen to a patient wtih chronic hypercapnia.

In addition, an increasingly recognized cause of intermittent hypercapnia is the *sleep apnea syndrome* [17,18]. This disorder is characterized by multiple (up to several hundred) apneic episodes per night associated with short periods of arousal (which are not apparent to the patient) due to hypoxemia and hypercapnia. Three different types of sleep apnea have been recognized [17,19]: central, in which rare brainstem disorders interfere with the medullary control of ventilation; obstructive, in which there is abnormal passive collapse of the pharyngeal muscles during inspiration such that the airway becomes occluded from the apposition of the tongue and soft palate against the posterior oropharynx; and a mixed central and obstructive picture. Most patients have at least some obstructive component which is typically manifested by loud snoring. Obesity, hypothyroidism, tonsillar enlargement, and nasal obstruction all may contribute to the development of inspiratory obstruction [20].

The sleep apnea syndrome is associated with a variety of occasionally subtle clinical manifestations which are due both to the repeated episodes of hypoxemia and the lack of uninterrupted sleep [17,18]. These include headaches, daytime somnolence and fatigue, morning confusion with difficulty in concentration, personality changes, depression, persistent pulmonary and systemic hypertension, and potentially life-threatening cardiac arrhythmias. Serious job-related and familial problems frequently ensue. The diagnosis should be suspected from the clinical history and can be confirmed by appropriate evaluation while the patient is asleep. Holter monitoring is also performed at this time to check for the presence of serious arrhythmias. Effective treatment can rapidly reverse almost all of the clinical findings (see "Treatment" below).

Chronic hypercapnia is unusual in the sleep apnea syndrome since the CO_2 retained during apneic episodes can be excreted when the patient is awake and ventilation is relatively normal. In rare cases, the combination of underlying lung disease and repetitive apneic episodes (including those due to daytime somnolence) can lead to a low total daily alveolar ventilation and persistent CO_2 retention [21].

Mechanical ventilation may be associated with hypercapnia if the rate of effective alveolar ventilation is inadequate. These patients, in whom ventilation is *fixed,* may also retain CO_2 if the rate of CO_2 production is increased. This sequence can occur with the administration of either $NaHCO_3$ to treat metabolic acidosis [22,23] or a large glucose load as part of a hyperalimentation regimen [24].

Chronic Respiratory Acidosis

The acid-base picture is different with chronic hypercapnia because of the compensatory renal response. The persistent elevation in the P_{CO_2} stimulates renal H^+ secretion over 3 to 5 days, resulting in the addition of HCO_3^- to the extracellular fluid (see page 237) [6]. The net effect is that patients with chronic respiratory acidosis have roughly a 3.5-meq/L increase in the plasma HCO_3^- concentration for every 10-mmHg increment in the P_{CO_2} (Fig. 20-3) [25,26]. If, for example, the P_{CO_2} were chronically increased to 80 mmHg, the plasma HCO_3^- concentration should increase by approximately 14 meq/L to 38 meq/L. This response is extremely effective since the pH only falls to 7.30, in contrast to 7.17 as seen above with a similar degree of acute hypercapnia. The efficiency of the renal compensation has allowed some patients to tolerate a P_{CO_2} as high as 90 to 110 mmHg without a fall in the arterial pH to less than 7.25 and without symptoms as long as adequate oxygenation is maintained [27].

The extent of the rise in the plasma HCO_3^- concentration in chronic respiratory acidosis is determined *solely by the increase in renal H^+ secretion.* Exogenous $NaHCO_3$ is not only unnecessary (since the pH is so well protected), but will be rapidly excreted in the urine without affecting the final plasma HCO_3^- concentration [28].

Chronic respiratory acidosis is a relatively common clinical disturbance that is most often due to chronic obstructive lung disease (bronchitis and emphysema) in smokers. Despite the presence of severe intrinsic pulmonary dysfunction, it is not completely understood why some patients become hypercapnic and hypoxemic rel-

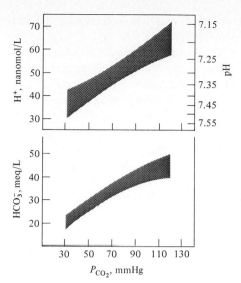

Figure 20-3 Ninety-five percent significance bands for plasma pH and H^+ and HCO_3^- concentrations in chronic hypercapnia. Note that, per change in P_{CO_2}, there is much less change in H^+ concentration and pH than in acute hypercapnia (Fig. 20-2). This reflects the effect of increased renal HCO_3^- generation. *(From W. B. Schwartz, N. C. Brackett, Jr., and J. J. Cohen, J. Clin. Invest., 44:291, 1965, by copyright permission of the American Society for Clinical Investigation.)*

atively early ("blue bloaters") whereas others do not ("pink puffers"). It has been observed that unaffected family members of patients with chronic hypercapnia have a reduced ventilatory response to hypoxemia and, to a lesser degree, hypercapnia, presumably due to some inherited abnormality in the respiratory center [29].

This finding suggests the following sequence: Lung disease initially impairs alveolar gas exchange, resulting in hypoxemia and eventually hypercapnia, which then stimulate ventilation and return the arterial P_{O_2} and P_{CO_2} toward normal. Persistent hypercapnia will occur relatively early (as in blue bloaters) when the ventilatory response to these stimuli is impaired. If, on the other hand, the central control of respiration is normal, persistent hypercapnia will not occur until pulmonary dysfunction is more severe (as in pink puffers).

A similar problem may be present when chronic respiratory acidosis occurs in extremely obese patients (called the Pickwickian syndrome) [30,31]. It had been assumed that the increased work of breathing and inspiratory muscle weakness induced by the excess weight of the chest wall were the primary problems in this disorder [31]. Reversal of the hypercapnia with weight loss in some patients is consistent with this hypothesis. However, the following observations suggest that factors other than obesity play a contributory role. First, most morbidly obese patients do not become hypercapnic and, in those who do, there is no correlation between the degree of obesity and the ventilatory abnormalities [32]. Second, a more normal ventilatory pattern frequently can be attained with progesterone (a direct respiratory stimulant) [33,34], which indicates that these patients can increase alveolar ventilation and raise the possibility of an associated central defect. An abnormality in respiratory control is also suggested by the demonstration that obese hypoventilators have a markedly impaired hypoxic drive and a lesser impairment in the hypercapnic drive [32,35]. In contrast, obese patients with normal ventilation have normal responses to hypoxemia and hypercapnia [36]. Once again, an inherited defect in ventilatory regulation may select out those obese patients who will develop chronic hypercapnia.

Obese hypoventilators may also have a component of obstructive sleep apnea [20]. The symptoms are somewhat similar since many Pickwickian patients complain of excessive daytime somnolence. However, there are important differences between these disorders since obstructive sleep apnea alone is only rarely associated with chronic CO_2 retention [21].

SYMPTOMS

Severe acute respiratory acidosis can produce a variety of neurologic abnormalities [37]. The initial symptoms include headache, blurred vision, restlessness, and anxiety which can progress to tremors, asterixis, delirium, and somnolence (called CO_2 narcosis). The cerebrospinal fluid (CSF) pressure is often elevated, and papilledema may be seen. These latter effects may be due to an elevation in cerebral blood flow induced by acidemia [38].

Both the neurologic symptoms and the increase in cerebral blood flow appear to be related to changes in the CSF (or cerebral interstitial) pH, not to the arterial pH or the P_{CO_2} [27,38,39]. CO_2 is lipid-soluble and rapidly equilibrates across the blood-brain barrier; HCO_3^- is a polar compound that crosses this barrier very slowly. Thus, acute hypercapnia produces a greater fall in CSF pH than does acute metabolic acidosis [40]; this probably explains why neurologic abnormalities are more prominent in respiratory acidosis. Symptoms are also less common with chronic hypercapnia since the renal compensation returns the arterial pH and ultimately the CSF pH toward normal [27].

If the systemic pH is reduced below 7.10 (most commonly seen when respiratory acidosis is complicated by metabolic acidosis), arrhythmias and peripheral vasodilation may combine to produce severe hypotension (see Chap. 19). Since acidemia also diminishes the inotropic response to catecholamines, elevation of the blood pressure may be difficult to achieve prior to raising the pH to greater than 7.15 to 7.20.

Chronic respiratory acidosis is also commonly associated with cor pulmonale and peripheral edema. The cardiac output and GFR are normal or near normal in this disorder, which usually occurs only in those patients with severe lung disease who are hypercapnic (see page 320) [41]. These findings suggest a direct role for CO_2 in the renal Na^+ retention that produces the edema; both the compensatory increase in $NaHCO_3$ reabsorption and a reduction in renal blood flow may contribute to this process [41].

DIAGNOSIS

The presence of an acid pH and hypercapnia is diagnostic of respiratory acidosis. Since the responses to acute and chronic respiratory acidosis are different, identifying the underlying disorder is more complicated than in metabolic acidosis or alkalosis. The following examples will illustrate how the confidence bands in Figs. 20-2 and 20-3 can be used in the evaluation of patients with respiratory acidosis. As will be demonstrated, *there is no substitute for an accurate and complete history* since a given set of arterial blood values can be interpreted in several different ways.

In acute hypercapnia, the plasma HCO_3^- concentration should be between 24 and 29 meq/L (Fig. 20-2). Values above or below this range indicate superimposed metabolic disorders. For example:

Case history 20-1 A previously well patient is brought into the emergency room in a moribund state. Physical examination and chest x-ray suggest acute pulmonary edema. The results of laboratory tests are:

$$pH = 7.02$$

$$P_{CO_2} = 60 \text{ mmHg}$$

$$[HCO_3^-] = 15 \text{ meq/L}$$

$$P_{O_2} = 40 \text{ mmHg}$$

COMMENT Since the plasma HCO_3^- concentration should rise 1 meq/L for each 10-mmHg increment in P_{CO_2} in acute respiratory acidosis, an acute elevation of the P_{CO_2} to 60 mmHg should increase the plasma HCO_3^- concentration to 26 meq/L (pH of 7.24). Therefore, these results represent a combined respiratory and metabolic acidosis, a life-threatening combination not infrequently seen in severe acute pulmonary edema [16]

The difficulties in interpretation in chronic respiratory acidosis are illustrated in Figs. 20-4 to 20-6. Consider the following set of arterial blood tests:

$$pH = 7.27$$

$$P_{CO_2} = 70 \text{ mmHg}$$

$$[HCO_3^-] = 31 \text{ meq/L}$$

$$P_{O_2} = 35 \text{ mmHg}$$

This 30-mmHg increase in the P_{CO_2} should be associated with a 3-meq/L increase in the plasma HCO_3^- concentration to 27 meq/L in acute hypercapnia and an 11-meq/L increment in the plasma HCO_3^- concentration (3.5 meq/L per 10-mmHg increase in the P_{CO_2}) to 35 meq/L in chronic hypercapnia. The observed plasma HCO_3^- concentration of 31 meq/L falls *between* the bands for acute and chronic respiratory acidosis (Fig. 20-4a, point A). This can represent (1) metabolic acidosis complicating chronic hypercapnia (Fig. 20-4b); (2) acute, superimposed on chronic, respiratory acidosis (Fig. 20-4c); or (3) metabolic alkalosis and acute hypercapnia (Fig. 20-4d). *These possibilities cannot be distinguished without the respective histories:*

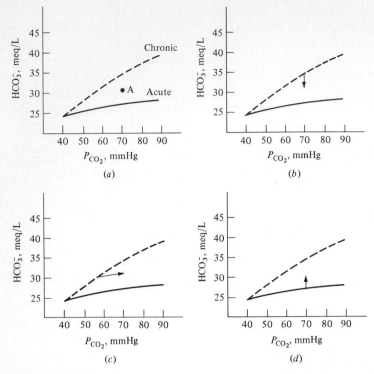

Figure 20-4 Confidence bands for acute and chronic hypercapnia have been transposed from Figs. 20-2 and 20-3. (*a*) Point A lies between the curves and can represent three different disorders. (*b*) Metabolic acidosis complicating chronic respiratory acidosis. (*c*) Acute, superimposed on chronic, hypercapnia. (*d*) Acute respiratory acidosis and metabolic alkalosis. (*Adapted from J. J. Cohen and W. B. Schwartz, Am. J. Med., 41:163, 1966.*)

1. A patient with chronic bronchitis develops persistent diarrhea.
2. A patient with chronic hypercapnia complains of fever and increased sputum production. The chest x-ray is consistent with pneumonia.
3. A patient with a history of extrinsic asthma has 5 days of vomiting due to theophylline toxicity and then develops an acute asthmatic attack after the theophylline is discontinued.

A different problem in interpretation is present in the following example. The initial laboratory data were:

$$pH = 7.53$$

$$P_{CO_2} = 50 \text{ mmHg}$$

$$[HCO_3^-] = 40 \text{ meq/L}$$

$$P_{O_2} = 55 \text{ mmHg}$$

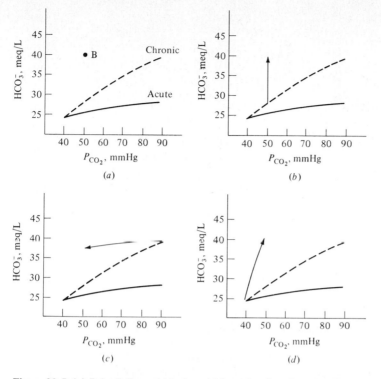

Figure 20-5 (*a*) Point B lies outside the confidence bands and can be due to one of three disorders. (*b*) Metabolic alkalosis complicating chronic hypercapnia. (*c*) An acute reduction in the P_{CO_2} in a patient with chronic respiratory acidosis. (*d*) Primary metabolic alkalosis with compensatory CO_2 retention.

Oxygen was started and repeat tests were obtained:

$$pH = 7.47$$

$$P_{CO_2} = 57 \text{ mmHg}$$

$$[HCO_3^-] = 40 \text{ meq/L}$$

$$P_{O_2} = 80 \text{ mmHg}$$

Because of the increase in P_{CO_2}, oxygen was discontinued for fear of further hypercapnia and CO_2 narcosis. After appropriate therapy with NaCl, the following values were obtained:

$$pH = 7.41$$

$$P_{CO_2} = 39 \text{ mmHg}$$

$$[HCO_3^-] = 24 \text{ meq/L}$$

$$P_{O_2} = 68 \text{ mmHg}$$

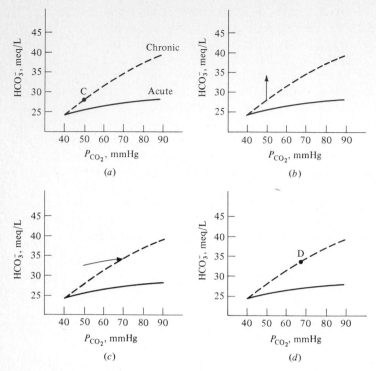

Figure 20-6 Patient with stable chronic hypercapnia (*a*, point C) develops metabolic alkalosis (*b*) due to vomiting. During one episode he aspirates some vomitus and has an acute increase in P_{CO_2} (*c*). At this point, his blood values lie within the confidence band and are indistinguishable from a patient with uncomplicated chronic respiratory acidosis (*d*, point D).

Despite the high P_{CO_2} on admission, the pH was alkaline. Most commonly, this degree of hypercapnia in an alkalemic patient is due to metabolic alkalosis complicating chronic hypercapnia (Fig. 20-5*b*). However, this can also represent acute hypocapnia superimposed on chronic respiratory acidosis (Fig. 20-5*c*) or hypercapnia as part of the normal respiratory compensation to metabolic alkalosis (Fig. 20-5*d*). The specific diagnosis cannot be made directly from the laboratory data, and the respective clinical histories are required:

1. A patient with chronic obstructive pulmonary disease develops pedal edema due to cor pulmonale and is started on diuretics.
2. Tracheal intubation and mechanical ventilation are begun in a patient with severe CO_2 retention. This entity is referred to as posthypercapnic alkalosis (see Chap. 18).
3. A patient has 5 days of persistent vomiting.

The correction of the alkalemia and hypercapnia with NaCl indicates that this patient's primary problem was metabolic alkalosis due to vomiting. Although a high P_{CO_2} which increases after the administration of oxygen is most often seen with chronic respiratory acidosis, it may also occur with metabolic alkalosis since the development of hypoxemia can limit the degree of compensatory hypoventilation (see page 379). In this patient, the further hypercapnia that followed oxygen therapy was beneficial as the arterial pH decreased toward normal (from 7.53 to 7.47). There was no danger of CO_2 narcosis, and oxygen could safely have been continued.

Finally, *one cannot assume that values within the confidence bands connote an uncomplicated disorder*. Suppose a patient with stable chronic hypercapnia (Fig. 20-6a, point C) develops recurrent vomiting (Fig. 20-6b) and then aspiration pneumonia (Fig. 20-6c). Although this represents a triad of chronic respiratory acidosis, metabolic alkalosis, and acute respiratory acidosis, the final blood values lie within the band and cannot be distinguished from pure, severe chronic hypercapnia (Fig. 20-6d).

In summary, the confidence bands are useful guides in the interpretation of acid-base measurements. However, this interpretation cannot proceed in a vacuum and must be correlated with a complete history and physical examination.

Use of the Alveolar-Arterial Oxygen Gradient

Calculation of the alveolar-arterial (A-a) oxygen gradient can be used to help differentiate intrinsic pulmonary disease from extrapulmonary disorders as the cause of the hypercapnia. The derivation of the formula used to estimate this gradient requires a brief review of the physiology of alveolar gas exchange [42]. At a barometric pressure of 1 atm (P_B equals 760 mmHg) in the inspired air at sea level, water vapor accounts for approximately 47 mmHg, nitrogen for 563 mmHg, and oxygen for the remaining 150 mmHg (Fig. 20-7). Since the pressure remains at 1 atm in the alveolus and there is no net movement of nitrogen or water vapor across the alveolar capillary, the P_{N_2} and P_{H_2O} in the alveolus will be the same as in the inspired air and will equal 610 mmHg. Thus, the sum of the partial pressures of the other

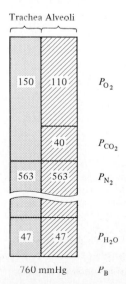

Trachea Alveoli

P_{O_2}

P_{CO_2}

P_{N_2}

P_{H_2O}

760 mmHg P_B

Figure 20-7 Composition of gas in the trachea at end inspiration (the same as that in the inspired air) and composition of alveolar gas. Total pressure and the partial pressures of nitrogen and water vapor are the same in both compartments; alveolar P_{O_2} goes down proportionately to the increase in alveolar P_{CO_2}. It should be noted that these values are for a patient breathing room air. If a patient is given supplemental oxygen, for example, P_{iO_2} equals 300 mmHg, then the alveolar P_{O_2} will be proportionately increased. *(Adapted from G. L. Snider, Chest, 63:801, 1973.)*

gases in the alveolus must be equal to 150 mmHg, i.e., to the partial pressure of oxygen in the inspired air (P_{iO_2}) (Fig. 20-7). In the alveolus, inspired O_2 enters the blood while CO_2 leaves the blood and enters the alveolus. If the amount of CO_2 added were equal to the amount of O_2 taken up, then the alveolar $P_{O_2}(P_{AO_2})$ would be less than the inspired P_{O_2} by an amount equal to the alveolar P_{CO_2} (P_{ACO_2}):

$$P_{AO_2} = P_{iO_2} - P_{ACO_2} \tag{20-5}$$

However, more O_2 is usually taken up than CO_2 produced. The respiratory quotient is defined as the amount of CO_2 produced divided by the amount of O_2 utilized and, on a normal diet, is approximately 0.8. This means that each molecule of CO_2 produced and entering the alveolus represents the utilization of 1.25 molecules of O_2. To account for this. Eq. (20-5) can be rewritten:

$$P_{AO_2} = P_{iO_2} - 1.25 P_{ACO_2} \tag{20-6}$$

Since CO_2 rapidly diffuses across the alveolar capillary (20 times as fast as O_2), the P_{ACO_2} is essentially equal to the arterial P_{CO_2} (P_{aCO_2}). Therefore

$$P_{AO_2} = P_{iO_2} - 1.25 P_{aCO_2} \tag{20-7}$$

In a subject inspiring room air (P_{iO_2} equals 150 mmHg) with a P_{aCO_2} of 40 mmHg,

$$P_{AO_2} = 150 - (1.25 \times 40) = 100 \text{ mmHg}$$

Not all this oxygen enters the blood since there is an alveolar-arterial (A-a) oxygen gradient averaging, on room air, 5 to 10 mmHg in subjects under the age of 30 and gradually increasing to 15 to 20 mmHg in the elderly. This gradient probably is due both to pulmonary arteriovenous shunts and to the perfusion of underventilated areas of the lung. Since

$$(\text{A-a}) \text{ } O_2 \text{ gradient} = P_{AO_2} - P_{aO_2} \tag{20-8}$$

by substituting Eq. (20-7) for P_{AO_2}, we have

$$(\text{A-a}) \text{ } O_2 \text{ gradient} = P_{iO_2} - 1.25 P_{CO_2} - P_{aO_2} \tag{20-9}$$

The (A-a) oxygen gradient is always increased in hypercapnic patients with intrinsic pulmonary disease and may be increased in some patients with extrapulmonary disorders [42]. However, a normal gradient essentially excludes pulmonary disease and suggests some form of central alveolar hypoventilation or an abnormality of the chest wall or inspiratory muscles.

TREATMENT

Acute Respiratory Acidosis

Patients with acute respiratory acidosis are at risk from both hypercapnia and hypoxemia. Although the P_{O_2} can usually be raised by the administration of supplemental oxygen, reversal of the hypercapnia requires an increase in effective alveolar ventilation. This can be achieved by control of the underlying disease (e.g., with the use of bronchodilators and, if necessary, corticosteroids in asthma) or by tracheal intubation and mechanical ventilation. Indications for the latter include refractory severe hypoxemia, symptomatic or progressive hypercapnia, and depression of the respiratory center due, for example, to a drug overdose.

The role of $NaHCO_3$ in the treatment of acute respiratory acidosis (without concomitant metabolic acidosis) is not well defined. Although the primary aim of therapy is to restore normal ventilation, small doses of $NaHCO_3$ (44 to 88 meq) can be infused over 5 to 10 min if the P_{CO_2} cannot be promptly controlled in a severely acidemic patient (pH less than 7.15). This regimen may have particular benefit in severe status asthmaticus requiring mechanical ventilation [43]. In this setting, elevating the plasma HCO_3^- concentration allows the pH to be controlled at a higher P_{CO_2} and, therefore, at a lower ventilatory rate and a lower peak inspiratory pressure. The latter change may minimize the incidence of potentially serious complications such as pneumothorax or pneumomediastinum [43].

There are, however, several potential hazards in the use of $NaHCO_3$ in acute respiratory acidosis. First, the administration of $NaHCO_3$ should be avoided, if possible, in patients with pulmonary edema because it can increase the degree of pulmonary congestion. In general, most patients can be managed without $NaHCO_3$ since correction of the pulmonary edema and hypoxemia is usually sufficient to restore acid-base balance [16,44]. Second, metabolic alkalosis (due to the excess HCO_3^-) may ensue after the P_{CO_2} has returned to normal. This is usually not a major problem. Third, the infusion of $NaHCO_3$ can result in an increase in the P_{CO_2} by the following reaction:

$$HCO_3^- + H^+ \rightarrow H_2CO_3 \rightarrow H_2O + CO_2$$

Normally, the CO_2 that is generated is rapidly eliminated by the lungs. However, in patients with inadequate ventilation (as occurs during a cardiac arrest), the CO_2 may be retained and the resultant elevation in the P_{CO_2} may prevent an increase in, and may actually reduce, the pH [22,23]. If a patient with respiratory acidosis demonstrates these changes, further $NaHCO_3$ administration should be avoided until ventilation can be enhanced.

Sleep apnea syndrome Treatment is required in this disorder to control the severe daytime symptoms and, when present, the complex arrhythmias which occur during the apneic episodes [17,18,45]. In some patients with *central* sleep apnea (diagnosed by monitoring during sleep), progesterone or acetazolamide may effectively stimulate ventilation and reduce the number of apneic episodes [19,33,34]. The latter is a carbonic anhydrase inhibitor that induces a $NaHCO_3$ diuresis by reducing proximal HCO_3^- reabsorption (see page 230). The ensuing metabolic acidosis may be responsible for the enhanced ventilation [19]. If these medications fail, diaphragmatic pacing or, in severe cases, tracheostomy with mechanical ventilation at night may be required [19].

Patients with *obstructive* sleep apnea should first be evaluated for a surgically correctible lesion such as enlarged tonsils and adenoids [46]. Respiratory stimulation with progesterone can be tried, although it is infrequently effective since it does not reverse the primary problem [47,48]. However, the antidepressant protriptyline can improve the daytime symptoms and nocturnal oxygenation in many patients [49]. The mechanism by which this occurs may be related to a reduction in the time spent in the rapid eye movement stage of sleep, during which apneic episodes are generally most severe. A nonspecific "alerting" effect may also contribute to the relief of daytime symptoms.

If drug therapy is ineffective or not tolerated, passive collapse of the pharyngeal muscles and airway obstruction can be directly prevented by the use of continuous positive airway pressure through a face mask while the patient is asleep [21]. If all else fails in a patient with severe symptoms or arrhythmias, tracheostomy (in which the tracheostomy tube is closed during the day and open only at night) can result in a rapid and permanent reversal of the apneic episodes and essentially all clinical abnormalities [18,45].

Chronic Respiratory Acidosis

Because of the effectiveness of the renal compensation, it is usually not necessary to treat the pH even in patients with severe hypercapnia [27]. The primary goal of therapy is to maintain adequate oxygenation and, if possible, to improve effective alveolar ventilation.

The appropriate treatment varies with the underlying disease. As a general rule, excessive oxygen and sedatives should be avoided since they can act as respiratory depressants, producing further hypoventilation. For patients with chronic obstructive lung disease, bronchodilators, expectorants, corticosteroids, and antibiotics may help to reverse the bronchoconstrictive and inflammatory components of the disease [50]. If severe hypoxemia persists (arterial P_{O_2} below 45 to 50 mmHg), continuous low-flow oxygen therapy may prolong survival and diminish the severity of cor pulmonale [51]. Although respiratory stimulants have not been generally effective in this disorder [50], some patients may have a beneficial response to progesterone [52]. Since hypoxemia and hypercapnia should maximally stimulate ventilation, the effect of progesterone suggests that these patients may have an associated defect in the central control of respiration. This possibility is also supported by the finding of abnormal respiratory regulation in unaffected family members of hypercapnic patients [29]. Progesterone may also be effective in massively obese patients [33,34] in whom weight reduction is another important component of therapy [30,31].

Mechanical ventilation may be required when there is an acute exacerbation of chronic hypercapnia (as with the development of pneumonia). In this setting, care must be taken *to lower the P_{CO_2} gradually*. A rapid reduction in the P_{CO_2} to near normal can lead to a marked rise in the pH of the extracellular fluid and the central nervous system (CNS) because of the persistence of the compensatory elevation in the plasma HCO_3^- concentration. The increase in CNS pH may be responsible for the development of severe neurologic abnormalities such as seizures and coma [53,54]. These findings may improve if the P_{CO_2} is allowed to rise toward its previous level.

Effect of superimposed metabolic alkalosis As described above, the apparent respiratory insensitivity to CO_2 in chronic hypercapnia is due at least in part to the compensatory rise in the plasma HCO_3^- concentration, which limits the fall in pH [8,10]. Consistent with this hypothesis is the observation that alveolar ventilation can be enhanced and the P_{CO_2} lowered in a patient with chronic hypercapnia if the plasma HCO_3^- concentration and, therefore, the arterial pH are reduced by the administration of acetazolamide [55].

On the other hand, raising the plasma HCO_3^- concentration further will increase the pH (occasionally to alkalemic levels), which then depresses ventilation, leading to more hypoxemia and hypercapnia [56,57]. In this setting, which is most often due to diuretic therapy for edema in cor pulmonale, lowering the plasma HCO_3^- concentration will improve ventilation and may produce a subjective sense of feeling better [56].

Correction of a superimposed metabolic alkalosis can be achieved by discontinuing the diuretics and administering NaCl. This is not practical, however, in the patient who is still significantly edematous. In this circumstance, acetazolamide (250 to 375 mg once or twice a day) can both lower the plasma HCO_3^- concentration and increase the urine output by inhibiting proximal $NaHCO_3$ reabsorption. Despite its efficacy, there are two potential problems with the use of acetazolamide. First, the plasma HCO_3^- concentration should be lowered to the level appropriate for the degree of hypercapnia (see Fig. 20-2). Returning the plasma HCO_3^- concentration to the normal value of 24 meq/L can lead to severe acidemia in the presence of marked hypercapnia [55]. Second, acetazolamide can produce a transient *elevation* in the P_{CO_2} (usually 3 to 7 mmHg) prior to its diuretic effect [58]. This complication, which is generally not clinically important, may be due to partial inhibition of carbonic anhydrase in the red blood cell [55,58]. This enzyme catalyzes the hydration of CO_2 to H_2CO_3, a reaction that is essential for CO_2 transport by the red cell [see Eqs. (20-1) and (20-2)] and, therefore, the elimination of CO_2 by the lungs.

PROBLEMS

20-1 Match the clinical histories with the appropriate arterial blood values:

	pH	P_{CO_2}, mmHg	$[HCO_3^-]$, meq/L
(a)	7.37	65	37
(b)	7.22	60	26
(c)	7.35	60	32

1. A 60-year-old man with chronic bronchitis develops persistent diarrhea.
2. A markedly obese 24-year-old man.
3. A 14-year-old girl with a severe acute asthmatic attack.
4. A 56-year old woman with chronic bronchitis is started on diuretic therapy for peripheral edema, resulting in a 3-kg weight loss.

20-2 A 54-year-old man with a history of chronic obstructive pulmonary disease has a 2-day episode of increasing shortness of breath and sputum production. The chest x-ray reveals a left lower-lobe pneumonia. The following laboratory data are obtained with the patient breathing room air:

$$\text{Arterial pH} = 7.25$$

$$P_{CO_2} = 70 \text{ mmHg}$$

$$[HCO_3^-] = 30 \text{ meq/L}$$

$$P_{O_2} = 30 \text{ mmHg}$$

$$\text{Urine } [Na^+] = 4 \text{ meq/L}$$

The patient is started on intravenous aminophylline and nasal oxygen and becomes less responsive. Repeat blood tests are obtained:

$$\text{Arterial pH} = 7.18$$

$$P_{CO_2} = 86 \text{ mmHg}$$

$$[HCO_3^-] = 31 \text{ meq/L}$$

$$P_{O_2} = 58 \text{ mmHg}$$

(*a*) What was the probable acid-base disturbance on admission?
(*b*) What was responsible for the increase in the P_{CO_2} in the hospital?
(*c*) What further therapy would you recommend?
(*d*) If the P_{CO_2} is rapidly lowered to 40 mmHg, what will happen to the arterial pH? If the patient is then maintained on a low-sodium diet, how long will it take for the plasma HCO_3^- concentration to return to normal?

20-3 A 65-year-old man has a history of smoking and hypertension, which is treated with a diuretic. The following arterial blood values are obtained on room air:

$$\text{pH} = 7.48$$

$$P_{CO_2} = 51 \text{ mmHg}$$

$$P_{O_2} = 73 \text{ mmHg}$$

$$[HCO_3^-] = 36 \text{ meq/L}$$

(*a*) What is the most likely acid-base disturbance?
(*b*) Does the patient have significant underlying lung disease?

REFERENCES

1. Lambertsen, C. J.: Chemical control of respiration at rest, in *Medical Physiology,* 14th ed., V. B. Mountcastle (ed.), Mosby, St. Louis, 1980.
2. Berger, A. J., R. A. Mitchell, and J. W. Severinghaus: Regulation of respiration, N. Engl. J. Med., 297:92, 138, 194, 1977.
3. Fencl, V., T. B. Miller, and J. R. Pappenheimer: Studies on the respiratory response to disturbances of acid-base balance, with deductions concerning the ionic composition of cerebral interstitial fluid, Am. J. Physiol., 210:459, 1966.
4. Lugliani, R., B. J. Whipp, C. Seard, and K. Wasserman: Effect of bilateral carotid-body resection on ventilatory control at rest and during exercise in man, N. Engl. J. Med., 285:1105, 1971.

5. West, J. B.: Causes of carbon dioxide retention in lung disease, N. Engl. J. Med. 284:1232, 1971.
6. Polak, A., G. D. Haynie, R. M. Hays, and W. B. Schwartz: Effects of chronic hypercapnia on electrolyte and acid-base equilibrium. I. Adaptation, J. Clin. Invest., 40:1223, 1961.
7. Hirshman, C. A., R. E. McCullough, and J. V. Weil: Normal values for hypoxic and hypercapnic ventilatory drives in man, J. Appl. Physiol., 38:1095, 1975.
8. Lourenco, R. V., and J. M. Miranda: Drive and performance of the ventilatory apparatus in chronic obstructive lung disease, N. Engl. J. Med., 279:53, 1968.
9. Lane, D. J., and J. B. L. Howell: Relationship between sensitivity to carbon dioxide and clinical features in patients with chronic airway obstruction. Thorax, 25:150, 1970.
10. Heinemann, H. O., and R. M. Goldring: Bicarbonate and the regulation of ventilation, Am. J. Med., 57:361, 1974.
11. Eldridge, F., and G. Gherman: Studies of oxygen administration in respiratory failure, Ann. Intern. Med., 68:569, 1968.
12. Franklin, W.: Treatment of severe asthma, N. Engl. J. Med., 290:1469, 1974.
13. Giebisch, G., L. Berger, and R. F. Pitts: The extrarenal response to acute acid-base disturbances of respiratory origin, J. Clin. Invest., 34:231, 1955.
14. Arbus, G. S., L. A. Hevert, P. R. Levesqué, B. E. Etsten, and W. B. Schwartz: Characterization and clinical application of the "significance band" for acute respiratory alkalosis, N. Engl. J. Med., 280:117, 1969.
15. Brackett, N. C., J. J. Cohen, and W. B. Schwartz: Carbon dioxide titration curve of normal man: effect of increasing degrees of acute hypercapnia on acid-base equilibrium, N. Engl. J. Med., 272:6, 1965.
16. Aberman, A., and M. Fulop: The metabolic and respiratory acidosis of acute pulmonary edema, Ann. Intern. Med., 76:173, 1972.
17. Guilleminault, C., J. Cummiskey, and W. C. Dement: Sleep apnea syndrome: recent advances, Adv. Intern. Med., 26:347, 1980.
18. Guilleminault, C., F. B. Simons, J. Motta, J. Commiskey, M. Rosekind, J. S. Schroeder, and W. C. Dement: Obstructive sleep apnea syndrome and tracheostomy: long term followup experience, Arch. Intern. Med., 141:985, 1981.
19. White, D. P., C. W. Zwillich, C. K. Pickett, N. J. Douglas, L. J. Findley, and J. V. Weil: Central sleep apnea: improvement with acetazolamide therapy, Arch. Intern. Med., 142:1816, 1982.
20. Stradling, J. R.: Obstructive sleep apnoea syndrome, Br. Med. J., 2:528, 1982.
21. Rapoport, D. M., B. Sorkin, S. M. Garay, and R. M. Goldring: Reversal of the "Pickwickian syndrome" by long-term use of nocturnal nasal-airway pressure, N. Engl. J. Med., 307:931, 1982.
22. Bishop, R. L., and M. L. Weisfeldt: Sodium bicarbonate administration during cardiac arrest: effect on arterial pH, P_{CO_2}, and osmolality, J. Am. Med. Assoc., 235:506, 1976.
23. Ostrea, E. M., and G. B. Odell: The influence of bicarbonate administration on blood pH in a "closed system": clinical implications, J. Pediatr., 80:671, 1972.
24. Covelli, H. D., J. W. Black, M. S. Olson, and J. F. Beekman: Respiratory failure precipitated by high carbohydrate loads, Ann. Intern. Med., 95:579, 1981.
25. Schwartz, W. B., N. C. Brackett, and J. J. Cohen: The response of extracellular hydrogen ion concentration to graded degrees of chronic hypercapnia: the physiologic limits of the defense of pH, J. Clin. Invest., 44:291, 1965.
26. van Ypersele de Strihou, C., L. Brasseur, and J. de Coninck: "Carbon dioxide response curve" for chronic hypercapnia in man, N. Engl. J. Med., 275:117, 1966.
27. Neff, T. A., and T. L. Petty: Tolerance and survival in severe chronic hypercapnia, Arch. Intern. Med., 129:591, 1972.
28. van Ypersele de Strihou, C., P. F. Gulyassy, and W. B. Schwartz: Effects of chronic hypercapnia on electrolyte and acid-base equilibrium. II. Characteristics of the adaptive and recovery process as evaluated by provision of alkali, J. Clin. Invest., 41:2246, 1962.
29. Mountain, R., C. Zwillich, and J. V. Weil: Hypoventilation in obstructive lung disease: the role of familial factors, N. Engl. J. Med., 298:521, 1978.
30. Fishman, A., G. M. Turino, and E. H. Bergofsky: The syndrome of alveolar hypoventilation, Am. J. Med., 23:333, 1957.

31. Rochester, D. F., and Y. Enson: Current concepts in the pathogenesis of the obesity-hypoventilation syndrome, Am. J. Med., 57:402, 1974.

32. Zwillich, C. W., F. D. Sutton, D. J. Pierson, E. M. Creagh, and J. V. Weil: Decreased hypoxic ventilatory drive in the obesity-hypoventilation syndrome, Am. J. Med., 59:343, 1975.

33. Lyons, H. A., and C. T. Huang: Therapeutic use of progesterone in alveolar hypoventilation associated with obesity, Am. J. Med., 44:881, 1968.

34. Sutton, F. D., C. W. Zwillich, E. Creagh, D. J. Pierson, and J. V. Weil: Progesterone for outpatient treatment of Pickwickian syndrome, Ann. Intern. Med., 83:476, 1975.

35. Kronenberg, R. S., C. W. Drage, and J. E. Stevenson: Acute respiratory failure and obesity with normal ventilatory response to carbon dioxide and absent hypoxic ventilatory drive, Am. J. Med., 62:773, 1977.

36. Kronenberg, R. S., R. A. Gabel, and J. W. Severinghaus: Normal chemoreceptor function in obesity before and after ileal bypass surgery to force weight reduction, Am. J. Med., 59:349, 1975.

37. Kilburn, K.: Neurologic manifestations of respiratory failure, Arch. Intern. Med., 116:409, 1965.

38. Fencl, V., J. R. Vale, and J. A. Broch: Respiration and cerebral blood flow in metabolic acidosis and alkalosis in humans, J. Appl. Physiol., 27:67, 1969.

39. Posner, J., and F. Plum: Spinal-fluid pH and neurologic symptoms in systemic acidosis, N. Engl. J. Med., 277:605, 1967.

40. Posner, J. B., A. G. Swanson, and F. Plum: Acid-base balance in cerebrospinal fluid, Arch. Neurol., 12:479, 1965.

41. Farber, M. O., L. R. Roberts, M. R. Weinberger, G. L. Robertson, N. S. Fineberg, and F. Manfredi: Abnormalities of sodium and H_2O handling in chronic obstructive lung disease, Arch. Intern. Med., 142:1326, 1982.

42. Snider, G. L.: Interpretation of the arterial oxygen and carbon dioxide partial pressures, Chest, 63:801, 1973.

43. Menitove, S. M., and R. M. Goldring: Combined ventilator and bicarbonate strategy in the management of status asthmaticus, Am. J. Med., 74:898, 1983.

44. Fulop, M., M. Horowitz, A. Aberman, and E. Jaffé: Lactic acidosis in pulmonary edema due to left ventricular failure, Ann. Intern. Med., 79:180, 1973.

45. Guilleminault, C., S. J. Connolly, and R. A. Winkle: Cardiac arrhythmia and conduction disturbances during sleep in 400 patients with sleep apnea syndrome, Am. J. Cardiol., 52:490, 1983.

46. Guilleminault, C., F. L. Eldridge, A. Tilkian, F. B. Simmons, and W. C. Dement: Sleep apnea syndrome: a review of 25 cases, Arch. Intern. Med., 137:296, 1977.

47. Orr, W. C., N. K. Imes, and R. J. Martin: Progesterone therapy in obese patients with sleep apnea, Arch. Intern Med., 139:109, 1979.

48. Strohl, K. P., M. J. Hensley, N. A. Saunders, S. M. Scarf, R. Brown, and R. H. Ingram: Progesterone administration and progressive sleep apneas, J. Am. Med. Assoc., 245:1230, 1981.

49. Brownell, L. G., P. West, P. Sweatmen, J. C. Acres, and M. H. Kryger: Protriptyline in obstructive sleep apnea: a double-blind trial, N. Engl. J. Med., 307:1037, 1982.

50. Bone, R. C.: Treatment of respiratory failure due to advanced chronic obstructive lung disease, Arch. Intern. Med., 140:1018, 1980.

51. Anthonisen, N. R.: Long-term oxygen therapy, Ann. Intern. Med., 99:519, 1983.

52. Skatrud, J. B., J. A. Dempsey, P. Bhansali, and C. Irvin: Determinants of chronic carbon dioxide retention and its correction in humans, J. Clin. Invest., 65:813, 1980.

53. Rotheram, E. B., Jr., P. Safar, and E. D. Robin: CNS disorder during mechanical ventilation in chronic pulmonary disease, J. Am. Med. Assoc., 189:993, 1964.

54. Kilburn, K. H.: Shock, seizures and coma with alkalosis during mechanical ventilation, Ann. Intern. Med., 65:977, 1966.

55. Dorris, R., J. V. Olivia, and T. Rodman: Dichlorphenamide, a potent carbonic anhydrase inhibitor: effect on alveolar ventilation, ventilation-perfusion relationships and diffusion in patients with chronic lung disease, Am. J. Med., 36:79, 1964.

56. Bear, R., M. Goldstein, E. Phillipson, M. Ho, M. Hammeke, R. Feldman, S. Handelsman, and M. Halperin: Effect of metabolic alkalosis on respiratory function in patients with chronic obstructive lung disease, Can. Med. Assoc. J., 117:900, 1977.

57. Miller, P. D., and A. S. Berns: Acute metabolic alkalosis perpetuating hypercapnia: a role for acet-azolamide in chronic obstructive pulmonary disease, J. Am. Med. Assoc., 238:2400, 1977.

58. Bell, A. L. L., C. N. Smith, and E. Andreae: Effects of the carbonic anhydrase inhibitor "6063" (Diamox) on respiration and electrolyte metabolism of patients with respiratory acidosis, Am. J. Med., 18:536, 1955.

TWENTY-ONE

RESPIRATORY ALKALOSIS

The introduction to acid-base disorders presented in Chap. 17 should be read before proceeding with this discussion. Respiratory alkalosis is a clinical disturbance characterized by an elevated arterial pH (or a decreased H^+ concentration), a low P_{CO_2} (hypocapnia), and a variable reduction in the plasma HCO_3^- concentration. This must be differentiated from metabolic acidosis in which the plasma HCO_3^- concentration and P_{CO_2} also are diminished but the pH is reduced rather than increased.

PATHOPHYSIOLOGY

A primary decrease in the P_{CO_2} occurs when effective alveolar ventilation is increased to a level beyond that needed to eliminate the daily load of metabolically produced

CO_2. Before discussing the different disorders that can cause a respiratory alkalosis, it is helpful to first review how the body responds to hypocapnia. From the law of mass action,

$$[H^+] = 24 \times \frac{P_{CO_2}}{[HCO_3^-]}$$

it can be seen that the reduction in the extracellular H^+ concentration induced by hypocapnia can be minimized by lowering the HCO_3^- concentration. This protective response involves two steps: rapid cell buffering and a later decrease in net renal acid excretion. As a result of the time differential between the cellular and renal effects, the changes in acute and chronic respiratory alkalosis are different.

Acute Respiratory Alkalosis

Within 10 min after the onset of respiratory alkalosis, H^+ ions move from the cells into the extracellular fluid where they combine with HCO_3^-, resulting in a fall in the plasma HCO_3^- concentration:

$$H^+ + HCO_3^- \rightarrow H_2CO_3 \rightarrow CO_2 + H_2O$$

These H^+ ions are derived from the protein, phosphate, hemoglobin buffers in the cells ($HBuf \rightarrow H^+ + Buf^-$) and, to a lesser degree, from increased cellular lactic acid production induced by the alkalosis [1].

In general, enough H^+ ions enter the extracellular fluid to lower the plasma HCO_3^- concentration 2 meq/L for each 10-mmHg decrease in the P_{CO_2} (see Fig. 20-2) [2]. For example, if the P_{CO_2} were reduced to 20 mmHg (20 mmHg less than normal), the plasma HCO_3^- concentration should fall by 4 meq/L to 20 meq/L (pH equals 7.63). This is not very efficient since the pH would have been only slightly greater, at 7.70, if there were no cell buffering and the plasma HCO_3^- concentration had remained at 24 meq/L.

Chronic Respiratory Alkalosis

In the presence of persistent hypocapnia, there is a decrease in renal H^+ secretion that begins within 2 h but is not complete for 2 to 3 days [3–5]. This response, which is directly induced by the fall in the P_{CO_2} [6], is manifested by HCO_3^- loss in the urine and by decreased urinary ammonium excretion [4,5]. Both of these effects lower the plasma HCO_3^- concentration, the latter by preventing the excretion of the daily H^+ load, thereby resulting in H^+ retention.

On the average, the combined effects of the cell buffers and the renal compensation will result in a new steady state in which the plasma HCO_3^- concentration has fallen 5 meq/L for each 10-mmHg reduction in the P_{CO_2} (Fig. 21-1) [3]. If the P_{CO_2} were chronically reduced to 20 mmHg, the plasma HCO_3^- concentration should fall by 10 meq/L to 14 meq/L. This response is extremely effective since the pH is only increased to 7.47 as compared with 7.63 as seen above with a similar degree of acute hypocapnia.

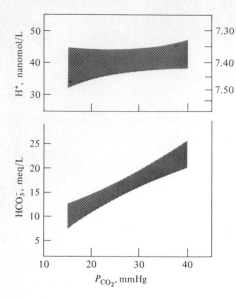

Figure 21-1 Significance bands of 95 percent probability for plasma pH and H^+ and HCO_3^- concentrations in chronic hypocapnia. Note that there is only a minimal change in the H^+ concentration and pH as the P_{CO_2} is reduced. *(Reproduced from J. F. Gennari, M. B. Goldstein, and W. B. Schwartz, J. Clin. Invest., 51:1722, 1972, by copyright permission of the American Society for Clinical Investigation.)*

ETIOLOGY

Respiration is physiologically governed by two sets of chemoreceptors: those in the respiratory center in the brainstem and those in the carotid and aortic bodies located at the bifurcation of the carotid arteries and in the aortic arch, respectively [7,8]. The central chemoreceptors are stimulated by an increase in the P_{CO_2} or metabolic acidosis, both of which appear to be sensed as a fall in the pH of the surrounding cerebral interstitial fluid [9]. The peripheral chemoreceptors, on the other hand, are primarily stimulated by hypoxemia, although they also contribute to the acidemic response [7,8,10]. Thus, *primary* hyperventilation resulting in respiratory alkalosis can be produced by hypoxemia, a reduction in the cerebral pH (an apparently rare event since the cerebrospinal fluid pH is usually elevated in respiratory alkalosis) [11,12], or a nonphysiologic stimulus to ventilation, as may happen with pulmonary disease or direct stimulation of the central respiratory center (Table 21-1).

Hypoxemia

Hypoxemic hyperventilation can occur when there is a fall in the arterial P_{O_2} to below 60 to 70 mmHg or when oxygen delivery is reduced by severe hypotension or anemia (Table 21-1). This hypoxemic response occurs in two stages which illustrate the interaction between the peripheral and central chemoreceptors [13,14]. Hypoxemia activates the peripheral chemoreceptors, resulting in hyperventilation, hypocapnia, and increases in the arterial and cerebral pH. However, the cerebral alkalosis inhibits the central respiratory center, initially limiting the degree of hyperventilation. This limitation is then partially removed as the renal compensation lowers the systemic

Table 21-1 Causes of respiratory alkalosis

A. Hypoxemia
  1. Pulmonary disease: pneumonia, interstitial fibrosis, emboli, edema
  2. Congestive heart failure
  3. High-altitude residence
  4. Hypotension
  5. Severe anemia
B. Pulmonary disease
C. Direct stimulation of the medullary respiratory center
  1. Psychogenic or voluntary hyperventilation
  2. Hepatic failure
  3. Gram-negative septicemia
  4. Salicylate intoxication
  5. Postcorrection of metabolic acidosis
  6. Pregnancy and the luteal phase of the menstrual cycle (progesterone)
  7. Neurologic disorders: cerebrovascular accidents, pontine tumors
D. Mechanical ventilation

pH toward normal. This allows further hyperventilation, which both raises the P_{O_2} and lowers the P_{CO_2} [12,14].

These interrelationships can also be appreciated from Fig. 21-2. If normal subjects are made hypoxemic by lowering the partial pressure of O_2 in the inspired air, alveolar ventilation increases very slowly (with only a slight fall in the arterial P_{CO_2}) until the arterial P_{O_2} falls below 40 to 50 mmHg. If, however, the depressant effect

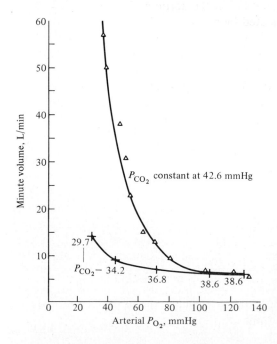

Figure 21-2 Influence of arterial P_{CO_2} (in mmHg) on the ventilatory response to hypoxemia in normal subjects. As the partial pressure of oxygen in the inspired air is reduced, the ensuing hypoxemia increases ventilation and lowers the arterial P_{CO_2} to only a minor degree until the arterial P_{O_2} is less than 40 to 50 mmHg (lower curve). The earlier and greater degree of hyperventilation seen when the arterial P_{CO_2} is held constant (upper curve) indicates that the development of hypocapnic alkalosis normally limits the hypoxemic drive. *(Adapted from H. H. Loeschcke and K. H. Gertz, Arch. Ges. Physiol., 267:460, 1958.)*

of acute hypocapnia is prevented by keeping the P_{CO_2} constant, then ventilation rises rapidly when the arterial P_{O_2} is less than 70 to 80 mmHg [7,13].

Pulmonary Disease

Respiratory alkalosis is a common finding in a variety of pulmonary diseases, including pneumonia, pulmonary emboli, and interstitial fibrosis [15–17]. It may also occur in pulmonary edema, but metabolic and respiratory acidosis are much more common in this disorder [18]. Although hyperventilation in pulmonary disease may be due in part to hypoxemia, it is frequently not corrected by the administration of oxygen [15,16].

This finding suggests that some other factor must be responsible for the increase in ventilation. In particular, intrapulmonary receptors may stimulate the respiratory center via afferent signals sent through the vagus nerves [15,16,19]. Two different receptors may participate in this response: juxtacapillary receptors in the interstitium of the alveolar wall which can be activated by interstitial edema, fibrosis, or pulmonary vascular congestion; and irritant receptors in the epithelial lining of the airways which are activated by the inhalation of irritants and perhaps local inflammatory processes such as pneumonia [15,19]. Although direct confirmation of the importance of these receptors in humans is not generally available, vagal blockade can reverse the hyperventilation associated with pulmonary diseases in experimental animals [15,19,20].

It should be noted that these receptors play no role in the control of ventilation in normal subjects and that their effect in pulmonary disease is somewhat maladaptive. For example, dyspnea and breathlessness are common complaints in diffuse pulmonary interstitial fibrosis, even in patients who are not very hypoxemic. These symptoms are probably due at least in part to the increased ventilatory drive [15].

Direct Stimulation of the Medullary Respiratory Center

Primary hyperventilation due to stimulation of the respiratory center is a common finding in a variety of disorders (Table 21-1). The possible mechanisms by which this occurs are variable and include the primary effect of cortical centers in psychogenic hyperventilation [21], retained amines in hepatic failure [22,23], bacterial toxins in gram-negative septicemia [24], salicylates in salicylate intoxication [25,26], progesterone in pregnancy, and, to a lesser degree, the luteal phase of the menstrual cycle [27,28], and a persistently acid cerebrospinal fluid (CSF) pH following the rapid correction of a metabolic acidosis [29,30]. In the last situation, the administration of NaHCO₃ raises the plasma HCO_3^- concentration and the arterial pH. As the increase in pH is sensed by the peripheral chemoreceptors, there is a decrease in the degree of compensatory hyperventilation and a moderate elevation in the P_{CO_2}. Since CO_2 but not HCO_3^- rapidly crosses the blood-brain barrier, the brain initially senses only the higher P_{CO_2}. This produces a paradoxical fall in the CSF pH [30], which tends to prolong the hyperventilatory state.

Respiratory alkalosis is also an occasional finding in neurologic disorders. With pontine tumors, a reduction in the cerebral pH due to local lactic acid production may be responsible for increased ventilation [31]. Hypocapnia also may be seen with acute cerebrovascular accidents. Although the mechanism by which this occurs is uncertain, the finding of an otherwise unexplained respiratory alkalosis is a poor prognostic sign in this setting [32].

Mechanical Ventilation

The use of mechanical ventilation not uncommonly leads to respiratory alkalosis. The imposition of forced hyperventilation in this setting usually results from an attempt to correct hypoxemia. If necessary, the respiratory alkalosis can be reversed by increasing the dead space or reducing either the tidal volume or respiratory rate.

SYMPTOMS

The symptoms produced by respiratory alkalosis are related to increased irritability of the central and peripheral nervous systems and include light-headedness, altered consciousness, paresthesias of the extremities and circumoral area, cramps, carpopedal spasm that is indistinguishable from that seen with hypocalcemia, and syncope [21,33]. A variety of supraventricular and ventricular arrhythmias may also occur, particularly in critically ill patients [34].

These abnormalities are thought to be related to the effect of CSF and cerebral interstitial fluid alkalosis on cerebral function as well as the direct effect of alkalosis to increase membrane excitability. Respiratory alkalosis also reduces cerebral blood flow (by as much as 35 to 40 percent if the P_{CO_2} falls by 20 mmHg) [35], and this may contribute to the neurologic symptoms. In addition, some symptoms may be unrelated to the change in pH. Patients with psychogenic hyperventilation, for example, frequently complain of headache, shortness of breath, chest pain or tightness, and other somatic symptoms which may be emotional in origin and not caused by the alkalemia.

The above problems primarily occur in acute respiratory alkalosis when the P_{CO_2} falls below 25 to 30 mmHg, a setting in which there is a substantial rise in cerebral pH. They are much less likely to be seen in chronic respiratory alkalosis (since the pH is so well protected) or in metabolic alkalosis, where there is a lesser elevation in CSF pH because of the relative inability of HCO_3^- to cross the blood-brain barrier [11,36].

An interesting finding in many patients with severe respiratory alkalosis is a reduction in the plasma phosphate concentration (measured in the laboratory as the plasma concentration of inorganic phosphorus) to as low as 0.5 to 1.5 mg/dL (normal is 2.5 to 4.5 mg/dL) [37]. This appears to be due to the movement of phosphate from the extracellular fluid into the cells since intracellular alkalosis stimulates glycolysis, resulting in the increased formation of phosphorylated compounds such as glucose 6-phosphate and fructose 1,6-diphosphate [37]. Although this phenomenon has not been reported to produce symptoms and does not require specific therapy, the physician should be aware of its existence since a workup for the other causes of hypophosphatemia is usually not indicated.

DIAGNOSIS

The physical finding of tachypnea may be an important clue to the presence of hypocapnia, due either to primary respiratory alkalosis or the respiratory compensation to metabolic acidosis. Once the presence of primary respiratory alkalosis has been confirmed by the measurement of the pH, P_{CO_2}, and HCO_3^- concentration, the next step is to identify the underlying disturbance responsible for the increase in ventilation, such as pulmonary disease or sepsis. For example, respiratory alkalosis is a relatively early finding in septicemia [24], and this diagnosis should be considered in the appropriate clinical setting when there is no other apparent cause for the hyperventilation.

Since the responses to acute and chronic hypocapnia are different, the determination of the correct acid-base disorder is more difficult than in metabolic acidosis or alkalosis. For example, suppose a patient has the following arterial blood values:

$$pH = 7.47$$

$$P_{CO_2} = 20 \text{ mmHg}$$

$$[HCO_3^-] = 14 \text{ meq/L}$$

The alkaline pH and hypocapnia are diagnostic of respiratory alkalosis. With a P_{CO_2} of 20 mmHg, the plasma HCO_3^- concentration should be roughly 20 meq/L in acute respiratory alkalosis (a reduction of 2 meq/L per 10-mmHg fall in the P_{CO_2}) and 14 meq/L in chronic respiratory alkalosis (a reduction of 5 meq/L per 10-mmHg fall in the P_{CO_2}). Thus, 14 to 19 meq/L describes the approximate range for the plasma HCO_3^- concentration in a patient with respiratory alkalosis and a P_{CO_2} of 20 mmHg. Values significantly above or below this range represent superimposed metabolic alkalosis or acidosis. In this patient, the plasma HCO_3^- concentration of 14 meq/L is consistent with uncomplicated chronic respiratory alkalosis. However, it must be emphasized that *the evaluation of the laboratory data must proceed in conjunction with the history and physical examination.* For example, the interpretation would be different if the following history had been obtained:

Case history 21-1 A 5-year-old child is brought into the emergency room in a stuporous condition. The only pertinent history is that he had been playing with a bottle of aspirin tablets earlier that day.

COMMENT In this setting, the most likely diagnosis is a salicylate overdose in which the plasma HCO_3^- concentration is lower than normally seen with acute hypocapnia due to the concurrent presence of a salicylate-induced metabolic acidosis [26].

TREATMENT

In general, treatment of the alkalemia is not necessary, and therapy should be aimed at the diagnosis and correction of the underlying disorder. There is no rationale for the use of respiratory depressants or the administration of acid, such as HCl, in an effort to normalize the pH. In severely symptomatic patients with acute respiratory alkalosis, rebreathing into a paper bag, i.e., increasing the P_{CO_2} in the inspired air, may partially correct the hypocapnia and relieve the symptoms. The arterial pH should be monitored since the compensatory decrease in the plasma HCO_3^- concentration will persist and may result in metabolic acidosis as the P_{CO_2} is increased toward normal. This is usually mild but rarely may require small amounts of $NaHCO_3$.

REFERENCES

1. Giebisch, G. E., L. Berger, and R. F. Pitts: The extrarenal response to acute acid-base disturbances of respiratory origin, J. Clin. Invest., 34:231, 1955
2. Arbus, G. S., L. A. Hevert, P. R. Levesque, B. E. Etsten, and W. B. Schwartz: Characterization and clinical application of the "significance band" for acute respiratory alkalosis, N. Engl. J. Med., 280:117, 1969.
3. Gennari, J. F., M. B. Goldstein, and W. B. Schwartz: The nature of the renal adaptation to chronic hypocapnia, J. Clin. Invest., 51:1722, 1972.
4. Gledhill, N., G. J. Beirne, and J. A. Dempsey: Renal response to short-term hypocapnia in man, Kidney Int., 8:376, 1975.
5. Gougoux, A., W. D. Kaehny, and J. J. Cohen: Renal adaptation to chronic hypocapnia: dietary constraints in achieving H^+ retention, Am. J. Physiol., 229:1330, 1975.
6. Madias, N. E., W. B. Schwartz, and J. J. Cohen: The maladaptive renal response to secondary hypocapnia during chronic HCl acidosis in the dog, J. Clin. Invest., 60:1393, 1977.
7. Lambertsen, C. J.: Chemical control of respiration at rest, in *Medical Physiology,* 14th ed., V. B. Mountcastle (ed.), Mosby, St. Louis, 1980.
8. Berger, A. J., R. A. Mitchell, and J. W. Severinghaus: Regulation of respiration, N. Engl. J. Med., 297.92, 138, 194, 1977.
9. Fencl, V., T. B. Miller, and J. R. Pappenheimer: Studies on the respiratory response to disturbances of acid-base balance, with deductions concerning the ionic composition of cerebral interstitial fluid, Am. J. Physiol., 210:459, 1966.
10. Lugliani, R., B. J. Whipp, C. Seard, and K. Wasserman: Effect of bilateral carotid-body resection on ventilatory control at rest and during exercise in man, N. Engl. J. Med., 285:1105, 1971.
11. Mitchell, R. A., C. T. Carman, J. W. Severinghaus, B. W. Richardson, M. M. Singer, and S. Shnider: Stability of cerebrospinal fluid pH in chronic acid-base disturbances in blood, J. Appl. Physiol., 20:443, 1965.
12. Dempsey, J. A., H. V. Foster, and G. A. DoPico: Ventilatory acclimatization to moderate hypoxemia in man: the role of spinal fluid $[H^+]$, J. Clin. Invest., 53:1091, 1974.
13. Weil, J. V., E. Byrne-Quinn, E. Sadal, W. O. Friesen, B. Underhill, G. F. Filley, and R. F. Grover: Hypoxic ventilatory drive in normal man, J. Clin. Invest., 49:1061, 1970.
14. Lenfant, C., and K. Sullivan: Adaptation to high altitude, N. Engl. J. Med., 284:1298, 1971.
15. Kornbluth, R. S., and G. M. Turino: Respiratory control in diffuse interstitial lung disease and diseases of the pulmonary vasculature, Clin. Chest Med., 1:91, 1980.
16. Lourenco, R. V., G. M. Turino, L. A. G. Davidson, and A. P. Fishman: The regulation of ventilation in diffuse pulmonary fibrosis, Am. J. Med., 38:199, 1965.

17. Szucs, M. M., H. L. Brooks, W. Grossman, J. S. Banas, S. G. Meister, L. Dexter, and J. E. Dalen: Diagnostic sensitivity of laboratory findings in acute pulmonary embolism, Ann. Intern. Med., 74:161, 1971.

18. Aberman, A., and M. Fulop: The metabolic and respiratory acidosis of acute pulmonary edema, Ann. Intern. Med., 76:173, 1972.

19. Trenchard, D., D. Gardner, and A. Guz: Role of pulmonary vagal afferent nerve fibres in the development of rapid shallow breathing in lung inflammation, Clin. Sci., 42:251, 1972.

20. Horres, A. D., and T. Bernthal: Localized multiple minute pulmonary embolism and breathing, J. Appl. Physiol., 16:842, 1961.

21. Rice, R. L.: Symptom patterns of the hyperventilation syndrome, Am. J. Med., 8:691, 1950.

22. Karetzky, M. S., and J. C. Mithoefer: The cause of hyperventilation and arterial hypoxia in patients with cirrhosis of the liver, Am. J. Med. Sci., 254:797, 1967.

23. Record, C. O., R. A. Iles, R. D. Cohen, and R. Williams: Acid-base and metabolic disturbances in fulminant hepatic failure, Gut, 16:144, 1975.

24. Simmons, D. H., J. Nicoloff, and L. B. Guze: Hyperventilation and respiratory alkalosis as signs of gram-negative bacteremia, J. Am. Med. Assoc., 174:2196, 1960.

25. Tenny, S. M. and R. M. Miller: The respiratory and circulatory actions of salicylate, Am. J. Med., 19:498, 1955.

26. Gabow, P. A., R. J. Anderson, D. E. Potts, and R. W. Schrier: Acid-base disturbances in the salicylate-intoxicated adult, Arch. Intern. Med., 138:1481, 1978.

27. Lim, V. S., A. I. Katz, and M. D. Lindheimer: Acid-base regulation in pregnancy, Am. J. Physiol., 231:1764, 1976.

28. Takano, N., and T. Kaneda: Renal contribution to acid-base regulation during the menstrual cycle, Am. J. Physiol., 244:F320, 1983.

29. Rosenbaum, B. J., J. W. Coburn, J. H. Shinaberger, and S. G. Massry: Acid-base status during the interdialytic period in patients maintained with chronic hemodialysis, Ann. Intern. Med., 71:1105, 1969.

30. Posner, J. B., and F. Plum: Spinal-fluid pH and neurologic symptoms in systemic acidosis, N. Engl. J. Med., 277:605, 1967.

31. Plum, F.: Mechanisms of central hyperventilation, Ann. Neurol., 11:636, 1982.

32. Rout, M. W., D. J. Lane, and L. Wollner: Prognosis in acute cerebrovascular accidents in relation to respiratory pattern and blood gas tensions, Br. Med. J., 3:7, 1971.

33. Saltzman, H., A. Heyman, and H. O. Sieker: Correlation of clinical and physiologic manifestations of sustained hyperventilation, N. Engl. J. Med., 268:1431, 1963.

34. Ayres, S. M., and W. J. Grace: Inappropriate ventilation and hypoxemia as causes of cardiac arrhythmias: the control of arrhythmias without antiarrhythmic drugs, Am. J. Med., 46:495, 1969.

35. Wasserman, A. J., and J. L. Patterson: The cerebral vascular response to reduction in arterial carbon dioxide tension, J. Clin. Invest., 40:1297, 1961.

36. Posner, J. B., A. G. Swanson, and F. Plum: Acid-base balance in cerebrospinal fluid, Arch. Neurol., 12:479, 1965.

37. Knochel, J. P.: The pathophysiology and clinical characteristics of severe hypophosphatemia, Arch. Intern. Med., 137:203, 1977.

INTRODUCTION TO DISORDERS OF OSMOLALITY

Hyponatremia and hypernatremia are common clinical problems. Although the plasma Na^+ concentration is abnormal, these disorders reflect abnormalities in osmolality and not necessarily in Na^+ balance itself. The review presented below is discussed in more detail in Chaps. 1 and 10.

WATER DISTRIBUTION AND OSMOTIC PRESSURE

The total body water (TBW) (about 60 percent of body weight in males, and 50 percent in females) is distributed between the intracellular (60 percent of body water) and extracellular (40 percent of body water) spaces. In addition, roughly one-fifth of the extracellular fluid is confined to the intravascular space (the plasma water). For example, in an average 70-kg male, the total body water is approximately 42 liters, or which 25 liters is intracellular and 17 liters is extracellular. Within the extracellular compartment, 3 liters is in the vascular space.

Osmotic forces are important in determining the distribution of water between these compartments. Each compartment has one major solute which, because it is restricted primarily to that compartment, acts to hold water within the compartment. Therefore, Na^+ salts (extracellular osmoles), K^+ salts (intracellular osmoles), and the plasma proteins (intravascular osmoles) help to maintain the volumes of the extracellular, intracellular, and intravascular spaces. In contrast to Na^+ and K^+, urea rapidly crosses cell membranes and equilibrates throughout the total body water. As a result, urea does not affect the distribution of water between the cells and the extracellular fluid, i.e., urea is an ineffective osmole.

Since the cell membranes are permeable to water, the extracellular and intracellular fluids (ECF and ICF) are in osmotic equilibrium. (The renal medulla is one of the minor exceptions.) *If an osmotic gradient is established, water will flow from the compartment of low osmolality to that of high osmolality until the osmotic pressures are equalized.*

PHYSIOLOGIC EFFECTS OF CHANGES IN PLASMA OSMOLALITY

The effects of variations in the effective plasma osmolality on internal water distribution can be illustrated by the responses to NaCl, water, and an isotonic solution of NaCl and water (Fig. 22-1). (The methods used to calculate the new steady states are discussed on page 27.) Since Na^+ is essentially limited to the ECF, the administration of NaCl without water augments ECF osmolality, resulting in water movement *out of* the cells (Fig. 22-1b). Equilibrium is characterized by hypernatremia and equal increases in the osmolality of the ECF (due to the excess NaCl) and the ICF (due to water loss). In addition, the redistribution of water enhances the extracellular volume and reduces the intracellular volume. Thus, *the osmolal effects of the administration of NaCl are distributed throughout the total body water even though NaCl itself is largely restricted to the ECF.* In this example, one might have expected the addition of 210 meq of Na^+ to 17 liters of ECF to increase the plasma Na^+ concentration by 12.5 meq/L ($\frac{210}{17} = 12.5$). However, the plasma Na^+ concentration rose by only 5 meq/L because of the movement of intracellular water into the ECF.

The results are different when only water is given. In this setting, there is an initial fall in ECF osmolality, thereby promoting water movement *into* the cells (Fig. 22-1c). The new steady state is characterized by a reduction in ECF and ICF osmolality, hyponatremia, and expansion of both the extracellular and intracellular volumes.

In contrast, the effect of an isotonic NaCl solution is limited to expansion of the ECF volume (Fig. 22-1d). Since there is no change in osmolality, there is no shifting of water and the composition of the ICF is unchanged.

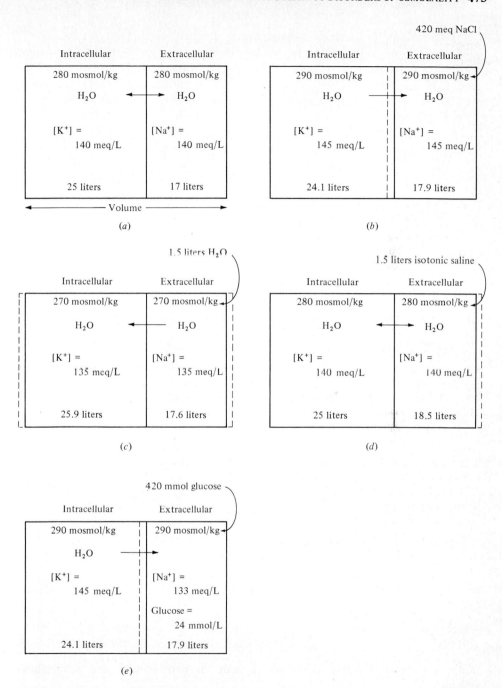

Figure 22-1 Osmolality of the body fluids and the distribution of water between the intracellular fluid and the extracellular fluid in the control state (*a*) and after the addition of NaCl (*b*), H_2O (*c*), isotonic NaCl and H_2O (*d*), or glucose (*e*) to the extracellular fluid. For simplicity, it is assumed that the only extracellular and intracellular osmoles are Na^+ salts and K^+ salts, respectively. See text for details.

These examples illustrate two important clinical points. First, an increase in effective ECF osmolality results in cellular dehydration (Fig. 22-1*b*), and a decrease in effective ECF osmolality results in cellular overhydration (Fig. 22-1*c*). As will be seen, it is this *flow of water out of and into cells, particularly brain cells, that is responsible for the symptoms produced by hyper- and hypoosmolality.* These water shifts do not occur and the symptoms of hyperosmolality are absent when the P_{osm} is elevated by an increase in the plasma urea concentration, as seen with renal failure. Urea, in contrast to Na^+, readily crosses the cell membrane, and osmotic equilibrium is reached by urea entry into cells rather than water movement out of cells.

Second, it can be seen that the plasma Na^+ concentration, which is a function of the *ratio* of the amounts of solute and water present, does not necessarily correlate with volume, which is a function of the *total* amount of Na^+ and water present. In each of the examples in Fig. 22-1, the extracellular volume is increased yet the plasma Na^+ concentration is high, low, and normal, respectively.

MEANING OF PLASMA SODIUM CONCENTRATION

An understanding of what the plasma Na^+ concentration represents, including its difference from the extracellular volume, is essential in the approach to patients with hyponatremia or hypernatremia. Although it may appear logical to consider alterations in the plasma Na^+ concentration as indicating abnormal Na^+ balance, they are almost always *a reflection of abnormal water balance.*

Plasma Sodium Concentration and Plasma Osmolality

The osmolality of a solution is determined by the number of solute particles per kilogram of water. Since Na^+ salts (particularly NaCl and $NaHCO_3$), glucose, and urea (measured as the blood urea nitrogen, or BUN) are the primary extracellular (and plasma) osmoles, the plasma osmolality (P_{osm}) can be approximated from (see page 29)

$$P_{osm} \cong 2 \times \text{plasma } [Na^+] + \frac{[\text{glucose}]}{18} + \frac{BUN}{2.8} \qquad (22\text{-}1)$$

where 2 reflects the osmotic contribution of the anion accompanying Na^+ and 18 and 2.8 represent the conversion of the plasma glucose concentration and BUN from units of milligrams per deciliter (mg/dL) into millimoles per liter (mmol/L).

Although urea contributes to the absolute value of the P_{osm}, it does not act to hold water within the extracellular space because of its membrane permeability. Therefore, urea is an ineffective osmole and does not contribute to the effective P_{osm}:

$$\text{Effective } P_{osm} \cong 2 \times \text{plasma } [Na^+] + \frac{[\text{glucose}]}{18} \qquad (22\text{-}2)$$

In humans, the normal values for these parameters are

$$P_{osm} = 275\text{--}290 \text{ mosmol/kg}$$

$$\text{Effective } P_{osm} = 270\text{--}285 \text{ mosmol/kg}$$

$$\text{Plasma } [Na^+] = 137\text{--}145 \text{ meq/L}$$

$$\text{Plasma } [glucose] = 60\text{--}100 \text{ mg/dL (fasting)}$$

$$BUN = 10\text{--}20 \text{ mg/dL}$$

Under normal conditions, glucose and urea contribute less than 10 mosmol/kg, and the plasma Na^+ concentration is the main determinant of the P_{osm}:

$$P_{osm} \cong 2 \times \text{plasma } [Na^+] \tag{22-3}$$

Thus, *hypernatremia represents hyperosmolality and, in most instances, hyponatremia reflects hypoosmolality.* One exception to this general relationship occurs with hyperglycemia. The elevation in the plasma glucose concentration raises the effective P_{osm}, pulling water out of the cells and lowering the plasma Na^+ concentration by dilution (Fig. 22-1e). This is clinically important because therapy should be directed toward hyperosmolality and not, as suggested by the reduced plasma Na^+ concentration, hypoosmolality.

Plasma Sodium Concentration and Total Body Osmolality

If the plasma Na^+ concentration is a reflection of the P_{osm} and the P_{osm} is in equilibrium with the total body osmolality, then (see page 30):

$$\text{Plasma } [Na^+] \propto \text{total body osmolality} \tag{22-4}$$

Since

$$\text{Total body osmolality} = \frac{\text{extracellular} + \text{intracellular solutes}}{TBW} \tag{22-5}$$

and Na^+ and K^+ salts are the primary extracellular and intracellular solutes, then Eq. (22-4) can be converted to

$$\text{Plasma } [Na^+] \cong \frac{Na_e^+ + K_e^+}{TBW} \tag{22-6}$$

where Na_e^+ and K_e^+ refer to the total "exchangeable" quantities of these ions (Fig. 22-2) [1]. The exchangeable portion is used since about 30 percent of the body Na^+ and a small fraction of the body K^+ are bound in areas such as bone where they are "nonexchangeable" and *osmotically inactive* (see Table 1-5).

The relationship in Eq. (22-6) indicates the settings in which hyponatremia or hypernatremia can occur. It is not suprising that changes in either Na^+ or water balance can affect the plasma Na^+ concentration. The effect of K^+ is less apparent than that of Na^+ or water but can be clinically important. If K^+ is lost from the extracellular fluid (due to renal or gastrointestinal losses), the extracellular K^+ con-

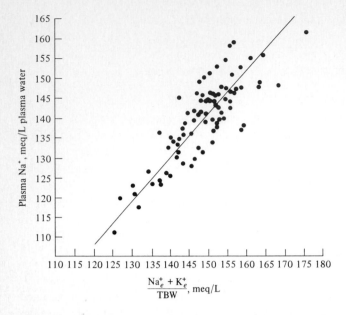

Figure 22-2 Relation between the plasma water Na^+ concentration and the ratio of $(Na_e^+ + K_e^+)/$ TBW. *(Adapted from I. Edelman, J. Leibman, M. O'Meara, and L. Birkenfeld. J. Clin. Invest., 37:1236, 1958, by copyright permission of the American Society for Clinical Investigation.)*

centration will fall. As a result, K^+ will move from the cells into the ECF down a concentration gradient. To maintain electroneutrality, Na^+ (and H^+) will enter the cells, thereby lowering the plasma Na^+ concentration [2]. In some patients with diuretic-induced hyponatremia, for example, it is the reduction in exchangeable K^+ not Na^+ that is primarily responsible for the fall in the plasma Na^+ concentration [2].

Hyponatremia and Hypernatremia

From Eq. (22-6), it can be seen that hyponatremia or hypernatremia can be induced by alterations in solute or water balance. In the clinical setting, however, these disorders are *almost always due to changes in water balance.* Hyponatremia, for example, almost always results from the retention of ingested or administered water. Although solute loss in excess of water also can lower the plasma Na^+ concentration, this is an uncommon event, occurring most often with the administration of thiazide diuretics (see Chap. 23). Similarly, hypernatremia usually results from water loss in excess of solute (see Chap. 24). In contrast, the loss of an isosmotic fluid *will cause volume depletion but will not affect body osmolality and therefore will not change the plasma Na^+ concentration.*

A clinical illustration of these principles occurs in patients with diarrhea. Diarrheal fluid is roughly isosmotic to plasma [3], so that the loss of this fluid will not directly alter either the P_{osm} or the plasma Na^+ concentration. However, diarrheal syndromes have other effects on water balance as they may be associated with fever, metabolic acidosis, and volume depletion. Fever increases loss of water as sweat, and metabolic acidosis produces a compensatory hyperventilation which enhances water loss from the lungs. On the other hand, volume depletion is a potent stimulus to thirst and antidiuretic hormone (ADH) secretion, resulting in water retention due to the combined effects of increased intake and reduced excretion (see below). Most commonly, the increments in water loss and water retention are of the same magnitude, and there is little change in the plasma Na^+ concentration. If, however, water intake is not increased, as in the case of infants who do not have access to water, enteric infections with fever can lead to negative water balance and hypernatremia [4]. Conversely, in the hypovolemic adult who is able to satisfy thirst, the attempt to restore normovolemia by augmenting water intake can produce positive water balance with consequent hyponatremia.

REGULATION OF PLASMA OSMOLALITY

The relationship of the plasma Na^+ concentration to water balance is also illustrated by the manner in which the plasma Na^+ concentration and P_{osm} are normally regulated: by alterations in the intake and excretion of water, not Na^+.

Each day, there is a variable degree of water intake and loss which can lead to changes in the P_{osm} (see Chap. 10). Water intake is derived from three sources: drinking, the water content of food, and water of oxidation (e.g., carbohydrates are metabolized to carbon dioxide and water) (Table 22-1). The retention of this water tends to lower the P_{osm}. On the other hand, water is lost in the urine and feces as well as from the skin and respiratory tract as insensible and sweat losses. This loss of water tends to raise the P_{osm}.

Under normal circumstances there is a balance between net water intake and excretion such that the P_{osm} is maintained within narrow limits. This regulatory response is mediated by osmoreceptors in the hypothalamus which sense changes in the P_{osm} of as little as 1 percent and which affect both water intake via thirst and

Table 22-1 Typical daily water balance in a normal human, assuming a low rate of sweat production†

Water intake, mL/day		Water output, mL/day	
Source		Source	
Ingested water	1400	Urine	1500
Water content of food	850	Skin	500
Water of oxidation	350	Respiratory tract	400
		Stool	200
Total	2600		2600

†Under conditions of increased sweat production, the water losses from the skin can increase markedly, occasionally exceeding 5 L/day. When this occurs, thirst is stimulated, resulting in an appropriate increase in the volume of ingested water.

water excretion via the secretion of ADH from the posterior lobe of the pituitary (see Chap. 8). In the kidney, ADH augments the water permeability of the collecting tubules, resulting in increased water reabsorption and the excretion of a hyperosmotic urine (high U_{osm} and specific gravity). When ADH is absent, water reabsorption falls, and a dilute urine is excreted (low U_{osm} and specific gravity) since the collecting tubules are now relatively impermeable to water.

The osmoreceptors regulate the P_{osm} in the following manner. After a water load, there is a fall in the P_{osm} which inhibits both ADH secretion and thirst. This promotes the urinary excretion of the excess water, which returns the P_{osm} to normal. If, on the other hand, a patient becomes hyperosmolal (as with hypernatremia due to insensible water losses), thirst and ADH release are stimulated. The combination of enhanced water intake and renal water conservation results in water retention and an appropriate reduction in the P_{osm}. (In contrast, the osmoreceptors are not stimulated by hyperosmolality due to an elevation in the concentration of urea, which is an ineffective osmole.)

This regulatory system can be disrupted either by neurologic disorders, which interfere with hypothalamic or posterior pituitary function, or by renal disorders, which affect concentrating or diluting ability. In addition, there are nonosmolal factors which can affect hypothalamic function and override the effects of osmolality (see Chap. 8). For example, volume depletion is a potent stimulus to ADH release and thirst. As a result, patients who are hypovolemic may have persistent thirst and ADH secretion even in the presence of hypoosmolality (see Chap. 8) [5–7]. In this setting, volume and tissue perfusion are maintained at the expense of osmolality.

VOLUME AND OSMOREGULATION

It is important to understand the differences between volume and osmoregulation. The latter maintains the effective P_{osm} and the plasma Na^+ concentration by varying the rate of *water* intake and excretion. This again illustrates that hyponatremia and hypernatremia are primarily problems of abnormal water balance.

In contrast, the effective circulating volume, which may be independent of the plasma Na^+ concentration as shown in Fig. 22-1, is maintained by alterations in the rate of urinary *sodium* excretion. An increase in volume enhances Na^+ excretion, whereas volume depletion induces renal Na^+ retention. This homeostatic mechanism involves sensors (in the cardiopulmonary circulation and the kidney) and effectors (including aldosterone, angiotensin II, natriuretic hormone, and the sympathetic nervous system) that are different from those involved in osmoregulation† (see Chap. 9). Thus, urinary Na^+ excretion is *affected by changes in the plasma Na^+ concentration only if there are associated changes in volume.* For example, the urine Na^+ concentration should be less than 10 meq/L when hyponatremia is due to net Na^+ loss (volume depletion) and greater than 20 meq/L when due to primary water retention

---

†The hypovolemia-induced increase in ADH secretion contributes to volume regulation by enhancing water reabsorption. It has no direct effect on Na^+ reabsorption.

(volume expansion). As a result, measurement of the urine Na^+ concentration is an important component of the diagnostic approach to hyponatremia (see Chap. 23).

The interaction of volume and osmoregulatory systems can be illustrated by the different responses elicited by NaCl, water, and isotonic NaCl and water (Fig. 22-1): (1) Isotonic NaCl and water enhances volume without a change in the P_{osm}. Thus, the volume receptors are activated, resulting in NaCl (and water) loss in the urine. (2) A water load lowers the P_{osm}, and, as ADH release is inhibited, a dilute urine is formed and the water is rapidly excreted in the urine. Normally this is so efficient that volume is only transiently increased and there is little change in NaCl excretion. However, if water excretion is impaired, the water load will be retained, resulting in hypervolemia and an increase in urinary Na^+ excretion. (3) The administration of NaCl without water enhances the extracellular volume and leads to renal NaCl loss. In addition, the increase in osmolality stimulates ADH release and thirst. This results in water retention, which both reduces osmolality toward normal and augments volume, further promoting the renal excretion of the NaCl load.

URINE OSMOLALITY AND SPECIFIC GRAVITY

Estimating the ability to concentrate or dilute the urine can be helpful in the diagnosis of patients with hypernatremia or hyponatremia. This can be done by measuring the osmolality or, if an osmometer is not available, the specific gravity of the urine. In general, the urinary specific gravity correlates reasonably well with the U_{osm} (see Fig. 1-8) according to the following approximate relationship:

Specific gravity	Osmolality
1.000	0
1.010	350
1.020	700
1.030	1050

However, when larger molecules are present in the urine, e.g., during an osmotic diuresis with glucose or after the administration of radiopaque dyes, the specific gravity can be elevated out of proportion to any change in the U_{osm}. In this situation, using the specific gravity for diagnosis can be misleading.

RELATION BETWEEN INTAKE AND OUTPUT

In the treatment of patients with hyponatremia or hypernatremia, attention is appropriately paid to comparing net fluid intake to urinary output since changing the state of water balance can return the plasma Na^+ concentration toward normal. For example, hyponatremic patients who are not volume-depleted can be treated with fluid restriction. If intake is kept below output, there will be a net loss of water and an elevation in the plasma Na^+ concentration.

It must be emphasized, however, that *the composition of the fluids given and excreted is often markedly different*. Therefore, merely comparing intake and output may be insufficient to accurately predict the effects of therapy. If, for example, urinary NaCl and water loss is induced by a diuretic and the fluid losses are replaced by an equal volume of water, the patient will be in *water* balance. However, the loss of the unreplaced solute will induce hypoosmolality and hyponatremia.

A more complex evaluation of fluid balance can be illustrated by the following case history:

Case history 22-1 A 58-year-old woman with an oat-cell carcinoma of the lung is admitted for progressive lethargy and confusion. The physical examination shows no focal neurologic findings and a weight of 60 kg. Laboratory data reveal:

$$\text{Plasma } [Na^+] = 102 \text{ meq/L}$$

$$P_{osm} = 230 \text{ mosmol/kg}$$

$$\text{Urine } [Na^+] = 70 \text{ meq/L}$$

$$U_{osm} = 420 \text{ mosmol/kg}$$

A diagnosis of inappropriate ADH secretion due to the lung tumor is made (see Chap. 23). In view of the severe hyponatremia, the patient is treated with water restriction, hypertonic saline (Na^+ concentration equals 513 meq/L; osmolality equals 1026 mosmol/kg) and furosemide.

Overnight, the patient is given 1700 mL of hypertonic saline and excretes 3300 mL of urine with an osmolality of 300 mosmol/kg. Repeat blood tests in the morning reveal a plasma Na^+ concentration of 123 meq/L and a plasma osmolality of 271 mosmol/kg.

COMMENT At first glance, it seems unlikely that a negative fluid balance of only 1600 mL can result in such a marked rise in the plasma Na^+ concentration and osmolality. However, a more complete evaluation of intake and output shows how this change occurred.† The patient weighed 60 kg on admission, approximately one-half of which was water. Thus, her total body water on admission was 30 L. Since the osmolality in all fluid compartments is equal and the plasma osmolality was 230 mosmol kg:

$$\text{Total body osmoles} = \text{total body water} \times \text{plasma osmolality} \qquad (22\text{-}7)$$

$$= 30 \times 230$$

$$= 6900 \text{ mosmol}$$

With the loss of 1600 mL of water, her total body water fell to 28.4 L. If her total body osmoles were still 6900, then, by rearranging Eq. (22-7):

$$\text{New plasma osmolality} = \text{total body osmoles} \div \text{total body water}$$

$$= \frac{6900}{28.4}$$

$$= 243 \text{ mosmol/kg}$$

This is clearly much different from the measured value of 271 mosmol/kg. The error lies in the assumption that the patient's total body osmoles were unchanged. The total osmolar intake was 1745 mosmol (1700 mL at 1026 mosmol/kg) and total osmolar excretion was 990 mosmol (3300

---

†The method used in this example is similar to that involved in the calculation of the new steady states in Fig. 22-1 (see page 27).

mL at 300 mosmol/kg). Thus, there was a 755-mosmol *increase* in total body osmoles from 6900 to 7655 mosmol. As a result:

$$\text{New plasma osmolality} = \frac{7655}{28.4}$$

$$= 270 \text{ mosmol/kg}$$

This value is essentially identical to the measured value of 271 mosmol/kg. Furthermore, the 41-mosmol/kg elevation in the P_{osm} is composed of increases in the plasma Na^+ and Cl^- concentrations. Thus, the plasma Na^+ concentration should have risen by 20.5 meq/L ($\frac{41}{2} = 20.5$), a change similar to that observed (from 102 to 123 meq/L).

PROBLEMS

22-1 A patient has the following laboratory data:

$$\text{Plasma } [Na^+] = 125 \text{ meq/L}$$

$$[\text{Glucose}] = 108 \text{ mg/dL}$$

$$\text{BUN} = 140 \text{ mg/dL}$$

Calculate the plasma osmolality. Would this patient have the symptoms of hyperosmolality?

22-2 Suppose a patient can only excrete urine that is isosmotic to plasma. If the patient's intake were limited to the administration of isotonic saline (Na^+ concentration equals 154 meq/L, the same as the Na^+ concentration in the plasma water; see page 22), what would happen to the plasma osmolality and Na^+ concentration? Would the slow infusion of half-isotonic saline (Na^+ concentration of 77 meq/L) supplemented with 77 meq/L of K^+ (as KCl) have different effects on the plasma osmolality and Na^+ concentration?

REFERENCES

1. Edelman, I. S., J. Leibman, and M. P. O'Meara: Interrelations between serum sodium concentration, serum osmolarity and total exchangeable sodium, total exchangeable potassium and total body water. J. Clin. Invest., 37:1236, 1958.
2. Fichman, M. P., H. Vorherr, C. R. Kleeman, and N. Telfer: Diuretic-induced hyponatremia, Ann. Intern. Med., 75:853, 1971.
3. Teree, T., E. Mirabal-Font, A. Ortiz, and W. Wallace: Stool losses and acidosis in diarrheal disease of infancy, Pediatrics, 36:704, 1965.
4. Bruck, E., G. Abal, and T. Aceto: Pathogenesis and pathophysiology of hypertonic dehydration with diarrhea, Am. J. Dis. Child., 115:122, 1968.
5. Leaf, A., and A. R. Mamby: An antidiuretic mechanism not regulated by extracellular fluid tonicity, J. Clin. Invest., 31:60, 1952.
6. Robertson, G. L.: Thirst and vasopressin function in normal and disordered states of water balance, J. Lab. Clin. Med., 101:351, 1983.
7. Schrier, R. W., and D. G. Bichet: Osmotic and nonosmotic control of vasopressin release and the pathogenesis of impaired water excretion in adrenal, thyroid, and edematous disorders, J. Lab. Clin. Med., 98:1, 1981.

TWENTY-THREE

HYPOOSMOLAL STATES—HYPONATREMIA

The introduction to disorders of water balance presented in Chap. 22 should be read before proceeding with this discussion.

PATHOPHYSIOLOGY

Since the plasma Na^+ concentration is the main determinant of the plasma osmolality (P_{osm}), *hyponatremia* (plasma Na^+ concentration below 135 meq/L) *usually reflects hypoosmolality*. This is an important relationship because the low P_{osm} results in water movement into the cells; it is this cellular overhydration, particularly in brain cells, that is primarily responsible for the symptoms of hyponatremia (see below).

The basic mechanisms by which hyponatremia and hypoosmolality occur can be most easily understood if we ask two separate questions:

1. How do patients develop hyponatremia?
2. Why do they stay hyponatremic?

Generation of Hyponatremia

From the relationship between the plasma Na^+ concentration and the osmolality of the body fluids (see Fig. 22-2),

$$\text{Plasma } [Na^+] \cong \frac{Na_e^+ + K_e^+}{\text{total body water}} \tag{23-1}$$

it can be seen that either solute (Na^+ or K^+) loss or water retention can produce hyponatremia. However, solute loss, as with vomiting or diarrhea, usually occurs in a fluid that is isosmotic to plasma [1]. Isosmotic fluid loss cannot directly produce hypoosmolality, but hyponatremia will ensue if these losses are replaced with ingested or administered water. Thus, *water retention leading to an excess of water in relation to solute is the common denominator in almost all hypoosmolal states.* The corollary of this is that hypoosmolality generally cannot be produced if there is no water intake.

Perpetuation of Hyponatremia

The primary response to a fall in the P_{osm}, as most often occurs after the ingestion of a water load, is to diminish ADH secretion. This results sequentially in decreased water reabsorption in the collecting tubules, the production of a dilute urine, and the rapid excretion of the excess water (more than 80 percent within 4 h in normal subjects). This is a dose-dependent effect, so the final urine osmolality (U_{osm}) is determined by how much ADH release is inhibited. As depicted in Fig. 23-1, ADH secretion is maximally suppressed when the P_{osm} falls below 275 to 280 mosmol/kg, a setting in which the plasma Na^+ concentration should be less than 135 meq/L. In

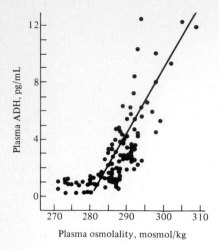

Figure 23-1 Relationship of plasma ADH concentration to plasma osmolality in normal humans in whom the plasma osmolality was changed by varying the state of hydration. ADH secretion is almost totally suppressed when the plasma osmolality falls below 275 to 280 mosmol/kg. *(Adapted from G. L. Robertson, P. Aycinena, and R. L. Zerbe, Am. J. Med., 72:339, 1982.)*

the relative absence of ADH, the U_{osm} can fall to 40 to 100 mosmol/kg (specific gravity equals 1.001 to 1.003) with a maximum water excretory capacity that can exceed 10 liters of solute-free water per day.

Since the capacity for water excretion is normally so great, *water retention resulting in hyponatremia essentially occurs only when there is a defect in renal water excretion,* i.e., in free-water clearance, C_{H_2O}. (This term is defined in detail in Chap 10 and represents the amount of solute-free water excreted by the kidney per unit time.) A rare exception to this rule occurs in patients with primary polydipsia who drink such large volumes that they overwhelm even the normal excretory capacity.

The excretion of free water is dependent upon two factors: (1) the generation of free water and a hypoosmotic urine by NaCl reabsorption without water in the diluting segments in the ascending limb of the loop of Henle and, to a lesser degree, the distal tubule; and (2) the excretion of this water by keeping the collecting tubules impermeable to water (see Chap. 6). Therefore, a *reduction in C_{H_2O} which promotes the development of hyponatremia must involve an abnormality in one or both of these steps* (Table 23-1).

This decrease in C_{H_2O} results in a U_{osm} that is inappropriately high (U_{osm} greater than 100 mosmol/kg) considering the presence of hypoosmolality. It should be noted that the impairment in water excretion does not have to be very severe. Suppose a

Table 23-1 Pathophysiologic factors which diminish renal water excretion

Diminished generation of free water in the loop of Henle and distal tubule
  A. Decreased delivery of fluid to these segments
    1. Effective circulating volume depletion
    2. Renal failure
  B. Inhibition of NaCl reabsorption by diuretics

Enhanced water permeability of the collecting tubules due to the presence of ADH
  A. Effective circulating volume depletion
  B. Syndrome of inappropriate ADH secretion
  C. Adrenal insufficiency or hypothyroidism

Table 23-2 Etiology of hypoosmolal states

Disorders in which renal water excretion is impaired
A. Effective circulating volume depletion
1. Gastrointestinal losses: vomiting, diarrhea, tube drainage, bleeding, intestinal obstruction
2. Renal losses: diuretics, hypoaldosteronism, Na^+-wasting renal disease
3. Skin losses: burns, cystic fibrosis
4. Edematous states: heart failure, hepatic cirrhosis, nephrotic syndrome
5. K^+ depletion
B. Renal failure
C. Diuretics
1. Thiazides (most commonly)
2. Furosemide
3. Ethacrynic acid
D. Presence of ADH
1. Syndrome of inappropriate ADH secretion
2. Effective circulating volume depletion
3. Cortisol deficiency
4. Hypothyroidism

Disorders in which renal water excretion is normal
A. Primary polydipsia
B. Reset osmostat

patient has a daily solute intake of 400 mosmol and a net water intake (intake minus insensible loss) of 2 liters. To excrete this load and remain in the steady state, the average U_{osm} will be 200 mosmol/kg. If this patient were unable to reduce the U_{osm} below 222 mosmol/kg (a level still hypoosmotic to plasma), the 400 mosmol of solute would be excreted in only 1800 mL of water, resulting in the daily retention of 200 mL of water and a gradual fall in the plasma Na^+ concentration.

ETIOLOGY

Since hyponatremia with hypoosmolality is caused by the retention of solute-free water, the differential diagnosis of this disturbance consists primarily of those conditions which limit water excretion (Table 23-2).

Effective Circulating Volume Depletion

The effective circulating volume refers to that fluid which is effectively perfusing the tissues (see Chap. 9). Effective volume depletion may be associated with either reduction or expansion of the extracellular volume. True volume depletion, i.e., depletion of both the intravascular and interstitial compartments, can be produced by Na^+ and water loss from the gastrointestinal tract, kidneys, or skin (Table 23-2). In contrast, a primary reduction in the cardiac output in heart failure or the extravascular sequestration of fluid in hepatic cirrhosis or the nephrotic syndrome can also lead to decreased tissue perfusion even though the total extracellular volume is increased (see Chap. 16).

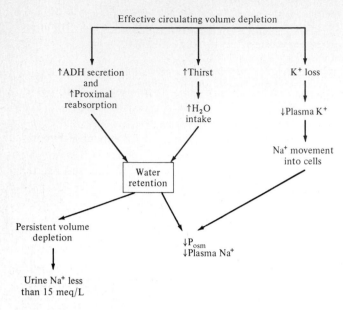

Figure 23-2 Pathophysiology of the development of hyponatremia in volume-depleted states.

Effective volume depletion predisposes toward the development of hyponatremia by its effects on renal water excretion, thirst, and K^+ balance (Fig. 23-2). Regardless of the underlying disorder, *volume depletion can impair C_{H2O} in two ways*. First, hypovolemia, acting via the carotid sinus baroreceptors, is a potent stimulus to ADH secretion (see Fig. 8-6), resulting in augmented water permeability of the collecting tubules. For example, almost all hyponatremic patients with heart failure or cirrhosis have elevated circulating ADH levels [2,3]. This can be called *appropriate ADH secretion* since the retained water attempts to restore normovolemia.

Second, the combination of a fall in GFR and an increase in proximal tubular Na^+ and water reabsorption, both induced by volume depletion, reduces the delivery of water to the diluting segments, thereby limiting the amount of free water that can be generated† [4,5]. The importance of reduced delivery has been shown by the response to exogenous mannitol. This osmotic diuretic enhances loop delivery by diminishing proximal reabsorption and, in hyponatremic patients with cirrhosis, heart failure, and true volume depletion, is able to significantly increase C_{H2O} [7,8]. In normal subjects, loop delivery is not limiting and mannitol has only a minor effect on C_{H2O}.

Not surprisingly, the tendency to increase ADH release and to reduce loop delivery is related to the severity of the effective volume depletion. Thus, a plasma Na^+

---

†An exception to the problem of decreased delivery may occur in hypoalbuminemic patients with the nephrotic syndrome. The concentration of albumin in the peritubular capillary is an important determinant of net proximal reabsorption (see page 95). Thus, hypoalbuminemia tends to reduce proximal reabsorption, thereby *increasing* loop delivery [6]. Hyponatremia in nephrotic patients with relatively normal function, therefore, results primarily from enhanced ADH secretion.

concentration below 125 meq/L usually does not occur in heart failure or hepatic cirrhosis until advanced disease is present [9,10]. If this degree of hyponatremia is seen with less severe disease, a superimposed problem such as diuretic therapy is likely to be responsible.

In the patient with reduced ability to excrete water, the volume of water that is retained is directly related to water intake. Since hypovolemia may also stimulate thirst [11], this further promotes the tendency toward hyponatremia. In addition to these effects on C_{H2O} and thirst, patients with gastrointestinal or renal losses can become K^+-depleted. This loss of K^+ can contribute to the development of hyponatremia by driving Na^+ into cells in exchange for cellular K^+, which moves into the extracellular fluid to maintain the plasma K^+ concentration. In this setting, the administration of KCl, without other efforts at volume repletion, can reverse this cation exchange and raise the plasma Na^+ concentration toward normal [12,13].

Renal Failure

Relative water excretion, as measured by C_{H2O}/GFR, is normal in most patients with renal failure [14]. However, when the GFR is very low, there is a proportionate fall in loop delivery and C_{H2O}; in other words, there are too few functioning nephrons to excrete the daily water load. In this setting, even normal levels of water intake can lead to hyponatremia.

Diuretics

Hyponatremia is a relatively common, though usually mild, complication of diuretic therapy. Severe hyponatremia, however, may occur, particularly in patients who drink large volumes of water [13,15,16]. Three mechanisms contribute to this tendency toward hyponatremia: volume depletion, K^+ depletion, and direct inhibition of urinary dilution by diminished NaCl reabsorption in the diluting segments [13,15]. In those patients in whom K^+ depletion is of primary importance, the administration of KCl alone is sufficient to raise the plasma Na^+ concentration toward normal [13,17].

Diuretic-induced hyponatremia usually *begins within the first two weeks of therapy* when urinary losses are maximal. After this time, a new steady state is established which prevents progressive hyponatremia unless a superimposed problem occurs such as vomiting, diarrhea, or increased water intake. When diuretics are given to patients with essential hypertension, for example, negative fluid balance is seen during the first week [18]. The ensuing activation of the renin-angiotensin-aldosterone system and other Na^+-retaining mechanisms then enhance proximal and collecting tubular reabsorption, counteracting the diuretic effect and preventing further volume depletion [18,19]. K^+ depletion also occurs within the first few weeks of therapy, and then a new steady state is achieved in which the K^+ deficit and the degree of hypokalemia remain relatively constant [18,20].

Similar considerations apply to the use of diuretics in edematous patients, except that steady states are not achieved so readily since the diuretic dose may be adjusted

Table 23-3 Disorders which cause SIADH, listed according to probable mechanism of action

Increased hypothalamic production of ADH
  A. Neuropsychiatric disorders†
    1. Infections: meningitis, encephalitis, abscess
    2. Vascular: thrombosis, subarachnoid hemorrhage, subdural hemorrhage
    3. Cerebral neoplasm: primary or metastatic
    4. Miscellaneous: Guillain-Barré syndrome, acute intermittent porphyria,
       acute psychosis, autonomic neuropathy, hypothalamic sarcoidosis
  B. Drugs
    1. Carbamazepine [22,23]
    2. Cyclophosphamide [24,25]
    3. Vincristine [26]
    4. Thiothixene [27]
    5. Thioridazine [28]
    6. Haloperidol [29]
    7. Amitriptyline [30]
    8. Monoamine oxidase inhibitors [31]
    9. Bromocriptine [31a]
  C. Pulmonary diseases†
    1. Pneumonia: viral, bacterial, or fungal [32,33]
    2. Tuberculosis [34]
    3. Acute respiratory failure [35]
    4. Other: asthma, atelectasis, pneumothorax
  D. Postoperative patient† [36,37]
  E. Idiopathic [38]

Ectopic (nonhypothalamic) production of ADH
  A. Carcinoma: oat-cell of lung,† bronchogenic, duodenum, pancreas, thymus [39,40]
  B. ?Pulmonary tuberculosis [41]

Potentiation of ADH effect
  A. Chlorpropamide† [42]
  B. Tolbutamide [43]
  C. ?Carbamazepine [43a]

Exogenous administration of ADH
  A. Vasopressin
  B. Oxytocin [44–46]

<hr>

†Common causes of SIADH.

upward to remove the excess fluid. In this setting, both the underlying disease (heart failure, hepatic cirrhosis) and the diuretics (which may further reduce the effective circulating volume; see page 339) contribute to the development of hyponatremia.

Although the loop diuretics (furosemide and ethacrynic acid) and the thiazides both cause volume and K^+ depletion and impair urinary dilution, *severe hyponatremia most often occurs with the thiazides* because of their lack of effect on urinary *concentration* [13,15,16,21]. A concentrated urine is produced by equilibration of the fluid in the collecting tubules with the hyperosmotic medullary interstitium.† The

<hr>

†The mechanisms responsible for concentration and dilution of the urine are discussed in detail in Chap. 6.

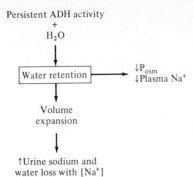

Persistent ADH activity
+
H_2O

Water retention → $\downarrow P_{osm}$ $\downarrow$Plasma Na^+

Volume
expansion

↑Urine sodium and
water loss with [Na^+]
greater than 20 meq/L

Figure 23-3 Pathophysiology of hyponatremia in the syndrome of inappropriate ADH secretion.

loop diuretics interfere with this process by inhibiting NaCl reabsorption in the medullary thick ascending limb, thereby diminishing NaCl accumulation in the medullary interstitium; the thiazides, in contrast, act in the cortex in the distal tubule and do not affect urinary concentration [21]. As a result, *the ability of ADH to increase collecting tubule water reabsorption and promote the development of hyponatremia is reduced by loop diuretics but remains intact with the thiazides.* The administration of a thiazide also represents the major clinical setting in which solute loss in excess of water can contribute to the development of hyponatremia: the combination of the diuretic-induced Na^+ and K^+ losses and the hypovolemia-induced increase in ADH secretion results in a concentrated urine in which the sum of the Na^+ and K^+ concentrations can exceed that in the plasma [16].

Syndrome of Inappropriate ADH Secretion (SIADH)

Pathogenesis Inappropriate secretion of ADH, i.e., not due to the physiologic stimuli of hyperosmolality or hypovolemia, is a common problem that occurs in a variety of clinical disorders (Table 23-3). The metabolic consequences of this persistent ADH activity are depicted in Fig. 23-3 [47]. Because of the hormonal effect to enhance renal water reabsorption, ingested water is retained, resulting in dilution (hyponatremia and hypoosmolality) and expansion of the body fluids. Edema does not occur because the increase in volume activates the volume receptors, producing an increase in urinary Na^+ and water excretion with the urine Na^+ concentration remaining above 20 meq/L. The net effect is that the hyponatremia is due both to a moderate degree of water retention and to urinary Na^+ loss† [see Eq. (23-1)] [48].

It is important to emphasize that the ingestion of water is an essential step in the development of hyponatremia in the SIADH. If water intake is restricted, water retention and Na^+ loss do not occur and *ADH excess has no effect on the plasma Na^+ concentration* [47,51].

If the levels of ADH release and water intake remain constant, a new steady

---

†It has also been suggested that movement of Na^+ into the cells, where it becomes bound and osmotically inactive, might contribute to the reduction in the plasma Na^+ concentration [49]. This thesis, however, is unproven, and it is likely that the combination of water retention and Na^+ loss is sufficient to explain the hyponatremia in most patients [48,50].

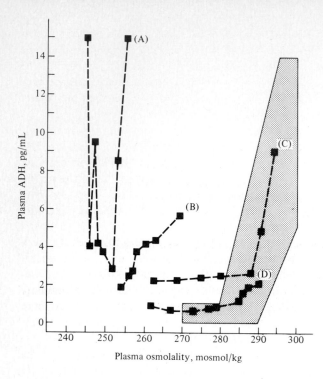

Figure 23-4 The relationship between plasma ADH levels and plasma osmolality in patients with the SIADH. The plasma osmolality is increased in these initially hypoosmolar patients by the administration of hypertonic saline. The shaded area represents the normal range. *(From G. L. Robertson, R. L. Shelton, and S. Athar, Kidney Int., 10:25, 1976. Reprinted by permission from Kidney International.)*

state will be reached within 1 or 2 weeks in which the U_{osm} falls, urinary Na^+ and water excretion match intake, and the plasma Na^+ concentration stabilizes at 105 to 130 meq/L [51,52]. A further reduction in the plasma Na^+ concentration will occur only if there is an increase in either the secretion of ADH or water intake. This stabilization appears to be due to resistance of the collecting tubule epithelium to ADH, an effect that may be mediated by an increase in cytosolic Ca^{2+} [53]. There is normally a passive Na^+-Ca^{2+} exchange occurring across the peritubular membrane. As hyponatremia develops, there is less of a gradient for Na^+ to enter the cell and consequently less movement of Ca^{2+} out of the cell. The ensuing increase in cell Ca^{2+} is known to be able to inhibit the effect of ADH [54].

ADH secretion Although it might be thought that ADH is secreted at random in the SIADH, this occurs in only a minority of patients, as four distinct patterns of ADH release have been identified (Fig. 23-4) [11]. Type A is found in about 20 percent of patients and is characterized by erratic changes in ADH secretion independent of the plasma osmolality. In this setting, ADH release is occurring randomly or in response to volume stimuli as osmotic regulation appears to be lost. Type B (35 percent) represents the "reset osmostat" in which ADH secretion varies appropriately with the plasma osmolality but the curve is shifted leftward (see below). In this situation, the plasma Na^+ is relatively stable (usually between 125 to 130 meq/L), the urine can be appropriately diluted after a water load, and progressive hyponatremia does not occur. Type C (35 percent) is characterized by normal ADH release when the plasma osmolality is normal or elevated but an inability to reduce ADH secretion below a certain level. This nonsuppressible "leak" of ADH results in water retention and hyponatremia. Type D is the least common (10 percent) and is associated with normal ADH release. Either increased sensitivity to ADH (as occurs, for example, with chlorpropamide) [55] or some other antidiuretic factor must be present in these patients. It should be noted that a strict correlation between the pattern of ADH release and the cause of SIADH cannot generally be made [11].

Acid-base and K^+ balance Although the retention of water lowers the plasma Na^+ concentration by dilution, it does not reduce either the plasma HCO_3^- concentration or, in most cases, the plasma K^+ concentration [47]. The maintenance of the plasma HCO_3^- concentration in the face of hypotonic volume expansion appears to be mediated by the movement of H^+ into cells [56] and by increased renal H^+ excretion [57], both of which prevent a fall in the plasma HCO_3^- concentration. The plasma K^+ concentration is maintained by the movement of K^+ out of the cells, some of which may be linked to the intracellular movement of H^+ [56]. However, when the P_{osm} is less than 240 mosmol/kg, renal K^+ excretion may increase and mild hypokalemia may ensue [57]. The mechanisms by which these processes are regulated have not been defined, but a hyponatremia-induced elevation of aldosterone secretion may contribute [58].

Etiology The SIADH can be produced by enhanced hypothalamic secretion, ectopic (nonhypothalamic) hormone production, the potentiation of ADH effect, or the administration of exogenous ADH or oxytocin (Table 23-3). A variety of *neuropsychiatric disorders* can promote ADH release, either directly or by activation of cortical neurons that can stimulate the hypothalamus [47,58a]. The mechanism by which *drugs* increase ADH secretion is uncertain [59]. Particular care must be taken with cyclophosphamide, an alkylating agent which can cause the SIADH when given intravenously in high doses but not when taken orally in low doses. A high fluid intake is generally recommended to limit drug contact with the bladder and prevent the development of hemorrhagic cystitis. However, the combination of ADH and a large volume of water can lead to severe, occasionally fatal hyponatremia within 24 h [24,25]. This complication can be minimized by using isosmotic saline rather than water to maintain a high urine output.

Chlorpropamide, an oral hypoglycemic agent, is the most common cause of drug-induced SIADH. When used in diabetes mellitus, it lowers the plasma Na^+ concentration in roughly 4 percent of patients [42]. Chlorpropamide is relatively unique in that it appears to act primarily by potentiating the effect of ADH, not by enhancing its secretion [55,60,61]. This effect is manifested by a greater than normal increment in ADH-induced adenyl cyclase activity and cyclic AMP generation (see Chap. 8), suggesting that chlorpropamide may increase the affinity of the receptor on the tubular membrane for ADH [55].

Although chlorpropamide increases the action of ADH, some circulating ADH must be present for this potentiation to occur [62]. As depicted in Fig. 23-1, some people maintain low basal levels of ADH (<2 pg/mL) despite the presence of hypoosmolality; it may be that it is these patients who are most likely to become hyponatremic with chlorpropamide [42]. The concurrent use of another drug which increases ADH secretion (such as a diuretic) will also increase the risk of hyponatremia [42].

Some other hypoglycemic drugs such as tolazamide and acetohexamide have opposite effects as they induce a small *increase* in water excretion (by an unknown mechanism) [63]. Thus, these agents can be substituted in patients who develop hyponatremia while on chlorpropamide.

Nonsteroidal anti-inflammatory drugs also can potentiate the effect of ADH. This is mediated by a reduction in renal prostaglandin synthesis as prostaglandins normally antagonize the action of ADH (see page 149). Despite this effect, spontaneous hyponatremia has not been reported, probably because ADH secretion is reduced due either to a direct drug effect [64] or the initial fall in P_{osm} if some water is retained. These agents may, however, exacerbate the tendency to hyponatremia in patients who are volume-depleted or have the SIADH. One exception is sulindac, which does not appear to inhibit renal prostaglandin synthesis, perhaps because no active drug or metabolite is excreted in the urine [65,66].

Pulmonary diseases, particularly pneumonia but including acute asthma, atelectasis, empyema, pneumothorax, tuberculosis, and acute respiratory failure, can lead to the SIADH [32–35,67,68]. The mechanism by which this occurs is uncertain, but a decrease in pulmonary venous return, activating the volume receptors, may be involved [67,69]. If present, hypercapnia also may contribute to the water retention, perhaps by increasing proximal $NaHCO_3$ reabsorption, thereby limiting distal delivery to the diluting segments (see page 93) [35,70]. It should be noted, however, that the urine Na^+ concentration is low in some patients, suggesting that hypovolemia also may be important [33,71].

In the *postoperative patient,* inappropriate ADH secretion may persist for several days [36,37]. This appears to be mediated by pain afferents which directly stimulate the hypothalamus [36]. Patients undergoing mitral commissurotomy to relieve mitral stenosis are particularly prone to develop the SIADH, perhaps because the acute reduction in left atrial pressure can activate the left atrial receptors and contribute to ADH release (see page 136) [37].

Ectopic tumor production of ADH has been reported with a variety of different neoplasms, particularly oat-cell carcinoma of the lung (Table 23-3). Direct evidence for tumor hormonal synthesis has come primarily from the study of lung carcinomas. In several of these tumors, ADH and its carrier, neurophysin, have been extracted from the tumor [39] and in vitro synthesis of ADH by the neoplastic tissue has been demonstrated [40]. Indirect evidence for ectopic production is also suggested by the finding that ethanol, an inhibitor of central ADH secretion, does not initiate a water diuresis in these patients, in contrast to its effect in other states in which ADH secretion is increased [47].

In one patient with *pulmonary tuberculosis* and hyponatremia, ADH was found in the tuberculous tissue [41]. It is not clear if this is another example of ectopic production or merely represents nonspecific hormone adsorption onto the diseased tissue. In most patients with pulmonary tuberculosis and hyponatremia, ethanol increases water excretion, suggesting a hypothalamic origin for the excess ADH [34].

Oxytocin is a second hormone synthesized in the hypothalamus and released from the neurohypophysis. Its primary effects are on uterine function and lactation, but oxytocin also possesses significant antidiuretic activity. The use of intravenous infusions of this hormone in dextrose and water to stimulate labor in pregnant women has resulted in water retention, severe hyponatremia, and seizures in both the mother and fetus [44–46]. This complication can be prevented by limiting the amount of water given and using isosmotic saline rather than dextrose and water. Hyponatremia can also be induced by the administration of *exogenous ADH* to patients with diabetes insipidus (see Chap. 24).

Rarely, no cause for SIADH can be identified [38]. Although some of these patients have remained idiopathic for many years, careful monitoring for the presence of an occult tumor (particularly pulmonary) is essential [38].

In summary, the SIADH can be produced by a variety of disorders and is characterized by (1) hyponatremia and hypoosmolality, (2) a U_{osm} that is inappropriately high (greater than 100 mosmol/kg), (3) a urine Na^+ concentration greater than 20 meq/L, (4) normovolemia, (5) normal renal, adrenal, and thyroid function, and (6) normal acid-base and K^+ balance.

Adrenal Insufficiency

Hyponatremia is a common complication of adrenal insufficiency. Although volume depletion due to diarrhea, vomiting, or renal Na^+ loss (resulting from lack of aldosterone) may contribute [72], hypocortisolism usually plays a major role since cortisol replacement rapidly increases the rate of water excretion and raises the plasma Na^+ concentration toward normal [73–75]. The deleterious effect of cortisol deficiency is primarily due to altered systemic hemodynamics as the blood pressure, cardiac output, and ultimately renal blood flow are reduced by an unknown mechanism [76]. The fall in blood pressure stimulates the nonosmotic release of ADH, which is primarily responsible for the development of hyponatremia [76,77]. Decreased loop delivery induced by the fall in renal perfusion may also play a contributory role [76].

Hypothyroidism

Similar factors may participate in the development of hyponatremia in moderate to severe hypothyroidism. The cardiac output and GFR are frequently reduced in this setting [78], and these changes can lead both to the release of ADH and to diminished delivery to the diluting segments [78–80]. Normal water balance can be restored by the administration of thyroid hormone.

Reset Osmostat

As described above, patients with a reset osmostat have normal osmoreceptor responses to changes in the P_{osm} but the threshold for ADH release (and in many cases thirst) is reduced (Fig. 23-4, pattern B) [11]. As a result, the plasma Na^+ concentration is below normal but stable (usually 125 to 135 meq/L) since the ability to excrete water is maintained. In addition to being present in approximately 35 percent of patients with SIADH, a reset osmostat and hyponatremia may occur in several other conditions. These include hypovolemic states (in which the baroreceptor stimulus to ADH release is superimposed upon normal osmoreceptor function; see Fig. 8-8) [11], pregnancy (in which vasodilation may induce effective volume depletion, with the plasma Na^+ concentration falling less than 5 meq/L) [11,81], psychosis [82], and certain chronic diseases such as tuberculosis and malnutrition [83]. In the last disorders, it has been suggested that defective cellular metabolism results in a decrease in cell osmolality and resetting of the osmostat. Correction of the underlying problem with antituberculous drugs or hyperalimentation has been effective in returning the plasma Na^+ concentration to normal [83].

Primary Polydipsia

Patients with primary polydipsia have a primary increase in water intake and may complain of polyuria or excessive thirst. This disorder is most often seen in anxious middle-aged women and in patients with psychiatric illnesses, particularly those taking phenothiazines in whom the sensation of a dry mouth stimulates thirst [82,84,85]. It may also occur with hypothalamic disorders (such as sarcoidosis) in which the regulation of thirst may be directly affected [85a].

There is usually no change in the plasma Na^+ concentration or P_{osm} in primary polydipsia since renal water excretion is not impaired [86]. In rare instances, however, water intake exceeds 10 to 15 L/day and overwhelms renal excretory capacity, resulting in potentially fatal hyponatremia [87,88]. For example, one patient with this disorder was able to lower her plasma Na^+ concentration to 84 meq/L even though her GFR was normal and her urine was maximally dilute (U_{osm} equal to 74 mosmol/kg, specific gravity of 1.001) [87]. More commonly, a marked reduction in the plasma Na^+ concentration in these patients is associated with some limitation in water excretion, as occurs, for example, with diuretic therapy [15].

An interesting example of this phenomenon has been described in excessive beer drinkers in whom the ability to excrete water is reduced by a poor dietary intake [89]. A normal subject may excrete 750 mosmol of solute per day consisting mostly of NaCl, KCl, and urea (which is derived from the metabolism of proteins). If the minimum U_{osm} is 50 mosmol/kg, then the maximum daily urine volume will be 15 liters [750 mosmol/day $\div$ 50 mosmol/kg = 15 L/day (or kg/day)]. However, the daily solute load can fall to 250 mosmol or less in beer drinkers who may ingest little or no NaCl or protein. In this setting, the maximum urine volume is only 5 liters ($\frac{250}{50} = 5$), and hyponatremia will ensue if more than this amount of fluid (primarily as beer) is ingested.

Pseudohyponatremia

In some patients, a decrease in the plasma Na^+ concentration is associated with a normal or increased effective P_{osm}, rather than hypoosmolality. This has been called *pseudohyponatremia* (Table 23-4). These disorders highlight the importance of mea-

Table 23-4 Etiology of pseudohyponatremia

A. Low plasma Na^+ concentration with normal P_{osm}
  1. Severe hyperlipemia
  2. Severe hyperproteinemia
  3. Posttransurethral resection of prostate or bladder
B. Low plasma Na^+ concentration with elevated P_{osm}
  1. Hyperglycemia
  2. Administration of hypertonic mannitol

suring the P_{osm} in patients with hyponatremia, since *therapy should not be directed toward the fall in the plasma Na^+ concentration.*

Hyponatremia with normal P_{osm} In each liter of plasma there is normally about 930 mL of water, with the plasma proteins and lipids occupying the remaining 70 mL (see page 22). However, the plasma water may fall to as low as 720 mL per liter of plasma in states of severe hyperlipemia (as in uncontrolled diabetes mellitus) or hyperproteinemia (as in multiple myeloma) [90,91]. Since the lipids exist in a separate phase and an osmometer measures only the activity of the plasma water, the P_{osm} will be unaffected†. However, the plasma Na^+ concentration, *measured per liter of plasma, not plasma water,* will be artifactually reduced to 110 meq/L (153 meq/L of plasma water × 0.72 liters of plasma water per liter of plasma). These patients do not have the symptoms of hypoosmolality and do not require therapy to normalize the plasma Na^+ concentration.

Isosmotic reduction in the plasma Na^+ concentration can also be produced by the acute addition of isosmotic but non-Na^+-containing fluid to the extracellular space. This is most often seen in patients undergoing transurethral resections of the bladder or prostate in whom as much as 20 to 30 liters of isotonic glycine may be used as a flushing solution. Variable quantities of this solution are absorbed, and the plasma Na^+ concentration may fall below 100 meq/L without change in the P_{osm} [92,93]. Confusion, disorientation, hypotension, and abnormalities on the electrocardiogram have been observed in these patients and are due either to the direct effect of hyponatremia (without hypoosmolality) or to glycine toxicity [93]. With time, renal excretion of the excess fluid will restore the plasma Na^+ concentration to normal and relieve any symptoms which may occur.

Hyponatremia with increased P_{osm} If a solute which penetrates cells poorly, such as glucose, is added to the extracellular fluid, the P_{osm} increases. This creates an osmotic gradient, and water moves from the cells into the extracellular fluid and lowers the plasma Na^+ concentration (see Fig. 22-1e). In general, every 62-mg/dL

†Since they are very large molecules, proteins make only a minimal contribution to the P_{osm}. For example, the normal plasma protein concentration of 7 g/dL (70 g/L) represents an osmolality of only 1.3 mosmol/kg (see page 34). Thus, doubling the plasma protein concentration, which will significantly reduce the fraction of the plasma that is water and lower the measured plasma Na^+ concentration, will have only a minor effect on the P_{osm}.

increment in the plasma glucose concentration will draw enough water out of the cells to reduce the plasma Na^+ concentration 1 meq/L [94]. Conversely, as insulin therapy drives glucose into the cells, water will follow, and the plasma Na^+ concentration will rise.

A similar situation can be produced by the administration of hypertonic mannitol to induce a diuresis in patients with renal failure [95]. Although measurement of the plasma mannitol concentration is not available in most laboratories, the presence of significant amounts of mannitol in the blood can be detected by the difference between the measured P_{osm}, which includes the effect of mannitol, and the calculated P_{osm}, which does not (calculated $P_{osm} \cong 2 \times$ plasma $[Na^+]$ + [glucose]/18 + BUN/2.8; see page 30).

Since there is a disparity between the P_{osm} and the plasma Na^+ concentration, an osmolal gap will also be seen in states of hyperlipemia, hyperproteinemia, or after the use of glycine solutions. In addition, drugs other than mannitol, such as ethanol and methanol, also can achieve significant osmolal concentrations in the blood, resulting in a gap between the measured and calculated P_{osm} (see page 410) [96,97]. However, methanol and ethanol are ineffective osmoles (like urea) and do not affect water distribution or the plasma Na^+ concentration [97].

Hyponatremia and azotemia Patients with renal failure have a high P_{osm} due to the increase in the BUN. However, urea is an ineffective osmole and the effective P_{osm} ($2 \times$ plasma $[Na^+]$ + [glucose]/18) is generally normal or reduced. For example, consider a patient with the following plasma values:

$$\text{Plasma } [Na^+] = 115 \text{ meq/L}$$

$$[\text{Glucose}] = 90 \text{ mg/dL}$$

$$\text{BUN} = 140 \text{ mg/dL}$$

$$P_{osm} = 285 \text{ mosmol/kg}$$

Despite the normal P_{osm}, the effective P_{osm} is markedly reduced at 235 mosmol/kg. Thus, this patient has true hyponatremia, not pseudohyponatremia, and may become symptomatic.

SYMPTOMS

The symptoms associated with hyponatremia are primarily due to the reduction in the P_{osm}. As the P_{osm} falls, an osmolal gradient is created across the blood-brain barrier, resulting in water movement into the brain (as well as other cells). This cerebral overhydration is probably responsible for the symptoms of hypoosmolality, which are primarily neurologic [98,99]. As depicted in Fig. 23-5, a rapid reduction in the plasma Na^+ concentration to 119 meq/L in rabbits induces a marked increase in brain water content, severe symptoms, and death. In comparison, a similar degree of hyponatremia produced slowly results in a lesser degree of cerebral edema and no

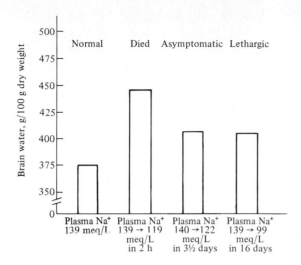

Figure 23-5 Brain water content in normal rabbits and in three groups of hyponatremic rabbits. In animals in whom the plasma Na$^+$ concentration is lowered to 119 meq/L in 2 h, brain water content is 17 percent above normal and is associated with severe symptoms and death. In contrast, when the plasma Na$^+$ concentration is lowered to almost the same value (122 meq/L) in 3½ days, brain water is only increased by 7 percent and the animals are asymptomatic. When the plasma Na$^+$ concentration is decreased to 99 meq/L in 16 days, brain water is again only 7 percent above normal but mild symptoms are present. *(From A. I. Arieff, F. Llach, and S. G. Massry, Medicine, 55:121, 1976.)*

neurologic changes. This return of brain volume toward normal appears to be due to the loss of osmoles (Na$^+$, K$^+$, and perhaps amino acids such as taurine) from the cell and/or to binding of these solutes to intracellular proteins, making them osmotically inactive [99–101]. The net decrease in cell osmolality then promotes water movement out of the brain cells [99,100]. Although it is not clear how these osmolal changes occur, they represent a specific response of the brain since the volume of other cells (muscle, liver, erythrocytes) remains elevated. Finally, a further, slow reduction in the plasma Na$^+$ concentration to 99 meq/L results in lethargy but brain water content remains stable (Fig. 23-5).

These findings indicate that the severity of the neurologic symptoms is *related to the rapidity, as well as the degree, of the reduction in the plasma Na$^+$ concentration* [98]. The changes associated with acute hyponatremia are primarily due to cerebral overhydration. With chronic hyponatremia, the low plasma Na$^+$ concentration is directly responsible, perhaps reflecting the importance of Na$^+$ in neural function.

The neurologic symptoms associated with hyponatremia are similar to those in other metabolic encephalopathies [98]. In general, the patient begins to complain of nausea and malaise as the plasma Na$^+$ concentration falls below 125 meq/L. Between 115 and 120 meq/L, headache, lethargy, and obtundation may appear. The more severe changes of seizures and coma are not usually seen until the plasma Na$^+$ concentration is less than 115 meq/L [19,98]. Focal neurologic findings are uncommon but may occur in patients with an underlying defect such as an old cerebral infarct† [102]. Hyponatremic encephalopathy is typically reversible, but *acute* marked reductions in the plasma Na$^+$ concentration can lead to permanent neurologic deficits and death [16,98,103].

†When the SIADH is due to a central nervous system disorder, it may be difficult to determine whether the neurologic disease or the hyponatremia is responsible for the symptoms. The inciting factor can be assessed more accurately by the response to correction of the hyponatremia.

Central pontine myelinolysis, a syndrome characterized by the rapid onset of paraparesis or quadriparesis, dysarthria, and dysphagia, has also been rarely associated with severe hyponatremia [104,105]. Although it has been suggested that this disorder may be related to an overly rapid correction of the plasma Na^+ concentration [106,106a], it is most likely due to the low P_{osm} [107–109]. There is no evidence in humans that there is a limit to how quickly severe hyponatremia can be safely corrected (see below) [107,109].

When volume depletion is present, patients may also complain of the symptoms of hypovolemia such as weakness, fatigue, muscle cramps, and postural dizziness (see Chap. 15). In contrast, signs of extracellular volume expansion such as edema usually are not seen in patients with water retention due to the SIADH or primary polydipsia since roughly two-thirds of the retained water is stored in the cells and persistent hypervolemia is prevented by increased Na^+ and water excretion [48].

DIAGNOSIS

As with other electrolyte and acid-base disorders, the history (?vomiting, diarrhea, diuretic therapy, or one of the causes of the SIADH) and physical examination (?physical findings of true volume depletion or edema) can provide important clues to the correct diagnosis. In addition, the initial laboratory evaluation should include measurement of the P_{osm}; the plasma concentrations of Na^+, K^+, Cl^-, HCO_3^-, urea, and glucose; the urine Na^+ concentration and osmolality; and, if the HCO_3^- concentration is abnormal, the plasma pH. Tables 23-5 to 23-7 illustrate how these tests can be used to identify the cause of the hyponatremia.

Plasma Osmolality

The first step in the approach to the patient with hyponatremia is to confirm the presence of hypoosmolality (Table 23-5). If the effective P_{osm} (measured P_{osm} minus BUN/2.8) is normal or elevated, evaluation for one of the causes of pseudohyponatremia should be carried out.

Urine Osmolality

Once it is demonstrated that the patient is hypoosmolal, measurement of the U_{osm} can be used to determine whether water excretion is normal or impaired. A value below 100 mosmol/kg (specific gravity ≤ 1.003) indicates that ADH secretion is almost completely and appropriately suppressed, a finding seen with either primary polydipsia or a reset osmostat (if the P_{osm} is below the new threshold for ADH release). These disorders can be distinguished by the response to water restriction. The urine will remain dilute until the plasma Na^+ concentration is normal in primary polydipsia. In contrast, the U_{osm} will rise progressively with a reset osmostat since a small elevation in the plasma Na^+ concentration will stimulate the release of ADH. By using these criteria, many psychotic patients initially thought to have primary polydipsia were shown to have a reset osmostat [82].

Table 23-5 Major steps in the initial evaluation of the hyponatremic patient

A. Plasma osmolality
   1. Low: true hyponatremia
   2. Normal or elevated: pseudohyponatremia or renal failure

B. Urine osmolality
   1. Less than 100 mosmol/kg: primary polydipsia or reset osmostat
   2. Greater than 100 mosmol/kg: other causes of true hyponatremia

C. Urine Na^+ concentration
   1. Less than 15 meq/L: effective circulating volume depletion
   2. Greater than 20 meq/L: those disorders in which volume is relatively normal or the ability to conserve Na^+ is impaired (see Table 23-6)

Table 23-6 Effective circulating volume and urine Na^+ concentration in the differential diagnosis of hyponatremia

Effective circulating volume		
Decreased		Normal or increased
Urine $[Na^+] < 15$ meq/L	Urine $[Na^+] > 20$ meq/L	Urine $[Na^+] > 20$ meq/L
Gastrointestinal losses	Diuretics (early)	SIADH
Burns	Adrenal insufficiency	Primary polydipsia††
Diuretics (late)	Salt-wasting renal disease	Renal failure
Edematous states	Osmotic diuretics†	Reset osmostat
Pure cortisol deficiency	Some patients with vomiting (see page 384)	

†In pseudohyponatremia due to glucose or mannitol, the urine Na^+ concentration is high due to the ability of these substances to act as osmotic diuretics.

††The urine $[Na^+]$ may be lower due to dilution in patients with a very high urine output. For example, 100 meq of Na^+ in 10 L of urine results in a $[Na^+]$ of only 10 meq/L.

Table 23-7 Disorders in acid-base and K^+ balance which may occur in patients with hyponatremia

Metabolic acidosis	Normal pH	Metabolic alkalosis
Plasma K^+ concentration may be normal or elevated	Plasma K^+ concentration usually normal	Plasma K^+ concentration may be normal or reduced
Renal failure	SIADH	
Adrenal insufficiency	Primary polydipsia (hypokalemia may occur)	Vomiting
		Nasogastric suction
Plasma K^+ concentration may be normal or reduced	Edematous states (no diuretics)	Diuretics
Diarrhea	Pure cortisol deficiency	
Tube drainage of intestinal secretions	Hypothyroidism	

In the vast majority of hyponatremic patients, however, water excretion is impaired and the U_{osm} exceeds 100 mosmol/kg†. It is important to emphasize that a U_{osm} of 100 to 230 mosmol/kg may be hypoosmotic to plasma but is still inappropriately high. This can be illustrated by a simple example. To raise the U_{osm} from a maximally dilute level of 70 mosmol/kg up to 210 mosmol/kg, for example, requires the removal of two-thirds of the water. Thus, a patient with a U_{osm} of 210 mosmol/kg will be able to excrete only one-third the normal amount of free water and, therefore, will be much more likely to develop hyponatremia.

Urine Sodium Concentration

The differential diagnosis of hyponatremia, hypoosmolality, and an inappropriately high U_{osm} frequently narrows down to effective circulating volume depletion, the SIADH, adrenal insufficiency, and hypothyroidism. In addition to assessing adrenal and thyroid function, the urine Na^+ concentration is usually helpful in differentiating between these disorders (Table 23-6). The urine Na^+ concentration should be less than 10 to 15 meq/L in hypovolemic states but greater than 20 meq/L in the SIADH, where the volume is somewhat expanded.

The response to the administration of NaCl can also be used in a patient with a borderline urine Na^+ concentration. This value should initially remain low in true volume depletion as Na^+ is retained in an attempt to replete the extracellular volume. In contrast, the exogenous Na^+ is rapidly and appropriately excreted in the SIADH. As a result, the urine Na^+ concentration rises quickly and may even exceed that in the plasma due to the combination of enhanced Na^+ loss and ADH-induced water reabsorption.

The meaning of the urine Na^+ concentration is more difficult to interpret if the patient is unable to conserve Na^+ normally. In disorders associated with renal Na^+ wasting, the urine Na^+ concentration may exceed 20 meq/L despite the presence of volume depletion. This may occur with diuretic therapy (if the diuretic is still acting), salt-wasting nephropathies (in which substantial renal failure is usually present; see page 281), and adrenal insufficiency. In the last condition, the inappropriate Na^+ excretion, as well as the hyperkalemia and metabolic acidosis, is due to aldosterone deficiency. In contrast, Na^+, K^+, and H^+ handling are normal in pure cortisol deficiency (as seen with hypopituitarism) and the urine Na^+ concentration may be low if the patient is hypovolemic [75]. The urine Na^+ concentration is also variable in hypothyroidism, depending primarily upon daily Na^+ intake.

Plasma pH and Potassium Concentration

Abnormalities in acid-base and K^+ homeostasis frequently are associated with hyponatremic disorders, and their presence can be helpful in establishing the correct

†A high U_{osm} is also seen with a reset osmostat if the P_{osm} is above the threshold for ADH release. This diagnosis should be suspected if the plasma Na^+ concentration is stable despite variations in Na^+ and water intake and can be confirmed by the ability to dilute the urine normally after a water load [83].

diagnosis (Table 23-7). For example, the combination of metabolic alkalosis and hypokalemia should lead one to suspect vomiting or diuretic therapy, whereas hyperkalemia and metabolic acidosis in a patient with relatively normal renal function is suggestive of adrenal insufficiency.

Hypokalemia may occur in patients with primary polydipsia and appears to be directly related to the high urine output [82]. In the presence of K^+ depletion, normal subjects can lower the urine K^+ concentration to a minimum of 5 to 10 meq/L [110]. If, however, the urine output is 10 L/day or more, then there will be an *obligatory* K^+ loss that can exceed 50 to 100 meq and promote the development of hypokalemia.

Clinical Example

The following case exemplifies the use of this method in the approach to the hyponatremic patient.

Case history 23-1 A 62-year-old man with a carcinoma of the lung develops nausea and vomiting after being given chemotherapy. Five days later, he is admitted to the hospital in an obtunded state. No other history is obtainable. The physical examination reveals reduced skin turgor and flat neck veins. The laboratory values include:

Plasma $[Na^+]$ = 114 meq/L   Arterial pH = 7.48

$[K^+]$ = 2.8 meq/L P_{osm} = 240 mosmol/kg

$[Cl^-]$ = 72 meq/L U_{osm} = 380 mosmol/kg

$[HCO_3^-]$ = 31 meq/L   Urine $[Na^+]$ = 7 meq/L

The patient is initially treated with NaCl, KCl, and water. Five days later, the plasma K^+ concentration and pH are normal, the plasma Na^+ concentration is stable at 124 meq/L, the U_{osm} is 310 meq/L, and the urine Na^+ concentration is 82 meq/L.

COMMENT The low P_{osm}, high U_{osm}, and low urine Na^+ concentration on admission indicate that the patient had true hyponatremia and impaired water excretion due to volume depletion. The hypokalemia and metabolic alkalosis suggest that vomiting (induced by chemotherapy) was the primary problem. However, volume repletion only partially corrected the plasma Na^+ concentration. The last laboratory values are consistent with the SIADH, presumably due to the underlying lung tumor. In this case, the SIADH was initially masked by the presence of volume depletion.

TREATMENT

General Principles

There are two basic principles involved in the treatment of hyponatremia: raising the plasma Na^+ concentration and treating the underlying cause (such as discontinuing chlorpropamide or giving cortisol in adrenal insufficiency). In general, hyponatremia is corrected acutely by giving Na^+ to patients who are volume-depleted and by restricting water intake in patients who are normovolemic or edematous (Table 23-8). However, more vigorous therapy (usually requiring hypertonic saline) is indicated

Table 23-8 Basic therapeutic regimen in the disorders which produce hyponatremia

NaCl	H_2O restriction
True volume depletion	Edematous states
Adrenal insufficiency	SIADH
Diuretics	Primary polydipsia
	Renal failure

when symptoms are present or the plasma Na^+ concentration is less than 110 to 115 meq/L since these are the settings in which irreversible neurologic damage and death can occur [16,98]. The first aim of treatment in this setting is to raise the plasma Na^+ concentration to 120 to 125 meq/L, a level at which the patient should be out of danger.

Sodium deficit The amount of Na^+ required to raise the plasma Na^+ concentration to a desired value can be derived from the following formula:

$$Na^+ \text{ deficit} = \text{volume of distribution of plasma } [Na^+]$$
$$\times \text{ plasma } [Na^+] \text{ deficit per liter} \qquad (23\text{-}2)$$

Although Na^+ itself is restricted to the extracellular fluid, changes in the plasma Na^+ concentration reflect changes in osmolality and are distributed through the total body water (see Fig. 22-1). The total body water is approximately 60 and 50 percent of *lean* body weight in men and women, respectively. Thus, from Eq. (23-2), the approximate amount of Na^+ (in meq) required to raise the plasma Na^+ concentration to 140 meq/L in men is equal to

$$Na^+ \text{ deficit} = 0.6 \times \text{lean body weight (kg)} \times (140 - \text{plasma } [Na^+]) \quad (23\text{-}3)$$

Suppose a 70-kg man is started on a thiazide diuretic and 5 days later presents with lethargy, confusion, decreased skin turgor, and a plasma Na^+ concentration of 108 meq/L. The initial aim of therapy is to raise the plasma Na^+ concentration to 120 meq/L. The amount of Na^+ required can be estimated by adapting Eq. (23-3):

$$Na^+ \text{ deficit} = 0.6 \times 70 \times (120 - 108) = 504 \text{ meq}$$

Since 3% saline contains 513 meq of Na^+ per liter, 1 liter of this solution will provide the required Na^+. The optimal rate at which this should be given is uncertain. It has been suggested that too rapid a correction may be harmful [106,106a], but it seems likely that persistent severe hyponatremia represents the greater danger [107–109]. In this example, it is reasonable to administer 1 liter of 3% saline over 5 h, a regimen that should raise the plasma Na^+ concentration by roughly 2.4 meq/L per hour (12 meq over 5 h).† The possible risk of overly rapid correction can be minimized by

---

†The increase in the plasma Na^+ concentration will actually be somewhat less than this because 1 liter of water is also being given. A similar overestimation will occur in the SIADH since the administered Na^+ will be relatively rapidly excreted in the urine (see below).

switching to isotonic saline and slowing the rate of infusion once the plasma Na^+ concentration reaches 120 meq/L, a level at which the patient should be out of danger.

Two further points about Eq. (23-3) deserve emphasis. First, it is only an estimate, and serial measurements of the plasma Na^+ concentration are necessary to assess the efficacy of treatment. Second, this formula does not include any isosmotic losses which may also be present. For example, a patient with diarrhea may lose 5 L of isosmotic fluid and then become hyponatremic by drinking and retaining 3 L of water. Equation (23-3) estimates the amount of Na^+ required to counteract the 3 L of free water; a 2-L isosmotic Na^+ and water deficit will still remain.

The adequacy of *volume* repletion can be determined by following the skin turgor, central venous pressure, and urine Na^+ concentration. If this patient had an initial urine Na^+ concentration of 2 meq/L and, after the administration of 400 meq of NaCl, the urine Na^+ concentration were only 7 meq/L, then hypovolemia persists and further replacement therapy is indicated. When the urine Na^+ concentration exceeds 30 to 40 meq/L, normovolemia probably has been achieved.

True Volume Depletion

True volume depletion due to gastrointestinal or renal losses represents the main indication for the use of Na^+ in the treatment of hyponatremia. Not only will Na^+ replace the lost solute, but volume repletion will increase fluid delivery to the diluting segments, inhibit ADH release, and allow renal excretion of the excess water.

With asymptomatic or mild hyponatremia, one-third of the Na^+ deficit estimated from Eq. (23-3) can be given as isotonic saline in the first 4 to 6 h, with the remainder being given over the ensuing 24 to 48 h. Oral NaCl, added to the diet or given in tablet form, is a simple, safe alternative in patients who are able to eat. Hypertonic saline is necessary only in severe cases.

Correction of K^+ depletion, if present, is another important component of therapy. The administration of K^+ in this setting will increase the P_{osm}. As this K^+ moves intracellularly to replace the cell deficit, Na^+ leaves the cells to preserve electroneutrality and water follows K^+ into the cells down an osmotic gradient; both of these changes will raise the plasma Na^+ concentration toward normal [12,13,17].

Edematous States

Treatment is different in edematous patients in whom the extracellular volume is increased and hyponatremia is truly a problem of water overload. Therapy must be aimed at water removal since the administration of Na^+ will increase the severity of the edema. If one assumes that the water content of food is approximately equal to the insensible water loss from the skin and respiratory tract (see Table 22-1), then negative water balance can be achieved by restricting water intake to a volume that is *less* than the output. If water intake is held to 800 mL but output is only 500 mL, the patient will continue to retain water.

In patients with severe or symptomatic hyponatremia, the plasma Na^+ concentration can be elevated more quickly by the use of loop diuretics in combination with hypertonic saline or by peritoneal dialysis or hemodialysis. With the former, Na^+ and water loss is induced by the diuretic and only the Na^+ loss is replaced; the net effect is negative water balance. This modality can also be used in patients in whom water restriction alone is ineffective.

A frequent concern involves the safety of the use of diuretics in the hyponatremic, edematous patient. Loop diuretics (furosemide and ethacrynic acid) act in the medulla and inhibit both diluting and concentrating ability [21]. The ensuing loss of a relatively isosmotic urine will not further lower the P_{osm} or the plasma Na^+ concentration. In contrast, the thiazides act in the cortex, do not impair concentrating ability, and can directly cause hyponatremia by causing Na^+ and K^+ loss in excess of water [16,21].

In some patients with severe heart failure or hepatic cirrhosis, the water-retaining tendency is so great that water restriction and loop diuretics are ineffective in maintaining a safe plasma Na^+ concentration. Since elevated ADH levels play an important role in this setting [2,3], inhibiting the effect of ADH with demeclocycline, a tetracycline derivative, may increase water excretion and raise the plasma Na^+ concentration (see below) [111,112]. This agent must be used with care, however, because hepatic metabolism is reduced, resulting in increased plasma drug levels and frequent nephrotoxicity [113].

Syndrome of Inappropriate ADH Secretion

Treatment of the SIADH is potentially more complicated and may be different in the acute and chronic settings (Table 23-9).

Acute Hyponatremia in the SIADH is due initially to water retention and then in part to Na^+ loss induced by the volume expansion [48]. The simplest therapy is restricting water intake while maintaining that of NaCl. If this is ineffective or severe hyponatremia is present, the combination of hypertonic saline and a loop diuretic can be used to raise the plasma Na^+ concentration [114]. To understand exactly how this regimen works in the SIADH, it is important to review the meaning of the U_{osm} as illustrated by the following case history:

Table 23-9 Treatment of the SIADH

Acute
  Water restriction
  Hypertonic saline or NaCl tablets
  Loop diuretic

Chronic
  Water restriction
  High-salt, high-protein diet
  Loop diuretic
  Urea
  Demeclocycline or lithium

Table 23-10 Effect of giving isotonic saline to a patient with SIADH and a U_{osm} of 720 mosmol/kg

	NaCl, mosmol	Water, mL
In	308	1000
Out	308	428
Net	0	+572

Case history 23-2 A 58-year-old woman with an oat-cell carcinoma of the lung is admitted because of slight obtundation. The following laboratory data are obtained which was consistent with the SIADH:

$$\text{Plasma } [Na^+] = 115 \text{ meq/L}$$

$$P_{osm} = 240 \text{ mosmol/kg}$$

$$U_{osm} = 720 \text{ mosmol/kg}$$

$$\text{Urine volume} = 1000 \text{ mL/day}$$

$$[Na^+] = 62 \text{ meq/L}$$

The U_{osm} in this patient is three times that of the plasma. This is achieved in the kidney by beginning with isosmotic urine and then reabsorbing two-thirds of the water in the collecting tubules (see Chap. 6). Thus, 1 liter of three-times-hyperosmotic urine is equivalent to 3 L of isosmotic urine from which 2 L of free water has been removed. In other words, the kidney is generating 2 L of water per day; this explains why dietary water restriction alone may be ineffective. In this setting, a loop diuretic tends to induce the excretion of an isosmotic urine [21], thereby eliminating most or all of this 2-L water load [115,116].

The U_{osm} also determines what kind of saline solution must be given. It is often stated that saline alone will produce only a transient rise in the plasma Na^+ concentration in the SIADH [47]. The reason for this can be appreciated if we return to the above patient with a U_{osm} of 720 mosmol/kg. If this patient is given 1000 mL of isotonic saline (Na^+ concentration equals 154 meq/L; osmolality equals 308 mosmol/kg), the plasma Na^+ concentration will initially increase because the solution has a higher osmolality than the patient. However, patients with the SIADH are relatively euvolemic and excrete Na^+ normally. Therefore, the patient will excrete the NaCl load but, since the U_{osm} is 720 mosmol/kg, the urine volume will only be 428 mL (308 mosmol in 428 mL equals 720 mosmol/kg) (Table 23-10). The net effect is *water retention and a further reduction in the plasma Na^+ concentration.*

To eliminate this problem, *the osmolality of the fluid given must be greater than that of the urine if the plasma Na^+ concentration is to rise.* Since the U_{osm} usually exceeds 300 mosmol/kg in the SIADH, there is *essentially no role for the use of isotonic saline†* in this disorder. Hypertonic saline (3% or 5%) in a volume calculated

---

†Isotonic saline can, however, be effectively used in hypovolemic patients since both the NaCl and water will be retained.

according to Eq. (23-3) is indicated for severe hyponatremia and, if fluids are to be given, even in mild to moderate hyponatremia. As an alternative, NaCl tablets can be used as this supplies Na^+ without water.

The combined effects of hypertonic saline and furosemide can be illustrated if we make calculations similar to those on page 480. If the woman in Case history 23-2 had a lean body weight of 60 kg, then:

$$\text{Total body water (TBW)} = 0.5 \times \text{body weight} = 30 \text{ L}$$

$$\text{Total body osmoles} = \text{TBW} \times P_{osm} = 30 \times 240 = 7200 \text{ mosmol}$$

Let us assume that on the first day the patient is given 1000 mL of 3% saline (containing 1026 mosmol) and puts out 1000 mL of urine containing 720 mosmol (U_{osm} equals 720 mosmol/kg). At this time, the total body water (TBW) is unchanged but the patient has gained 306 mosmol (1026 minus 720), resulting in a new total body osmoles of 7506. Thus,

$$\text{New } P_{osm} = \text{total body osmoles} \div \text{TBW}$$

$$= 7506/30 = 250 \text{ mosmol/kg}$$

This should increase the plasma Na^+ concentration by 5 meq/L (to 120 meq/L) since both Na^+ *and* Cl^- make up the 10-mosmol/kg elevation in the P_{osm}.

These results illustrate that 3% saline alone may not be very effective in correcting the hyponatremia, in large part because the kidney is continuing to generate a large amount of free water by excreting a three-times-hyperosmotic urine. The net effect is substantially different, however, if furosemide is also given to diminish this free-water generation. Suppose furosemide increases the urine output to 3 L per day while lowering the U_{osm} to an almost isosmotic level of 300 mosmol/kg (total osmolal excretion equals 900 mosmol). In this setting, the TBW will *fall* by 2 L (1 L in, 3 L out) to 28 L while the total body osmoles will *increase* by 126 mosmol (1026 mosmol in, 900 mosmol out) to a total of 7326 mosmol. Consequently,

$$\text{New } P_{osm} = 7326/28 = 263 \text{ mosmol/kg}$$

This 23-mosmol/kg increase in the P_{osm} will raise the plasma Na^+ concentration by almost 12 meq/L to 127 meq/L, a safe level that should not be associated with any symptoms. Thus, furosemide potentiates the effect of hypertonic saline when the urine is very concentrated.

Chronic The SIADH is frequently a transient phenomenon that resolves after discontinuation of the offending drug or recovery from the underlying disease process (such as meningitis or pneumonia). Normalization of the plasma Na^+ concentration does not necessarily indicate recovery, however, since inducing negative water balance will raise the plasma Na^+ concentration independent of ADH levels. Thus, the simplest regimen is to slowly increase water intake once normonatremia is achieved, particularly if it appears that the causative factor has been corrected. If hyponatremia recurs while the U_{osm} remains high, then the SIADH persists and water restriction should be reintroduced.

Occasionally, the SIADH does not resolve, particularly in patients with ectopic hormone production. These patients must be treated with chronic water restriction. If this is ineffective, therapy must be aimed at increasing water excretion either by enhancing solute excretion or antagonizing the effect of ADH (Table 23-9). If, for example, the U_{osm} is relatively fixed at 720 mosmol/kg, then *the urine output will be determined by the rate of solute excretion.* In this setting, the daily urine output will be 1000 mL if 720 mosmol of solute is excreted but 2000 mL if 1440 mosmol of solute is excreted. This increase in solute output can be achieved by putting the patient on a high-salt, high-protein diet (the unused protein will be excreted as urea) or, if available, by giving 30 to 60 g of urea per day [117,118]. As long as renal function is normal, urea has few side effects other than some gastrointestinal discomfort.

As an alternative, the U_{osm} can be lowered by antagonizing the effect of ADH. This can be achieved by the administration of 40 to 80 mg of furosemide per day in divided doses (with NaCl to prevent hypovolemia) [115,116] or of demeclocycline or lithium [119,120]. Furosemide impairs the countercurrent mechanism by decreasing NaCl reabsorption in the medullary loop of Henle, whereas demeclocycline and lithium directly interfere with the effect of ADH on the collecting tubules [121,122]. In general, demeclocycline (in a dose of 300 to 600 mg twice a day) is more effective and better tolerated than lithium [123], although lithium is preferred in children since tetracyclines can interfere with bone development [124].

The choice of which regimen to use is in part dependent upon the U_{osm} [116,118]. Patients with a U_{osm} of 300 to 400 mosmol/kg or less can usually be treated with dietary means alone. Once the U_{osm} exceeds 600 to 700 mosmol/kg, however, there is substantial generation of free water by the kidney, and furosemide (which is probably safer) or demeclocycline may be required to maintain the plasma Na^+ concentration at a safe level.

Other Disorders

Somewhat different considerations may apply to the treatment of other causes of hyponatremia. Patients with *primary adrenal insufficiency* are both cortisol- and aldosterone-deficient. The administration of cortisol will rapidly increase water excretion and return the plasma Na^+ concentration toward normal [73–75]. Mineralocorticoid (such as fludrocortisone) is frequently required to correct the urinary Na^+ wasting and hyperkalemia which are commonly present. Mineralocorticoid alone will not normalize renal water excretion and is not necessary in patients with hypopituitarism in whom aldosterone secretion is relatively normal. Hormone replacement will also correct the hyponatremia in patients with *hypothyroidism* [78–80].

Raising the plasma Na^+ concentration is not very feasible in patients with a *reset osmostat* since Na^+ and water excretion will appropriately vary with changes in intake [83]. Treatment must be aimed at correcting the underlying disorder such as tuberculosis or malnutrition.

The imposition of water restriction will rapidly return the plasma Na^+ concentration to normal in patients with *primary polydipsia* since water-excretory capacity is intact. It may also be helpful to alter the drug regimen when phenothiazines stimulate water intake by causing the sensation of a dry mouth [85].

Finally, no specific therapy to raise the plasma Na^+ concentration is required with *pseudohyponatremia*. If hypertonic saline infusions or water restriction are mistakenly used in a hyponatremic-hyperosmolal patient (due to hyperglycemia or mannitol), there will be a further elevation in the P_{osm} with a potentially deleterious result.

PROBLEMS

23-1 A 45-year-old woman is started on hydrochlorothiazide and a low-sodium diet for the treatment of hypertension. After 1 week she complains of weakness, muscle cramps, and postural dizziness. On physical examination, the patient is found to be alert and oriented. The blood pressure is 130/86 (previous was 150/100). The skin turgor is decreased, and the neck veins are flat. The laboratory data are:

Plasma $[Na^+]$ = 119 meq/L $\qquad$ P_{osm} = 252 mosmol/kg

$[K^+]$ = 2.1 meq/L $\qquad$ Urine $[Na^+]$ = 4 meq/L

$[Cl^-]$ = 71 meq/L $\qquad$ $[K^+]$ = 20 meq/L

$[HCO_3^-]$ = 34 meq/L $\qquad$ U_{osm} = 540 mosmol/kg

Which of the following has contributed to the development of hyponatremia?
  (*a*) Hydrochlorothiazide
  (*b*) Volume depletion
  (*c*) Increased ADH secretion
  (*d*) Water retention
  (*e*) K^+ depletion

The appropriate therapy should include which of the following?
  (*a*) Water restriction alone
  (*b*) Potassium citrate
  (*c*) Potassium chloride
  (*d*) Half-isotonic (0.45%) saline
  (*e*) Isotonic (0.9%) saline

23-2 A 52-year-old man with hypertension treated with unknown medications is admitted to the hospital in a comatose state, responding only to deep pain. On physical examination, the blood pressure is found to be 200/120. The skin turgor is reduced, and the neck veins are flat. After appropriate studies, the diagnosis of an intracerebral hemorrhage is made. To minimize the degree of brain swelling, the patient is given a total of 25 g of mannitol. Only 100 mL of other fluids is given. The laboratory data at this time include:

Plasma $[Na^+]$ = 120 meq/L $\qquad$ BUN = 15 mg/dL

$[K^+]$ = 3.3 meq/L $\qquad$ P_{osm} = 253 mosmol/kg

$[Cl^-]$ = 78 meq/L $\qquad$ Urine $[Na^+]$ = 46 meq/L

$[HCO_3^-]$ = 29 meq/L $\qquad$ U_{osm} = 240 mosmol/kg

$[Glucose]$ = 125 mg/dL

What is the most likely cause of the hyponatremia?
  (*a*) Pseudohyponatremia due to mannitol
  (*b*) Volume depletion
  (*c*) SIADH

23-3 Match the correct therapy with the appropriate clinical setting:

1. A 41-year-old man with end-stage renal failure, mild peripheral edema, and a plasma Na^+ concentration of 125 meq/L

   (a) Water restriction with normal sodium intake

2. A 53-year-old woman with an oat-cell carcinoma of the lung, a plasma Na^+ concentration of 107 meq/L, and a urine osmolality of 640 mosmol/kg

   (b) Water and sodium restriction

3. A 27-year-old woman with chronic diarrhea, decreased skin turgor, and a plasma Na^+ concentration of 126 meq/L

   (c) No therapy required

4. A 38-year-old man with multiple myeloma, a plasma Na^+ concentration of 127 meq/L, and a plasma osmolality of 286 mosmol/kg

   (d) Isotonic saline

5. A 58-year-old diabetic man with congestive heart failure, a plasma Na^+ concentration of 126 meq/L, and a plasma osmolality of 268 mosmol/kg

   (e) Hypertonic saline

6. A 49-year-old woman with carcinoma of the lung, a stable plasma Na^+ concentration of 118 meq/L, and a urine osmolality of 290 mosmol/kg

   (f) Hypertonic saline plus furosemide

23-4 A 60-year-old man weighing 70 kg has an oat-cell carcinoma of the lung and is admitted to the hospital with a 2-week history of progressive lethargy and obtundation. The physical examination is within normal limits except for the obtundation. The following laboratory studies are obtained:

$$\text{Plasma } [Na^+] = 105 \text{ meq/L} \qquad P_{osm} = 222 \text{ mosmol/kg}$$

$$[K^+] = 4 \text{ meq/L} \qquad \text{Urine } [Na^+] = 78 \text{ meq/L}$$

$$[Cl^-] = 72 \text{ meq/L} \qquad U_{osm} = 604 \text{ mosmol/kg}$$

$$[HCO_3^-] = 21 \text{ meq/L}$$

What is the most likely diagnosis? How would you raise the plasma Na^+ concentration? If NaCl is to be given, what solution should be used and at what approximate hourly rate to get the patient out of danger?

23-5 Uric acid reabsorption in the proximal tubule is related to Na^+ reabsorption (see Chap. 5). Considering the effects of volume on Na^+ reabsorption, how might the plasma uric acid concentration be used to help differentiate between hyponatremia due to the SIADH or volume depletion?

REFERENCES

1. Teree, T., E. Mirabal-Font, A. Ortiz, and W. Wallace: Stool losses and acidosis in diarrheal disease of infancy, Pediatrics, 36:704, 1965.
2. Szatalowicz, V. L., P. E. Arnold, C. Chaimovitz, D. Bichet, T. Berl, and R. W. Schrier: Radioimmunoassay of plasma arginine vasopressin in hyponatremic patients with congestive heart failure, N. Engl. J. Med., 305:263, 1981.
3. Bichet, D., V. Szatalowicz, C. Chaimovitz, and R. W. Schrier: Role of vasopressin in abnormal water excretion in cirrhotic patients, Ann. Intern. Med., 96:413, 1982.
4. Anderson, R. J., P. Cadnapaphornchai, J. A. Harbottle, K. M. McDonald, and R. W. Schrier: Mechanism of effect of thoracic inferior vena cava constriction on renal water excretion, J. Clin. Invest., 54:1473, 1974.
5. Schrier, R. W., and D. G. Bichet: Osmotic and nonosmotic control of vasopressin release and the

pathogenesis of impaired water excretion in adrenal, thyroid, and edematous disorders, J. Lab. Clin. Med., 98:1, 1981.

6. Grausz, H., R. Lieberman, and L. E. Earley: Effect of plasma albumin on sodium reabsorption in patients with nephrotic syndrome, Kidney Int., 1:47, 1972.

7. Schedl, H. P., and F. C. Bartter: An explanation for and experimental correction of the abnormal water diuresis in cirrhosis, J. Clin. Invest., 39:248, 1960.

8. Bell, N. H., H. P. Schedl, and F. C. Bartter: An explanation for abnormal water retention and hypoosmolality in congestive heart failure, Am. J. Med., 36:351, 1964.

9. Takasu, T., N. Lasker, and R. J. Shalhoub: Mechanisms of hyponatremia in chronic congestive heart failure, Ann. Intern. Med., 55:368, 1961.

10. Baldus, W. P., R. M. Feichter, W. H. Summerskill, J. C. Hunt, and K. G. Wakim: The kidney in cirrhosis. II. Disorders of renal function, Ann. Intern. Med., 60:366, 1964.

11. Robertson, G. L., P. Aycinena, and R. L. Zerbe: Neurogenic disorders of osmoregulation, Am. J. Med., 72:339, 1982.

12. Laragh, J. H.: The effect of potassium chloride on hyponatremia, J. Clin. Invest., 33:807, 1954.

13. Fichman, M. P., H. Vorherr, C. R. Kleeman, and N. Telfer: Diuretic-induced hyponatremia, Ann. Intern. Med., 75:853, 1971.

14. Kleeman, C. R., D. A. Adams, and M. H. Maxwell: An evaluation of maximal water diuresis in chronic renal disease. I. Normal solute intake, J. Lab. Clin. Med., 58:169, 1961.

15. Kennedy, R. M., and L. Earley: Profound hyponatremia resulting from a thiazide-induced decrease in urinary diluting capacity in a patient with primary polydipsia, N. Engl. J. Med., 282:1185, 1970.

16. Ashraf, N., R. Locksley, and A. I. Arieff: Thiazide-induced hyponatremia associated with death or neurologic damage in outpatients, Am. J. Med., 70:1163, 1981.

17. Hamburger, S., B. Koprivica, E. Ellerbeck, and J. O. Covinsky: Thiazide-induced syndrome of inappropriate secretion of antidiuretic hormone: time course of resolution, J. Am. Med. Assoc., 246:1235, 1981.

18. Maronde, R. F., M. Milgrom, N. D. Vlachakis, and L. Chan: Response of thiazide-induced hypokalemia to amiloride, J. Am. Med. Assoc., 249:237, 1983.

19. Martinez-Maldonado, M., G. Eknoyan, and W. N. Suki: Diuretics in nonedematous states: physiological basis for the clinical use, Arch. Intern. Med., 131:797, 1973.

20. Wilkinson, P. R., H. Issler, R. Hesp, and E. B. Raftery: Total body and serum potassium during prolonged thiazide therapy for essential hypertension, Lancet, 1:759, 1975.

21. Szatalowicz, V. L., P. D. Miller, J. W. Lacher, J. A. Gordon, and R. W. Schrier: Comparative effect of diuretics on renal water excretion in hyponatremic oedematous disorders, Clin. Sci., 62:235, 1982.

22. Smith, N. J., and M. L. E. Espir: Raised plasma arginine vasopressin concentration in carbamazepine-induced water intoxication, Br. Med. J., 2:804, 1977.

23. Flegel, K. M., and C. H. Cole: Inappropriate antidiuresis during carbamazepine treatment, Ann. Intern. Med., 87:722, 1977.

24. DeFronzo, R. A., H. Braine, O. M. Calvin, and P. J. Davis: Water intoxication in man after cyclophosphamide therapy: time course and relation to drug activation, Ann. Intern. Med., 78:861, 1973.

25. Harlow, P. J., Y. A. Declerck, N. A. Shore, J. A. Ortega, A. Carranza, and E. Heuser: A fatal case of inappropriate ADH secretion induced by cyclophosphamide therapy, Cancer, 44:896, 1979.

26. Robertson, G. L., N. Bhoopalam, and L. J. Zelkowitz: Vincristine neurotoxicity and abnormal secretion of antidiuretic hormone, Arch. Intern. Med., 132:717, 1973.

27. Ajlouni, K., M. W. Kern, J. F. Teres, G. B. Theil, and T. C. Hagen: Thiothixene-induced hyponatremia, Arch. Intern. Med., 134:1103, 1974.

28. Vincent, F. M., and S. Emery: Antidiuretic hormone syndrome and thioridazine, Ann. Intern. Med., 89:147, 1978.

29. Peck, V., and L. Shenkman: Haloperidol-induced syndrome of inappropriate secretion of antidiuretic hormone, Clin. Pharmacol. Therap., 26:442, 1979.

30. Luzecky, M. H., K. D. Burman, and E. R. Schultz: The syndrome of inappropriate antidiuretic hormone secretion associated with amitriptyline administration, South. Med. J., 67:495, 1974.

31. Peterson, J. C., R. W. Pollack, J. J. Mahoney, and T. J. Fuller: Inappropriate antidiuretic hormone secondary to a monoamine oxidase inhibitor, J. Am. Med. Assoc., 239:1422, 1978.

31a. Marshall, A. W., A. W. Jakobovitz, and M. Y. Morgan: Bromocriptine-associated hyponatremia in cirrhosis, Br. Med. J., 285:1534, 1982.

32. Mor, J., E. Ben-Galin, and A. Abrakomov: Inappropriate antidiuretic hormone secretion in an infant with severe pneumonia, Am. J. Dis. Child., 129:133, 1975.

33. Thomas, T. H., D. B. Morgan, R. Swaminathan, S. G. Ball, and M. R. Lee: Severe hyponatraemia: a study of 17 patients, Lancet, 1:621, 1978.

34. Shalhoub, R. J., and L. D. Antoniou: The mechanism of hyponatremia in pulmonary tuberculosis, Ann. Intern. Med., 70:943, 1969.

35. Szatalowicz, V. L., J. P. Goldberg, and R. J. Anderson: Plasma antidiuretic hormone in acute respiratory failure, Am. J. Med., 72:583, 1982.

36. Ukai, M., W. J. Moran, and B. Zimmerman: The role of visceral afferent pathways on vasopressin secretion and urinary excretory patterns during surgical stress, Ann. Surg., 168:16, 1968.

37. Bruce, R. A., K. A. Merendina, M. Dunning, B. H. Scribner, D. Donohue, E. Carlson, and J. Cummings: Observations on hyponatremia following mitral valve surgery, Surg. Gynecol. Obstet., 100:293, 1955.

38. Martinez-Maldonado, M.: Inappropriate antidiuretic hormone secretion of unknown origin, Kidney Int., 17:554, 1980.

39. Hamilton, B. P. B., G. V. Upton, and T. T. Amatruda: Evidence for the presence of neurophysin in tumors producing the syndrome of inappropriate antidiuresis, J. Clin. Endocrinol. Metab., 35:764, 1972.

40. George, J. M., C. C. Capen, and A. S. Phillips: Biosynthesis of vasopressin in vitro and ultrastructure of a bronchogenic carcinoma, J. Clin. Invest., 51:141, 1972.

41. Vorkerr, H., S. G. Massry, R. Fallet, L. Kaplan, and C. R. Kleeman: Antidiuretic principle in tuberculosis and hyponatremia, Ann. Intern. Med., 72:383, 1970.

42. Weissman, P., L. Shenkman, and R. I. Gregerman: Chlorpropamide hyponatremia: drug induced inappropriate antidiuretic-hormone activity, N. Engl. J. Med., 284:65, 1971.

43. Hagen, G. A., and T. F. Frawley: Hyponatremia due to sulfonylurea compounds, J. Clin. Endocrinol. Metab., 31:570, 1970.

43a. Meinder, A. E., V. Cejka, and G. L. Robertson: The antidiuretic effect of carbamazepine in man, Clin. Sci. Mol. Med., 47:289, 1974.

44. Pittman, J. G.: Water intoxication due to oxytocin, N. Engl. J. Med., 268:481, 1963.

45. Schwartz, R. H., and R. W. A. Jones: Transplacental hyponatraemia due to oxytocin, Br. Med. J., 1:152, 1978.

46. Feeney, J. G.: Water intoxication and oxytocin, Br. Med. J., 285:243, 1982.

47. Bartter, F. C., and W. B. Schwartz: The syndrome of inappropriate secretion of antidiuretic hormone, Am. J. Med., 42:790, 1967.

48. Cooke, C. R., M. D. Turin, and W. G. Walker: The syndrome of inappropriate antidiuretic hormone secretion: pathophysiologic mechanisms in solute and volume regulation, Medicine, 58:240, 1979.

49. Nolph, K. D., and R. W. Schrier: Sodium, potassium and water metabolism in the syndrome of inappropirate antidiuretic hormone secretion, Am. J. Med., 49:533, 1970.

50. Gross, P. A., and R. J. Anderson: Effects of DDAVP and AVP on sodium and water balance in conscious rat, Am. J. Physiol., 243:R512, 1982.

51. Leaf, A., F. C. Bartter, R. F. Santos, and O. Wrong: Evidence in man that urinary electrolyte loss induced by Pitressin is a function of water retention, J. Clin. Invest., 32:868, 1953.

52. Jaenike, J. R., and C. Waterhouse: The renal response to sustained administration of vasopressin and water in man, J. Clin. Endocrinol. Metab., 21:231, 1961.

53. Frindt, G., E. E. Windhager, and A. Taylor: Hydroosmotic resonse of collecting tubules to ADH or cAMP at reduced peritubular sodium, Am. J. Physiol., 243:F503, 1982.

54. Humes, H. D., C. F. Simmons, Jr., and B. M. Brenner: Effect of verapamil on the hydroosmotic response to antidiuretic hormone in toad urinary bladder, Am. J. Physiol., 239:F250, 1980.

55. Moses, A. M., R. Fenner, E. T. Schroeder, and R. Coulson: Further studies on the mechanism by which chlorpropamide alters the action of vasopressin, Endocrinology, 111:2025, 1982.

56. Garella, S., A. H. Tzamaloukas, and J. A. Chazan: Effect of isotonic volume expansion on extracellular bicarbonate stores in normal dogs, Am. J. Physiol., 225:628, 1973.

57. Lowance, D. C., H. B. Garfinkel, W. D. Mattern, and W. B. Schwartz: The effect of chronic hypotonic volume expansion on the renal regulation of acid-base equilibrium, J. Clin. Invest., 51:2928, 1972.

58. Cohen, J. J., H. N. Hulter, N. Smithline, J. C. Melby, and W. B. Schwartz: The critical role of the adrenal gland in the renal regulation of acid-base equilibrium during chronic hypotonic expansion: evidence that chronic hyponatremia is a potent stimulus to aldosterone secretion, J. Clin. Invest., 58:1201, 1976.

58a. Dubovsky, S. L., S. Grabon, T. Berl, and R. W. Schrier: Syndrome of inappropriate secretion of antidiuretic hormone with exacerbated psychosis, Ann. Intern. Med., 79:551, 1973.

59. Moses, A. M., and M. Miller: Drug-induced dilutional hyponatremia, N. Engl. J. Med., 291:1234, 1974.

60. Berndt, W. O., M. Miller, W. M. Kettyle, and H. Valtin: Potentiation of the antidiuretic effect of vasopressin by chlorpropamide, Endocrinology, 86:1028, 1970.

61. Moses, A. M., and R. Coulsen: Augmentation of chlorpropamide of 1-deamino-8-D-arginine vasopressin-induced antidiuresis and stimulation of renal medullary adenylate cyclase and accumulation of adenosine 3′, 5′-monophosphate, Endocrinology, 106:967, 1980.

62. Miller, M., and A. M. Moses: Mechanism of chlorpropamide action in diabetes insipidus, J. Clin. Endocrinol. Metab., 30:488, 1970.

63. Moses, A. M., J. Howanitz, and M. Miller: Diuretic action of three sulfonylurea drugs, Ann. Intern. Med., 78:541, 1973.

64. Ishikawa, S-E., T. Saito, and S. Yoshida: The effect of prostaglandins on the release of arginine vasopressin from the guinea pig hypothalamo-neurohypophyseal complex in organ culture, Endocrinology, 108:193, 1981.

65. Ciabattoni, G., F. Pugliese, G. A. Cinotti, and C. Patrono: Renal effects of anti-inflammatory drugs, Eur. J. Rheumatol. Inflam., 3:210, 1980.

66. Bunning, R. D., and W. F. Barth: Sulindac: a potentially renal-sparing nonsteroidal anti-inflammatory drug, J. Am. Med. Assoc., 248:2864, 1982.

67. Baker, J. W., S. Yerger, and W. E. Segar: Elevated plasma antidiuretic hormone levels in status asthmaticus, Mayo Clin. Proc., 51:31, 1976.

68. Paxson, C. L., J. W. Stoerner, S. E. Denson, E. W. Adcock, and F. H. Morriss, Jr.: Syndrome of inappropriate antidiuretic hormone secretion in neonates with pneumothorax or atelectasis, J. Pediatr., 91:459, 1977.

69. Benson, H., M. Akbarian, L. N. Adler, and W. H. Abelmann: Hemodynamic effects of pneumonia. I. Normal and hypodynamic responses, J. Clin. Invest., 49:791, 1970.

70. Farber, M. O., S. S. O. Kiblawi, R. A. Strawbridge, G. L. Robertson, M. H. Weinberger, and F. Manfredi: Studies on plasma vasopressin and the renin-angiotensin-aldosterone system in chronic obstructive lung disease, J. Lab. Clin. Med., 90:373, 1977.

71. Miller, A. C.: Hyponatraemia in legionnaires' disease, Br. Med. J., 284:558, 1982.

72. Gill, J. R., Jr., D. S. Gann, and F. C. Bartter: Restoration of water diuresis in Addisonian patients by expansion of the volume of extracellular fluid, J. Clin. Invest., 41:1078, 1962.

73. Ahmed, A. B., B. C. George, C. Conyalez-Auvert, and J. F. Dingman: Increased plasma arginine vasopressin in clinical adrenocortical insufficiency and its inhibition by glucosteroids, J. Clin. Invest., 46:111, 1967.

74. Green, H. H., A. R. Harrington, and H. Valtin: On the role of antidiuretic hormone in the inhibition of acute water diuresis in adrenal insufficiency and the effects of gluco- and mineralocorticoids in reversing the inhibition, J. Clin. Invest., 49:1724, 1970.

75. Bethune, J. E., and D. H. Nelson: Hyponatremia in hypopituitarism, N. Engl. J. Med., 272:771, 1965.

76. Linas, S. L., T. Berl, G. L. Robertson, G. A. Aisenbrey, R. W. Schrier, and R. J. Anderson: Role of vasopressin in impaired water excretion of glucocorticoid deficiency, Kidney Int., 18:58, 1980.

77. Ishikawa, S-E., and R. W. Schrier: Effect of arginine vasopressin antagonist on renal water excretion in glucocorticoid and mineralocorticoid deficient rats, Kidney Int., 22:587, 1982.

78. DeRubertis, F. R., Jr., M. F. Michelis, M. E. Bloom, D. H. Mintz, J. B. Field, and B. B. Davis: Impaired water excretion in myxedema, Am. J. Med., 51:41, 1971.

79. Skowsky, W. R., and T. A. Kikuchi: The role of vasopressin in the impaired water excretion of myxedema, Am. J. Med., 64:613, 1978.

80. Macaron, C., and O. Famuyiawa: Hyponatremia of hypothyroidism: appropriate suppression of antidiuretic hormone levels, Arch. Intern. Med., 138:820, 1978.

81. Durr, J. A., B. Stamoutsos, and M. D. Lindheimer: Osmoregulation during pregnancy in the rat: evidence for resetting of the threshold for vasopressin secretion during gestation, J. Clin. Invest., 68:337, 1981.

82. Hariprasad, M. K., R. P. Eisinger, I. M. Nadler, C. S. Padmanabhan, and B. D. Nidus: Hyponatremia in psychogenic polydipsia, Arch. Intern. Med., 140:1639, 1980.

83. DeFronzo, R. A., M. Goldberg, and Z. S. Agus: Normal diluting capacity in hyponatremic patients: reset osmostat or a variant of the syndrome of inappropriate antidiuretic hormone secretion, Ann. Intern. Med., 84:538, 1976.

84. Barlow, E. D., and H. E. deWardener: Compulsive water drinking, Q. J. Med., 28:235, 1959.

85. Rao, K. J., M. Miller, and A. Moses: Water intoxication and thioridazine, Ann. Intern. Med., 82:61, 1975.

85a. Stuart, C. A., F. A. Neelon, and H. E. Lebovitz: Disordered control of thirst in hypothalamic-pituitary sarcoidosis, N. Engl. J. Med., 303:1078, 1980.

86. Miller, M., T. Dalakos, A. M. Moses, H. Fellerman, and D. Streeten: Recognition of partial defects in antidiuretic hormone secretion, Ann. Intern. Med., 73:721, 1970.

87. Langgard, H., and W. O. Smith: Self-induced water intoxication without predisposing illness, N. Engl. J. Med., 266:378, 1962.

88. Rendell, M., D. McGrane, and M. Cuesta: Fatal compulsive water drinking, J. Am. Med. Assoc., 240:2557, 1978.

89. Hilden, T., and T. L. Svendsen: Electrolyte disturbances in beer drinkers, Lancet, 2:245, 1975.

90. Albrink, M. J., P. M. Hald, E. B. Man, and J. P. Peters: The displacement of serum water by the lipids of hyperlipemic serum: a new method for the rapid determination of serum water, J. Clin. Invest., 34:1481, 1955.

91. Tarail, R., K. W. Buchwald, J. F. Holland, and O. S. Selawry: Misleading reductions of serum sodium and chloride: association with hyperproteinemia in patients with multiple myeloma, Proc. Soc. Exp. Biol. Med., 110:145, 1962.

92. Hagstrom, R.: Studies on fluid absorption during transurethral prostatic resection, J. Urol., 73:852, 1955.

93. Osborn, D. E., P. N. Rao, M. J. Greene, and R. J. Barnard: Fluid absorption during transurethral resection, Br. Med. J., 2:1549, 1980.

94. Katz, M.: Hyperglycemia-induced hyponatremia: calculation of expected serum sodium depression, N. Engl. J. Med., 289:843, 1973.

95. Aviram, A., A. Pfau, J. W. Czaczkes, and T. D. Ullman: Hyperosmolality with hyponatremia caused by inappropriate administration of mannitol, Am. J. Med., 42:648, 1967.

96. Glasser, L., P. D. Sternglanz, J. Combie, and A. Robinson: Serum osmolality and its applicability to drug overdose, Am. J. Clin. Pathol., 60:695, 1973.

97. Robinson, A. G., and J. N. Loeb: Ethanol ingestion: commonest cause of elevated plasma osmolality?, N. Engl. J. Med., 284:1253, 1971.

98. Arieff, A. I., F. Llach, and S. G. Massry: Neurological manifestations and morbidity of hyponatremia: correlation with brain water and electrolytes, Medicine, 55:121, 1976.

99. Arieff, A. I., and R. Guisado: Effects on the central nervous system of hypernatremic and hyponatremic states, Kidney Int., 10:104, 1976.

100. Pollock, A. S., and A. I. Arieff: Abnormalities of cell volume regulation and their functional consequences, Am. J. Physiol., 239:F195, 1980.

101. Thurston, J. H., R. E. Hauhart, and J. A. Dirgo: Taurine: a role in osmotic regulation of mammalian brain and possible clinical significance, Life Sci., 26:1561, 1980.

102. Gilbert, G. J.: Neurologic manifestations of hyponatremia, N. Engl. J. Med., 274:1153, 1966.

103. Lipsmeyer, E., and G. L. Ackerman: Irreversible brain damage after water intoxication, J. Am. Med. Assoc., 196:286, 1966.

104. Burcar, P. J., M. D. Norenberg, and P. R. Yarnell: Hyponatremia and central pontine myelinolysis, Neurology, 27:223, 1977.

105. Schneck, S. A., J. S. Burks, and P. R. Yarnell: Antemortem diagnosis of central pontine myelinolysis, Neurology, 28:389, 1978.

106. Kleinschmidt-DeMasters, B. K., and M. D. Norenberg: Rapid correction of hyponatremia causes demyelination: relation to central pontine myelinolysis, Science, 211:1068, 1981.

106a. Laureno, R.: Central pontine myelinolysis following rapid correction of hyponatremia, Ann. Neurol., 13:232, 1983.

107. Laureno, R., and A. I. Arieff: Rapid correction of hyponatremia: cause of pontine myelinolysis?, Am. J. Med., 71:846, 1981.

108. Dubois, G. D., W. Leach, and A. I. Arieff: Hyponatremia untreated can cause central pontine myelinolysis (abstract), Kidney Int., 23:121, 1983.

109. Ayus, J. C., J. J. Olivero, and J. P. Frommer: Rapid correction of severe hyponatremia with intravenous hypertonic saline solution, Am. J. Med., 72:43, 1982.

110. Womersley, R. A., and J. H. Darragh: Potassium and sodium restriction in the normal human, J. Clin. Invest., 34:456, 1955

111. Cox, M. M., J. Guzzo, G. Morrison, and I. Singer: Demeclocycline and therapy of hyponatremia, Ann. Intern. Med., 86:113, 1977.

112. DeTroyer, A., W. Pilloy, I. Broeckaert, and J. C. Demanet: Demeclocycline treatment of water retention in cirrhosis, Ann. Intern. Med., 85:336, 1976.

113. Miller, P. D., S. L. Linas, and R. W. Schrier: Plasma demeclocycline levels and nephrotoxicity: correlation in hyponatremic cirrhotic patients, J. Am. Med. Assoc., 243:2513, 1980.

114. Hantman, D., B. Rossier, R. Zohlman, and R. W. Schrier: Rapid correction of hyponatremia in the syndrome of inappropriate secretion of antidiuretic hormone: an alternative to hypertonic saline, Ann. Intern. Med., 78:870, 1973.

115. Decaux, G., Y. Waterlot, F. Genette, and J. Mockel: Treatment of the syndrome of inappropriate secretion of antidiuretic hormone with furosemide, N. Engl. J. Med., 304:329, 1981.

116. Decaux, G., Y. Waterlot, F. Genette, R. Hallemans, and J. C. Demanet: Inappropriate secretion of antidiuretic hormone treated with frusemide, Br. Med. J., 285:89, 1982.

117. Decaux, G., S. Brimioulle, F. Genette, and J. Mockel: Treatment of the syndrome of inappropriate secretion of antidiuretic hormone by urea, Am. J. Med., 69:99, 1980.

118. Decaux, G., and F. Genette: Urea for long-term treatment of syndrome of inappropriate secretion of antidiuretic hormone, Br. Med. J., 2:1081, 1981.

119. DeTroyer, A., and J. C. Demanet: Correction of antidiuresis by demeclocycline, N. Engl. J. Med., 293:915, 1975.

120. White, M., and C. D. Fetner: Treatment of the syndrome of inappropriate secretion of antidiuretic hormone with lithium carbonate, N. Engl. J. Med., 292:390, 1975.

121. Singer, I., and D. Rotenberg: Demeclocycline-induced nephrogenic diabetes insipidus, Ann. Intern. Med., 79:679, 1973.

122. Cox, M., and I. Singer: Lithium and water metabolism, Am. J. Med., 59:153, 1975.

123. Forrest, J. N., Jr., M. Cox, C. Hong, G. Morrison, M. Bia, and I. Singer: Superiority of demeclocycline over lithium in the treatment of chronic syndrome of inappropriate secretion of antidiuretic hormone, N. Engl. J. Med., 298:173, 1978.

124. Baker, R. S., R. M. Hurley, and W. Feldman: Treatment of recurrent syndrome of inappropriate secretion of antidiuretic hormone with lithium, J. Pediatr., 90:480, 1977.

TWENTY-FOUR

HYPEROSMOLAL STATES—HYPERNATREMIA

The introduction to disorders of water balance presented in Chap. 22 should be read before proceeding with this discussion.

PATHOPHYSIOLOGY

Hyperosmolality

Hypernatremia represents hyperosmolality. Since Na^+ is an effective osmole, the increase in the plasma osmolality (P_{osm}) induced by hypernatremia creates an osmotic gradient that results in water movement out of the cells into the extracellular fluid. It is this cellular dehydration in the brain that is primarily responsible for the neurologic symptoms associated with hypernatremia (see below).

A similar syndrome can be produced when the P_{osm} is elevated by hyperglycemia. However, when hyperosmolality is due to the accumulation of a cell-permeable solute, such as urea (as in renal failure) or ethanol [1], there is no water shift in the steady state because osmotic equilibrium is reached by solute entry into cells; i.e., urea and ethanol are ineffective osmoles. These conditions do not produce the symptoms of hyperosmolality and will not be considered further here.

Since it is the effective P_{osm} that is clinically important, *the contribution of urea to the P_{osm} should be excluded.* In general, the effective P_{osm} can be calculated from (see page 30):

$$\text{Effective } P_{osm} \cong \text{measured } P_{osm} - \frac{\text{BUN}}{2.8} \qquad (24\text{-}1)$$

or estimated from

$$\text{Effective } P_{osm} \cong 2 \times \text{plasma } [Na^+] + \frac{[\text{glucose}]}{18} \qquad (24\text{-}2)$$

The normal value for the effective P_{osm} is 270 to 285 mosmol/kg.

Generation of Hypernatremia

From the relationship between the plasma Na^+ concentration and the osmolality of the body fluids (see Fig. 22-2),

$$\text{Plasma } [Na^+] \cong \frac{Na_e^+ + K_e^+}{\text{TBW}} \qquad (24\text{-}3)$$

it can be seen that *hypernatremia can result from water loss or Na^+ retention* (Table 24-1). The serious toxicity of hyperkalemia prevents the retention of enough K^+ to significantly raise the plasma Na^+ concentration.

To cause hypernatremia, water loss must occur in excess of solute. This can be produced by insensible losses from the skin and respiratory tract and by the excretion of a hypoosmotic urine. The latter requires either decreased secretion of antidiuretic hormone [central diabetes insipidus (CDI)] or end-organ resistance to its effect [nephrogenic diabetes insipidus (NDI)]. Gastrointestinal water loss, whether as vomiting or diarrhea, usually results in isosmotic dehydration and does not directly induce hypernatremia.

Table 24-1 Etiology of hypernatremia

A. Water loss
  1. Insensible loss
    a. Increased sweating: fever, exposure to high temperatures
    b. Burns
    c. Respiratory infections
  2. Renal loss
    a. Central diabetes insipidus
    b. Nephrogenic diabetes insipidus
    c. Osmotic diuresis
  3. Gastrointestinal loss
    a. Lactulose
  4. Hypothalamic disorders
    a. Hypodipsia (diminished thirst)
    b. Essential hypernatremia
    c. Resetting of the osmostat due to primary mineralocorticoid excess
  5. Water loss into the cells
    a. Seizures or severe exercise

B. Sodium retention
  1. Administration of hypertonic NaCl or NaHCO₃
  2. Ingestion of sodium

Thirst and the Maintenance of Hypernatremia

The normal defense against the development of hypernatremia is the stimulation of both antidiuretic hormone (ADH) release and thirst by the hypothalamic osmoreceptors. The combination of decreased water excretion and increased water intake results in water retention and return of the plasma Na^+ concentration to normal. The secretion of ADH generally begins when the P_{osm} exceeds 275 to 285 mosmol/kg (see Fig. 23-1), whereas the threshold for thirst is approximately 5 to 10 mosmol/kg higher [2]. Osmoregulation is normally so effective that the P_{osm} is maintained within a range of 1 to 2 percent (usually between 280 and 290 mosmol/kg) despite wide variations in sodium and water intake.

Although ADH release occurs earlier, *it is thirst that provides the ultimate protection against hypernatremia.* In patients with CDI who secrete no ADH, for example, renal water reabsorption falls and the urine output can exceed 10 to 15 L/day. Nevertheless, near normal water balance is maintained because water intake is augmented to match output. Conversely, even with maximum ADH secretion, the kidney may be unable to retain enough water to offset insensible losses from the skin and respiratory tract in a patient with hypodipsia (diminished thirst). Thus, *hypernatremia due to water loss occurs only in patients who have hypodipsia or, much more commonly, in infants and comatose patients who have an intact thirst mechanism but are unable to ask for water.* A plasma Na^+ concentration greater than 150 meq/L is virtually never seen in an alert patient with a normal thirst mechanism and access to water.

ETIOLOGY

Insensible Water Loss

Insensible fluid losses from the skin and respiratory tract are hypoosmotic to plasma and average 800 to 1000 mL/day in adults. Any condition which increases these losses, such as fever, respiratory infections, burns, or exposure to high temperatures, predisposes toward the development of hypernatremia. This complication occurs primarily in infants [3,4] and obtunded or comatose patients who are unable to increase their water intake to compensate for the enhanced water loss. In addition, hyperglycemia may be present in infants with enteric infections, further elevating the effective P_{osm} [4]. This increase in the plasma glucose concentration (to as high as 300 to 500 mg/dL) appears to be a stress response, perhaps mediated by catecholamines, and is corrected with rehydration. In recent years, the incidence of hypernatremic dehydration following gastroenteritis has fallen in infants, primarily because of the use of low-solute feedings, which supply more free water to replace the insensible losses [5].

An unusual cause of insensible *gastrointestinal* water loss is the administration of lactulose to treat hepatic encephalopathy [6]. This drug is given in a hyperosmotic solution, resulting in the osmotic flow of water into the gastrointestinal tract. When these losses are large, as manifested by watery diarrhea, hypernatremia can ensue.

Diabetes Insipidus

Diabetes insipidus is characterized by the complete or partial failure of ADH secretion (CDI) or of renal response to ADH (NDI). As a result, renal water reabsorption falls, and a diuresis of dilute urine ensues (3 to 20 L/day). It must be emphasized again that the majority of these patients maintain water balance with a near normal plasma Na^+ concentration because of the efficacy of the thirst mechanism [7,8]. Their major complaints are polyuria and polydipsia, not the symptoms of hypernatremia.

The outcome is different, however, if the hypothalamic disorder producing CDI also interferes with thirst. In this setting, even a partial defect in ADH release can lead to water loss and hypernatremia [2]. For example, a patient with partial CDI may be able to maximally concentrate his or her urine (U_{max}) to 400 mosmol/kg. This is hyperosmotic to plasma but less than the normal U_{max} of 800 to 1400 mosmol/kg. If this patient excreted 800 mosmol of solute per day (primarily Na^+ and K^+ salts and urea), the solute load would be excreted in the urine in a minimum of 2000 mL of water (800 mosmol of solute in 2000 mL of water equals 400 mosmol/kg). In contrast, the obligatory renal water loss would be only 800 mL if the U_{max} were normal at 1000 mosmol/kg (800 mosmol in 800 mL of water equals 1000 mosmol/kg). Thus, the reduction in renal concentrating ability can result in an extra 1200 mL of water lost in the urine per day. In a hypodipsic patient, this added loss might not be replaced, leading to the development of hypernatremia.

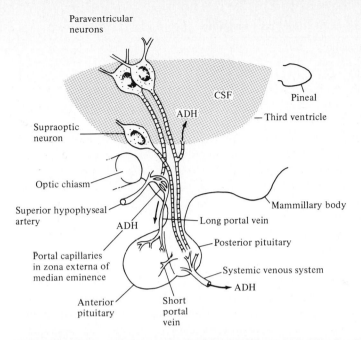

Figure 24-1 Diagram of the mammalian hypothalamus and pituitary gland depicting pathways for the secretion of ADH. The hormone is formed in the supraoptic and paraventricular nuclei, transported in granules along their axons, and then secreted at three sites: the posterior pituitary gland; the portal capillaries of the median eminence; and the cerebrospinal fluid (CSF) of the third ventricle. *(Adapted from E. A. Zimmerman and A. G. Robinson, Kidney Int., 10:12, 1976. Reprinted by permission from Kidney International.)*

Central diabetes insipidus ADH is synthesized in the supraoptic and paraventricular nuclei in the hypothalamus. It then streams down the axons of the supraopticohypophyseal tract and is stored in and subsequently released from the posterior lobe of the pituitary (neurohypophysis) (Fig. 24-1). Impaired secretion of ADH can be induced by a variety of clinical disorders which disrupt the osmoreceptors, the hypothalamic nuclei, or the supraopticohypophyseal tract (Table 24-2) [9,10]. In contrast, damage to the tract below the median eminence or removal of the posterior pituitary usually produces only a transient period of diabetes insipidus [10]. In these settings, ADH can still be secreted into the systemic circulation via the portal capillaries in the median eminence (Fig. 24-1).

Among the causes of CDI (Table 24-2), the most common are probably idiopathic, head trauma, and hypoxic encephalopathy [10]. Idiopathic CDI may be familial (with autosomal dominant or X-linked inheritance) and in at least some patients appears to result from degeneration of cells in the hypothalamic nuclei [11,12]. A possible role for immune mechanisms in this disorder is suggested by the finding of circulating antibodies directed against the ADH-secreting cells in about one-third of patients [13].

Table 24-2 Etiology of central diabetes insipidus

A. Idiopathic (may be familial)
B. Trauma
C. Hypoxic encephalopathy
D. Posthypophysectomy
E. Neoplastic
   1. Primary: craniopharyngioma, pinealoma, cyst
   2. Metastatic: breast, lung
F. Miscellaneous
   1. Histiocytosis X
   2. Sarcoidosis
   3. Aneurysm
   4. Encephalitis or meningitis

     Surgical hypophysectomy or trauma, in which the hypothalamus or tract is damaged, can produce a typical triphasic response (Fig. 24-2). Initially, there is a polyuric state which lasts 4 to 5 days and probably represents inhibition of ADH release due to hypothalamic dysfunction. From days 6 to 11, there is an antidiuretic phase which represents slow release of stored hormone from the degenerating posterior pituitary. During this time, excessive water intake can produce hyponatremia in a manner similar to the syndrome of inappropriate ADH secretion (see Chap. 23). This stage is followed by permanent CDI once the neurohypophyseal stores are depleted.

Nephrogenic diabetes insipidus Nephrogenic diabetes insipidus (NDI) is a congenital or acquired disorder in which hypothalamic function and ADH release are normal but the ability to concentrate the urine is reduced because of diminished or absent renal responsiveness to ADH [14,15]. Concentration of the urine normally

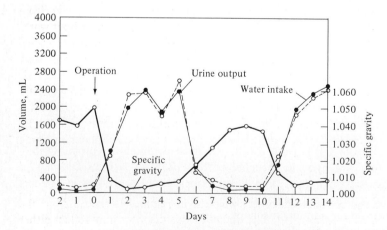

Figure 24-2 Typical triphasic cycle produced by section of the hypophyseal stalk and damage to the median eminence. The interphase of excess ADH secretion extends from day 6 to day 10 or 11. *(From W. H. Hollinshead, Mayo Clin. Proc., 39:92, 1964.)*

Table 24-3 Causes of nephrogenic diabetes insipidus listed according to probable mechanism of action

Disorders which decrease the water permeability of the collecting tubules
  A. Reduced cyclic AMP generation by ADH
     1. Congenital
     2. Hypercalcemia and hypokalemia†
     3. Drugs
        a. Lithium†
        b. Demeclocycline
  B. Reduced cyclic AMP effect
     1. Drugs
        a. Lithium
        b. Demeclocycline
     2. Hypokalemia
     3. Congenital
     4. ?Amyloidosis
     5. ?Sjögren's syndrome

Disorders which interfere with countercurrent function
  A. Osmotic diuresis†
  B. Diuretics
  C. Acute and chronic renal failure
  D. Hypokalemia
  E. Hypercalcemia
  F. Sickle cell anemia

Mechanism unknown
  A. Drugs
     1. Propoxyphene overdose
     2. Methoxyflurane

†Most common causes of NDI and polyuria in adults.

involves two basic steps: (1) creation of a hyperosmotic medullary interstitium (to a maximum to 800 to 1400 mosmol/kg) primarily by NaCl reabsorption without water in the ascending limb of the loop of Henle (a process called countercurrent multiplication); and (2) osmotic equilibration of the urine in the collecting tubules with the medullary interstitium (see Chap. 6). ADH, acting via the generation of cyclic AMP, is essential for the second step since it markedly increases the water permeability of the collecting tubules (see page 133). Thus, NDI must be associated with an abnormality in either countercurrent function or the ability to respond to ADH. The probable mechanisms of action of the various causes of NDI appear in Table 24-3.

    Congenital NDI is an uncommon condition that is transmitted in an X-linked fashion with varying degrees of penetrance in heterozygous females [16,17]. Thus, males tend to have the complete disorder, whereas in females it ranges from the carrier state to marked polyuria [16]. The defect in this disorder may reflect either impaired cyclic AMP generation or reduced cyclic AMP effect [16].

    The major causes of NDI and polyuria in adults are *lithium toxicity, hypercalcemia, hypokalemia,* and the osmotic diuresis associated with *uncontrolled diabetes mellitus.* Lithium appears to act by interfering with both cyclic AMP generation and effect [18]. Polyuria may appear within 8 to 12 weeks and ultimately occurs in 15 to 30 percent of patients [18–20]. Although generally reversible, the defect may be permanent after prolonged drug usage [20]. A similar effect occurs with the use of demeclocycline, a tetracycline derivative [21]. The duration of administration of this drug is usually too short for polyuria to be a serious problem. However, its antagonism of ADH has made demeclocycline useful in the treatment of some patients with hyponatremia due to the syndrome of inappropriate ADH secretion (see Chap. 23) [22].

Hypercalcemia and hypokalemia produce a form of NDI that is generally reversible within 1 to 12 weeks after the correction of the electrolyte disturbance [23–26]. With hypercalcemia, the concentrating defect may become clinically apparent when the plasma Ca^{2+} concentration exceeds 11 mg/dL [24]. Cyclic AMP generation is impaired in this setting, an effect that may involve the ADH receptor in the cell membrane or the ability of the hormone-receptor complex to activate adenyl cyclase [27,28].

With hypokalemia, the concentrating defect usually occurs with K^+ deficits of 200 to 400 meq, a setting in which the plasma K^+ concentration should be under 3.0 meq/L (see Fig. 13-1). [25]. Collecting tubule permeability to water is diminished by hypokalemia [29], an effect that may be mediated by reductions in ADH-induced cyclic AMP generation and in cyclic AMP effect [30].

In addition to antagonizing the effect of ADH on the tubular cell, two other factors may contribute to the polyuria seen with these electrolyte disorders. First, countercurrent function may be impaired, preventing the establishment of a normally hyperosmotic medullary interstitium [26,31]. Second, hypokalemia and possibly hypercalcemia may directly stimulate thirst [26,32,33]. How this might occur is not known.

An *osmotic diuresis* refers to the enhanced urinary water loss that is induced by the presence of large amounts of nonreabsorbed solute in the tubular lumen (see page 119). Both the increase in solute excretion and an impairment in countercurrent function may contribute to this effect. Water is generally lost in excess of solute in this disorder, thereby promoting the development of hyperosmolality if the losses are not replaced. Uncontrolled diabetes mellitus with glucosuria is the most common cause of an osmotic diuresis (see Chap. 25). However, this problem may also occur in patients given either high-protein tube feedings (resulting in the formation of urea from hepatic protein metabolism) [34] or prolonged infusions of mannitol [35].

Loop diuretics, such as *furosemide* and *ethacrynic acid,* impair urinary concentration by inhibiting NaCl reabsorption in the ascending limb of the loop of Henle. However, these agents also impair urinary dilution and are more commonly associated with water retention and hyponatremia (see Chap. 23). Hypernatremia is an unusual consequence of diuretic therapy.

An inability to concentrate the urine maximally is also an early finding in most forms of *renal failure.* Several factors contribute to this problem, including the osmotic diuresis induced by the necessity to increase solute excretion in the remaining functioning nephrons [36,37], decreased tubular responsiveness to ADH [38], and interference with the countercurrent mechanism in disorders affecting the renal medulla such as chronic pyelonephritis and analgesic abuse nephropathy [39]. The net effect is that, as the renal failure becomes more severe, the U_{max} falls, becoming isosmotic or even slightly hypoosmotic to plasma [40]. However, the degree of polyuria is usually limited by the reduction in functioning renal mass. Rarely, a more severe concentrating defect (U_{max} less than 150 mosmol/kg) with marked polyuria may transiently follow the relief of urinary tract obstruction [41].

Diminished concentrating ability is an early and uniform finding in *sickle cell anemia* [42,43]. The low partial pressure of oxygen and high osmolality of the renal medulla favor sickling in the vasa recta, thereby impairing countercurrent function [44]. In a child with the disease the net effect is that, by the age of 10, the U_{max} is only 400 to 500 mosmol/kg, less than one-half the normal value [42]. These changes occur later and are less severe in patients with sickle cell trait or hemoglobin SC disease [42]. Tranfusions with hemoglobin A can initially reverse the concentrating defect, presumably by restoring vasa recta flow [42,43]. However, this beneficial response is lost by age 15, at which time chronic medullary ischemia has produced irreversible interstitial fibrosis and tubular atrophy.

On rare occasions, *amyloidosis* [45] and *Sjögren's syndrome* [46] are associated with NDI and polyuria. Biopsy specimens reveal, respectively, amyloid deposits in, and lymphocytic infiltration around, the collecting tubules. These changes presumably interfere with tubular function and are responsible for the concentrating defect. Other unusual causes of NDI are a *propoxyphene* overdose [47] and fluoride toxicity to the tubular cells following administration of the general anesthetic *methoxyflurane* [48].

Polyuria in diabetes insipidus Several factors contribute to the degree of polyuria in CDI and NDI, including the severity of the concentrating defect, the rate of solute excretion, and the patient's volume status. The interrelationship between the U_{max} and solute excretion can be illustrated by the following examples. Suppose the daily rate of solute excretion is 750 mosmol (composed mostly of Na^+ and K^+ salts and urea). If the U_{max} is 300 mosmol/kg (similar to the P_{osm}), then the minimum urine output will be 2.5 L/day (750 mosmol/day ÷ 300 mosmol/kg = 2.5 kg (or liters) per day). In comparison, the minimum urine output will exceed 10 L/day if the U_{max} is 75 mosmol/kg or less. In general, such a severe concentrating defect is seen only with complete CDI or congenital NDI. Most cases of acquired NDI are associated with a U_{max} that is greater than 300 mosmol/kg. In this setting, nocturia may be the major complaint since the degree of polyuria is relatively mild.

When the U_{osm} is relatively fixed, as it is in diabetes insipidus, *the rate of solute excretion becomes the primary determinant of the urine output.*† If, for example, the U_{osm} is 100 mosmol/kg, then the daily urine volume will be 8 liters if 800 mosmol is excreted but only 4 liters if 400 mosmol is excreted. This has potential therapeutic

---

†This is in contrast to normal subjects in whom the major factor affecting the urine output is water intake via its effect on ADH secretion and, subsequently, the U_{osm}.

importance since a low-sodium, low-protein diet can limit the degree of polyuria by diminishing solute excretion.

Effective circulating volume depletion also can limit the urine volume. Since collecting tubule water reabsorption is diminished in diabetes insipidus, the urine output in this condition is directly related to the volume of water delivered to these segments. The kidney responds to volume depletion in part by increasing proximal Na^+ and water reabsorption (see Chap. 9). As a result, distal delivery and, consequently, the urine output will fall. This effect of hypovolemia constitutes the rationale for the use of diuretics and a low-sodium diet in the therapy of CDI or NDI (see "Treatment" below). It also explains why cortisol deficiency, which is associated with reductions in systemic blood pressure, cardiac output, and renal blood flow [49], limits the urine output in diabetes insipidus. Thus, patients with coexistent anterior and posterior pituitary insufficiency may not initially complain of polyuria. However, the underlying CDI will be unmasked and polyuria will ensue when cortisol replacement is given [50].

Hypothalamic Dysfunction

Chronic hypernatremia in an alert patient is indicative of hypothalamic disease affecting thirst. Two somewhat different syndromes have been described which are most often due to tumors (similar to those producing CDI in Table 24-2), granulomatous diseases such as sarcoidosis, and vascular disease [2,51]. In one disorder, there is a defect in thirst with or without concomitant CDI [2,51–53]. As a result, water lost in the urine and from the skin and lungs is not replaced by drinking. In general, forced water intake is sufficient to return the plasma Na^+ concentration to normal, although partial CDI, if present, may also have to be treated (see below).

In other hypodipsic patients, water loading is ineffective in lowering the plasma Na^+ concentration, as the administered water is excreted in the urine [54–58]. This diuretic response to water, presumably mediated by inhibition of ADH secretion, initially suggested that the osmoreceptors in the hypothalamus may have been reset to recognize the elevated plasma Na^+ concentration as normal. This syndrome has been called *essential hypernatremia.*

Although rare, essential hypernatremia affords an interesting clinical opportunity to study the independent effects of osmolality and volume on ADH secretion. If the osmostat has been reset, its characteristics should be similar to the normal osmostat: (1) the inhibition of ADH release and the excretion of a dilute urine after a water load; (2) the stimulation of ADH release and the excretion of a concentrated urine after water restriction; and (3) maintenance of the new "normal" plasma Na^+ concentration *within narrow limits* (± 1 to 2 percent). Patients with essential hypernatremia satisfy the first two criteria but usually display wide variations in the plasma Na^+ concentration, e.g., plasma Na^+ concentration ranging between 150 and 180 meq/L [54–58].

The latter finding suggests that the osmoreceptors are relatively insensitive, rather than having been reset at a higher level. If this is so, then the appropriate responses to variations in water intake might be mediated by the volume receptors and not the osmoreceptors. For example, water loading increases the effective cir-

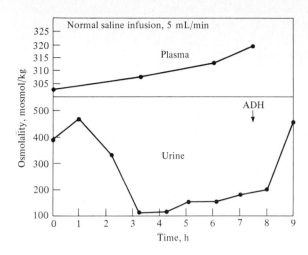

Figure 24-3 Response to saline infusion in a patient with essential hypernatremia. After overnight dehydration, an infusion of approximately 500 mL of saline resulted in urinary dilution (U_{osm} is 102 mosmol/kg of water). With continued saline infusion, a sustained water diuresis was observed despite a rising plasma osmolality (301 to 320 mosmol/kg). The ability of exogenous ADH to terminate the water diuresis suggests that endogenous ADH release was inhibited despite the rise in P_{osm}. *(From F. R. DeRubertis, M. F. Michelis, N. Beck, J. B. Field, and B. B. Davis, J. Clin. Invest., 50:97, 1971, by copyright permission of The American Society for Clinical Investigation.)*

culating volume, which could inhibit ADH secretion and allow the excess water to be excreted. Conversely, water restriction decreases volume, which could augment ADH secretion. To test this hypothesis, hypertonic saline can be administered. This increases both the P_{osm} and volume, respectively stimulating and inhibiting ADH release. In a normal subject, the osmolal effect predominates, causing ADH secretion and an increase in U_{osm}. However, in patients with essential hypernatremia, the U_{osm} may fall, indicating reduced ADH release despite the rise in P_{osm} (Fig. 24-3).

These findings suggest that essential hypernatremia represents *selective damage to the osmoreceptors* resulting in hypodipsia, hypernatremia, and volume-mediated ADH release [2,56–58]. This hypothesis has been directly confirmed in at least one patient in whom plasma ADH levels increased normally with the induction of hypotension but showed little change with an elevation in the P_{osm} [58]. The normal response to volume again illustrates that the osmoreceptor cells are distinct from the hormone-producing cells.

Hypernatremia due to true resetting of the osmostat has been reported only in states of *primary mineralocorticoid excess* such as primary hyperaldosteronism [2]. The chronic mild hypervolemia induced by the mineralocorticoid effect retards ADH secretion. This shifts the osmotic threshold for ADH release upward by 5 to 10 mosmol/kg (see Fig. 8-8). As a result, the normal plasma Na^+ concentration in these disorders is slightly elevated at about 145 meq/L. Normal osmoregulation can be restored by removing the source of hormone secretion or lowering the effective circulating volume with diuretics [2].

Water Loss into Cells

Transient hypernatremia (in which the plasma Na^+ concentration can exceed 150 meq/L) can be induced by severe exertion, as with exercise or seizures [59,60]. This effect is presumed to result from an increase in intracellular osmolality which promotes water movement into the cells. Lactic acidosis also occurs in this setting, and it may be the breakdown of glycogen into smaller molecules (such as lactate) that is responsible for the cellular hyperosmolality [59].

Sodium Overload

Severe hypernatremia can be produced by the acute ingestion or infusion of hypertonic Na^+ solutions. This problem can occur in infants given either high Na^+ feedings [61,62] or $NaHCO_3$ for the respiratory distress syndrome [63], after the use of $NaHCO_3$ to correct the lactic acidosis induced by cardiopulmonary arrest [64], or with the use of hypertonic saline to induce a therapeutic abortion [65]. For example, the inadvertent administration of only one tablespoon of NaCl to a newborn can raise the plasma Na^+ concentration by as much as 70 meq/L [62].

SYMPTOMS

The symptoms of hypernatremia (hyperosmolality) are primarily neurologic. Lethargy, weakness, and irritability are the earliest findings, which can then progress to twitching, seizures, coma, and death [66,67]. These symptoms are related less to the absolute level of the plasma Na^+ concentration than to the movement of water out of the brain cells down the osmotic gradient created by the rise in the effective P_{osm} (Fig. 24-4). Studies in experimental animals and in humans have revealed that this decrease in brain volume causes rupture of the cerebral veins, resulting in focal intracerebral and subarachnoid hemorrhages and neurologic dysfunction which may be irreversible [61,68–70]. A lumbar puncture at this time may reveal blood in the cerebrospinal fluid.

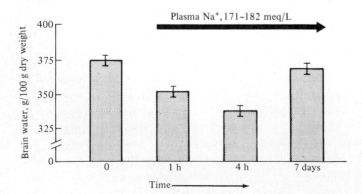

Figure 24-4 Effect of sustained hypernatremia on brain water content in rabbits. Brain water was significantly reduced by 1 to 4 h but returned to normal within 1 week *(From A. S. Pollock and A. I. Arieff, Am. J. Physiol., 239:F195, 1980.)*

A clinically significant water shift appears to require at least a 30- to 35-mosmol/kg osmolal gradient between the plasma and brain [71,72]. This gradient, which is derived from animal studies, correlates well with the findings in humans. In children with acute hypernatremia, for example, seizures and potentially permanent neurologic damage are most likely to occur when the plasma Na^+ concentration exceeds 158 meq/L [69]. This 17-meq/L elevation above normal represents, if the anion accompanying Na^+ is included, approximately a 34-mosmol/kg rise in the P_{osm}.

The cerebral dehydration induced by hypernatremia is transient. Within 24 h, the brain begins to adapt to the hyperosmolal state with an increase in brain cell osmolality, resulting in water movement into the brain and the return of brain volume toward normal (Fig. 24-4) [67,73]. This effect is mediated by increase in cellular Na^+, K^+, amino acids (particularly taurine), and other unidentified solutes (called idiogenic osmoles) [74–76]. Although it is not known how this adaptation occurs, it is specific for the brain since other cells such as those in skeletal muscle remain dehydrated [74,76].

This normalization of brain water content has two important clinical consequences. First, patients with chronic hypernatremia may be relatively asymptomatic despite a plasma Na^+ concentration of as high as 170 to 180 meq/L [77]. Thus, *the severity of the neurologic symptoms is related to both the degree and, more importantly, the rate of rise in the effective* P_{osm}. Although hypernatremia can directly alter cerebral energy metabolism [74], it appears that symptoms are primarily related to the cerebral dehydration. Second, overly rapid correction of the hyperosmolality can cause the now normal brain water to increase above normal, leading to cerebral edema and possible neurologic deterioration (see "Treatment" below).

It should be emphasized that underlying neurologic disease frequently precedes the onset of hypernatremia, and it may be difficult to tell initially whether the neurologic abnormalities are, in fact, due to the increase in the plasma Na^+ concentration. For example, patients with CDI, primary hypodipsia, and essential hypernatremia have hypothalamic lesions which may be due to tumors. Also, patients with decreased mentation due to cerebrovascular disease are particularly prone to develop hypernatremia because of their decreased access to water. The relative roles of hyperosmolality and the underlying disease can be evaluated more accurately after the restoration of a normal plasma Na^+ concentration.

In addition to the neurologic changes, hypernatremic patients may show signs of volume expansion or volume depletion. Patients with Na^+ overload may have peripheral and/or pulmonary edema, whereas those with an osmotic diuresis or enteric infections (who have lost both Na^+ and water) may have flat neck veins, decreased skin turgor, and postural hypotension. In contrast, in patients with pure water loss due to diabetes insipidus or insensible losses, roughly two-thirds of the water deficit comes from the cells. Thus, signs of extracellular volume depletion are usually absent unless marked water losses have occurred, a setting in which the plasma Na^+ concentration is usually greater than 165 to 170 meq/L.

As described above, hypernatremia is unusual in either form of diabetes insipidus because of the effectiveness of thirst. Thus, these patients complain of polyuria, nocturia, and polydipsia rather than symptoms of hyperosmolality. It is important to note that normal subjects form a highly concentrated urine only while sleeping at night since no water is taken in during this time. As a result, nocturia may be the only symptom in patients with a mild to moderate concentrating defect. The underlying partial diabetes insipidus may be masked during the day when there is no need for maximum ADH effect.

Untreated patients with chronic polyuria may also develop functional dilation of the bladder, hydroureter, and hydronephrosis because of voluntary suppression of urination in an attempt to minimize urinary frequency. This can result in a marked increase in bladder capacity such that the patient may void as much as 1000 mL at a time, leading to a decrease in urinary frequency and nocturia.

DIAGNOSIS

Hypernatremia

Hypernatremia generally occurs in adults with an altered mental status or in infants since these are the settings in which thirst is most often impaired. An awake, alert patient with hypernatremia, on the other hand, can be assumed to have a hypothalamic lesion affecting the thirst center. Although the history may be helpful (?polyuria, polydipsia, diabetes mellitus), neurologic abnormalities induced by the hyperosmolality or underlying cerebral disease frequently limit the information that can be attained. In this setting, measurement of the U_{osm} can be particularly helpful. To understand the meaning of this measurement, let us review the response of a normal

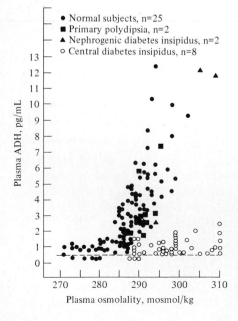

Figure 24-5 Relationship of plasma ADH levels to plasma osmolality in normal subjects and patients with polyuria of diverse etiologies. ADH secretion is reduced only in central diabetes insipidus. *(From G. L. Robertson, E. A. Mahr, S. Atkar, and T. Sinka, J. Clin. Invest., 52:2340, 1973, by copyright permission of the American Society for Clinical Investigation.)*

Table 24-4 U_{osm} and response to ADH in patients with hypernatremia

U_{osm}	Response to vasopressin
Less than 300 mosmol/kg	
CDI	+
NDI	−
300–800 mosmol/kg	
Volume depletion in CDI	+
Partial CDI	+
Partial NDI	−
Osmotic diuresis	−
Greater than 800 mosmol/kg	
Na$^+$ overload	−
Insensible water loss	−
Primary hypodipsia	−
Variable	
Essential hypernatremia	Variable

subject to the induction of hypernatremia by water restriction or the administration of hypertonic saline. As the P_{osm} rises, ADH release is stimulated (Fig. 24-5), resulting in enhanced renal water reabsorption and an elevation in the U_{osm} to a maximum value of 800 to 1400 mosmol/kg (specific gravity equals 1.023 to 1.035). This limit represents *maximum ADH effect on the kidney and is reached when the P_{osm} is 285 to 295 mosmol/kg.* The administration of exogenous ADH at this time will not induce a further increase in the U_{osm} [7].

Patients with hypernatremia (plasma Na$^+$ concentration above 150 meq/L) already have a P_{osm} greater than 295 mosmol/kg, the level at which the urine should be maximally concentrated. Thus, two conclusions can be drawn from the U_{osm} in this setting: (1) there is at least a partial defect in ADH release or effect if the U_{osm} is less than 800 mosmol/kg; and (2) exogenous ADH (given as 5 units of aqueous vasopressin† subcutaneously or 10 μg of dDAVP by nasal insufflation) will increase the U_{osm} only if endogenous secretion is impaired, as occurs in CDI [7].

These responses can be used to evaluate the hypernatremic patient (Table 24-4). Concentrating ability should be normal in subjects with Na$^+$ overload, enhanced insensible loss, and primary hypodipsia without CDI. In these conditions, the U_{osm} should exceed 800 mosmol/kg if there is no underlying concentrating defect, and will be unaffected by vasopressin. In contrast, either severe CDI or NDI is present if the urine is hypoosmotic to plasma (U_{osm} less than 300 mosmol/kg, specific gravity less than 1.010). These disorders can be differentiated by the administration of vasopressin, which will produce at least a 50 percent increase in the U_{osm} and a marked fall in urine volume in CDI but will be without effect in NDI [7].

Many patients fall in an intermediate area with the U_{osm} ranging from 300 to 800 mosmol/kg (specific gravity of 1.010 to 1.023). This can reflect volume depletion

†The chemical name for human ADH is arginine vasopressin.

in severe CDI†, partial CDI, partial NDI, or an osmotic diuresis. Vasopressin is effective only in the first two conditions, augmenting the U_{osm} by at least 60 mosmol/kg and frequently by much more [7].

A potential source of confusion may occur in patients with CDI who do not replace their water losses and become hypernatremic. Although the U_{osm} is low and polyuria may be present when the plasma Na^+ concentration is normal, two factors can combine to raise the U_{osm} (and lower the urine volume) as hypernatremia develops: (1) the osmoreceptor may be stimulated, resulting in a small (though subnormal) increase in ADH release in patients with partial CDI (see Fig. 24-5); and (2) volume depletion induced by the water loss limits fluid delivery to the collecting tubules [78,79]. As a result of these changes, polyuria may be absent, obscuring the presence of CDI. The correct diagnosis can be made in two ways: by observing the response to exogenous vasopressin; and by reversing the hypernatremia and volume depletion, which will unmask the ADH deficiency.

It should be emphasized that the excretion of urine with an osmolality exceeding 300 mosmol/kg cannot directly raise the plasma Na^+ concentration since water is not being lost in excess of solute. For example, many elderly patients with diminished mental status become hypernatremic because of insensible water losses that are not replaced because of impaired thirst. They may also have an inappropriately low U_{osm} since concentrating ability is reduced in the elderly due in large part to a fall in GFR [80]. Although not directly responsible for the hypernatremia, the low U_{osm} *indirectly* contributes by diminishing the ability of the kidney to conserve water.

The U_{osm} in essential hypernatremia is variable and depends upon the state of hydration: high if water-restricted, low if water-loaded. The presence of this rare syndrome of selective osmoreceptor dysfunction should be suspected in a persistently hypernatremic patient who is alert and in whom the administration of water is relatively ineffective in lowering the plasma Na^+ concentration.

Evaluation of Polyuria

Polyuria is a relatively common clinical complaint. The differential diagnosis includes *inappropriate* water loss due to CDI or NDI (particularly uncontrolled diabetes mellitus) or *appropriate* water loss due to increased water intake (primary polydipsia; see page 494) [7]. Once glucosuria has been excluded by testing the urine for glucose, there are several clues in the history and laboratory data that may aid in establishing the correct diagnosis.

---

†Effective volume depletion can raise the U_{osm} in severe CDI to as high as 400 mosmol/kg or more [78,79]. The ability to concentrate the urine in this setting is related to two factors: the collecting tubules have some permeability to water even in the absence of ADH; and the combination of a hypovolemia-induced reduction in GFR and increase in proximal reabsorption can markedly diminish water delivery to the collecting tubules. Since distal delivery is so low, the reabsorption of a small amount of water in the collecting tubules (occurring in the absence of ADH) can substantially raise the U_{osm}. Volume repletion will unmask the underlying CDI, producing a marked increase in the urine output and a fall in the U_{osm}.

1. Patients with CDI frequently have a predilection for very cold or iced water, a finding that does not seem to be present in other polyuric disorders. In addition, CDI typically begins abruptly, and the patient can date the exact onset of the disease. A more gradual onset suggests NDI or primary polydipsia.
2. Severe polyuria with a urine output exceeding 4 to 5 L/day is seen only with primary polydipsia or a severe concentrating defect (U_{max} less than 200 to 250 mosmol/kg). In general, the latter occurs primarily with CDI or congenital NDI. Marked polyuria in acquired NDI is unusual but may occasionally be found with lithium toxicity or hypercalcemia [20,24].
3. Measurement of the P_{osm} may be helpful. Rarely, patients with primary polydipsia can ingest enough water to overwhelm renal excretory capacity and develop water retention and hypoosmolality [8,81]. Thus, this disorder should be suspected if the P_{osm} is less than 275 mosmol/kg. Conversely, patients with primary renal water loss due to diabetes insipidus may be slightly hyperosmolal (P_{osm} greater than 290 mosmol/kg) [8]. More commonly, however, the baseline P_{osm} is within normal range and is of no diagnostic value.

The definitive diagnosis can be made by inducing hyperosmolality with complete water restriction, thereby stimulating endogenous ADH release (Fig. 24-5) and raising the U_{osm} [7]. The urine volume, U_{osm}, and body weight are measured hourly, and the P_{osm} every 4 h. Water restriction is continued until the U_{osm} reaches a plateau (defined as less than a 30-mosmol/kg increase in the U_{osm} in two consecutive hourly specimens) or until the P_{osm} reaches 295 mosmol/kg,† the level at which maximum endogenous ADH effect on the kidney should be present. At this point, exogenous ADH (5 units of aqueous vasopressin subcutaneously or 10 μg of dDAVP by nasal insufflation) is given and the hourly measurements continued.

The different patterns of response are illustrated in Fig. 24-6. In normal subjects, the urine becomes maximally concentrated, the urine volume falls to less than 0.5 mL/min, and vasopressin is without effect. Patients with complete or partial CDI or NDI respond to induced hyperosmolality and vasopressin in the same manner as when they spontaneously develop hypernatremia and hyperosmolality (Table 24-4). Urinary concentration is impaired and, therefore, the urine output does not fall to the level seen in normal subjects. A positive response to vasopressin is seen only with CDI.‡

As depicted in Fig. 24-6, the U_{max} achieved in CDI (after vasopressin) and primary polydipsia is less than that in normal subjects. In both of these disorders, ADH

---

†Accurate measurement of the P_{osm} is an essential component of this test. A potential error of as much as 8 mosmol/kg can occur if the blood specimen is stored at room temperature for 1 to 4 h after it has been obtained. In this setting, persistent glycolytic activity in erythrocytes and leukocytes results in the production of lactic acid and its release into the plasma. This problem can be prevented if the specimen is refrigerated at 0°C or the plasma is separated from the cells within 20 min [82].

‡Accurate interpretation of this test requires that vasopressin not be given before the U_{osm} has stabilized or the P_{osm} has reached 295 mosmol/kg. Below this level, maximum endogenous ADH effect may not be present and the administration of vasopressin is of little diagnostic benefit since it can raise the U_{osm} even in normal subjects.

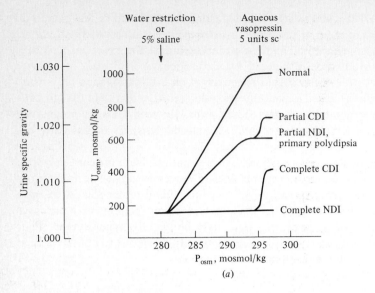

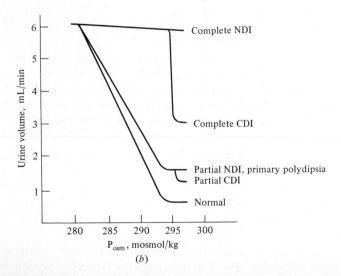

Figure 24-6 Effect of the induction of hyperosmolality, either by water restriction or 5% saline, and exogenous vasopressin on (a) urine osmolality and (b) urine volume in normal subjects and in polyuric states. In normal subjects, as the P_{osm} reaches 285 to 295 mosmol/kg, there will be maximum ADH effect on the kidney, resulting in a U_{osm} greater than 800 mosmol/kg and a urine volume less than 0.5 mL/min. Exogenous vasopressin will be without effect. In patients with complete central diabetes insipidus (CDI) or nephrogenic diabetes insipidus (NDI), the urine will remain hypoosmotic to plasma with a high urine volume. Vasopressin increases the U_{osm} and lowers the urine volume only in CDI. Patients with partial CDI (or complete CDI with volume depletion) or NDI show an intermediate response, and only the former will respond to vasopressin. Since primary polydipsia may induce a form of acquired NDI, these tests may not differentiate this condition from other mild forms of NDI.

secretion is reduced (due, in primary polydipsia, to the chronic water loading). Since the lack of ADH impairs urea accumulation in the medullary interstitium, interstitial osmolality and, therefore, the U_{max} are diminished (see Fig. 6-7). This defect is readily correctible by the chronic administration of ADH in CDI or the control of water intake in primary polydipsia. Thus, both conditions represent reversible forms of acquired NDI. Primary polydipsia can usually be differentiated from other causes of NDI by the history and laboratory data (?drugs, hypokalemia, hypercalcemia, glucosuria) and, since ADH secretion is normal (Fig. 24-5), from partial CDI by its lack of response to vasopressin.

Patients undergoing the water restriction test must be monitored carefully since there are complications that may occur. In some patients with complete CDI, the urine output can reach 700 to 800 mL/h. In this setting, severe volume depletion and vascular collapse can occur if water deprivation is allowed to continue beyond the above limits (stable U_{osm} or P_{osm} of 295 mosmol/kg). In general, the maximum weight loss should not exceed 3 to 5 percent of body weight. Usually, a loss of 1.0 to 2.5 kg is sufficient, requiring 4 to 12 h of water restriction. A longer period may be necessary in patients with primary polydipsia, who may be water-overloaded at the start of the test. These patients must be observed carefully, since they can discover bizarre ways in which to ingest water, e.g., from a flower pot.

Results similar to the water restriction test can be obtained by inducing hyperosmolality with an intravenous infusion of hypertonic saline. This has been called the Hickey-Hare test. Five percent saline (Na^+ concentration of 855 meq/L) is infused at the rate of 0.05 mL/(kg·min) for no more than 2 h and the urine volume and U_{osm} followed [83]. This method is used less frequently than water restriction because of the danger of circulatory overload. It may be particularly useful, however, in essential hypernatremia where hypertonic saline may produce a paradoxical fall in the U_{osm} since ADH secretion is governed primarily by volume, not osmolality (Fig. 24-3).

Although the response to water restriction or hypertonic saline has been the standard approach to patients with polyuria, these tests are indirect since the U_{osm} is used as an index of ADH secretion or effect. The accuracy of water restriction has recently been evaluated by concomitant measurement of the plasma ADH levels [84]. The water restriction test established the correct diagnosis in 80 percent of patients, with the major error occurring in the distinction between partial CDI and primary polydipsia. Some patients with partial CDI appear to have increased sensitivity to ADH (by an unknown mechanism). As a result, they are polyuric at the normal P_{osm} of 280 to 290 mosmol/kg, when ADH secretion is low, but have a maximally concentrated urine at a P_{osm} of 295 mosmol/kg, when their ADH levels are higher but still subnormal. Since exogenous vasopressin will have no further effect, a diagnosis of primary polydipsia will be made. Establishing the correct diagnosis may be important in this setting because treatment for CDI will relieve the polyuria and polydipsia [84]. This may be difficult, however, since measurement of plasma ADH levels is expensive and not widely available. In appropriate patients, in whom primary polydipsia seems less likely from the history, it may be reasonable to institute a trial of dDAVP (see below). This must be done in the hospital because the administration of dDAVP to a patient with primary polydipsia can lead to marked water retention and severe hyponatremia.

Polyuria in hospitalized patients The above approach applies primarily to polyuria in ambulatory patients. In this setting, there is usually a problem in water balance involving defective ADH release or effect or increased water intake. In contrast, polyuria developing in the hospital frequently reflects a solute (or osmotic) diuresis due to the administration of large amounts of saline solutions, the use of hypertonic, high-protein enteral feedings [34], or the relief of urinary tract obstruction. It is useful to evaluate such patients by asking two questions:

1. Does the polyuria reflect a solute or a water diuresis?
2. Is the diuresis appropriate or inappropriate?

If polyuria is arbitrarily defined as a urine output exceeding 3 to 4 L/day, the distinction between a solute and a water diuresis can usually be made by measuring the U_{osm}. A U_{osm} below 250 mosmol/kg indicates a water diuresis, and the patient should undergo a water restriction test to differentiate diabetes insipidus from increased water intake (including fluids administered intravenously).

An isosmotic or hyperosmotic urine (U_{osm} greater than 300 mosmol/kg), on the other hand, is generally indicative of a solute diuresis. Although partial CDI or NDI can also result in a similar U_{osm}, *they will not lead to marked polyuria since solute excretion is normal*. For example, the normal range of solute excretion is 600 to 900 mosmol/day. If the U_{max} is 300 mosmol/kg, then the maximum urine output will be 2 to 3 L/day (900 mosmol/day $\div$ 300 mosmol/kg $=$ 3 L/day). A value greater than this can be achieved only if solute excretion is increased or the U_{osm} is reduced.

In patients with a solute diuresis, hyperglycemia and high-protein feedings can be excluded by the history and laboratory data. In the absence of these disorders, polyuria due to a solute diuresis is almost always *appropriate* and may result from the inadvertent administration of excessive amounts of saline or the release of bilateral urinary tract obstruction. Some patients may have a urine output in excess of 10 L/day because of the initial infusion of 1 to 2 liters of saline followed by orders to replace the urine output with an equivalent volume of saline. As a result, the urine output gradually increases since the patient remains volume-expanded. Similarly, a postobstructive diuresis is almost always appropriate as it represents the excretion of fluid retained during the period of obstruction [85,86]. The correct therapy in either of these settings is to limit fluid intake to a maintenance level, thereby allowing the patient to develop negative fluid balance. The polyuria will cease when the excess fluid has been excreted.

An *inappropriate* solute diuresis is less common and should be suspected if the patient develops hypotension, reduced skin turgor, or decreased renal function. Although true salt wasting can occur with renal insufficiency, polyuria is an unusual complaint in these disorders since the obligatory urine output is generally less than 2 L/day [87].

Clinical example The sequential approach to the polyuric patient can be illustrated by the following case history.

> **Case history 24-1** A 60-year-old man has a cardiac arrest and is resuscitated. Although circulatory function is restored, the patient remains comatose. The urine output is noticed to increase to 300 to 400 mL/h within the first day. At this time, the laboratory data reveal:
>
> $$\text{Plasma } [Na^+] = 144 \text{ meq/L}$$
>
> $$\text{Osmolality} = 290 \text{ meq/L}$$
>
> $$\text{Urine osmolality} = 120 \text{ mosmol/kg}$$
>
> A water restriction test is begun. When the P_{osm} is 296 mosmol/kg, the U_{osm} is only 130 mosmol/kg but rises to 370 mosmol/kg after 5 units of aqueous vasopressin. A diagnosis of CDI is made, and the patient is begun on vasopressin tannate in oil and enteral hyperalimentation. The urine output is initially well controlled but increases to over 150 mL/h on the fourth day and is refractory to vasopressin. At this time, examination of the urine reveals:
>
> $$\text{Urine osmolality} = 500 \text{ mosmol/kg}$$
>
> $$[Na^+] = 30 \text{ meq/L}$$
>
> $$[K^+] = 33 \text{ meq/L}$$
>
> $$[\text{Glucose}] = 0$$
>
> $$[\text{Urea nitrogen}] = 840 \text{ mg/dL}$$

COMMENT The low U_{osm} when the patient is initially polyuric indicates the presence of a water diuresis. The subsequent response to the water restriction test is consistent with CDI, presumably induced by hypoxic encephalopathy. The recurrent polyuria was refractory to vasopressin, an unusual finding in CDI. However, the high U_{osm} of 500 mosmol/kg indicates that this is a solute diuresis, probably due to the high-protein enteral alimentation. This diagnosis was confirmed by analysis of the urine, which demonstrated that urea was the major urinary osmole; a urea nitrogen concentration of 840 mg/dL represents a concentration of 300 mmol/L or 300 mosmol/kg.† Treatment of the polyuria now requires a decrease in protein intake, not administration of higher doses of vasopressin.

TREATMENT

General Principles

Rapid correction of hypernatremia can induce cerebral edema, seizures, permanent neurologic damage, and death [74,88]. This potential complication is a direct result of the beneficial increase in brain volume to normal that initially protects against the symptoms of hypernatremia (see Fig. 24-4). A reduction in the P_{osm} causes water to enter the brain down an osmotic gradient. Once the cerebral adaptation to hyper-

---

†The urea nitrogen concentration in milligrams per deciliter (mg/dL) must be divided by 2.8 to convert to millimoles per liter (mmol/L) or mosmoles per kilogram (mosmol/kg) (see page 15). Thus,

$$[\text{Urea nitrogen}] = 840 \text{ mg/dL} \div 2.8 = 300 \text{ mosmol/kg}$$

natremia has occurred, this water movement augments brain volume above normal and can lead to cerebral edema.

A clinically significant alteration in brain size appears to require an acute change in P_{osm} of approximately 30 to 35 mosmol/kg, which represents a 15-meq/L change in the plasma Na^+ concentration [71,72]. Thus, to minimize the risk of cerebral edema in the treatment of hypernatremia, the plasma Na^+ concentration should be lowered by no more than 15 meq/L in any 8-h period, i.e., a *maximum reduction of 2 meq/L per hour*. The potential danger of more rapid correction of hypernatremia is illustrated in the following example:

> **Case history 24-2** A slightly somnolent 16-year-old female had a plasma Na^+ concentration of 183 meq/L. Her past history revealed 3 years of progressive panhypopituitarism which was being treated with cortisol and thyroid hormone. Because of the hypernatremia, the patient was treated with large volumes of dextrose and water. During the first 6 h the plasma Na^+ concentration was reduced to 154 meq/L but the patient had become unresponsive. A lumbar puncture revealed an opening pressure of 30 cmH$_2$O (normal is 10 to 20), clear fluid, and no cells. Over the next 36 h the patient gradually improved.

Water Loss

Water deficit When hypernatremia is due to water loss (Table 24-1), therapy is aimed at lowering the plasma Na^+ concentration by water replacement. The formula to estimate the amount of water required can be derived in the following way. The quantity of osmoles in the body is equal to the osmolal space [the total body water (TBW)] times the osmolality of the body fluids:

$$\text{Total body osmoles} = \text{TBW} \times P_{osm} \qquad (24\text{-}4)$$

Since the P_{osm} is primarily determined by the plasma Na^+ concentration,

$$\text{Total body osmoles} \propto \text{TBW} \times \text{plasma } [Na^+] \qquad (24\text{-}5)$$

If hypernatremia results only from water loss, then

$$\text{Current body osmoles} = \text{Normal body osmoles}$$

or

Current body water (CBW) $\times$ plasma $[Na^+]$

$$= \text{normal body water (NBW)} \times \text{normal plasma } [Na^+]$$

$$\text{NBW} = \text{CBW} \times \frac{\text{plasma } [Na^+]}{140} \qquad (24\text{-}6)$$

The water deficit can then be estimated from

$$\text{Water deficit} = \text{NBW} - \text{CBW} \qquad (24\text{-}7)$$

$$= \left(\text{CBW} \times \frac{\text{plasma } [Na^+]}{140} \right) - \text{CBW} \qquad (24\text{-}8)$$

$$= \text{CBW} \times \left(\frac{\text{plasma } [Na^+]}{140} - 1 \right) \qquad (24\text{-}9)$$

The total body water is normally about 60 and 50 percent of lean body weight in men and women, respectively. However, it is probably reasonable to use values about 10 percent lower in hypernatremic patients who are water-depleted. Thus, in women, Eq. (24-9) becomes:

$$\text{Water deficit} = 0.4 \times \text{lean body weight} \times \left(\frac{\text{plasma } [Na^+]}{140} - 1 \right) \quad (24\text{-}10)$$

This formula estimates the amount of positive water balance required to return the plasma Na^+ concentration to 140 meq/L. It does not include an additional isosmotic fluid deficit that is frequently present when both Na^+ and water have been lost, as occurs with an osmotic diuresis.

The patient described in Case history 24-2 can be used as an example of the approach to therapy. She weighed 50 kg, and, therefore, her total water deficit was approximately 6 liters [$0.4 \times 50 \times (\frac{183}{140} - 1)$]. The initial aim of therapy is to lower the plasma $Na^{\text{ }}$ concentration no more than 15 meq/L in the first 8 h. This requires the administration of 1.8 liters of water [$0.4 \times 50 \times (\frac{183}{168} - 1)$] or about 225 mL/h. Since the aim is to induce *positive water balance,* estimated insensible losses (usually 30 to 40 mL/h) should be replaced, raising the infusion rate to about 260 mL/h. Urinary losses should also be added to this figure in patients with diabetes insipidus excreting a dilute urine since this is another source of free-water loss.

The type of fluid administered to replace these losses is variable, depending upon the patient's clinical state and the cause of the hypernatremia:

1. Free water can be given orally or intravenously (as dextrose in water)† in patients with hypernatremia due to pure water loss.
2. An infusion of quarter-isotonic saline is preferable if Na^+ depletion is also present, as typically occurs with an osmotic diuresis or concurrent diuretic use. One liter of this solution is a combination of 750 mL of free water and 250 mL of isotonic saline. Thus, about 350 mL/h of quarter-isotonic saline must be administered to provide 260 mL/h of free water.
3. Isotonic saline should be used initially if the patient is hypotensive. In this setting, restoration of tissue perfusion is the most urgent requirement, and this can be best achieved with isotonic saline. This solution may also lower the plasma Na^+ concentration since it is hypoosmotic to the hypernatremic patient. More dilute solutions can be substituted once tissue perfusion is adequate.
4. The contribution of K^+ salts must be taken into account when calculating the tonicity of the fluid that is to be given. For example, quarter-isotonic saline to which 40 meq of K^+ has been added is osmotically equivalent to half-isotonic saline. On the other hand, the contribution of glucose usually can be ignored since it is rapidly metabolized in nondiabetics to carbon dioxide and water. Thus,

---

†The intravenous administration of large volumes of dextrose in water can, in some patients, lead to marked hyperglycemia since the quantity of glucose given can exceed the maximum amount that can normally be metabolized [89]. This problem can be avoided by giving free water orally, by minimizing the urine output in CDI with an ADH preparation (see below), or by careful monitoring of the plasma glucose concentration.

5% dextrose in water has an osmolality of 278 mosmol/kg but is equivalent to free water in the body.

This process of a gradual correction of the hypernatremia can be repeated until the plasma Na^+ concentration is normal. It must be emphasized, however, that Eq. (24-10) is only an approximation and that serial measurements of the plasma Na^+ concentration are required to ascertain that the desired effect is being achieved. For example, the total body water may be substantially less than 40 to 50 percent of lean body weight in elderly patients who are very cachectic. In this setting, the calculated TBW and water deficit will be falsely elevated, possibly leading to an overly rapid reduction in the plasma Na^+ concentration.

Central diabetes insipidus The most physiologic therapy of CDI is to give exogenous ADH (Table 24-5). Acutely, aqueous vasopressin can be given subcutaneously (20 units/mL, 0.25 to 0.50 mL every 3 to 4 h). However, this preparation is too short-acting to be useful in the long-term control of the disease. Vasopressin tannate in oil given by intramuscular injection (0.3 to 1.0 mL every 1 to 3 days) and a lysine-vasopressin nasal spray (one to two sprays in each nostril, four to six times a day) have been used for chronic therapy, but they have recently been supplanted by dDAVP (Desmopressin), a two amino acid substitute of arginine vasopressin that is also administered as a nasal spray. In contrast to lysine vasopressin, dDAVP has more antidiuretic activity, has no vasopressor effect, and has to be taken only once or twice a day (in 5- to 20-µg doses) because of a longer duration of action [90].

It is important to be aware that there is some potential risk to the administration of dDAVP in CDI. Patients with this disorder are polyuric but not in danger as long as their thirst mechanism is intact. However, once dDAVP is given, the patient has nonsuppressible ADH activity and is at risk of developing water retention and hypo-

Table 24-5 Drugs which reduce renal water excretion and are effective in the treatment of CDI,† listed according to probable mechanism of action

ADH preparations
  A. Aqueous vasopressin
  B. Vasopressin tannate in oil
  C. Lysine-vasopressin nasal spray
  D. dDAVP nasal spray (Desmopressin)

Drugs which increase ADH secretion
  A. Clofibrate
  B. ?Carbamazepine

Drugs which potentiate ADH
  A. Chlorpropamide
  B. ?Carbamazepine

Drugs not requiring ADH
  A. Diuretics

† Only diuretics are effective in NDI.

natremia.† As a result, the minimum dose must be used to allow the maintenance of an adequate urine output. This can be achieved by giving the first dose (5 to 10 μg) in the late evening to control the most troubling symptom of nocturia. The size of and necessity for a daytime dose can then be determined by the effectiveness of the evening dose. If, for example, polyuria does not recur until noon, then one-half the evening dose may be sufficient at that time.

Aside from the occasional patient who has an incomplete response to dDAVP [90], the major problem with this medication is its cost, which can exceed $100 per month. As a result, it may be desirable to try some of the other drugs listed in Table 24-5, which are substantially less expensive. For example, hydrochlorothiazide or chlorpropamide may cost less than $3 and $12 per month, respectively.

The induction of mild volume depletion with a low-sodium diet and *diuretics* (such as hydrochlorothiazide, 25 mg once or twice daily) may be extremely effective in diabetes insipidus. Although it seems paradoxical to treat polyuria with a diuretic, as little as a 1.0- to 1.5-kg weight loss can reduce the urine output by more than 50 percent, from, for example, almost 10 L to below 3.5 L per day [92].

Two mechanisms may contribute to this response. Volume depletion is associated with enhanced proximal NaCl and water reabsorption (see Chap. 9). As a result, less water is delivered to the collecting tubules (the site of ADH action) and therefore less water is excreted [92]. Thiazide-induced hypovolemia also can lead to an increase in medullary interstitial osmolality by an unknown mechanism. This will increase the osmotic gradient between the tubular fluid and the interstitium, promoting further water reabsorption in the less than normally permeable collecting tubules [93].

The addition of moderate dietary protein restriction can also contribute to control of the polyuria. The combination of sodium and protein restriction reduces the rate of solute excretion, which, as described above, will diminish the urine output in diabetes insipidus.

The action of diuretics and diet is independent of ADH. In contrast, the other drugs used in the treatment of CDI act by potentiating the effect of ADH or by increasing its secretion (Table 24-5). Since some ADH must be present, *these drugs are effective only in partial CDI but not in complete CDI* [94–97]. In the latter condition, at least some exogenous ADH is generally required.

Chlorpropamide, the most widely used of these drugs, is an oral hypoglycemic agent which primarily enhances the action of ADH [96–98]. This effect is manifested by a greater than normal increment in ADH-induced adenyl cyclase activity, suggesting that chlorpropamide may enhance the affinity of the receptor on the tubular membrane for ADH [98,99]. This unique action of chlorpropamide allows it to

---

†Because of this risk, as well as the difficulty of nasal administration, it is recommended that children up to the age of 2 to 3 be treated with frequent, low-solute feedings (to provide free water to replace the urinary losses) rather than an ADH preparation [91]. The addition of a thiazide diuretic (see below) may also be helpful in controlling the polyuria. Once nasal administration becomes possible, a low dose of dDAVP or a single dose of short-acting lysine vasopressin can be given at bedtime. This will minimize the degree of nocturia and thirst, thereby allowing both the child and parents at least some uninterrupted sleep [91].

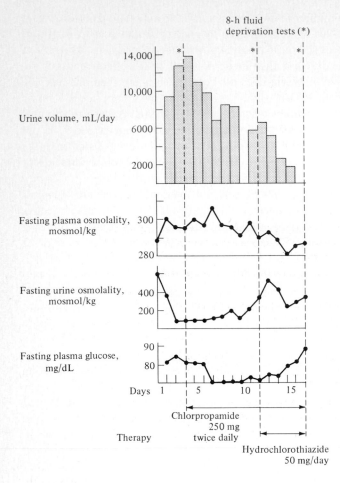

Figure 24-7 Serial observations in control and treatment periods in a patient with partial CDI. Note the additive antidiuretic effects of chlorpropamide and hydrochlorothiazide and the ability of the latter drug to protect against a fall in the plasma glucose concentration induced by chlorpropamide. *(From B. Webster and J. Bain, J. Clin. Endocrinol. Metab., 30:215, 1970. Used by permission from the Journal of Clinical Endocrinology and Metabolism and from Lippincott.)*

be given alone or in conjunction with dDAVP since it will potentiate the antidiuretic response.

The usual dose of chlorpropamide is 125 to 250 mg, once or twice a day. Higher doses (up to 1250 mg/day) can increase the antidiuresis [95], but they also increase the risk of hypoglycemia and should not be used. If hypoglycemia does develop, a problem particularly likely to occur in patients with associated anterior pituitary insufficiency, the dose can be reduced or a thiazide can be added since the latter tends to raise the plasma glucose concentration (Fig. 24-7) [100]. Occasionally, chlorpropamide has to be discontinued because of severe or recurrent hypoglycemia.

Both clofibrate (used in the treatment of hyperlipemia) (500 mg every 6 h) and carbamazepine (used for tic douloureux) (100 to 300 mg, twice daily) also can effectively lower the urine output in partial CDI [94,95,101–103]. Clofibrate appears to enhance ADH secretion [94] whereas carbamazepine seems to increase ADH effect [102,102a] and perhaps augment its secretion [103].

As with diuretics, these drugs can produce more than a 50 percent reduction in urine output in responsive patients. If, however, this represents a decrease in output from 12 to 6 liters daily, the patient, although much improved, will still complain of polyuria. In this situation, combination therapy can be used. By choosing agents which have different mechanisms of action, one can achieve additive antidiuretic effects. For example, diuretics and chlorpropamide can be used together or in conjunction with dDAVP with excellent results (Fig. 24-7) [96]. Carbamazepine and chlorpropamide also can be an effective combination [95].

As discussed in Chap. 23, the antidiuretic properties of diuretics, chlorpropamide, and carbamazepine can lead to water retention and hyponatremia. This is particularly true in patients with primary polydipsia. Therefore, one must be certain to exclude primary polydipsia as the cause of polyuria before beginning therapy for CDI (or NDI). Less commonly, chlorpropamide can induce symptomatic hyponatremia in patients with partial CDI [96]. Since the symptoms of hyponatremia can mimic those of hypernatremia and the therapy is diametrically opposite (water restriction versus water loading), it is important to establish the correct diagnosis and not to assume the presence of hypernatremia because of the history of CDI.

Nephrogenic diabetes insipidus Chronic therapy for NDI should be reserved for those patients with symptomatic polyuria in whom the renal defect is not rapidly correctable. This represents a very small group of patients such as those with congenital NDI or lithium toxicity [16,20]. Specific treatment is not required when the concentrating defect is reversible (drugs, osmotic diuresis, hypercalcemia, hypokalemia) or when polyuria is not a problem (renal failure, sickle cell anemia).

Since patients with NDI do not respond to ADH, dDAVP or drugs depending upon ADH for their action are ineffective. The major form of therapy in this disorder is the use of diuretics and a low-sodium, low-protein diet [92,104]. There is some risk, however, in using diuretics in patients who must continue to take lithium. Although the ensuing volume depletion can reduce the urine output [20], it also increases the proximal reabsorption of Na^+ *and* secondarily of lithium, resulting in decreased excretion and potential lithium toxicity [105]. Careful monitoring of the plasma lithium concentration is required in this setting.

One other possibility that has been effective in some patients with congenital NDI or lithium toxicity is the administration of a nonsteroidal anti-inflammatory drug (NSAID) which inhibits renal prostaglandin synthesis [106–108]. Renal prostaglandins have a variety of effects which impair concentrating ability, including a reduction in ADH-induced stimulation of adenyl cyclase (see page 149). As a result, prostaglandin synthesis inhibitors enhance the effect of ADH and can even raise the U_{osm} in the absence of ADH [109]. Although it is not clear how the latter effect occurs, it can be therapeutically useful in NDI. In one child with congenital NDI, for example, the U_{osm} increased from 94 mosmol/kg on a thiazide diuretic to 174 mosmol/kg with a NSAID [106]. This almost twofold elevation in the U_{osm} means that the urine volume will fall by nearly 50 percent. Thus, it is reasonable to try a NSAID if diuretics and dietary modification do not sufficiently control the polyuria. One exception is sulindac (Clinoril), which should not be used in this setting since it is the one NSAID that does not appear to inhibit renal prostaglandin production [110,111].

Hypothalamic dysfunction The proper treatment of patients with hypothalamic dysfunction depends upon the pattern of ADH secretion. Patients with primary hypodipsia but without CDI can be simply treated with forced water intake. The patient can be instructed to drink 1500 to 2000 mL of water per day, regardless of thirst. There is no risk of water overload in this setting since the ability to excrete water is not impaired.

Correction of the hypernatremia is somewhat more difficult in hypodipsic patients who also have partial CDI or essential hypernatremia. In these conditions, water loading will diminish ADH secretion, resulting in polyuria, excretion of the ingested water, and persistent elevation of the plasma Na^+ concentration. Chlorpropamide has been effective in many of these patients [53,55–57], probably by enhancing the action of the small amount of ADH being released [55]. There is, however, some risk of hyponatremia due to the combination of chlorpropamide and the osmoreceptor defect, which may prevent complete suppression of ADH secretion [2].

In addition to these regimens, an integral part of the therapeutic approach to patients with CDI or hypothalamic disease is a neurologic evaluation to determine the underlying cause. Some tumors, for example, may respond to radiotherapy.

Sodium Overload

Patients with primary Na^+ overload are volume-expanded, and the administration of water to lower the plasma Na^+ concentration can precipitate heart failure. Therefore, therapy is best aimed at removing the excess Na^+. When renal function is normal, the Na^+ load may be excreted rapidly in the urine. This can be facilitated by inducing a Na^+ and water diuresis with diuretics and replacing the urine output solely with water. In patients with poor renal perfusion, e.g., after cardiopulmonary arrest, or in infants, peritoneal dialysis with an electrolyte-free, hypertonic (8%) dextrose and water solution can initially be used to remove the excess Na^+ [62]. Water retention is minimized, since the dialysis solution is hyperosmotic to plasma. The rate of dialysis should be adjusted to prevent an overly rapid fall in the plasma Na^+ concentration and the development of cerebral edema.

PROBLEMS

24-1 A 45-year-old woman with sarcoidosis complains of drinking 8 to 10 liters of water per day. The results of laboratory studies are:

$$Plasma\ [Na^+] = 134\ meq/L$$

$$P_{osm} = 274\ mosmol/kg$$

$$U_{osm} = 80\ mosmol/kg$$

What is the most likely diagnosis? How would you establish the correct diagnosis?

24-2 An 80-year-old partially senile woman treated with hydrochlorothiazide for hypertension is admitted from a nursing home with a 4-day history of a viral-like illness, diarrhea, and increasing confusion. Physical examination reveals a 50-kg female with decreased skin turgor and mentation but a normal blood pressure. The plasma sodium concentration is found to be 174 meq/L, the urine sodium concentration 5 meq/L, and the urine osmolality 606 mosmol/kg. Which of the following are the most important factors in the development of the hypernatremia?

    (*a*) Diarrhea
    (*b*) Decreased thirst
    (*c*) Diabetes insipidus
    (*d*) Diuretic therapy
    (*e*) Insensible losses

What is the most appropriate initial therapy for the hypernatremia?

    (*a*) Isotonic saline at 100 mL/h
    (*b*) Five percent dextrose in water at 200 mL/h
    (*c*) Quarter-isotonic saline at 100 mL/h
    (*d*) Quarter-isotonic saline at 300 mL/h
    (*e*) Five percent dextrose in water at 500 mL/h

24-3 A 40-year-old male alcoholic is brought into the hospital in a comatose state. He is found to have a skull fracture. It is noted that his urine output is 175 mL/h. The following laboratory data are obtained:

$$\text{Plasma } [Na^+] = 168 \text{ meq/L} \qquad \text{Plasma } [HCO_3^-] = 25 \text{ meq/L}$$

$$[K^+] = 4 \text{ meq/L} \qquad\qquad P_{osm} = 350 \text{ mosmol/kg}$$

$$[Cl^-] = 130 \text{ meq/L} \qquad\qquad U_{osm} = 80 \text{ mosmol/kg}$$

The diagnosis of central diabetes insipidus is considered. How would you confirm this diagnosis? If his weight is 70 kg, what is the approximate water deficit? To lower the plasma Na^+ concentration by 15 meq/L in the first 8 h, how much free water should be given?

    The diagnosis of central diabetes insipidus is made and the patient started on vasopressin tannate in oil with an appropriate reduction in the urine output. Two days later, the plasma Na^+ concentration has fallen to 124 meq/L. What is responsible for this?

REFERENCES

1. Robinson, A. G., and J. N. Loeb: Ethanol ingestion: commonest cause of elevated plasma osmolality?, N. Engl. J. Med., 284:1253, 1971.
2. Robertson, G. L., P. Aycinena, and R. L. Zerbe: Neurogenic disorders of osmoregulation, Am. J. Med., 72:339, 1982.
3. Bruck, E., G. Abal, and T. Aceto: Pathogenesis and pathophysiology of hypertonic dehydration with diarrhea, Am. J. Dis. Child., 115:122, 1968.
4. Finberg, L.: Hypernatremic (hypertonic) dehydration in infants, N. Engl. J. Med., 289:196, 1973.
5. Finberg, L.: Dehydration and osmolality, Am. J. Dis. Child., 135:997, 1981.
6. Nelson, D. C., W. R. G. McGrew, and A. M. Hoyumpa: Hypernatremia and lactulose therapy, J. Am. Med. Assoc., 249:1295, 1983.
7. Miller, M., T. Kalkos, A. M. Moses, H. Fellerman, and D. Streeten: Recognition of partial defects in antidiuretic hormone secretion, Ann. Intern. Med., 73:721, 1970.
8. Barlow, E. D., and H. E. de Wardener: Compulsive water drinking, Q. J. Med., 28:235, 1959.
9. Baylis, P. H., M. B. Gaskill, and G. L. Robertson: Vasopressin secretion in primary polydipsia and cranial diabetes insipidus, Q. J. Med., 50:345, 1981.
10. Leaf, A.: Neurogenic diabetes insipidus, Kidney Int., 15:572, 1979.
11. Braverman, L. E., J. P. Mancini, and D. M. McGoldrick: Hereditary idiopathic diabetes insipidus, Ann. Intern. Med., 63:503, 1965.

12. Kaplowitz, P. B., A. J. D'Ercole, and G. L. Robertson: Radioimmunoassay of vasopressin in familial central diabetes insipidus, J. Pediatr., 100:76, 1982.

13. Scherbaum, W. A., and G. F. Botazzo: Autoantibodies to vasopressin cells in idiopathic diabetes insipidus: evidence for an autoimmune variant, Lancet, 1:897, 1983.

14. Jamison, R. L., and R. E. Oliver: Disorders of urinary concentration and dilution, Am. J. Med., 72:308, 1982.

15. Singer, I., and J. N. Forrest: Drug-induced states of nephrogenic diabetes insipidus, Kidney Int., 10:82, 1976.

16. Culpepper, R. M., S. C. Hebert, and T. E. Andreoli: Nephrogenic diabetes insipidus, in *The Metabolic Basis of Inherited Disease,* 5th ed., J. B. Stanbury, J. B. Wyngaarden, D. S. Frederickson, J. L. Goldstein, and M. S. Brown (eds.), McGraw-Hill, New York, 1983.

17. Bode, H. H., and J. D. Crawford: Nephrogenic diabetes insipidus in North America: the Hopewell hypothesis, N. Engl. J. Med., 280:750, 1969.

18. Cox, M., and I. Singer: Lithium and water metabolism, Am. J. Med., 59:153, 1975.

19. Baylis, P. H., and D. A. Heath: Water disturbances in patients treated with oral lithium carbonate, Ann. Intern. Med., 88:607, 1978.

20. Simon, N. M., E. Garber, and A. J. Arieff: Persistent nephrogenic diabetes insipidus after lithium carbonate, Ann. Intern. Med., 86:446, 1977.

21. Singer, I., and D. Rotenberg: Demeclocycline-induced nephrogenic diabetes insipidus, Ann. Intern. Med., 79:679, 1973.

22. Forrest, J. N., M. Cox, C. Hong, G. Morrison, M. Bia, and I. Singer: Superiority of demeclocycline over lithium in the treatment of chronic syndrome of inappropriate secretion of antidiuretic hormone, N. Engl. J. Med., 298:173, 1978.

23. Schwartz, W. B., and A. S. Relman: Effects of electrolyte disorders on renal structure and function, N. Engl. J. Med., 276:383, 452, 1967.

24. Zeffren, J. L., and H. O. Heinemann: Reversible defect in renal concentrating mechanism in patients with hypercalcemia, Am. J. Med., 33:54, 1962.

25. Rubini, M.: Water excretion in potassium-deficient man, J. Clin. Invest., 40:2215, 1961.

26. Levi, M., L. Peterson, and T. Berl: Mechanism of concentrating defect in hypercalcemia. Role of polyuria and prostaglandins, Kidney Int., 23:489, 1983.

27. Beck, N., H. Singh, S. W. Reed, H. V. Murdaugh, and B. B. Davis: Pathogenic role of cyclic AMP in the impairment of urinary concentrating ability in acute hypercalcemia, J. Clin. Invest., 54:1049, 1974.

28. Wiesmann, W., S. Sinha, and S. Klahr: Effects of ionophore A23187 on base-line and vasopressin-stimulated sodium transport in the toad bladder, J. Clin. Invest., 59:418, 1977.

29. Manitius, A., H. Levitin, D. Beck, and F. H. Epstein: On the mechanism of impairment of renal concentrating ability in potassium deficiency, J. Clin. Invest., 39:684, 1960.

30. Beck, N., and S. K. Webster: Impaired urinary concentrating ability and cyclic AMP in K$^+$-depleted rat kidney, Am. J. Physiol. 231:1204, 1976.

31. Bennett, C. M.: Urine concentration and dilution in hypokalemic and hypercalcemic dogs, J. Clin. Invest., 49:1447, 1970.

32. Berl, T., S. L. Linas, G. A. Aisenbrey, and R. J. Anderson: On the mechanism of polyuria in potassium depletion. The role of polydipsia, J. Clin. Invest., 60:620, 1977.

33. Fourman, P., and P. M. Leeson: Thirst and polyuria, Lancet, 1:268, 1959.

34. Gault, M. H., M. E. Dixon, M. Doyle, and W. M. Cohen: Hypernatremia, azotemia, and dehydration due to high-protein tube feeding, Ann. Intern. Med., 68:778, 1968.

35. Gipstein, R. M., and J. D. Boyle: Hypernatremia complicating prolonged mannitol diuresis, N. Engl. J. Med., 272:1116, 1965.

36. Dorhout-Mees, E. J.: Relation between maximal urine concentration, maximal water reabsorption capacity, and mannitol clearance in patients with renal disease, Br. Med. J., 1:1159, 1959.

37. Kleeman, C. R., D. A. Adams, and M. H. Maxwell: An evaluation of maximal water diuresis in chronic renal disease. I. Normal solute intake, J. Lab. Clin. Med., 58:169, 1961.

38. Fine, L. G., D. Schlondorff, W. Trizna, and R. M. Gilbert: Functional profile of the isolated uremic nephron: impaired water permeability and adenylate cyclase responsiveness of the cortical collecting tubule to vasopressin, J. Clin. Invest., 61:1519, 1978.

39. Gilbert, R. M., H. Weber, L. Turchin, L. G. Fine, J. J. Bourgoignie, and N. S. Bricker: A study of the intrarenal recycling of urea in the rat with chronic experimental pyelonephritis, J. Clin. Invest., 58:1348, 1976.

40. Tannen, R. L., E. M. Regal, M. J. Dunn, and R. W. Schrier: Vasopressin-resistant hyposthenuria in advanced chronic renal failure, N. Engl. J. Med., 280:1135, 1969.

41. Earley, L. E.: Extreme polyuria in obstructive uropathy, N. Engl. J. Med., 255:600, 1956.

42. Statius van Eps, L. W., and L. E. Earley: The kidney in sickle cell disease, in *Strauss and Welt's Diseases of the Kidney,* 3rd ed., L. E. Earley and C. W. Gottschalk (eds.), Little, Brown, Boston, 1979.

43. Keitel, H. G., D. Thompson, and H. A. Itano: Hyposthenuria in sickle cell anemia: a reversible renal defect, J. Clin. Invest., 35:998, 1956.

44. Statius van Eps, L. W., C. E. Pinedo-Veels, G. H. de Vries, and J. de Koning: Nature of concentrating defect in sickle-cell nephropathy: microradioangiographic studies, Lancet, 1:450, 1970.

45. Carone, F. A., and F. H. Epstein: Nephrogenic diabetes insipidus caused by amyloid disease. Evidence in man of the role of the collecting ducts in concentrating urine, Am. J. Med., 29:539, 1960.

46. Shearn, M. A., and W. Tu: Nephrogenic diabetes insipidus and other defects of renal tubular function in Sjögren's syndrome, Am. J. Med., 39:312, 1965.

47. McCarthy, W. H., and R. L. Keenan: Propoxyphene hydrochloride poisoning, J. Am. Med. Assoc., 187: 460, 1964.

48. Mazze, R. I., J. R. Trudell, and M. J. Cousins: Methoxyflurane metabolism and renal dysfunction: clinical correlation in man, Anesthesiology, 35:247, 1971.

49. Linas, S. L., T. Berl, G. L. Robertson, G. A. Aisenbrey, R. W. Schrier, and R. J. Anderson: Role of vasopressin in impaired water excretion of glucocorticoid deficiency, Kidney Int., 18:58, 1980.

50. Martin, M. M.: Coexisting anterior pituitary and neurohypophyseal insufficiency, Arch. Intern. Med., 123:409, 1969.

51. Miller, P. D., R. A. Krebs, B. J. Neal, and D. O. McIntyre: Hypodipsia in geriatric patients, Am. J. Med., 73:354, 1982.

51a. Robertson, G. L.: Thirst and vasopressin function in normal and disordered states of water balance, J. Lab. Clin. Med., 101:351, 1983.

52. Hays, R. M., P. R. McHugh, and H. E. Williams: Absence of thirst in association with hydrocephalus, N. Engl. J. Med., 269:227, 1963.

53. Bode, H. H., B. M. Harley, and J. D. Crawford: Restoration of normal drinking behavior by chlorpropamide in patients with hypodipsia and diabetes insipidus, Am. J. Med., 51:304, 1971.

54. Alford, F. P., and B. Scoggins: Symptomatic normovolemic essential hypernatremia, Am. J. Med., 54:359, 1973.

55. Sridhar, D. B., G. D. Calvert, and H. K. Ibbertson: Syndrome of hypernatremia, hypodipsia, and partial diabetes insipidus: a new interpretation, J. Clin. Endocrinol. Metab., 38:890, 1974.

56. DeRubertis, F. R., M. F. Michelis, N. Beck, J. B. Field, and B. B. Davis: Essential hypernatremia due to ineffective osmotic and intact volume regulation of vasopressin secretion, J. Clin. Invest., 50:97, 1971.

57. DeRubertis, F. R., M. F. Michelis, and B. B. Davis: Essential hypernatremia, Arch. Intern. Med., 134:889, 1974.

58. Halter, J. B., A. P. Goldberg, G. L. Robertson, and D. Porte, Jr.: Selective osmoreceptor dysfunction in the syndrome of chronic hypernatremia, J. Clin. Endocrinol. Metab., 44:609, 1977.

59. Felig, P., C. Johnson, M. Levitt, J. Cunningham, F. Keefe, and B. Boglioli: Hypernatremia induced by maximal exercise, J. Am. Med. Assoc., 248:1209, 1982.

60. Welt, L. G., J. Orloff, M. Kydd, and J. E. Oltman: An example of cellular hyperosmolarity, J. Clin. Invest., 29:935, 1950.

61. Finberg, L., J. Kiley, and C. N. Luttrell: Mass accidental salt poisoning in infancy, J. Am. Med. Assoc., 184:187, 1963.

62. Miller, N. L., and L. Finberg: Peritoneal dialysis for salt poisoning, N. Engl. J. Med., 263:1347, 1960.

63. Simmons, M. A., E. Adcock III, H. Bard, and F. Battageia: Hypernatremia, intracranial hemorrhage and $NaHCO_3$ administration in neonates, N. Engl. J. Med., 291:6, 1974.

64. Mattar, J. A., M. H. Weil, H. Shubin, and L. Stein: Cardiac arrest in the critically ill. II. Hyperosmolal states following cardiac arrest, Am. J. Med., 56:162, 1974.

65. De Villota, E. D., J. M. Cavanilles, L. Stein, H. Shubin, and M. H. Weil: Hyperosmolal crisis following infusion of hypertonic sodium chloride for purposes of therapeutic abortion, Am. J. Med., 55:116, 1973.

66. Ross, E. J., and S. B. M. Christie: Hypernatremia, Medicine, 48:441, 1969.

67. Arieff, A. I., and R. Guisado: Effects on the central nervous system of hypernatremic and hyponatremic states, Kidney Int., 10:104, 1976.

68. Finberg, L., E. Luttrell, and H. Redd: Pathogenesis of lesions in the nervous system in hypernatremic states. II. Experimental studies of gross anatomic changes and alterations of chemical composition of the tissues, Pediatrics, 23:46, 1959.

69. Morris-Jones, P. H., I. B. Houston, M. B. Lord, and M. D. Manc: Prognosis of the neurological complications of acute hyponatraemia, Lancet, 2:1385, 1967.

70. Macaulay, D., and M. Watson: Hypernatraemia in infants as a cause of brain damage, Arch. Dis. Child., 42:485, 1967.

71. Stern, W. E., and R. V. Coxon: Osmolality of brain tissue and its relation to brain bulk, Am. J. Physiol., 206:1, 1964.

72. Guisado, R., A. I. Arieff, and S. G. Massry: Effects of glycerol infusions on brain water and electrolytes, Am. J. Physiol., 227:865, 1974.

73. Holliday, M. A., M. N. Kalyci, and J. Harrah: Factors that limit brain volume changes in response to acute and sustained hyper-and hyponatremia, J. Clin. Invest., 47:1916, 1968.

74. Pollock, A. S., and A. I. Arieff: Abnormalities of cell volume regulation and their functional consequences, Am. J. Physiol., 239:F195, 1980.

75. Thurston, J. H., R. E. Hauhart, and J. A. Dirgo: Taurine: a role in osmotic regulation of mammalian brain and possible clinical significance, Life Sci., 26:1561, 1980.

76. Bradbury, M. W. B., and C. R. Kleeman: The effect of chronic osmotic disturbance on the concentrations of cations in cerebrospinal fluid, J. Physiol. (London), 204:181, 1969.

77. Kastin, A. J., M. B. Lipsett, A. K. Ommaya, and J. M. Moser, Jr.: Asymptomatic hypernatremia, Am. J. Med., 38:306, 1965.

78. Berliner, R. W., and D. G. Davidson: Production of hypertonic urine in the absence of pituitary antidiuretic hormone, J. Clin. Invest., 36:1416, 1957.

79. Edwards, B. R., M. Gellai, and H. Valtin: Concentration of urine in the absence of ADH with minimal or no decrease in GFR, Am. J. Physiol., 239:F84, 1980.

80. Sporn, I. N., R. G. Lancestremere, and S. Papper: Differential diagnosis of oliguria in aged patients, N. Engl. J. Med., 267:130, 1962.

81. Rendell, M., D. McGrane, and M. Cuesta: Fatal compulsive water drinking, J. Am. Med. Assoc., 240:2557, 1978.

82. Redetzki, H. M., J. R. Hughes, and J. E. Redetzki: Differences between serum and plasma osmolalities and their relationship to lactic acid values, Proc. Soc. Exp. Biol. Med., 139:315, 1972.

83. Moses, A. M., and D. H. P. Streeten: Differentiation of polyuric states by measurement of responses to changes in plasma osmolality induced by hypertonic saline infusions, Am. J. Med., 42:368, 1967.

84. Zerbe, R. L., and G. L. Robertson: A comparison of plasma vasopressin measurements with a standard indirect test in the differential diagnosis of polyuria, N. Engl. J. Med., 305:1539, 1981.

85. Howards, S. S.: Post-obstructive diuresis: a misunderstood phenomenon, J. Urol., 110:537, 1973.

86. Rose, B. D., *Pathophysiology of Renal Disease,* McGraw-Hill, New York, 1981, p. 361.

87. Coleman, A. J., M. Arias, N. W. Carter, F. C. Rector, Jr., and D. W. Seldin: The mechanism of salt-wasting in chronic renal disease, J. Clin. Invest., 45:1116, 1966.

88. Hogan, G., P. R. Dodge, S. Gill, S. Master, and J. Sotos: Pathogenesis of seizures occurring during restoration of plasma tonicity to normal in animals previously chronically hypernatremic, Pediatrics, 43:54, 1969.

89. Freidenberg, G. R., E. J. Kosnik, and J. F. Sotos: Hyperglycemic coma after suprasellar surgery, N. Engl. J. Med., 303:863, 1980.

90. Cobb, W. E., S. Spare, and S. Reichlin: Neurogenic diabetes insipidus: management with dDAVP (1-desamino-8-D-arginine vasopressin), Ann. Intern. Med., 88:183, 1978.

91. Bode, H. H.: Disorders of the posterior pituitary, in *Clinical Pediatric and Adolescent Endocrinology,* S. A. Kaplan (ed.), Saunders, Philadelphia, 1982.
92. Earley, L. E., and J. Orloff: The mechanism of antidiuresis associated with the administration of hydrochlorothiazide to patients with vasopressin-resistant diabetes insipidus, J. Clin. Invest., 41:1988, 1962.
93. Shirley, D. G., S. J. Walter, and J. F. Laycock: The antidiuretic effect of chronic hydrochlorothiazide treatment in rats with diabetes insipidus: renal mechanisms, Clin. Sci., 63:533, 1982.
94. Moses, A. M., J. Howanitz, M. van Gemert, and M. Miller: Clofibrate-induced antidiuresis, J. Clin. Invest., 52:535, 1973.
95. Radó, J.: Combination of carbamazepine and chlorpropamide in the treatment of "hyporesponder" pituitary diabetes insipidus, J. Clin. Endocrinol. Metab. 38:1, 1974.
96. Webster, B., and J. Bain: Antidiuretic effect and complications of chlorpropamide therapy in diabetes insipidus, J. Clin. Endocrinol. Metab., 30:215, 1970.
97. Miller, M., and A. M. Moses: Mechanism of chlorpropamide action in diabetes insipidus, J. Clin. Endocrinol. Metab., 30:488, 1970.
97a. Murase, T., and S. Yoshida: Mechanism of chlorpropamide action in patients with diabetes insipidus, J. Clin. Endocrinol. Metab., 36:174, 1973.
98. Moses, A. M., R. Fenner, E. T. Schroeder, and R. Coulson: Further studies on the mechanisms by which chlorpropamide alters the action of vasopressin, Endocrinology, 111:2025, 1982.
99. Moses, A. M., and R. Coulson: Augmentation by chlorpropamide of 1-deamino-8-D-arginine vasopressin-induced antidiuresis and stimulation of renal medullary adenylate cyclase and accumulation of adenosine 3',5'-monophosphate, Endocrinology, 106:967, 1980.
100. Helderman, J. H., D. Elahi, D. K. Anderson, G. S. Raizes, J. D. Tobin, D. Shocken, and R. Andres: Prevention of the glucose intolerance of thiazide diuretics by maintenance of body potassium, Diabetes, 32:106, 1983.
101. Hamuth, Y. A., and M. Gelb: Clofibrate treatment of idiopathic diabetes insipidus, J. Am. Med. Assoc., 224:1041, 1973.
102. Wales, J. K.: Treatment of diabetes insipidus with carbamazepine, Lancet, 2:948, 1975.
102a. Meinders, A. E., V. Cejka, and G. L. Robertson: The antidiuretic action of carbamazepine in man, Clin. Sci. Mol. Med., 47:289, 1974.
103. Kimura, T., K. Matsui, T. Sato, and K. Yoshinaga: Mechanism of carbamazepine (Tegretol)-induced antidiuresis: evidence for release of antidiuretic hormone and impaired excretion of a water load, J. Clin. Endocrinol. Metab., 38:356, 1974.
104. Cutler, R. E., C. R. Kleeman, M. H. Maxwell, and J. T. Dowling: Physiologic studies in nephrogenic diabetes insipidus, J. Clin. Endocrinol. Metab., 22:827, 1962.
105. Petersen, V., S. Hvidt, K. Thomsen, and M. Schou: Effect of prolonged thiazide treatment on renal lithium clearance, Br. Med. J., 3:143, 1974.
106. Usberti, M., M. Dechaux, M. Guillot, R. Seligmann, H. Pavlovitch, C. Loirat, C. Sachs, and M. Broyer: Renal prostaglandin E_2 in nephrogenic diabetes insipidus: effects of inhibition of prostaglandin synthesis by indomethacin, J. Pediatr., 97:476, 1980.
107. Chevalier, R. L., and A. D. Rogol: Tolmetin sodium in the management of nephrogenic diabetes insipidus, J. Pediatr., 101:787, 1982.
108. Gross, P. A., R. W. Schrier, and R. J. Anderson: Prostaglandins and water metabolism: a review with emphasis on in vivo studies, Kidney Int., 19:839, 1981.
109. Walker, R. M., R. S. Brown, and J. S. Stoff: Role of renal prostaglandins during antidiuresis and water diuresis in man, Kidney Int., 21:365, 1981.
110. Ciabattoni, G., F. Pugliese, G. A. Cinotti, and C. Patrono: Renal effects of anti-inflammatory drugs, Eur. J. Rheumatol. Inflam., 3:210, 1980.
111. Bunning, R. D., and W. F. Barth: Sulindac: a potentially renal-sparing nonsteroidal anti-inflammatory drug, J. Am. Med. Assoc., 248:2864, 1982.

TWENTY-FIVE

HYPEROSMOLAL STATES—HYPERGLYCEMIA

PATHOPHYSIOLOGY

Hyperglycemia is a common clinical problem. In addition to producing a hyperosmolal state similar to hypernatremia, hyperglycemia may be associated with severe metabolic acidosis (ketoacidosis) and volume depletion, both of which can jeopardize the life of the patient. Although a complete discussion of the regulation of carbohydrate and fat metabolism is beyond the scope of this chapter, a review of the role of insulin, glucagon, and other hormones is essential to understanding the pathophysiology of this disorder. As will be seen, both *insulin deficiency and glucagon excess* play a major role in the development of these biochemical changes: glucagon by altering hepatic metabolism to promote glucose and ketoacid production; and insulin deficiency primarily by increasing the supply of substrates to the liver to allow these processes to occur [1–4].

Glucose Metabolism

The glucose concentration in the extracellular fluid is determined by the relationship between production and utilization. Net glucose production is influenced by three factors: dietary intake, glycogenolysis, and hepatic gluconeogenesis using lactate, amino acids (primarily alanine), and glycerol as substrates (Fig. 25-1) [4]. The glucose that is produced can then be utilized for energy or stored, for future use, in the liver and skeletal muscle as glycogen.

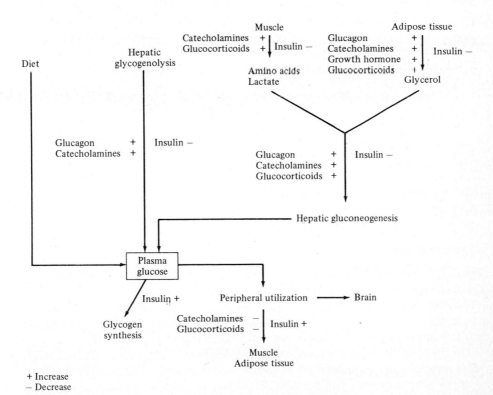

+ Increase
− Decrease

Figure 25-1 Hormonal regulation of carbohydrate metabolism. Insulin and glucagon are most important in this process: the former by increasing the utilization of glucose; the latter by enhancing its production.

In the normal subject, the extracellular supply of glucose is carefully regulated, with the plasma glucose concentration being maintained within narrow limits (60 to 100 mg/dL, fasting). Insulin and glucagon, which are secreted from the pancreas, play a central role in this process by affecting both glucose production and utilization (Fig. 25-1) [3,4]. Insulin is released when the plasma glucose concentration is elevated and acts to restore normoglycemia by reducing glycogenolysis and gluconeogenesis and by enhancing glucose uptake by the liver, muscle, and adipose tissue. The fall in gluconeogenesis is primarily due to the antiproteolytic and antilipolytic effects of insulin, which diminish alanine and glycerol delivery to the liver. Insulin also has an inhibitory effect on glucagon secretion, further reducing hepatic glucose production [1,3]. In contrast, the diminished release of insulin in conditions of glucose lack allows increased secretion of glucagon. Glucagon then appropriately increases the availability of glucose by promoting hepatic glycogenolysis and gluconeogenesis.

In the fasting state, the liver produces glucose to meet basal energy requirements, particularly for the brain. This process is primarily mediated by glucagon [3,4]. If a carbohydrate meal is then ingested, the ensuing rise in the plasma glucose concentration increases the secretion of insulin and reduces that of glucagon. The major effect of these changes is on the liver: hepatic glucose production is markedly reduced while approximately 85 percent of the ingested glucose is initially taken up by the liver and stored as glycogen [5,6]. Only 15 percent of the exogenous glucose is delivered to the extrahepatic tissues. This predominant hepatic effect is probably related to the release of insulin into the portal vein. As a result, the insulin concentration in the portal vein (which perfuses the liver) is 3 to 10 times that in the peripheral circulation [7].

These hormonal changes are reversed with hypoglycemia as diminished release of insulin directly promotes the secretion of glucagon [9]. Catecholamines (particularly epinephrine) and, to a lesser degree, glucocorticoids and growth hormone also may contribute to this protective response (Fig. 25-1) [8,9]. The ensuing increases in glycogenolysis and gluconeogenesis restore the plasma glucose concentration toward normal.

Hyperglycemia

Hyperglycemia generally requires the presence of insulin deficiency and/or resistance, changes which most often occur in idiopathic diabetes mellitus [5,10,11]. In addition to its direct effects on glucose metabolism, the lack of insulin also contributes to the development of hyperglycemia by promoting the secretion of glucagon and, to a lesser degree, catecholamines and growth hormone [3,12–14].

In the early stages of the disease when the hormonal changes are relatively mild, the plasma glucose concentration is normal in the fasting state since low insulin secretion is appropriate in this setting. However, the insulin response to a glucose load is subnormal, resulting in decreased hepatic and peripheral utilization and an excessive postprandial elevation in the plasma glucose concentration [5]. Fasting hyperglycemia occurs with more severe insulin deficiency; it is due initially to enhanced glycogenolysis and ultimately (since glycogen stores are limited) to an increase in gluconeogenesis, most of which occurs from amino acids derived from protein breakdown [5]. Thus, fasting hyperglycemia may represent a catabolic state since protein stores can become depleted. In patients with severe uncontrolled diabetes mellitus (plasma glucose concentration greater than 400 mg/dL), for example, hepatic glucose production rises to a level more than twice that seen in normal fasting subjects [5]. The magnitude of this change becomes more apparent when it is noted that even minimal hyperglycemia in normal subjects increases insulin secretion and almost abolishes hepatic glucose production [6].

Although insulin deficiency plays a major role, the elevation in hepatic gluconeogenesis in diabetes mellitus is also strongly dependent upon the hypersecretion of glucagon [1]. The importance of glucagon is illustrated by the observation that fasting hyperglycemia in insulin-deficient subjects can be markedly attenuated if glucagon release is prevented either by infusing somatostatin or by the presence of a somatostatin-producing tumor [3,15]. However, excess glucagon alone does not lead to hyperglycemia since its effects can be counteracted by normal insulin secretion [16].

The increases in catecholamine and growth hormone release induced by insulin deficiency can also contribute to the development of hyperglycemia, primarily by increasing the availability of alanine and glycerol for gluconeogenesis (Fig. 25-1) and, with catecholamines, by directly increasing glucagon secretion [9,17,18]. In addition, catecholamines and cortisol are released in response to stress. This may explain why stresses such as infection or volume depletion are the most common precipitating factors of uncontrolled diabetes mellitus [19,20]. Although these hormones can moderately raise the plasma glucose concentration in normal subjects [18], the concurrent presence of insulin deficiency can lead to severe hyperglycemia.

Ketoacidosis

When glucose utilization is impaired by fasting or insulin deficiency (as in diabetes mellitus), ketones are produced by the liver from free fatty acids to supply an alternate source of energy [2,21]. This adaptation also conserves body protein stores [21]. If ketones were not available, there would be a greater requirement for gluconeogenesis, much of which is derived from the metabolism of amino acids [22].

Two basic steps are required for ketogenesis to occur (Fig. 25-2) [1,2,23]. First, lipolysis must be increased to enhance the delivery of free fatty acids to the liver. This is achieved by the same hormonal changes that are responsible for hyperglycemia—insulin deficiency and an excess of glucagon, and, to a lesser degree, catecholamines, growth hormone, and cortisol [3,17,23,24]. The last three hormones act in part by enhancing the release of glucagon [24,25].

Second, hepatic metabolism must be altered to allow ketone formation to occur. Excess fatty acids alone are insufficient since they are normally metabolized in the cytosol into triglycerides [1,2]. The rate-limiting step in hepatic ketogenesis is the entry of fatty acyl CoA into the mitochondria, a process regulated by the cytosolic enzyme carnitine acyltransferase (CAT) (Fig. 25-2). CAT activity in turn appears to vary inversely with the level of malonyl CoA [26]. In the fed state, malonyl CoA

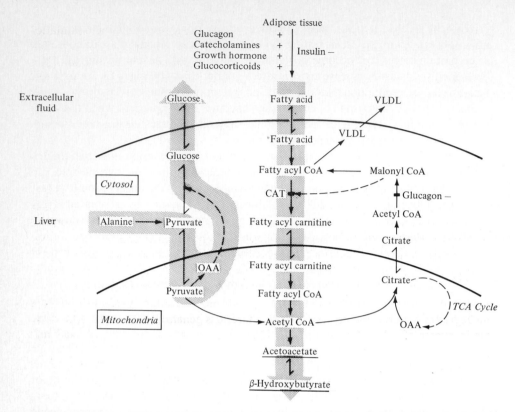

Figure 25-2 Summary of hepatic fatty acid and carbohydrate metabolism in uncontrolled diabetes mellitus, a low-insulin, high-glucagon state. In this setting, malonyl CoA levels are low, in part because glucagon decreases the activity of the enzyme leading to its formation. As a result, carnitine acyltransferase (CAT) activity is high and fatty acids presented to the hepatocyte are able to enter the mitochondria and be metabolized to ketones. Note also that pyruvate is preferentially utilized for gluconeogenesis in this setting. The metabolic pathways are different in the fed state since malonyl CoA is abundant and CAT activity is low. In this setting, any free fatty acids delivered to the liver will remain in the cytosol and be converted to very low-density lipoprotein (VLDL) triglycerides. *(Adapted from G. F. Cahill, Jr., Kidney Int., 20:461, 1982. Reprinted by permission from Kidney International.)*

is relatively abundant, CAT activity is low, and ketone synthesis will not occur even if free fatty acids are available.

These changes are reversed in diabetes mellitus as malonyl CoA levels are low, CAT activity is increased, and ketogenesis can occur. Glucagon excess appears to play a major role in this hepatic response whereas the primary role of insulin deficiency in ketogenesis is to increase the supply of free fatty acids [1,3,23]. At least some of the effect of glucagon is mediated by diminished activity of acetyl CoA carboxylase, the enzyme that converts acetyl CoA into malonyl CoA (Fig. 25-2) [1,27].

Acetoacetic acid is the initial ketone formed. It may then be reduced to β-hydroxybutyric acid or nonenzymatically decarboxylated to acetone [28]. Acetone is chemically neutral, but the other ketones are organic acids and their accumulation will lead to metabolic acidosis. The degree of ketoacidosis is limited in fasting (usually to less than 10 meq/L) [29] because ketones stimulate insulin production [30], thereby limiting the availability of free fatty acids. In contrast, severe metabolic acidosis can occur in insulin-deficient patients with diabetes mellitus. Although overproduction is the primary factor responsible for the development of ketoacidosis in this setting, insulin deficiency may also impair peripheral ketone utilization [31,32].

Hypovolemia

Hypovolemia is an almost invariable finding with marked hyperglycemia and is primarily induced by the associated glucosuria. As the filtered load of glucose (GFR times plasma glucose concentration) rises, it eventually exceeds tubular reabsorptive capacity (see Chap. 5). As a result, glucose remains in the tubular lumen and acts as an osmotic diuretic, increasing the urinary loss of electrolytes and water [33]. In severely hyperglycemic patients, the fluid loss may reach 8 to 10 liters (almost 25 percent of the total body water) and produce circulatory insufficiency [34].

In addition to reducing tissue perfusion, the osmotic diuresis tends to increase the degree of hyperosmolality. An osmotic diuretic is generally associated with water loss in excess of solute (see page 119) [35], thereby raising the plasma Na^+ and glucose concentrations and the plasma osmolality [19,36].

Renal Insufficiency

Hypovolemia also reduces renal function. This may be an extremely important change since the kidney normally acts to minimize the severity of uncontrolled diabetes mellitus by excreting some of the excess glucose and by increasing net acid excretion. As β-hydroxybutyrate (the major ketoacid anion) appears in the urine, it can buffer secreted H^+ to form β-hydroxybutyric acid or it can be excreted with a cation, including NH_4^+ (another source of H^+ excretion). In one study of patients with diabetic ketoacidosis, ketone production averaged 51 meq/h while net acid excretion with the ketoacid anions increased by 15 meq/h [37]. Thus, 30 percent of the generated ketoacids were excreted in the urine, limiting the severity of the metabolic acidosis. (A second protective mechanism is the conversion of acetoacetic acid to acetone, which neutralizes another 15 to 25 percent of the acid load) [28,37].

Plasma Sodium Concentration

Variable changes in the plasma sodium concentration occur with hyperglycemia. Since glucose penetrates cells slowly, an increase in the plasma glucose concentration raises the effective P_{osm} and causes water to move from the cells into the ECF. By dilution, this lowers the plasma Na^+ concentration. In theory, every 62-mg/dL increment in the plasma glucose concentration should draw enough water out of the cells to reduce the plasma Na^+ concentration 1 meq/L [38]. For example, if the plasma glucose concentration is 720 mg/dL (620 mg/dL greater than normal), the plasma

Na^+ concentration should fall 10 meq/L, from 140 to 130 meq/L. It is important to be aware of this situation because therapy must be directed toward hyperosmolality and not, as suggested by the presence of hyponatremia, toward hypoosmolality.

In many patients, the plasma Na^+ concentration differs from the value predicted from the increment in the plasma glucose concentration. Ketoacidosis may be associated with marked hyperlipemia, which can lead to a further reduction in the plasma Na^+ concentration (see Chap. 23). On the other hand, the preferential loss of water induced by the osmotic diuresis can result in a plasma Na^+ concentration that is normal or even elevated, a frequent finding in patients with severe hyperglycemia [34].

ETIOLOGY

Hyperglycemia

There are two major symptomatic hyperglycemic syndromes in humans: *diabetic ketoacidosis* (DKA) and *nonketotic hyperglycemia* (NKH).† The absence of ketoacidosis in NKH may be due to the differential sensitivity of fat and glucose metabolism to the effects of insulin. Studies in humans indicate that the concentration of insulin necessary to suppress lipolysis is only one-tenth that required to promote glucose utilization [40]. Consequently, with moderate insulin deficiency, there might be enough insulin available to block lipolysis but not enough to enhance glucose utilization. The result would be hyperglycemia without ketoacidosis. With more severe insulin deficiency, lipolysis would be increased and ketoacidosis would accompany the increase in the plasma glucose concentration. Consistent with this model are the findings of some observers of a higher plasma free fatty acid concentration and a lower plasma insulin concentration in patients with DKA than in those with NKH [34,41,42]. In addition, DKA tends to occur in patients with type 1 (usually juvenile-onset) diabetes mellitus who produce little or no endogenous insulin. In contrast, NKH occurs most commonly in older patients (average age 60 to 65) in whom insulin levels are reduced, but not absent, and who have either no history or one of mild, type 2 (maturity-onset) diabetes mellitus [34,43,44].

It should be noted, however, that other observers have been unable to demonstrate these differences in the plasma insulin concentration [25,45,46]. If the free fatty acid levels are similar, then the liver may be more efficient in making ketones in DKA than in NKH. This could be due to an increase in the glucagon concentration or perhaps the glucagon/insulin ratio in the portal vein. Alternatively, the free fatty acid levels could be elevated with the same insulin and glucagon concentrations if the secretion of lipolytic hormones such as cortisol and growth hormone were enhanced [44].

---

†These disorders are most often due to idiopathic diabetes mellitus but can also occur in other conditions, including severe pancreatic disease or total pancreatectomy (see below) [25,39].

Both DKA and NKH are most often precipitated by various stresses (infection, hypovolemia, surgery, emotional trauma), which act in part by increasing the secretion of catecholamines, glucagon, and cortisol [19,34,44]. Omission of insulin (or oral hypoglycemic) therapy or failure to augment insulin dosage when the plasma glucose concentration is poorly controlled are less often responsible. NKH can also be induced in diabetics by glucose loading, as occurs with peritoneal dialysis [36]. This complication is more common when hypertonic glucose solutions (such as 4.25% glucose) are used and can be prevented by careful monitoring of the plasma glucose concentration and the addition of insulin to the dialysis fluid [47].

Both DKA and NKH can also be induced in patients who do not have diabetes mellitus. This may occur with the administration of glucocorticoids, which antagonize insulin action [48], or drugs which inhibit insulin release such as diazoxide, phenytoin, or thiazide and loop diuretics [49–53]. NKH may also follow marked glucose loading in acutely ill stressed patients [54,55] or the administration of dextrose in water to replace the high urine output in diabetes insipidus [56]. In the latter setting, the amount of glucose infused may be so large that it exceeds the maximum metabolic capacity.

Mannitol

Mannitol is a nonreabsorbable sugar which is given as a hypertonic solution to augment the urine output in oliguric patients and to treat cerebral edema. In patients with renal failure, however, the mannitol may not be excreted in the urine. The retention of mannitol in the extracellular fluid will increase the plasma osmolality and lower the plasma Na^+ concentration in a manner similar to hyperglycemia [57].

SYMPTOMS

The patient with hyperglycemia may suffer from symptoms due to hyperosmolality, volume depletion, and, in DKA, metabolic acidosis. The severity of these symptoms is generally proportional to both the degree and the duration of the hyperglycemia. In some patients, however, these findings may be masked by the symptoms associated with the acute illness that precipitated the hyperglycemic state.

The earliest complaints associated with hyperglycemia are polyuria, polydipsia, and weight loss. This characteristic triad is due to the combination of the glucose osmotic diuresis and hypovolemia. In more severely affected patients, focal or generalized neurologic abnormalities may be seen, including lethargy, twitching, obtundation, motor or sensory defects, seizures, and coma [25,58].

These symptoms and the response of the brain are similar to those seen when hyperosmolality is due to hypernatremia (see Chap. 23) [58]. The increase in plasma osmolality initially causes water movement out of the cells, leading to cellular dehydration, most importantly in the brain. Within 4 to 6 h, however, brain cell volume returns toward normal due to the generation of new (or idiogenic) osmoles which pull water back into the cell.† Despite this adaptation, the severity of neurologic symptoms is roughly proportional to the degree of hyperosmolality [59]. This suggests either that hyperglycemia acts in some way other than by causing cerebral dehydration or that the impairment in carbohydrate metabolism increases the sensitivity to osmotic injury [58].

Coma is not usually seen unless the plasma osmolality is greater than 340 to 350 mosmol/kg [34,36,59]. In NKH, the plasma glucose concentration frequently exceeds 1000 mg/dL, the plasma osmolality may reach 380 mosmol/kg, and neurologic abnormalities may be the reason that the patient is brought to medical care. This degree of hyperosmolality occurs less often in DKA, where the plasma glucose concentration is generally below 800 mg/dL [1,34].

The more severe hyperglycemia in NKH may be due to at least two factors. First, patients with NKH are older and frequently have impaired renal function. In contrast, DKA occurs primarily in young patients with type 1 diabetes mellitus who have a GFR that, in the first 5 years of the disease, may be as much as 50 percent above normal [59a]. As a result, these patients generally have a much greater capacity to excrete glucose than those with NKH, a protective mechanism that will limit the degree of hyperglycemia. Second, patients with DKA may present early with the symptoms of metabolic acidosis (nausea, vomiting, abdominal pain, dyspnea) rather than late with those of hyperosmolality.

The water losses induced by the glucose osmotic diuresis also appear to play an important role in the development of severe hyperosmolality. If these losses do not occur, as in diabetic patients with end-stage renal failure, the plasma glucose concentration can reach 1000 to 1500 mg/dL but the effective plasma osmolality (see page 30) usually remains below 325 mosmol/kg because water movement out of the cells lowers the plasma Na^+ concentration. Severe neurologic symptoms are not seen since the degree of hyperosmolality is relatively mild [36].

The symptoms and signs produced by hypovolemia and metabolic acidosis are discussed in Chaps. 15 and 19. Circulatory insufficiency with hypotension or shock is not uncommon due to the marked fluid losses.

DIAGNOSIS

The history and physical examination may provide important clues to the presence of uncontrolled diabetes mellitus. Hyperglycemia should be suspected in any patient complaining of polyuria, polydipsia, and weight loss. The findings of hyperventilation and the fruity odor of acetone on the patient's breath suggest that ketoacidosis is also present. In addition, hyperglycemia (and hypoglycemia) must be included in the differential diagnosis of the comatose patient.

Once suspected, the diagnosis can be easily confirmed by measuring the plasma glucose concentration directly or by using reagent sticks. Tablets or reagent sticks

---

†This response of the brain is relatively specific. Other cells such as those in skeletal muscle do not generate new osmoles and remain contracted [58].

can also be used to detect glucosuria, ketonuria, and ketonemia. If serum or plasma that is undiluted or diluted 1:1 with normal saline has a 4+ reaction for ketones, then ketoacidosis can be assumed to be present. It should be noted, however, that the nitroprusside tablets used in this test react with acetoacetate and acetone but not β-hydroxybutyrate. The ratio of β-hydroxybutyrate to acetoacetate is about 3:1 in DKA but may be as high as 8:1 with alcoholic ketoacidosis or when there is coexistent lactic acidosis (see Chap. 19) [60]. In the latter disorder, for example, the altered redox state that promotes pyruvate conversion to lactate also favors acetoacetate conversion to β-hydroxybutyrate. Thus, nitroprusside can underestimate the severity of the ketoacidosis. An indirect way around this problem (since an assay for β-hydroxybutyrate is not routinely available) is to add a few drops of hydrogen peroxide to the urinary specimen. This will convert β-hydroxybutyrate to acetoacetate, which will then be detectable by nitroprusside [61].

The presence of ketoacidosis should also be suspected in any patient with a high anion gap metabolic acidosis (see Chap. 19). In this setting, the H^+ ion is responsible for the acidemia and the retained ketoacid anion (primarily β-hydroxybutyrate) for the elevated anion gap. These findings, however, may be modified by the loss of the ketoacid anions in the urine. Patients who are able to excrete large quantities of ketones can present with only a slightly increased or even a normal anion gap [62]. The acidemia will persist since these anions are excreted with Na^+ and K^+, not exclusively with H^+ (as NH_4^+).

Although the presence of uncontrolled diabetes mellitus may be suggested by many of the above findings, hypoglycemic therapy with insulin should not be begun until the presence of hyperglycemia has been demonstrated. *The isolated finding of glucosuria or ketonemia does not necessarily mean that hyperglycemia is present, and the administration of insulin to such a patient may be dangerous* [63]. Diabetics, particularly those on insulin, can be in a coma due to hypoglycemia, rather than hyperglycemia. If the patient has not emptied his or her bladder for several hours, the urine may contain glucose, reflecting a period of hyperglycemia that may have existed several hours previously and is not the current state. The administration of insulin in this setting will further reduce the plasma glucose concentration and can produce neurologic deterioration and death. Whenever there is a question as to the presence of hypoglycemia or hyperglycemia, it is always safer to give glucose (50 mL of 50% glucose intravenously) immediately after blood has been drawn for measurement of the plasma glucose concentration. This will dramatically improve the status of the hypoglycemic patient but will not be harmful to the patient with hyperglycemia.

In addition to diabetes mellitus, alcoholism can also produce severe ketoacidosis (see page 406) [64,65]. In this condition, the plasma glucose concentration is less than 250 mg/dL and may be less than 100 mg/dL. The treatment of this disorder is to give glucose, which will stimulate endogenous insulin secretion, resulting in the correction of the ketoacidosis. Since the plasma glucose concentration frequently is normal, the administration of exogenous insulin can precipitate severe hypoglycemia.

The diagnosis of hyperosmolality due to mannitol must be made indirectly since an assay for measuring mannitol is not routinely available. The presence of this condition should be suspected in an oliguric patient who is given hypertonic mannitol and then becomes hyponatremic as water is drawn out of the cells [57]. The measured plasma osmolality will be elevated in this setting (not reduced as with most causes of hyponatremia) and will be significantly greater than the value calculated from the following (see Chap. 1):

$$\text{Plasma osmolality} \cong 2 \times \text{plasma } [Na^+] + \frac{[\text{glucose}]}{18} + \frac{BUN}{2.8}$$

The gap between the measured and calculated plasma osmolality is due to the retained mannitol (see page 410 for a review of the other causes of such an osmolal gap).

TREATMENT

Therapy must be directed toward each of the metabolic disturbances that may be present in the hyperglycemic patient: hyperosmolality, ketoacidosis, hypovolemia, and potassium and phosphate depletion [1,10]. Since absolute or relative insulin deficiency is responsible for most of these problems, the administration of insulin and volume repletion are the mainstays of therapy. To assess the effect of treatment, the plasma glucose concentration should be measured every 2 h and the plasma electrolytes and arterial pH every 2 to 4 h until the patient is out of danger.

Insulin

Insulin acts to correct the hyperglycemia, diminish ketone production (by diminishing both lipolysis and glucagon secretion), and increase ketone utilization. The major effect of insulin on glucose metabolism is to increase hepatic uptake, with peripheral utilization being quantitatively less important [66]. The net effect is that the plasma glucose concentration usually falls at a maximum rate of 65 to 125 mg/dL per hour [66–68].

The promotion of ketone utilization by insulin will also lead to correction of the acidemia since metabolism of the ketoacid anions results in the regeneration of HCO_3^- [33]. Although this occurs relatively rapidly over the first 5 to 10 h, acetone is cleared more slowly (in part via the lungs) and may remain in the blood for more than 36 h [69]. Thus, the persistence of ketonemia and ketonuria is not necessarily indicative of a failure of insulin therapy.

Regular (crystalline) insulin should be used initially. Long-acting insulins given subcutaneously or intramuscularly are released too slowly and may not achieve circulating levels high enough to correct severe hyperglycemia or ketoacidosis [19].

Most patients are currently treated with a low-dose insulin regimen in which the total amount of insulin administered is usually between 40 and 100 units [66–68,70–72]. Insulin requirements do not appear to be substantially different in DKA and NKH [34]. A loading dose of 15 to 20 units should be given initially (in part to saturate any anti-insulin antibodies that may be present in patients previously treated with animal insulin) followed by 8 to 15 units every hour until the biochemical abnormalities are under reasonable control. Higher doses (such as 50 to 100 units every 2 h) can be used but do not usually lower the plasma glucose concentration more rapidly [67,68,71]. Although diabetics are frequently insulin-resistant, the equivalent effectiveness of the low- and high-dose regimens suggests that both doses saturate the cell membrane receptors and that the insulin resistance is generally due to a postreceptor defect [1,72a]. Some patients, however, do not respond to low-dose insulin, presumably due to circulating anti-insulin antibodies or a defect at the membrane receptor. In this setting, the dose must be increased to a level adequate to control the plasma glucose concentration.

Regular insulin can be given by the intravenous, subcutaneous, or intramuscular routes. When administered as an intravenous bolus, one is sure of the dose reaching the bloodstream, but the half-life of the insulin is only 4 to 5 min, and it is completely cleared from the plasma within 30 min [73]. Subcutaneous insulin begins to act within 30 min, reaching a peak effect at 2 to 3 h. Intramuscular insulin acts slightly faster than subcutaneous insulin. However, absorption from either of these sites may be erratic, particularly in patients with circulatory insufficiency, and late release of the insulin may result in hypoglycemia. Despite these potential advantages and disadvantages, each route of administration appears to be equally effective in correcting the hyperglycemia and ketoacidosis [72]. If an intravenous infusion is chosen, relatively concentrated solutions should be used to minimize insulin absorption onto the glass and plastic tubing. If, for example, 40 units of regular insulin is added to 100 mL of isotonic saline, an infusion at 20 to 30 mL/h will deliver 8 to 12 units of insulin per hour to the patient.

Fluids

The average fluid loss is 3 to 6 liters in DKA but can reach 8 to 10 liters in NKH [1,34]. With each liter of water, about 70 meq of monovalent cation, primarily Na^+, is lost. This reflects the enhanced water loss relative to solute induced by the glucose osmotic diuresis. In most cases, volume repletion can be achieved with half-isotonic saline (Na^+ concentration of 77 meq/L), a fluid similar in composition to that lost in the urine. When severe hypovolemia or shock is present, initial rehydration should be begun with isotonic saline (Na^+ concentration equal to 154 meq/L), which will still be hypotonic to plasma. In general, 4 to 5 liters of fluid is given in the first 12 h, with the remainder of the deficit replaced in the ensuing 24 to 36 h. If necessary, a central venous pressure catheter can be used to monitor the efficacy of volume repletion and to prevent overhydration.

In addition to correcting hypovolemia, the administration of fluids alone can lower the plasma glucose concentration by as much as 35 to 70 mg/dL per hour [74]. This effect is due to both dilution of the extracellular fluid and improved renal function, which allows enhanced glucose excretion.

The increase in renal perfusion is associated with reductions in the BUN and plasma creatinine concentration. Although the plasma creatinine concentration generally varies inversely with GFR, this value may be misleading in DKA since acetoacetate is a noncreatinine chromogen that is measured as creatinine in the standard colorimetric assay. As a result, the plasma creatinine concentration may be falsely elevated by as much as 2 mg/dL or more, leading to a marked underestimation of the GFR [75]. Metabolism of the acetoacetate following the administration of insulin will rapidly lower the measured plasma creatinine concentration toward its true value.

Rarely, patients develop potentially fatal cerebral edema within the first day after therapy has been initiated [76–79]. The mechanism by which this occurs is uncertain. The combination of insulin and fluids can lower the plasma glucose concentration by a maximum of 200 mg/dL or 11 mosmol/kg per hour. This rapid reduction in the plasma osmolality could promote water movement into the brain, as occurs in hypernatremia (see Chap. 23). This explanation, however, does not seem to apply to most patients with this disorder.

Studies in diabetic animals with marked hyperglycemia suggest that insulin itself may play an important role in the genesis of cerebral edema [80]. Lowering the plasma glucose concentration below 250 to 300 mg/dL with insulin can result in the generation of new (idiogenic) osmoles within the brain cells. This increase in brain cell osmolality can draw water into the brain and produce cerebral edema. The importance of insulin in this phenomenon is suggested by the absence of idiogenic osmole formation when the plasma glucose concentration is reduced with peritoneal dialysis. This model, however, is not entirely applicable to humans since cerebral edema has been described at a time when more marked hyperglycemia was still present [77]. Nevertheless, it seems prudent to discontinue regular insulin and to use dextrose-containing saline solutions when the plasma glucose concentration falls below 300 mg/dL [34]. Further insulin therapy can be ordered according to fluctuations in the plasma glucose concentration or if ketoacidosis persists. If cerebral edema does develop, acutely raising the plasma osmolality with hypertonic mannitol may be beneficial [79].

Bicarbonate

The potential problems associated with HCO_3^- replacement in ketoacidosis are discussed in detail in Chap. 19. Reviewed briefly, the metabolism of ketoacid anions following the administration of insulin results in the generation of HCO_3^- and spontaneous correction of the acidemia. Therefore, treatment with alkali should generally be restricted to patients with severe acidemia (arterial pH less than 7.10 to 7.15) and HCO_3^- given only in small doses sufficient to maintain the pH above 7.20 [81]. In a patient with normal respiratory compensation, an arterial pH above 7.20 will be reached when the plasma HCO_3^- concentration is raised to about 10 meq/L; this may require only 45 to 90 meq of $NaHCO_3$. Overzealous alkali administration can result in a paradoxical fall in CSF pH and possible neurologic deterioration [81,82].

There are two other situations in which HCO_3^- can be given in DKA. First, patients presenting with a normal or only slightly elevated anion gap have excreted most of their ketoacid anions in the urine [62]. In this setting, insulin will not correct the acidemia and exogenous HCO_3^- is more likely to be required. Second, HCO_3^-

should be administered to patients with potentially life-threatening hyperkalemia (see Chap. 28). HCO_3^- drives K^+ into the cells, resulting in a reduction in the plasma K^+ concentration.

Potassium

Patients with DKA or NKH are markedly K^+-depleted. The average K^+ deficit is 3 to 5 meq/kg but can exceed 10 meq/kg in some patients [10]. Several factors contribute to this problem, including vomiting, increased urinary losses due to the osmotic diuresis, and the loss of K^+ from the cells due to glycogenolysis and proteolysis [10,83].

Despite this loss of K^+, the plasma K^+ concentration is usually normal or, in about one-third of patients, elevated [10,34,83]. This paradoxical finding results from a change in the internal distribution of K^+ as K^+ moves from the cells into the extracellular fluid due both to insulin deficiency and hyperglycemia (see Chap. 13) [84–86]. The latter acts by solvent drag since the increase in plasma osmolality pulls water and secondarily K^+ out of the cells [87]. Diabetic patients also may develop hypoaldosteronism, another abnormality that can cause hyperkalemia [88]. It is important to note that acidemia itself does *not* play a major role since organic acids do not appear to affect the distribution of K^+ (see page 254) [83,89]. Thus, hyperkalemia is probably as likely to occur in NKH, where the pH is relatively normal, as it is in DKA [34].

Although masked initially, the K^+ depletion becomes rapidly apparent as insulin therapy drives K^+ into the cells (both directly and by causing glucose utilization). To prevent the development of potentially severe hypokalemia, 20 to 40 meq/L of KCl should be added to the intravenous infusions once the plasma K^+ concentration falls below 5.0 meq/L. The need for K^+ repletion is more urgent in those patients who are hypokalemic prior to therapy. In this setting, K^+ replacement must be begun immediately at an initial rate of up to 20 to 40 meq/h (it is more safely given orally, if possible) if serious cardiac arrhythmias are present (see Chap. 27). This should be followed by careful monitoring of the plasma K^+ concentration, muscle strength, and the electrocardiogram.

Phosphate

Cellular phosphate depletion with an initially normal plasma phosphate concentration is another common finding in DKA or NKH [10,33,90]. Multiple factors contribute to this effect: decreased intake; enhanced release from cells due to glycogenolysis, proteolysis, and an acidemia-induced decomposition of organic acids; and increased urinary losses produced by the osmotic diuresis and acidemia. These processes are rapidly reversed with the administration of insulin as glucose and phosphate enter the cells. The net effect is a reduction in the plasma phosphate concentration to as low as 1.0 mg/dL or less (normal equals 2.5 to 4.5 mg/dL) within the first 5 to 12 h [33,90].

The routine use of phosphate supplements has been recommended in this setting because of the multiple organ failure that can be induced by severe hypophosphatemia [90]. However, two controlled studies have shown that although phosphate administration can minimize the fall in the plasma phosphate concentration, there is no effect on morbidity or the rate of correction of the electrolyte disturbances [91,92]. Furthermore, phosphate administration is not without risk since hyperphosphatemia and hypocalcemia may ensue [93]. It seems prudent, therefore, to reserve phosphate supplements for the occasional patient who develops severe, symptomatic hypophosphatemia.

PROBLEMS

25-1 A 68-year-old woman with adequately controlled diabetes mellitus and previously normal renal function presents with fever, dysuria, nausea, recurrent vomiting, flank pain, and polyuria which have become progressively more severe over 4 days. The physical examination reveals a temperature of 39.6°C, reduced skin turgor, flat neck veins, postural hypotension, and marked tenderness over the right costovertegral angle. The urine shows pyuria and bacteriuria, and a diagnosis of acute pyelonephritis is made. Other laboratory data reveal:

$$\text{Plasma [glucose]} = 570 \text{ mg/dL} \qquad \text{[BUN]} = 32 \text{ mg/dL}$$

$$\text{[Na}^+\text{]} = 135 \text{ meq/L} \qquad \text{[Creatinine]} = 4.0 \text{ mg/dL}$$

$$\text{[K}^+\text{]} = 2.6 \text{ meq/L} \qquad \text{[Ketones]} = 4+, \text{ diluted } 1:1$$

$$\text{[Cl}^-\text{]} = 87 \text{ meq/L} \qquad \text{Arterial pH} = 7.36$$

$$\text{[HCO}_3^-\text{]} = 20 \text{ meq/L} \qquad P_{CO_2} = 37 \text{ mmHg}$$

The electrocardiogram shows prominent U waves in the precordial leads and occasional multifocal premature ventricular beats.

    (*a*) What is the acid-base disturbance on admission?

    (*b*) What factors account for the elevations in the BUN and plasma creatinine concentration?

    (*c*) What would be your initial therapeutic regimen?

25-2 A 27-year-old woman with type 1, insulin-dependent diabetes is admitted to the hospital with a soft-tissue infection of the palate. The initial laboratory data included:

$$\text{Plasma [glucose]} = 147 \text{ mg/dL}$$

$$\text{[Na}^+\text{]} = 140 \text{ meq/L}$$

$$\text{[K}^+\text{]} = 3.8 \text{ meq/L}$$

$$\text{[Cl}^-\text{]} = 110 \text{ meq/L}$$

$$\text{[HCO}_3^-\text{]} = 23 \text{ meq/L}$$

The patient ate sparingly because of pain on swallowing. To minimize the risk of hypoglycemia, her insulin was withheld. Repeat blood tests were obtained 36 h later:

$$\text{Plasma [glucose]} = 270 \text{ mg/dL} \qquad \text{Anion gap} = 15$$

$$\text{[Na}^+\text{]} = 135 \text{ mg/dL} \qquad \text{Ketones} = 4+, \text{ undiluted}$$

$$\text{[K}^+\text{]} = 5.0 \text{ meq/L} \qquad \text{Arterial pH} = 7.32$$

$$\text{[HCO}_3^-\text{]} = 15 \text{ meq/L} \qquad P_{CO_2} = 30 \text{ mmHg}$$

A diagnosis of diabetic ketoacidosis was made.

(*a*) Why is the anion gap only slightly elevated despite the presence of ketoacidosis?

(*b*) How would you treat the patient at this time?

REFERENCES

1. Foster, D. W., and J. D. McGarry: The metabolic derangements and treatment of diabetic ketoacidosis, N. Engl. J. Med., 309:159, 1983.
2. Cahill, G. F., Jr.: Ketosis, Kidney Int., 20:416, 1981.
3. Unger, R. H., and L. Orci: Glucagon and the A cell. Physiology and pathophysiology, N. Engl. J. Med., 304:1518, 1575, 1981.
4. Ensinck, J. W., and R. H. Williams: Disorders causing hypoglycemia, in *Textbook of Endocrinology,* 6th ed., R. H. Williams (ed.), Saunders, Philadelphia, 1981.
5. Felig, P., and J. Wahren: The liver as site of insulin and glucagon action in normal, diabetic, and obese humans, Isr. J. Med. Sci., 11:528, 1975.
6. Felig, P., and J. Wahren: Influence of endogenous insulin secretion on splanchnic glucose and amino acid metabolism in man, J. Clin. Invest., 50:1702, 1971.
7. Blackard, W. G., and N. C. Nelson: Portal and peripheral vein immunoreactive insulin concentrations before and after glucose, Diabetes, 19:302, 1970.
8. Cryer, P. E.: Glucose counterregulation in man, Diabetes, 30:261, 1981.
9. Unger, R. H.: Insulin-glucagon relationships in the defense against hypoglycemia, Diabetes, 32:575, 1983.
10. Kreisberg, R. A.: Diabetic ketoacidosis: new concepts and trends in pathogenesis and treatment, Ann. Intern. Med., 88:681, 1978.
11. DeFronzo, R. A., and E. Ferrannini: The pathogenesis of non-insulin-dependent diabetes. An update, Medicine, 61:125, 1982.
12. Felig, P., and R. S. Sherwin: Glucagon and blood glucose: insights from artificial pancreas studies, Ann. Intern. Med., 92:856, 1980.
13. Raskin, P., A. Pietri, and R. Unger: Changes in glucagon levels after four to five weeks of glucoregulation by portable insulin infusion pumps, Diabetes, 28:1033, 1979.
14. Tamborlane, W. V., R. S. Sherwin, V. Koivisto, R. Hendler, M. Genel, and P. Felig: Normalization of the growth hormone and catecholamine response to exercise in juvenile-onset diabetic subjects treated with a portable insulin infusion pump, Diabetes, 28:785, 1979.
15. Unger, R. H.: Somatostatinoma, N. Engl. J. Med., 296:998, 1977.
16. Sherwin, R. S., M. Fisher, R. Hendler, and P. Felig: Hyperglucagonemia and blood glucose regulation in normal, obese, and diabetic subjects, N. Engl. J. Med., 294:455, 1976.
17. Barnes, A. J., E. M. Kohner, S. R. Bloom, D. J. Johnston, K. G. M. M. Alberti, and P. Smythe: Importance of pituitary hormones in aetiology of diabetic ketoacidosis, Lancet, 1:1171, 1978.
18. Shamoon, H., R. Hendler, and R. S. Sherwin: Synergistic interactions among antiinsulin hormones in the pathogenesis of stress hyperglycemia in humans, J. Clin. Endocrinol. Metab., 52:1235, 1981.
19. Schade, D. S., and R. P. Eaton: Prevention of diabetic ketoacidosis, J. Am. Med. Assoc., 242:2455, 1979.
20. Gerich, J. E., M. M. Martin, and L. Recant: Clinical and metabolic characteristics of hyperosmolar nonketotic coma, Diabetes, 20:228, 1971.
21. Cahill, G.: Starvation in man, N. Engl. J. Med., 282:668, 1970.
22. Chiasson, J. L., R. L. Atkinson, A. D. Cherrington, U. Keller, B. C. Sinclair-Smith, W. W. Lacy, and J. E. Liljenquist: Effects of fasting on gluconeogenesis from alanine in diabetic man, Diabetes, 28:56, 1979.
23. Miles, J. M., M. W. Haymond, S. L. Nissen, and J. E. Gerich: Effects of free fatty acid availability, glucagon excess, and insulin deficiency on ketone body production in postabsorptive man, J. Clin. Invest., 71:1554, 1983.
24. Schade, D. S., and R. P. Eaton: The regulation of plasma ketone body concentration by counterregulatory hormones in man. I. Effects of norepinephrine in diabetic man, Diabetes, 26:989, 1977.

25. Porte, D., Jr., and J. B. Halter: The endocrine pancreas and diabetes mellitus, in *Textbook of Endocrinology,* 6th ed., R. H. Williams (ed.), Saunders, Philadelphia, 1981.

26. McGarry, J. D., G. P. Mannaerts, and D. W. Foster: A possible role for malonyl-CoA in the regulation of hepatic fatty acid oxidation and ketogenesis, J. Clin. Invest., 60:265, 1977.

27. Cook, G. A., R. C. Nielson, R. A. Hawkins, M. A. Mehlman, M. R. Lakshmanan, and R. L. Veech: Effect of glucagon on hepatic malonyl coenzyme A concentration and on lipid synthesis, J. Biol. Chem., 252:4421, 1977.

28. Owen, O. E., V. E. Trapp, C. L. Skutches, M. A. Mozzoli, R. D. Hoeldtke, G. Boden, and G. A. Reichard, Jr.: Acetone metabolism during diabetic ketoacidosis, Diabetes, 31:242, 1982.

29. Reichard. G. A., Jr., O. E. Owen, A. C. Haff, P. Paul, and W. M. Bortz: Ketone body production and oxidation in fasting obese humans, J. Clin. Invest., 53:508, 1974.

30. Miles, J. M., M. W. Haymond, and J. E. Gerich: Suppression of glucose production and stimulation of insulin secretion by physiological concentrations of ketone bodies in man, J. Clin. Endocrinol. Metab., 52:34, 1981.

31. Ruderman, N. B., and M. N. Goodman: Inhibition of muscle acetoacetate utilization during diabetic ketoacidosis, Am. J. Physiol., 226:136, 1974.

32. Sherwin, R. S., R. G. Hendler, and P. Felig: Effect of diabetes mellitus and insulin on the turnover and metabolic response to ketones in man, Diabetes, 25:776, 1976.

33. Seldin, D. W., and R. Tarail: The metabolism of glucose and electrolytes in diabetic acidosis, J. Clin. Invest., 29:552, 1950.

34. Arieff, A. I., and H. J. Carroll: Nonketotic hyperosmolar coma with hyperglycemia: clinical features, pathophysiology, renal function, acid-base balance, plasma-cerebrospinal fluid equilibria and the effects of therapy in 37 cases, Medicine, 51:73, 1972.

35. Seely, J. F., and J. H. Dirks: Micropuncture study of hypertonic mannitol diuresis in the proximal and distal tubule of the dog kidney, J. Clin. Invest., 48:2330, 1969.

36. Al-Kudsi, R. R., J. T. Daugirdas, T. S. Ing, A. O. Kheirbek, S. Popli, J. E. Hano, and V. C. Gandhi: Extreme hyperglycemia in dialysis patients, Clin. Nephrol., 17:228, 1982.

37. Owen, O. E., J. H. Licht, and D. G. Sapir: Renal function and effects of partial rehydration during diabetic ketoacidosis, Diabetes, 30:510, 1981.

38. Katz, M.: Hyperglycemia-induced hyponatremia: calculation of expected serum sodium depression, N. Engl. J. Med., 289:843, 1973.

39. Barnes, A. J., S. R. Bloom, K. G. M. M. Alberti, P. Smythe, F. P. Alford, and D. J. Chisholm: Ketoacidosis in pancreatectomized man, N. Engl. J. Med., 296:1250, 1977.

40. Zierler, K. L., and D. Rabinowitz: Effect of very small concentrations of insulin on forearm metabolism. Persistence of its action on potassium and free fatty acids without its effect on glucose, J. Clin. Invest., 43:950, 1964.

41. Johnson, R. D., J. C. Conn, C. J. Dykman, S. Pek, and J. Starr: Mechanisms and management of hyperosmolar coma without ketoacidosis in the diabetic, Diabetes, 18:111, 1969.

42. Charles, M. A., and E. Danforth: Nonketoacidotic hyperglycemia and coma during intravenous diazoxide therapy in uremia, Diabetes, 20:501, 1971.

43. DiBenedetto, R. J., J. A. Crocco, and J. L. Soscia: Hyperglycemic nonketotic coma, Arch. Intern. Med., 116:74, 1965.

44. Gerich, J. E., M. M. Martin, and L. Recant: Clinical and metabolic characteristics of hyperosmolar nonketotic coma, Diabetes, 20:228, 1971.

45. Joffe, B. I., H. C. Seftel, R. Goldberg, M. Van As, L. Krut, and I. Bersohn: Factors in the pathogenesis of experimental nonketotic and ketoacidotic diabetic stupor, Diabetes, 22:653, 1973.

46. Joffe, B. I., R. B. Goldberg, L. H. Krut, and H. C. Seftel: Pathogenesis of nonketotic hyperosmolar diabetic coma, Lancet, 1:1069, 1975.

47. Amair, P., R. Khanna, B. Leibel, A. Pierratos, S. Vas, E. Meema, G. Blair, L. Chisholm, M. Vas, W. Zingg, G. Digenis, and D. Oreopoulos: Continuous ambulatory peritoneal dialysis in diabetics with end-stage renal disease, N. Engl. J. Med., 306:625, 1982.

48. Boyer, M.: Hyperosmolar anacidotic coma in association with glucocorticoid therapy, J. Am. Med. Assoc., 202:1007, 1967.

49. Charles, M. A., and E. Danforth: Nonketoacidotic hyperglycemia and coma during intravenous diazoxide therapy in uremia, Diabetes, 20:501, 1971.

50. Updike, S. J., and A. R. Harrington: Acute diabetic ketoacidosis: a complication of intravenous diazoxide treatment for refractory hypertension, N. Engl. J. Med., 280:768, 1969.
51. Goldberg, E. M., and S. S. Sanbar: Hyperglycemic, nonketotic coma following administration of Dilantin (diphenylhydantoin), Diabetes, 18:101, 1969.
52. Rapoport, M. E., and H. F. Hurd: Thiazide-induced glucose intolerance treated with potassium, Arch. Intern. Med., 113:405, 1964.
53. Fonseca, V., and D. N. Phear: Hyperosmolar non-ketotic diabetic syndrome precipitated by treatment with diuretics, Br. Med. J., 1:36, 1982.
54. Wyrick, W. J., W. J. Rea, and R. M. McClelland: Rare complications with intravenous hyperosmotic alimentation, J. Am. Med. Assoc., 211:1697, 1970.
55. Rosenberg, S. A., D. K. Brief, J. Kinney, M. Herrera, R. E. Wilson, and F. D. Moore: The syndrome of dehydration, coma, and severe hyperglycemia without ketosis in patients convalescing from burns, N. Engl. J. Med., 272:931, 1965.
56. Freidenberg, G. R., E. J. Kosnik, and J. F. Sotos: Hyperglycemic coma after suprasellar surgery, N. Engl. J. Med., 303:863, 1980.
57. Aviram, A., A. Pfau, J. W. Czaczkes, and T. D. Ullman: Hyperosmolality with hyponatremia caused by inappropriate administration of mannitol, Am. J. Med., 42:648, 1967.
58. Pollack, A. S., and A. I. Arieff: Abnormalities of cell volume regulation and their functional consequences, Am. J. Physiol., 239:F195, 1980.
59. Fulop, M., H. Tannenbaum, and N. Dreyer: Ketotic hyperosmolar coma, Lancet, 2:635, 1973.
59a. Mogensen, C. E.: Glomerular filtration rate and renal plasma flow in short-term and long-term juvenile diabetes mellitus, Scand. J. Clin. Lab. Invest., 28:91, 1971
60. Marliss, E. B., J. L. Ohman, Jr., T. T. Aoki, and G. P. Kozak: Altered redox state obscuring ketoacidosis in diabetic patients with lactic acidosis, N. Engl. J. Med., 283:978, 1970.
61. Narins, R. G., E. R. Jones, M. C. Stom, M. R. Rudnick, and C. P. Bastl: Diagnostic strategies in disorders of fluid, electrolyte and acid-base homeostasis, Am. J. Med., 72:496, 1982.
62. Adrogué, H. J., H. Wilson, A. E. Boyd, W. N. Suki, and G. Eknoyan: Plasma acid-base patterns in diabetic ketoacidosis, N. Engl. J. Med., 307:1603, 1982.
63. Felig, P.: Diabetic ketoacidosis, N. Engl. J. Med., 290:1360, 1974.
64. Levy, L. J., J. Duga, M. Girgis, and E. E. Gordon: Ketoacidosis associated with alcoholism in nondiabetic subjects, Ann. Intern. Med., 78:213, 1973.
65. Miller, P. D., R. E. Heinig, and C. Waterhouse: Treatment of alcoholic acidosis. The role of dextrose and phosphorus, Arch. Intern. Med., 138:67, 1978.
66. Brown, P. M., C. V. Tompkins, S. Juul, and P. H. Sönksen: Mechanism of action of insulin in diabetic patients: a dose-related effect on glucose production and utilisation, Br. Med. J., 1:1239, 1978.
67. Genuth, S. M.: Constant intravenous insulin infusion in diabetic ketoacidosis, J. Am. Med. Assoc., 223:1348, 1973.
68. Padilla, A. J., and J. N. Loeb: "Low dose" versus "high dose" insulin regimens in the management of uncontrolled diabetes. A survey, Am. J. Med., 63:843, 1977.
69. Sulway, M. J., and J. M. Malins: Acetone in diabetic ketoacidosis, Lancet, 2:736, 1970.
70. Soler, N. G., A. D. Wright, M. G. Fitzgerald, and J. M. Malins: Comparative study of different insulin regimens in management of diabetic ketoacidosis, Lancet, 2:1221, 1975.
71. Sönksen, P. H., M. C. Srivastava, C. V. Tompkins, and J. D. N. Nabarro: Growth-hormone and cortisol responses to insulin infusion in patients with diabetes mellitus, Lancet, 2:155, 1972.
72. Fisher, J. N., M. M. Shahshahani, and A. E. Kitabshi: Diabetic ketoacidosis: low-dose insulin therapy by various routes, N. Engl. J. Med., 297:238, 1977.
72a. Scarlett, J. A., O. G. Kolterman, T. P. Ciaraldi, M. Kao, and J. M. Olefsky: Insulin treatment reverses the postreceptor defect in adipocyte 3-0-methyl-glucose transport in type II diabetes mellitus, J. Clin. Endocrinol. Metab., 56:1195, 1983.
73. Turner, R. C., J. A. Grayburn, G. B. Newman, and J. D. N. Nabarro: Measurement of the insulin delivery rate in man, J. Clin. Endocrinol. Metab., 33:279, 1971.
74. Page, M. McB., K. G. M. M. Alberti, R. Greenwood, K. A. Gumau, T. D. R. Hockaday, C. Lowry, J. D. N. Nabarro, D. A. Pyke, P. H. Sönksen, P. J. Watkins, and T. E. T. West: Treatment of diabetic coma with continuous low-dose insulin infusion, Br. Med. J., 2:687, 1974.

75. Molitch, M. E., E. Roidman, C. A. Hirsch, and E. Dubinsky: Spurious serum creatinine elevations in ketoacidosis, Ann. Intern. Med., 93:280, 1980.

76. Young, E., and R. F. Bradley: Cerebral edema with irreversible coma in severe diabetic ketoacidosis, N. Engl. J. Med., 276:665, 1967.

77. Rosenbloom, A. L., W. J. Riley, F. T. Weber, J. I. Malone, and W. H. Donnelly: Cerebral edema complicating diabetic ketoacidosis in childhood, J. Pediatr., 96:357, 1980.

78. Editorial: Crystalloid infusions in diabetic ketoacidosis, Lancet, 2:308, 1982.

79. Franklin, B., J. Liu, and F. Ginsberg-Fellner: Cerebral edema and ophthalmoplegia reversed by mannitol in a new case of insulin-dependent diabetes mellitus, Pediatrics, 69:87, 1982.

80. Arieff, A. I., and C. R. Kleeman: Studies on mechanisms of cerebral edema in diabetic comas, J. Clin. Invest., 52:571, 1973.

81. Kaye, R.: Diabetic ketoacidosis: the bicarbonate controversy, J. Pediatr., 87:156, 1975.

82. Posner, J., and F. Plum: Spinal-fluid pH and neurologic symptoms in systemic acidosis, N. Engl. J. Med., 277:605, 1967.

83. Perez, G. O., J. R. Oster, and C. A. Vaamonde: Serum potassium concentration in acidemic states, Nephron, 27:233, 1981.

84. Cox, M., R. H. Sterns, and I. Singer: The defense against hyperkalemia: the roles of insulin and aldosterone, N. Engl. J. Med., 299:525, 1978.

85. Nicolis, G. L., T. Kahn, A. Sanchez, and J. L. Gabrilove: Glucose-induced hyperkalemia in diabetic subjects, Arch. Intern. Med., 141:48, 1981.

86. Goldfarb, S., M. Cox, I. Singer, and M. Goldberg: Acute hyperkalemia induced by hyperglycemia: hormonal mechanisms, Ann. Intern. Med., 84:426, 1976.

87. Moreno, M., C. Murphy, and C. Goldsmith: Increase in serum potassium resulting from the administration of hypertonic mannitol and other solutions, J. Lab. Clin. Med., 73:291, 1969.

88. DeFronzo, R. A.: Hyperkalemia in hyporeninemic hypoaldosteronism, Kidney Int., 17:118, 1980.

89. Adrogué, H. J., and N. E. Madias: Changes in plasma potassium concentration during acute acid-base disturbances, Am. J. Med., 71:456, 1981.

90. Knochel, J. P.: The pathophysiology and clinical characteristics of severe hypophosphatemia, Arch. Intern. Med., 137:203, 1977.

91. Keller, U., and W. Berger: Prevention of hypophosphatemia by phosphate infusion during treatment of diabetic ketoacidosis and hyperosmolar coma, Diabetes, 29:87, 1980.

92. Wilson, H. K., S. P. Kever, A. S. Lea, A. E. Boyd, and G. Eknoyan: Phosphate therapy in diabetic ketoacidosis, Arch. Intern Med., 142:517, 1982.

93. Winter, R. J., C. J. Harris, L. S. Phillips, and O. C. Green: Diabetic ketoacidosis. Induction of hypocalcemia and hypomagnesemia by phosphate therapy, Am. J. Med., 67:897, 1979.

TWENTY-SIX

INTRODUCTION TO DISORDERS OF POTASSIUM BALANCE

The maintenance of K^+ balance is essential for a variety of cellular functions. This chapter will review the physiologic effects of K^+ and the factors governing K^+ homeostasis, topics which are generally discussed in greater detail in Chap. 13. The application of these principles to the common clinical problems of K^+ depletion and K^+ excess will then be presented in Chaps. 27 and 28.

PHYSIOLOGIC EFFECTS OF POTASSIUM

The total body K^+ stores in a normal adult are approximately 3000 to 4000 meq (50 to 55 meq/kg body weight). Roughly 98 percent of the body K^+ is located in the cells; this is in contrast to Na^+, which is primarily limited to the extracellular fluid. The localization of Na^+ and K^+ to the different fluid compartments is maintained by the Na^+-K^+-ATPase pump in the cell membrane, which transports Na^+ out of and K^+ into the cells in a 3:2 ratio (see Chap. 1) [1,2]. The net effect is that the K^+

concentration is about 150 meq/L in the cells but only 4 to 5 meq/L in the extracellular fluid (including the plasma).

Cell Function

Potassium plays an important role in cell function and neuromuscular transmission. In the cells, K^+ participates in the regulation of such processes as protein and glycogen synthesis [3,4]. As a result, conditions of K^+ imbalance are associated with a variety of signs and symptoms. For example, patients with K^+ depletion frequently complain of polyuria and polydipsia (increased urine output and thirst) [5]. These problems, which are reversed with K^+ repletion, appear to be due to both a reduced ability to concentrate the urine (resulting from decreased tubular responsiveness to antidiuretic hormone; see Fig. 13-1) and a primary stimulation of water intake [6–9].

Resting Membrane Potential

In addition to the importance of the *absolute* amount of K^+ present, the *ratio* of the K^+ concentration in the cells to that in the extracellular fluid is the major determinant of the resting membrane potential (E_m) across the cell membrane. This relationship can be expressed by the following formula

$$E_m = -61 \log \frac{r[K^+]_c + 0.01[Na^+]_c}{r[K^+]_e + 0.01[Na^+]_e} \qquad (26\text{-}1)$$

where r is the 3:2 active transport ratio of the Na^+-K^+-ATPase pump, 0.01 is the relative membrane permeability of Na^+ to K^+, and the subscripts c and e refer to the cellular and extracellular concentrations, respectively [2].

If the normal concentrations for K^+ and Na^+ are substituted in Eq. 26-1 (see Table 1-6),

$$E_m = -61 \log \frac{\frac{3}{2}(150) + 0.01(12)}{\frac{3}{2}(4.4) + 0.01(145)}$$

$$= -88 \text{ mV}$$

As described in Chap. 1, this resting potential is generated largely by the diffusion of K^+ out of the cell down its concentration gradient (Na^+ diffusion in the opposite direction being less prominent because of the lower membrane permeability). This movement of positively charged K^+ ions makes the interior of the cell electrically negative with respect to the extracellular fluid.

It is the resting membrane potential that sets the stage for the generation of the action potential that is essential for normal neural and muscular function [10]. During excitation, the release of acetylcholine at synapses and motor end plates produces an increase in Na^+ permeability. This change has three consequences: Na^+ enters the cell down its concentration gradient; this Na^+ movement causes the membrane potential to be depolarized—i.e., it declines toward zero; and K^+ then moves out of the cell since the interior of the cell is now less electronegative. The net effect depends

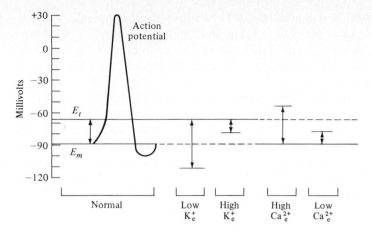

Figure 26-1 Relationship between the extracellular concentrations of K^+ ($[K^+]_e$) and Ca^{2+} ($[Ca^{2+}]_e$) on the resting (E_m) and threshold (E_t) potentials of a normal skeletal muscle. When the threshold potential is reached during depolarization, an action potential is seen. The height of the arrows is equal to the difference between the resting and threshold potentials and represents the excitability of the cell membrane. Changes in the $[K^+]_e$ affect the resting potential whereas those in the $[Ca^{2+}]_e$ affect the threshold potential. (*Adapted from A. Leaf and R. Cotran, Renal Pathophysiology, Oxford University Press, New York, 1976.*)

upon the degree of depolarization. When the depolarizing stimulus is relatively small, there is only a minor increment in sodium permeability. As a result, the initial Na^+ movement into the cell is followed by a period in which K^+ exit exceeds further Na^+ entry because of the greater K^+ permeability. This flux of K^+ raises the membrane potential back toward its baseline value, and generation of an action potential does not occur.

The *threshold potential* (E_t) is that potential at which the sodium permeability is sufficiently enhanced so that Na^+ entry continues to exceed K^+ exit. This induces a self-perpetuating cycle characterized by more depolarization (since the entry of Na^+ makes the cell interior less electronegative), a further increase in sodium permeability (up to 1000 times the basal value, as depolarization itself increases the sodium permeability), more Na^+ entry, more depolarization, etc. [10]. The net effect is the generation of an action potential in which the cell interior becomes electro*positive* due to the massive influx of Na^+ (Fig. 26-1). The propagation of these changes to adjacent cells is responsible for the transmission of neural impulses and the initiation of muscle contraction.†

†The action potential is followed by repolarization and recovery. During repolarization, the permeability to Na^+ returns to its baseline value whereas that to K^+ is slightly increased [10]. In this setting, the high cell K^+ concentration, high K^+ permeability, and the electrical gradient (cell interior now electrically positive) all favor the passive movement of K^+ out of the cell, returning the potential to its negative resting level. In the recovery phase, the Na^+-K^+ pump extrudes the Na^+ that entered the cell during depolarization and pumps in the K^+ that left the cell during repolarization, resulting in the normalization of cell composition. It should be noted that the quantity of ions that must cross the cell membrane to produce these changes in potential is extremely small. To generate a resting potential of 90 mV requires the separation of only 10^{-9} meq of K^+ per square centimeter of membrane [2,10].

Membrane excitability The excitability (or irritability) of neuromuscular tissue is defined as the difference between the resting and threshold potentials ($E_m - E_t$). Thus, any factor which alters either of these potentials affects excitability. In particular, small changes in the extracellular K^+ concentration (which is much lower than that in the cells) can produce relatively large changes in the $[K^+]_c/[K^+]_e$ ratio and consequently in the resting membrane potential (Fig. 26-1) [11,12].

Hypokalemia increases the magnitude of the resting potential,† thereby hyperpolarizing the cell membrane. As a result, the $E_m - E_t$ is increased, and the cell is less sensitive to exciting stimuli. If severe hypokalemia is present, flaccid paralysis may ensue. On the other hand, hyperkalemia reduces the magnitude of the membrane potential, initially making the cell more excitable. With severe hyperkalemia, however, the resting potential may become less than the threshold potential. If this occurs, the cell is unable to repolarize after a single action potential and is no longer excitable. Thus, hyperkalemia can also produce muscle paralysis, which can be fatal if the respiratory muscles are involved. In addition to these changes in skeletal muscle, the conducting fibers in the heart are affected, producing characteristic changes in the electrocardiogram and potentially fatal arrhythmias (see Chaps. 27 and 28).

Despite the effects of K^+ on the resting potential, the changes associated with hypokalemia or hyperkalemia are variable. For example, one patient may have severe muscle weakness with a plasma K^+ concentration of 1.8 meq/L whereas another patient may be asymptomatic at the same level. Two factors appear to be responsible for this individual variation. First, the effect of a change in the plasma K^+ concentration on the resting potential is dependent upon the degree to which the cellular stores can "buffer" the changes in the extracellular fluid and minimize the alteration in the $[K^+]_c/[K^+]_e$ ratio. In hypokalemic periodic paralysis, for example, extracellular K^+ acutely moves into the cell. This results in a relatively large change in the $[K^+]_c/[K^+]_e$ ratio, and paralysis may ensue (see Chap. 27). In contrast, with chronic K^+ depletion due to gastrointestinal or renal losses, K^+ can move from the cells into the extracellular fluid down the concentration gradient created by the fall in the plasma K^+ concentration [13]. As a result, there is less change in the ratio and, therefore, in the resting potential since the K^+ concentration falls in both the cells and the extracellular fluid. Similarly, chronic hyperkalemia may be better tolerated than acute hyperkalemia, although the capacity of the cells to store excess K^+ is relatively limited [13].

Second, membrane excitability is determined by factors other than K^+, including the plasma Ca^{2+} concentration and pH. In contrast to K^+, Ca^{2+} acts on the threshold potential, not the resting potential (Fig. 26-1) [10,14]. This appears to be mediated by an effect of calcium on the relationship between the membrane potential and sodium permeability. As described above, threshold occurs when the sodium permeability is high enough to allow Na^+ entry to exceed K^+ exit; this is normally

---

†Since the membrane potential is stated in terms of the cell being electronegative, what is meant by an increase or decrease in potential may be confusing. An increase in potential refers to an increase in electronegativity across the cell membrane (or hyperpolarization) whereas a decrease in potential refers to a fall in electronegativity toward zero (or depolarization).

seen at a membrane potential of about -65 mV. Hypocalcemia appears to increase membrane excitability by allowing the threshold sodium permeability to be reached with a lesser degree of depolarization, e.g., at -75 mV [10]. On the other hand, the membrane becomes less excitable with hypercalcemia since more depolarization is now required to attain the threshold sodium permeability. These relationships become clinically important in certain settings. For example, an increase in the plasma Ca^{2+} concentration tends to counteract the effect of an increase in the plasma K^+ concentration by returning membrane excitability ($E_m - E_t$) toward normal. This constitutes the rationale for the administration of calcium salts in the treatment of severe hyperkalemia (see Chap. 28).

Membrane excitability is also influenced by changes in the arterial pH, being directly increased by alkalemia and decreased by acidemia. In addition, alterations in pH affect the plasma K^+ concentration. For example, metabolic acidosis can induce the membrane changes of hyperkalemia because of the release of intracellular K^+ into the extracellular fluid (see page 254) [15].

In summary, the degree to which a change in K^+ balance affects neuromuscular excitability is dependent upon a variety of factors, including the rapidity with which it occurs, the plasma Ca^{2+} concentration, and the pH. As a result, the severity of symptoms does not necessarily correlate with the magnitude of the alteration in the plasma K^+ concentration. Since the electrocardiogram and muscle strength reflect the *functional consequences* of K^+ excess or depletion, the monitoring of these parameters as well as the plasma K^+ concentration is essential in the management of patients with severe K^+ imbalance.

REGULATION OF POTASSIUM BALANCE

The maintenance of K^+ balance involves two functions: (1) the normal distribution of K^+ between the cells and extracellular fluid and (2) the renal excretion of the K^+ added to the extracellular fluid from dietary intake and endogenous cellular breakdown.

Distribution between Extracellular Fluid and Cells

Regulation of the internal distribution of K^+ must be extremely efficient since the movement of as little as 1.5 to 2 percent of the cell K^+ into the extracellular fluid can result in a potentially fatal increase in the plasma K^+ concentration to as high as 8 meq/L or more. In the basal state, normal K^+ distribution is achieved primarily by the Na^+-K^+-ATPase pump. In addition, the ability of K^+ to move between the cells and the extracellular fluid is also important. As described above, the fall in the plasma K^+ concentration following gastrointestinal or urinary K^+ losses is minimized by the movement of K^+ out of the cells.

Conversely, K^+ can enter the cells after a K^+ load. The importance of this response can be appreciated from the following example. Suppose a normal 70-kg subject drinks three glasses of orange juice containing 40 meq of K^+. If this K^+

Table 26-1 Factors influencing the distribution of K^+ between the cells and the extracellular fluid

Physiologic
  Na^+-K^+-ATPase
  Catecholamines
  Insulin
  Plasma K^+ concentration
  Exercise

Pathologic
  Chronic diseases
  Arterial pH
  Rate of cell breakdown
  Hyperosmolality

remained in the extracellular fluid (the extracellular volume being approximately 17 liters), there would be a potentially dangerous 2.4-meq/L increase in the plasma K^+ concentration. This is prevented by the entry of most of the K^+ load into the cells [16] until the excess K^+ can be excreted in the urine [17,18].

The physiologic and pathologic factors that influence the distribution of K^+ are listed in Table 26-1. The role of these factors in hypokalemic and hyperkalemic states will be discussed in the following two chapters. Nevertheless, it is useful at this time to briefly review the physiologic roles of catecholamines, insulin, and the plasma K^+ concentration [16,19,20].

Catecholamines and insulin Catecholamines and insulin can affect K^+ distribution as both β_2-adrenergic stimuli (primarily epinephrine) and insulin promote the cellular uptake of K^+ [16,19–20]. The physiologic importance of these effects has been demonstrated by the response to the administration of β-adrenergic blockers or somatostatin (which impairs insulin secretion). In these settings, the increment in the plasma K^+ concentration after a K^+ load is greater and more prolonged than normal (see Figs. 13-2 and 13-3) [20–22].

It appears that it is the *basal* levels of epinephrine and insulin that enhance K^+ uptake since a K^+ load does not promote the release of these hormones [22,23]. If, however, the availability of epinephrine or insulin is increased (as with a glucose load for insulin), there will be a further tendency for K^+ to move into the cells [21,24]. This effect lasts only several hours because other factors (perhaps the plasma K^+ concentration itself) then cause K^+ to move back into the extracellular fluid [24].

Thus, the primary physiologic effect of epinephrine and insulin is *to facilitate the disposition of an acute K^+ load, not to regulate the baseline plasma K^+ concentration*. Although a deficiency of these hormones may cause mild hyperkalemia [20], this effect is transient since the excess K^+ can reenter the cells or be excreted in the urine. As a result, the fasting plasma K^+ concentration is typically normal in patients treated with β-adrenergic blockers (for hypertension or coronary artery disease) and in patients with diabetes mellitus given enough insulin to prevent marked hyperglycemia [20–22].

Plasma potassium concentration The internal distribution of K^+ is also affected by the plasma K^+ concentration. For example, a K^+ load leads to an elevation in the plasma K^+ concentration, which can then promote K^+ entry into the cells, perhaps by passive mechanisms. The observation that the combination of insulin deficiency and sympathetic blockade impairs but does not prevent the intracellular movement of K^+ in this setting is consistent with this hypothesis [19,25].

The net effect is that in most situations, the plasma K^+ concentration varies directly with body K^+ stores, decreasing with K^+ depletion and increasing with K^+ retention. In general, a reduction in the plasma K^+ concentration from 4.0 to 3.0 meq/L is associated with a 100- to 400-meq deficit in total body K^+ [13,26]. Conversely, an elevation in the plasma K^+ concentration from 4.0 to 5.0 meq/L is usually associated with the retention of 100 to 200 meq of K^+ [13].

There are some exceptions to this rule, and they primarily occur with disorders that affect K^+ distribution. For example, insulin deficiency with hyperglycemia, metabolic acidosis, severe exercise, and excess tissue breakdown all induce K^+ movement out of the cells. In these settings, hyperkalemia can occur even though total body K^+ stores are normal or even reduced. These problems are discussed in detail in the next two chapters. It is important, however, to note that the effect of exercise can interfere with the routine measurement of the plasma K^+ concentration. After a tourniquet is applied to obtain a blood sample, the patient is frequently instructed to repeatedly clench and unclench his or her fist in an attempt to increase local blood flow and make the veins more apparent. This can result in K^+ movement out of the cells and an elevation in the plasma K^+ concentration of as much as 2 meq/L, leading to erroneous evaluation of the state of K^+ balance [27].

Renal Excretion

Although small amounts of K^+ are lost each day in the feces and sweat, the urine is the major route by which the K^+ derived from the diet and endogenous cellular breakdown is eliminated from the body. The primary event in urinary K^+ excretion is the *secretion* of K^+ from the tubular cell into the lumen in the distal nephron, particularly the cortical collecting tubule (see Chap. 13) [28–30]. Although a substantial amount of K^+ is filtered, almost all of this is reabsorbed prior to the distal secretory site. The amount of K^+ secreted varies appropriately with the state of K^+ balance: it is enhanced by a K^+ load and reduced by a low-K^+ diet. In addition, net distal reabsorption rather than secretion can occur in states of K^+ depletion [30].

The secretion of K^+ from the cell into the lumen is primarily passive and therefore is a function of luminal membrane permeability and the concentration and electrical gradients across the luminal membrane [28,29]. *Aldosterone* and the *plasma K^+ concentration* are the major physiologic determinants of K^+ secretion as they vary directly with the state of K^+ balance (Fig. 26-2). The *flow rate to the distal nephron* and the *potential difference* generated by Na^+ reabsorption are also important, but they generally play a permissive rather than a regulatory role in that they do not necessarily change with alterations in K^+ balance.

After a K^+ load, the small increase in the plasma K^+ concentration stimulates the release of aldosterone [31], both of which then promote distal K^+ secretion (see

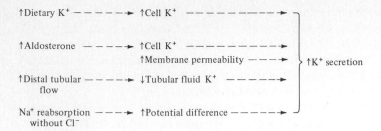

Figure 26-2 Summary of the factors which increase distal K^+ secretion by altering the electrochemical gradient favoring the movement of K^+ from the cell into the lumen. Conversely, K^+ secretion is decreased by reductions in dietary intake, aldosterone secretion or effect, distal tubular flow, or potential difference.

Fig. 13-7) [32–34]. Aldosterone has two major effects on K^+ handling in the distal nephron as it increases both the activity of the Na^+-K^+-ATPase pump in the peritubular membrane (thereby enhancing the cell K^+ concentration) and the K^+ permeability of the luminal membrane, which directly favors K^+ secretion (see Chap. 8) [35,36]. The elevation in the plasma K^+ concentration potentiates the effect of aldosterone [33], probably by increasing the cell K^+ concentration, a change which may be mediated in part by enhanced Na^+-K^+-ATPase activity [34,37]. These changes are reversed with a low-K^+ diet or K^+ depletion [30,38]. The net effect is that K^+ excretion can range from a low of 5 to 15 meq/day with marked K^+ depletion [39] to a high of more than 500 meq/day with chronic K^+ loading [40].

The distal flow rate affects K^+ secretion in a different manner—by influencing the tubular fluid, not the cell K^+ concentration. The secretion of K^+ raises the tubular fluid K^+ concentration, thereby limiting the concentration gradient for further diffusion out of the cell. This effect can be minimized and K^+ secretion enhanced if the flow rate is relatively high, since the secreted K^+ will be washed away by fluid with a very low K^+ concentration due to the reabsorption of filtered K^+ in the proximal tubule and loop of Henle (see Fig. 13-8) [41].

The distal flow rate plays an important role in both normal and disease states. In particular, it allows aldosterone to regulate Na^+ balance without interfering with that of K^+ (see Table 8-2). In hypovolemic states such as congestive heart failure, enhanced secretion of aldosterone contributes to the retention of Na^+. These patients (if untreated) are not usually hypokalemic, however, since the associated increase in proximal reabsorption reduces distal flow, thereby counteracting the effect of aldosterone on K^+ secretion [42]. Conversely, an elevation in K^+ secretion does not occur in normal subjects if distal flow rate is enhanced by high Na^+ intake [43]. In this setting, the ensuing volume expansion suppresses the secretion of aldosterone, allowing the excess Na^+ to be excreted without wasting K^+.

Inappropriate K^+ loss and hypokalemia will be seen, however, if distal flow is enhanced while aldosterone secretion is normal or elevated. This sequence may follow the administration of Na^+ to a patient with an aldosterone-producing adenoma (in whom aldosterone secretion is not suppressible by volume expansion) [43] or the use of diuretics such as furosemide or a thiazide [44]. In the latter setting, flow to the secretory site is increased because tubular reabsorption is impaired in the loop of Henle or distal tubule (see Chap. 16); this is frequently accompanied by hyperaldosteronism due to the diuretic-induced volume loss.

Since K^+ is a charged particle, its secretion is also affected by the transepithelial potential difference across the tubular cell. The normal potential difference in the K^+-secreting cells in the cortical collecting tubule is approximately -48 mV (lumen negative); this potential is generated by the reabsorption of Na^+ (which is positively charged) from the lumen into the peritubular capillary (see Fig. 13-9). This luminal negativity favors K^+ secretion into the lumen [45].

The importance of Na^+ transport in this process can be illustrated by the response to the diuretic amiloride [46]. This drug impairs the entry of luminal Na^+ into the cells of the distal nephron, apparently by decreasing the Na^+ permeability of the luminal membrane. The net effect is diminished Na^+ reabsorption, a reduction in the transepithelial potential difference, and a marked fall in K^+ secretion. Since amiloride has no known direct effect on K^+ handling, it is likely that the decrease in the potential difference is responsible for the change in K^+ secretion [46].

On the other hand, an increase in potential difference promotes K^+ secretion. The normal electrical gradient favors the passive reabsorption of Cl^-, which partially dissipates the Na^+-generated potential. If, however, Na^+ is presented to the distal tubule with a nonreabsorbable anion such as sulfate or the antimicrobial carbenicillin, the anion is unable to follow. As a result, the potential difference rises [45] as does K^+ secretion [47,48]. Thus, the administration of carbenicillin can lead to excess K^+ losses and hypokalemia [49].

Hypokalemia and Hyperkalemia

In summary, K^+ enters the body by dietary ingestion (normal K^+ intake is 40 to 120 meq/day) or intravenous infusion, is stored primarily in the cells, and is excreted in the urine and to a lesser degree in the feces and sweat (Fig. 26-3). Thus, changes in the plasma K^+ concentration must involve changes in one or more of these processes. For example, hypokalemia can be produced by increased K^+ entry into cells or by increased losses. Decreased dietary intake can contribute to these disorders; it will not, however, cause significant K^+ depletion in normal subjects unless intake is severely restricted, since the kidney can reduce K^+ losses to less than 15 to 20 meq/day [39]. Conversely, hyperkalemia is most often due to enhanced release of K^+ from cells or to decreased urinary excretion. Unless given acutely, increased intake alone will not lead to hyperkalemia in normal subjects since the excess K^+ will be excreted in the urine [40].

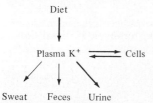

Figure 26-3 Schematic illustration of the factors involved in K^+ homeostasis.

PROBLEMS

26-1 A patient with recurrent diarrhea complains of severe muscle weakness. There is no history, e.g., carpopedal spasm, or physical findings, e.g., Trousseau's or Chvostek's signs, consistent with hypocalcemia. The electrocardiogram reveals ST segment and T wave changes with premature ventricular beats which are felt to be compatible with hypokalemia. The following laboratory data are obtained:

$$\text{Plasma } [Na^+] = 140 \text{ meq/L}$$

$$[K^+] = 1.3 \text{ meq/L}$$

$$[Cl^-] = 117 \text{ meq/L}$$

$$[HCO_3^-] = 10 \text{ meq/L}$$

$$[\text{Albumin}] = 4.0 \text{ g/dL (normal} = 3.5\text{--}5.0 \text{ g/dL)}$$

$$[Ca^{2+}] = 6.3 \text{ mg/dL (normal} = 8.8\text{--}10.5 \text{ mg/dL)}$$

$$\text{Arterial pH} = 7.26$$

$$P_{CO_2} = 23 \text{ mmHg}$$

(a) What effect would correction of the metabolic acidosis have on the plasma K^+ concentration?

(b) Would correction of hypocalcemia be part of your initial therapeutic regimen?

REFERENCES

1. Sweadner, K. J., and S. M. Goldin: Active transport of sodium and potassium ions: mechanism, function, and regulation, N. Engl. J. Med., 302:777, 1980.
2. DeVoe, R. D., and P. C. Maloney: Principles of cell homeostasis, in *Medical Physiology,* 14th ed., V. B. Mountcastle (ed.), Mosby, St. Louis, 1980.
3. Cannon, P. J., L. E. Frazier, and R. H. Hughes: Sodium as a toxic ion in potassium deficiency, Metabolism, 2:297, 1953.
4. Lubin, M.: Intracellular potassium and control of protein synthesis, Fed. Proc., 23:994, 1964.
5. Rubini, M.: Water excretion in potassium-deficient man, J. Clin. Invest., 40:2215, 1961.
6. Manitius, A., H. Levitin, D. Beck, and F. H. Epstein: On the mechanism of impairment of renal concentrating ability in potassium deficiency, J. Clin. Invest., 39:684, 1960.
7. Beck, N., and S. K. Webster: Impaired urinary concentrating ability and cyclic AMP in K^+-depleted rat kidney, Am. J. Physiol., 231:1204, 1976.
8. Bennett, C. M.: Urine concentration and dilution in hypokalemic and hypercalcemic dogs, J. Clin. Invest., 49:1447, 1970.
9. Berl, T., S. L. Linas, G. A. Aisenbrey, and R. J. Anderson: On the mechanisms of polyuria in potassium depletion. The role of polydipsia, J. Clin. Invest., 60:620, 1977.
10. Kuffler, S. W., and J. G. Nicholls: *From Neuron to Brain,* Sinauer Assoc., Sunderland, Mass., 1976, chap. 6.
11. Adrian, R. H.: The effect of internal and external potassium concentration on the membrane potential of frog muscle, J. Physiol. (London), 133:631, 1956.
12. Shanes, A. M.: Electrochemical aspects of physiological and pharmacological action in excitable cells. Part II. The action potential and excitation, Pharmacol. Rev., 10:165, 1958.
13. Sterns, R. H., M. Cox, P. U. Feig, and I. Singer: Internal potassium balance and the control of the plasma potassium concentration, Medicine, 60:339, 1981.
14. Frankenhaeuser, B., and A. L. Hodgkin: The action of calcium on the electrical properties of squid axons, J. Physiol. (London), 137:218, 1957.

15. Abrams, W. B., D. W. Lewis, and S. Bellet: The effect of acidosis and alkalosis on the plasma potassium concentration and the electrocardiogram of normal and potassium depleted dogs, Am. J. Med. Sci., 222:506, 1951.

16. Bia, M. J., and R. A. DeFronzo: Extrarenal potassium homeostasis, Am. J. Physiol., 240:F257, 1981.

17. Winkler, A. W., H. E. Hoff, and P. K. Smith: The toxicity of orally administered potassium salts in renal insufficiency, J. Clin. Invest., 20:119, 1941.

18. DeFronzo, R. A., P. A. Taufield, H. Black, P. McPhedran, and C. R. Cooke: Impaired renal tubular potassium secretion in sickle cell disease, Ann. Intern. Med., 90:310 1979.

19. Silva, P., and K. Spokes: Sympathetic system in potassium homeostasis, Am. J. Physiol., 241:F151, 1981.

19a. Brown, M. J., D. C. Brown, and M. B. Murphy: Hypokalemia from beta$_2$-receptor stimulation by circulating epinephrine, N. Engl. J. Med., 309:1414, 1983.

20. DeFronzo, R. A., R. S. Sherwin, M. Dillingham, R. Hendler, W. V. Tamborlane, and P. Felig: Influence of basal insulin and glucagon secretion on potassium and sodium metabolism: studies with somatostatin in normal dogs and in normal and diabetic human beings, J. Clin. Invest., 61:472, 1978.

21. Rosa, R. M., P. Silva, J. B. Young, L. Landsberg, R. S. Brown, J. W. Rowe, and F. H. Epstein: Adrenergic modulation of extrarenal potassium disposal, N. Engl. J. Med., 302:431, 1980.

22. DeFronzo, R. A., M. Bia, and G. Birkhead: Epinephrine and potassium homeostasis, Kidney Int., 20:83, 1981.

23. Cox, M., R. H. Sterns, and I. Singer: The defense against hyperkalemia: the roles of insulin and aldosterone, N. Engl. J. Med., 299:525, 1978.

24. Minaker, K. L., and J. W. Rowe: Potassium homeostasis during hyperinsulinemia: effect of insulin level, β-blockade, and age, Am. J. Physiol., 242:E373, 1982.

25. DeFronzo, R. A., R. Lee, A. Jones, and M. Bia: Effect of insulinopenia and adrenal hormone deficiency on acute potassium tolerance, Kidney Int., 17:586, 1980.

26. Scribner, B. H., and J. M. Burnell: Interpretation of the serum potassium concentration, Metabolism, 5:468, 1956.

27. Brown, J. J., R. H. Chinn, D. L. Davies, R. Fraser, A. F. Lever, R. J. Rae, and J. I. S. Robertson: Falsely high plasma potassium values in patients with hyperaldosteronism, Br. Med. J., 2:18, 1970.

28. Wright, F. S.: Sites and mechanisms of potassium transport along the renal tubule, Kidney Int., 11:415, 1977.

29. Wright, F. S.: Potassium transport by successive segments of the mammalian nephron, Fed. Proc., 40:2398, 1981.

30. Stanton, B. A., D. Biemesderfer, J. B. Wade, and G. Giebisch: Structural and functional study of the rat distal nephron: effects of potassium adaptation and depletion, Kidney Int., 19:36, 1981.

31. Himathongham, T., R. Dluhy, and G. Williams: Potassium-aldosterone-renin interrelationships, J. Clin. Endocrinol. Metab., 41:153, 1975.

32. Stokes, J. B., M. J. Ingram, A. D. Williams, and D. Ingram: Heterogeneity of the rabbit collecting tubule: localization of mineralocorticoid hormone action to the cortical portion, Kidney Int., 20:340, 1981.

33. Young, D. B., and A. W. Paulsen: Interrelated effects of aldosterone and plasma potassium on potassium excretion, Am. J. Physiol., 244:F28, 1983.

34. Silva, P., B. D. Ross, A. N. Charney, A. Besarab, and F. H. Epstein: Potassium transport by the isolated perfused kidney, J. Clin. Invest., 56:862, 1975.

35. Hierholzer, K., and M. Wiederholt: Some aspects of distal tubular solute and water transport, Kidney Int., 9:198, 1976.

36. Petty, K. J., J. P. Kokko, and D. Marver: Secondary effect of aldosterone on Na-K-ATPase activity in the rabbit cortical collecting tubule, J. Clin. Invest., 68:1514, 1981.

37. Doucet, A., and A. I. Katz: Renal potassium adaptation: Na-K-ATPase activity along the nephron after chronic potassium loading, Am. J. Physiol., 238:F380, 1980.

38. Linas, S. L., L. N. Peterson, R. J. Anderson, G. A. Aisenbrey, F. R. Simon, and T. Berl: Mechanism of renal potassium conservation in the rat, Kidney Int., 15:601, 1979.

39. Squires, R. D., and E. J. Huth: Experimental potassium depletion in normal human subjects. I. Relation of ionic intakes to the renal conservation of potassium, J. Clin. Invest., 38:1134, 1959.

40. Talbott, J. H., and R. S. Schwab: Recent advances in the biochemistry and therapeusis of potassium salts, N. Engl. J. Med., 222:585, 1940.

41. Khuri, R. M., M. Wiederholt, N. Strieder, and G. Giebisch: Effects of flow rate and potassium intake on distal tubular potassium transfer, Am. J. Physiol., 228:1249, 1975.

42. Seldin, D., L. Welt, and J. Cort: The role of sodium salts and adrenal steroids in the production of hypokalemic alkalosis, Yale J. Biol. Med., 29:229, 1956.

43. George, J. M., L. Wright, N. H. Bell, and F. C. Bartter: The syndrome of primary aldosteronism, Am. J. Med., 48:343, 1970.

44. Duarte, C. G., F. Chomety, and G. Giebisch: Effect of amiloride, ouabain, and furosemide on distal tubular function in the rat, Am. J. Physiol., 221:632, 1971.

45. Giebisch, G., G. Malnic, R. M. Klose, and E. E. Windhager: Effect of ionic substitutions on distal potential differences in rat kidney, Am. J. Physiol., 211:560, 1966.

46. Garcia-Filho, E., G. Malnic, and G. Giebisch: Effects of changes in electrical potential difference on tubular potassium transport, Am. J. Physiol., 238:F235, 1980.

47. Schwartz, W. B., R. L. Jenson, and A. S. Relman: Acidification of the urine and increased ammonium excretion without change in acid-base equilibrium: sodium reabsorption as a stimulus to the acidifying process, J. Clin. Invest., 34:673, 1955.

48. Lipner, H. I., F. Ruzany, M. Dasgupta, P. D. Lief, and N. Bank: The behavior of carbenicillin as a nonreabsorbable anion, J. Lab. Clin. Med., 86:183, 1975.

49. Klastersky, J., B. Vanderkelen, D. Daneua, and M. Mathieu: Carbenicillin and hypokalemia (letter), Ann. Intern. Med., 78:774, 1973.

The introduction to disorders of potassium balance presented in Chap. 26 should be read before proceeding with this discussion.

ETIOLOGY

Potassium enters the body by dietary intake or intravenous infusion, is primarily stored in the cells, and is then excreted in the urine and, to a lesser degree, in the stool and in sweat. An abnormality in any one or more of these processes can lead to hypokalemia (Table 27-1). This section will review the causes of K^+ depletion as well as some aspects of the diagnosis and treatment of certain disorders. The general

Table 27-1 Etiology of hypokalemia

Decreased net intake
  A. Low-K^+ diet or K^+-free intravenous fluids
  B. Clay ingestion

Increased entry into cells
  A. Elevation in arterial pH
  B. Increased availability of insulin
  C. Acute elevation in β-adrenergic activity
  D. Periodic paralysis—hypokalemic form
  E. Treatment of anemia
  F. Pseudohypokalemia
  G. Delirium tremens
  H. Hypothermia

Increased gastrointestinal losses†

Increased urinary losses
  A. Primary mineralocorticoid excess, e.g., hyperaldosteronism
  B. Enhanced distal flow
    1. Diuretics†
    2. Salt-wasting nephropathies
    3. Hypercalcemia
    4. Acute leukemia
  C. Na^+ reabsorption with a nonreabsorbable anion
    1. Vomiting or nasogastric suction (early)†
    2. Some forms of metabolic acidosis†
    3. Carbenicillin or other penicillin derivative
  D. Miscellaneous
    1. Hypomagnesemia†
    2. Polyuric states
    3. L-dopa

Increased sweat losses

Dialysis

Potassium depletion without hypokalemia

† Most common causes.

principles involved in the approach to the hypokalemic patient will be discussed separately later in the chapter.

Decreased Net Intake

The normal range of dietary K^+ intake is approximately 40 to 120 meq/day, with most of this K^+ then being excreted in the urine. If intake is diminished, urinary K^+ excretion can be appropriately reduced to a minimum of 5 to 25 meq/day [1,2]. Consequently, a low K^+ diet (or a low K^+ content of intravenous feedings) will not lead to significant K^+ depletion unless intake is severely limited. Since K^+ is present in meat, fruit, and some vegetables, marked K^+ restriction is difficult to sustain and is a rare cause of hypokalemia in otherwise normal subjects. However, reduced intake can contribute to other causes of K^+ depletion. For example, poor people living in rural areas may have an average K^+ intake of only 25 meq/day because of the relatively high cost of K^+-containing foods [3]. These patients are more likely to become hypokalemic if treated with diuretics for hypertension.

Net K^+ intake can also be limited by chronic clay ingestion,† a not uncommon practice in some rural areas in the southeastern United States [4]. The clay appears to bind dietary K^+ and iron directly, diminishing their ability to be absorbed. Hypokalemia and iron-deficiency anemia may ensue if the ingestion is continued for a prolonged period.

Increased Entry into Cells

Elevation in arterial pH Alkalemia, either metabolic or respiratory, can promote K^+ entry into the cells. In alkalemic states, H^+ ions are released from the cellular buffers and move into the extracellular fluid to minimize the change in pH. To preserve electroneutrality, extracellular K^+ (and Na^+) move into the cells. In general, the plasma K^+ concentration falls less than 0.4 meq/L per 0.1-unit increase in arterial pH [6]; as a result, the degree of hypokalemia induced by the alkalemia is typically mild. A similar reduction in the plasma K^+ concentration may occur when $NaHCO_3$ is administered to correct a metabolic acidosis. In this setting, both the elevation in pH and a direct effect of the increased plasma HCO_3^- concentration appear to contribute to the entry of K^+ into the cells [7]. However, the fall in the plasma K^+ concentration may be greater than that seen with primary alkalemia [6].

Although the effect of alkalemia is relatively small, a substantial degree of hypokalemia is frequently present in metabolic alkalosis. Perhaps the major reason for this association is that the causative factor (diuretics, vomiting, hyperaldosteronism) causes *both* K^+ and H^+ loss. In addition, the development of hypokalemia may be an important factor in the genesis and maintenance of the metabolic alkalosis (see

---

†The effect of clay on K^+ balance varies with the type of clay ingested. Red clay, for example, contains a relatively large amount of K^+. The associated K^+ load can lead to hyperkalemia in patients with renal failure who are unable to excrete the excess K^+ [5].

Chap. 18) [8–10]. As the plasma K^+ concentration falls due to renal or extrarenal losses, intracellular K^+ diffuses into the extracellular fluid in exchange for extracellular H^+ (and Na^+). The net effect is an extracellular alkalosis with an intracellular acidosis [11,12]. When the latter occurs in renal tubular cells, H^+ secretion is stimulated, resulting in H^+ loss and more alkalosis [10]. The importance of these reciprocal cation shifts has been demonstrated by the ability of KCl to correct both the hypokalemia and metabolic alkalosis; most of the administered K^+ enters the cells, allowing H^+ to move back into the extracellular fluid [13,14].

Increased availability of insulin Insulin promotes the entry of K^+ into skeletal muscle and hepatic cells [15,16]. The major setting in which this leads to hypokalemia is during the treatment of severe hyperglycemia due to uncontrolled diabetes mellitus. In this setting, the patients are markedly K^+-depleted but the initial plasma K^+ concentration is usually normal or elevated because the combination of insulin deficiency and hyperglycemia promotes the movement of intracellular K^+ into the extracellular fluid (see Chap. 25) [17]. These abnormalities are corrected by insulin, which then unmasks the underlying K^+ depletion.

Mild hypokalemia can also be induced by a carbohydrate load or the administration of exogenous insulin [16,18,19]. This effect is transient unless there is persistent hypersecretion of insulin, as may occur with intravenous hyperalimentation. If insulin administration is associated with hypoglycemia, the ensuing increase in epinephrine release can also contribute to the fall in the plasma K^+ concentration [19].

Elevated β-adrenergic activity Catecholamines promote K^+ entry into the cells, a response that is mediated by the β_2-adrenergic receptors [20,21]. As a result, transient hypokalemia can be induced when β-adrenergic activity is increased, as with enhanced epinephrine release (due to hypoglycemia or the stress of an acute illness) [19,20,21a] or with the excessive use of salbutamol (albuterol), a β_2-adrenergic agonist used to treat bronchospasm or to inhibit uterine contractions [22,23].

Periodic paralysis Periodic paralysis is a rare disorder characterized by recurrent episodes of muscle weakness or paralysis which can be fatal if the respiratory muscles are involved. The severity of individual attacks is variable, ranging from weakness in a single muscle group to diffuse paralysis.

Hypokalemic, hyperkalemic, and normokalemic forms have been described [24,25]. The hypokalemic form may be familial with autosomal dominant inheritance or may be acquired as a result of thyrotoxicosis (particularly in Chinese males) [24–26]. In either disorder, episodes can be precipitated by rest after exercise, a carbohydrate meal, stress, or the administration of insulin or epinephrine. These attacks are associated with the sudden movement of K^+ into the cells, resulting in an acute reduction in the plasma K^+ concentration (which is normal between attacks) to as low as 1.5 to 2.5 meq/L. If untreated, muscle strength returns after 6 to 48 h as K^+ moves back into the extracellular fluid. A similar acute form of paralysis can be induced by barium poisoning, which usually results from contaminated foods [25].

The possible mechanisms responsible for the membrane defect in this rare disorder are generally beyond the scope of this chapter and have been reviewed elsewhere [25]. In thyrotoxicosis, there is an increased sensitivity to catecholamines, and the administration of β-adrenergic blockers can minimize the severity and number of attacks and, usually, the fall in the plasma K^+ concentration [27,28]. These findings suggest an important role for increased sympathetic activity, although how it leads to an exaggerated fall in the plasma K^+ concentration is uncertain [25].

The diagnosis of periodic paralysis should be suspected from the history (including a possible familial incidence), the severity of the hypokalemia in the absence of any obvious cause, and the rapid normalization of the plasma K^+ concentration and relief of symptoms following the administration of K^+. Thyroid function studies should be obtained in patients with a negative family history.

Treatment Treatment of the acute episode involves the oral administration of 60 to 120 meq of KCl. This should lead to increased muscle strength within 15 to 20 min. If no improvement is observed, another 60 meq can be given. The presence of hypokalemia must be confirmed *prior* to the initiation of therapy since K^+ can exacerbate both the hyperkalemic and normokalemic forms [24].

Therapy must also be directed toward preventing further paralytic episodes [24]. β-Adrenergic blockers followed by treatment of the underlying disease leads to a cure in thyrotoxic patients. In the familial form, the following, used singly or in combination, have been effective: K^+ supplementation, K^+-sparing diuretics, a low-carbohydrate diet, and acetazolamide [24,25,29]. It is unclear how acetazolamide works since it increases urinary K^+ and $NaHCO_3$ loss. A direct effect on the cell membrane or the ensuing metabolic acidosis (which promotes K^+ movement out of the cells) may be involved.

Treatment of anemia Within 48 h after the administration of folic acid or vitamin B_{12} to patients with megaloblastic anemia, there is frequently a reduction in the plasma K^+ concentration to about 3.0 meq/L and occasionally to as low as 2.1 meq/L with the possible development of cardiac arrhythmias [30]. This response is presumably due to the uptake of K^+ by the newly produced red cells and platelets. This effect is less prominent with other anemias (such as that due to iron deficiency) since treatment of these anemias results in a much slower rate of new cell production.

The plasma K^+ concentration may also fall below 3.0 meq/L following multiple transfusions with frozen, washed red cells [31]. These cells, but not those stored in acid-citrate-dextran, lose up to 50 percent of their K^+ during storage. In the recipient, K^+ rapidly moves into the cells to repair the deficit.

Pseudohypokalemia Metabolically active cells can also take up K^+ after blood has been drawn. In this setting, which has been described in cases of acute myeloid leukemia that have a very high white cell count, the patient may have a relatively normal plasma K^+ concentration but the measured value may be below 1.0 meq/L (without any symptoms) if the blood is first allowed to stand for a prolonged period at room temperature [32]. This problem can be avoided if the plasma or serum is rapidly separated from the cells or if the blood is stored at $4°C$.

Delirium tremens An acute reduction in the plasma K^+ concentration to about 3.0 meq/L may also occur in patients with delirium tremens [33]. This problem is presumably due to the movement of K^+ into the cells. Although the mechanism by which it occurs is not known, increased secretion of epinephrine may play a contributory role.

Hypothermia Accidental or induced hypothermia can lower the plasma K^+ concentration to below 3.0 meq/L, apparently as a result of K^+ entry into the cells [34,34a]. This effect is readily reversible during rewarming, and may be associated with "overshoot" *hyper*kalemia, particularly if K^+ has been given during the period of hypothermia [34].

Increased Gastrointestinal Losses

In normal subjects, approximately 3 to 6 liters of gastric, pancreatic, biliary, and intestinal secretions is secreted into the gastrointestinal lumen each day. Almost all these fluids are then reabsorbed, as only 100 to 200 mL of water and 5 to 10 meq of K^+ are lost in the stool. Since each of these secretions contains K^+, the loss of any of them (because of decreased absorption or increased secretion) can lead to K^+ depletion. This can be seen with vomiting (although urinary losses also contribute; see below), diarrhea, intestinal fistulas or tube drainage, or the loss of colonic secretions from a villous adenoma or from chronic laxative abuse [35–41].

In some patients, massive fluid and electrolyte losses can occur. In cholera, for example, daily stool losses may average 8 liters of water, 1000 meq of Na^+, and 130 meq of K^+ [36]. Similarly, daily fluid losses in excess of 6 liters and K^+ losses in excess of 300 meq have been reported in patients with the Verner-Morrison syndrome (severe, watery diarrhea and histamine-fast achlorhydria usually but not always due to a non-β-cell, islet cell tumor) [37–39]. The diarrhea in this disorder is most often due to the secretion of vasoactive intestinal peptide (which increases fluid secretion in Brunner's glands) [42], but prostaglandins may be involved in some patients [43].

Therapy must be directed toward minimizing further fluid loss by treating the underlying disease. In the watery diarrhea syndrome, for example, surgical removal of the tumor, if possible, or the administration of the chemotherapeutic agent streptozotocin may be beneficial [37–39]. If these modalities are ineffective, clonidine (an α_2-adrenergic agonist), trifluoperazine (which interferes with the action of calmodulin), or prostaglandin-synthesis inhibitors have reduced the degree of diarrhea in selected patients [39,43,44].

Increased Urinary Losses

Urinary K^+ excretion is primarily determined by K^+ secretion in the distal nephron, particularly the cortical collecting tubule. Inappropriate urinary K^+ loss leading to hypokalemia is most often due to conditions associated with mineralocorticoid excess, increased urinary flow to the distal secretory site, or the reabsorption of Na^+ in the presence of a nonreabsorbable anion (Table 27-1).

Table 27-2 Causes of primary mineralocorticoid excess

Primary hyperaldosteronism
  Adenoma
  Hyperplasia
  Carcinoma

Cushing's syndrome

Congenital adrenal hyperplasia
  17α-hydroxylase deficiency
  11β-hydroxylase deficiency

Chronic ingestion of exogenous mineralocorticoid
  Licorice
  Carbenoxolone
  Fludrocortisone

Bartter's syndrome

Hyperreninism
  Renin-secreting tumor
  Renal artery stenosis

Hypersecretion of deoxycorticosterone
  or other mineralocorticoids

Mineralocorticoid excess Aldosterone, the primary endogenous mineralocorticoid, stimulates the reabsorption of Na^+ and the secretion of K^+ and H^+. Consequently, the excessive secretion of aldosterone (or any other mineralocorticoid) can lead to hypokalemia and metabolic alkalosis [8–10]. Edema, however, does not usually occur in otherwise normal subjects since the initial Na^+ retention is followed by a spontaneous natriuresis, a phenomenon referred to as *aldosterone escape*† (see page 144).

For hypokalemia to occur, there must be adequate delivery of Na^+ and water to the distal nephron. When distal flow is reduced because of effective circulating volume depletion, K^+ secretion may be relatively unchanged despite the presence of hyperaldosteronism [45]. Thus, patients with uncomplicated heart failure or hepatic cirrhosis typically have a normal plasma K^+ concentration. However, hypokalemia may rapidly ensue if distal delivery is enhanced by the administration of diuretics.

Primary mineralocorticoid excess occurs in a variety of uncommon conditions (Table 27-2) [46–48]. In addition to hypokalemia and metabolic alkalosis, these disorders (with the exception of Bartter's syndrome) are also associated with hypertension and mild hypernatremia. Volume expansion initiates the elevation in blood pressure [49] and also accounts for the rise in the plasma Na^+ concentration by causing an upward resetting of the osmostat (see page 525) [50].

---

† Hyperaldosteronism can contribute to fluid retention in edematous states such as heart failure and hepatic cirrhosis. Aldosterone escape does not occur in these disorders since the patient remains effectively volume-depleted due, for example, to a low cardiac output in heart failure (see Chap. 16).

Primary hyperaldosteronism The autonomous hypersecretion of aldosterone may result from a unilateral adenoma or carcinoma or bilateral hyperplasia [46–48,51]. Because of the mild hypervolemia, the plasma renin activity is typically (but not always) reduced in these disorders [52].

The mechanism responsible for adrenal hyperplasia is not well understood. Cyproheptadine, a serotonin antagonist, markedly reduces aldosterone secretion in some patients, suggesting involvement of central serotoninergic pathways which might mediate the secretion of a hypothalamic secretagogue for aldosterone [53]. In another disorder with autosomal dominant inheritance, the hypersecretion of aldosterone is reversed by the administration of a glucocorticoid such as dexamethasone [51,54]. Increased sensitivity of the zona glomerulosa to even small amounts of ACTH appears to be present in this disorder and may contribute to the hyperaldosteronism [51,54].

Cushing's syndrome (glucocorticoid excess) Cortisol, the most active glucocorticoid, is synthesized in the zona fasciculata under the influence of ACTH. Since cortisol also has weak mineralocorticoid activity, cortisol excess can lead to hypokalemia. In addition, the oversecretion of other mineralocorticoids such as deoxycorticosterone and corticosterone may also contribute to the K^+ loss [55].

Hypercortisolism can result from hypersecretion of ACTH (due to a pituitary adenoma or a nonendocrine ACTH-producing tumor) or from primary adrenal diseases (adenoma or carcinoma)† [56]. The degree of hypokalemia is directly related to the level of cortisol secretion, being most severe in those states with the highest hormone production, namely, adrenal carcinoma and nonendocrine tumors [57].

In contrast to primary hyperaldosteronism, in which the elevation in blood pressure can be prevented by Na^+ restriction, the hypertension in Cushing's syndrome is not as directly related to volume expansion [49]. As a result, the plasma renin activity is generally normal or elevated, not reduced as with aldosterone excess [48,58,59]. Cortisol-induced production of renin substrate by the liver may contribute to this response [59].

Congenital adrenal hyperplasia Deoxycorticosterone (DOC) and corticosterone are synthesized in the adrenal cortex and possess significant mineralocorticoid activity. Unlike aldosterone, the secretion of these hormones is regulated by ACTH. When cortisol secretion is reduced because of an adrenal enzyme deficiency, ACTH release is persistently elevated. In two forms of congenital adrenal hyperplasia, 17α-hydroxylase and 11β-hydroxylase deficiency (Fig. 27-1), this hypersecretion of ACTH results in the overproduction of DOC and corticosterone [46,48]. The former is non-virilizing since androgen production is also dependent upon 17-hydroxylation. K^+ balance can be restored in these disorders by the administration of cortisol which lowers ACTH secretion to normal.

---

†Exogenous synthetic glucocorticoids such as prednisone and dexamethasone have little mineralocorticoid activity and are unlikely to produce hypokalemia.

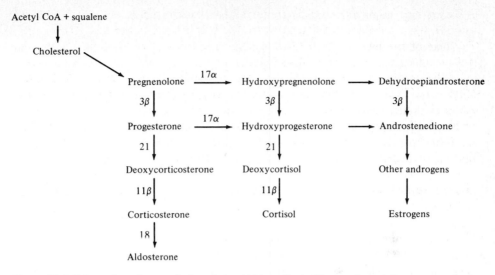

Figure 27-1 Schematic pathways of adrenal steroid biosynthesis. The numbers at the arrows refer to specific enzymes: 17α = 17α-hydroxylase: 3β = 3β-hydroxysteroid dehydrogenase; 21 = 21-hydroxylase; 11β = 11β-hydroxylase; 18 refers to a two-step process resulting in the addition of an aldehyde at the 18-carbon position. Deficiencies in any of these enzymes can lead to abnormal mineralocorticoid (as well as glucocorticoid and androgen) production.

Exogenous mineralocorticoid   Licorice contains a steroid, glycyrrhizic acid, which has mineralocorticoid activity. As a result, subjects chronically ingesting large amounts of licorice (or licorice-containing chewing tobacco) can develop a reversible syndrome indistinguishable from primary hyperaldosteronism [60,61]. Similar abnormalities can be induced by fludrocortisone, a synthetic mineralocorticoid used in the treatment of hypoaldosteronism and also present in some nasal sprays [62].

Bartter's syndrome   Bartter's syndrome is a rare disorder characterized by hyperreninemia, hyperaldosteronism, hyperplasia of the juxtaglomerular apparatus (the source of renin in the kidney), and hypokalemic alkalosis [63]. Increased secretion of vasodilatory prostaglandins (prostaglandin E and prostacyclin) is also present in this condition [63,64] and may partially explain why the blood pressure remains normal [65].

Recent studies suggest that impaired NaCl reabsorption in the loop of Henle or the distal tubule may be the primary abnormality in this disorder [66,67]. The ensuing volume depletion enhances renin secretion, resulting in the formation of angiotensin II, which then stimulates the release of aldosterone. This combination of increased distal flow (due to the reabsorptive defect) and hyperaldosteronism promotes K^+ secretion and the development of hypokalemia. Prostaglandins, which directly increase renin secretion (see Chap. 3), also contribute to this process as evidenced by the ability of prostaglandin-synthesis inhibitors to reverse most of the biochemical and hormonal changes (but not the impairment in NaCl reabsorption) [66,68].

The mechanism by which prostaglandin production is increased in this disorder is uncertain. Experimental studies suggested that hypokalemia might be responsible, but this does not appear to apply generally to humans [63,69]. An alternate possibility is that the salt wasting in this disorder is similar to that induced by the loop diuretics since these agents also increase prostaglandin production [70]. Decreased reabsorption at the macula densa may mediate this response since diuretics which act distal to this site (such as the thiazides) do not augment prostaglandin release [70].

Hyperreninism The primary hypersecretion of renin can result in a syndrome that mimics the clinical findings of primary hyperaldosteronism except for the elevated plasma renin activity. This may occur with rare, renin-secreting tumors of the juxtaglomerular apparatus [71] or, more commonly, with renovascular hypertension (renal artery stenosis, malignant hypertension, vasculitis, scleroderma) in which renal ischemia induces nonsuppressible renin release [72].

Other mineralocorticoids On rare occasions, the secretion of other mineralocorticoids can lead to hypokalemia and hypertension. Included in this group are patients with DOC-producing adenomas [73,74] and patients with hyperplasia who secrete an as yet unidentified, ACTH-sensitive mineralocorticoid [75,76].

Diagnosis Primary mineralocorticoid excess should be suspected in any patient with hypertension and *unexplained* hypokalemia. Although occasional patients are normokalemic [52,77], it is not realistic to screen every hypertensive patient for hyperaldosteronism. Even more rare are those patients who are hypokalemic but normotensive [78]. In this setting, the unexplained persistent reduction in the plasma K^+ concentration warrants evaluation for mineralocorticoid excess.

A stepwise approach to the hypokalemic patient with hypertension is illustrated in Fig. 27-2 [79]. A 24-h urine collection for K^+ will differentiate inappropriate urinary losses (greater than 25 to 30 meq/day) from extrarenal or previous urinary losses. Diuretics must be discontinued prior to the collection since they directly increase K^+ excretion. It is also important that the patient not be volume-depleted (as evidenced by low urinary sodium excretion) since the associated decrease in distal delivery can lower K^+ excretion even in patients with hyperaldosteronism [45]. On the other hand, the degree of K^+ wasting, and therefore the diagnostic accuracy, can be increased by a high-sodium diet as the combination of augmented flow and elevated aldosterone secretion enhances K^+ secretion [80]. A high-Na^+ diet can also be given to patients with a borderline plasma K^+ concentration since Na^+-induced hypokalemia is strongly indicative of nonsuppressible hyperaldosteronism [52,80]. In contrast, K^+ wasting and hypokalemia are not induced in normal subjects by Na^+ loading because the ensuing volume expansion lowers renin and aldosterone release [80].

In patients with persistent hypokalemia and inappropriately high urinary K^+

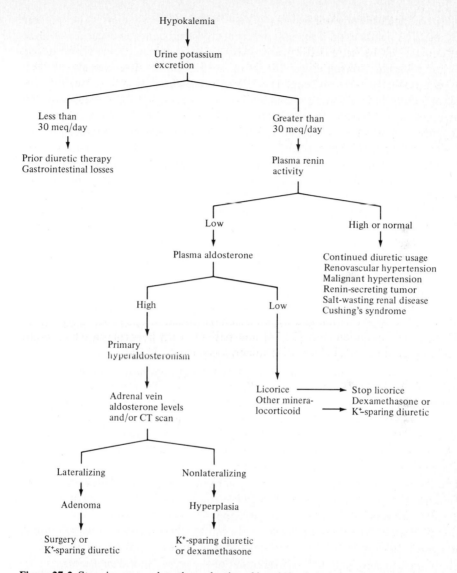

Figure 27-2 Stepwise approach to the evaluation of hypokalemia in hypertensive patients.

excretion, the plasma renin activity should be determined [47,52]. An elevated value is most often due to renovascular or malignant hypertension or diuretic usage.† A low value, however, is highly suggestive of some form of mineralocorticoid excess.

The disorders associated with hypersecretion of mineralocorticoids can be differentiated from one another by measuring the plasma aldosterone concentration or the urinary excretion of aldosterone metabolites [47,52]. A low plasma level indi-

†The diagnostic and therapeutic approach to Cushing's syndrome is reviewed in Refs. 56 and 81.

cates some nonaldosterone mineralocorticoid, whereas a clearly high value (greater than 30 ng/dL) points toward hyperaldosteronism. The diagnostic accuracy of these measurements can be increased by attempting to suppress endogenous aldosterone production by administering either 2 liters of isotonic saline intravenously over 4 h (while recumbent) or 0.6 to 1.2 mg/day of fludrocortisone (a synthetic mineralocorticoid) for 3 days. Patients without primary hyperaldosteronism should suppress their plasma aldosterone concentration to below 10 ng/dL; values above this level are consistent with hyperaldosteronism [47,52,82]. To prevent possible misinterpretation, the plasma aldosterone concentration should be measured with the patient recumbent, off K^+ supplements (both standing and K^+ loading can increase aldosterone secretion), and relatively normokalemic (since hypokalemia can reduce aldosterone release) [83].

Once the diagnosis of primary hyperaldosteronism is established, a unilateral adenoma or carcinoma must be distinguished from bilateral hyperplasia because of their differing responses to therapy (see below). Although a variety of indirect tests have been proposed [48,52,84], this distinction can be most accurately made by measurement of adrenal vein aldosterone levels [47,52]. To be certain that the samples are from the adrenal veins, the cortisol and/or epinephrine concentration, which should be roughly the same on both sides but much greater than that from a peripheral vein, should also be measured [47,85]. Unilateral disease is associated with a marked increase in the aldosterone concentration on the side of the tumor, whereas there should be little difference with bilateral hyperplasia.

A possible and simpler alternative is CT scanning; a unilateral adrenal mass is indicative of an adenoma (or carcinoma), but the absence of a mass is not diagnostic of hyperplasia since small tumors may be missed [52,84]. Adrenal venography and scintillation scanning with ^{131}I-iodocholesterol (a precursor of aldosterone) can also miss some small tumors and will not detect hyperplasia [47,86,87]. In addition, the latter procedure is not widely available, and venography may be associated with complications such as adrenal hemorrhage [86].

The diagnosis of Bartter's syndrome, a normotensive state, is basically one of exclusion [63]. Most of the biochemical and hormonal findings in this disorder can be replicated by vomiting or diuretic usage, either apparent or surreptitious [88–90]. In general, Bartter's syndrome presents early in life (the diagnosis is usually made by the age of 25) and is frequently associated with growth and mental retardation [63,91]. Even in young patients, measurement of the urine chloride concentration (a value below 10 to 15 meq/L suggests vomiting) [88] and an assay of the urine for diuretics [90] should be performed to exclude these other, much more common causes.

Treatment The mode of treatment varies with the underlying disease. Surgery is the preferred treatment in patients with an adenoma or carcinoma since unilateral adrenalectomy results in a fall in blood pressure, a marked reduction in aldosterone secretion, and correction of the hypokalemia in virtually all patients [46,48,92]. However, despite the decrease in blood pressure, a lesser degree of hypertension persists in as many as 40 percent of patients [93]. Surgery (subtotal adrenalectomy) is much less successful with adrenal hyperplasia, as only a minority of patients become normotensive [46,48,92].

The reasons for the failure of surgery to control the blood pressure are incompletely understood. The administration of a mineralocorticoid plus a high-salt diet produces hypertension in subjects with labile hypertension but not in normal subjects [92]. Thus, it may be that in many cases primary hyperaldosteronism leads to an earlier expression of an underlying hypertensive state, particularly in adrenal hyperplasia, where the degree of hyperaldosteronism and hypokalemia is often relatively mild [46,52,92]. In addition, if the elevation in blood pressure has persisted for a period of years, the development of nephrosclerosis may contribute to the persistence of hypertension even after the endocrine abnormality has been corrected [93].

As a result of these observations, *surgery should be considered only in patients with a unilateral tumor* as determined by measurement of adrenal vein aldosterone concentrations or CT scanning. In patients who are not surgical candidates or who have hyperplasia, a K^+-sparing diuretic (spironolactone, triamterene, or amiloride) can lower the blood pressure and correct the hypokalemia [94–96]. Amiloride (10 to 40 mg/day) may be the best tolerated since it can be given once daily and avoids the gastrointestinal and endocrine side effects (menstrual irregularities, gynecomastia, and hypogonadism) associated with spironolactone [96].

As an alternative, dexamethasone can be used in those patients with bilateral hyperplasia in whom aldosterone secretion appears to be mediated by ACTH or a related compound [51,54]. This disorder should be suspected when children with a positive family history are affected. Glucocorticoid therapy to lower ACTH levels is also beneficial in patients with congenital adrenal hyperplasia or with low aldosterone and DOC levels in whom some other ACTH-dependent mineralocorticoid must be present [75,76].

Bartter's syndrome may be effectively treated with a K^+-sparing diuretic (such as amiloride) and a prostaglandin-synthesis inhibitor [63,68,97]. Sulindac, however, should not be used since it is the one prostaglandin-synthesis inhibitor that does not appear to reduce *renal* prostaglandin production, perhaps because no drug or active metabolite is excreted in the urine [98,99].

Increased flow to the distal nephron Since water reabsorption follows that of NaCl in the proximal tubule and loop of Henle, increased distal flow and possible K^+ wasting occurs in those states in which proximal or loop NaCl reabsorption is decreased (Table 27-1).

Diuretics The loop diuretics and the thiazides enhance distal flow and consequently K^+ secretion by inhibiting NaCl reabsorption proximal to the K^+-secretory site (see Chap. 16) [100]. Other factors may also contribute to the K^+ loss in this setting, including hyperaldosteronism (due to diuretic-induced hypovolemia and/or the underlying disease such as heart failure), decreased loop K^+ reabsorption with loop diuretics since K^+ and NaCl reabsorption are linked at this site [101], and diuretic-induced hypomagnesemia (see below) [102].

In hypertensive patients, most of the K^+ loss occurs during the first 2 weeks of therapy, after which a new steady state† is established [103,104]. The combination of a decrease in distal flow induced by the associated hypovolemia and the direct effect of hypokalemia prevents continued K^+ wasting. It should also be noted that overtreatment contributes to the hypokalemia in many patients. Daily doses above 50 mg of hydrochlorothiazide and 25 mg of chlorthalidone produce increased K^+ losses but generally little further reduction in blood pressure [105,106].

Salt-wasting nephropathies In a variety of renal diseases, particularly tubulointerstitial disorders such as pyelonephritis or urinary tract obstruction, Na^+ reabsorption is impaired. If this occurs in the proximal tubule or loop of Henle, there will be concomitant increases in distal flow and K^+ secretion, similar to that induced by diuretics [107–109]. In severe cases, K^+ losses can exceed 200 meq/day‡ [107].

This mechanism may also account for the development of hypokalemia in hypercalcemic states since calcium-induced tubular damage may impair Na^+ reabsorption [111]. Similarly, lysozyme-induced tubular damage may be at least in part responsible for the K^+ wasting and relatively marked hypokalemia that may be associated with leukemia, particularly acute monocytic or myelomonocytic leukemia [112–116]. K^+ entry into the metabolically active leukemic cells may also contribute to the fall in the plasma K^+ concentration in this disorder [113].

Sodium reabsorption with a nonreabsorbable anion The presentation of sodium to the distal secretory site with a nonreabsorbable anion results in Na^+ reabsorption in exchange for K^+ and H^+. This effect also tends to be associated with increased distal delivery and is most prominent when there is a stimulus to conserve Na^+ due to hyperaldosteronism.

Vomiting or nasogastric suction Hypokalemia is a common finding with persistent loss of gastric secretions. Although some K^+ is present in this fluid (about 5 to 10 meq/L), most of the K^+ deficit is initially due to urinary losses [117]. The mechanism by which this occurs is as follows. Gastric juice contains a high concentration of HCl. Thus, its removal leads to elevations in the plasma HCO_3^- concentration and the filtered HCO_3^- load (GFR times plasma HCO_3^- concentration) as well as hypovolemia and enhanced aldosterone release. However, a comparable increase in proximal HCO_3^- reabsorptive capacity cannot occur acutely, leading to augmented $NaHCO_3$ and water delivery to the distal nephron. The reabsorption of Na^+ at this site is accompanied by K^+ loss since HCO_3^- acts as a nonreabsorbable anion. Within 48 to 72 h, proximal Na^+ and HCO_3^- reabsorption rises in response to the hypovolemia, less HCO_3^- is delivered distally, and K^+ excretion falls [117]. Further K^+ loss at this time is primarily due to continued loss of gastric secretions.

---

†A new steady state is less readily established in edematous patients since the diuretic dose may be increased to remove more fluid. As a result, K^+ loss is more likely to increase with time.

‡Na^+-wasting states can also lead to *hyper*kalemia if associated with hypoaldosteronism or reduced renal perfusion due to lack of replacement of urinary Na^+ losses [110,110a].

Metabolic acidosis Increased urinary K$^+$ loss also occurs in several forms of metabolic acidosis, generally by mechanisms similar to that in vomiting [118]. In diabetic ketoacidosis, for example, increased quantities of Na$^+$ and water (due to the glucose osmotic diuresis) are presented to the distal nephron with β-hydroxybutyrate and acetoacetate. In type 2 (proximal) renal tubular acidosis, Na$^+$ is delivered with HCO$_3^-$ because of a primary reduction in proximal HCO$_3^-$ reabsorption (see Fig. 19-5) [119].

A somewhat different mechanism is operative in type 1 (distal) renal tubular acidosis. In this condition, distal H$^+$ secretion is reduced. As a result, Na$^+$ reabsorption must occur in exchange for K$^+$ if Na$^+$ balance is to be maintained (see page 421) [119]. Severe K$^+$ depletion with a plasma K$^+$ concentration below 2.0 meq/L may occur in this disorder. Although relatively rare, one cause of this condition that is becoming more common in certain groups is the sniffing of toluene, a component of airplane glues and paint thinners [120,121]

In addition to these factors, distal flow may be directly increased in metabolic acidosis. As described in Chap. 5, approximately one-third of proximal Na$^+$ reabsorption is passive, occurring down gradients created primarily by HCO$_3^-$ reabsorption. In metabolic acidosis, less HCO$_3^-$ is reabsorbed proximally (since less is filtered), thereby decreasing passive NaCl and water transport and augmenting distal delivery [122].

Penicillin derivatives Penicillin and its synthetic derivatives are frequently given as the Na$^+$ salts. In this setting, Na$^+$ is presented to the distal nephron with the non-reabsorbable penicillin anion, leading to increased K$^+$ and H$^+$ secretion, hypokalemia, and metabolic alkalosis [123]. This problem may be seen with any of these drugs but is most likely to occur with intravenous carbenicillin since it is given in large doses (24 to 36 g) and contains 4.7 meq of Na$^+$ per gram [124–126].

Hypomagnesemia Hypomagnesemia of any cause can lead to K$^+$ depletion and hypokalemia [127]. Hypocalcemia, due to both diminished secretion of parathyroid hormone and skeletal resistance to its effect, is also commonly present [127–129]. If otherwise unexplained, this combination of hypokalemia and hypocalcemia is suggestive of magnesium depletion.

Increased urinary and fecal K$^+$ excretion contribute to the loss of K$^+$ [127,130]. Although the mechanism by which this occurs is incompletely understood, enhanced secretion of aldosterone appears to play an important role [131].

Polyuria A marked increase in urine volume can induce K$^+$ loss by an unusual mechanism. Normal subjects can lower the urine K$^+$ concentration to a minimum of 5 to 10 meq/L in the presence of K$^+$ depletion [1]. Although this generally leads to adequate K$^+$ conservation, the obligatory K$^+$ loss can exceed 50 to 100 meq/day if the urine output is 10 L/day or more. This degree of K$^+$ loss is most likely to occur with primary polydipsia [132] since patients with complete diabetes insipidus or severe hyperglycemia usually seek medical care soon after the polyuria has begun.

L-**Dopa** Increased urinary loss and mild hypokalemia can be induced by the administration of L-dopa [133]. The mechanism by which this occurs is uncertain, although dopamine formation may be important.

Increased Sweat Losses

Only small amounts of K^+ are normally lost in the sweat each day since the volume is low and the K^+ concentration is only 5 to 10 meq/L. However, substantial K^+ losses can occur when sweat production is chronically increased. For example, 10 L/day or more may be produced in subjects exercising in a hot climate [134]. Unless intake is appropriately increased, the ensuing K^+ depletion plays an important role in the rhabdomyolysis that can occur in this setting (see below). Urinary losses also contribute to this problem since aldosterone secretion is enhanced by both exercise (via catecholamine-induced renin secretion) and volume loss and is not immediately shut off when salt is ingested to restore normovolemia [134,134a].

Dialysis

Patients on chronic dialysis are dialyzed against a low K^+ concentration since K^+ retention is commonly present. Patients on chronic peritoneal dialysis, for example, may lose 30 meq of K^+ per day. This loss is generally well tolerated, but it can lead to K^+ depletion if intake is reduced or there are concurrent gastrointestinal losses [135].

A somewhat different mechanism may be operative in patients with underlying K^+ depletion in whom severe acidemia results in the movement of K^+ out of the cells and therefore a relatively normal baseline plasma K^+ concentration. In this setting, acute hemodialysis can rapidly correct the acidemia, resulting in K^+ entry into cells and a potentially large fall in the plasma K^+ concentration even though little or no K^+ is lost by dialysis [136].

Potassium Depletion without Hypokalemia

In most conditions, the plasma K^+ concentration varies directly with body K^+ stores. Thus, hypokalemia is generally associated with K^+ depletion as a 100- to 400-meq K^+ deficit is required to lower the plasma K^+ concentration from 4 to 3 meq/L [137]. However, this relationship is disturbed in disorders which affect the distribution of K^+ between the cells and the extracellular fluid. As described above, an elevation in arterial pH, increased availability of insulin, and periodic paralysis are all associated with K^+ movement into the cells and hypokalemia without K^+ depletion. Conversely, acidemia can mask underlying K^+ depletion by maintaining a normal plasma K^+ concentration [6,136].

Another example of this dissociation occurs in a variety of chronic diseases such as heart failure, renal failure, hepatic cirrhosis, and malnutrition. Patients with these disorders may have a normal plasma K^+ concentration despite a 10 to 15 percent fall in body K^+ stores [138–141]. The preferential loss of K^+ from the cells is asso-

Table 27-3 Abnormalities induced by hypokalemia

Muscle weakness or paralysis (including intestinal ileus)
Cardiac arrhythmias (especially with digitalis)
Rhabdomyolysis
Renal dysfunction
  Renal insufficiency
  Impaired urinary concentration (leading to polyuria and polydipsia)
  Increased ammonia production (can induce hepatic coma in cirrhosis)
  Impaired urinary acidification
  Increased bicarbonate reabsorption
  Abnormal NaCl reabsorption
Hyperglycemia

ciated with an increase in the cell Na^+ concentration and a reduction in the magnitude of the resting membrane potential, suggesting impaired function of the Na^+-K^+-ATPase pump in the cell membrane [141–143].

The mechanism by which the underlying disease induces this defect in cell metabolism and the clinical significance of these changes are not well understood. Since the plasma K^+ concentration is normal (the K^+ that had been in the cells being excreted in the urine), these patients do not have the symptoms of hypokalemia.† Normalization of K^+ balance will occur only with reversal of the underlying disease [140]. K^+ supplements alone are ineffective as they are excreted in the urine, not taken up by the cells [139,140].

SYMPTOMS

A variety of abnormalities may be associated with hypokalemia (Table 27-3). Although the severity of these changes is generally related to the degree of hypokalemia, there is substantial individual variability. Nevertheless, marked symptoms are unusual unless the plasma K^+ concentration is below 2.5 to 3.0 meq/L [145,146].

Muscle Weakness

Hypokalemia can induce muscle weakness and paralysis. The mechanism by which this occurs is complex. Initially, hypokalemia hyperpolarizes the cell membrane by increasing the ratio of the intracellular K^+ concentration to that in the extracellular fluid (see Fig. 26-1) [147]. As a result, membrane excitability is reduced since the resting potential is further away from the threshold that must be reached before an action potential can be generated. The net effect is decreased responsiveness of the membrane to exciting stimuli (such as acetylcholine).

---

†One exception may occur with the K^+ depletion commonly seen with the liquid protein diets used for rapid weight loss. It is possible that the K^+ deficit contributes to the potentially fatal cardiac arrhythmias that can develop in this setting [144].

However, severe hypokalemia leading to marked muscle weakness or paralysis appears to be associated with *depolarization* (not hyperpolarization) of the cell membrane, as the resting potential falls from the normal of -88 mV to -50 to -55 mV [25,147,148]. In this setting, the muscle cell is unable to sustain an action potential because the resting potential is below the threshold level of about -65 mV. The cause for this depolarization is uncertain but could be due to an increase in Na^+ permeability (resulting in Na^+ entry into the cell) or a decrease in K^+ permeability (limiting K^+ exit, which generates the potential; see page 12) [25].

Muscle weakness usually does not begin until the plasma K^+ concentration is less than 2.5 meq/L. There is, however, variability in this relationship because of the effects of the plasma Ca^{2+} concentration, arterial pH, and the rapidity and manner in which hypokalemia develops (see page 570). For example, patients with chronic K^+ loss may be relatively asymptomatic [41] because part of the K^+ lost from the extracellular fluid is replaced by K^+ movement out of the cells. The deficit in both compartments minimizes the change in the ratio of the K^+ concentrations and, therefore, in the resting potential. In contrast, acute K^+ entry into cells can lead to paralysis [25,136], in part because the cell K^+ concentration has *increased,* resulting in a much larger change in the concentration ratio.

The pattern of muscle weakness is relatively characteristic [24,149]. The lower extremities are most commonly involved first, particularly the quadriceps. In more severe cases, the muscles of the trunk, of the upper extremities, and eventually of respiration become affected. Death may ensue from respiratory failure. The cranial nerves are rarely involved. Cramps, paresthesias, tetany, muscle tenderness, and atrophy also may occur [149]. In addition, involvement of smooth muscle in the gastrointestinal tract can produce a paralytic ileus and the symptoms of abdominal distention, anorexia, nausea, vomiting, and constipation.

Cardiac Arrhythmias

A variety of cardiac arrhythmias can be induced by hypokalemia. These include premature atrial and ventricular beats, sinus bradycardia, paroxysmal atrial or junctional tachycardia, atrioventricular block, and even ventricular tachycardia or fibrillation [105,150–153]. Potentially life-threatening complex ventricular arrhythmias usually do not occur until the plasma K^+ concentration is less than 3.0 meq/L [146,153]. An exception to this rule, however, occurs in patients taking digitalis who are very sensitive to hypokalemia [151,154,155]. Digitalis-induced arrhythmias typically occur with toxic serum levels if K^+ balance is normal but may occur with normal levels if hypokalemia is present [155].

Electrocardiographic changes The electrocardiogram is a reflection of the electrical events in the heart. The P wave represents *atrial depolarization,* the QRS complex represents *ventricular depolarization,* and the ST segment and T and U waves represent *ventricular repolarization.* The atrial repolarization wave is lost in the QRS complex.

In addition to arrhythmias, hypokalemia produces characteristic changes in the electrocardiogram that are primarily due to delayed ventricular repolarization

| Plasma K⁺, meq/L | 4.0 | 3.0 | 2.0 | 1.0 |

Figure 27-3 Electrocardiogram in hypokalemia. As the plasma K^+ concentration falls, the initial changes are decreased amplitude of the T wave, ST-segment depression, and increased height of the U wave. With more severe hypokalemia, the P-wave amplitude is increased, as is the duration of the QRS complex. The approximate relationship between these changes and the plasma K^+ concentration is indicated. *(Adapted from B. Surawicz, Am. Heart J., 73:814, 1967.)*

[150]†. The result is ST-segment depression, decreased amplitude or inversion of the T wave, increased height of the U wave to greater than 1 mm, and prolongation of the QU interval (Fig. 27-3). With more severe K^+ depletion, increased amplitude of the P wave, prolongation of the PR interval, and widening of the QRS complex may occur [150]. These changes begin to be seen when the plasma K^+ concentration is less than 3.0 meq/L and are present in approximately 90 percent of patients with a plasma K^+ concentration under 2.7 meq/L [150,157]. They are rapidly reversible with K^+ repletion.

Rhabdomyolysis

Muscle cramps, rhabdomyolysis, and myoglobinuria, possibly leading to renal failure, may be seen in patients with severe K^+ depletion (plasma K^+ concentration less than 2.5 meq/L) [120,158–160]. The role of K^+ in the regulation of skeletal muscle blood flow appears to play an important role in the development of this problem [161]. During exercise, there is normally an appropriate increase in muscle perfusion to meet enhanced energy demands. This hyperemic response is mediated in part by the release of K^+ from skeletal muscle cells. The ensuing local elevation in the K^+ concentration causes vasodilation, which enhances blood flow. However, the cellular release of K^+ is impaired by K^+ depletion. As a result, there is a lesser increase in blood flow, and cramps, ischemic necrosis, and rhabdomyolysis may ensue [161]. In addition to hypoperfusion, hypokalemia-induced impairment in glycogen accumulation may also contribute to the muscle dysfunction [159].

Renal Dysfunction

K^+ depletion can interfere with a variety of renal functions (see Table 27-3) [162–164]. Each of these changes is usually reversible with K^+ repletion. One important

---

†During repolarization, the primary ionic event is the movement of K^+ out of the cell (see page 569). The rapidity with which repolarization occurs is therefore proportional to the K^+ permeability of the membrane, which appears to vary directly with the plasma K^+ concentration [150,156]. Consequently, the reduction in K^+ permeability with hypokalemia slows K^+ movement out of the cell and delays repolarization. Conversely, K^+ permeability is enhanced by hyperkalemia, resulting in an increased rate of repolarization. This can be detected on the electrocardiogram as tall, narrow T waves with a shortened QT interval (see Fig. 28-2).

function that is maintained, however, is the ability to conserve K^+ [162], an adaptive response mediated by decreased secretion and increased reabsorption in the collecting tubules [165,166]. Thus, patients with extrarenal causes of K^+ depletion should excrete less than 25 meq of K^+ per day in the urine [1,2].

Nephropathy of potassium depletion In humans, K^+ depletion produces a characteristic vacuolar lesion in the epithelial cells of the proximal tubule and occasionally the distal tubule [162,163]. This lesion occurs primarily with chronic K^+ depletion and in most patients requires at least 1 month to develop. In addition to this change, which is clearly linked to K^+ depletion, interstitial fibrosis and tubular atrophy and dilation may be seen. It is not clear if these lesions also are due to K^+ depletion or are related to concurrent renal disease such as nephrosclerosis or pyelonephritis.

Within weeks to months after K^+ repletion, the vacuolar changes are reversed. However, the secondary changes of interstitial fibrosis and tubular dilation may be irreversible. Although the GFR may be reduced when the patient is hypokalemic, it usually is improved with the restoration of normal K^+ balance [162]. Persistent renal insufficiency following chronic K^+ depletion may occur but is rare [166a].

Impaired urinary concentration Polyuria and polydipsia are common complaints in hypokalemic patients [167]. These symptoms may be due to both a primary stimulation of thirst and decreased urinary concentrating ability [168]. The latter is associated with diminished collecting tubule permeability to water, an effect that may be mediated by reductions in antidiuretic hormone–induced cyclic AMP generation and in cyclic AMP effect [169,170]. In addition, countercurrent function may be impaired, preventing the establishment of a normally hyperosmotic medullary interstitium [171]. Diluting ability, in contrast to concentrating ability, is usually well maintained with hypokalemia [167,171].

In general, the maximum urine osmolality (which in normal subjects is 900 to 1400 mosmol/kg) remains above 300 mosmol/kg with K^+ depletion (Fig. 27-4). This is in contrast to values that may be below 150 mosmol/kg in patients with central diabetes insipidus (see Chap. 24). Consequently, with hypokalemia the degree of polyuria is relatively mild, as the urine output is usually less than 3 to 4 L/day.

The concentrating defect is both dose- and time-related [167]. The maximum urine osmolality begins to fall when the K^+ deficit exceeds 200 meq and reaches its minimum at a deficit of 400 meq, a level at which the plasma K^+ concentration should be below 3.0 meq/L. The fall in the urine osmolality occurs slowly over 2 to 3 weeks (Fig. 27-4). Acute hypokalemia, on the other hand, does not appear to affect urinary concentration [167]. The reason for this difference is not known.

Ammonia production and urinary acidification Hypokalemia results in an increase in NH_3 production by the renal tubular cells [172]. The NH_3 that is produced enters both the tubular lumen and the peritubular capillary, resulting in increases in urinary NH_4^+ excretion and in the NH_3 concentration in the renal vein [41,172–174].

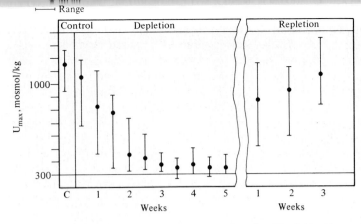

Figure 27-4 Ability to maximally concentrate the urine (U_{max}) in patients with progressive potassium depletion. The average K^+ deficit was 350 meq or about 10 percent of the total body K^+. *(From M. Rubini, J. Clin. Invest., 40:2215, 1961, by copyright permission of The American Society for Clinical Investigation.)*

This effect may be related to the movement of K^+ out of and, to maintain electroneutrality, H^+ into the cells that occurs with hypokalemia [11,12]. The intracellular acidosis could then stimulate NH_3 production and H^+ secretion [175], a mechanism similar to that thought to be responsible for the appropriate increase in NH_4^+ excretion with metabolic acidosis that allows the excess acid to be excreted (see Chap. 12).

The increase in NH_3 production may be clinically important in patients with severe hepatic disease in whom hypokalemia can precipitate hepatic coma [174,176]. In addition to the plasma NH_3 levels, the commonly associated metabolic alkalosis (which promotes the movement of NH_3 and other toxic amines into the brain; see page 349) may also contribute to this problem. Correction of these electrolyte disturbances with KCl may reverse the encephalopathy without any other therapy [174,176].

Bicarbonate reabsorption The increases in the cell H^+ concentration and H^+ secretion induced by hypokalemia also promote HCO_3^- reabsorption (see Fig. 12-11) [177–179]. In patients who have metabolic alkalosis, this increase in HCO_3^- reabsorption contributes to the perpetuation of the alkalosis since it prevents the urinary excretion of the excess HCO_3^- (see Chap. 18) [10].

NaCl reabsorption Hypokalemia has a dual effect on Na^+ excretion, impairing both the ability to excrete a Na^+ load and the ability to conserve Na^+ maximally. For example, K^+ depletion can lead to Na^+ retention and edema in subjects on a high-sodium diet [1,41,180]. This effect could be mediated by increased Na^+ reabsorption in the proximal tubule or the loop of Henle [172,181]. How this occurs is not well understood.

Conversely, the ability to lower the urine concentration of NaCl to below 15 meq/L in the presence of volume depletion may also be impaired by moderate-to-severe hypokalemia [163,182]. Impaired Cl^- reabsorption in the distal nephron (perhaps due to a decrease in Cl^- permeability) has been demonstrated in this setting [183,184]. This finding can explain the defect in NaCl conservation and also the importance of hypokalemia in the genesis of metabolic alkalosis in primary hyperaldosteronism [9,184]. If Na^+ is reabsorbed but Cl^- cannot follow, then H^+ (and K^+) secretion must increase to maintain electroneutrality [184].

Other

Mild hyperglycemia may be induced by hypokalemia even in patients without a history of diabetes mellitus. This appears to be due primarily to impaired insulin secretion [185–187].

Neurologic symptoms such as lethargy and confusion may also occur. This is a relatively uncommon finding, perhaps because of the maintenance of normal cerebrospinal fluid and cerebral interstitial concentrations of K^+ despite the reduction in the plasma K^+ concentration [188].

DIAGNOSIS

The etiology of hypokalemia can usually be determined from the history—for example, complaints of vomiting or diarrhea, the use of diuretics, or recurring acute episodes of muscle weakness in periodic paralysis. When the diagnosis is not readily apparent, measurement of *urinary K^+ excretion* and assessment of the *acid-base status* may be helpful.

Urinary Potassium Excretion

As described above (see "Mineralocorticoid Excess: Diagnosis"), patients with extrarenal losses (or diuretics after the drug effect has worn off) should excrete less than 25 to 30 meq/day in the urine whereas values above this level suggest at least a contribution from urinary K^+ wasting. Random measurement of the urine K^+ concentration is simpler to perform but may be less accurate than a 24-h collection. If the urine K^+ concentration is below 15 meq/L, it is likely that extrarenal losses are present (unless the patient is polyuric). Somewhat higher values, however, do not necessarily imply K^+ wasting. For example, a patient excreting 20 meq/day will have a urine K^+ concentration of 40 meq/L if the urine volume is only 500 mL.

Hypovolemia (as evidenced by a low urine Na^+ concentration) can interfere with the interpretation of urinary K^+ excretion: raising it in patients with extrarenal losses because of the associated secondary hyperaldosteronism or lowering it in

Table 27-4 Acid-base disorders in hypokalemia

Metabolic acidosis may be seen	Metabolic alkalosis may be seen
Loss of intestinal secretions (diarrhea, laxative abuse)	Diuretics
Renal tubular acidosis	Vomiting or nasogastric suction
Ketoacidosis	Mineralocorticoid excess
Salt-wasting nephropathies	Penicillin derivatives

patients with urinary losses (as with primary mineralocorticoid excess) because of the decrease in distal flow. In this setting, the response to volume repletion may be helpful. Aldosterone secretion and K^+ excretion will fall in patients with extrarenal losses; in contrast, distal flow and K^+ excretion will increase in patients with non-suppressible mineralocorticoid excess [80].

Acid-Base Status

Some hypokalemic states are associated with acid-base disturbances that may aid in establishing the correct diagnosis (Table 27-4). In a patient with metabolic acidosis, for example, a salt-wasting nephropathy (in which renal insufficiency is typically present and accounts for the acidemia), renal tubular acidosis (see Chap. 19), and ketoacidosis should be excluded. If these disorders are not present, then surreptitious diarrhea perhaps due to laxative abuse is likely to be present [41].

In contrast, the combination of hypokalemia and metabolic alkalosis is usually due to diuretic use, vomiting (both of which may be surreptitious), or, less often, one of the causes of mineralocorticoid excess.

Vomiting and diuretics may be associated with volume depletion. Thus, the findings of decreased skin turgor, flat neck veins, and postural hypotension are suggestive of one of these problems since primary mineralocorticoid excess leads to mild volume expansion. The urine Cl^- concentration also may be helpful since a value below 15 meq/L is strongly suggestive of vomiting (or possibly diuretics after the drug effect has dissipated). If, on the other hand, the urine Cl^- concentration exceeds 15 to 20 meq/L and the patient is *normotensive,* a urinary assay for diuretics should be obtained [90]. If this is negative and hypokalemia persists, then Bartter's syndrome may be present. If the patient is *hypertensive,* then the evaluation outlined in Fig. 27-2 should be initiated.

Clinical Examples

The following case histories illustrate how this approach can be utilized.

Case history 27-1 A 36-year-old woman is started on diuretics twice weekly for mild pedal edema. During a routine follow-up, she is found to have a mild elevation in blood pressure (duration unknown). Also noted are flat neck veins and a moderate decrease in skin turgor. The following laboratory tests are obtained:

Plasma $[Na^+]$ = 137 meq/L	Arterial pH = 7.47
$[K^+]$ = 3.0 meq/L	Urine $[Na^+]$ = 12 meq/L
$[Cl^-]$ = 98 meq/L	$[K^+]$ = 90 meq/L
$[HCO_3^-]$ = 29 meq/L	Volume = 600 mL/day

COMMENT The combination of hypertension, hypokalemia, hyperkaliuria, and metabolic alkalosis suggests the diagnosis of primary hyperaldosteronism, e.g., due to an adrenal adenoma. It was initially felt that, although diuretics also can produce this electrolyte picture, the patient was taking the drug too infrequently. However, the physical findings and low urine Na^+ concentration pointed to the presence of volume depletion. As a result, the diuretic was discontinued and the patient given sodium and potassium. As volume repletion occurred, the urine Na^+ concentration rose to 40 meq/L with a concomitant reduction in urinary K^+ excretion. This response is consistent with diuretic-induced hypokalemia and not with primary hyperaldosteronism. The hypokalemia did not recur, and the elevation in blood pressure was shown to represent essential hypertension.

Case history 27-2 A 22-year-old woman complains of persistent weakness but denies all other symptoms. The physical examination is unremarkable; blood pressure is normal. The following laboratory data are obtained:

Plasma $[Na^+]$ = 136 meq/L	Arterial pH = 7.30
$[K^+]$ = 2.7 meq/L	Urine $[Na^+]$ = 7 meq/L
$[Cl^-]$ = 108 meq/L	$[K^+]$ = 12 meq/L
$[HCO_3^-]$ = 17 meq/L	

COMMENT The low urine K^+ concentration suggests extrarenal losses. Although she denies vomiting or diarrhea, the low urine Na^+ concentration indicates that she is volume-depleted and must have some unadmitted source of fluid loss. The presence of metabolic acidosis points toward diarrhea, perhaps induced by laxative abuse.

TREATMENT

The initial step in therapy must be the assessment of the physiologic effects of the K^+ deficit. As described above, there is a wide variation in the degree to which a given reduction in the plasma K^+ concentration will produce symptoms. Thus, monitoring of the electrocardiogram and muscle strength, *which reflect the functional consequences of K^+ depletion,* is an essential part of the management of patients with severe hypokalemia.

As with other electrolyte disorders, the first aim of therapy is to get the patient out of danger, not to immediately correct the entire K^+ deficit. The potential risk of the rapid administration of K^+ is illustrated in Fig. 27-5. In this patient, who had hypokalemia and flaccid paralysis, the intravenous infusion of 80 meq of K^+ over 15 min resulted in a change in the electrocardiogram from one typical of hypokalemia to one with the findings of severe hyperkalemia.

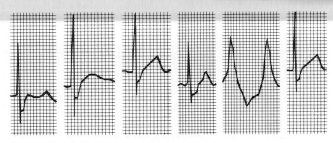

Time P.M.

(a) 12:10   (b) 12:18   (c) 12:22   (d) 12:24   (e) 12:25   (f) 12:30

Figure 27-5 Electrocardiogram (ECG) during the intravenous administration of 80 meq of K^+ over 15 min. *(a)* ECG showing hypokalemia with absent T wave and prominent U wave. *(b)* After 45 meq, the T wave is returning. *(c)* After 65 meq, the T wave is normal. *(d* and *e)* After 80 meq, the appearance of signs of hyperkalemia: peaked T wave and widening of the QRS complex. *(f)* Slowing of K^+ infusion with return of the tracing to normal. *(From H. C. Seftel and M. C. Kew, Diabetes, 15:694, 1966.)*

Potassium Deficit

The potassium deficit can only be approximated since there is no definite correlation between the plasma K^+ concentration and body K^+ stores. In general, a reduction in the plasma K^+ concentration from 4.0 to 3.0 meq/L requires the loss of 100 to 400 meq of K^+ [137,189]. An additional 200- to 400-meq deficit will lower the plasma K^+ concentration to 2.0 meq/L. However, continued K^+ losses may not produce much more hypokalemia as the release of K^+ from the cells is usually able to maintain the plasma K^+ concentration near 2.0 meq/L [189].

These estimates of the K^+ deficit assume that there is a normal distribution of K^+ between the cells and the extracellular fluid. In patients with periodic paralysis, for example, body K^+ stores are normal since the hypokalemia is due to K^+ movement into the cells. In this setting, K^+ is given to normalize the plasma K^+ concentration, not to repair a K^+ deficit.

The effects of pH are also important in evaluating the K^+ status of the patient. In particular, acidemia frequently raises the plasma K^+ concentration and masks the severity of the K^+ depletion. This has important therapeutic implications since correction of the acidemia may worsen the hypokalemia.

Use of Potassium Chloride

A variety of K^+ preparations are available for oral and intravenous use, including the Cl^-, HCO_3^-,† phosphate, and gluconate salts. There are two major advantages to the use of KCl in K^+-depleted patients. First, metabolic alkalosis is commonly

---

†The acetate and citrate salts also are available. These organic anions are rapidly metabolized into HCO_3^- in the body.

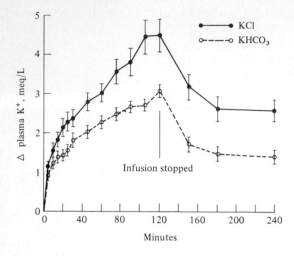

Figure 27-6 Changes in plasma potassium concentration of potassium-depleted dogs during infusion with KCl and KHCO$_3$. *(From M. F. Villamil, E. C. DeLand, P. Henney, and J. V. Maloney, Jr., Am. J. Physiol., 229:161, 1975.)*

associated with hypokalemia. In these patients, who tend to be Cl$^-$-depleted as well, e.g., because of vomiting or diuretics, the administration of Cl$^-$ is essential for the correction of both the alkalosis and the K$^+$ deficit. The use of other K$^+$ salts is less effective in inducing positive K$^+$ balance and may increase the severity of the alkalemia (see page 387) [190,191].

Second, the $[K^+]_c/[K^+]_e$ ratio (the main determinant of the resting membrane potential) is affected primarily by changes in $[K^+]_e$ since the extracellular K$^+$ concentration is so much less than that in the cells. Consequently, in a severely hypokalemic patient with muscle weakness or advanced electrocardiographic changes, the aim of therapy is to rapidly increase the plasma (and extracellular) K$^+$ concentration. If equal doses of KCl and KHCO$_3$ are given, there will be a significantly greater increase in the plasma K$^+$ concentration with KCl than with KHCO$_3$ (Fig. 27-6). This difference is probably related to the ability of HCO$_3^-$ to enter the cells in comparison with that of Cl$^-$, which is mostly limited to the extracellular fluid [192]. As a result, K$^+$ follows HCO$_3^-$ into the cells to maintain electroneutrality,† producing a lesser increase in the plasma K$^+$ concentration.

Although KHCO$_3$ is not as effective as KCl in metabolic alkalosis or severe hypokalemia, it may be the preferred K$^+$ salt in certain patients with mild degrees of hypokalemia and metabolic acidosis. For example, patients with renal tubular acidosis tend to waste K$^+$ in the urine and become K$^+$-depleted (see Chap. 19). In this setting, chronic therapy with KHCO$_3$ (or K$^+$ citrate, which may be more palatable) tends to correct both the K$^+$ depletion and the acidemia.

Oral KCl can be given orally in crystalline form (salt substitutes), as a liquid, or in a slow-release tablet or capsule. Enteric-coated tablets also are available, but their

---

†Rather than HCO$_3^-$ entry into the cells, intracellular H$^+$ ions may be released into the extracellular fluid where they buffer the excess HCO$_3^-$ ions. In either case, K$^+$ tends to move into the cells to maintain electroneutrality.

use *should be avoided* because of the relatively high incidence of small-bowel ulceration and/or stenosis seen with these agents [193,194]. This effect appears to be due to the local accumulation of high concentrations of K^+ as the enteric coating dissolves rapidly at the alkaline intestinal pH. This complication can be largely avoided by the use of the other K^+ preparations, which generally disperse the K^+ more widely, thereby avoiding local accumulation [195]. However, the slow-release tablets can cause lesions, particularly when there is underlying obstruction due, for example, to an enlarged left atrium pressing on the esophagus or decreased intestinal motility [195–197].

The optimum form of oral K^+ supplementation is uncertain. KCl solutions are safe but poorly palatable, whereas slow-release preparations are better-tolerated but have some (although small) risk [198]. It may be that the use of a salt substitute (which contains between 50 and 65 meq per level teaspoon) is the ideal oral therapy since it is much cheaper than the other supplements (by 85 to 95 percent) and is as safe as the KCl solutions but better-tolerated [199]. The traditional therapy of treating chronic hypokalemia (most often due to diuretics; see below) with K^+-rich foods such as orange juice or bananas is less desirable. This regimen contains less chloride, is somewhat more expensive than KCl solutions or tablets, and involves an increase in caloric intake in excess of 350 kcal/day. The last problem is a particular disadvantage with obese hypertensives in whom weight reduction would be beneficial.

Intravenous In patients who are unable to eat, K^+ must be given intravenously. The standard intravenous KCl solution contains 2 meq each of K^+ and Cl^- per milliliter. In most circumstances, 20 to 40 meq of K^+ (10 to 20 mL) is added to each liter of dextrose or saline solution. However, if this quantity of K^+ is added to 1 liter of a dextrose solution, there may acutely be a small reduction in the plasma K^+ concentration of 0.2 to 1.4 meq/L, particularly when only 20 meq/L is added [18]. This effect is probably due to enhanced insulin secretion stimulated by the infusion of glucose. Although normal subjects can tolerate this decrease in the plasma K^+ concentration, arrhythmias may be precipitated in patients who are hypokalemic or taking digitalis [18]. Consequently, K^+ supplementation should be given in a non-dextrose-containing solution, usually in a concentration of 40 meq/L. In general, no more than 60 meq/L should be given through a peripheral vein since higher concentrations of K^+ are very irritating, resulting in pain and sclerosis of the vein.

Rate of Potassium Repletion

The majority of patients have mild to moderate hypokalemia, with the plasma K^+ concentration ranging from 2.7 to 3.5 meq/L. This degree of K^+ depletion is usually well tolerated in the absence of digitalis therapy or severe hepatic disease [154,155, 174,176]. Treatment is not urgent in this setting and must be directed toward both repair of the K^+ deficit and prevention of further K^+ loss by correcting the underlying disorder (such as diarrhea). These patients can usually be treated with oral KCl at an initial dose of 60 to 80 meq/day. However, larger amounts will be required if there is continued K^+ loss. In primary hyperaldosteronism, for example, oral K^+

therapy is generally ineffective and a K^+-sparing diuretic is required to maintain a normal plasma K^+ concentration [94–96].

In patients with severe symptoms or marked hypokalemia, K^+ must be given more rapidly. This is more easily done orally as the plasma K^+ concentration will acutely rise by as much as 1.0 to 1.5 meq/L after 40 to 60 meq and by 2.5 to 3.5 meq/L after 135 to 160 meq [200,201]. These maximum effects are transient since most of the administered K^+ will enter the cells to repair the cell deficit [202]. As a result, the plasma K^+ concentration must be carefully monitored and more K^+ given as necessary. A patient with a plasma K^+ concentration of 2.0 meq/L, for example, may have a total K^+ deficit of as much as 400 to 800 meq [137].

Large doses of K^+ are much more difficult to give intravenously since the limit of 60 meq/L means that a large volume of fluid also must be given. This tendency to fluid overload is enhanced by the preferential use of saline-containing solutions (such as one-quarter–isotonic saline) because of the desire to avoid dextrose administration [18]. Despite these obstacles, some patients must be treated intravenously. In general, intravenous K^+ is given at a maximum rate of 10 to 20 meq/h although as much as 40 to 100 meq/h has been given to patients with paralysis or life-threatening arrhythmias [203–205]. In the latter setting, solutions containing as much as 150 to 180 meq of K^+ per liter have been used [206,207]. This must be given into a large vein (such as the femoral vein); infusions through a central venous line should probably be avoided since the local increase in the K^+ concentration could have deleterious effects on the heart.

The necessity for such aggressive therapy has been reported primarily in patients with diabetic ketoacidosis who are also hypokalemic (see Chap. 25) [204,205]. In this setting, the administration of insulin and the consequent reduction in the plasma glucose concentration will drive K^+ into the cells, further reducing the plasma K^+ concentration if KCl is withheld. Since the average fluid deficit in these patients is 3 to 6 liters, rapid K^+ replacement can be achieved by adding 60 meq of K^+ to each liter of fluid. This example also illustrates that the factors that influence the internal distribution of K^+ (arterial pH, availability of insulin) must be taken into account when evaluating the hypokalemic patient.

It must be emphasized that the rapid administration of K^+ is *potentially dangerous even in severely K^+-depleted patients* and should be used only in life-threatening situations. A rate in excess of 80 meq/h can result in the electrocardiographic changes of hyperkalemia (Fig. 27-5) or complete heart block [208]. Thus, continuous monitoring of the electrocardiogram is essential in this setting.

Diuretic-Induced Hypokalemia

The necessity for treating the mild hypokalemia induced by diuretic therapy for hypertension is uncertain [145,146]. In most patients, the plasma K^+ concentration remains above 3.0 meq/L [103–106,145,146] a level at which symptoms are unlikely to occur. Furthermore, the administration of KCl or a K^+-sparing diuretic can result in possibly fatal *hyper*kalemia, particularly in patients with underlying renal insufficiency [145,146,209]. Thus, it may be reasonable to conclude that routine K^+ supplementation is not required in asymptomatic patients on a regular diet who are not taking digitalis and who have a plasma K^+ concentration above 3.0 meq/L. If KCl is given, 60 meq/day (one level teaspoon of a salt substitute, for example) is usually necessary to normalize K^+ balance [210]. KCl should not be given with a K^+-sparing diuretic since the incidence of hyperkalemia with this combination is substantially increased, reaching 42 percent in patients with renal insufficiency (blood urea nitrogen greater than 50 mg/dL) [211].

It should be noted, however, that asymptomatic ventricular arrhythmias (at rest and after exertion) occur with greater frequency even in mildly hypokalemic patients [105,153]. Although it has been assumed that these arrhythmias are benign, patients with mild hypertension (diastolic pressure between 90 and 94 mmHg) treated with a diuretic may have an increased incidence of cardiac complications [212], an effect that could be related in part to K^+ depletion. Furthermore, there is some evidence that a low-K^+ diet may predispose to the development of hypertension and that K^+ supplements may be desirable to lower the blood pressure, if not to correct the hypokalemia [3,213,214]. As a result, the benignity of mild K^+ depletion is not definitively proved and the role of K^+ supplementation remains uncertain.

PROBLEMS

27-1 A 22-year-old woman complains of easy fatigability and weakness. She has no other complaints. The physical examination is unremarkable, including a normal blood pressure. The following laboratory data are obtained:

Plasma [Na^+] = 141 meq/L	Urine [Na^+] = 80 meq/day
[K^+] = 2.1 meq/L	[K^+] = 170 meq/day
[Cl^-] = 85 meq/L	
[HCO_3^-] = 45 meq/L	

(*a*) What is the differential diagnosis of the hypokalemia?

(*b*) What test would you order next?

27-2 A patient is noted to have a plasma K^+ concentration of 2.7 meq/L. Match the other laboratory changes that are present with the likely diagnosis.

(*a*) Plasma [HCO_3^-] = 27 meq/L   1. Renal tubular acidosis

Arterial pH = 7.43

Urine [K^+] = 10 meq/L

Osmolality = 102 mosmol/kg   2. Hypomagnesemia

(*b*) Plasma [HCO_3^-] = 27 meq/L

[Ca^{2+}] = 7.3 mg/dL   3. Primary polydipsia

[Albumin] = 4.1 g/dL

Arterial pH = 7.46

Urine [K^+] = 45 meq/day   4. Laxative abuse

(c) Plasma $[HCO_3^-]$ = 14 meq/L

   Arterial pH = 7.28

   Urine $[K^+]$ = 52 meq/day

   pH = 6.0

(d) Plasma $[HCO_3^-]$ = 14 meq/L

   Arterial pH = 7.28

   Urine $[K^+]$ = 18 meq/day

   pH = 4.9

5. Primary hyperaldosteronism

6. Vomiting

REFERENCES

1. Womersley, R. A., and J. H. Darragh: Potassium and sodium restriction in the normal human, J. Clin. Invest., 34:456, 1955.
2. Squires, R. D., and E. J. Huth: Experimental potassium depletion in normal human subjects: I. Relation of ionic intakes to the renal conservation of potassium, J. Clin. Invest., 38:1134, 1959.
3. Langford, H. G.: Dietary potassium and hypertension: epidemiologic data, Ann. Intern. Med., 98(part 2):770, 1983.
4. Gonzalez, J. J., W. Owens, P. C. Ungaro, E. E. Werk, Jr., and P. W. Wentz: Clay ingestion: a rare cause of hypokalemia, Ann. Intern. Med., 97:65, 1982.
5. Gelfand, M. C., A. Zarate, and J. H. Knepshield: Geophagia: a cause of life-threatening hyperkalemia in patients with chronic renal failure, J. Am. Med. Assoc., 234:738, 1975.
6. Adrogué, H. J., and N. E. Madias: Changes in plasma potassium concentration during acute acid-base disturbances, Am. J. Med., 71:456, 1981.
7. Fraley, D. S., and S. Adler: Correction of hyperkalemia by bicarbonate despite constant blood pH, Kidney Int., 12:354, 1977.
8. Hulter, H. N., J. F. Sigala, and A. Sebastian: K^+ deprivation potentiates the renal alkalosis-producing effect of mineralocorticoid, Am. J. Physiol., 235:F298, 1978.
9. Kassirer, J. P., A. M. London, D. M. Goldman, and W. B. Schwartz: On the pathogenesis of metabolic alkalosis in hyperaldosteronism, Am. J. Med., 49:306, 1970.
10. Kurtzman, N. A., M. G. White, and P. W. Rogers: Pathophysiology of metabolic alkalosis, Arch. Intern. Med., 131:702, 1973.
11. Wilson, A. F., and D. H. Simmons: Relationships between potassium, chloride, intracellular and extracellular pH in dogs, Clin. Sci., 39:731, 1970.
12. Adler, S., B. Anderson, and B. Zett: The effect of acute potassium depletion on muscle cell pH in vitro, Kidney Int., 2:159, 1972.
13. Cooke, R. E., W. Segar, D. B. Cheek, F. Coville, and D. Darrow: The extrarenal correction of alkalosis associated with potassium deficiency, J. Clin. Invest., 31:798, 1952.
14. Orloff, J., T. Kennedy, Jr., and R. W. Berliner: The effect of potassium in nephrectomized rats with hypokalemic alkalosis, J. Clin. Invest., 32:538, 1953.
15. Zierler, K. L., and D. Rabinowitz: Effect of very small concentrations of insulin on forearm metabolism: persistence of its action on potassium and free fatty acids without its effect on glucose, J. Clin. Invest., 43:950, 1964.
16. Minaker, E. L., and J. W. Rowe: Potassium homeostasis during hyperinsulinemia: effect of insulin level, β-blockade, and age, Am. J. Physiol., 242:E373, 1982.
17. Kreisberg, R. A.: Diabetic ketoacidosis: new concepts and trends in pathogenesis and treatment, Ann. Intern. Med., 88:681, 1978.

18. Kunin, A. S., B. Surawicz, and E. A. H. Sims: Decrease in serum potassium concentration and appearance of cardiac arrhythmias during infusion of potassium with glucose in potassium-depleted patients, N. Engl. J. Med., 266:228, 2962.

19. Petersen, K-G., K. J. Schlüter, and L. Kerp: Regulation of serum potassium during insulin-induced hypoglycemia, Diabetes, 31:615, 1982.

20. Brown, M. J., D. C. Brown, and M. B. Murphy: Hypokalemia from beta$_2$-receptor stimulation by circulating epinephrine, N. Engl. J. Med., 309:1414, 1983.

21. DeFronzo, R. A., M. Bia, and G. Birkhead: Epinephrine and potassium homeostasis, Kidney Int., 20:83, 1981.

21a. Struthers, A. D., R. Whitesmith, and J. L. Reid: Prior thiazide diuretic treatment increases adrenalin-induced hypokalaemia, Lancet, 1:1358, 1983.

22. O'Brien, I. A. D., J. Fitzgerald, I. G. Lewin, and R. J. M. Corrall: Hypokalemia due to salbutamol overdosage, Br. Med. J., 1:1515, 1981.

23. Chew, W. C., and L. C. Lew: Ventricular ectopics after salbutamol infusion for preterm labor, Lancet, 2:1383, 1979.

24. Pearson, C. M., and K. Kalyanaraman: Periodic paralysis, in *The Metabolic Basis of Inherited Disease,* 3d ed., J. B. Stanbury, J. B. Wyngaarden, and D. S. Frederickson (eds.), McGraw-Hill, New York, 1972.

25. Layzer, R. B.: Periodic paralysis and the sodium-potassium pump, Ann. Neurol., 11:547, 1982.

26. McFadzean, A. J. S., and R. Yeung: Periodic paralysis complicating thyrotoxicosis in Chinese, Br. Med. J., 1:451, 1967.

27. Conway, M. J., J. A. Seibel, and R. P. Eaton: Thyrotoxicosis and periodic paralysis: improvement with beta blockade, Ann. Intern. Med., 81:332, 1974.

28. Yeung, R. T. T., and T. F. Tse: Thyrotoxic periodic paralysis: effect of propranolol, Am. J. Med., 57:584, 1974.

29. Resnick, J. S., W. L. Engel, R. C. Griggs, and A. C. Stam: Acetazolamide prophylaxis in hypokalemic periodic paralysis, N. Engl. J. Med., 278:582, 1968.

30. Lawson, D. H., R. M. Murray, and J. L. W. Parker: Early mortality in the megaloblastic anaemias, Q. J. Med., 41:1, 1972.

31. Rao, T. L. K., M. Mathru, M. R. Salem, and A. A. El-Etr: Serum potassium levels following transfusion of frozen erythrocytes, Anesthesiology, 52:170, 1980.

32. Adams, P. C., K. W. Woodhouse, M. Adela, and A. Parnham: Exaggerated hypokalemia in acute myeloid leukaemia, Br. Med. J., 1:1034, 1981.

33. Wadstein, J., and G. Skude: Does hypokalaemia precede delirium tremens?, Lancet, 2:549, 1978.

34. Koht, A., L. J. Cerullo, P. C. Land, and H. W. Linde: Serum potassium levels during prolonged hypothermia, Anesthesiology, 51(suppl).S203, 1979.

34a. Bruining, H. A., and R. U. Boelhouwer: Acute transient hypokalemia and body temperature (letter), Lancet, 2:1283, 1982.

35. Kassirer, J. P., and W. B. Schwartz: The response of normal man to selective depletion of hydrochloric acid: factors in the genesis of persistent gastric alkalosis, Am. J. Med., 40:10, 1966.

36. Watten, R. H., F. M. Morgan, Y. N. Songkhla, B. Vanikiati, and R. A. Phillips: Water and electrolyte studies in cholera, J. Clin. Invest., 38:1879, 1959.

37. Verner, J. V., and A. B. Morrison: Endocrine pancreatic islet disease with diarrhea: report of a case due to diffuse hyperplasia of nonbeta islet tissue with a review of 54 additional cases, Arch. Intern. Med., 133:492, 1974.

38. Kahn, C. R., A. G. Levy, J. D. Garner, J. V. Miller, P. Gordon, and P. S. Schein: Pancreatic cholera: beneficial effects of treatment with streptozotocin, N. Engl. J. Med., 292:941, 1975.

39. McArthur, K. E., D. S. Anderson, T. E. Durbin, M. J. Orloff, and K. Dharmsathaphorn: Clonidine and lidamidine to inhibit watery diarrhea in a patient with lung cancer, Ann. Intern. Med., 96:323, 1980.

40. Crane, C. W.: Observations on the sodium and potassium content of mucus from the large intestine, Gut, 6:439, 1965.

41. Schwartz, W. B., and A. S. Relman: Metabolic and renal studies in chronic potassium depletion resulting from overuse of laxatives, J. Clin. Invest., 32:258, 1953.

42. Kirkegaard, P., J. M. Lundberg, S. S. Poulsen, and J. Christiansen: Vasoactive intestinal polypeptidergic nerves and Brunner's gland secretion in the rat, Gastroenterology, 81:872, 1981.

43. Jaffe, B. M., D. F. Kopen, K. DeSchryver-Kecskemeti, R. L. Gingerich, and M. Greider: Indomethacin-responsive pancreatic cholera, N. Engl. J. Med., 297:817, 1977.

44. Donowitz, M., G. Elta, S. R. Bloom, and L. Nathanson: Trifluoperazine reversal of secretory diarrhea in pancreatic cholera, Ann. Intern. Med., 93:284, 1980.

45. Seldin, D., L. G. Welt, and J. Cort: The role of sodium salts and adrenal steroids in the production of hypokalemic alkalosis, Yale J. Biol. Med., 29:229, 1956.

46. Biglieri, E. G., J. R. Stockigt, and M. Schambelan: Adrenal mineralocorticoids causing hypertension, Am. J. Med., 52:623, 1972.

47. Weinberger, M. H., C. E. Grim, J. W. Hollifield, D. C. Kem, A. Ganguly, N. J. Kramer, H. Y. Yune, H. Wellman, and J. P. Donohue: Primary aldosteronism, Ann. Intern. Med., 90:386, 1979.

48. Biglieri, E. G.: Adrenocortical components in hypertension, Cardiovasc. Rev. Rep., 3(5):734, 1982.

49. Haack, D., J. Mohring, B. Mohring, M. Petri, and E. Hackenthal: Comparative study on development of corticosterone and DOCA hypertension in rats, Am. J. Physiol., 233:F403, 1978.

50. Robertson, G. L., P. Aycinena, and R. L. Zerbe: Neurogenic disorders of osmoregulation, Am. J. Med., 72:339, 1982.

51. Ganguly, A.: New insights and questions about glucocorticoid-suppressible hyperaldosteronism, Am. J. Med., 72:851, 1982.

52. Bravo, E. L., R. C. Tarazi, H. P. Dustan, F. M. Fouad, S. C. Textor, R. W. Gifford, and D. G. Vidt: The changing clinical spectrum of primary aldosteronism, Am. J. Med., 74:641, 1983.

53. Gross, M. D., R. J. Grekin, T. C. Gniadek, and J. Z. Villareal: Suppression of aldosterone by cyproheptadine in idiopathic aldosteronism, N. Engl. J. Med., 305:181, 1981.

54. Mulrow, P. J.: Glucocorticoid-suppressible hyperaldosteronism: a clue to the missing hormone?, N. Engl. J. Med., 305:1012, 1981.

55. Biglieri, E. G., P. E. Slaton, M. Schambelan, and S. J. Kronfield: Hypermineralocorticoidism, Am. J. Med., 45:170, 1968.

56. Gold, E. M.: The Cushing's syndromes: changing views of diagnosis and treatment, Ann. Intern. Med., 90:829, 1979.

57. Christy, N. P., and J. H. Laragh: Pathogenesis of hypokalemic alkalosis in Cushing's syndrome, N. Engl. J. Med., 265:1083, 1961.

58. Elijovich, F., and L. R. Krakoff: Effect of converting enzyme inhibition on glucocorticoid hypertension in the rat, Am. J. Physiol., 238:H844, 1980.

59. Dalakos, T. G., A. N. Elias, G. H. Anderson, Jr., D. H. P. Streeten, and E. T. Schroeder: Evidence for an angiotensinogenic mechanism of the hypertension of Cushing's syndrome, J. Clin. Endocrinol. Metab., 46:114, 1978.

60. Conn, J. W., D. R. Roune, and E. L. Cohen: Licorice-induced pseudoaldosteronism, J. Am. Med. Assoc., 205:492, 1968.

61. Blachley, J. D., and J. P. Knochel: Tobacco chewer's hypokalemia: licorice revisited, N. Engl. J. Med., 302:784, 1980.

62. Mantero, F., D. Armanini, G. Opocher, F. Fallo, L. Sampieri, B. Cuspidi, C. Ambrosi, and G. Faglia: Mineralocorticoid hypertension due to a nasal spray containing α-fluroprednisolone, Am. J. Med., 71:352, 1981.

63. Dunn, M. J.: Prostaglandins and Bartter's syndrome, Kidney Int., 19:86, 1981.

64. Güllner, H. G., C. Cerlette, F. C. Bartter, J. B. Smith, and J. R. Gill, Jr.: Prostacyclin overproduction in Bartter's syndrome, Lancet, 2:767, 1979.

65. Fujita, T., K. Ando, Y. Sato, K. Yamashita, M. Nomura, and T. Fukui: Independent roles of prostaglandins and the renin-angiotensin system in abnormal vascular reactivity in Bartter's syndrome, Am. J. Med., 73:71, 1982.

66. Gill, J. R., Jr., and F. C. Bartter: Evidence for a prostaglandin-independent defect in chloride reabsorption in the loop of Henle as a proximal cause of Bartter's syndrome, Am. J. Med., 65:766, 1978.

67. Carmine, Z., B. Ettore, C. Giuseppe, and M. Quirino: The renal tubular defect of Bartter's syndrome, Nephron, 32:140, 1982.

68. Vinci, J. M., J. R. Gill, Jr., R. E. Bowden, J. J. Pisano, J. L. Izzo, Jr., N. Radfar, A. A. Taylor, R. M. Zusman, F. C. Bartter, and H. R. Keiser: The kallikrein-kinin system in Bartter's syndrome and its response to prostaglandin synthetase inhibition, J. Clin. Invest., 61:1671, 1978.

69. Zipser, R. D., R. K. Rude, P. K. Zia, and M. P. Fichman: Regulation of urinary prostaglandins in Bartter's syndrome, Am. J. Med., 67:263, 1979.

70. Patak, R. V., S. Z. Fadem, S. G. Rosenblatt, M. D. Lifschitz, and J. H. Stein: Diuretic-induced changes in renal blood flow and prostaglandin E excretion in the dog, Am. J. Physiol., 236:F494, 1979.

71. Conn, J. W., E. L. Cohen, C. P. Lucas, W. J. McDonald, G. H. Mayor, W. M. Blough, W. C. Evenland, J. J. Bookstein, and J. Lapides: Primary reninism, Arch. Intern. Med., 130:682, 1972.

72. Rose, B. D.: *Pathophysiology of Renal Disease,* McGraw-Hill, New York, 1981, pp. 592–599.

73. Kondo, K., T. Saruta, I. Saito, R. Yoshida, H. Maruyama, and S. Matsuki: Benign desoxycorticosterone-producing adrenal tumor, J. Am. Med. Assoc., 236:1042, 1976.

74. Hogan, M. J., M. Schambelan, and E. G. Biglieri: Concurrent hypercortisolism and hypermineralocorticoidism, Am. J. Med., 62:777, 1977.

75. New, M. I., and L. S. Levine: Mineralocorticoid hypertension in childhood, Mayo Clin. Proc., 52:323, 1977.

76. New, M. I., L. S. Levine, E. G. Biglieri, J. Pareira, and S. Ulick: Evidence for an unidentified steroid in a child with apparent mineralocorticoid hypertension, J. Clin. Endocrinol. Metab., 44:924, 1977.

77. Conn, J. W., E. L. Cohen, D. R. Rovner, and R. M. Nesbit: Normokalemic primary aldosteronism: a detectable cause of curable "essential" hypertension, J. Am. Med. Assoc., 193:200, 1965.

78. Kono, T., F. Ikeda, F. Oseko, H. Imura, and H. Tanimura: Normotensive primary aldosteronism: report of a case, J. Clin. Endocrinol. Metab., 52:1009, 1981.

79. Rose, B. D.: *Pathophysiology of Renal Disease,* McGraw-Hill, New York, 1981, pp. 603–606.

80. George, J. M., L. Wright, N. H. Bill, and F. C. Bartter: The syndrome of primary aldosteronism, Am. J. Med., 48:343, 1970.

81. Rose, B. D.: *Pathophysiology of Renal Disease,* McGraw-Hill, New York, 1981, pp. 607–609.

82. Streeten, D. H. P., N. Tomycz, and G. H. Anderson, Jr.: Reliability of screening methods for the diagnosis of primary aldosteronism, Am. J. Med., 67:403, 1979.

83. Kaplan, N. M.: Hypokalemia in the hypertensive patient: with observations on the incidence of primary aldosteronism, Ann. Intern. Med., 66:1079, 1967.

84. White, E. A., M. Schambelan, C. R. Roist, E. G. Biglieri, A. A. Moss, and M. Korobkin: Use of computed tomography in diagnosing the cause of primary aldosteronism, N. Engl. J. Med., 303:1503, 1980.

85. Levinson, P. D., Z. Zadik, B. P. M. Hamilton, J. H. Mersey, R. I. White, and A. A. Kowarski: Adrenal vein epinephrine levels: a useful aid for venous sampling in primary aldosteronism, Ann. Intern. Med., 97:690, 1982.

86. Melby, J. C.: Identifying the adrenal lesion in primary aldosteronism, Ann. Intern. Med., 76:1039, 1972.

87. Conn, J. W., E. L. Cohen, and K. R. Herwig: The dexamethasone-modified adrenal scintiscan in hyporeninemic aldosteronism (tumor versus hyperplasia): a comparison with adrenal venography and adrenal venous aldosterone, J. Lab. Clin. Med., 88:841, 1976.

88. Veldhuis, J. D., C. W. Bardin, and L. M. Demers: Metabolic mimickry of Bartter's syndrome by covert vomiting: utility of urinary chloride determinations, Am. J. Med., 66:361, 1979.

89. Jamison, R. L., J. C. Ross, R. L. Kempson, C. R. Sufit, and T. E. Parker: Surreptitious diuretic ingestion and pseudo-Bartter's syndrome, Am. J. Med., 73:142, 1982.

90. Rodman, J. S., and M. M. Reidenberg: Symptomatic hypokalemia resulting from surreptitious diuretic ingestion, J. Am. Med. Assoc., 246:1687, 1981.

91. Chan, J. C. M.: Bartter's syndrome, Nephron, 26:155, 1980.

92. Biglieri, E. G., M. Schambelan, P. E. Slaton, and J. R. Stockigt: The intercurrent hypertension of primary aldosteronism, Circ. Res., 27 (suppl 1):195, 1970.

93. O'Neil, L. W., J. M. Kissane, and P. M. Hartroft: The kidney in endocrine hypertension, Arch. Surg., 100:498, 1970.

94. Brown, J. J., D. L. Davies, J. B. Ferriss, R. Fraser, F. Haywood, A. F. Lever, and J. I. S. Robertson: Comparison of surgery and prolonged spironolactone therapy in patients with hypertension, aldosterone excess, and low plasma renin, Br. Med. J., 2:729, 1972.

95. Ganguly, A., and M. H. Weinberger: Triamterene-thiazide combination: alternative therapy for primary aldosteronism, Clin. Pharmacol. Therap., 30:246, 1981.

96. Griffing, G. T., A. G. Cole, S. A. Aurecchia, B. H. Sindler, P. Komanicky, and J. C. Melby: Amiloride in primary hyperaldosteronism, Clin. Pharmacol. Therap., 31:57, 1982.

97. Griffing, G. T., P. Komanicky, S. A. Aurecchia, B. H. Sindler, and J. C. Melby: Amiloride in Bartter's syndrome, Clin. Pharmacol. Therap., 31:713, 1982.

98. Ciabattoni, G., F. Pugliese, G. A. Cinotti, and C. Patrono: Renal effects of anti-inflammatory drugs, Eur. J. Rheumatol. Inflam., 3:210, 1980.

99. Bunning, R. D., and W. F. Barth: Sulindac: a potentially renal-sparing nonsteroidal anti-inflammatory drug, J. Am. Med. Assoc., 248:2864, 1982.

100. Duarte, C. G., F. Chomety, and G. Giebisch: Effect of amiloride, ouabain, and furosemide on distal tubular function in the rat, Am. J. Physiol., 221:632, 1971.

101. Burg, M. B., and N. Green: Effect of ethacrynic acid on the thick ascending limb of Henle's loop, Kidney Int., 4:301, 1973.

102. Swales, J. D.: Magnesium deficiency and diuretics, Br. Med. J., 2:1377, 1982.

103. Maronde, R. F., M. Milgrom, N. D. Vlachakis, and L. Chan: Response of thiazide-induced hypokalemia to amiloride, J. Am. Med. Assoc., 249:237, 1983.

104. Morgan, D. B., and C. Davidson: Hypokalaemia and diuretics: an analysis of publications, Br. Med. J., 1:905, 1980.

105. Hollifield, J. W., and P. E. Slaton: Thiazide diuretics, hypokalemia and cardiac arrhythmias, Acta Med. Scand., 209(suppl. 647):67, 1981.

106. Materson, B. J., J. R. Oster, U. F. Michael, S. M. Bolton, Z. C. Burton, J. E. Stambaugh, and J. Morledge: Dose response to chlorthalidone in patients with mild hypertension: efficacy of a lower dose. Clin. Pharmacol. Therap., 24:192, 1978.

107. Bricker, N. S., E. S. Shwayri, J. B. Reardon, D. Kellogg, J. P. Merrill, and J. H. Holmes: An abnormality in renal function resulting from urinary tract obstruction, Am. J. Med., 23:554, 1957.

108. Potter, W. Z., C. W. Trygstad, O. M. Helmer, W. E. Nance, and W. E. Judson: Familial hypokalemia associated with renal interstitial fibrosis, Am. J. Med., 57:971, 1974.

109. Jones, N. F., and I. H. Mills: Reversible renal potassium loss with urinary tract infection, Am. J. Med., 37:305, 1964.

110. Popovtzer, M. M., F. H. Katz, W. F. Pinggera, J. Robinette, C. C. Halgrimson, and D. E. Butkus: Hyperkalemia in salt-wasting nephropathy: study of the mechanism, Arch. Intern. Med., 132:203, 1973.

110a. Uribarri, J., M. S. Oh, and H. J. Carroll: Salt-losing nephropathy. Clinical presentation and mechanisms, Am. J. Nephrol., 3:193, 1983.

111. Aldinger, K. A., and N. A. Samaan: Hypokalemia with hypercalcemia: prevalence and significance in treatment, Ann. Intern. Med., 87:571, 1977.

112. Muggia, F. M., H. O. Heinemann, M. Farhangi, and E. F. Osserman: Lysozymuria and renal tubular dysfunction in monocytic and myelomonocytic leukemia, Am. J. Med., 47:351, 1969.

113. Mir, M. A., B. Brabin, M. J. Leyland, and I. W. Delamore: Hypokalaemia in acute myeloid leukaemia, Ann. Intern. Med., 82:54, 1975.

114. Skarin, A. T., Y. Matsuo, and W. C. Moloney: Muramidase in myeloproliferative disorders terminating in acute leukemia, Cancer, 29:1336, 1972.

115. Evans, J. J., and M. J. Bozdech: Hypokalemia in nonblastic chronic myelogenous leukemia, Arch. Intern. Med., 141:786, 1981.

116. Greenberger, J. S., D. S. Rosenthal, and W. C. Moloney: Studies on hypermuramidasemia in the normal and chloroleukemic rat: the role of the kidney, J. Lab. Clin. Med., 81:116, 1973.

117. Kassirer, J. P., and W. B. Schwartz: The response of normal man to selective depletion of hydrochloric acid: factors in the genesis of persistent gastric alkalosis, Am. J. Med., 40:10, 1966.

118. Gennari, F. J., and J. J. Cohen: Role of the kidney in potassium homeostasis: lessons from acid-base disturbances, Kidney Int., 8:1, 1975.

119. Sebastian, A., E. McSherry, and R. C. Morris, Jr.: Renal potassium wasting in renal tubular acidosis (RTA): its occurrence in types 1 and 2 RTA despite sustained correction of systemic acidosis, J. Clin. Invest., 50:667, 1971.

120. Streicher, H. Z., P. A. Gabow, A. H. Moss, D. Kono, and W. D. Kaehny: Syndromes of toluene sniffing in adults, Ann. Intern. Med., 94:758, 1981.

121. Taher, S. M., R. J. Anderson, R. McCartney, M. M. Popovtzer, and R. W. Schrier: Renal tubular acidosis associated with toluene "sniffing", N. Engl. J. Med., 290:765, 1974.

122. Cogan, M. G., and F. C. Rector: Proximal reabsorption during metabolic acidosis in the rat, Am. J. Physiol., 242:F499, 1982.

123. Lipner, H. I., F. Ruzany, M. Dasgupta, P. D. Leif, and N. Bank: The behavior of carbenicillin as a nonreabsorbable anion, J. Lab. Clin. Med., 86:183, 1975.

124. Klastersky, J., B. Vanderkelen, D. Daneua, and M. Mathieu: Carbenicillin and hypokalemia (letter), Ann. Intern. Med., 78:774, 1973.

125. Brunner, F. P., and P. G. Frick: Hypokalemia, metabolic alkalosis, and hypernatremia due to "massive" sodium penicillin therapy, Br. Med. J., 4:550, 1968.

126. Mohr, J. A., R. M. Clark, T. C. Waack, and R. Whang: Nafcillin-associated hypokalemia, J. Am. Med. Assoc., 242:544, 1979.

127. Shils, M. E.: Experimental human magnesium depletion, Medicine, 48:61, 1969.

128. Rude, R. K., S. B. Oldham, C. F. Sharp, Jr., and F. R. Singer: Parathyroid hormone secretion in magnesium deficiency, J. Clin. Endocrinol. Metab., 47:800, 1978.

129. Estep, H., W. A. Shaw, C. Watlington, R. Hoke, W. Holland, and St. G. Tucker: Hypocalcemia due to hypomagnesemia and reversible parathyroid hormone unresponsiveness, J. Clin. Endocrinol. Metab., 29:842, 1969.

130. Petersen, V. P.: Potassium and magnesium turnover in magnesium deficiency, Acta Med. Scand., 174:595, 1963.

131. Francisco, L. L., L. L. Sawin, and G. F. DiBona: Mechanism of negative potassium balance in the magnesium-deficient rat, Proc. Soc. Exp. Biol. Med., 168:382, 1982.

132. Hariprasad, M. K., R. P. Eisinger, I. M. Nadler, C. S. Padmanabhan, and B. D. Nidus: Hyponatremia in psychogenic polydipsia, Arch. Intern. Med., 140:1639, 1980.

133. Granérus, A-K., R. Jagenburg, and A. Svanborg: Kaliuretic effect of L-dopa treatment in Parkinsonian patients, Acta Med. Scand., 201:291, 1977.

134. Knochel, J. P., L. N. Dotin, and R. J. Hamburger: Pathophysiology of intense physical conditioning in hot climate: I. Mechanism of potassium depletion, J. Clin. Invest., 51:242, 1972.

134a. Kosunen, K. J., and A. J. Pakarinen: Plasma renin, angiotensin II, and plasma and urinary aldosterone in running exercise, J. Appl. Physiol., 41:26, 1976.

135. Rostand, S. G.: Profound hypokalemia in continuous ambulatory peritoneal dialysis, Arch. Intern. Med., 143:377, 1983.

136. Wiegand, C. F., T. D. Davin, L. Raij, and C. M. Kjellstrand: Severe hypokalemia induced by hemodialysis, Arch. Intern. Med., 141:167, 1981.

137. Sterns, R. H., M. Cox, P. U. Feig, and I. Singer: Internal potassium balance and the control of the plasma potassium concentration, Medicine, 60:339, 1981.

138. Flear, C. T. G., W. T. Cooke, and A. Quinton: Serum-potassium levels as an index of body content, Lancet, 1:458, 1957.

139. Nagant de Deuxchaisnes, C., R. A. Collet, R. Busset, and R. S. Mach: Exchangeable potassium in wasting, amyotrophy, heart disease, and cirrhosis of the liver, Lancet, 1:681, 1961.

140. Casey, T. H., W. J. H. Summerskill, and A. L. Orvis: Body and serum potassium in liver disease: I. Relationship to hepatic function and associated factors, Gastroenterology, 48:198, 1965.

141. Bilbrey, G. L., N. W. Carter, M. G. White, J. F. Schilling, and J. P. Knochel: Potassium deficiency in chronic renal failure, Kidney Int., 4:423, 1973.

142. Edmundson, R. P. S., R. D. Thomas, R. J. Hilton, J. Patrick, and N. F. Jones: Leucocyte electrolytes in cardiac and non-cardiac patients receiving diuretics, Lancet, 1:12, 1974.

143. Cunningham, J. N., Jr., N. W. Carter, F. C. Rector, Jr., and D. W. Seldin: Resting transmembrane potential difference of skeletal muscle in normal subjects and severely ill patients, J. Clin. Invest., 50:49, 1971.

144. Amatruda, J. M., T. L. Biddle, M. L. Patton, and D. H. Lockwood: Vigorous supplementation of a hypocaloric diet prevents cardiac arrhythmias and mineral depletion, Am. J. Med., 74: 1016, 1983.

145. Kassirer, J. P., and J. T. Harrington: Diuretics and potassium metabolism: a reassessment of the need, effectiveness and safety of potassium therapy, Kidney Int., 11:505, 1977.

146. Harrington, J. T., J. M. Isner, and J. P. Kassirer: Our national obsession with potassium, Am. J. Med., 73:155, 1982.

147. Bilbrey, G. L., L. Herbin, N. W. Carter, and J. P. Knochel: Skeletal muscle resting membrane potential in potassium deficiency, J. Clin. Invest., 52:3011, 1973.

148. Hoffman, W. W., and R. A. Smith: Hypokalaemic periodic paralysis studied in vitro, Brain, 93:445, 1970.

149. Epstein, F. H.: Signs and symptoms of electrolyte disorders, in *Clinical Disorders of Fluid and Electrolyte Metabolism,* 3d ed., M. H. Maxwell and C. R. Kleeman (eds.), McGraw-Hill, New York, 1980.

150. Surawicz, B.: Relationship between electrocardiogram and electrolytes, Am. Heart J., 73:814, 1967.

151. Davidson, S., and B. Surawicz: Incidence of supraventricular and ventricular ectopic beats and rhythms and of atrioventricular conduction disturbances in patients with hypopotassemia, Circulation, 34(suppl. 3):85, 1966.

152. Curry, P., D. Fitchett, W. Stubbs, and D. Krikler: Ventricular arrhythmias and hypokalemia, Lancet, 2:231, 1976.

153. Holland, O. B., J. V. Nixon, and L. Kuhnert: Diuretic-induced ventricular ectopic activity, Am. J. Med., 70:762, 1981.

154. Lown, B., H. Salzberg, C. D. Enselberg, and R. E. Weston: Interrelationship between potassium metabolism and digitalis toxicity in heart failure, Proc. Soc. Exp. Biol. Med., 76:797, 1951.

155. Shapiro, W., and K. Taubert: Hypokalaemia and digoxin-induced arrhythmias, Lancet, 2:604, 1975.

156. Weidmann, S.: Membrane excitation in cardiac muscle, Circulation, 24:499, 1961.

157. Dreifus, L. S., and A. Pick: A clinical correlative study of the electrocardiogram in electrolyte imbalance, Circulation, 14:815, 1956.

158. Gross, E. G., J. D. Dexter, and R. G. Roth: Hypokalemic myopathy with myoglobinuria associated with licorice ingestion, N. Engl. J. Med., 274:602, 1966.

159. Knochel, J. P.: Neuromuscular manifestations of electrolyte disorders, Am. J. Med., 72:521, 1982.

160. Dominic, J. A., M. Koch, G. P. Guthrie, and J. H. Galla: Primary aldosteronism presenting as myoglobinuric acute renal failure, Arch. Intern. Med., 138:1433, 1978.

161. Knochel, J. P., and E. M. Schlein: On the mechanism of rhabdomyolysis in potassium depletion, J. Clin. Invest., 51:1750, 1972.

162. Relman, A. S., and W. B. Schwartz: The nephropathy of potassium depletion: a clinical and pathological entity, N. Engl. J. Med., 255:195, 1956.

163. Schwartz, W. B., and A. S. Relman: Effects of electrolyte disorders on renal structure and function, N. Engl. J. Med., 276:383, 1967.

164. Massry, S. G.: Effects of electrolyte disorders on the kidney, in *Strauss and Welt's Diseases of the Kidney,* 3d ed., L. E. Earley and C. W. Gottschalk (eds.), Little, Brown, Boston, 1979.

165. Linas, S. L., L. N. Peterson, R. J. Anderson, G. A. Aisenbrey, F. R. Simon, and T. Berl: Mechanism of renal potassium conservation in the rat, Kidney Int., 15:601, 1979.

166. Stetson, D. L., J. B. Wade, and G. Giebisch: Morphologic alterations in the rat medullary collecting duct following potassium depletion, Kidney Int., 17:45, 1980.

166a. Riemenschneider, Th., and A. Bohle: Morphologic aspects of low-potassium and low-sodium nephropathy, Clin. Nephrol., 19: 271, 1983.

167. Rubini, M.: Water excretion in potassium deficient man, J. Clin. Invest., 40:2215, 1961.

168. Berl, T., S. L. Linas, G. A. Aisenbrey, and R. J. Anderson: On the mechanism of polyuria in potassium depletion: the role of polydipsia, J. Clin. Invest., 60:620, 1977.

169. Manitius, A., H. Levitan, D. Beck, and F. H. Epstein: On the mechanism of impairment of renal concentrating ability in potassium deficiency, J. Clin. Invest., 39:684, 1960.

170. Beck, N., and S. K. Webster: Impaired urinary concentrating ability and cyclic AMP in K⁺-depleted rat kidney, Am. J. Physiol., 231:1204, 1976.

171. Bennett, C. M.: Urine concentration and dilution in hypokalemic and hypercalcemic dogs, J. Clin. Invest., 49:1447, 1970.

172. Tannen, R. L.: Relationship of renal ammonia production and potassium homeostasis, Kidney Int., 11:453, 1977.

173. Baertle, J. M., S. M. Sancelta, and G. J. Gabuzda: Relation of acute potassium depletion to renal ammonium metabolism in patients with cirrhosis, J. Clin. Invest., 42:696, 1963.

174. Gabuzda, G. J., and P. W. Hall III: Relation of potassium depletion to renal ammonium metabolism and hepatic coma, Medicine, 45:481, 1966.

175. Kamm, D. E., and G. L. Strope: Glutamine and glutamate metabolism in renal cortex from potassium-depleted rats, Am. J. Physiol., 224:1241, 1973.

176. Artz, S. A., I. C. Paes, and W. W. Faloon: Hypokalemia-induced hepatic coma in cirrhosis: occurrence despite neomycin therapy, Gastroenterology, 51:1046, 1966.

177. Fuller, G. R., M. B. MacLeod, and R. F. Pitts: Influence of administration of potassium salts on the renal tubular reabsorption of bicarbonate, Am. J. Physiol., 182:111, 1955.

178. Giebisch, G., M. B. MacLeod, and R. F. Pitts: Effect of adrenal steroids on renal tubular reabsorption of bicarbonate, Am. J. Physiol., 183:377, 1955.

179. Chan, Y. L., B. Biagi, and G. Giebisch: Control mechanisms of bicarbonate transport across the rat proximal convoluted tubule, Am. J. Physiol., 242:F532, 1982.

180. Lennon, E. J., and J. Lemann, Jr.: The effect of a potassium-deficient diet on the pattern of recovery from experimental metabolic acidosis, Clin. Sci., 34:365, 1968.

181. Stokes, J. B.: Consequences of potassium recycling in the renal medulla: effects on ion transport by the medullary thick ascending limb of Henle's loop, J. Clin. Invest., 70:219, 1982.

182. Garella, S., J. A. Chazan, and J. J. Cohen: Saline-resistant metabolic alkalosis or "chloride-wasting nephropathy," Ann. Intern. Med., 73:31, 1970.

183. Luke, R. G., F. S. Wright, N. Fowler, M. Kashgarian, and G. H. Giebisch: Effects of potassium depletion on renal tubular chloride transport in the rat, Kidney Int., 14:414, 1978.

184. Hulter, H. N., J. F. Sigala, and A. Sebastian: K⁺ deprivation potentiates the renal alkalosis-producing effect of mineralocorticoid, Am. J. Physiol., 235:F298, 1978.

185. Conn, J.: Hypertension, the potassium ion and impaired carbohydrate tolerance, N. Engl. J. Med., 273:1135, 1965.

186. Helderman, J. H., D. Elahi, D. K. Andersen, G. S. Raizes, J. D. Tobin, D. Shocken, and R. Andres: Prevention of the glucose intolerance of thiazide diuretics by maintenance of body potassium, Diabetes, 32:106, 1983.

187. Gorden, P., B. M. Sherman, and A. P. Simopoulos: Glucose intolerance with hypokalemia: an increased proportion of circulating proinsulin-like component, J. Clin. Endocrinol. Metab., 34:235, 1972.

188. Bradbury, M. W. B., and C. R. Kleeman: Stability of the potassium content of cerebrospinal fluid and brain, Am. J. Physiol., 213:519, 1967.

189. Scribner, B. H., and J. M. Burnell: Interpretation of the serum potassium concentration, Metabolism, 5:468, 1956.

190. Schwartz, W. B., C. E. van Ypersele de Strihou, and J. P. Kassirer: Role of anions in metabolic alkalosis and potassium deficiency, N. Engl. J. Med., 279:630, 1968.

191. Bleich, H. L., R. L. Tannen, and W. B. Schwartz: The induction of metabolic alkalosis by correction of potassium deficiency, J. Clin. Invest., 45:573, 1966.

192. Villamil, M. F., E. C. DeLand, R. P. Henney, and J. V. Maloney: Anion effects on cation movements during correction of potassium depletion, Am. J. Physiol., 229:161, 1975.

193. Boley, S. J., A. C. Allen, L. Schultz, and S. Schwartz: Potassium-induced lesions of the small bowel: I. Clinical aspects, J. Am. Med. Assoc., 193:997, 1965.

194. Allen, A. C., S. J. Boley, L. Schultz, and S. Schwartz: Potassium-induced lesions of the small bowel: II. Pathology and pathogenesis, J. Am. Med. Assoc., 193:1001, 1965.

195. Hutcheon, D. E.: Benefit-risk factors associated with supplemental potassium therapy, J. Clin. Pharmacol., 16:85, 1976.

196. McMahon, F. G., and K. Akadamar: Gastric ulceration after "Slow-K," N. Engl. J. Med., 295:733, 1976.

197. Weiss, S. M., H. L. Rutenberg, D. L. Paskin, and H. A. Zaren: Gut lesions due to slow-release KCl tablets, N. Engl. J. Med., 296:111, 1977.

198. Aselton, P. J., and H. Jick: Short-term follow-up study of wax matrix potassium chloride in relation to gastrointestinal bleeding, Lancet, 1:184, 1983.

199. Sopko, J. A., and R. M. Freeman: Salt substitutes as a source of potassium, J. Am. Med. Assoc., 238:608, 1977.

200. DeFronzo, R. A., P. A. Taufield, H. Black, P. McPhedran, and C. R. Cooke: Impaired renal tubular potassium secretion in sickle cell disease, Ann. Intern. Med., 90:310, 1979.

201. Keith, N. M., A. E. Osterberg, and H. B. Burchell: Some effects of potassium salts in man, Ann. Intern. Med., 16:879, 1942.

202. Sterns, R. H., P. U. Feig, M. Pring, J. Guzzo, and I. Singer: Disposition of intravenous potassium in anuric man: a kinetic analysis, Kidney Int., 15:651, 1979.

203. Pullen, H., A. Doig, and A. T. Lambie: Intensive intravenous potassium replacement therapy, Lancet, 2:809, 1967.

204. Seftel, H. C., and M. C. Kew: Early and intensive potassium replacement in diabetic acidosis, Diabetes, 15:694, 1966.

205. Abramson, E., and R. Arky: Diabetic acidosis with initial hypokalemia, J. Am. Med. Assoc., 196:401, 1966.

206. Clementsen, H. J.: Potassium therapy: a break with tradition, Lancet, 2:175, 1962.

207. Fisch, C., J. P. Shields, S. A. Ridolfo, and H. Feigenbaum: Effect of potassium on conduction and ectopic rhythms in atrial fibrillation treated with digitalis, Circulation, 18:98, 1958.

208. Swales, J. D.: Hypokalemia and the electrocardiogram, Lancet, 2:1365, 1964.

209. Lawson, D. H.: Adverse reactions to potassium chloride, Q. J. Med., 43:433, 1974.

210. Schwartz, A. B., and C. D. Swartz: Dosage of potassium chloride elixir to correct thiazide-induced hypokalemia, J. Am. Med. Assoc., 230:702, 1974.

211. Greenblatt, D. J., and J. Koch-Weser: Adverse reactions to spironolactone: a report from the Boston Collaborative Drug Surveillance Program, J. Am. Med. Assoc., 225:40, 1973.

212. Multiple Risk Factor Intervention Trial: Risk factor changes and mortality results, MRFIT Research Group, National Heart, Lung, and Blood Institute, National Institutes of Health, Bethesda, Md., J. Am. Med. Assoc., 248:1465, 1982.

213. MacGregor, G. A., N. D. Markandu, S. J. Smith, R. A. Banks, and G. A. Sagnella: Moderate potassium supplementation in essential hypertension, Lancet, 2:567, 1982.

214. Tannen, R. L.: Effects of potassium on blood pressure control, Ann. Intern. Med., 98(part 2):773, 1983.

TWENTY-EIGHT
HYPERKALEMIA

The introduction to disorders of K^+ balance presented in Chap. 26 should be read before proceeding with this discussion.

DEFENSE AGAINST HYPERKALEMIA

Hyperkalemia is a rare occurrence in normal subjects because the body is extremely effective in preventing the accumulation of excess K^+ in the extracellular fluid. As described in Chap. 26, this protective response involves two steps: initial K^+ uptake by the cells followed by the urinary excretion of most of the excess K^+ within 6 to 8 h [1–3]. Both the elevation in the plasma K^+ concentration and enhanced secretion of aldosterone contribute to the appropriate increase in K^+ excretion [4].

In addition to these acute responses, the ability to tolerate a K^+ load is increased by the chronic ingestion of a high-potassium diet. As a result, normal subjects can maintain K^+ balance even if K^+ intake is *slowly* increased from the normal of about 80 meq/day to 500 meq/day or more [5]. This ability to handle what might be a lethal K^+ load if given acutely is called *K^+ adaptation*. This phenomenon is due both to enhanced K^+ entry into the cells and, more importantly, to more rapid K^+ excretion in the urine [6–8]. Increased release of aldosterone appears to be responsible, by an unknown mechanism, for the cellular adaptation [7] and may contribute to, but does not explain all of, the renal response [9,10].

The increase in K^+ excretion during adaptation results from enhanced K^+ secretion throughout the distal nephron, particularly in the cortical and medullary collecting tubules [8,11]. Na^+-K^+-ATPase activity in these segments, but not in other segments, is elevated in this setting [10], probably due both to aldosterone (which acts primarily in the cortical collecting tubule) and to a direct effect of a small elevation in the plasma K^+ concentration [10,12]. This elevation in Na^+-K^+-ATPase activity promotes K^+ entry from the peritubular capillary into the tubular cell, thereby augmenting cellular K^+ stores and facilitating K^+ secretion into the lumen.

The major clinical example of K^+ adaptation is chronic renal failure. In this disorder, the combination of a constant K^+ intake and fewer functioning nephrons requires an increase in K^+ excretion per nephron [13,14]. This response allows K^+ balance to be maintained even in advanced renal failure as long as the urine output and therefore the distal flow rate is adequate [15]. Aldosterone contributes to this effect as evidenced by a fall in K^+ excretion if an aldosterone antagonist is administered [16]. In addition, Na^+-K^+-ATPase activity in the distal nephron cells is increased when K^+ intake is normal but not when intake is restricted in proportion to the fall in GFR, a situation in which enhanced K^+ excretion per nephron is not required [17]. This finding suggests that the increment in Na^+-K^+-ATPase activity (which favors K^+ secretion) is appropriate and specific, not incidentally induced by the renal disease.

K^+ adaptation in renal failure is also associated with aldosterone-induced increases in Na^+-K^+-ATPase activity and K^+ secretion in the colon [18]. This response becomes physiologically important in patients with end-stage renal failure on chronic dialysis in whom enhanced fecal losses may account for the excretion of as much as 30 to 50 percent of dietary K^+ intake [19].

ETIOLOGY

Potassium enters the body by oral intake or intravenous infusion, is stored in the cells, and is then excreted primarily in the urine. Thus, an abnormality in any one

Table 28-1 Etiology of hyperkalemia

Increased intake†
  A. Oral
  B. Intravenous

Movement from cells into extracellular fluid
  A. Pseudohyperkalemia†
  B. Metabolic acidosis
  C. Insulin deficiency and hyperglycemia†
  D. Tissue catabolism†
  E. β-adrenergic blockade
  F. Severe exercise
  G. Digitalis overdose
  H. Periodic paralysis—hyperkalemic form
  I. Cardiac surgery
  J. Succinylcholine
  K. Arginine

Decreased urinary excretion
  A. Renal failure†
  B. Effective circulating volume depletion†
  C. Hypoaldosteronism†
  D. Type 1 (distal) renal tubular acidosis
  E. Selective potassium secretory defect

†Most common causes.

or more of these processes can lead to hyperkalemia (Table 28-1). It should be noted, however, that *chronic hyperkalemia is always associated with an impairment in urinary K^+ excretion* since the elevation in the plasma K^+ concentration would not persist if excretory capacity were normal.

This section will review the major causes of hyperkalemia as well as some aspects of the diagnosis and treatment of specific disorders. The general principles involved in the approach to the hyperkalemic patient will be considered separately later in the chapter.

Increased Intake

In normal subjects, an acute K^+ load produces a dose-dependent elevation in the plasma K^+ concentration. For example, 135 to 160 meq of oral K^+ can transiently raise the plasma K^+ concentration by 2.5 to 3.5 meq/L, a change that generally is well tolerated [2]. Ingesting more than 160 meq, however, can produce a potentially fatal increase in the plasma K^+ concentration to above 8.0 meq/L, even in patients with normal renal function [20]. This problem is more likely to occur with a rapid intravenous infusion (see Fig. 27-5) or in infants because of their small size. Severe hyperkalemia and even cardiac arrest have been reported in infants after the administration of potassium penicillin as an intravenous bolus [21], the accidental ingestion of a KCl-containing salt substitute [22], or the use of stored blood† for exchange transfusions [23].

Hyperkalemia is also more common when K^+ is given to patients with any of the causes of impaired K^+ excretion listed in Table 28-1. In this setting, K^+ loads that would normally be well tolerated can lead to substantial elevations in the plasma K^+ concentration. In addition to dietary K^+, K^+ supplements, or the use of a salt substitute [26,27], unexpected sources of K^+ include low-sodium soups [28], red clay (clay ingestion is relatively frequent in certain rural areas in the southeastern United States) [29], and multiple transfusions of stored whole blood [25].

Movement from Cells into Extracellular Fluid

Pseudohyperkalemia Pseudohyperkalemia refers to those disorders in which the elevation in the measured K^+ concentration is due to K^+ movement out of the cells *after* the blood specimen has been drawn. The major cause of this problem is mechanical trauma during venipuncture, resulting in the release of K^+ from red cells. Since hemoglobin is also released in this setting, the serum will have a characteristic red tint.

Another cause of pseudohyperkalemia results from measurement of the serum (in which the extracellular fluid is separated from the red cells *after* clotting has occurred) rather than the plasma K^+ concentration. In normal subjects, a small amount of K^+ moves out of white cells and platelets during coagulation. Consequently, the measured serum K^+ concentration exceeds the true level in the plasma by as much as 0.5 meq/L, a difference that is clinically unimportant [30]. However, much more K^+ may be released in patients with marked leukocytosis or thrombocytosis (white cell or platelet count greater than 100,000 per mm³ or 500,000 per mm³, respectively). In these conditions, there may be a spurious elevation in the serum K^+ concentration to as high as 9 meq/L [30–32].

A rare familial condition has also been described in which K^+ leaks out of abnormally permeable red cells [33,34]. True hyperkalemia does not occur in vivo since the excess K^+ is excreted in the urine.

The presence of pseudohyperkalemia should be suspected when there is no apparent cause for the elevations in the plasma K^+ concentration and when there are no changes in muscle strength or the electrocardiogram since the true K^+ concentration is normal. Careful venipuncture to avoid hemolysis and measurement of the plasma (not the serum) K^+ concentration usually establishes the correct diagnosis. In the familial disorder, in vitro K^+ leakage can be prevented by rapid centrifugation to separate the red cells from the plasma.

Metabolic acidosis Metabolic acidosis (other than organic acidoses such as lactic acidosis or ketoacidosis) results in K^+ movement out of the cells as some of the excess

---

†As blood is stored, K^+ is gradually released from the red cells into the plasma. By 21 days, the K^+ concentration in the plasma can reach 30 meq/L [24,25]. The risk of K^+ overload can be minimized by the use of fresh blood or by removing the plasma and administering packed red cells.

H^+ ions are buffered intracellularly (see page 254) [35]. As a result, severe acidemia can induce hyperkalemia even in patients who are K^+-depleted. The true state of K^+ balance will become apparent as the acidemia is corrected. Although similar K^+ movement can occur with respiratory acidosis, the elevation in the plasma K^+ concentration is generally small [35].

Insulin deficiency and hyperglycemia Hyperkalemia is a common finding in patients with diabetic ketoacidosis or nonketotic hyperglycemia even though total body K^+ stores are almost invariably depleted (see Chap. 25) [36,37]. Insulin deficiency contributes to this response since insulin normally promotes K^+ entry into cells [38]. However, the associated hyperglycemia may be of greater importance [39–41] as the elevation in the effective plasma osmolality pulls water out of the cells with K^+ following because of solvent drag [42]. In contrast, ketoacidosis itself does not appear to contribute since it is one form of metabolic acidosis that does not usually cause K^+ movement out of the cells [35].

Several other factors also may promote the development of hyperkalemia in diabetic patients even if they are relatively well controlled. These include renal failure due to diabetic nephropathy, hyporeninemic hypoaldosteronism (see below) [41], and decreased sympathetic activity due either to diabetic autonomic neuropathy [43] or the use of β-adrenergic blockers to treat hypertension.

Tissue catabolism When the rate of tissue breakdown is increased, large amounts of K^+ may be released into the extracellular fluid; hyperkalemia may occur in this setting, particularly if renal failure is present. Clinical examples of hypercatabolism may be seen after trauma [44], the administration of cytotoxic agents to patients with malignant lymphomas [45], or massive hemolysis [46]. Since proteins (metabolized in part into urea), phosphates, and nucleic acids (metabolized into uric acid) are also released from the cells, increases in the BUN and plasma phosphate and uric acid concentrations may be seen.

β-Adrenergic blockade β-Adrenergic blockers such as propranolol are widely used in the treatment of hypertension, coronary artery disease, and other disorders. These drugs can produce a mild, transient elevation in the plasma K^+ concentration because they interfere with the β-adrenergic facilitation of K^+ entry into cells (see Fig. 13-2) [47]. A more substantial degree of hyperkalemia may occur, however, if a K^+ load, severe exercise, diabetes mellitus with hypoaldosteronism, or cardiac surgery is superimposed (see below) [47–51]. In these settings, the treatment of hypertension with a central adrenergic inhibitor such as methyldopa or clonidine may be safer since these agents do not interfere with K^+ homeostasis [52,53].

Severe exercise Potassium is normally released from muscle cells during exercise. This is an appropriate response since the *local* increase in the plasma K^+ concentration has a vasodilatory effect which contributes to the enhanced blood flow (and therefore energy delivery) to the exercising muscle [54]. The elevation in the *systemic* plasma K^+ concentration is less pronounced and related to the degree of exercise: 0.3 to 0.4 meq/L with slow walking [55]; 0.7 meq/L with moderate exertion (including prolonged aerobic exercise as with marathon running) [48,55a]; and as much as 2.0 meq/L with electrocardiographic changes following severe exercise to exhaustion [49,56]. The plasma K^+ concentration returns to normal after 3 to 4 min of rest as K^+ moves back into the cells [56].

Although these changes are generally well tolerated, it is possible that hyperkalemia may be responsible for some of the cases of sudden death that occur during exercise [56a]. This may be more likely if exercise is superimposed upon some other abnormality in K^+ handling. For example, the plasma K^+ concentration can approach 8.0 meq/L with severe exercise in patients taking a β-adrenergic blocker [49].

Digitalis overdose The Na^+-K^+-ATPase pump in the cell membrane that is responsible for the maintenance of a high cell K^+ concentration is inhibited in a dose-related manner by digitalis [57]. Thus, the administration of digitalis tends to increase the plasma K^+ concentration because of the release of K^+ from the cells. When used in therapeutic doses, this effect of digitalis is relatively small, although there may be some impairment in the ability to handle a large K^+ load [58]. However, severe hyperkalemia (plasma K^+ concentration up to 13 meq/L) has resulted from the ingestion of massive amounts of digitalis following a suicide attempt [59,60].

Periodic paralysis The hyperkalemic form of periodic paralysis is a familial disorder with autosomal dominant inheritance that is characterized by recurrent attacks of muscle weakness or paralysis [61,62]. Most patients with this disorder also have myotonic symptoms, particularly in the cold. Episodes are precipitated by rest after exercise or the ingestion of K^+ and are associated with an increase in the plasma K^+ concentration due either to K^+ release from the cells or to an inability of ingested K^+ to enter the cells [61–63]. The degree of hyperkalemia is frequently mild (less than 5.5 meq/L), although the plasma K^+ concentration can exceed 7 meq/L in some patients [62].

The pathogenesis of this rare disorder is uncertain. The observations that the plasma Na^+ concentration falls and the plasma protein concentration rises (due to hemoconcentration) during an attack suggest that Na^+ and water enter the cell as K^+ leaves [63]. These changes could result from depolarization of the cell membrane, although the mechanism by which this might occur is not known.

The diagnosis of periodic paralysis should be suspected from the personal and family history of recurrent episodes of muscle weakness and the elevated plasma K^+ concentration during an attack. In contrast to *hypo*kalemic periodic paralysis in which the muscle weakness may be profound and last for up to 48 h, the attacks are usually mild in the hyperkalemic form with a duration of less than 1 to 2 h [61]. The diagnosis can be confirmed by the induction of muscle weakness and hyperkalemia after a relatively small oral K^+ load (0.5 to 1.0 meq/kg) [61–63].

Treatment is aimed at correcting the hyperkalemia and then attempting to prevent further episodes. Salbutamol (albuterol), a β_2-adrenergic agonist used to treat bronchoconstriction, may be particularly effective in reversing the acute symptoms by driving K^+ into the cells [64]. Modalities used chronically include limiting exercise (if this precipitates attacks), prescribing a low-K^+–high-carbohydrate diet (carbohydrates promote K^+ entry into cells via increased insulin secretion), and inducing mild K^+ depletion with a thiazide diuretic or a mineralocorticoid (such as fludrocortisone) [61]. The addition of the carbonic anhydrase inhibitor acetazolamide may also be beneficial [61,65]. Although this drug is a diuretic which enhances urinary K^+ excretion, its effectiveness in the hypokalemic form as well (see Chap. 27) suggests that it may directly affect the cell membrane. Salbutamol may also be effective in prophylactic therapy in this disorder [63].

Cardiac surgery Patients on cardiac bypass may develop a mild elevation in the plasma K^+ concentration as normal circulation is restored [66], particularly if they have been taking a β-adrenergic blocker [51]. Two factors may contribute to this problem: washout of ischemic areas underperfused during bypass, and rewarming (since the surgery is performed under hypothermic conditions). The induction of hypothermia causes K^+ to move into the cells by an unknown mechanism [67,68]. This effect is reversed with rewarming and can be associated with "overshoot" hyperkalemia, particularly if K^+ has been given during the period of hypothermia [67].

Succinylcholine Succinylcholine is a muscle relaxant used in general anesthesia. It acts by depolarizing the cell membrane, i.e., it reduces the magnitude of the resting membrane potential. Since the cell interior becomes less electronegative, this favors the movement of positively charged K^+ ions out of the cells into the extracellular fluid. In normal subjects, the result is a small rise in the plasma K^+ concentration of 0.5 meq/L or less [69]. However, in patients with burns, extensive trauma, tetanus, or neuromuscular diseases, succinylcholine can induce an increase in the plasma K^+ concentration of as much as 6 meq/L, leading to cardiac arrhythmias and even cardiac arrest [69,70]. Although it is unclear why these patients are at such high risk, the increase in the plasma K^+ concentration (which usually occurs within 5 min) can be minimized by the prior administration of tubocurarine [69].

Arginine Arginine hydrochloride is metabolized in part to hydrochloric acid and has been used in the treatment of refractory metabolic alkalosis. Marked hyperkalemia is a potential complication with this drug, and is presumably due to the movement of K^+ out of cells as the cation arginine enters the cells [71,72].

Decreased Urinary Excretion

There are three major conditions in which urinary K^+ excretion is reduced: renal failure, effective circulating volume depletion, and hypoaldosteronism. It is likely that one of these disorders is present in any patient with persistent hyperkalemia since excretory capacity must be diminished to perpetuate the elevation in the plasma K^+ concentration.

Renal failure As described above, K^+ balance is maintained in renal failure by increased excretion per functioning nephron [13–15]. This adaptation, mediated in part by aldosterone and enhanced Na^+-K^+-ATPase activity, is effective as long as the urine output remains adequate. However, the ability to excrete K^+ falls once oliguria develops, probably because of the decrease in flow to the distal secretory site [15,73]. In this setting, some of the K^+ derived from dietary intake is likely to be retained, resulting in a persistent elevation in the plasma K^+ concentration.

When hyperkalemia develops in a nonoliguric patient, some other factor is usually superimposed, such as enhanced tissue breakdown, hypoaldosteronism (see below), or increased K^+ intake. For example, subjects who are in balance on a regular diet may have an exaggerated rise in the plasma K^+ concentration following a K^+ load [2,14,15,74]. Despite a low absolute rate of K^+ excretion in this setting, the rate divided by the GFR (an index of K^+ excretion per functioning nephron) is similar to that in normal subjects given a K^+ load [14]. This finding suggests that the retention of K^+ is due to too few nephrons, not a specific defect in K^+ secretion [15,74].

In addition to the diminished kaliuresis, K^+ entry into the cells is impaired in renal failure. This is manifested by a low cell K^+ concentration in the basal state (despite a normal or elevated plasma level) [75] and diminished cellular uptake of K^+ after a K^+ load [76,77]. Although the mechanism by which this occurs is not well understood, renal failure is associated with insulin resistance and decreased Na^+-K^+-ATPase activity,† both of which can impair K^+ entry into the cells [75,77,78].

Volume depletion Effective circulating volume depletion can be produced by fluid loss from the body, sequestration into a noncirculating space, or a primary decrease in cardiac output as in heart failure (see Chap. 9). When due to fluid loss, volume depletion is frequently associated with hypokalemia since K^+ is lost as well as Na^+ and water. However, the ability to handle a K^+ load is impaired by hypovolemia, and this can lead to an elevation in the plasma K^+ concentration in some patients. Reductions in both K^+ excretion and entry into cells contribute to this problem [79,80]. The combination of a low GFR and increased proximal Na^+ and water reabsorption induced by volume depletion decreases distal delivery and may be responsible for the reduction in K^+ excretion. Why cell entry is diminished is not understood [80].

A clinical example of the effect of volume depletion may be seen in renal failure. One of the changes that occurs in this setting is an inability to maximally conserve Na^+ [81,82]. In most patients, the obligatory Na^+ loss is relatively small and not clinically important on a regular diet. If, however, intake is reduced (as in the treatment of hypertension) or extrarenal losses are enhanced, volume depletion will ensue. The resultant fall in renal perfusion can lead to decreased K^+ excretion and hyperkalemia [83]. Na^+ wasting and hyperkalemia can also result from decreased secretion of, or tubular responsiveness to, aldosterone [84,85].

---

†One exception is the tubular cells of the distal nephron where Na^+-K^+-ATPase activity is increased because of the need to enhance K^+ excretion per nephron [17].

Table 28-2 Causes of hypoaldosteronism

Associated with decreased activity of the renin-angiotensin system
  Hyporeninemic hypoaldosteronism
  Nonsteroidal anti-inflammatory drugs
  Converting enzyme inhibition
  Hypervolemia in chronic dialysis patients

Decreased adrenal synthesis
  Primary adrenal insufficiency
  Enzyme deficiencies
    Congenital adrenal hyperplasia
    Isolated hypoaldosteronism
  Heparin
  Post-removal of adrenal adenoma

Aldosterone resistance
  Potassium-sparing diuretics
  Pseudohypoaldosteronism

Hypoaldosteronism Hypoaldosteronism can be induced by a variety of conditions which interfere with either the production or effect of aldosterone (Table 28-2). The most common causes are hyporeninemic hypoaldosteronism and K^+-sparing diuretics in adults and adrenal enzyme deficiencies (particularly 21-hydroxylase deficiency) in children.

In addition to hyperkalemia, varying degrees of Na^+ wasting and metabolic acidosis are also present in this disorder since aldosterone normally promotes Na^+ reabsorption and H^+ as well as K^+ secretion (see Chap. 8) [85,86]. The metabolic acidosis (called type 4 renal tubular acidosis; see Chap. 19) is also due in large part to the hyperkalemia since lowering the plasma K^+ concentration frequently corrects the acidosis [87,88]. This effect of hyperkalemia may be due to a transcellular cation exchange as some of the excess K^+ moves into the cells and H^+ (and Na^+) move out to maintain electroneutrality [89]. The ensuing intracellular alkalosis then reduces NH_3 production and H^+ secretion by the renal tubular cells (see page 240). As a result, less HCO_3 is reabsorbed (see Fig. 12-11) and less H^+ is excreted (as NH_4^+), both of which promote the development of metabolic acidosis [87,90].

Hyporeninemic hypoaldosteronism In the absence of an obvious case (oliguric renal failure, K^+ supplements, K^+-sparing diuretics), the syndrome of hyporeninemic hypoaldosteronism may account for 50 to 75 percent of cases of initially unexplained hyperkalemia in adults [91,92]. This disorder has the following characteristics [85]:

1. Most patients have mild to moderate renal insufficiency with a creatinine clearance of 20 to 75 mL/min.
2. Approximately 50 percent have diabetes mellitus.
3. About 85 percent have a reduced plasma renin activity.
4. Patients typically present with asymptomatic hyperkalemia.

Since angiotensin II and K^+ are the major physiologic stimuli of aldosterone secretion (see Chap. 8), it was initially thought that the low angiotensin II levels induced by hyporeninism were responsible for the decreased aldosterone release. However, several observations suggest that there must also be an *intraadrenal* defect. These include a normal plasma renin activity in some patients, an inability of infused angiotensin II to stimulate aldosterone secretion [93,94], and the demonstration that nephrectomized patients (who have no renin) still have normal aldosterone production that is directly stimulated by hyperkalemia [95].

It now appears that an acquired adrenal insensitivity to angiotensin II and K^+ is largely responsible for the hypoaldosteronism in this disorder [85,93,94]. How this might occur is not known. One possibility (although unproven) is that a defect in prostaglandin production might be involved since prostaglandins promote renin secretion (see Chap. 3) and appear to facilitate aldosterone release by angiotensin II [96]. Furthermore, nonsteroidal anti-inflammatory drugs which inhibit prostaglandin synthesis can induce hyporeninemic hypoaldosteronism in some patients [97,98].

The reason why diabetics are so prone to developing this disorder is uncertain. Decreased conversion of inactive to active renin or an autonomic neuropathy (with decreased β-adrenergic activity) has been described in diabetics and could account for the decrease in renin secretion [85,99]. These abnormalities, however, do not necessarily explain the hypoaldosteronism. Regardless of the mechanism, the defect in renin and aldosterone release may be relatively common since it can be demonstrated (although to a lesser degree) in some diabetics who are nonazotemic and normokalemic [100,101]. The superimposition of renal insufficiency may be required for hyperkalemia to occur.

Two other settings in which a syndrome similar to hyporeninemic hypoaldosteronism can occur are volume expansion in patients on maintenance dialysis and the administration of a converting enzyme inhibitor such as captopril [102,103]. The former is an appropriate physiologic response that is reversible with removal of the excess fluid. It is of uncertain clinical importance since these patients generally have a low urine output. Captopril, used in the treatment of hypertension and heart failure, acts by reducing the formation of angiotensin II and therefore the release of aldosterone. In this setting, the plasma renin activity is *increased* (not reduced) since angiotensin II normally decreases renin secretion by feedback inhibition. The degree of hyperkalemia is variable, being most prominent in patients with renal insufficiency [103].

Primary adrenal insufficiency Patients with primary adrenal insufficiency (or bilateral adrenalectomy) have diminished glucocorticoid as well as mineralocorticoid secretion [104]. In contrast, patients with secondary hypoadrenalism due to reduced ACTH secretion have relatively normal aldosterone secretion since ACTH does not have a major role in the regulation of aldosterone release [105].

Enzyme deficiencies The pathways involved in adrenal steroid synthesis are illustrated in Fig. 28-1. In addition to aldosterone, deoxycorticosterone (DOC) and corticosterone also have mineralocorticoid activity. Their secretion, however, is deter-

Acetyl CoA + squalene

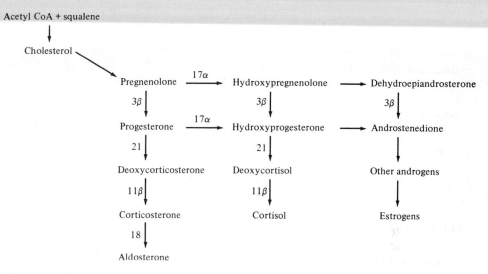

Figure 28-1 Schematic pathways of adrenal steroid biosynthesis. The numbers at the arrows refer to specific enzymes: 17α = 17α-hydroxylase: 3β = 3β-hydroxysteroid dehydrogenase; 21 = 21-hydroxylase; 11β = 11β-hydroxylase; 18 refers to a two-step process catalyzed by the enzymes corticosterone methyl oxidase I and II that results in the addition of an aldehyde at the 18-carbon position. This conversion of corticosterone to aldosterone occurs only in the zona glomerulosa whereas cortisol and androgen production occur in the zone fasciculata and reticularis. Deficiencies in any of these enzymes can lead to abnormal mineralocorticoid (as well as glucocorticoid and androgen) production.

mined primarily by ACTH not, as with aldosterone, by angiotensin II or the plasma K^+ concentration [104]. Signs of mineralocorticoid deficiency may be seen with reduced activity of enzymes involved in steps prior to the formation of DOC (3β-hydroxysteroid dehydrogenase or 21-hydroxylase) or in the conversion of corticosterone to aldosterone (corticosterone methyl oxidase I and II).†

21-Hydroxylase deficiency is the most common form of congenital adrenal hyperplasia. Lack of this enzyme also impairs the conversion (primarily in the zona fasciculata) of 17-hydroxyprogesterone to deoxycortisol, the precursor of cortisol (Fig. 28-1). The ensuing cortisol deficiency stimulates the release of ACTH, resulting in enhanced adrenal androgen synthesis and virilization‡ [106]. The associated accumulation of 17-hydroxyprogesterone (since its metabolism is blocked) may also

†Although impaired conversion of DOC to corticosterone (11β-hydroxylase deficiency) also decreases aldosterone production, there is a buildup of DOC. This typically results in signs of mineralocorticoid excess (hypokalemia and hypertension), not mineralocorticoid deficiency (see Chap. 27).

‡Some children with this disorder do not have Na^+ wasting or hyperkalemia and have normal aldosterone levels [107–109]. These patients appear to have relatively adequate 21-hydroxylase activity in the zona glomerulosa but are still cortisol-deficient because enzyme activity is very low in the zona fasciculata [108,109]. The reason for this difference is uncertain, but it is possible that 21-hydroxylase activity in the two zones is under somewhat separate genetic control [108]. Similar findings may occur with 3β-hydroxysteroid dehydrogenase deficiency where enzyme activity may be adequate only in the zona glomerulosa, resulting in normal aldosterone production [109a].

contribute to the hypomineralocorticoidism since this compound antagonizes the effect of aldosterone [107].

Two much rarer disorders are deficiencies of 3β-hydroxysteroid dehydrogenase (which is nonvirilizing since androgen production is also limited; Fig. 28-1) or one of the enzymes involved in the conversion of corticosterone to aldosterone [104]. The latter represents isolated hypoaldosteronism [110,111] because corticosterone methyl oxidase I and II are not involved in cortisol synthesis and, therefore, are not associated with enhanced ACTH release.

Heparin Heparin reduces aldosterone secretion by a direct action on the adrenal gland [112]. This effect is seen within 4 to 7 days but is likely to produce hyperkalemia only if some superimposed problem is present, such as renal insufficiency in a diabetic [112,113]. The aldosterone deficiency is readily reversible with discontinuation of the drug.

Post-removal of adrenal adenoma In patients with primary *hyper*aldosteronism due to an adrenal adenoma, the overproduction of aldosterone suppresses the normal tissue in the zona glomerulosa. As a result, surgical removal of the tumor leads to a transient period of hypoaldosteronism that can last up to 6 months or more [114,115].

Potassium-sparing diuretics The K^+-sparing diuretics impair distal K^+ secretion: spironolactone by antagonizing the effect of aldosterone, and triamterene and amiloride by directly interfering with tubular function (see Chap. 16). As a result, the use of any of these drugs can produce hyperkalemia, particularly in patients with renal failure or taking K^+ supplements [27,116].

Pseudohypoaldosteronism Reduced aldosterone effect may rarely be induced by end-organ resistance (called pseudohypoaldosteronism). This disorder is associated with volume depletion, Na^+ wasting, hyperkalemia, and markedly elevated levels of renin and aldosterone, findings similar to that with K^+-sparing diuretics [84,117–119]. The congenital form presents in infancy and is associated in at least some children with generalized aldosterone resistance as evidenced by inappropriately elevated Na^+ concentrations in the feces and sweat as well as the urine [119]. In comparison, the acquired form is limited to the kidney and is seen primarily with tubulointerstitial diseases such as chronic pyelonephritis, medullary cystic disease, methicillin nephritis, and amyloidosis [84,120–122].

Type 1 renal tubular acidosis Type 1 (distal) renal tubular acidosis (RTA) is an uncommon disorder characterized by impaired distal H^+ secretion, resulting in metabolic acidosis with an inappropriately high urine pH above 5.3 (see Chap. 19). *Hypo*kalemia frequently occurs since the decrease in H^+ secretion requires that Na^+ reabsorption occur in exchange for K^+ [123]. However, hyperkalemia may be seen when the underlying mechanism is a primary decrease in distal Na^+ reabsorption. This impairs the generation of the lumen negative potential difference that promotes

both H^+ and K^+ secretion [124]. This form of hyperkalemic renal tubular acidosis has been associated with obstructive uropathy and sickle cell disease [125–127]. These disorders, however, may also be associated with hypoaldosteronism [125,126], a condition that is treated differently from RTA. The correct diagnosis can be established by measuring the plasma aldosterone level (see "Diagnosis" below), which is normal in RTA, and the urine pH, which in the presence of metabolic acidosis is appropriately below 5.3 with aldosterone deficiency [126].

Selective potassium secretory defect In some patients, hyperkalemia occurs with inappropriately low urinary K^+ excretion, normal renin and aldosterone levels, and no Na^+ wasting. This rare syndrome of an apparently selective defect in K^+ secretion which does not respond to exogenous mineralocorticoid has been described with a renal transplant and lupus nephritis [128,129]. The absence of Na^+ wasting and the presence of a normal degree of Na^+ retention following the administration of exogenous mineralocorticoid indicate that this is not simple aldosterone resistance.

The pathogenesis of this disorder is not understood, but a process similar to type 1 RTA may be involved since about one-half of the patients have a metabolic acidosis which, in at least one case, was associated with an elevated urine pH [128,129]. Alternatively, selective impairment of K^+ secretion may be responsible—with hyperkalemia then inducing the metabolic acidosis by, as described above, reducing NH_3 production and net acid excretion [87,88,90].

Somewhat similar findings have been described in other patients who also have hypertension [130–132]. It has been proposed that distal Cl^- reabsorption is enhanced in these patients (by an unknown mechanism) [130]. As a result, Na^+ is reabsorbed distally with Cl^- (leading to volume expansion and hypertension) *not* in exchange for K^+ and H^+ [130]. The major finding in support of this hypothesis is the demonstration that K^+ secretion does not increase when aldosterone is given with NaCl (the latter to increase distal flow) but rises *normally* when Na^+ is administered when a nonreabsorbable anion such as SO_4^{2-} or HCO_3^- [130].

SYMPTOMS

The changes induced by hyperkalemia are essentially limited to muscle weakness and abnormal cardiac conduction which can lead to potentially fatal arrhythmias. Patients may also complain of symptoms related to the underlying disease such as polyuria and polydipsia in uncontrolled diabetes mellitus or weight loss and failure to thrive in infants with hypoaldosteronism.

Muscle Weakness

The muscle weakness associated with hyperkalemia appears to be due to the concomitant changes in the resting membrane potential. The increase in the plasma K^+ concentration reduces the ratio of the intracellular K^+ concentration to that in the extracellular fluid, resulting in a decrease in the magnitude of the resting membrane

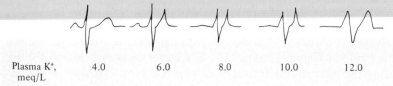

| Plasma K$^+$, meq/L | 4.0 | 6.0 | 8.0 | 10.0 | 12.0 |

Figure 28-2 Electrocardiogram in hyperkalemia. The initial change is peaking and narrowing of the T wave with a short QT interval. With more severe hyperkalemia, widening of the QRS complex, decreased amplitude and eventual loss of the P wave, and a sine wave pattern as the QRS complex merges with the T wave may be seen. The approximate relationship between these changes and the plasma K$^+$ concentration is indicated. *(Adapted from B. Surawicz, Am. Heart J., 73:814, 1967.)*

potential (see Chap. 26). If the resting potential falls to or below the threshold potential, the cell is unable to sustain an action potential and muscle weakness or paralysis ensues. These symptoms most often begin in the lower extremities and ascend to the trunk and upper extremities [133]. The respiratory muscles and those supplied by the cranial nerves are usually spared.

Muscle weakness typically does not develop until the plasma K$^+$ concentration exceeds 8 meq/L [133–135]. However, patients with periodic paralysis may become symptomatic at a plasma K$^+$ concentration less than 5.5 meq/L [62], indicating that they are particularly sensitive to even minor degrees of hyperkalemia.

Cardiac Arrhythmias

Disturbances in cardiac conduction, which can lead to ventricular fibrillation or standstill, pose the greatest danger to the patient with hyperkalemia [133,136]. Consequently, monitoring of the electrocardiogram is an essential part of the management of this disorder. As the plasma K$^+$ concentration is increased, there is a characteristic sequence of changes in the electrocardiogram that is due to the effects of hyperkalemia on atrial and ventricular depolarization (represented by the P wave and QRS complex, respectively) and repolarization (represented by the T wave for ventricular repolarization, the atrial repolarization wave being lost in the QRS complex) [136–139]. The earliest changes are peaked narrow T waves and a shortened QT interval, which reflect more rapid repolarization (Fig. 28-2).† On occasion, this may be confused with the tall T waves seen with myocardial ischemia. However, the QT interval is usually normal or prolonged during ischemic episodes [136].

The alteration in T-wave configuration typically becomes prominent when the plasma K$^+$ concentration exceeds 6 meq/L. At a plasma K$^+$ concentration above 7 to 8 meq/L, further changes in the electrocardiogram occur which are primarily due

---

†Repolarization is characterized by the movement of K$^+$ out of the cells (see page 569). The rapidity with which repolarization occurs therefore is proportional to the K$^+$ permeability of the cell membrane, which varies directly with the plasma K$^+$ concentration. Thus, hyperkalemia increases the permeability to K$^+$, resulting in more rapid repolarization which is manifested by peaked, narrow T waves on the electrocardiogram [136,137].

to delayed *depolarization* [136,140].† The result is widening of the QRS complex with no change in configuration (the electrocardiographic manifestation of slowed ventricular depolarization) and decreased amplitude, widening, and eventual loss of the P wave (Fig. 28-2). The final changes are a sine wave pattern as the widened QRS complex merges with the T wave, followed by ventricular fibrillation or standstill. Despite these changes in conduction, the contractility of cardiac muscle seems to be unaffected [142].

The plasma K^+ concentration at which the electrocardiographic changes are seen is somewhat variable because of the influence of other factors. For example, the cardiac toxicity of hyperkalemia is enhanced by hypocalcemia [143], hyponatremia [144], acidemia [145], and a rapid elevation in the plasma K^+ concentration [138,146]. Thus, patients with renal failure may be particularly sensitive to hyperkalemia since hypocalcemia (due to phosphate retention and vitamin D deficiency), hyponatremia (due to water retention), and metabolic acidosis (due to reduced NH_3 formation) all may be present [147].

DIAGNOSIS

The initial evaluation of the hyperkalemic patient should include: a complete history (including questions about dietary intake or a history of kidney disease, diabetes mellitus, the use of K^+-sparing diuretics, or recurrent episodes of muscle weakness); physical examination (for muscle weakness or the signs of volume depletion or edema); an electrocardiogram; and measurement of arterial pH and the BUN, plasma creatinine, glucose, Na^+, and Ca^{2+} concentrations.

With this information, the approach to diagnosis can be simplified by considering separately the three groups of conditions associated with hyperkalemia: increased intake, K^+ release from the cells, and reduced urinary K^+ excretion. As described above, increased intake plays only a contributory role unless a massive load has been given acutely. Nevertheless, a careful dietary history should be obtained including the possible use of salt substitute or K^+ supplements. The diagnosis of one of the disorders associated with K^+ release from the cells can usually be made from the history or laboratory data, as, for example, with marked hyperglycemia. Pseudohyperkalemia should be suspected if there is no apparent cause for the elevation of the plasma K^+ concentration or if there are no electrocardiographic changes at a plasma K^+ concentration greater than 6.5 to 7.0 meq/L. Once hemolysis has been excluded, the diagnosis of one of the other forms of pseudohyperkalemia can be confirmed by the findings of a normal plasma (not serum) K^+ concentration and a marked elevation in either the white blood cell or platelet count.

---

†Depolarization, in contrast to repolarization, is primarily due to a marked increase in the Na^+ permeability of the cell membrane, resulting in the movement of Na^+ into the cell. The rate at which the Na^+ permeability increases is related, in an unknown way, to the magnitude of the resting membrane potential [140,141]. Since hyperkalemia lowers the resting potential by decreasing the $[K^+]_c/[K^+]_e$ ratio, there is a slower increase in Na^+ permeability and therefore a slower rate of depolarization [136,140].

If none of these disorders is present or the patient has persistent hyperkalemia, then decreased K^+ excretion in the urine must be at least contributing to the rise in the plasma K^+ concentration. Severe renal failure is characterized by marked elevations in the BUN and plasma creatinine concentration or, in patients with acute renal failure, a progressive increase in these parameters. Hyperkalemia is particularly likely to occur if there is an increased K^+ load. For example, approximately 50 percent of patients with post-traumatic renal failure, in whom tissue catabolism is increased, develop a plasma K^+ concentration greater than 7 meq/L [44].

Effective volume depletion can usually be excluded by the history (?vomiting, diarrhea, diuretics, heart disease) and physical examination. True volume depletion may be associated with decreased skin turgor, flat neck veins, tachycardia, and a postural fall in blood pressure. In contrast, peripheral and/or pulmonary edema is characteristically present in effectively hypovolemic patients with heart failure, cirrhosis, or the nephrotic syndrome. If underlying renal function is normal, the urine Na^+ concentration should be less than 10 to 15 meq/L in these disorders as Na^+ reabsorption is enhanced in an attempt to restore normovolemia. This is an important finding since it also tends to exclude hypoaldosteronism, where the ability to conserve Na^+ maximally is impaired because of the decrease in distal Na^+ reabsorption.

Hypoaldosteronism

If renal function is normal or only moderately impaired and no other etiology is apparent (such as type 1 renal tubular acidosis in which the urine pH is inappropriately elevated), the patient should be evaluated for one of the causes of mineralocorticoid deficiency. Any potential offending drug (nonsteroidal anti-inflammatory drug, converting enzyme inhibitor, heparin, or K^+-sparing diuretic) should be discontinued, if possible. If these agents are not being used, measurement of the morning plasma renin activity and aldosterone and cortisol concentrations should establish the correct diagnosis (Table 28-3). To minimize the incidence of confusing borderline values, the patient should be given 40 mg of furosemide at 6 P.M. and 6 A.M. before the blood specimen is obtained. This regimen enhances aldosterone secretion (via stimulation of renin) in normal subjects but not in patients with hypoaldosteronism [85].

Table 28-3 Plasma renin activity, aldosterone, and cortisol levels and responsiveness to aldosterone therapy in major causes of idiopathic mineralocorticoid deficiency

Disorder	Plasma renin activity	Plasma aldosterone	Plasma cortisol	Response to aldosterone
Hyporeninism	↓ −nl	↓	nl	+
Primary adrenal insufficiency	↑	↓	↓	+ −
Enzyme deficiencies				
Congenital adrenal hyperplasia	↑	↓	↓	+
Isolated hypoaldosteronism	↑	↓	nl	+
Pseudohypoaldosteronism	↑	↑	nl	−

The age of the patient is also important. In adults, hyporeninemic hypoaldosteronism and primary adrenal insufficiency are most common. The latter diagnosis should be suspected clinically if signs of cortisol deficiency are present, such as salt craving, fasting hypoglycemia, or hyperpigmentation of the skin and mucous membranes (due to the hypersecretion of ACTH).

In comparison, enzyme deficiencies or pseudohypoaldosteronism begin in infancy or childhood. Salt wasting may be relatively severe as the child presents with volume depletion, hyperkalemia, hyponatremia, metabolic acidosis, and an elevated urine Na^+ concentration. Plasma 17-hydroxyprogesterone levels should also be measured since this compound accumulates in 21-hydroxylase deficiency, the most common cause of hypoaldosteronism in children.

Distinction between pseudohypoaldosteronism and a selective K^+ secretory defect may be somewhat difficult since aldosterone levels are normal or elevated in both. However, the former is primarily a disease of infants [118] or patients with underlying renal disease [84] whereas the latter has a late onset, is not characterized by Na^+ wasting, and has a normal reduction in Na^+ excretion (without change in that of K^+) after the administration of a mineralocorticoid such as fludrocortisone [128–132].

TREATMENT

Treatment should be aimed toward correcting both the hyperkalemia and the underlying disorder. In addition to following the plasma K^+ concentration, the electrocardiogram and muscle strength must be monitored carefully since they reflect the *functional consequences* of the elevation in the plasma K^+ concentration. For example, periodic paralysis [62] and metabolic acidosis [145] potentiate the toxicity of hyperkalemia. As a result, a patient with these disorders may be symptomatic at a plasma K^+ concentration well tolerated by others.

The therapeutic measures that should be used are determined by the degree of K^+ intoxication. A plasma K^+ concentration greater than 8 meq/L, muscle weakness, or electrocardiographic changes are considered signs of severe hyperkalemia and require *immediate* treatment.

Hyperkalemia

Reversing the effects of hyperkalemia can be achieved by direct antagonism of its membrane actions and by lowering the plasma K^+ concentration either by driving K^+ into the cells (which will increase the $[K^+]_c/[K^+]_e$ ratio toward normal) or removing K^+ from the body (Table 28-4). In addition, further K^+ intake should be limited. Commonly ignored sources of K^+ are salt substitutes, stored blood, and K^+ penicillin, which contains 1.6 meq of K^+ per million units of penicillin.

Calcium Membrane excitability is determined by the difference between the resting membrane potential and the threshold potential that must be reached before an action potential is fired. The threshold potential varies inversely with the plasma

Table 28-4 Treatment of hyperkalemia

Antagonism of membrane actions
  A. Calcium
  B. Hypertonic Na^+ solution

K^+ entry into the cells
  A. Glucose and insulin
  B. $NaHCO_3$
  C. Hypertonic Na^+ solution

K^+ removal from the body
  A. Diuretics
  B. Cation-exchange resin
  C. Peritoneal dialysis or hemodialysis

Ca^{2+} concentration (see Fig. 26-1) [141]. Since an elevation in the plasma K^+ concentration reduces the magnitude of the *resting potential,* lowering the *threshold potential* by the administration of Ca^{2+} returns membrane excitability toward normal and protects against hyperkalemia [148]. Conversely, a decrease in the plasma Ca^{2+} concentration raises the threshold potential and enhances the effects of hyperkalemia [143].

The protective effect of Ca^{2+} begins within minutes but is relatively short-lived. Thus, Ca^{2+} is transiently effective pending the onset of action of other therapeutic modalities. In general, Ca^{2+} is used only in patients with severe K^+ intoxication who cannot wait the 30 to 60 min before glucose and insulin or $NaHCO_3$ begin to act. The usual dose is 10 mL (1 ampul) of a 10% calcium gluconate solution infused slowly over 2 to 3 min under electrocardiographic monitoring.† This dose can be repeated after 5 min if the electrocardiographic changes persist. Ca^{2+} should be used only when absolutely necessary, e.g., with loss of P waves or widening of the QRS complex, in patients taking digitalis since hypercalcemia, like hypokalemia, can precipitate digitalis toxicity.

Glucose and insulin Increasing the availability of insulin lowers the plasma K^+ concentration by driving K^+ into cells [149]. In nondiabetics, this can be achieved by the administration of glucose (for example, 500 mL of 10% dextrose and water solution infused over 15 to 30 min), which will rapidly enhance endogenous insulin secretion. In patients with diabetes mellitus, the concurrent subcutaneous or intravenous administration of 1 unit of regular insulin per 3 to 5 g of glucose will achieve the same result. This regimen usually lowers the plasma K^+ concentration 1 to 2 meq/L [150]. The effect begins within 1 h and may last for many hours. If necessary, the glucose and insulin infusions can be repeated.

In patients with uncontrolled diabetes mellitus and marked hyperglycemia, insulin alone is sufficient. These patients, however, are typically total body K^+–depleted

---

†Ca^{2+} should not be infused in HCO_3^--containing solutions because it can result in the precipitation of the insoluble salt, $CaCO_3$.

and must be watched carefully for the development of *hypo*kalemia as the plasma glucose concentration is reduced [36].

Sodium bicarbonate Whereas metabolic acidosis can result in the release of K^+ from the cells, raising the pH with $NaHCO_3$ drives K^+ into the cells [35]. In addition to the pH change, the elevation in the plasma HCO_3^- concentration appears to directly contribute to this effect [151]. Consequently, the infusion of $NaHCO_3$ will lower the plasma K^+ concentration and alleviate the signs of hyperkalemia, particularly in a patient with metabolic acidosis [151,152]. This effect begins within 30 to 60 min and may persist for many hours. Forty-five meq of $NaHCO_3$ (1 ampul of a 7.5% $NaHCO_3$ solution) can be infused slowly over 5 min and repeated within 30 min. Alternatively, $NaHCO_3$ can be added to a glucose and saline solution (see below).

Hypertonic sodium solutions Hyponatremia increases the toxicity of hyperkalemia on the heart [144]. Although the mechanism by which this occurs is unclear, it may be related in part to impaired cell uptake of K^+. When the plasma Na^+ concentration is reduced, correction of the hyponatremia with hypertonic Na^+ solutions may reverse the electrocardiographic changes of hyperkalemia [144]. This is associated with a fall in the plasma K^+ concentration, which may be due both to K^+ entry into the cells and to dilution since increasing the plasma Na^+ concentration and therefore the plasma osmolality causes water to move from the cells into the extracellular fluid (see Fig. 22-1). In addition, raising the plasma Na^+ concentration may directly counteract the membrane effects of hyperkalemia by increasing the rate of depolarization [153].

For appropriate patients, a therapeutic "cocktail" can be devised to lower the plasma K^+ concentration. For example, for a patient who is hyponatremic, 90 meq of $NaHCO_3$ can be added to 1 liter of 10% dextrose in isotonic saline (total Na^+ concentration of 154 (as NaCl) + 90 (as $NaHCO_3$) equals 244 meq/L). Five hundred milliliters of this solution can be infused over 30 min, with the remainder given over 2 to 3 h. If the plasma Na^+ concentration is normal, the infusion should consist of 90 meq of $NaHCO_3$ added to 1 liter of 10% dextrose in water. However, these solutions should be used only in patients who are able to tolerate a Na^+ and water load. In patients with underlying cardiac disease or oliguric renal failure, the infusion of large quantities of $NaHCO_3$ or NaCl can precipitate pulmonary edema. To minimize the volume of fluid given to these patients, 50 mL of a 50% glucose solution (and insulin if necessary) can be given over 5 min. Small doses of $NaHCO_3$ (45 to 90 meq) also can be infused slowly if the patient has no signs of pulmonary congestion.

The effects of Ca^{2+}, insulin, and $NaHCO_3$ are usually transient. For example, the K^+ that has been driven into the cell by insulin may come back into the extracellular fluid after several hours. Therefore, these measures are effective for a short period but usually must be followed by the removal of the excess K^+ from the body by the use of diuretics, cation-exchange resins, or dialysis. In addition, the short-acting measures may not be necessary in patients with mild hyperkalemia. A cation-exchange resin alone may be sufficient in a patient who has a plasma K^+ concentration less than 6.5 meq/L without muscle or electrocardiographic abnormalities.

Diuretics Diuretics which inhibit NaCl reabsorption in the ascending limb of the loop of Henle or distal tubule (furosemide, ethacrynic acid, and the thiazides) increase K^+ excretion primarily by enhancing the flow of Na^+ and water to the K^+ secretory site in the distal tubule [154]. In general, diuretics are not widely used in the treatment of hyperkalemia since patients with renal failure or hypoaldosteronism are unlikely to respond to these agents with a significant increase in K^+ excretion. However, diuretics may have a dual benefit in patients with heart failure and other edematous states, removing both the edema fluid and the excess K^+.

Cation-exchange resins The major cation-exchange resin available is sodium polystyrene sulfonate (Kayexalate), prepared in the sodium phase. In the gut, the resin takes up K^+ (and, to lesser degrees, Ca^{2+} and Mg^{2+}) and releases Na^+. Each gram of resin may bind as much as 1 meq of K^+ and release 1 to 2 meq of Na^+ [155,156]. When administered orally, 20 g of resin should be given with 100 mL of a 20% sorbitol solution to prevent constipation. This can be repeated every 4 to 6 h as necessary.

In patients who cannot take oral fluids, the resin can be given as a retention enema. In this situation, 50 g of resin is mixed with 50 mL of 70% sorbitol plus 100 to 150 mL of tap water and kept in the colon for at least 30 to 60 min and preferably 2 to 3 h [150]. Each enema can lower the plasma K^+ concentration by as much as 0.5 to 1.0 meq/L. The enemas can be repeated every 2 to 4 h if necessary.

The major side effects of sodium polystyrene sulfonate are nausea, constipation, hypokalemia (due to excessive use), and retention of the Na^+ that has been exchanged for K^+. In patients with oliguric renal failure or those with cardiac disease, enough Na^+ may be retained to precipitate pulmonary edema [155]. Consequently, the resin should be used with care in these conditions.

Dialysis In almost all patients, the conservative measures described above will control hyperkalemia. However, when these measures are ineffective or severe hyperkalemia is present, either peritoneal dialysis or hemodialysis can be used. For example, patients with acute renal failure who are hypercatabolic because of trauma can release large quantities of K^+ into the extracellular fluid [44] and are best treated by dialysis. Hemodialysis is preferred because the rate of K^+ removal is many times faster than with peritoneal dialysis [157].

Hypoaldosteronism

The proper therapy of hypoaldosteronism varies with the underlying cause. If drug-induced, the offending agent should be discontinued, if possible. With the nonsteroidal anti-inflammatory drugs, it may be sufficient to switch to sulindac, which does not appear to interfere with renal prostaglandin or renin production [158,159].

Fludrocortisone, a synthetic mineralocorticoid, is generally used in patients with deficient aldosterone production. Although the amount required is variable, care must be taken to avoid excessive replacement since this can lead to fluid retention, hypokalemia, and hypertension. Adults with primary adrenal insufficiency can usually be treated with 0.05 to 0.10 mg/day of fludrocortisone, as well as with cortisol replacement [104]. Patients with mild mineralocorticoid deficiency may not even need fludrocortisone to prevent hyperkalemia since cortisol has some mineralocorticoid activity.

Patients with hyporeninemic hypoaldosteronism, on the other hand, may require as much as 0.4 to 1.0 mg/day to normalize K^+ balance [85,88]. This supraphysiologic requirement may reflect aldosterone resistance induced by the associated renal insufficiency. In patients with an underlying edematous state, however, mineralocorticoid replacement can lead to further fluid overload. In this setting, use of a low-K^+ diet plus a diuretic such as furosemide or a thiazide to increase urinary K^+ losses is generally effective in controlling the hyperkalemia [85,160].

Children with an adrenal enzyme deficiency can also be treated with fludrocortisone. This is particularly important with 21-hydroxylase deficiency since inadequate mineralocorticoid replacement leads to volume depletion and activation of the renin–angiotensin II system. The ensuing elevation in angiotensin II appears to stimulate ACTH secretion, resulting in the necessity for a higher cortisol dose to suppress ACTH release and virilization [161,162]. This increase in the cortisol requirement can then impair growth. Thus, enough fludrocortisone should be given to normalize the plasma renin activity; the cortisol dose, on the other hand, is equal to that amount required to normalize the plasma 17-hydroxyprogesterone level since it is this compound that accumulates with excess ACTH [161].

Pseudohypoaldosteronism can usually be treated with a high-Na^+ diet which prevents volume depletion and maintains distal flow at a level that allows adequate K^+ excretion [118]. Fludrocortisone is ineffective since these patients are resistant to aldosterone. Infants should be followed carefully since the aldosterone resistance typically, but not always, resolves within a few years [118,163].

Other

Patients with type 1 RTA or a selective K^+ secretory defect can be treated with $NaHCO_3$ (which both corrects the acidemia and maintains a high distal flow), a low-K^+ diet, and, if necessary, a diuretic. Thiazides may be more effective than furosemide in this setting, although the mechanism by which this might occur is uncertain [128,131]. In patients who appear to have a primary increase in distal Cl^- reabsorption, $NaHCO_3$ has the added benefit of increasing K^+ excretion since Na^+ is presented to the distal secretory site with an anion other than Cl^- [130].

Patients with oliguric renal failure can also be treated with $NaHCO_3$ and a low-K^+ diet. If this is ineffective, chronic use of a cation-exchange resin can be tried, although the institution of dialysis is usually necessary in this setting.

PROBLEMS

28-1 A 62-year-old man with mild chronic renal failure (plasma creatinine concentration equals 2.1 mg/dL) and normokalemia is started on a low-sodium diet for hypertension. Two weeks later, he notices that he is unable to lift himself out of a chair. On physical examination, slightly decreased skin turgor and marked proximal muscle weakness are found. The electrocardiogram reveals peaked T waves and some widening of the P wave and QRS complex. The following blood test results are obtained:

$$\text{Plasma [creatinine]} = 2.7 \text{ mg/dL}$$

$$[Na^+] = 130 \text{ meq/L}$$

$$[K^+] = 9.8 \text{ meq/L}$$

$$[Cl^-] = 98 \text{ meq/L}$$

$$[HCO_3^-] = 17 \text{ meq/L}$$

$$\text{Arterial pH} = 7.32$$

(*a*) What are the most likely factors responsible for the elevation in the plasma K^+ concentration?

(*b*) How do you know this is not pseudohyperkalemia?

(*c*) How would you treat the hyperkalemia?

(*d*) If the plasma K^+ concentration were only 6.4 meq/L and there were no changes in muscle strength or the electrocardiogram, how would you lower the plasma K^+ concentration?

28-2 A 54-year-old man with no prior medical history complains of chronic fatigue. The positive physical findings include a blood pressure of 100/60 and increased skin pigmentation. The skin turgor is relatively normal. The laboratory data are:

$$BUN = 28 \text{ mg/dL}$$

$$\text{Plasma [creatinine]} = 1.0 \text{ mg/dL}$$

$$[Na^+] = 130 \text{ meq/L}$$

$$[K^+] = 6.8 \text{ meq/L}$$

$$[Cl^-] = 100 \text{ meq/L}$$

$$[HCO_3^-] = 20 \text{ meq/L}$$

$$[\text{Glucose}] = 90 \text{ mg/dL}$$

$$\text{Urine } [Na^+] = 50 \text{ meq/L}$$

$$[K^+] = 18 \text{ meq/L}$$

The electrocardiogram shows mild peaking of the T waves in the precordial leads. An infusion of glucose and insulin in appropriate proportions results in an episode of hypoglycemia.

(*a*) What does the urine K^+ concentration of 18 meq/L mean?

(*b*) What is the most likely diagnosis?

(*c*) How would you treat this patient?

REFERENCES

1. Bia, M. J., and R. A. DeFronzo: Extrarenal potassium homeostasis, Am. J. Physiol., 240:F257, 1981.
2. Winkler, A. W., H. E. Hoff, and P. K. Smith: The toxicity of orally administered potassium salts in renal insufficiency, J. Clin. Invest., 20:119, 1941.

3. DeFronzo, R. A., P. A. Taufield, H. Black, P. McPhedran, and C. R. Cooke: Impaired renal tubular potassium secretion in sickle cell disease, Ann. Intern. Med., 90:310, 1979.
4. Young, D. B., and A. W. Paulsen: Interrelated effects of aldosterone and plasma potassium on potassium excretion, Am. J. Physiol., 244:F28, 1983.
5. Talbott, J. H., and R. S. Schwab: Recent advances in the biochemistry and therapeusis of potassium salts, N. Engl. J. Med., 222:585, 1940.
6. Hayslett, J. P., and H. J. Binder: Mechanism of potassium adaptation, Am. J. Physiol., 243:F103, 1982.
7. Alexander, E. A., and N. G. Levinsky: An extrarenal mechanism of potassium adaptation, J. Clin. Invest., 47:740, 1968.
8. Schon, D. A., K. A. Backman, and J. P. Hayslett: Role of the medullary collecting duct in potassium excretion in potassium-adapted animals, Kidney Int., 20:655, 1981.
9. Silva, P., J. P. Hayslett, and F. H. Epstein: The role of Na-K-activated adenosine triphosphatase in potassium adaptation: stimulation of enzymatic activity by potassium loading, J. Clin. Invest., 52:2665, 1973.
10. Doucet, A., and A. I. Katz: Renal potassium adaptation: Na-K-ATPase activity along the nephron after chronic potassium loading, Am. J. Physiol., 238:F380, 1980.
11. Stanton, B. A., D. Biemesderfer, J. B. Wade, and G. Giebisch: Structural and functional study of the rat distal nephron: effects of potassium adaptation and depletion, Kidney Int., 19:36, 1981.
12. Garg, L. C., M. A. Knepper, and M. B. Burg: Mineralocorticoid effects on Na-K-ATPase in individual nephron segments, Am. J. Physiol., 240:F536, 1981.
13. Schultze, R. G., D. D. Taggart, H. Shapiro, J. P. Pennell, S. Caglar, and N. S. Bricker: On the adaptation in potassium excretion associated with nephron reduction in the dog, J. Clin Invest., 50:1061, 1971.
14. Bourgoignie, J. J., M. Kaplan, J. Pincus, G. Gavellas, and A. Rabinovitch: Renal handling of potassium in dogs with chronic renal insufficiency, Kidney Int., 20:482, 1981.
15. Gonick, H. C., C. R. Kleeman, M. E. Rubini, and M. H. Maxwell: Functional impairment in chronic renal disease: III. Studies of potassium excretion, Am. J. Med. Sci., 261:281, 1971.
16. Schrier, R. W., and E. M. Regal: Physiological role of aldosterone in sodium, water and potassium metabolism in chronic renal disease, Kidney Int., 1:156, 1972.
17. Schon, D. A., P. Silva, and J. P. Hayslett: Mechanism of potassium excretion in renal insufficiency, Am. J. Physiol., 227:1323, 1974.
18. Bastl, C., J. P. Hayslett, and H. J. Binder: Increased large intestinal secretion of potassium in renal insufficiency, Kidney Int., 12:9, 1977.
19. Hayes, C. P., Jr., and R. R. Robinson: Fecal potassium excretion in patients on chronic intermittent hemodialysis, Trans. Am. Soc. Artif. Intern. Organs, 11:242, 1965.
20. Illingworth, R. N., and A. T. Proudfoot: Rapid poisoning with slow-release potassium, Br. Med. J., 2:485, 1980.
21. Moss, M. H., and A. R. Rasen: Potassium toxicity due to intravenous penicillin therapy, Pediatrics, 29:1032, 1962.
22. Kallen, R. J., C. H. L. Rieger, H. S. Cohen, M. A. Sutter, and R. Ong: Near-fatal hyperkalemia due to ingestion of salt substitute by an infant, J. Am. Med. Assoc., 235:2125, 1976.
23. Scanlon, J. W., and R. Krakaur: Hyperkalemia following exchange transfusion, J. Pediatr., 96:108, 1980.
24. Perkins, H. A., M. R. Rolfs, and D. J. Acra: Studies on bank blood collected and stored under various conditions with particular reference to its use in open heart surgery, Transfusion, 1:151, 1961.
25. LeVeen, H. H., H. S. Posternack, I. Lustrin, R. B. Shapiro, and E. Beeker: Hemorrhage and transfusion as the major cause of cardiac arrest, J. Am. Med. Assoc., 173:770, 1960.
26. Lawson, D. H.: Adverse reactions to potassium chloride, Q. J. Med., 43:433, 1974.
27. Greenblatt, D. J., and J. Koch-Weser: Adverse reactions to spironolactone: a report from the Boston Collaborative Drug Surveillance Program, J. Am. Med. Assoc., 225:40, 1973.
28. Bay, W. H., and J. A. Hartman: High potassium in low sodium soups (letter), N. Engl. J. Med., 308:1166, 1983.

29. Gelfand, M. C., A. Zarate, and J. H. Knepshield: Geophagia: a cause of life-threatening hyperkalemia in patients with chronic renal failure, J. Am. Med. Assoc., 234:738, 1975.

30. Chumbley, L. C.: Pseudohyperkalemia in acute myelocytic leukemia, J. Am. Med. Assoc., 211:1007, 1970.

31. Bronson, W. R., V. T. DeVita, P. P. Carbone, and E. Cotlove: Pseudohyperkalemia due to release of potassium from white blood cells during clotting, N. Engl. J. Med., 274:369, 1966.

32. Hartmann, R. C., J. V. Auditore, and D. P. Jackson: Studies on thrombocytosis: I. Hyperkalemia due to release of potassium from platelets during coagulation, J. Clin. Invest., 37:699, 1958.

33. Stewart, G. W., J. A. Fyffe, R. J. M. Corrall, G. Stockdill, and J. A. Strong: Familial pseudohyperkalaemia: a new syndrome, Lancet, 2:175, 1979.

34. Luciani, J-C., T. Lavabre-Bertrand, J. Fourcade, P. Bayon, A. Mimran, and A. Callis: Familial pseudohyperkalaemia, Lancet, 1:491, 1980.

35. Adrogué, H. J., and N. E. Madias: Changes in plasma potassium concentration during acute acid-base disturbances, Am. J. Med., 71:456, 1981.

36. Kreisberg, R. A.: Diabetic ketoacidosis: new concepts and trends in pathogenesis and treatment, Ann. Intern. Med., 88:681, 1978.

37. Arieff, A., and H. J. Carroll: Nonketotic hyperosmolar coma with hyperglycemia: clinical features, pathophysiology, renal function, acid-base balance, plasma-cerebrospinal fluid equilibria and the effects of therapy in 37 cases, Medicine, 51:73, 1972.

38. DeFronzo, R. A., R. S. Sherwin, M. Dillingham, R. Hendler, W. V. Tamborlane, and P. Felig: Influence of basal insulin and glucagon secretion on potassium and sodium metabolism: studies with somatostatin in normal dogs and in normal and diabetic human beings, J. Clin. Invest., 61:472, 1978.

39. Viberti, G. C.: Glucose-induced hyperkalaemia: a hazard for diabetics?, Lancet, 1:690, 1978.

40. Nichols, G. L., T. Kahn, A. Sanchez, and L. Gabrilove: Glucose-induced hyperkalemia in diabetic subjects, Arch. Intern. Med., 141:49, 1981.

41. Goldfarb, S., M. Cox, I. Singer, and M. Goldberg: Acute hyperkalaemia induced by hyperglycemia: hormonal mechanisms, Ann. Intern. Med., 84:426, 1976.

42. Makoff, D. L., J. A. Da Silva, B. J. Rosenbaum, S. E. Levy, and M. H. Maxwell: Hypertonic expansion: acid-base and electrolyte changes, Am. J. Physiol., 218:1201, 1970.

43. Hoeldtke, R. D., G. Boden, C. R. Shuman, and O. E. Owen: Reduced epinephrine secretion and hypoglycemia unawareness in diabetic autonomic neuropathy, Ann. Intern. Med., 96:459, 1982.

44. Lordon, R. E., and J. R. Burton: Post-traumatic renal failure in military personnel in Southeast Asia, Am. J. Med., 53:137, 1972.

45. Arseneau, J. C., C. M. Bagley, T. Anderson, and G. P. Canellos: Hyperkalaemia, a sequel to chemotherapy of Burkitt's lymphoma, Lancet, 1:10, 1973.

46. Fortner, R. W., A. Nowakowski, C. B. Carter, L. H. King, and J. H. Knepshield: Death due to overheated dialysate during dialysis, Ann. Intern. Med., 73:443, 1970.

47. Rosa, R. M., P. Silva, J. B. Young, L. Landsberg, R. S. Brown, J. W. Rowe, and F. H. Epstein: Adrenergic modulation of extrarenal potassium disposal, N. Engl. J. Med., 302:431, 1980.

48. Carlsson, E., E. Fellenius, P. Lundborg, and L. Svensson: β-adrenoceptor blockers, plasma potassium, and exercise (letter), Lancet, 2:424, 1978.

49. Lim, M., R. A. F. Linton, C. B. Wolff, and D. M. Band: Propranolol, exercise and arterial plasma potassium, Lancet, 2:591, 1981.

50. Silva, P., and K. Spokes: Sympathetic system in potassium homeostasis, Am. J. Physiol., 241:F151, 1981.

51. Bethune, D. W., and R. McKay: Paradoxical changes in serum-potassium during cardiopulmonary bypass in association with non-cardioselective beta blockade, Lancet, 2:380, 1978.

52. Rosenthal, L. S., D. T. Lowenthal, M. B. Affrime, B. Falkner, and A. B. Gould: The renin-aldosterone-potassium response to methyl-dopa during dynamic physical activity, Clin. Pharmacol. Therap., 31:264, 1982.

53. Lowenthal, D. T., M. B. Affrime, L. Rosenthal, A. B. Gould, J. Borruso, and B. Falkner: Dynamic and biochemical responses to single and repeated doses of clonidine during dynamic physical activity, Clin. Pharmacol. Therap., 32:18, 1982.

54. Knochel, J. P., and E. M. Schlein: On the mechanism of rhabdomyolysis in potassium depletion, J. Clin. Invest., 51:1750, 1972.

55. Sessard, J., M. Vincent, G. Annat, and C. A. Bizollon: A kinetic study of plasma renin and aldosterone during changes of posture in man, J. Clin. Endocrinol. Metab., 42:20, 1976.

55a. Rose, L. I., D. R. Carroll, S. L. Lowe, E. W. Peterson, and K. H. Cooper: Serum electrolyte changes after marathon running, J. Appl. Physiol., 29:449, 1970.

56. Coester, N., J. C. Elliott, and U. C. Luft: Plasma electrolytes, pH, and ECG during and after exhaustive exercise, J. Appl. Physiol., 34:677, 1973.

56a. Ledingham, I. McA., S. MacVicar, I. Watt, and G. A. Weston: Early resuscitation after marathon collapse, Lancet, 2:1096, 1982.

57. Smith, T. W., and E. Haber: Digitalis, N. Engl. J. Med., 289:945, 1973.

58. Lown, B., H. Black, and F. D. Moore: Digitalis, electrolytes and the surgical patient, Am. J. Cardiol., 6:309, 1960.

59. Asplund, J., O. Edhag, L. Mongenson, O. Nyquist, E. Orinius, and A. Sjögren: Four cases of massive digitalis poisoning, Acta Med. Scand., 189:293, 1971.

60. Reza, M. J., R. B. Kovick, K. I. Shine, and M. L. Pearce: Massive intravenous digoxin overdosage, N. Engl. J. Med., 291:777, 1974.

61. Pearson, C. M., and K. Kalyanaraman: Periodic paralysis, in *The Metabolic Basis of Inherited Disease,* 3d ed., J. B. Stanbury, J. B. Wyngaarden, and D. S. Fredcrickson (eds.), McGraw-Hill, New York, 1972.

62. Gamstorp, I., M. Hauge, H. F. Helweg-Larsen, and H. Mjones: Adynamia episodica hereditaria: a disease clinically resembling familial periodic paralysis but characterized by increasing serum potassium during the paralytic attacks, Am. J. Med., 23:385, 1957.

63. Clausen, T., P. Wang, H. Ørskov, and O. Kristen: Hyperkalemic periodic paralysis: relationship between changes in plasma water, electrolytes, insulin, and catecholamines during attacks, Scand. J. Clin. Lab. Invest., 40:211, 1980.

64. Wang, P., and T. Clausen: Treatment of attacks in hyperkalaemic familial periodic paralysis by inhalation of salbutamol, Lancet, 1:221, 1976.

65. Streeten, D. H., T. G. Dalakos, and H. Fellerman: Studies on hyperkalemic periodic paralysis: evidence of changes in plasma Na and Cl and induction of paralysis by adrenal glucocorticoids, J. Clin. Invest., 50:142, 1971.

66. Lim, M., R. A. F. Linton, and D. M. Band: Rise in plasma potassium during rewarming in open-heart surgery, Lancet, 1:241, 1983.

67. Koht, A., L. J. Cerullo, P. C. Land, and H. W. Linde: Serum potassium levels during prolonged hypothermia, Anesthesiology, 51(suppl):S203, 1979.

68. Bruining, H. A., and R. U. Boelhouwer: Acute transient hypokalemia and body temperature (letter), Lancet, 2:1283, 1982.

69. Birch, A. A., G. D. Mitchell, G. A. Playford, and C. A. Lang: Changes in serum potassium response to succinylcholine following trauma, J. Am. Med. Assoc., 210:490, 1969.

70. Cooperman, L. H.: Succinylcholine-induced hyperkalemia in neuro-muscular disease, J. Am. Med. Assoc., 213:1867, 1970.

71. Hertz, P., and J. A. Richardson: Arginine-induced hyperkalemia in renal failure patients, Arch. Intern. Med., 130:778, 1972.

72. Bushinsky, D. A., and F. J. Gennari: Life-threatening hyperkalemia induced by arginine, Ann. Intern. Med., 89:632, 1978.

73. Elkinton, J. R., R. Tarail, and J. P. Peters: Transfers of potassium in renal insufficiency, J. Clin Invest., 28:378, 1949.

74. Kleeman, C. R., R. Okun, and R. J. Heller: The renal regulation of potassium in patients with chronic renal failure and the effect of diuretics on the excretion of these ions, Ann. N.Y. Acad. Sci., 139:520, 1966.

75. Bilbrey, G. L., N. W. Carter, M. G. White, J. F. Schilling, and J. P. Knochel: Potassium deficiency in chronic renal failure, Kidney Int., 4:423, 1973.

76. Kahn, T., M. Kaji, G. Nichols, L. R. Krakoff, and R. M. Stein: Factors related to potassium transport in chronic stable renal disease in man, Clin. Sci., 54:661, 1978.

77. Bia, M. J., and R. A. DeFronzo: Extrarenal potassium homeostasis, Am. J. Physiol., 240:F257, 1981.

78. Smith, E. K. M., and L. G. Welt: The red blood cell as a model for the study of uremic toxins, Arch. Intern. Med., 126:827, 1970.

79. Malnic, G., R. M. Klose, and G. Giebisch: Micropuncture study of distal tubular potassium and sodium transport in rat nephron, Am. J. Physiol., 211:529, 1966.

80. Anderson, H. M., and J. M. Laragh: Renal excretion of potassium in normal and sodium depleted dogs, J. Clin Invest., 37:323, 1958.

81. Coleman, A. J., M. Arias, N. W. Carter, F. C. Rector, Jr., and D. W. Seldin: The mechanism of salt-wasting in chronic renal disease, J. Clin. Invest., 45:1116, 1966.

82. Danovitch, G. M., J. J. Bourgoignie, and N. S. Bricker: Reversibility of the "salt-losing" tendency of chronic renal failure, N. Engl. J. Med., 296:14, 1977.

83. Popovtzer, M. M., F. H. Katz, W. F. Pinggera, J. Robinette, C. C. Halgrimson, and D. E. Butkus: Hyperkalemia in salt-wasting nephropathy: study of the mechanism, Arch. Intern. Med., 132:203, 1973.

84. Uribarri, J., M. S. Oh, and H. J. Carroll: Salt-losing nephropathy. Clinical presentation and mechanisms Am. J. Nephrol., 3:193, 1983.

85. DeFronzo, R. A.: Hyperkalemia in hyporeninemic hypoaldosteronism, Kidney Int., 17:118, 1980.

86. Gabow, P. A., S. Moore, and R. W. Schrier: Spironolactone-induced hyperchloremic acidosis in cirrhosis, Ann. Intern. Med., 90:338, 1979.

87. Szylman, P., O. S. Better, C. Chaimowitz, and A. Rosler: Role of hyperkalemia in the metabolic acidosis of isolated hypoaldosteronism, N. Engl. J. Med., 294:361, 1976.

88. Sebastian, A., M. Schambelan, S. Lindenfeld, and R. C. Morris, Jr.: Amelioration of metabolic acidosis with fludrocortisone therapy in hyporeninemic hypoaldosteronism, N. Engl. J. Med., 297:576, 1977.

89. Kamm, D. E., and G. L. Strope: Glutamine and glutamate metabolism in renal cortex from potassium-depleted rats, Am. J. Physiol., 224:1241, 1973.

90. Tannen, R. L., E. Wedell, and R. Moore: Renal adaptation to a high potassium intake. The role of hydrogen ion, J. Clin. Invest., 52:2089, 1973.

91. Schambelan, M., A. Sebastian, and E. Biglieri: Prevalence, pathogenesis, and functional significance of aldosterone deficiency in hyperkalemic patients with chronic renal insufficiency, Kidney Int., 17:89, 1980.

92. Tan. S. Y., and M. Burton: Hyporeninemic hypoaldosteronism: an overlooked cause of hyperkalemia, Arch. Intern. Med., 141:30, 1981.

93. Tuck, M. L., and D. M. Mayes: Mineralocorticoid biosynthesis in patients with hyporeninemic hypoaldosteronism, J. Clin Endocrinol. Metab., 50:341, 1980.

94. Morimoto, S., K. S. Kim, I. Yamamoto, K. Uchida, R. Takeda, and L. Kornel: Selective hypoaldosteronism with hyperreninemia in a diabetic patient, J. Clin. Endocrinol. Metab., 49:742, 1979.

95. Bayard, F., C. Cooke, D. Tiller, I. Beitins, A. Kovorski, W. Walker, and C. Migeon: The regulation of aldosterone secretion in anephric man, J. Clin. Invest., 50:1585, 1971.

96. Campbell, W. B., C. E. Gomez-Sanchez, B. V. Adams, J. M. Schmitz, and H. D. Itskovitz: Attenuation of angiotensin II- and III-induced aldosterone released by prostaglandin synthesis inhibitors, J. Clin. Invest., 64:1552, 1979.

97. Tan, S. Y., R. Shapiro, R. Franco, H. Stockard, and P. J. Mulrow: Indomethacin-induced prostaglandin inhibition with hyperkalemia: a reversible cause of hyporeninemic hypoaldosteronism, Ann. Intern. Med., 90:783, 1979.

98. Kutyrina, I. M., S. O. Androsova, and I. E. Tareyeva: Indomethacin-induced hyporeninaemic hypoaldosteronism, Lancet, 1:785, 1979.

99. Hsueh, W. A., E. J. Carlson, J. A. Leutscher, and G. Grislis: Activation and characterization of inactive big renin in plasma of patients with diabetic nephropathy and unusual active renin, J. Clin. Endocrinol. Metab., 51:535, 1980.

100. Perez, G. O., L. Lespier, J. Jacobi, J. R. Oster, F. H. Katz, C. A. Vaamonde, and L. M. Fishman: Hyporeninemia and hypoaldosteronism in diabetes mellitus, Arch. Intern. Med., 137:852, 1977.

101. Beretta-Piccoli, P. Weidmann, and G. Keusch: Responsiveness of plasma renin and aldosterone in diabetes mellitus, Kidney Int., 20:259, 1981.

102. Krause, J. Z., R. A. Matarese, P. M. Zabetakis, and M. F. Michelis: Reversibility of hyporeninemia and hypoaldosteronemia in chronic hemodialysis patients by correction of fluid excess, J. Lab. Clin. Med., 96:734, 1980.

103. Textor, S. C., E. L. Bravo, F. M. Fouad, and R. C. Tarazi: Hyperkalemia in azotemic patients during angiotensin-converting enzyme inhibition and aldosterone reduction with captopril, Am. J. Med., 73:719, 1982.

104. Liddle, G. W.: The adrenals, in *Textbook of Endocrinology,* 6th ed., R. H. Williams (ed.), Saunders, Philadelphia, 1981.

105. Ganong, W. F., L. Coultan, A. B. Alpert, and T. C. Less: ACTH and the regulation of adrenocortical secretion, N. Engl. J. Med., 290:1006, 1974.

106. Kohn, B., L. S. Levine, M. S. Pollack, S. Pang, T. Lorenzen, D. Levy, A. J. Lerner, G. F. Rondanini, B. Dupont, and M. I. New: Late-onset steroid 21-hydroxylase deficiency: a variant of classical congenital adrenal hyperplasia, J. Clin. Endocrinol. Metab., 55:817, 1982.

107. Bartter, F. C., R. I. Henkin, and G. R. Bryan: Aldosterone hypersecretion in "nonsalt-losing" congenital adrenal hyperplasia, J. Clin. Invest., 47:1742, 1968.

108. Kuhnle, U., D. Chow, R. Rapaport, S. Pang, L. S. Levine, and M. I. New: The 21-hydroxylase activity in the glomerulosa and fasciculata of the adrenal cortex in congenital adrenal hyperplasia, J. Clin. Endocrinol. Metab., 52:534, 1981.

109. Biglieri, E. G., B. L. Wajchenberg, D. A. Malerbi, H. Okada, C. E. Leme, and C. E. Kater: The zonal origins of the mineralocorticoid hormones in the 21-hydroxylation deficiency of congenital adrenal hyperplasia, J. Clin. Endocrinol. Metab., 53:964, 1981

109a. Pang, S., L. S. Levine, E. Stoner, J. M. Opitz, M. S. Pollack, B. Dupont, and M. I. New: Nonsalt-losing congenital adrenal hyperplasia due to 3β-hydroxysteroid dehydrogenase deficiency with normal glomerulosa function, J. Clin. Endocrinol. Metab., 56:808, 1983.

110. Ulick, S.: Diagnosis and nomenclature of the disorders of the terminal portion of the aldosterone biosynthetic pathway, J. Clin. Endocrinol. Metab., 43:92, 1976.

111. Veldhuis, J. D., H. E. Kulin, R. J. Santen, T. E. Wilson, and J. C. Melby: Inborn error in the terminal step of aldosterone biosynthesis: corticosterone methyl oxidase type II deficiency in a North American pedigree, N. Engl. J. Med., 303:117, 1980.

112. O'Kelly, R., F. Magee, and T. J. McKenna: Routine heparin therapy inhibits adrenal aldosterone production, J. Clin. Endocrinol. Metab., 56:108, 1983.

113. Phelps, K. R., M. S. Oh, and H. J. Carroll: Heparin-induced hyperkalemia: report of a case, Nephron, 25:254, 1980.

114. Biglieri, E. G., P. E. Slaton, Jr., W. S. Silen, M. Galante, and P. H. Forsham: Postoperative studies of adrenal function in primary aldosteronism, J. Clin. Endocrinol. Metab., 26:553, 1966.

115. Bravo, E. L., H. P. Dustan, and R. C. Tarazi: Selective hypoaldosteronism despite prolonged pre- and postoperative hyperreninemia in primary aldosteronism, J. Clin. Endocrinol. Metab., 41:611, 1975.

116. Cohen, A. B.: Hyperkalemic effects of triamterene, Ann. Intern. Med., 65:521, 1966.

117. Donnell, G. N., N. Litman, and M. Roldan: Pseudohypo-adrenocorticism, Am. J. Dis. Child., 97:813, 1959.

118. Proesmans, W., H. Geussens, L. Corbeel, and R. Eeckels: Pseudohypoaldosteronism, Am. J. Dis. Child., 126:510, 1973.

119. Oberfield, S. E., L. S. Levine, R. M. Carey, R. Bejar, and M. I. New: Pseudohypoaldosteronism: multiple target organ unresponsiveness to mineralocorticoid hormones, J. Clin. Endocrinol. Metab., 48:228, 1979.

120. Daughaday, W. H., and D. Rendleman: Severe symptomatic hyperkalemia in an adrenalectomized woman due to enhanced mineralocorticoid requirement, Ann. Intern. Med., 66:1197, 1967.

121. Cogan, M. C., and A. I. Arieff: Sodium wasting, acidosis and hyperkalemia induced by methicillin interstitial nephritis: evidence for selective distal tubular dysfunction, Am. J. Med., 64:500, 1978.

122. Luke, R. G., M. E. Allison, J. F. Davidson, and W. P. Duquid: Hyperkalemia and renal tubular acidosis due to renal amyloidosis, Ann. Intern. Med., 70:1211, 1969.

123. Sebastian, A., E. McSherry, and R. C. Morris, Jr.: Renal potassium wasting in renal tubular acidosis (RTA): its occurrence in types 1 and 2 RTA despite sustained correction of systemic acidosis, J. Clin. Invest., 50:667, 1971.

124. Arruda, J. A. L., and N. A. Kurtzman: Mechanisms and classification of deranged distal urinary acidification, Am. J. Physiol., 239:F515, 1980.
125. Batlle, D. C., J. A. L. Arruda, and N. A. Kurtzman: Hyperkalemic distal renal tubular acidosis associated with obstructive uropathy, N. Engl. J. Med., 304:373, 1981.
126. Batlle, D., K. Itsarayoungyuen, J. A. L. Arruda, and N. A. Kurtzman: Hyperkalemic hyperchloremic metabolic acidosis in sickle cell hemoglobinopathies, Am. J. Med., 72:188, 1982.
127. DeFronzo, R. A., P. A. Taufield, H. Black, P. McPhedran, and C. R. Cooke: Impaired renal tubular potassium secretion in sickle cell disease, Ann. Intern. Med., 90:310, 1979.
128. DeFronzo, R. A., M. Goldberg, C. R. Cooke, C. Barker, R. A. Grossman, and Z. S. Agus: Investigations into mechanisms of hyperkalemia following renal transplantation, Kidney Int., 11:357, 1977.
129. DeFronzo, R. A., C. R. Cooke, M. Goldberg, M. Cox, A. R. Myers, and Z. S. Agus: Impaired renal tubular potassium secretion in systemic lupus erythematosus, Ann. Intern. Med., 86:268, 1977.
130. Schambelan, M., A. Sebastian, and F. C. Rector, Jr.: Mineralocorticoid-resistant renal hyperkalemia without salt-wasting (type II pseudohypoaldosteronism): role of increased renal chloride reabsorption, Kidney Int., 19:716, 1981.
131. Arnold, J. E., and J. K. Healy: Hyperkalemia, hypertension and systemic acidosis without renal failure associated with a tubular defect in potassium excretion, Am. J. Med., 47:461, 1969.
132. Brautbar, N., J. Levi, A. Rosler, E. Leitesdorf, M. Djaldeti, M. Epstein, and C. R. Kleeman: Familial hyperkalemia, hypertension, and hyporeninemia with normal aldosterone levels. A tubular defect in potassium handling, Arch. Intern. Med., 138:607, 1978.
133. Epstein, F. H.: Signs and symptoms of electrolyte disorders, in *Clinical Disorders of Fluid and Electrolyte Metabolism,* 3d ed., M. H. Maxwell, and C. R. Kleeman (eds.), McGraw-Hill, New York, 1980.
134. Finch, C. A., C. G. Sawyer, and J. M. Flynn: Clinical syndrome of potassium intoxication, Am. J. Med., 1:337, 1946.
135. Bele, H., W. L. Hayes, and J. Vosburgh: Hyperkalemic paralysis due to adrenal insufficiency, Arch. Intern. Med., 115:418, 1965.
136. Surawicz, B.: Relationship between electrocardiogram and electrolytes, Am. Heart J., 73:814, 1967.
137. Weidmann, S.: Membrane excitation in cardiac muscle, Circulation, 24:499, 1961.
138. Surawicz, B., H. Chlebus, and A. Mazzoleni: Hemodynamic and electrocardiographic effects of hyperpotassemia. Differences in response to slow and rapid increases in concentration of plasma K, Am. Heart J., 73:647, 1967.
139. Gettes, L. S., B. Surawicz, and J. C. Shiue: Effect of high K, low K, and quinidine on QRS duration and ventricular action potential, Am. J. Physiol., 203:1135, 1962.
140. Weidmann, S.: The effect of the cardiac membrane potential on the rapid availability of the sodium carrying system, J. Physiol. (London), 127:213, 1955.
141. Kuffler, S. W., and J. G. Nicholls: *From Neuron to Brain,* Sinauer Assoc., Sunderland, Mass., 1976, chap. 6.
142. Goodyer, A. V. N., M. J. Goodkind, and E. J. Stanley: The effects of abnormal concentrations of the serum electrolytes on left ventricular function in the intact animal, Am. Heart J., 67:779, 1964.
143. Braun, H. A., R. Van Horne, C. Bettinger, and S. Bellet: The influence of hypocalcemia induced by sodium ethylenediamine acetate on the toxicity of potassium: an experimental study, J. Lab. Clin. Med., 46:544, 1955.
144. Garcia-Palmieri, M. R.: Reversal of hyperkalemic cardiotoxicity with hypertonic saline, Am. Heart J., 64:483, 1962.
145. Abrams, W. B., D. W. Lewis, and S. Bellet: The effect of acidosis and alkalosis on the plasma potassium concentration and the electrocardiogram in normal and potassium depleted dogs, Am. J. Med. Sci., 222:506, 1951.
146. Surawicz, B., and L. S. Gettes: Two mechanisms of cardiac arrest produced by potassium, Circ. Res., 12:415, 1963.
147. Rose, B. D.: *Pathophysiology of Renal Disease,* McGraw-Hill, New York, 1981, chap. 9.

148. Winkler, A. W., H. E. Hoff, and P. K. Smith: Factors affecting the toxicity of potassium, Am. J. Physiol., 127:430, 1939.

149. Zierler, K. L., and D. Rabinowitz: Effect of very small concentrations of insulin on forearm metabolism: persistence of its action on potassium and free fatty acids without its effect on glucose, J. Clin. Invest., 43:950, 1964.

150. Levinsky, N.: Management of emergencies: VI. Hyperkalemia, N. Engl. J. Med., 274:1076, 1966.

151. Fraley, D. S., and S. Adler: Correction of hyperkalemia by bicarbonate despite constant blood pH, Kidney Int., 12:354, 1977.

152. Schwarz, K. C., B. D. Cohen, G. D. Lubash, and A. L. Rubin: Severe acidosis and hyperpotassemia treated with sodium bicarbonate infusion, Circulation, 19:215, 1959.

153. Ballantyne, F., III, L. D. Davis, and E. W. Reynolds, Jr.: Cellular basis for reversal of hyperkalemic electrocardiographic changes by sodium, Am. J. Physiol., 229:935, 1975.

154. Duarte, C. G., F. Chomety, and G. Giebisch: Effect of amiloride, ouabain and furosemide on distal tubular function in the rat, Am. J. Physiol., 221:632, 1971.

155. Berlyne, G. M., K. Janab, M. B. Manc, M. B. Baghdad, and A. B. Shaw: Dangers of resonium A in the treatment of hyperkalaemia in renal failure, Lancet, 1:167, 1966.

156. Steinmetz, P. R., and J. E. Kiley: Hyperkalemia in renal failure: the effectiveness of treatment depends on the gastrointestinal tract as a locus of exchange, J. Am. Med. Assoc., 175:689, 1961.

157. Rose, B. D.: *The Pathophysiology of Renal Disease,* McGraw-Hill, New York, 1981, chap. 10.

158. Ciabattoni, G., F. Pugliese, G. A. Cinotti, and C. Patrono: Renal effects of anti-inflammatory drugs, Eur. J. Rheumatol. Inflam., 3:210, 1980.

159. Bunning, R. D., and W. F. Barth: Sulindac: a potentially renal-sparing nonsteroidal anti-inflammatory drug, J. Am. Med. Assoc., 248:2864, 1982.

160. Sebastian, A., and M. Schambelan: Amelioration of type 4 renal tubular acidosis in chronic renal failure with furosemide, Kidney Int., 12:534, 1977.

161. Winter, J. S. D.: Current approaches to the treatment of congenital adrenal hyperplasia, J. Pediatr., 97:81, 1980.

162. Schaison, G., B. Couzinet, M. Gourmelen, F. Elkik, and P. Bougneres: Angiotensin and adrenal steroidogenesis: study of 21-hydroxylase-deficient congenital adrenal hyperplasia, J. Clin. Endocrinol. Metab., 51:1390, 1980.

163. Satayaviboon, S., F. Dawgert, P. L. Monteleone, and J. A. Monteleone: Persistent pseudohypoaldosteronism in a 7-year-old boy, Pediatrics, 69:458, 1982.

TWENTY-NINE

ANSWERS TO THE PROBLEMS

CHAPTER 1

1-1 An increase in the plasma Na^+ concentration will stimulate thirst since Na^+ is an effective osmole. In comparison, urea is an ineffective osmole. Thus, an elevation in the BUN will raise the total P_{osm} but not the effective P_{osm}. As a result, thirst will not be affected.

1-2 The plasma Na^+ concentration is the main determinant of the P_{osm} since Na^+ salts are the major extracellular osmoles. There is, however, *no direct relationship* between the plasma Na^+ concentration and the extracellular volume: the latter is determined by the *total* amount of Na^+ and water present whereas the former is determined by the *relative* amount of Na^+ and water. For example, a low plasma Na^+ concentration can be produced by Na^+ loss (volume depletion) or water retention (volume expansion).

1-3 The addition of glucose to the extracellular fluid will initiate the following sequence:

(*a*) Elevate the effective P_{osm}.

(*b*) Increase the extracellular volume since the rise in P_{osm} will pull water out of the cells.

(*c*) Reduce the intracellular volume due to water loss.

(*d*) Lower the plasma Na^+ concentration by dilution. Thus, hyperglycemia results in a dissociation of the usually direct relationship between the plasma Na^+ concentration and P_{osm}.

1-4

$$P_{osm} \cong 2 \times \text{plasma } [Na^+] + \frac{BUN}{2.8} + \frac{[glucose]}{18}$$

$$290 = 2 \times 125 + \frac{28}{2.8} + \frac{[glucose]}{18}$$

$$[\text{Glucose}] = 540 \text{ mg/dL}$$

1-5 Due to the Gibbs-Donnan equilibrium, the concentration of anions in the protein-free glomerulate filtrate is 1.05 times that in the plasma. In addition, each liter of protein-free glomerular filtrate contains almost 1000 mL of water. This is in contrast to each liter of plasma, which contains only 930 mL of water (and therefore less Cl⁻) because of the volume occupied by proteins and, to a lesser degree, fats. As a result of these two factors,

$$\text{Glomerular filtrate } [Cl^-] = \text{plasma } [Cl^-] \times \frac{1.05}{0.93} = 117 \text{ meq/L}$$

1-6 (*a*) An increase in arterial pressure has little effect on filtration, because of autoregulation of capillary hydrostatic pressure by the precapillary sphincter.

(*b*) A decrease in venous pressure reduces the capillary hydrostatic pressure, promoting the movement of interstitial fluid into the capillary.

(*c*) An increase in the plasma albumin concentration also favors the movement of fluid into the capillary.

1-7 The osmolality of a solution is determined by the total number of particles in the solution. Thus, both effective and ineffective osmoles contribute to the osmolality. In contrast, the osmotic pressure is affected only by the concentration of effective osmoles.

1-8 The capillary membrane is freely permeable to Na⁺. In comparison, the cell membrane is effectively impermeable to Na^+. Thus, Na^+ is an ineffective osmole across the capillary wall but an effective osmole across the cell membrane.

CHAPTER 2

2-1 Excretion equals filtration minus reabsorption. This simple relationship is often overlooked clinically. For example, acute renal failure is characterized by a low glomerular filtration rate. If the urine output increases, it is frequently assumed that there has been an increase in filtration and therefore improved renal function. However, a decrease in reabsorption with no change in filtration also may be responsible for the enhanced output. These possibilities can be differentiated by following the plasma creatinine concentration. If filtration has risen, more creatinine will be filtered and excreted, resulting in a reduction in the plasma creatinine concentration (see Chap. 3). In contrast, this parameter will remain stable or continue to rise if there has only been less reabsorption.

CHAPTER 3

3-1 The GFR *cannot* be estimated from the plasma creatinine concentration since the patient is not in a steady state.

3-2 Constricting the left renal artery will do the following:

(*a*) Augment renin secretion from the left kidney, because of decreased renal perfusion pressure.

(*b*) Increase blood pressure, because of increased angiotensin II production.

(*c*) Reduce renin secretion from the normal right kidney since the elevated blood pressure increases the perfusion pressure to this kidney.

The clinical correlate of this question is seen in patients with hypertension due to renal artery stenosis, in whom measurement of renal vein renin levels is used to evaluate the physiologic significance of the stenosis. The criteria for a positive test predicting surgical cure of the hypertension are (1) the renal vein renin concentration from the affected kidney is more than 1.5 times that from the normal kidney and (2) renin secretion by the normal kidney is suppressed as evidenced by the renin concentration in the renal vein being the same as that in the renal artery [1].†

3-3 (*a*)

$$C_{er} = \frac{U_{er} \times V}{P_{er}} \text{ (in mL/min)}$$

$$= \frac{125 \text{ mg/dL}}{3.5 \text{ mg/dL}} \times \frac{800 \text{ mL/day}}{1440 \text{ min/day}}$$

$$= 20 \text{ mL/min}$$

(*b*) The total creatinine excretion is 1000 mg or 12.5 mg/kg. Since a normal male should excrete 20 to 25 mg/kg, this probably represents an incomplete collection and underestimates the GFR.

3-4 Angiotensin II primarily constricts the efferent arteriole. As a result:

(*a*) The renal plasma flow (RPF) will fall due to increased renal vascular resistance.

(*b*) The change in GFR will be variable since the effect of reduced flow is counteracted by the increase in glomerular capillary hydrostatic pressure induced by efferent arteriolar constriction. Thus, the GFR may rise, be unchanged, or fall (although by a lesser degree than the RPF).

(*c*) The filtration fraction, GFR/RPF, will rise.

(*d*) The peritubular capillary albumin concentration will increase since a greater fraction of the RPF has been filtered. This change may contribute to the increased proximal tubular reabsorption seen with hypovolemia (see page 95) [2].

CHAPTER 4

4-1 (*a*) The reabsorption of water will increase the tubular fluid K^+ concentration and therefore the TF/P [K^+] ratio.

(*b*) The C_{K+}/C_{in} ratio represents the percentage of the filtered K^+ remaining in the lumen. Therefore, the reabsorption of water will not affect the C_{K+}/C_{in}. Although the tubular fluid K^+ concentration will be increased, there will be a similar elevation in the tubular fluid inulin concentration.

†Reference citations refer to references found at the end of this chapter.

4-2 The TF/P [K^+] ratio can be increased by K^+ secretion or water reabsorption. Since the C_{K+}/C_{in} rises in the cortical distal nephron (primarily in the initial cortical collecting tubule), K^+ secretion must be occurring. In contrast, the C_{K+}/C_{in} does not change in the late cortical and medullary collecting tubules since the ratio in the end distal sample is the same as that in the ureteral urine. Thus, water reabsorption (not K^+ secretion) is responsible for the elevation in the TF/P [K^+] ratio in these segments.

CHAPTER 5

5-1 Because of glomerulotubular balance, an increase in the GFR will do the following:

(*a*) Have no effect on fractional proximal Na^+ reabsorption.

(*b*) Increase *absolute* proximal Na^+ reabsorption since a constant fraction of a larger quantity is reabsorbed.

(*c*) Increase absolute proximal HCO_3^- reabsorption.

These responses prevent the loss of Na^+ and HCO_3^- in the urine.

5-2 The peritubular capillary oncotic pressure is determined by (1) the plasma protein concentration and (2) the filtration fraction. A reduction in the peritubular capillary oncotic pressure reduces capillary uptake and therefore net proximal Na^+ reabsorption.

5-3 (*a*) In response to volume depletion, fractional proximal Na^+ reabsorption is increased. The enhanced removal of water raises the tubular fluid urea concentration, resulting in more urea reabsorption, a fall in urea excretion, and an elevation in the BUN.

(*b*) The GFR is probably unchanged since the plasma creatinine concentration has remained constant.

(*c*) Both increased uric acid reabsorption (due to enhanced Na^+ reabsorption) and reduced secretion (due to the ketoacids) results in a fall in excretion, leading to hyperuricemia.

5-4 Proximal NaCl reabsorption should fall since the reduction in bicarbonate transport decreases the generation of concentration and osmotic gradients that promote *passive* NaCl reabsorption.

CHAPTER 6

6-1 NaCl reabsorption without water in the medullary ascending limb, and urea movement from the medullary collecting tubule into the interstitium contribute to the production of medullary hyperosmolality. The hairpin configuration of the vasa recta plays an important role in its maintenance by minimizing the removal of the excess solute.

6-2 NaCl reabsorption without water in the medullary ascending limb has two effects: (1) the medullary interstitium becomes hyperosmotic and (2) the tubular fluid becomes dilute. Thus, this step plays an important role in both concentration and dilution.

NaCl reabsorption without water in the cortical ascending limb and distal tubule further lowers the urine osmolality and contributes to urinary dilution. Since cortical blood flow is so high, however, the cortical interstitium does not become hyperosmotic and urinary concentration is not affected.

6-3 From the answer to Prob. 6-2, a diuretic which inhibits both concentration and dilution acts in the medullary ascending limb. In comparison, a diuretic which impairs dilution but not concentration probably acts in the cortical ascending limb or distal tubule. Thus, the effects of diuretics on concentration and dilution can be used to determine their sites of action within the nephron (see Chap. 16).

6-4 The tubular fluid leaving the proximal tubule is isosmotic to plasma. Water reabsorption in the descending limb occurs down an osmotic gradient between the tubular fluid and the hyperosmotic interstitium. Thus, any factor which decreases interstitial tonicity will reduce descending limb water reabsorption. This occurs with an osmotic diuresis because the associated increase in medullary blood flow washes out some of the medullary solute. A similar effect should occur in central diabetes insipidus where the absence of ADH leads to diminished medullary accumulation of urea (Fig. 6-7).

6-5 On a low-protein diet, less urea is excreted. Thus, less urea enters the medullary interstitium, and interstitial tonicity is reduced. The result is a mild decrease in concentrating ability [3].

CHAPTER 8

8-1 Renin secretion is reduced by an autonomous adrenal adenoma, probably as a result of the volume expansion produced by Na^+ and water retention. In contrast, renin secretion is increased by volume depletion, and the enhanced formation of angiotensin II is responsible for the increase in aldosterone secretion.

8-2 The secretion of ADH is affected by changes in the effective P_{osm}. Since urea is an ineffective osmole, an elevation in the BUN does not increase the effective P_{osm}.

8-3 As water is lost in the urine, the effective P_{osm} increases, resulting in a potent stimulus to thirst. Although patients who lack ADH may excrete more than 10 liters of urine per day, they remain in near-normal water balance because water intake is increased via the thirst mechanism.

8-4 With licorice, as with the administration of aldosterone, there is a transient period of Na^+ and water retention followed by a spontaneous diuresis. This is the phenomenon of aldosterone escape.

8-5 A variety of factors contribute to the development of hypocalcemia in renal failure, including phosphate retention, decreased 1,25-D formation, and resistance to the effects of both PTH and 1,25-D. The administration of vitamin D (or a metabolite such as 1,25-D) can correct the hypocalcemia but may also raise the plasma phosphate concentration. The resultant increase in the calcium phosphate product can cause metastatic calcification due to the precipitation of calcium phosphate. Thus, vitamin D should not be used when hyperphosphatemia is present. The initial aim of therapy should be to lower the plasma phosphate concentration with a phosphate-binding antacid such as aluminum hydroxide. In many patients, reducing the plasma phosphate concentration results in normalization of the plasma Ca^{2+} concentration. If hypocalcemia persists after the plasma phosphate concentration has returned to normal levels, vitamin D may then be added.

8-6 Hypercalciuria may transiently lower the plasma Ca^{2+} concentration. However, this stimulates the secretion of PTH, which then returns the plasma Ca^{2+} concentration to normal by increasing both bone resorption and, via enhanced 1,25-D formation, intestinal Ca^{2+} absorption. This example illustrates the importance of PTH in the regulation of calcium balance.

8-7 ADH enhances renal water reabsorption. The result is the retention of ingested water and volume expansion. However, this is not always a beneficial response since the retained water can lead to symptomatic hyponatremia and hypoosmolality (see Chap. 23). ADH has no physiologically important effect on renal Na^+ reabsorption.

8-8 Renal prostaglandin production is increased when there are high levels of vasoconstrictors such as angiotensin II and norepinephrine. This is most likely to occur with effective volume depletion, as with a low-salt diet, heart failure, or severe vomiting (*a*, *d*, and *e*). In these settings, a nonsteroidal anti-inflammatory drug can lead to renal ischemia and a fall in the GFR.

CHAPTER 9

9-1 (*a*) If the cardiac output is acutely reduced, there will be no change in the plasma or extracellular volumes (assuming pulmonary edema does not develop). However, the effective circulating volume and urinary Na^+ excretion will be decreased.

(*b*) A high-sodium diet will increase the plasma, extracellular, and effective circulating volumes and Na^+ excretion.

(*c*) The retention of water will also increase the plasma, extracellular, and effective circulating volumes and Na^+ excretion.

(*d*) Hypoalbuminemia favors the movement of fluid from the vascular space into the interstitium. Acutely, this redistribution of fluid will reduce the plasma volume without affecting the total extracellular volume. The fall in the plasma volume will lower both the effective circulating volume and urinary Na^+ excretion. With time, ingested Na^+ will be retained (because of the decrease in Na^+ excretion), resulting in an increase in the total extracellular volume and edema.

9-2 Diuretics increase the secretion of renin. Since this results in the formation of the vasoconstrictor angiotensin II, the fall in blood pressure will be minimized. Thus, agents such as a β-adrenergic blocker (which inhibits renin secretion) or captopril (a converting enzyme inhibitor which impairs the generation of angiotensin II) can potentiate the hypotensive action of diuretics.

9-3 The rate of urinary Na^+ excretion is generally the best measure of the effective circulating volume since it reflects the physiologic assessment by the kidney of systemic hemodynamics. A low rate of Na^+ excretion is diagnostic of effective hypovolemia unless there is selective renal or glomerular hypoperfusion, as with bilateral renal artery stenosis or acute glomerulonephritis. The cardiac output, plasma volume, and systemic blood pressure are less accurate. For example, the blood pressure tends to fall with hypovolemia, but this finding may be masked by the compensatory increase in sympathetic tone. On the other hand, the cardiac output may be misleadingly elevated if there are arteriovenous fistulas, as occurs in hepatic cirrhosis.

9-4 If the patient is stable, he or she is in a steady state and Na^+ intake and urinary excretion will be roughly equal. Although diuretics initially increase Na^+ excretion above intake, the ensuing volume loss will activate counteracting mechanisms (such as hyperaldosteronism) that result in increased Na^+ reabsorption at sites in the nephron not affected by the diuretics. This new steady state, in which the extracellular volume is reduced due to fluid loss during the period when excretion exceeded intake, is usually established within 7 days [4].

CHAPTER 10

10-1 The loss of isosmotic diarrheal fluid will have the following results:

(*a*) Reduce the effective circulating volume.

(*b*) Lower urinary Na^+ excretion because of the decrease in the effective circulating volume.

(*c*) Have no effect of the plasma osmolality since the loss of isosmotic fluid does not alter the osmolality of the body fluids.

(*d*) Have no effect on the plasma Na^+ concentration. This is a frequently misunderstood concept. Suppose diarrheal fluid has a Na^+ concentration of 90 meq/L and a K^+ concentration of 60 meq/L. It is tempting to say that the loss of fluid with a Na^+ concentration less than that of the plasma (and extracellular fluid) should lead to an increase in the plasma Na^+ concentration. This would be true if the diarrheal fluid were derived only from the extracellular fluid. However, since the plasma K^+ concentration is only 4 meq/L, most of the K^+ that is lost must come from the cells which have a high K^+ concentration. Each liter of diarrheal fluid with the above Na^+ and K^+ concentrations can be viewed as being composed of 600 mL of an isosmotic Na^+ solution (Na^+ concentration equals 150 meq/L) and 400 mL of an isosmotic K^+ solution (K^+ concentration equals 150 meq/L). The Na^+ solution represents extracellular fluid loss, and the K^+ solution represents intracellular fluid loss. Since the fluid lost from the extracellular fluid has a Na^+ concentration similar to that of the plasma, there will be no change in the plasma Na^+ concentration.

(e) Increase ADH secretion because of volume depletion.

(f) Increase the urine osmolality due to the effect of ADH.

(g) Increase thirst because of volume depletion.

If such a patient ingested water, it would be retained because of the increased ADH release, resulting in hyponatremia and hypoosmolality. Thus, hyponatremia in patients with diarrhea is not due directly to the loss of diarrheal fluid but to the subsequent retention of ingested water (see Chap. 23).

10-2 As described in Chap. 9, urinary Na^+ excretion is determined primarily by the effective circulating volume. However, the plasma Na^+ concentration is a measure of concentration, not volume (see the answer to Prob. 1-2) and therefore is not necessarily related to urinary Na^+ excretion. For example, the above patient with diarrhea and hyponatremia is hypovolemic and has a low urine Na^+ concentration. In contrast, hyponatremia due to water retention is associated with volume expansion and a high urine Na^+ concentration. Thus, measurement of the urine Na^+ concentration can be useful in the differential diagnosis of hyponatremia (see Chap. 23).

10-3 Water excretion is impaired by volume depletion via two mechanisms: increased ADH secretion; and reduced fluid delivery to the diluting segment in the ascending limb of the loop of Henle due to enhanced proximal Na^+ and water reabsorption.

10-4

$$T^C_{H_2O} = \frac{1040 \times 0.5}{286} - 0.5 = 1.3 \text{ mL/min}$$

$$C_{H_2O} = 10.4 - \frac{50 \times 10.4}{282} = 8.6 \text{ mL/min}$$

After a water load, there is a reduction in the P_{osm}, which suppresses ADH release. As a result, the U_{osm} falls and the urine volume increases. The maximum diuresis is seen after 90 to 120 min, the time required for the metabolism of the previously circulating ADH.

10-5 The minimum U_{osm} is unaffected by the change in diet. However, the maximum C_{H_2O} varies directly with the rate of solute excretion, which will fall dramatically with the switch to beer drinking. If, for example, the minimum U_{osm} is 75 mosmol/kg, then the maximum urine volume will be 10 L/day if 750 mosmol of solute is excreted (regular diet) but only 4 L/day if 300 mosmol of solute is excreted (beer drinking). If the P_{osm} is 280 mosmol/kg, the respective free water clearances will be:

$$C_{H_2O} = 10 - \frac{75 \times 10}{280} = 7.3 \text{ L/day} \qquad \text{(regular diet)}$$

$$C_{H_2O} = 4 - \frac{75 \times 4}{280} = 2.9 \text{ L/day} \qquad \text{(beer drinking)}$$

The combination of a reduced ability to excrete free water and increased fluid intake makes the beer drinker susceptible to water retention and hyponatremia [5].

CHAPTER 11

11-1 Buffers minimize changes in the free H^+ concentration by appropriately taking up ($H^+ + Buf^- \rightarrow HBuf$) or releasing ($HBuf \rightarrow H^+ + Buf^-$) H^+ ions. The efficacy of a buffer is determined by the quantity of the buffer present and the relation of the pK_a of the buffer to the pH of the solution. In addition, the ability to excrete CO_2 increases the effectiveness of the $HCO_3^- - CO_2$ buffer system.

11-2 The fall in the plasma HCO_3^- concentration is due to the different rates with which the administered HCO_3^- enters the different fluid compartments. Initially, the added HCO_3^- is limited to the vascular space, resulting in a large increase in the plasma HCO_3^- concentration. Within 15 min, the HCO_3^- equilibrates throughout the total extracellular fluid, reducing the plasma HCO_3^- concentration. The entry of HCO_3^- into cells and the release of H^+ ions from the intracellular buffers require 2 to 4 h to reach completion and produce a further decrease in the plasma HCO_3^- concentration. As discussed in Chap. 12, acid-base balance is restored by the excretion of the excess HCO_3^- in the urine.

The time-related effect of exogenous HCO_3^- on the plasma HCO_3^- concentration is clinically important when HCO_3^- is administered to treat metabolic acidosis (see Chap. 19). The increase in the plasma HCO_3^- concentration and therefore in the arterial pH will be greater if measured within 15 to 30 min than after 2 to 4 h when equilibrium with the intracellular buffers has been reached. Thus, it should not be assumed that metabolic acidosis has been corrected on the basis of measurements made shortly after the administration of HCO_3^-.

11-3 The quantity of available extracellular and intracellular buffers will determine how much of a reduction in pH will occur. Buffering capacity is best estimated from the initial plasma HCO_3 concentration. Patients with a low baseline level due to an underlying metabolic acidosis are more prone to having a major reduction in pH following an acid load.

CHAPTER 12

12-1 The primary response of the kidney to a H^+ load is increased NH_3 production, allowing increased H^+ excretion as NH_4^+. This response is relatively slow, occurring over several days. To minimize the reduction in the pH prior to this renal response, alveolar ventilation is increased, resulting in a fall in the P_{CO_2} and an increase in the arterial pH toward normal.

12-2 (*a*) Reduced titratable acid excretion has little effect on acid-base balance since NH_3 production and therefore NH_4^+ excretion can increase to compensate for the fall in titratable acid excretion.

(*b*) In contrast, the ability to augment titratable acid excretion is limited. As a result, a decrease in NH_4^+ excretion can lead to a reduction in net H^+ excretion and metabolic acidosis. This is the primary mechanism responsible for the metabolic acidosis in patients with renal failure (see Chap. 19).

12-3 The buffering of HCl and H_2SO_4 by $NaHCO_3$ results in the formation of NaCl and Na_2SO_4, respectively. When NaCl is presented to the distal nephron, the reabsorption of Na^+ will be followed by the reabsorption of Cl^-. With Na_2SO_4, on the other hand, the reabsorption of Na^+ will be accompanied by enhanced H^+ and K^+ secretion since SO_4^{2-} is poorly reabsorbed. Because of the increase in H^+ excretion, there will be a lesser elevation in the arterial H^+ concentration after H_2SO_4 than after HCl [6].

12-4 At a pH of 5.80 with 60 mmol of phosphate:

$$5.80 = 6.80 + \log \frac{x}{60 - x}$$

$$[HPO_4^{2-}] = x = 5.45 \text{ mmol}$$

$$[H_2PO_4^-] = 60 - x = 54.55 \text{ mmol}$$

In the filtrate, however, the initial pH was 7.40, similar to that in the plasma:

$$7.40 = 6.80 + \log \frac{x}{60 - x}$$

$$[HPO_4^{2-}] = 48 \text{ mmol}$$

$$[H_2PO_4^-] = 12 \text{ mmol}$$

Thus, *42.55 mmol* of HPO_4^{2-} (48 − 5.45) has been converted to $H_2PO_4^-$ by buffering—this is the quantity of titratable acidity excreted as $H_2PO_4^-$.

Titratable acidity is measured by the number of milliequivalents of NaOH that must be added to the urine to return the pH to 7.40. Ammonium excretion is not included in this measurement since the pK_a of the NH_3/NH_4^+ system is 9.3. Thus, raising the pH of the urine from 5.80 to 7.40 will have little effect on the NH_3/NH_4^+ ratio.

12-5 A low P_{CO_2} has the following results:
   (*a*) Elevation of the arterial pH.
   (*b*) Decrease in HCO_3^- reabsorption.
   (*c*) Decrease in plasma HCO_3^- concentration, which will appropriately reduce the arterial pH toward normal.

12-6 The loss of gastric secretions results in volume depletion as well as metabolic alkalosis. The former enhances HCO_3^- reabsorption, thereby preventing correction of the alkalosis.

CHAPTER 13

13-1 Distal K^+ secretion varies inversely with the tubular fluid K^+ concentration. The degree to which the tubular fluid K^+ concentration is increased after K^+ secretion is determined by the distal flow rate. For example, increased flow washes away the secreted K^+, maintaining a favorable concentration gradient for further K^+ secretion.

13-2 Increased Na^+ intake increases distal flow, resulting in enhanced K^+ secretion and hypokalemia in patients with primary hyperaldosteronism [7]. This does not happen in normal subjects, in whom a high-sodium diet reduces aldosterone secretion (see Fig. 8-12) [7].

13-3 In the presence of K^+ depletion due to extrarenal causes, urinary K^+ excretion usually is reduced to less than 25 meq/day. In comparison, K^+ excretion exceeds this level with renal K^+ wasting (see Chap. 27).

13-4 Spontaneous K^+ wasting and hypokalemia do not typically occur with effective hypovolemia because the effect of hyperaldosteronism is counteracted by the reduction in distal flow that results from enhanced proximal reabsorption. If, however, distal flow is augmented by furosemide (which inhibits NaCl and water reabsorption in the loop of Henle), urinary K^+ losses will increase and hypokalemia may occur.

13-5 K^+ excretion is not significantly affected by the absence of ADH. Although the urine output is markedly increased, this is due to decreased collecting tubule water reabsorption which occurs distal to the K^+ secretory site.

CHAPTER 15

15-1 The shock state is probably due to the sequestration of fluid in the infarcted bowel. Fluid replacement should proceed primarily by using isotonic saline. Blood is probably not necessary since the hematocrit of 53 percent suggests hemoconcentration due to the loss of extracellular fluid from the vascular space.

When faced with a polyuric patient, one must determine whether the high urine output is *appropriate or inappropriate*. In this patient who had a *positive* balance of 7 liters prior to surgery, it is more likely that the polyuria is an appropriate response to volume expansion than an inappropriate response, as might occur during the diuretic phase of acute tubular necrosis. Thus, the correct therapy would be to allow the patient to develop negative fluid balance while being carefully monitored. If the diuresis is appropriate, the urine output will eventually fall without the patient developing signs of volume depletion, such as poor skin turgor or hypotension. If the latter develop, then the fluid loss is inappropriate and replacement therapy should be started. If the urinary losses are replaced when the patient is polyuric and volume-expanded, the hypervolemia and therefore the high urine output will persist indefinitely.

15-2 For each liter of water lost, approximately 60 percent comes from the cells and 40 percent from the extracellular fluid. Although the water is initially lost from the extracellular fluid, the ensuing elevation in the plasma osmolality pulls a proportionate volume of water out of the cells. In contrast, each liter of isotonic Na^+ loss comes entirely from the extracellular fluid since there is no change in osmolality to affect the intracellular fluid. Consequently, there is a greater reduction in extracellular volume and blood pressure with Na^+ and water loss than with water loss alone.

15-3 Pure dextrose solutions have no role in the treatment of hypovolemic shock since only 40 percent of the fluid will remain in the extracellular space. In addition, the retention of free water can cause symptomatic hyponatremia. Even in patients who are hypernatremic, isotonic saline (which is hypoosmotic to plasma in this setting) should be infused until adequate tissue perfusion is restored.

15-4 (*a*) Volume expansion is responsible for the increases in urine Na^+ concentration and urine volume.

(*b*) Since there is a relatively wide range of normal values, the central venous pressure is only an estimate of the volume status of the patient. For example, 3 cmH_2O may be normal in some subjects and low in others. In contrast, in a patient who is able to conserve Na^+ normally, as this patient was on admission, an increase in the urine Na^+ concentration to 75 meq/L suggests that the kidney is sensing an adequate effective circulating volume.

(*c*) The BUN elevation on admission reflects urea accumulation over 10 days due to reduced urea excretion. Although normovolemia was restored over 18 h, a longer period is necessary for the renal excretion of the excess urea.

15-5 This patient is depleted of Na^+ and water (physical findings) and K^+ (low plasma K^+ concentration). The elevation in the plasma Na^+ concentration indicates that water has been lost in excess of solute. Thus, the replacement fluid should be hypotonic and contain Na^+ and K^+, for example, quarter- or half-isotonic saline plus 20 to 40 meq of K^+ per liter. This solution can be given at 50 to 100 mL/h to slowly and safely rehydrate the patient.

15-6 (*a*) Any definition of hypotension must be made in relation to the patient's baseline blood pressure. Although 110/70 appears normal, it is probably low in this patient because of the history of hypertension.

(*b*) If a patient becomes volume-depleted from insensible losses, the plasma Na^+ concentration must rise because relatively solute-free water has been lost. The normal plasma Na^+ concentration of 138 meq/L in this patient indicates that *Na*$^+$ *and water must have been lost in proportion*. Therefore, a source of Na^+ loss must be present. Diuretic-induced losses are likely in this setting considering the history of hypertension (which is typically treated with diuretics) and the presence of hypokalemia and probably metabolic alkalosis.

CHAPTER 16

16-1 Diuretics may lower the effective circulating volume in heart failure, nephrotic syndrome, and the adult respiratory distress syndrome, disorders in which effective hypovolemia may be present prior to therapy. In contrast, renal failure and hepatic cirrhosis with overflow are associated with primary volume expansion. Fluid removal in this setting will reduce the effective volume toward but not (unless excessive) below normal.

The simplest way to detect a reduction in the effective circulating volume in these conditions is to follow the BUN and plasma creatinine concentration: stable values indicate perfusion is being maintained, whereas increasing values are usually indicative of decreased renal perfusion secondary to effective volume depletion.

16-2 The oliguria and increase in BUN are due to effective circulating volume depletion which may be due both to the diuretics and to a primary decrease in cardiac output following the myocardial infarction. Since the patient was previously well (and presumably normovolemic), the infarct resulted in an acute redistribution of some of the plasma into the pulmonary interstitium and alveoli. Thus, the administration of diuretics will reduce the total extracellular volume below normal while having the beneficial effect of relieving the pulmonary edema.

Therapy should be aimed at increasing the effective circulating volume. If there is no evidence of pulmonary congestion, carefully liberalizing Na^+ intake can be tried. However, if pulmonary congestion persists or develops during Na^+ repletion, therapy must be aimed at improving the cardiac output by using digitalis or vasodilators.

16-3 The metabolic alkalosis is due to volume contraction and enhanced renal H^+ excretion, both of which are produced by the diuretics. The hyperuricemia is also diuretic-induced as the ensuing volume depletion increases proximal Na^+ reabsorption. Since proximal urate transport is indirectly linked to that of Na^+ (see Chap. 5), urate reabsorption is also enhanced in this setting.

These abnormalities can be reversed by expanding the effective circulating volume. This may be difficult to achieve, however, since the patient cannot be given Na^+ (because pulmonary congestion persists) and is already on digitalis. In addition to increasing the cardiac output with afterload reduction, both the alkalosis and the edema can be treated by inducing a $NaHCO_3$ diuresis with acetazolamide. The elevation in the plasma urate concentration probably does not require specific therapy unless gouty arthritis develops. Uric acid nephropathy due to uric acid precipitation in the distal nephron is not likely to occur in this setting since the hyperuricemia is due to a *decrease* in uric acid excretion.

CHAPTER 17

17-1 (*a*) The H^+ concentration at a pH of 7.60 is 26 nanoeq/L ($40 \times 0.8 \times 0.8$).

(*b*) The H^+ concentration at a pH of 7.20 is 63 nanoeq/L ($40 \times 1.25 \times 1.25$). The H^+ concentration at a pH of 7.10 is 80 meq/L (63×1.25). Thus, the H^+ concentration at a pH of 7.15 is 72 nanoeq/L [$63 + 0.5 \times (80 - 63)$].

(*c*) The H^+ concentration at a pH of 7.30 is 50 nanoeq/L (40×1.25). Thus, the H^+ concentration at a pH of 7.24 is 59 nanoeq/L [$50 + 0.6 \times (63 - 50)$].

17-2 (*a*) Metabolic acidosis (low pH, low HCO_3^- concentration, compensatory reduction in P_{CO_2}).

(*b*) Respiratory alkalosis (high pH, low P_{CO_2}, compensatory reduction in HCO_3^- concentration). Note that a low HCO_3^- concentration does not necessarily imply a metabolic acidosis.

(*c*) Combined respiratory and metabolic acidosis (low pH, high P_{CO_2}, low HCO_3^- concentration).

(*d*) Metabolic alkalosis (high pH, high HCO_3^- concentration, compensatory increase in P_{CO_2}).

17-3 The acid-base disorder is metabolic acidosis. Assuming the P_{CO_2} remains constant, the plasma HCO_3^- concentration must be raised to 5.0 meq/L to increase the pH to 7.20 (H^+ concentration equals 63 nanoeq/L):

$$63 = 24 \times \frac{13}{[HCO_3^-]}$$

$$[HCO_3^-] = 5.0 \text{ meq/L}$$

If the P_{CO_2} increased to 18 mmHg with correction of the acidemia, the plasma HCO_3^- concentration would have to be raised to 6.9 meq/L:

$$63 = 24 \times \frac{18}{[HCO_3^-]}$$

$$[HCO_3^-] = 6.9 \text{ meq/L}$$

These examples illustrate that, in patients who are able to hyperventilate in response to metabolic acidosis, only a small increase in the plasma HCO_3^- concentration is usually required to get the patient out of danger.

CHAPTER 18

18-1 The acute metabolic alkalosis is due to the citrate load given with the blood transfusions. Since cirrhosis with ascites is associated with effective circulating volume depletion, the urine Na^+ concentration should be less than 10 meq/L and the urine pH acid. In view of the ascites, volume expansion with saline is not indicated. In this setting, acetazolamide can be used to induce a $NaHCO_3$ diuresis.

18-2 The physical findings and low urine Na^+ concentration point to volume depletion. The patient should be treated with both NaCl and KCl, for example, half-isotonic saline to which 40 meq of K^+ (as KCl) has been added to each liter.

During therapy, volume expansion reduces Na^+ reabsorption, thereby allowing correction of the alkalosis by excretion of the excess HCO_3^-. Thus, the anion gap between the high urine Na^+ concentration and the low urine Cl^- concentration is primarily due to HCO_3^-. The degree of HCO_3^- wasting seen with an alkaline diuresis is often not appreciated. If the urine pH is 7.8 (H^+ concentration equals 16 nanoeq/L) and the urine P_{CO_2} is 46 mmHg (similar to the renal venous P_{CO_2}), then

$$16 = 24 \times \frac{46}{[HCO_3^-]}$$

$$[HCO_3^-] = 69 \text{ meq/L}$$

Since the urine Cl^- concentration remains low, Cl^- depletion persists and further Cl^- replacement is necessary. This illustrates the advantage in measuring the urine Cl^- concentration in patients with metabolic alkalosis since the high urine Na^+ concentration could be interpreted as showing adequate NaCl replacement.

CHAPTER 19

19-1 (*a*) No. The arterial pH has not been measured, so it is not certain that the patient has metabolic acidosis. The low plasma HCO_3^- concentration could also be due to chronic respiratory alkalosis.

(*b*) Yes. Although the arterial pH is only slightly reduced, this is maintained by marked hyperventilation (P_{CO_2} equals 14 mmHg), which is probably symptomatic. The administration of $NaHCO_3$ will correct the acidemia, thereby decreasing the stimulus to ventilation.

19-2 (*a*) This is a combined respiratory acidosis (elevated P_{CO_2}) and high anion gap metabolic acidosis. The latter is probably due to seizure-induced lactic acidosis.

(*b*) $NaHCO_3$ is not required since the cessation of seizure activity will allow lactate to be metabolized back to bicarbonate.

(*c*) The plasma K^+ concentration will be unchanged since neither lactic acidosis nor its spontaneous correction affects the internal distribution of K^+. If this had been a nonorganic acidosis, then the administration of $NaHCO_3$ would lead to a reduction in the plasma K^+ concentration.

19-3 (*a*) 3. Type 1 RTA in adults is characterized by a progressive decline in the plasma HCO_3^- concentration, which occurs slowly as some of the dietary H^+ load is retained each day.

(*b*) 2. Type 1 RTA in children may be associated with a higher urine pH and, therefore, fixed HCO_3^- wasting. As a result, the plasma HCO_3^- concentration will fall rapidly and progressively.

(*c*) 1. The plasma HCO_3^- concentration falls rapidly in type 2 RTA and then stabilizes at 12 to 20 meq/L once the new level of bicarbonate reabsorptive capacity has been reached.

At a normal plasma HCO_3^- concentration, these disorders can be distinguished by calculating the fractional excretion of HCO_3^-: less than 3 percent in adult type 1 RTA, 5 to 10 percent in childhood type 1 RTA with HCO_3^- wasting, and greater than 15 percent in type 2 RTA (see Fig. 19-4).

19-4 (*a*) This patient has a combined metabolic and respiratory acidosis. At a plasma HCO_3^- concentration of 9 meq/L (15 meq/L less than normal), the respiratory compensation should lower the P_{CO_2} to approximately 22 mmHg (15 × 1.2 = 18 mmHg less than normal). Thus, a P_{CO_2} of 40 mmHg is inappropriately high.

(*b*) HCO_3^- deficit = 0.7 × 80 × (24 − 9) = 840 meq.

(*c*) The H^+ concentration at a pH of 7.20 is 63 nanoeq/L (40 × 1.25 × 1.25). Thus

$$63 = 24 \times \frac{40}{[HCO_3^-]}$$

$$[HCO_3^-] = 15.2 \text{ meq/L}$$

To raise the plasma HCO_3^- concentration from 9 to 15 meq/L would require approximately 336 meq of HCO_3^- [0.7 × 80 × (15 − 9)].

(d) The formula for the calculation of the HCO_3^- deficit is dependent upon a steady state. Since this patient is losing 1 liter of diarrheal fluid per hour, there is a continuing HCO_3^- loss that has not been replaced.

(e) A normal plasma K^+ concentration at a pH of 6.97 usually indicates K^+ depletion which is masked because acidemia results in the movement of K^+ out of the cells into the extracellular fluid. Thus, the plasma K^+ concentration is inappropriately high in relation to the reduced total body K^+ stores.

19-5 (a) The underlying disorder is chronic respiratory alkalosis, a common finding in patients with cirrhosis (see Chap. 21). The low plasma HCO_3^- concentration represents the renal compensation to a low P_{CO_2}.

(b) The reduction in the plasma HCO_3^- concentration is due to the induction of hypoaldosteronism by spironolactone [8]. The concurrent increase in the plasma K^+ concentration is more than would be expected from the fall in pH and is also due to the decreased aldosterone effect induced by spironolactone.

(c) Since type 1 (distal) RTA is defined as an inability to reduce the urine pH below 5.3, the urine pH of 5.0 rules out this disorder. In addition, type 1 RTA is a chronic condition which would not be expected to develop acutely in the hospital.

19-6 The metabolic acidosis of renal failure results from retention of some of the daily H^+ load because of decreased NH_3 production and NH_4^+ excretion. In some patients, HCO_3^- wasting also contributes to the acidosis.

The second arterial pH was drawn only 30 min after the administration of HCO_3^-, before equilibration of the HCO_3^- with the intracellular buffers had occurred. Thus, the reductions in the plasma HCO_3^- concentration and arterial pH on the next day probably reflected the effect of the intracellular buffers. In addition, continued retention of acid may have contributed to the fall in pH.

CHAPTER 20

20-1 It is easiest to answer this problem by first determining the acid-base disorders represented by the three sets of blood values.

(a) The low pH and high P_{CO_2} indicate respiratory acidosis. In chronic respiratory acidosis, a P_{CO_2} of 65 mmHg (25 mmHg greater than normal) should be associated with a plasma HCO_3^- concentration of approximately 33 meq/L (3.5-meq/L increase in the plasma HCO_3^- concentration for each 10-mmHg elevation in the P_{CO_2}). Thus, the HCO_3^- concentration of 37 meq/L represents a superimposed metabolic alkalosis.

(b) At a P_{CO_2} of 60 mmHg, the plasma HCO_3^- concentration should be roughly 26 meq/L in acute respiratory acidosis (1-meq/L increase per 10-mmHg elevation in the P_{CO_2}) and 31 meq/L in chronic respiratory acidosis. Therefore, these values may represent acute respiratory acidosis or metabolic acidosis (lowering the HCO_3^- concentration from 31 to 26 meq/L) superimposed upon chronic respiratory acidosis.

(c) Chronic respiratory acidosis *or* metabolic alkalosis (raising the HCO_3^- concentration from 26 to 32 meq/L) superimposed upon acute respiratory acidosis.

From the history:

1. Chronic bronchitis plus diarrhea suggests combined chronic respiratory acidosis and metabolic acidosis, or (*b*).
2. Marked obesity suggests chronic hypercapnia, or (*c*).
3. Severe acute asthma suggests acute respiratory acidosis, or (*b*).
4. Chronic bronchitis plus diuretics suggests chronic hypercapnia with superimposed metabolic alkalosis, or (*a*).

20-2 (*a*) The probable acid-base disturbance on admission was acute, superimposed upon chronic, respiratory acidosis. This conclusion is based upon the history since other acid-base disorders may also be associated with the same values (see Fig. 20-4).

(*b*) Patients with chronic hypercapnia rely on the hypoxemic drive to ventilation. This is removed by the administration of oxygen, resulting in a decrease in alveolar ventilation and an increase in the P_{CO_2}.

(*c*) The patient cannot tolerate the administration of oxygen *nor* can he tolerate the P_{O_2} of 30 mmHg in room air. Thus, some form of mechanical ventilation and probably endotracheal intubation are required.

(*d*) Rapidly reducing the P_{CO_2} will produce metabolic alkalosis since the plasma HCO_3^- concentration will remain elevated. In this volume-depleted patient (urine Na^+ concentration equals 4 meq/L), HCO_3^- reabsorption is increased (see Chap. 18). Thus, the excess HCO_3^- will not be excreted in the urine until normovolemia is restored with a high-sodium diet.

20-3 (*a*) Metabolic alkalosis, with the elevated P_{CO_2} representing the appropriate respiratory compensation.

(*b*) The (A-a) O_2 gradient is 13 mmHg, making underlying lung disease and chronic hypercapnia unlikely.

CHAPTER 22

22-1 $P_{osm} = 2 \times 125 + (108/18) + (140/2.8) = 306$ mosmol/kg. However, urea is an ineffective osmole. The effective P_{osm} is actually reduced to 256 mosmol/kg because of the fall in the plasma Na^+ concentration. Thus, the patient may have symptoms of hypoosmolality, not hyperosmolality.

22-2 If isotonic fluids are given to a patient who can excrete only an isosmotic urine, the insensible free-water losses from the skin and respiratory tract will not be replaced, resulting in increases in the plasma osmolality and the plasma Na^+ concentration

Since half-isotonic saline plus 77 meq/L of K^+ per liter is also an isotonic fluid, it will have the *same effect* on the P_{osm} and the plasma Na^+ concentration as the administration of isotonic saline. At first glance, it might seem that the addition of fluid to the extracellular space with a Na^+ concentration less than that of the plasma (and extracellular fluid) should lower the plasma Na^+ concentration. However, not all of this fluid remains in the extracellular space. This can be appreciated if each liter of the added fluid is viewed as having two components: 500 mL of isotonic NaCl and 500 mL of isotonic KCl. The NaCl component stays in the extracellular fluid (and therefore does not lower the plasma Na^+ concentration), but most of the KCl component must either enter the cells or be excreted in the urine. If not, fatal hyperkalemia would ensue. Thus, the effect of half-isotonic saline plus K^+ on the extracellular Na^+ concentration is similar to that of isotonic saline. (See also the answer to Prob. 10-1*d*). In contrast, the administration of half-isotonic saline alone tends to lower the P_{osm} and the plasma Na^+ concentration since it is a hypotonic fluid. This concept is clinically important because the osmotic contribution of K^+ in intravenous solutions is frequently ignored.

CHAPTER 23

23-1 All these factors contributed to the reduction in the plasma Na^+ concentration. Hydrochlorothiazide-induced volume depletion (physical findings plus urine Na^+ concentration of 4 meq/L), which stimulated ADH secretion (U_{osm} equals 540 mosmol/kg), resulting in the retention of ingested water and hyponatremia. The loss of K^+ resulted in Na^+ movement into the cells, further reducing the plasma Na^+ concentration.

Therapy should include the administration of Na^+ and K^+ in a hypertonic solution. KCl added to isotonic saline would be an appropriate solution. There is little justification for water restriction since the patient is volume-depleted. Half-isotonic saline should not be used because it is hypotonic. In view of the metabolic alkalosis, potassium chloride, not potassium citrate (citrate being converted to HCO_3^- in the body), should be used for K^+ replacement.

23-2 The hyponatremia is probably due to volume depletion, perhaps the result of diuretics used in the treatment of hypertension. Consistent with this are the physical findings, low plasma K^+ concentration, and high plasma HCO_3^- concentration. The inappropriately high urine Na^+ concentration is probably due to the osmotic diuretic action of mannitol.

Pseudohyponatremia due to mannitol is not present since the calculated P_{osm} [2 $\times$ 120 + (125/18) + (15/2.8) = 252 mosmol/kg] is similar to the measured P_{osm}, indicating no retention of mannitol in the extracellular fluid. The fall in the plasma Na^+ concentration in the SIADH is due to the retention of ingested water. Although an intracerebral hemorrhage can cause the SIADH, this patient did not receive enough water after the hemorrhage to reduce the plasma Na^+ concentration to 120 meq/L. Therefore, the hyponatremia must have been present before the neurologic symptoms began.

23-3 1. (*b*) This patient is both water- and sodium-overloaded.

2. (*f*) The combination of severe hyponatremia and a very concentrated urine in a patient with the SIADH is best treated with hypertonic saline and furosemide.

3. (*d*) True volume depletion with mild hyponatremia can be corrected with isotonic saline.

4. (*c*) This case represents pseudohyponatremia (normal P_{osm}) due to hyperproteinemia.

5. (*b*) The patient is both water- and sodium-overloaded.

6. (*a*) or (*e*) Either water restriction alone or the use of hypertonic saline would be reasonable in this patient with presumed SIADH and asymptomatic hyponatremia. Furosemide would add little in this setting since the U_{osm} is already almost isosmotic to plasma.

23-4 The most likely diagnosis is SIADH due to an oat-cell carcinoma of the lung. Since the plasma Na^+ concentration is very low and the U_{osm} high, hypertonic saline and furosemide should be given. The aim should be to raise the plasma Na^+ concentration to about 120 meq/L in the first 12 h. The amount of Na^+ required can be estimated from:

$$Na^+ \text{ deficit} = 0.6 \times \text{lean body weight} \times (120 - 105)$$

$$= 0.6 \times 70 \times 15$$

$$= 630 \text{ meq}$$

Three percent saline contains 513 meq/L of Na^+. Therefore, 1200 mL over 12 h (100 mL/h) should raise the plasma Na^+ concentration to near the desired level. Furosemide will enhance this effect by making the urine relatively isosmotic to plasma, thereby reducing free-water generation by the kidney.

23-5 Volume depletion increases Na^+ reabsorption and secondarily that of uric acid. The result is an increase in the plasma uric acid concentration. In contrast, Na^+ and uric acid reabsorption are reduced by volume expansion, as in the SIADH. Thus, the plasma uric acid concentration tends to be reduced (usually below 4 mg/dL) in this setting [9].

CHAPTER 24

24-1 Polyuria with a dilute urine is due to either increased water intake (primary polydipsia) or an inability to concentrate the urine (central or nephrogenic DI). The low plasma Na^+ concentration and P_{osm} in this patient suggest that she is water-overloaded due to primary polydipsia. The diagnosis can be established by a water restriction test followed by the administration of vasopressin after the maximum urine osmolality has been achieved or the P_{osm} reaches 295 mosmol/kg. It is interesting to note that sarcoidosis can produce each of the disorders considered here: primary polydipsia and central DI by hypothalamic involvement, and nephrogenic DI by hypercalcemia [10].

24-2 Two factors were of primary importance in the development of hypernatremia in this patient: insensible water losses and decreased thirst (due to the depressed mental status), which prevented correction of the water deficit. Diarrhea and diuretics cause relatively isosmotic water losses, and the U_{osm} of 606 mosmol/kg makes diabetes insipidus unlikely since water is not being lost in excess of solute.

The aim of treatment is to lower the plasma Na^+ concentration by 15 meq/L in the first 8 to 10 h. The amount of water required can be calculated from:

$$\text{Water deficit} = 0.4 \times \text{lean body weight} \times (\tfrac{174}{159} - 1)$$

$$= 0.4 \times 50 \times 0.1$$

$$= 2 \text{ liters}$$

If insensible losses are 25 mL/h (or 250 mL/10 h), then 2250 mL of water should be given over the first 10 h, that is, 225 mL/h. Since the patient is also Na^+-depleted from diarrhea, quarter-isotonic saline would be a reasonable replacement solution. However, this fluid is only three-quarters free water. Therefore, 300 mL/h ($\tfrac{4}{3} \times 225$) must be given to provide the necessary free water. Thus, (*d*) is the correct answer.

24-3 At a plasma Na^+ concentration of 168 meq/L, endogenous ADH effect should be maximal. Therefore, the diagnosis of CDI can be confirmed by the administration of aqueous vasopressin, which should lower the urine volume and increase the U_{osm}.

The water deficit is approximately 7.0 liters [$0.5 \times 70 \times (\tfrac{168}{140} - 1)$]. To lower the plasma Na^+ concentration by 15 meq/L in the first 8 h, 3.5 liters [$0.5 \times 70 \times (\tfrac{168}{153} - 1)$] of free water should be given *in addition* to replacing the urine output and estimated insensible losses.

The reduction in the plasma Na^+ concentration probably represents an iatrogenic form of SIADH. Vasopressin tannate in oil is a long-acting ADH preparation. Consequently, administered water will be retained, resulting in hyponatremia.

CHAPTER 25

25-1 (*a*) The patient has both diabetic keotacidosis and metabolic alkalosis (due to vomiting). Although the laboratory data are consistent with a mild, simple metabolic acidosis (low pH and plasma HCO_3^- concentration), the anion gap of 28 is elevated to a much greater degree than the minor fall in plasma HCO_3^- concentration. This suggests that the effect of ketoacidois (which on its own might have lowered the plasma HCO_3^- concentration to 10 meq/L) was balanced in part by vomiting, which raised the plasma HCO_3^- concentration back toward normal without affecting the retention of ketoacid anions responsible for the high anion gap.

(*b*) Dehydration undoubtedly is responsible for much of the decline in renal function. In addition, acetoacetate is measured as creatinine in the standard assay, resulting in a further apparent elevation in the plasma creatinine concentration.

(*c*) The major electrolyte problem in this patient is hypokalemia. The hyperglycemia and acidemia are relatively mild, and correction of these problems with insulin will drive K^+ into the cells, possibly inducing arrhythmias. Therefore, the initial therapy should consist of fluids and K^+, for example, half-isotonic saline plus 40 meq of KCl per liter. This regimen will correct the hypokalemia and ameliorate the hyperglycemia, both by improving renal function (which will allow glucose to be excreted) and by dilution. Insulin should be withheld until the plasma K^+ concentration nears normal.

The patient should also be started on antimicrobial therapy (pending culture results) to treat the apparent acute pyelonephritis. This infection was presumably responsible for the loss of diabetic control.

25-2 (*a*) The acidemia is due to the retention of H^+ ions from the ketoacids; the associated anions (β-hydroxybutyrate and acetoacetate) presumably were excreted in the urine, resulting in only a small elevation in the anion gap.

(*b*) The patient should be given insulin with glucose. This will allow partial reversal of the ketoacidosis without the risk of hyperglycemia.

CHAPTER 26

26-1 (*a*) Correction of the acidemia will drive K^+ into the cells, further reducing the plasma K^+ concentration. In this setting, in which the acidemia is not severe, increasing the plasma K^+ concentration with K^+ supplements should be the first aim of therapy.

(*b*) Hypokalemia increases the magnitude of the *resting* potential whereas hypocalcemia increases the magnitude of the *threshold* potential. Thus, hypocalcemia protects against the effects of hypokalemia by returning membrane excitability ($E_m - E_t$) toward normal. Thus, correction of the hypokalemia should precede correction of the hypocalcemia.

It should be noted that, for the same reasons, hypokalemia protects against hypocalcemia. Therefore, increasing the plasma K^+ concentration may precipitate hypocalcemic tetany [11]. However, this is less serious than the muscle paralysis or cardiac arrhythmias that may be induced by severe hypokalemia.

CHAPTER 27

27-1 (*a*) The findings of hypokalemia, metabolic alkalosis, a high rate of urinary K^+ excretion, and a normal blood pressure suggest vomiting, diuretics, or Bartter's syndrome. The other forms of mineralocorticoid excess are excluded by the normal blood pressure.

(*b*) The urine Cl^- concentration should be ordered next. A value below 15 meq/L is virtually diagnostic of vomiting, which would probably be self-induced in view of the negative history. A higher value is consistent with diuretic use or Bartter's syndrome, disorders which can be differentiated by a urinary assay for diuretics.

27-2 (*a*) 3. The low urine osmolality is consistent with primary water overload, which shuts off ADH secretion. Although the urine K^+ concentration is appropriately low, the urine volume is probably very high, resulting in an inappropriately high absolute rate of K^+ excretion.

(*b*) 2. The major clue suggesting hypomagnesemia is the presence of hypocalcemia.

(*c*) 1. Metabolic acidosis with an inappropriately high urine pH is diagnostic of renal tubular acidosis.

(*c*) 4. Metabolic acidosis (with a normally acid urine pH) and a low rate of K^+ excretion in the urine suggest extrarenal losses of K^+ and perhaps HCO_3^-, as occurs with laxative abuse.

CHAPTER 28

28-1 (*a*) The underlying renal disease certainly is a contributing factor but cannot explain the acute increase in the plasma K^+ concentration. Since patients with renal insufficiency are unable to conserve Na^+ normally, a low-sodium diet can lead to volume depletion, decreased renal perfusion, and a reduced ability to excrete K^+. The findings of decreased skin turgor and an increase in the plasma creatinine concentration are consistent with this possibility. In addition, since it is presumed that the hyperkalemia is due to an inability to excrete K^+, the patient should always be questioned about dietary intake. This patient gave a history of using large quantities of a KCl-containing salt substitute after being placed on the low-sodium diet. It is likely that this was the primary factor responsible for the hyperkalemia. The mild metabolic acidosis in this patient is playing only a minor role in the increase in the plasma K^+ concentration.

(*b*) By definition, pseudohyperkalemia produces no signs of K^+ intoxication.

(*c*) This patient has both severe muscle and electrocardiographic changes. Therefore, therapy should be initiated with Ca^{2+} gluconate followed by glucose and $NaHCO_3$; for example, 500 mL of 10% dextrose in saline plus 45 meq of $NaHCO_3$ infused over 30 min will lower the plasma K^+ concentration, raise the plasma sodium concentration, and produce volume expansion. Sodium polystyrene sulfonate should be given orally and repeated as necessary to remove the excess K^+. Dialysis should not be required since the patient does not have severe renal failure.

(*d*) With mild asymptomatic hyperkalemia, sodium polystyrene sulfonate alone should be sufficient.

28-2 (*a*) The low urine K^+ concentration in the presence of hyperkalemia indicates that the kidney is unable to excrete K^+ appropriately.

(*b*) Of the causes of reduced renal K^+ excretion, neither oliguric renal failure nor significant volume depletion is present. The findings of low blood pressure and increased pigmentation of the skin and the development of hypoglycemia after glucose and insulin point to the probable diagnosis of primary adrenal insufficiency. This can be confirmed by measuring plasma cortisol and aldosterone levels.

(*c*) Acutely, sodium polystyrene sulfonate can be used to lower the plasma K^+ concentration. Chronically, both glucocorticoid and mineralocorticoid replacement should be given.

REFERENCES

1. Vaughan, E. D., Jr., F. R. Bühler, J. H. Laragh, J. E. Sealey, L. Baer, and R. H. Bard: Renovascular hypertension: renin measurements to indicate hypersecretion and contralateral suppression, estimate renal plasma flow, and score for surgical curability, Am. J. Med., 55:402, 1973.
2. Stein, J. H., R. W. Osgood, S. Boonjarern, J. W. Cox, and T. F. Ferris: Segmental sodium reabsorption in rats with mild and severe volume depletion, Am. J. Physiol., 227:351, 1974.
3. Epstein, F. H., C. R. Kleeman, S. Pursel, and A. Hendrikx: The effect of feeding protein and urea on the renal concentrating process, J. Clin. Invest., 36:635, 1957.
4. Maronde, R. F., M. Milgrom, N. D. Vlachakis, and L. Chan: Response of thiazide-induced hypokalemia to amiloride, J. Am. Med. Assoc., 249:237, 1983.
5. Hilden, T., and T. L. Svendsen: Electrolyte disturbances in beer drinkers, Lancet, 2:245, 1975.
6. DeSousa, R. C., J. T. Harrington, E. S. Ricanati, J. W. Shelkrot, and W. B. Schwartz: Renal regulation of acid-base equilibrium during chronic administration of mineral acid, J. Clin. Invest., 53:465, 1974.
7. George, J. M., L. Wright, N. H. Bell, and F. C. Bartter: The syndrome of primary aldosteronism, Am. J. Med., 48:343, 1970.
8. Gabow, P. A., S. Moore, and R. W. Schrier: Spironolactone-induced hyperchloremic acidosis in cirrhosis, Ann. Intern. Med., 90:338, 1979.
9. Beck, L. H.: Hypouricemia in the syndrome of inappropriate secretion of antidiuretic hormone, N. Engl. J. Med., 301:528, 1979.
10. Stuart, C. A., F. A. Neelon, and H. E. Lebovitz: Disordered control of thirst in hypothalamic-pituitary sarcoidosis, N. Engl. J. Med., 303:1078, 1980.
11. Engel, F. L., S. P. Martin, and H. Taylor: On the relation of potassium to the neurologic manifestations of hypocalcemic tetany, Bull. Johns Hopkins Hosp., 84:285, 1949.

THIRTY

SUMMARY OF EQUATIONS AND FORMULAS

UNITS OF MEASUREMENT

$$mmol/L = \frac{mg/dL \times 10}{molecular\ weight}$$

$$meq/L = mmol/L \times valence$$

$$mosmol/kg = n \times mmol/L \qquad n = number\ of\ dissociable\ particles\ per\ molecule$$

EVALUATION OF RENAL FUNCTION

Glomerular Filtration Rate

$$GFR = creatinine\ clearance = \frac{U_{cr} \times V}{P_{cr}}$$

The creatinine clearance, in the steady state, can be estimated from

$$C_{cr}(mL/min) = \frac{(140 - age) \times lean\ body\ weight\ (kg)}{P_{cr} \times 72}$$

The value obtained from this formula should be multiplied by 0.85 in women who have less muscle mass than men.

Tubular Function

Variations in tubular reabsorption are the major way that the kidneys alter solute and water excretion. There are, however, *no absolute normal values* for the urine

Na^+ or K^+ concentration, osmolality, or pH since these parameters vary with intake. For example, the urine Na^+ concentration should be less than 10 meq/L in volume depletion but may exceed 100 meq/L after a Na^+ load. Similarly, the U_{osm} should be less than 100 mosmol/kg in a hyponatremic (hypoosmolal) patient but more than 800 mosmol/kg in a hypernatremic patient. Thus, a U_{osm} of 200 mosmol/kg is *inappropriately high* (although still hypoosmotic to plasma) in a patient with a plasma Na^+ concentration of 115 meq/L.

ACID-BASE

$$pH = 6.10 + \log \frac{[HCO_3^-]}{0.03 P_{CO_2}}$$

$$[H^+] = 24 \times \frac{P_{CO_2}}{[HCO_3^-]}$$

Conversion of pH into Hydrogen Concentration

$$pH \text{ of } 7.40 = [H^+] \text{ of } 40 \text{ nanoeq/L}$$

For each 0.1-unit increase in pH, multiply $[H^+]$ by 0.8:

$$pH \text{ of } 7.60 = 40 \times 0.8 \times 0.8 = 26 \text{ nanoeq/L}$$

For each 0.1-unit decrease in pH, multiply $[H^+]$ by 1.25:

$$pH \text{ of } 7.30 = 40 \times 1.25 = 50 \text{ nanoeq/L}$$

Renal and Respiratory Compensations in Acid-Base Disorders

Metabolic acidosis:
   1.2-mmHg fall in P_{CO_2} per 1-meq/L decrease in plasma $[HCO_3^-]$
Metabolic alkalosis:
   0.6-mmHg increase in P_{CO_2} per 1-meq/L elevation in plasma $[HCO_3^-]$
Respiratory acidosis:
   Acute: 1-meq/L increase in plasma $[HCO_3^-]$ per 10-mmHg rise in P_{CO_2}
   Chronic: 3.5-meq/L elevation in plasma $[HCO_3^-]$ per 10-mmHg increase in P_{CO_2}
Respiratory alkalosis:
   Acute: 2-meq/L fall in plasma $[HCO_3^-]$ for each 10-mmHg decrease in P_{CO_2}
   Chronic: 5-meq/L reduction in plasma $[HCO_3^-]$ for each 10-mmHg fall in P_{CO_2}

Calculation of Bicarbonate Deficit and Excess

HCO_3^- deficit (meq) in metabolic acidosis

$$\cong 0.7 \times \text{lean body weight (kg)} \times (24 - \text{plasma } [HCO_3^-])$$

HCO_3^- excess (meq) in metabolic alkalosis

$$\cong 0.5 \times \text{lean body weight (kg)} \times (\text{plasma } [HCO_3^-] - 24)$$

OSMOLALITY

$$P_{osm} \cong 2 \times \text{plasma } [Na^+] + \frac{[\text{glucose}]}{18} + \frac{BUN}{2.8}$$

$$\text{Effective } P_{osm} \cong 2 \times \text{plasma } [Na^+] + \frac{[\text{glucose}]}{18}$$

$$\text{Plasma } [Na^+] \propto \frac{Na_e^+ + K_e^+}{TBW}$$

Plasma Sodium Concentration and Plasma Glucose Concentration

For each 62-mg/dL increase in the plasma glucose concentration, there will be a reciprocal 1-meq/L reduction in the plasma Na^+ concentration, because of the movement of water into the extracellular fluid down an osmotic gradient. Thus, hyperglycemia can result in a dissociation between the P_{osm} (which is increased) and the plasma Na^+ concentration (which may be reduced).

Hyponatremia

$$Na^+ \text{ deficit (meq)} \cong 0.6\dagger \times \text{lean body weight} \times (140 - \text{plasma } [Na^+])$$

This formula estimates the amount of Na^+ required to raise the plasma Na^+ concentration back up to 140 meq/L. It may not represent the total Na^+ deficit, however, since there may be an additional isosmotic Na^+ and water loss (due, for example, to vomiting or diuretics).

Hypernatremia

$$\text{Water deficit (liters)} \cong 0.5 \times \text{lean body weight} \times \left(\frac{\text{plasma } [Na^+]}{140} - 1 \right)$$

---

†The percent of lean body weight (in kg) used to estimate the total body water in this and the following equation refers to men. In women, a 10 percent lower value should be used, for example, 0.5 not 0.6 times the lean body weight for the sodium deficit.

MISCELLANEOUS

Alveolar-Arterial Oxygen Gradient

$$\text{(A-a) } O_2 \text{ gradient} = P_{iO_2}(150 \text{ mmHg on room air}) - 1.25 \times P_{aCO_2} - P_{aO_2}$$

Plasma Calcium Concentration and Plasma Albumin Concentration

For every 1.0-g/dL fall in the plasma albumin concentration, there will be approximately a 0.8-mg/dL reduction in the plasma Ca^{2+} concentration. Since this does not reflect a change in the physiologically important, free Ca^{2+} concentration, this does not represent true hypocalcemia.

INDEX

Acetate metabolism into bicarbonate, 383, 389
Acetazolamide, 331–332
  effect on bicarbonate reabsorption, 231–232, 331–332
  effect on glomerular filtration rate, 69
  in metabolic alkalosis, 331, 388, 457
  in periodic paralysis, 583, 623
  in refractory edema, 346
  in respiratory acidosis, 457
  in sleep apnea, 455
Acetoacetic acid:
  and measurement of plasma creatinine, 559–560
  (*See also* Ketoacidosis)
Acetone (*see* Ketoacidosis)
Acid-base balance:
  regulation of, 225–245
    by alveolar ventilation, 242–243
    by renal acid excretion, 227–241
    (*See also* Hydrogen, tubular secretion of)
  role of buffers in, 209–220
  (*See also* pH)
Acid-base disorders:
  compensatory responses in, 366, 368
  definition of, 216, 365–366
  introduction to, 361–371
  mixed, 367–369
    in metabolic acidosis, 397
    in respiratory acidosis, 449–453
  (*See also* Acidosis; Alkalosis)
Acid-base map, 369–370
Acidemia, 216, 365
Acidosis, 216, 366
  metabolic, 366, 394–430
    acid excretion in, 237, 398

Acidosis, metabolic (*Cont.*):
  alveolar ventilation in, 242–243, 244, 366, 397
  ammonium excretion in, 234–235, 398
  anion gap in, 399–403, 665
  bicarbonate deficit in, 427
  bone disease in, 421, 425
  buffering in, 217–218, 244, 395–397, 427
    time course of, 218, 427–428
  calcium excretion in, 417
  citrate excretion in, 417
  etiology and diagnosis of, 399, 403–424
    ammonium chloride, 412
    cholestyramine, 413
    diarrhea, 281, 293, 413
    ethylene glycol ingestion, 411–412
    formaldehyde ingestion, 410
    hyperalimentation fluids, 413
    hyperkalemia, 240–241, 414, 421, 625
    hypoaldosteronism, 241, 293, 338, 421, 625
    ketoacidosis (*see* Ketoacidosis)
    lactic acidosis, 403–406
    methanol ingestion, 410–411
    paraldehyde, 412
    potassium-sparing diuretics, 338
    renal failure, 293, 407–408, 414
    renal tubular acidosis, 414–424
    rhabdomyolysis, 424
    salicylate intoxication, 408–410
    sulfur, 412
    toluene, 412
    ureterosigmoidostomy, 413
  growth in, 408, 423
  hyperchloremic, 399–400
  in hypokalemia, 601